Springer Tracts in Civil Engineering

Springer Tracts in Civil Engineering (STCE) publishes the latest developments in Civil Engineering - quickly, informally and in top quality. The series scope includes monographs, professional books, graduate textbooks and edited volumes, as well as outstanding PhD theses. Its goal is to cover all the main branches of civil engineering, both theoretical and applied, including:

- Construction and Structural Mechanics
- Building Materials
- Concrete, Steel and Timber Structures
- Geotechnical Engineering
- Earthquake Engineering
- Coastal Engineering; Ocean and Offshore Engineering
- Hydraulics, Hydrology and Water Resources Engineering
- Environmental Engineering and Sustainability
- Structural Health and Monitoring
- Surveying and Geographical Information Systems
- Heating, Ventilation and Air Conditioning (HVAC)
- Transportation and Traffic
- Risk Analysis
- Safety and Security

Indexed by Scopus

To submit a proposal or request further information, please contact:
Pierpaolo Riva at Pierpaolo.Riva@springer.com (Europe) Mengchu Huang
at mengchu.huang@springer.com (China)

More information about this series at http://www.springer.com/series/15088

Horst Werkle

Finite Elements in Structural Analysis

Theoretical Concepts and Modeling
Procedures in Statics and Dynamics
of Structures

 Springer

Horst Werkle
Department of Civil Engineering
Konstanz University of Applied Sciences
Konstanz, Germany

ISSN 2366-259X   ISSN 2366-2603   (electronic)
Springer Tracts in Civil Engineering
ISBN 978-3-030-49842-9   ISBN 978-3-030-49840-5   (eBook)
https://doi.org/10.1007/978-3-030-49840-5

English translation of the original 4th German edition published by Springer Fachmedien Wiesbaden GmbH, 2021

This Springer imprint is published by the registered company Springer Nature Switzerland AG
The registered company address is: Gewerbestrasse 11, 6330 Cham, Switzerland

Preface

The finite element method has established itself worldwide as the standard method in structural design. As with any structural analysis method, a basic understanding of the method and knowledge of its areas of application and limits are required to avoid potentially serious errors in practice. In the case of the finite element method, this means that the structural engineer should have an understanding of the calculation processes taking place in the computer, i.e. of the mechanical and mathematical principles of the method. In addition the modelling, i.e. the transfer of a specific building construction into a structural model, is of importance. The enormously increased possibilities of the finite element method compared to a conventional analysis result in a higher complexity of the models and mean a high responsibility of the structural engineer. After the calculation has been carried out, its results must be interpreted in the context of the structural model. For this purpose, again a good knowledge of the method itself as well as of the model behavior are required. The book is limited to linear problems of statics and dynamics. It therefore covers the major part of the structural design of buildings. Non-linear problems—apart from stability problems—have a rather special character in structural design and are reserved for a later edition.

The successful application of finite element software requires sound knowledge of its fundamentals as well as a good understanding of the underlying models. The book is intended to convey both to the reader. The presentation of the theory follows a bottom-up approach. Theoretical fundamentals are presented "as simple as possible, but not simpler" (A. Einstein) and illustrated with a large number of examples.

The outline of the book follows a structured approach. It starts with the mathematical and mechanical basics, moves on to the finite element analysis of truss and beam structures, and leads on to the theoretically more demanding surface and solid structures. Finally, the last chapter provides a detailed introduction to the computational methods of structural dynamics with finite elements.

Chapter 1 deals with the mathematical basics of matrix calculation and solution methods for linear systems of equations and eigenvalue problems that are found in finite element systems. In Chap. 2, the basic equations of the theory of elasticity are compiled as far as they are needed in the following chapters. The presentation of the finite element method starts in Chap. 3 using bar systems. This chapter deals with trusses and beams subjected to bending loads. However, the methods for creating the underlying systems of equations are of general scope and can be applied to surface and solid structures as well. Chapter 4 is the heart of the book. It begins by discussing the approximate character of the finite element method as it is characteristic for surface and solid structures. This is followed by a presentation of finite element formulations for plates in plane stress and in bending as well as for shells and solid models. Chapter 5 deals with the dynamics of structures. It covers the computational methods for finite element models with a special focus on earthquake-excited vibrations.

Chapters 3 to 5 each contain a detailed discussion of the modeling of structures for static and dynamic investigations. Exercises accompanying the individual chapters allow the reader to explore the subject in greater depth.

The book as a whole is a comprehensive introduction to the finite element method in structural design. However, the reader may also confine himself to reading individual sections. For example one may start with Sect. 3.2 which deals with the structure of the system stiffness matrix of a finite element model. Sections 4.3 and 4.4 provide an introduction to the approximate character of the finite element method based on plates in plane stress. However, it should be noted that Sects. 4.3 to 4.9, which deal with the theoretical principles of surface and solid models, are based on the explanations in Sect. 3.2. Sections 3.7 and 4.11 deal with the modeling of beam and shell structures, and can be considered as stand-alone sections, just like Section 5.8 for dynamic modeling.

The book originates from many years of teaching in the Faculty of Civil Engineering at the HTWG Konstanz, Germany. It is intended for students of civil engineering as well as for practical structural engineers. My thanks for numerous inspiring technical discussions go to my colleague Prof. Dr. Detlef Rothe, University of Applied Sciences Darmstadt, Germany, who also contributed several worked examples. For his support in reading the manuscript and numerous hints I would like to thank Dr.-Ing. Andrei Firus. I would also like to thank my life partner, Mrs. Charlotte Jäkel, for her understanding but also for her tireless active support in preparing the drawings.

The book has already been published in several editions on the German book market. I wish the new English edition the same success which enjoyed the previous editions in German.

Contents

Chapter 1
Mathematical Background

Abstract The first chapter deals with the fundamentals of mathematics as they are required to formulate and apply the finite element method. Basics of matrix calculation are introduced and explained in examples. Solution methods for linear systems of equations and conditions to ensure their resolvability are treated. Chapters 3 and 4 will come back to it as all linear static finite element calculations lead to linear systems of equations. The next section deals with eigenvalue problems, their properties and some fundamental solution methods. Chapter 5 will make extensive reference to this section in the context of vibrations of structures modeled with finite elements.

1.1 Introduction

Each physical or other theory requires a "language" in which it can be expressed in an appropriate way. The "language" in which the finite element method is expressed is the matrix analysis and its application in linear algebra. Matrix formulations are particularly suitable for computer implementation and the analysis of large systems which are characteristic of computer-based methods in mechanics.

Hence, when studying the finite element method, the concepts of matrix analysis should be well understood. In this chapter, some basics of matrix analysis are given, and mathematical methods used in finite element analysis are explained.

1.2 Matrices and Vectors

Matrices allow the clear and effective representation of linear relationships with any number of variables. The clarity of matrix notation becomes evident when looking at a linear system of equations. For example, four equations with the unknowns x_1, x_2, x_3 and x_4 may be given as

$$5x_1 - 3x_2 - 6x_3 = 0$$
$$-3x_1 + 8x_2 + 2x_3 + x_4 = 2$$

© Springer Nature Switzerland AG 2021
H. Werkle, *Finite Elements in Structural Analysis*, Springer Tracts
in Civil Engineering, https://doi.org/10.1007/978-3-030-49840-5_1

$$4x_1 - 9x_2 + 5x_3 + 7x_4 = 1$$
$$3x_2 + 4x_3 + 5x_4 = -2.$$

In matrix notation, the system of equations is written as

$$
\begin{bmatrix}
5 & -3 & -6 & 0 \\
-3 & 8 & 2 & 1 \\
4 & -9 & 5 & 7 \\
0 & 3 & 4 & 5
\end{bmatrix}
\cdot
\begin{bmatrix}
x_1 \\ x_2 \\ x_3 \\ x_4
\end{bmatrix}
=
\begin{bmatrix}
0 \\ 2 \\ 1 \\ -2
\end{bmatrix}
$$

where the coefficients of the unknowns on the left side are written in a "condensed" form as a matrix. Similarly, the unknowns on the left and the values on the right side are shortly written as vectors. Using matrix notation, the equation can be written simply as

$$\underline{A} \cdot \underline{x} = \underline{b} \tag{1.0}$$

where $\underline{A}$ is the coefficient matrix of the linear system of equations, $\underline{x}$ is the vector of the unknowns, and $\underline{b}$ is the vector of the known quantities on the right side. Hence, it is

$$
\underline{A} =
\begin{bmatrix}
5 & -3 & -6 & 0 \\
-3 & 8 & 2 & 1 \\
4 & -9 & 5 & 7 \\
0 & 3 & 4 & 5
\end{bmatrix},
\quad
\underline{x} =
\begin{bmatrix}
x_1 \\ x_2 \\ x_3 \\ x_4
\end{bmatrix},
\quad
\underline{b} =
\begin{bmatrix}
0 \\ 2 \\ 1 \\ -2
\end{bmatrix}.
$$

The definition of a matrix is as follows:

Definition

A matrix is a system of $m \times n$ elements arranged in an array with m rows and n columns.

$$
\underline{A} =
\begin{bmatrix}
a_{11} & a_{12} & \cdot & \cdot & \cdot & a_{1n} \\
a_{21} & a_{22} & \cdot & \cdot & \cdot & a_{2n} \\
\cdot & \cdot & & a_{ki} & & \cdot \\
\cdot & & \cdot & a_{ik} & & \cdot \\
\cdot & \cdot & & & & \cdot \\
a_{m1} & a_{m2} & \cdot & \cdot & \cdot & a_{mn}
\end{bmatrix}.
$$

The matrix has the order $m \times n$. A matrix with one column only is also denoted as a vector.

The term vector is not defined geometrically, but it is understood in a more general way as a system of ordered elements. Hence, quantities with different units such as sectional forces M, N, V (bending moment, normal force and shear force) may be summed up to a vector.

Special forms of matrices are given in Table 1.1. Thereafter, matrix $\underline{A}$ in the example given above is a nonsymmetric square matrix of order 4. Matrices can be divided into submatrices. In this case, the elements of the matrices themselves represent matrices.

Since (square) matrices can be understood as coefficients of a linear system of equations, the linear independence of the rows (or columns) is decisive for the

Table 1.1 Special matrices

Denomination	Properties	
Rectangular matrix	Number of columns $n \neq$ number of rows m	
Square matrix	Number of columns $n =$ number of rows m	
Symmetric matrix	Square matrix where all off-diagonal elements are reflected at the diagonal, i.e. $a_{ik} = a_{ki}$ and $\underline{A} = \underline{A}^T$	
Diagonal matrix	$\begin{bmatrix} a_{11} & & & \\ & a_{22} & 0 & \\ & & \ddots & \\ & 0 & & \ddots \\ & & & & a_{nn} \end{bmatrix}$	Square matrix with all elements outside the diagonal equal to zero: $a_{ik} = 0$ for $i \neq k$
Unit matrix (Identity matrix)	$\begin{bmatrix} 1 & & & \\ & 1 & 0 & \\ & & 1 & \\ & 0 & & 1 \\ & & & & 1 \end{bmatrix} = \underline{I}$	Diagonal matrix with all elements on the diagonal equal to 1, i.e. $a_{ik} = 1$ for $i = k$ $a_{ik} = 0$ for $i \neq k$
Zero matrix	All elements are zero: $a_{ik} = 0$ for all i, k	
Triangular matrix	Square matrix where all elements above the diagonal (lower triangular) or below the diagonal (upper triangular) are equal to zero	
Transpose of a matrix	$\underline{A}^T$ is obtained by interchanging the columns and rows of the original matrix $\underline{A}$. Therewith, $(\underline{A} \cdot \underline{B})^T = \underline{B}^T \cdot \underline{A}^T$ and $(\underline{A}^T)^T = \underline{A}$	
Positive definiteness	A quadratic matrix $\underline{A}$ is positive definite, if $\underline{x}^T \cdot \underline{A} \cdot \underline{x} > 0$ and positive semidefinite, if $\underline{x}^T \cdot \underline{A} \cdot \underline{x} \geq 0$ for all vectors $\underline{x} \neq \underline{0}$	
Trace of a matrix	$\mathrm{tr}(\underline{A}) = a_{11} + a_{22} + \cdots + a_{nn} = \sum_{k=1}^{n} a_{kk}$	

(continued)

Table 1.1 (continued)

Denomination	Properties
Determinant of a matrix	2×2 *matrices: second-order determinant*
	$\det\left(\begin{bmatrix} a_{11} & a_{12} \\ a_{21} & a_{22} \end{bmatrix}\right) = a_{11} \cdot a_{22} - a_{21} \cdot a_{12}$
	3×3 *matrices: third-order determinant*
	$\det\left(\begin{bmatrix} a_{11} & a_{12} & a_{13} \\ a_{21} & a_{22} & a_{23} \\ a_{31} & a_{32} & a_{33} \end{bmatrix}\right)$
	$= a_{11} \cdot a_{22} \cdot a_{33} + a_{12} \cdot a_{23} \cdot a_{31} + a_{13} \cdot a_{21} \cdot a_{32}$
	$- a_{31} \cdot a_{22} \cdot a_{13} - a_{32} \cdot a_{23} \cdot a_{11} - a_{33} \cdot a_{21} \cdot a_{12}$
	Determinants of higher order see [1]

possible solution of the system of equations. A square matrix with linear dependent columns and rows is called singular. In this case, the corresponding system of equations possesses no unique solution. If the rows (or columns) are linearly independent, the matrix is called regular or non-singular.

As for real numbers, for matrices, mathematical operations can be defined.

1.3   Matrix Algebra

1.3.1   Addition and Subtraction

In order to add or subtract two matrices, their corresponding elements are added or subtracted, respectively. Hence, the addition or substraction of two matrices is defined only for two matrices possessing the same size as

$$\underline{A} + \underline{B} = \underline{C} \text{ where } a_{ij} + b_{ij} = c_{ij}. \tag{1.1}$$

As for real numbers the following calculation rules apply:

$$\underline{A} + \underline{B} = \underline{B} + \underline{A} \quad \text{(commutative law)} \tag{1.1a}$$

$$\underline{A} + (\underline{B} + \underline{C}) = (\underline{A} + \underline{B}) + \underline{C} \quad \text{(associative law)} \tag{1.1b}$$

1.3.2 Multiplication

In order to multiply a matrix with a scalar, each element of the matrix has to be multiplied with the scalar, that is

$$c \cdot \underline{A} = \underline{B} \quad \text{where } b_{ij} = c \cdot a_{ij}. \tag{1.2}$$

The product of two matrices $\underline{A}$ and $\underline{B}$ is defined only if the number of columns of $\underline{A}$ equals the number of rows of $\underline{B}$. The element c_{ij} of the matrix $\underline{C}$ is obtained by multiplying the elements of the ith row of $\underline{A}$ with the corresponding elements of the $j$the column of $\underline{B}$, and adding the products as

$$\underline{A} \cdot \underline{B} = \underline{C} \quad \text{where } c_{ij} = \sum_{k=1}^{n} \left(a_{ik} \cdot b_{jj} \right). \tag{1.3}$$

The matrix $\underline{C}$ has as many rows as matrix $\underline{A}$ and as many columns as matrix $\underline{B}$.

Matrix multiplication can be understood as replacing the "solution vector" of one system of equations by the right-hand-side vector of another system of equations (cf. Example 1.3). For calculations by hand, Falk's scheme is useful [1].

Example 1.1 The matrices $\underline{A}$ and $\underline{B}$ are specified as

$$\underline{A} = \begin{bmatrix} \underline{A}_1 & \underline{A}_2 \end{bmatrix}, \quad \underline{A}_1 = \begin{bmatrix} 4 & -3 \\ -3 & 8 \\ 4 & -9 \end{bmatrix}, \quad \underline{A}_2 = \begin{bmatrix} -3 & 0 \\ 2 & 1 \\ 0 & 7 \end{bmatrix}$$

$$\underline{B} = \begin{bmatrix} \underline{B}_1 \\ \underline{B}_2 \end{bmatrix}, \quad \underline{B}_1 = \begin{bmatrix} 1 & 3 \end{bmatrix}, \quad \underline{B}_2 = \begin{bmatrix} 0 & 2 \\ 4 & -1 \\ -1 & 2 \end{bmatrix}.$$

Verify the calculation rule $(\underline{A} \cdot \underline{B})^{\mathrm{T}} = \underline{B}^{\mathrm{T}} \cdot \underline{A}^{\mathrm{T}}$ for the two matrices given.

$$\underline{A} = \begin{bmatrix} 4 & -3 & -3 & 0 \\ -3 & 8 & 2 & 1 \\ 4 & -9 & 0 & 7 \end{bmatrix}, \quad \underline{B} = \begin{bmatrix} 1 & 3 \\ 0 & 2 \\ 4 & -1 \\ -1 & 2 \end{bmatrix}, \quad \underline{A} \cdot \underline{B} = \begin{bmatrix} -8 & 9 \\ 4 & 7 \\ -3 & 8 \end{bmatrix}$$

$$\underline{B}^{\mathrm{T}} = \begin{bmatrix} 1 & 0 & 4 & -1 \\ 3 & 2 & -1 & 2 \end{bmatrix}, \quad \underline{A}^{\mathrm{T}} = \begin{bmatrix} 4 & -3 & 4 \\ -3 & 8 & -9 \\ -3 & 2 & 0 \\ 0 & 1 & 7 \end{bmatrix}, \quad \underline{B}^{\mathrm{T}} \cdot \underline{A}^{\mathrm{T}} = \begin{bmatrix} -8 & 4 & -3 \\ 9 & 7 & 8 \end{bmatrix}.$$

The transpose of the product $\underline{A} \cdot \underline{B}$ is obtained to be the product of $\underline{B}^{\mathrm{T}}$ and $\underline{A}^{\mathrm{T}}$. ◄

The following calculation rules apply:

$$\underline{A} \cdot (\underline{B} \cdot \underline{C}) = (\underline{A} \cdot \underline{B}) \cdot \underline{C} \quad \text{(associative law)} \tag{1.4a}$$

$$\underline{A} \cdot (\underline{B} + \underline{C}) = \underline{A} \cdot \underline{B} + \underline{A} \cdot \underline{C} \quad \text{(distributive law).} \tag{1.4b}$$

The commutative law, however, does not apply to matrix multiplication, which means that $\underline{A} \cdot \underline{B} \neq \underline{B} \cdot \underline{A}$.

Example 1.2 By means of the matrix products $(\underline{B} \cdot \underline{A})$ and $(\underline{A} \cdot \underline{B})$, it has to be shown that the commutative law does not hold up.

$$\underline{A} = \begin{bmatrix} 1 & 1 \\ -1 & -1 \end{bmatrix}, \quad \underline{B} = \begin{bmatrix} 1 & 1 \\ 1 & 1 \end{bmatrix}.$$

One obtains

$$\underline{A} \cdot \underline{B} = \begin{bmatrix} 2 & 2 \\ -2 & -2 \end{bmatrix}, \quad \underline{B} \cdot \underline{A} = \begin{bmatrix} 0 & 0 \\ 0 & 0 \end{bmatrix}.$$

The example also shows that the product of two matrices may be obtained as a zero matrix even if none of the matrices to be multiplied is a zero matrix. In case of scalars, however, one of the scalars to be multiplied must be zero if the product is zero.   ◄

Example 1.3 Two equations with the unknowns x and y are given as

$$\underline{A} \cdot \underline{x} = \underline{f}$$

$$\underline{x} = \underline{B} \cdot \underline{y}.$$

Replacing $\underline{x}$ in the equation $\underline{A} \cdot \underline{x} = \underline{f}$ by $\underline{B} \cdot \underline{y}$ gives

$$\underline{A} \cdot \underline{B} \cdot \underline{y} = \underline{f}.$$

It is to be shown that the matrix product $(\underline{A} \cdot \underline{B})$ gives the same coefficient matrix of the unknowns y as by replacing x_1 and x_2 in the first system of equations by the second system of equations.

Solution in matrix notation with $\underline{A} = \begin{bmatrix} 3 & 2 \\ 4 & 1 \end{bmatrix}$, $\underline{B} = \begin{bmatrix} 2 & 4 \\ 3 & 2 \end{bmatrix}$:

$$\underline{A} \cdot \underline{x} = \underline{f} \qquad\qquad \underline{x} = \underline{B} \cdot \underline{y}$$

$$\begin{bmatrix} 3 & 2 \\ 4 & 1 \end{bmatrix} \cdot \begin{bmatrix} x_1 \\ x_2 \end{bmatrix} = \begin{bmatrix} 4 \\ 7 \end{bmatrix} \qquad \begin{bmatrix} x_1 \\ x_2 \end{bmatrix} = \begin{bmatrix} 2 & 4 \\ 3 & 2 \end{bmatrix} \cdot \begin{bmatrix} y_1 \\ y_2 \end{bmatrix}$$

$$\underline{A} \cdot \underline{B} \cdot \underline{y} = \underline{f}$$

$$\underline{A} \cdot \underline{B} = \begin{bmatrix} 2 \cdot 3 + 3 \cdot 2 & 4 \cdot 3 + 2 \cdot 2 \\ 2 \cdot 4 + 3 \cdot 1 & 4 \cdot 4 + 2 \cdot 1 \end{bmatrix} = \begin{bmatrix} 12 & 16 \\ 11 & 18 \end{bmatrix}.$$

Solution by scalar operations:

$$
\begin{array}{lll}
3x_1 + 2x_2 = 4 & x_1 = 2y_1 + 4y_2 & (2 \cdot 3 + 3 \cdot 2)y_1 + (4 \cdot 3 + 2 \cdot 2)y_2 = 4 \\
4x_1 + x_2 = 7 & x_2 = 3y_1 + 2y_2 & (2 \cdot 4 + 3 \cdot 1)y_1 + (4 \cdot 4 + 2 \cdot 1)y_2 = 7
\end{array}
$$

or written in matrix notation

$$\begin{bmatrix} 2 \cdot 3 + 3 \cdot 2 & 4 \cdot 3 + 2 \cdot 2 \\ 2 \cdot 4 + 3 \cdot 1 & 4 \cdot 4 + 2 \cdot 1 \end{bmatrix} \cdot \begin{bmatrix} y_1 \\ y_2 \end{bmatrix} = \begin{bmatrix} 4 \\ 7 \end{bmatrix} \quad \text{or} \quad \begin{bmatrix} 12 & 16 \\ 11 & 18 \end{bmatrix} \cdot \begin{bmatrix} y_1 \\ y_2 \end{bmatrix} = \begin{bmatrix} 4 \\ 7 \end{bmatrix}$$

which is identical to $\underline{A} \cdot \underline{B} \cdot \underline{y} = \underline{f}$. ◀

1.3.3 Inverse of a Matrix

The inversion of a matrix corresponds to the division of numbers in arithmetic. The division $a/b = c$ of two numbers, for example, may be written as

$$a \cdot b^{-1} = c.$$

Formally this expression may be translated into matrix notation as

$$\underline{A} \cdot \underline{B}^{-1} = \underline{C},$$

where $\underline{B}^{-1}$ is the inverse of the square matrix $\underline{B}$. If a matrix is multiplied with its inverse, the unit matrix is obtained as

$$\underline{A} \cdot \underline{A}^{-1} = \underline{I}. \tag{1.5}$$

This equation can be taken as the initial equation for the computation of the inverse $\underline{A}^{-1}$ of a given matrix $\underline{A}$. This will be shown in an example. In the matrix product

$$\underline{A} \cdot \underline{B} = \underline{I}$$

the matrix $\underline{B} = \underline{A}^{-1}$ is the inverse of the matrix $\underline{A}$. Assuming $\underline{A}$ and $\underline{B}$ to be two 3×3 matrices, the equation may be written as

$$\begin{bmatrix} a_{11} & a_{12} & a_{13} \\ a_{21} & a_{22} & a_{23} \\ a_{31} & a_{32} & a_{33} \end{bmatrix} \cdot \begin{bmatrix} b_{11} & b_{12} & b_{13} \\ b_{21} & b_{22} & b_{23} \\ b_{31} & b_{32} & b_{33} \end{bmatrix} = \begin{bmatrix} 1 & 0 & 0 \\ 0 & 1 & 0 \\ 0 & 0 & 1 \end{bmatrix}.$$

This equation may be subdivided into three matrix equations:

$$\begin{bmatrix} a_{11} & a_{12} & a_{13} \\ a_{21} & a_{22} & a_{23} \\ a_{31} & a_{32} & a_{33} \end{bmatrix} \cdot \begin{bmatrix} b_{11} \\ b_{21} \\ b_{31} \end{bmatrix} = \begin{bmatrix} 1 \\ 0 \\ 0 \end{bmatrix}$$

$$\begin{bmatrix} a_{11} & a_{12} & a_{13} \\ a_{21} & a_{22} & a_{23} \\ a_{31} & a_{32} & a_{33} \end{bmatrix} \cdot \begin{bmatrix} b_{12} \\ b_{22} \\ b_{32} \end{bmatrix} = \begin{bmatrix} 0 \\ 1 \\ 0 \end{bmatrix}$$

$$\begin{bmatrix} a_{11} & a_{12} & a_{13} \\ a_{21} & a_{22} & a_{23} \\ a_{31} & a_{32} & a_{33} \end{bmatrix} \cdot \begin{bmatrix} b_{13} \\ b_{23} \\ b_{33} \end{bmatrix} = \begin{bmatrix} 0 \\ 0 \\ 1 \end{bmatrix}.$$

These are three linear systems of equations with the 3 columns of the matrix $\underline{B}$ as unknowns. In general, the inversion of an $n \times n$ matrix requires the solution of n systems of equations. The columns of the unity matrix $\underline{I}$ are the right-hand sides of the equations. Therewith, the individual columns of the matrix $\underline{A}^{-1}$ are computed as the solution of a linear system of equations according to Sect. 1.4.3.

The inverse of a matrix does exist only if the matrix is square and regular (non-singular). For the inverse of a matrix product, the following rule applies:

$$\left(\underline{A} \cdot \underline{B}\right)^{-1} = \underline{B}^{-1} \cdot \underline{A}^{-1} \tag{1.5a}$$

1.4   Linear Systems of Equations

1.4.1   Inhomogeneous and Homogeneous Systems of Equations

A system of equations is called inhomogeneous if the right-hand side possesses values unequal to zero.

For example, the following system of equations is given:

$$
\begin{bmatrix}
a_{11} & a_{12} & \cdots & a_{1n} \\
a_{21} & a_{22} & \cdots & a_{2n} \\
\vdots & \vdots & & \vdots \\
a_{n1} & a_{n2} & \cdots & a_{nn}
\end{bmatrix}
\cdot
\begin{bmatrix}
x_1 \\ x_2 \\ \vdots \\ x_n
\end{bmatrix}
=
\begin{bmatrix}
b_1 \\ b_2 \\ \vdots \\ b_n
\end{bmatrix},
\tag{1.6}
$$

$$
\underline{A} \cdot \underline{x} = \underline{b}
$$

with the coefficients a_{ik}, the unknowns x_i, and the values b_i at the right-hand side. $\underline{A}$ is the coefficient matrix, $\underline{x}$ the solution vector, $\underline{b}$ the vector of the right-hand side, and n the number of equations. The system of equations is inhomogeneous if at least one value b_i is unequal to zero.

Systems of equations with the right side being zero are called homogeneous. In this section, only inhomogeneous systems of equations are considered. Homogeneous systems of equations are discussed in Sect. 1.5.

1.4.2 Existence of Solutions

Linear systems of equations do not necessarily possess a solution. Consider, for example, the system of equations

$$
\begin{aligned}
1000 \cdot x_1 + 500 \cdot x_2 &= 500 \\
2000 \cdot x_1 + 1000 \cdot x_2 &= 1000.
\end{aligned}
$$

It can be seen that the second equation is obtained by multiplying the first equation with a factor of 2. Hence, the second equation doesn't contain new "information" compared to the first equation. The rows of the matrix are called linearly dependent. Considering the rows of the matrix as vectors, the following definition applies:

Definition

A system of vectors $\underline{a}_i$ is called linearly dependent, if there exist constants c_i (with at least one $c_i \neq 0$), so that $\sum c_i \cdot \underline{a}_i = 0$, i.e. one vector can be represented as a linear combination of the other vectors.

In the example given above, the vectors (of the rows) are

$$
\underline{a}_1 = \begin{bmatrix} 1000 & 500 \end{bmatrix}, \quad \underline{a}_2 = \begin{bmatrix} 2000 & 1000 \end{bmatrix}.
$$

With $c_1 = -2$ and $c_2 = 1$, one obtains

$$c_1 \cdot \underline{a}_1 + c_2 \cdot \underline{a}_2 = \underline{0} \quad \text{or}$$

$$-2 \cdot \begin{bmatrix} 1000 & 500 \end{bmatrix} + 1 \cdot \begin{bmatrix} 2000 & 1000 \end{bmatrix} = \begin{bmatrix} 0 & 0 \end{bmatrix}.$$

Hence, the rows of the matrix are linearly dependent.

Regularity/singularity of a matrix

The matrix of a system of equations with linearly dependent rows or columns is called singular, otherwise, the matrix is regular (or non-singular). If a matrix is singular or regular, it can be examined by means of its determinant. If the determinant is zero, i.e.

$$\det(\underline{A}) = 0,$$

the matrix is singular, otherwise, it is regular.

The determinant of 2×2 and 3×3 matrices can be computed for 2×2 matrices as

$$\det(\underline{A}) = a_{11} \cdot a_{22} - a_{21} \cdot a_{12}$$

and for 3×3 matrices as

$$\det(\underline{A}) = a_{11} \cdot a_{22} \cdot a_{33} + a_{12} \cdot a_{23} \cdot a_{31} + a_{13} \cdot a_{21} \cdot a_{32}$$
$$- a_{31} \cdot a_{22} \cdot a_{13} - a_{32} \cdot a_{23} \cdot a_{11} - a_{33} \cdot a_{21} \cdot a_{12}.$$

Determinants of higher order for $n \times n$ matrices ($n > 3$) can be computed by a reduction to determinants of lower order with a Laplace expansion or by the Leibnitz formula (permutation rule) [1].

In the example given above, the determinant is obtained as

$$\det(\underline{A}) = 1000 \cdot 1000 - 2000 \cdot 500 = 0.$$

Hence, the coefficient matrix of the system of equations is singular.

For the solution, the following condition applies.

> **Existence of a unique solution of a linear inhomogeneous system of equations**
>
> Inhomogeneous linear systems of equations with n equations and n unknowns possess a unique solution only if the n rows are linearly independent, i.e. the coefficient matrix is regular.

If the matrix of an inhomogeneous system of equations is singular, a unique solution is no more possible. Depending on the values on the right side, there exists either an infinite number of solutions or there is no solution at all.

In the example, all pairs of x_1, x_2, fulfilling the condition

$$1000 \cdot x_1 + 500 \cdot x_2 = 500$$

are a solution of the system of equations. Hence there exist an infinite number of solutions. If the system of equations would be

$$1000 \cdot x_1 + 500 \cdot x_2 = 1000$$
$$2000 \cdot x_1 + 1000 \cdot x_2 = 1000$$

there would be a contradiction between the first and the second equation, and consequently, no solution would exist.

1.4.3 Numerical Solution Methods

The solution of large systems of equations is an essential part of a linear static finite element analysis. Their size may easily comprise 10000–100000 equations or even more. This section deals with some basic numerical solution methods. More sophisticated methods are given in textbooks as [2, 3].

Inhomogeneous linear systems of equations can be solved by direct or by iterative methods. Iterative methods as the *Gauss-Seidel method* give a solution for a single right-hand side vector (as in Eq. 1.0) whereas with direct methods, the solutions for multiple right-hand side vectors—corresponding to load cases of static analysis—can be easily obtained simultaneously. For this reason, they are preferred in programs for finite element analysis.

The *Gauss algorithm* and its variation called the Cholesky method are classical elimination methods. The practical analysis takes advantage of the properties of the systems of equations of the finite element method: the coefficient matrix is symmetric and often sparse (most of the elements are zero). In addition, it is positive definite which means especially that the elements on the diagonal are positive and larger than off-diagonal elements (Table 1.1) [1]. Particularly, the positive definiteness simplifies the solution procedure since a pivot search is not necessary, and the symmetry of the system of equations is preserved during the solution procedure.

In the Gauss method, the unknowns of the system of equations are eliminated successively. In the first step, the unknown x_1 is eliminated. To eliminate x_1 from equation i, the first row of the coefficient matrix and of the right-hand vector are multiplied with the factor

$$l_{i1()} = \frac{a_{i1}}{a_{11}} \quad (i = 2, 3, \ldots, n) \tag{1.7}$$

and the result is subtracted from the ith row:

$$a_{ik}^{(1)} = a_{ik} - l_{i1} \cdot a_{1k} \qquad b_i^{(1)} = b_i - l_{i1} \cdot b_1$$
$$(i = 2, 3, \ldots, n; \quad k = 2, 3, \ldots, n). \tag{1.7a}$$

Hence, after the first elimination step, the ith row does no more contain the unknown x_1. The coefficients of the matrix and the right-side vector are now denoted as $a_{ik}^{(1)}$ and $b_i^{(1)}$, respectively. In the same manner, x_1 is eliminated from the rows 2 to n.

The other unknowns x_i are eliminated by proceeding similarily. To eliminate the next unknown x_2, for example, the multiplication factors of the second row are obtained as

$$l_{i2} = \frac{a_{i2}^{(1)}}{a_{22}^{(1)}} \quad (i = 2, 3, \ldots, n) \tag{1.7b}$$

and the coefficients of the second reduction step as

$$a_{ik}^{(2)} = a_{ik}^{(1)} - l_{i2} \cdot a_{2k}^{(1)}, \qquad b_i^{(2)} = b_i^{(1)} - l_{i2} \cdot b_2^{(1)}$$
$$(i = 2, 3, \ldots, n; \quad k = 2, 3, \ldots, n). \tag{1.7c}$$

After $(n - 1)$ reduction steps, the system of equations is reduced to a single equation with the unknown x_n. The symmetry of the system of equations is preserved during the solution procedure.

Example 1.4 Perform the Gauss elimination steps for the equation system

$$\begin{bmatrix} 1.35 & -0.35 & -1.00 & 0 & -0.35 \\ -0.35 & 1.35 & 0 & 0 & 0.35 \\ -1.00 & 0 & 1.35 & 0.35 & 0 \\ 0 & 0 & 0.35 & 1.35 & 0 \\ -0.35 & 0.35 & 0 & 0 & 1.35 \end{bmatrix} \cdot \begin{bmatrix} x_1 \\ x_2 \\ x_3 \\ x_4 \\ x_5 \end{bmatrix} = \begin{bmatrix} 0 \\ 0 \\ 10 \\ -10 \\ 0 \end{bmatrix}.$$

$$\underline{A} \cdot \underline{x} = \underline{b}.$$

To reduce a system with 5 unknowns, in total 4 elimination steps are required.
Reduction step 1:

$$
l_1 = \begin{bmatrix} 1 \\ -0.259 \\ -0.741 \\ 0 \\ -0.259 \end{bmatrix},
$$

$$
\begin{bmatrix} 0 & 0 & 0 & 0 & 0 \\ 0 & 1.259 & -0.259 & 0 & 0.259 \\ 0 & -0.259 & 0.609 & 0.350 & -0.259 \\ 0 & 0 & 0.350 & 1.350 & 0 \\ 0 & 0.259 & -0.259 & 0 & 1.259 \end{bmatrix} \cdot \begin{bmatrix} 0 \\ x_2 \\ x_3 \\ x_4 \\ x_5 \end{bmatrix} = \begin{bmatrix} 0 \\ 0 \\ 10 \\ -10 \\ 0 \end{bmatrix}.
$$

Reduction step 2:

$$
l_2 = \begin{bmatrix} 0 \\ 1 \\ -0.206 \\ 0 \\ 0.206 \end{bmatrix},
\begin{bmatrix} 0 & 0 & 0 & 0 & 0 \\ 0 & 0 & 0 & 0 & 0 \\ 0 & 0 & 0.556 & 0.350 & -0.206 \\ 0 & 0 & 0.350 & 1.350 & 0 \\ 0 & 0 & -0.206 & 0 & 1.206 \end{bmatrix} \cdot \begin{bmatrix} 0 \\ 0 \\ x_3 \\ x_4 \\ x_5 \end{bmatrix} = \begin{bmatrix} 0 \\ 0 \\ 10 \\ -10 \\ 0 \end{bmatrix}.
$$

Reduction step 3:

$$
l_3 = \begin{bmatrix} 0 \\ 0 \\ 1 \\ 0.630 \\ -0.370 \end{bmatrix},
\begin{bmatrix} 0 & 0 & 0 & 0 & 0 \\ 0 & 0 & 0 & 0 & 0 \\ 0 & 0 & 0 & 0 & 0 \\ 0 & 0 & 0 & 1.130 & 0.130 \\ 0 & 0 & 0 & 0.130 & 1.130 \end{bmatrix} \cdot \begin{bmatrix} 0 \\ 0 \\ 0 \\ x_4 \\ x_5 \end{bmatrix} = \begin{bmatrix} 0 \\ 0 \\ 0 \\ -16.296 \\ 3.704 \end{bmatrix}.
$$

Reduction step 4:

$$
l_4 = \begin{bmatrix} 0 \\ 0 \\ 0 \\ 1 \\ 0.115 \end{bmatrix},
\begin{bmatrix} 0 & 0 & 0 & 0 & 0 \\ 0 & 0 & 0 & 0 & 0 \\ 0 & 0 & 0 & 0 & 0 \\ 0 & 0 & 0 & 0 & 0 \\ 0 & 0 & 0 & 0 & 1.115 \end{bmatrix} \cdot \begin{bmatrix} 0 \\ 0 \\ 0 \\ 0 \\ x_5 \end{bmatrix} = \begin{bmatrix} 0 \\ 0 \\ 0 \\ 0 \\ 5.574 \end{bmatrix}.
$$

◄

Based on the reduced system of equations, the unknowns can easily be determined by back substitution. The unknown x_n follows directly from the last reduction step. With x_n known, the unknown x_{n-1} is obtained from the second to last equation in the second to last reduction step, and so on. Finally, the unknown x_1 is obtained from the first equation of the original system of equations.

Example 1.5 Determine the unknowns of the system of equations in Example 1.4 by back substitution.

Solution:

$$
\begin{aligned}
x_5 &= & 5.574/1.115 &= & 5.0 \\
x_4 &= & (-0.130 \cdot 5.0 - 16.296)/1.130 &= & -15.0 \\
x_3 &= & (-0.350 \cdot (-15.0) + 0.206 \cdot 5.0 + 10)/0.556 &= & 29.3 \\
x_2 &= & (0.259 \cdot 29.3 - 0.259 \cdot 5.0)/1.259 &= & 5.0 \\
x_1 &= & (0.35 \cdot 5.0 + 29.3 + 0.35 \cdot 5.0)/1.350 &= & 24.3.
\end{aligned}
$$
◀

As can be shown, the system of equations (1.6) can also be written by the matrix product (see [2]):

$$\underline{L} \cdot \underline{D} \cdot \underline{L}^{\mathrm{T}} \cdot \underline{x} = \underline{b}. \tag{1.8}$$

The coefficients of the matrices $\underline{L}$ and $\underline{D}$ follow from the Gaussian solution procedure. For a 5×5 matrix, for example, they are given by

$$
\underline{L} =
\begin{bmatrix}
1 & 0 & 0 & 0 & 0 \\
l_{21} & 1 & 0 & 0 & 0 \\
l_{31} & l_{32} & 1 & 0 & 0 \\
l_{41} & l_{42} & l_{43} & 1 & 0 \\
l_{51} & l_{52} & l_{53} & l_{54} & 1
\end{bmatrix}
\tag{1.8a}
$$

$$
\underline{D} =
\begin{bmatrix}
a_{11} & 0 & 0 & 0 & 0 \\
0 & a_{22}^{(1)} & 0 & 0 & 0 \\
0 & 0 & a_{33}^{(2)} & 0 & 0 \\
0 & 0 & 0 & a_{44}^{(3)} & 0 \\
0 & 0 & 0 & 0 & a_{55}^{(4)}
\end{bmatrix}.
\tag{1.8b}
$$

Introducing the auxiliary vectors $\underline{y}$ and $\underline{c}$ as

$$\underline{c} = \underline{L}^{\mathrm{T}} \cdot \underline{x} \tag{1.9a}$$

$$\underline{y} = \underline{D} \cdot \underline{L}^{\mathrm{T}} \cdot \underline{x} = \underline{D} \cdot \underline{c} \tag{1.9b}$$

the following solution procedure is obtained as

Gauss' elimination method
1. Gauss decomposition of $\underline{A}$ in order to determine $\underline{L}$ and $\underline{D}$.
2. $\underline{L} \cdot \underline{y} = \underline{b}$ (forward substitution)
   $\underline{D} \cdot \underline{c} = \underline{y}$.
3. $\underline{L}^{\mathrm{T}} \cdot \underline{x} = \underline{c}$ (back substitution).

Since $\underline{L}$ is a triangular matrix and $\underline{D}$ a diagonal, the numerical computations can be performed easily.

Example 1.6 Determine the matrices $\underline{L}$ and $\underline{D}$ for the system of equations in Example 1.4 and solve the system of equations by forward and back substitution.

$$
\underline{L} = \begin{bmatrix} 1 & 0 & 0 & 0 & 0 \\ l_{21} & 1 & 0 & 0 & 0 \\ l_{31} & l_{32} & 1 & 0 & 0 \\ l_{41} & l_{42} & l_{43} & 1 & 0 \\ l_{51} & l_{52} & l_{53} & l_{54} & 1 \end{bmatrix} = \begin{bmatrix} 1 & 0 & 0 & 0 & 0 \\ -0.259 & 1 & 0 & 0 & 0 \\ -0.741 & -0.206 & 1 & 0 & 0 \\ 0 & 0 & 0.630 & 1 & 0 \\ -0.259 & 0.206 & -0.370 & 0.115 & 1 \end{bmatrix}
$$

$$
\underline{D} = \begin{bmatrix} a_{11} & 0 & 0 & 0 & 0 \\ 0 & a_{22}^{(1)} & 0 & 0 & 0 \\ 0 & 0 & a_{33}^{(2)} & 0 & 0 \\ 0 & 0 & 0 & a_{44}^{(3)} & 0 \\ 0 & 0 & 0 & 0 & a_{55}^{(4)} \end{bmatrix} = \begin{bmatrix} 1.350 & 0 & 0 & 0 & 0 \\ 0 & 1.259 & 0 & 0 & 0 \\ 0 & 0 & 0.556 & 0 & 0 \\ 0 & 0 & 0 & 1.130 & 0 \\ 0 & 0 & 0 & 0 & 1.115 \end{bmatrix}
$$

$$
\underline{A} = \underline{L} \cdot \underline{D} \cdot \underline{L}^{\mathrm{T}}.
$$

Forward substitution:

$$
\underline{L} \cdot \underline{y} = \underline{b}
$$

$$
\begin{bmatrix} 1 & 0 & 0 & 0 & 0 \\ -0.259 & 1 & 0 & 0 & 0 \\ -0.741 & -0.206 & 1 & 0 & 0 \\ 0 & 0 & 0.630 & 1 & 0 \\ -0.259 & 0.206 & -0.370 & 0.115 & 1 \end{bmatrix} \cdot \begin{bmatrix} y_1 \\ y_2 \\ y_3 \\ y_4 \\ y_5 \end{bmatrix} = \begin{bmatrix} 0 \\ 0 \\ 10 \\ -10 \\ 0 \end{bmatrix}
$$

$$
y_1 = 0 \quad y_2 = 0 \quad y_3 = 10.00 \quad y_4 = -16.30 \quad y_5 = 5.58
$$

$$\underline{D} \cdot \underline{c} = \underline{y}$$

$$\begin{bmatrix} 1.35 & 0 & 0 & 0 & 0 \\ 0 & 1.259 & 0 & 0 & 0 \\ 0 & 0 & 0.556 & 0 & 0 \\ 0 & 0 & 0 & 1.130 & 0 \\ 0 & 0 & 0 & 0 & 1.115 \end{bmatrix} \cdot \begin{bmatrix} c_1 \\ c_2 \\ c_3 \\ c_4 \\ c_5 \end{bmatrix} = \begin{bmatrix} 0 \\ 0 \\ 10 \\ -16.30 \\ 5.58 \end{bmatrix}$$

$$c_1 = 0 \quad c_2 = 0 \quad c_3 = 17.99 \quad c_4 = -14.43 \quad c_5 = 5.00.$$

Back substitution:

$$\underline{L}^{\mathrm{T}} \cdot \underline{x} = \underline{c}$$

$$\begin{bmatrix} 1 & -0.259 & -0.741 & 0 & -0.259 \\ 0 & 1 & -0.206 & 0 & 0.206 \\ 0 & 0 & 1 & 0.630 & -0.370 \\ 0 & 0 & 0 & 1 & 0.115 \\ 0 & 0 & 0 & 0 & 1 \end{bmatrix} \cdot \begin{bmatrix} x_1 \\ x_2 \\ x_3 \\ x_4 \\ x_5 \end{bmatrix} = \begin{bmatrix} 0 \\ 0 \\ 17.99 \\ -14.43 \\ 5.00 \end{bmatrix}$$

$$\begin{aligned} x_5 &= & = & \ 5.0 \\ x_4 &= & -14.43 - 0.115 \cdot 5.00 = & -15.0 \\ x_3 &= & 17.99 - 0.630 \cdot (-15.0) + 0.370 \cdot 5.00 = & \ 29.3 \\ x_2 &= & 0.206 \cdot 29.3 - 0.206 \cdot 5.00 = & \ 5.0 \\ x_1 &= 0.259 \cdot 5.00 + 0.741 \cdot 29.3 + 0.259 \cdot 5.0 = & & \ 24.3. \end{aligned}$$ ◀

When programming Gauss elimination, numerical values of the factors l_{ik} can be stored at the corresponding matrix entries at the end of a reduction step only. This drawback can be overcome by the complete symmetric decomposition according to Cholesky. Whereas Gauss elimination is applicable to any system of equations, Cholesky's method presumes symmetric, positive definite coefficient matrices. The systems of equations found in finite element analyses fulfill this requirement.

According to Cholesky, the matrix product $\underline{L} \cdot \underline{D} \cdot \underline{L}^{\mathrm{T}}$ can be written as

$$\underline{A} = \underline{L} \cdot \underline{D} \cdot \underline{L}^{\mathrm{T}} = \underline{L} \cdot \underline{D}^{1/2} \cdot \underline{D}^{1/2} \cdot \underline{L}^{\mathrm{T}} = \underline{L}^* \cdot \underline{L}^{*\mathrm{T}} \quad \text{with} \quad \underline{L}^* = \underline{L} \cdot \underline{D}^{1/2}.$$

With this substitution, the following alternative version of Gauss elimination is obtained as:

Cholesky method
1. Gauss decomposition of $\underline{A}$ and computation of $\underline{L}$, $\underline{D}$, and $\underline{L}^*$.
2. $\underline{L}^* \cdot \underline{c} = \underline{b}$ (forward substitution).
3. $\underline{L}^{*T} \cdot \underline{x} = \underline{c}$ (back substitution).

If the system of equations has to be solved for multiple right-hand sides $\underline{b}$, the time-consuming decomposition of the matrix $\underline{A}$ has to be done only once. This situation is typical in finite element analysis where the vector $\underline{b}$ contains the nodal loads of the system and the system of equations has to be solved for several load cases.

Example 1.7 Compute the Cholesky matrix $\underline{L}^*$ of the system of equations in Example 1.4, and solve the system of equations with the Cholesky method.

$$\underline{L} \cdot \underline{D} \cdot \underline{L}^T = \underline{L} \cdot \underline{D}^{1/2} \cdot \underline{D}^{1/2} \cdot \underline{L}^T = \underline{L}^* \cdot \underline{L}^{*T}$$

$$\underline{L}^* = \begin{bmatrix} 1 & 0 & 0 & 0 & 0 \\ l_{21} & 1 & 0 & 0 & 0 \\ l_{31} & l_{32} & 1 & 0 & 0 \\ l_{41} & l_{42} & l_{43} & 1 & 0 \\ l_{51} & l_{52} & l_{53} & l_{54} & 1 \end{bmatrix} \cdot \begin{bmatrix} \sqrt{a_{11}} & 0 & 0 & 0 & 0 \\ 0 & \sqrt{a_{22}^{(1)}} & 0 & 0 & 0 \\ 0 & 0 & \sqrt{a_{33}^{(2)}} & 0 & 0 \\ 0 & 0 & 0 & \sqrt{a_{44}^{(3)}} & 0 \\ 0 & 0 & 0 & 0 & \sqrt{a_{55}^{(4)}} \end{bmatrix}$$

$$\underline{L}^* = \begin{bmatrix} \sqrt{a_{11}} & 0 & 0 & 0 & 0 \\ \sqrt{a_{11}} \cdot l_{21} & \sqrt{a_{22}^{(1)}} & 0 & 0 & 0 \\ \sqrt{a_{11}} \cdot l_{31} & \sqrt{a_{22}^{(1)}} \cdot l_{32} & \sqrt{a_{33}^{(2)}} & 0 & 0 \\ \sqrt{a_{11}} \cdot l_{41} & \sqrt{a_{22}^{(1)}} \cdot l_{42} & \sqrt{a_{33}^{(2)}} \cdot l_{43} & \sqrt{a_{44}^{(3)}} & 0 \\ \sqrt{a_{11}} \cdot l_{51} & \sqrt{a_{22}^{(1)}} \cdot l_{52} & \sqrt{a_{33}^{(2)}} \cdot l_{53} & \sqrt{a_{44}^{(3)}} \cdot l_{54} & \sqrt{a_{55}^{(4)}} \end{bmatrix}$$

$$= \begin{bmatrix} 1.162 & 0 & 0 & 0 & 0 \\ -0.301 & 1.122 & 0 & 0 & 0 \\ -0.861 & -0.231 & 0.746 & 0 & 0 \\ 0 & 0 & 0.470 & 1.063 & 0 \\ -0.301 & 0.231 & -0.276 & 0.122 & 1.056 \end{bmatrix}$$

Forward substitution:

$$\underline{L}^* \cdot \underline{c} = \underline{b}$$

$$
\begin{bmatrix}
1.162 & 0 & 0 & 0 & 0 \\
-0.301 & 1.122 & 0 & 0 & 0 \\
-0.861 & -0.231 & 0.746 & 0 & 0 \\
0 & 0 & 0.470 & 1.063 & 0 \\
-0.301 & 0.231 & -0.276 & 0.122 & 1.056
\end{bmatrix}
\cdot
\begin{bmatrix}
c_1 \\ c_2 \\ c_3 \\ c_4 \\ c_5
\end{bmatrix}
=
\begin{bmatrix}
0 \\ 0 \\ 10 \\ -10 \\ 0
\end{bmatrix}
$$

$$c_1 = 0 \quad c_2 = 0 \quad c_3 = 13.41 \quad c_4 = -15.33 \quad c_5 = 5.28.$$

Back substitution:

$$\underline{L}^{\mathrm{T}} \cdot \underline{x} = \underline{c}$$

$$
\begin{bmatrix}
1.162 & -0.301 & -0.861 & 0 & -0.301 \\
0 & 1.122 & -0.231 & 0 & 0.231 \\
0 & 0 & 0.746 & 0.470 & -0.276 \\
0 & 0 & 0 & 1.063 & 0.122 \\
0 & 0 & 0 & 0 & 1.056
\end{bmatrix}
\cdot
\begin{bmatrix}
x_1 \\ x_2 \\ x_3 \\ x_4 \\ x_5
\end{bmatrix}
=
\begin{bmatrix}
0 \\ 0 \\ 13.41 \\ -15.33 \\ 5.28
\end{bmatrix}
$$

$$
\begin{aligned}
x_5 &= & 5.28/1.056 &= & 5.0 \\
x_4 &= & (-15.33 - 0.122 \cdot 5.0)/1.063 &= & -15.0 \\
x_3 &= & (13.41 - 0.470 \cdot (-15.0) + 0.276 \cdot 5.0)/0.746 &= & 29.3 \\
x_2 &= & (0.231 \cdot 29.3 - 0.231 \cdot 5.0)/1.122 &= & 5.0 \\
x_1 &= & (-0.301 \cdot 5.0 - 0.861 \cdot 29.3 - 0.301 \cdot 5.0)/1.162 &= & 24.3.
\end{aligned}
$$
◀

1.4.4   Norms and Condition Number

Sometimes it is useful or necessary to define a positive scalar number in order to measure the size of a vector or a matrix. The length of a two-dimensional vector, e.g. describing a displacement of a point, may be such a number. For general purposes, norms of vectors and matrices are defined.

There are several definitions of norms. All of them have to fulfill some mathematical conditions [3]. The absolute value norm is defined for vectors as

$$\|\underline{x}\|_1 = \sum_{i=1}^{n} |x_i| \quad \text{(Absolute value norm)}. \tag{1.10}$$

It is also denoted as 1-norm. Mostly for vectors, the so-called Euclidean norm (or 2-norm) and for matrices the Frobenius norm are used. They are defined for vectors as

$$\|\underline{x}\| = \sqrt{\sum_{i=1}^{n} x_i^2} \quad \text{(Euclidean norm)} \tag{1.10a}$$

and for matrices as

$$\|\underline{A}\| = \sqrt{\sum_{i=1}^{n} \sum_{k=1}^{m} a_{js}^2} \quad \text{(Frobenius norm)}. \tag{1.10b}$$

For two- and three-dimensional vectors in a plane or in space, respectively, the Euclidean norm corresponds to the its length.

Another norm often used is the infinity norm which is defined as for vectors as

$$\|\underline{x}\|_\infty = \max_i |x_i| \tag{1.10c}$$

and for matrices as

$$\|\underline{A}\|_\infty = \max_i \sum_{j=1}^{n} |a_{ij}|. \tag{1.10d}$$

The condition number of a matrix $\underline{A}$ is obtained as the product of two matrix norms as

$$\kappa(\underline{A}) = \|\underline{A}\| \cdot \|\underline{A}^{-1}\|. \tag{1.11}$$

It can be defined for any norm. The condition number describes the sensitivity of the solution of a linear system of equations to small variations of the coefficient matrix $\underline{A}$ as they may arise from numerical errors. If the condition number is very large, very small relative errors in the coefficient matrix may cause very large errors in the solution vector. Such systems of equations are to be avoided in finite element analyses (see Example 3.24). In iterative methods, a large condition number results in a wrong convergence behavior.

1.5   Eigenvalue Problems

1.5.1   General Eigenvalue Problem

Eigenvalue problems are special systems of equations typical in stability analyses and structural dynamics. The following two equations may be given as example:

$$(100 - \lambda \cdot 0.250) \cdot x_1 \quad -40 \cdot x_2 \qquad\qquad = 0$$
$$-40 \cdot x_1 \qquad\qquad (40 - \lambda \cdot 0.125) \cdot x_2 \; = 0$$

or

$$\left(\begin{bmatrix} 100 & -40 \\ -40 & 40 \end{bmatrix} - \lambda \cdot \begin{bmatrix} 0.250 & 0 \\ 0 & 0.125 \end{bmatrix} \right) \cdot \begin{bmatrix} x_1 \\ x_2 \end{bmatrix} = \begin{bmatrix} 0 \\ 0 \end{bmatrix}.$$

They describe a homogeneous linear system of equations with the unknowns x_1 and x_2 and the parameter λ. As a homogeneous system of equations, it posesses either no solution except the trivial solution $x_1 = x_2 = 0$ or an infinite number of solutions. Now the free parameter λ is chosen such that the system of equations possesses other solutions than $x_1 = x_2 = 0$. For this, the condition for the solvability of the homogeneous system of equations, i.e. that its determinant is equal to zero, is written as

$$\det \left(\begin{bmatrix} 100 & -40 \\ -40 & 40 \end{bmatrix} - \lambda \cdot \begin{bmatrix} 0.250 & 0 \\ 0 & 0.125 \end{bmatrix} \right)$$
$$= \det \begin{bmatrix} 100 - \lambda \cdot 0.250 & -40 \\ -40 & 40 - \lambda \cdot 0.125 \end{bmatrix} = 0.$$

This condition leads to the so-called "characteristic polynomial"

$$\det \begin{bmatrix} 100 - \lambda \cdot 0.250 & -40 \\ -40 & 40 - \lambda \cdot 0.125 \end{bmatrix}$$
$$= ((100 - \lambda \cdot 0.250) \cdot (40 - \lambda \cdot 0.125) - (-40) \cdot (-40))$$
$$= 2400 - 22.5 \cdot \lambda + 0.03125 \cdot \lambda^2.$$

The roots of the characteristic polynomial are obtained to be

$$\lambda_1 = 130.21 \quad \lambda_2 = 589.78.$$

They are called the *eigenvalues* of the eigenvalue problem. With the two eigenvalues λ_1 and λ_2, two solutions of the homogeneous system of equations are obtained. Introducing the first eigenvalue into the first (for example) equation of the system of equations gives

$$(100 - 130.21 \cdot 0.250) \cdot x_1 - 40 \cdot x_2 = 0 \quad \text{or} \quad x_1 = 0.593 \cdot x_2.$$

For the first eigenvalue $\lambda_1 = 130.21$, all pairs of values (x_1, x_2) fulfilling the equation above are solutions of the system of equations. This holds, for example, for the vector

$$\begin{bmatrix} x_1 \\ x_2 \end{bmatrix} = \begin{bmatrix} 0.593 \\ 1.000 \end{bmatrix}.$$

But, due to the homogeneity of the system of equations, all vectors obtained by multiplying the vector with a constant factor are solutions of the eigenvalue problem too.

Similarily, with the second eigenvalue $\lambda_2 = 589.78$, one obtains

$$(100 - 589.78 \cdot 0.250) \cdot x_1 - 40 \cdot x_2 = 0 \quad \text{or} \quad x_1 = -0.843 \cdot x_2$$

and

$$\begin{bmatrix} x_1 \\ x_2 \end{bmatrix} = \begin{bmatrix} -0.843 \\ 1.000 \end{bmatrix}.$$

The vectors are denoted as the two *eigenvectors* of the eigenvalue problem. They are determined except for a multiplication factor. Each eigenvector is associated with an eigenvalue λ_i.

In general, the following definition applies:

Definition

Mathematical problems described by a homogeneous system of equations as

$$\left(\underline{A} - \lambda_i \cdot \underline{B} \right) \cdot \underline{x}_i = 0 \tag{1.12}$$

are called general eigenvalue problems. $\underline{A}$ and $\underline{B}$ are square matrices.

The values λ_i are called eigenvalues of the pair of matrices $\underline{A}, \underline{B}$. The solution vectors $x_i \neq \underline{0}$ associated with the eigenvalues λ_i are called eigenvectors.

This formulation of an eigenvalue problem is called the general eigenvalue problem.

The general eigenvalue problem can be transformed, in general, into the special eigenvalue problem

$$\underline{C} \cdot \underline{x}_i = \lambda_i \cdot \underline{x}_i$$

or

$$\left(\underline{C} - \lambda_i \cdot \underline{I}\right) \cdot \underline{x}_i = \underline{0}. \tag{1.12a}$$

Multiplying (1.12) with $\underline{B}^{-1}$ gives

$$\left(\underline{B}^{-1} \cdot \underline{A} - \lambda_i \cdot \underline{B}^{-1} \cdot \underline{B}\right) \cdot \underline{x}_i = 0$$

or

$$\left(\underline{B}^{-1} \cdot \underline{A} - \lambda_i \cdot \underline{I}\right) \cdot \underline{x}_i = 0.$$

Hence, the standard eigenvalue problem is related to the general eigenvalue problem by

$$\underline{C} = \underline{B}^{-1} \cdot \underline{A}. \tag{1.12b}$$

The eigenvalues are generally obtained by the condition that the determinant of the coefficient matrix equals zero, i.e. for the general eigenvalue problem

$$\det\left(\underline{A} - \lambda_i \cdot \underline{B}\right) = 0. \tag{1.13}$$

The solutions of this equation are the roots of the characteristic polynomial. In the example given above, a 2×2 matrix has been considered whose determinant leads to a second-order polynomial. In general, two $n \times n$ matrices $\underline{A}$ and $\underline{B}$ lead to a characteristic polynomial of order n with exactly n roots. Hence, an eigenvalue problem with n equations posesses n eigenvalues and n corresponding eigenvectors. The eigenvalues are ordered according to their size as

$$\lambda_1 \leq \lambda_2 \leq \cdots \leq \lambda_i \leq \lambda_{i+1} \ldots \leq \lambda_n. \tag{1.13a}$$

In structural dynamics and for stability analyses, often, only the lowest eigenvalues and eigenvectors are of practical interest.

The eigenvalues and eigenvectors of matrices $\underline{A}$ and $\underline{B}$ with arbitrary coefficients can be complex numbers. However, for real, symmetric matrices appearing in finite element analyses, the following rules apply [1]:

Properties of eigenvalues and eigenvectors of symmetric matrices

(a)   All eigenvalues and eigenvectors are real if $\underline{A}$ and $\underline{B}$ are real, symmetric matrices.

(b)   All eigenvalues are positive if $\underline{A}$ and $\underline{B}$ are real, positive-definite matrices.

(c) Eigenvectors of symmetric matrices $\underline{A}$ and $\underline{B}$ corresponding to different eigenvalues are orthogonal to each other. They fulfill the orthogonality conditions

$$\begin{aligned} \underline{x}_i^T \cdot \underline{A} \cdot \underline{x}_j = 0, \quad \underline{x}_i^T \cdot \underline{B} \cdot \underline{x}_j = 0 \quad \text{for } i \neq j \\ \underline{x}_i^T \cdot \underline{A} \cdot \underline{x}_j \neq 0, \quad \underline{x}_i^T \cdot \underline{B} \cdot \underline{x}_j \neq 0 \quad \text{for } i = j. \end{aligned} \tag{1.14}$$

Because of their importance in structural dynamics, the proof of the orthogonality conditions will be discussed here. λ_i and λ_j are presumed to be two distinct eigenvalues of two symmetric matrices $\underline{A}$ and $\underline{B}$, and $\underline{x}_i$ and $\underline{x}_j$ are the corresponding eigenvectors. To begin with, the eigenvalue equations of both eigenvalues are multiplied from the left side by $\underline{x}_j^T$ and $\underline{x}_i^T$, respectively:

$$\begin{aligned} \underline{x}_j^T \cdot \underline{A} \cdot \underline{x}_i - \lambda_i \cdot \underline{x}_j^T \cdot \underline{B} \cdot \underline{x}_i = 0 \\ \underline{x}_i^T \cdot \underline{A} \cdot \underline{x}_j - \lambda_j \cdot \underline{x}_i^T \cdot \underline{B} \cdot \underline{x}_j = 0. \end{aligned}$$

For symmetric matrices $\underline{A}$ and $\underline{B}$ according to the calculation rules given in Sect. 1.3, one gets

$$\underline{x}_j^T \cdot \underline{A} \cdot \underline{x}_i = \underline{x}_i^T \cdot \underline{A} \cdot \underline{x}_j \qquad \underline{x}_j^T \cdot \underline{B} \cdot \underline{x}_i = \underline{x}_i^T \cdot \underline{B} \cdot \underline{x}_j.$$

Substraction of the two scalar equations gives

$$\underline{x}_i^T \cdot \underline{A} \cdot \underline{x}_j - \underline{x}_i^T \cdot \underline{A} \cdot \underline{x}_j - (\lambda_i - \lambda_j) \cdot \underline{x}_i^T \cdot \underline{B} \cdot \underline{x}_j = 0$$

or

$$(\lambda_i - \lambda_j) \cdot \underline{x}_i^T \cdot \underline{B} \cdot \underline{x}_j = 0.$$

Since according to the premise $\lambda_i \neq \lambda_j$, the second term of the product must be zero, i.e. $\underline{x}_i^T \cdot \underline{B} \cdot \underline{x}_j = 0$. From the equation $\underline{x}_i^T \cdot \underline{A} \cdot \underline{x}_j - \lambda_j \cdot \underline{x}_i^T \cdot \underline{B} \cdot \underline{x}_j = 0$, it follows that $\underline{x}_i^T \cdot \underline{A} \cdot \underline{x}_j = 0$. Herewith, the orthogonality condition of the vectors $\underline{x}_i$ and $\underline{x}_j$ is proven.

Eigenvectors are defined except for a "scaling" factor, i.e. if x_i is an eigenvector corresponding to the eigenvalue λ_i, the vector $\alpha \cdot \underline{x}_i$ is also an eigenvector. The scaling factor can be chosen in order to fulfill some scaling requirements. For example, the maximum absolute element of a vector can be chosen to be 1 in order to improve the comparability of the eigenvectors. Often, the eigenvectors are normalized to the $\underline{B}$ matrix as

$$\underline{x}_i^T \cdot \underline{B} \cdot \underline{x}_i = 1. \tag{1.14a}$$

With $\underline{x}_i^T \cdot \underline{A} \cdot \underline{x}_i - \lambda_i \cdot \underline{x}_i^T \cdot \underline{B} \cdot \underline{x}_i = 0$ for the matrix $\underline{A}$, the relationships

$$\underline{x}_i^T \cdot \underline{A} \cdot \underline{x}_i = \lambda_i \qquad (1.14b)$$

hold.

Example 1.8 The eigenvectors of the introductory example normalized to the absolute maximum value 1 as

$$\underline{x}_1 = \begin{bmatrix} 0.593 \\ 1.000 \end{bmatrix}, \quad \underline{x}_2 = \begin{bmatrix} -0.843 \\ 1.000 \end{bmatrix}$$

are to be normalized on the matrix $\underline{B}$. In addition, the orthogonality of both eigenvectors has to be checked.

The representation of the eigenvectors normalized to the matrix $\underline{B}$ is obtained with

$$\begin{bmatrix} 0.593 & 1.000 \end{bmatrix} \cdot \begin{bmatrix} 0.250 & 0 \\ 0 & 0.125 \end{bmatrix} \cdot \begin{bmatrix} 0.593 \\ 1.000 \end{bmatrix} = 0.213$$

and

$$\begin{bmatrix} -0.843 & 1.000 \end{bmatrix} \cdot \begin{bmatrix} 0.250 & 0 \\ 0 & 0.125 \end{bmatrix} \cdot \begin{bmatrix} -0.843 \\ 1.000 \end{bmatrix} = 0.302$$

after multiplication with $1/\sqrt{0.213} = 2.167$ and $1/\sqrt{0.302} = 1.819$, respectively, as

$$\underline{x}_1 = \begin{bmatrix} 1.285 \\ 2.167 \end{bmatrix}, \quad \underline{x}_2 = \begin{bmatrix} -1.532 \\ 1.818 \end{bmatrix}.$$

These vectors fulfill Eqs. (1.14a) and (1.14b) as can be shown easily:

$$\underline{x}_1^T \cdot \underline{B} \cdot \underline{x}_1 = \begin{bmatrix} 1.285 & 2.167 \end{bmatrix} \cdot \begin{bmatrix} 0.250 & 0 \\ 0 & 0.125 \end{bmatrix} \cdot \begin{bmatrix} 1.285 \\ 2.167 \end{bmatrix} = 1.000$$

$$\underline{x}_2^T \cdot \underline{B} \cdot \underline{x}_2 = \begin{bmatrix} -1.532 & 1.818 \end{bmatrix} \cdot \begin{bmatrix} 0.250 & 0 \\ 0 & 0.125 \end{bmatrix} \cdot \begin{bmatrix} -1.532 \\ 1.818 \end{bmatrix} = 1.000$$

$$\underline{x}_1^T \cdot \underline{A} \cdot \underline{x}_1 = \begin{bmatrix} 1.285 & 2.167 \end{bmatrix} \cdot \begin{bmatrix} 100 & -40 \\ -40 & 40 \end{bmatrix} \cdot \begin{bmatrix} 1.285 \\ 2.167 \end{bmatrix} = 130.19$$

$$\underline{x}_2^T \cdot \underline{A} \cdot \underline{x}_2 = \begin{bmatrix} -1.532 & 1.818 \end{bmatrix} \cdot \begin{bmatrix} 100 & -40 \\ -40 & 40 \end{bmatrix} \cdot \begin{bmatrix} -1.532 \\ 1.818 \end{bmatrix} = 589.72$$

Furthermore with (1.14), one obtains

$$\underline{x}_1^T \cdot \underline{B} \cdot \underline{x}_2 = \begin{bmatrix} 1.285 & 2.167 \end{bmatrix} \cdot \begin{bmatrix} 0.250 & 0 \\ 0 & 0.125 \end{bmatrix} \cdot \begin{bmatrix} -1.532 \\ 1.818 \end{bmatrix} = 0$$

$$\underline{x}_1^T \cdot \underline{A} \cdot \underline{x}_2 = \begin{bmatrix} 1.285 & 2.167 \end{bmatrix} \cdot \begin{bmatrix} 100 & -40 \\ -40 & 40 \end{bmatrix} \cdot \begin{bmatrix} -1.532 \\ 1.818 \end{bmatrix} = 0$$

Herewith, it is shown that the orthogonality conditions are fulfilled. ◀

An eigenvalue can be single or multiple. Multiple eigenvectors occur if the characteristic polynomial has multiple (repeated) roots. In the case of multiple roots, the eigenvalues are not separated clearly, i.e. several succeeding values of λ_i in (1.13, 1.13a) are equal. An eigenvalue with multiplicity m also corresponds to m different eigenvectors. However, the m eigenvectors are not unique as in the case of single eigenvalues. Moreover, linear combinations of m eigenvectors fulfilling the orthogonality condition (1.14) are also eigenvectors corresponding to the eigenvalue λ_i with multiplicity m.

Example 1.9 For the eigenvalue problem

$$\left(\begin{bmatrix} 40 & 0 & 0 \\ 0 & 40 & 0 \\ 0 & 0 & 100 \end{bmatrix} - \lambda \cdot \begin{bmatrix} 0.2 & 0 & 0 \\ 0 & 0.2 & 0 \\ 0 & 0 & 0.2 \end{bmatrix} \right) \cdot \begin{bmatrix} x_1 \\ x_2 \\ x_3 \end{bmatrix} = \begin{bmatrix} 0 \\ 0 \\ 0 \end{bmatrix}$$

it has to be shown that multiple eigenvalues exist and that the eigenvectors are not unique.

In this example, the eigenvalues can be found easily since $\underline{A}$ and $\underline{B}$ are diagonal matrices. Writing the three equations of the matrix equation as three scalar equations gives

$$(40 - \lambda \cdot 0.2) \cdot x_1 = 0 \quad \text{and with } x_1 \neq 0: \lambda_1 = 200$$
$$(40 - \lambda \cdot 0.2) \cdot x_2 = 0 \quad \text{and with } x_2 \neq 0: \lambda_2 = 200$$
$$(100 - \lambda \cdot 0.2) \cdot x_3 = 0 \quad \text{and with } x_3 \neq 0: \lambda_3 = 500.$$

Hence, the first eigenvalue is double.

In order to determine the eigenvectors, the eigenvalues are inserted into the system of equations. With the third eigenvalue $\lambda_3 = 500$, one obtains

$$\left(\begin{bmatrix} 40 & 0 & 0 \\ 0 & 40 & 0 \\ 0 & 0 & 100 \end{bmatrix} - 500 \cdot \begin{bmatrix} 0.2 & 0 & 0 \\ 0 & 0.2 & 0 \\ 0 & 0 & 0.2 \end{bmatrix}\right) \cdot \begin{bmatrix} x_1 \\ x_2 \\ x_3 \end{bmatrix} = \begin{bmatrix} 0 \\ 0 \\ 0 \end{bmatrix}$$

$$\rightarrow \underline{x}_3 = \begin{bmatrix} x_1 \\ x_2 \\ x_3 \end{bmatrix} = \begin{bmatrix} 0 \\ 0 \\ 1 \end{bmatrix}.$$

Apart from a constant multiplication factor this eigenvector is unique.

Introducing the first or second eigenvalue $\lambda_1 = \lambda_2 = 200$ into the system of equations gives

$$\left(\begin{bmatrix} 40 & 0 & 0 \\ 0 & 40 & 0 \\ 0 & 0 & 100 \end{bmatrix} - 200 \cdot \begin{bmatrix} 0.2 & 0 & 0 \\ 0 & 0.2 & 0 \\ 0 & 0 & 0.2 \end{bmatrix}\right) \cdot \begin{bmatrix} x_1 \\ x_2 \\ x_3 \end{bmatrix} = \begin{bmatrix} 0 \\ 0 \\ 0 \end{bmatrix}$$

$$\rightarrow \underline{x}_{1,2} = \begin{bmatrix} x_1 \\ x_2 \\ x_3 \end{bmatrix} = \begin{bmatrix} x_1 \\ x_2 \\ 0 \end{bmatrix}.$$

The vectors $\underline{x}_1$ and $\underline{x}_2$ are not unique. However, they must be chosen such that the orthogonality conditions of the eigenvectors $\underline{x}_1$ and $\underline{x}_2$ are fulfilled. This condition is satisfied, for example, by the vectors

$$\underline{x}_1 = \begin{bmatrix} 1 \\ 0 \\ 0 \end{bmatrix} \text{ and } \underline{x}_2 = \begin{bmatrix} 0 \\ 1 \\ 0 \end{bmatrix} \text{ with}$$

$$\underline{x}_1^T \cdot \underline{B} \cdot \underline{x}_2 = \begin{bmatrix} 1 & 0 & 0 \end{bmatrix} \cdot \begin{bmatrix} 0.2 & 0 & 0 \\ 0 & 0.2 & 0 \\ 0 & 0 & 0.2 \end{bmatrix} \cdot \begin{bmatrix} 0 \\ 1 \\ 0 \end{bmatrix} = 0.$$

However, there are other pairs of vectors as

$$\underline{x}_1 = \begin{bmatrix} 1 \\ 1 \\ 0 \end{bmatrix} \text{ and } \underline{x}_2 = \begin{bmatrix} 1 \\ -1 \\ 0 \end{bmatrix}, \text{ or}$$

$$\underline{x}_1 = \begin{bmatrix} 1 \\ 1 \\ 0 \end{bmatrix} \text{ and } \underline{x}_2 = \begin{bmatrix} -1 \\ 1 \\ 0 \end{bmatrix},$$

which are possible solutions of the eigenvalue problem. They are obtained as linear combinations of the eigenvectors given above and fulfill the orthogonality conditions (1.14). ◄

1.5.2 Numerical Solution of Eigenvalue Problems

The computation of eigenvalues and eigenvectors of large matrices typical for finite element analyses is a demanding and (computer) time-consuming task. Some mathematical methods have been established for practical application. Basically, there are vector iteration methods and transformation methods. In addition, methods have been developed based on the determination of the roots of the characteristic polynomial, and methods which are a combination of the methods mentioned above. Essentially, all methods for the computation of eigenvalues are iterative since their aim is the determination of the roots of a polynomial. In the following, a short overview of the methods used in finite element programs is given. The *inverse iteration* is a simple numerical method being presented in the following. It is the basis for the development of more sophisticated methods used in practice. For a more detailed discussion of numerical solution methods of eigenvalue problems, see [2–4].

The transformation methods are based on the orthogonality conditions (1.14a) and (1.14b) which can be written including all eigenvectors as

$$\underline{X}^{\mathrm{T}} \cdot \underline{B} \cdot \underline{X} = \underline{I} \tag{1.15a}$$

$$\underline{X}^{\mathrm{T}} \cdot \underline{A} \cdot \underline{X} = \underline{\Lambda} \tag{1.15b}$$

with

$$\underline{X} = [\underline{x}_1 \ \underline{x}_2 \ldots \underline{x}_n], \quad \underline{\Lambda} = \begin{bmatrix} \lambda_1 & 0 & \ldots & 0 \\ 0 & \lambda_2 & \ldots & 0 \\ \ldots & \ldots & \ldots & \ldots \\ 0 & 0 & \ldots & \lambda_n \end{bmatrix}, \quad \underline{I} = \begin{bmatrix} 1 & 0 & \ldots & 0 \\ 0 & 1 & \ldots & 0 \\ \ldots & \ldots & \ldots & \ldots \\ 0 & 0 & \ldots & 1 \end{bmatrix}. \tag{1.15c}$$

According to (1.15b), the diagonal matrix $\underline{\Lambda}$ of the eigenvalues can be understood as the matrix which is obtained by the transformation of the matrix $\underline{A}$ with the matrix $\underline{X}$ of the eigenvectors observing (1.15a). Herewith, all eigenvalues and eigenvectors are obtained simultaneously. Transformation methods are the basis of the *Jacobi method*, the *Householder triangularization*, and the *QR algorithm*.

Vector iteration methods start with Eq. (1.12) of the eigenvalue problem. It can be written as

$$\underline{A} \cdot \underline{x}_i = \lambda_i \cdot \underline{B} \cdot \underline{x}_i \tag{1.16}$$

or

$$\underline{F} = \lambda_i \cdot \underline{B} \cdot \underline{x}_i \tag{1.16a}$$

$$\underline{A} \cdot \underline{x}_i = \underline{F}. \tag{1.16b}$$

Equation (1.16b) can be understood as a linear system of equations to determine the unknown vector $\underline{x}_i$ with a vector $\underline{F}$ on the right-hand side. To specify $\underline{F}$ according to (1.16a), as a start, the vector $\underline{x}_i$ is estimated and λ_i is assumed as 1, for example. In the next iteration steps, $\underline{F}$ is computed with the vector $\underline{x}_i$ obtained from (1.16b) in the last step. Suchlike, vector iteration methods determine an eigenvector iteratively by solving repeatedly linear systems of equations.

The Rayleigh–Ritz method, inverse iteration, the forward iteration method, and the Lanczos method are vector iteration methods. They are very effective if only a few eigenvalues and eigenvectors are searched for. Vector iteration methods are also suited for sparse matrices characteristic of finite element models.

As an example for a simple vector iteration method, inverse iteration is explained in the following. The matrices $\underline{A}$ and $\underline{B}$ are assumed to be symmetric and $\underline{A}$ should be regular. According to (1.16), the following iteration rule is obtained:

$$\underline{A} \cdot \hat{\underline{x}}_i^{(k+1)} = \underline{B} \cdot \underline{x}_i^{(k)}. \tag{1.17}$$

The resulting vector will be normalized to $\underline{B}$ as

$$\underline{x}_i^{(k+1)} = \frac{1}{\sqrt{\hat{\underline{x}}_i^{(k+1)} \cdot \underline{B} \cdot \hat{\underline{x}}_i^{(k+1)}}} \cdot \hat{\underline{x}}_i^{(k+1)}. \tag{1.17a}$$

The starting vector $\hat{\underline{x}}_i^{(1)}$ can be chosen arbitrarily. However, it should not be orthogonal in the sense of (1.14) to the eigenvector $\underline{x}_i$ to be determined. For example, a vector fully occupied with ones can be taken as a starting vector. The solution of the linear system of Eq. (1.17) gives the vector $\hat{\underline{x}}_i^{(k+1)}$. After normalization with (1.17a), the next approximation $\underline{x}_i^{(k+1)}$ of the eigenvector is obtained. It can be shown that the vectors $\underline{x}_i^{(k)}$ converge to the vector corresponding to the smallest eigenvalue.

The iteration requires the repeated solution of the linear system of Eq. (1.17). Therefore, to begin with, a triangularization of $\underline{A}$ according to Gauss with $\underline{A} = \underline{L} \cdot \underline{D} \cdot \underline{L}^{\mathrm{T}}$ is recommended. With the lower triangular matrix $\underline{L}$, the system of equations $\underline{A} \cdot \underline{x} = \underline{b}$ can be solved very efficiently in two steps, that is, by forward substitution with $\underline{L} \cdot \underline{y} = \underline{b}$, $\underline{D} \cdot \underline{c} = \underline{y}$, followed by back substitution with $\underline{L}^{\mathrm{T}} \cdot \underline{x} = \underline{c}$ (see Sect. 1.4.3).

The eigenvalue is determined by means of the Rayleigh quotient. Multiplying the eigenvalue Eq. (1.12) from left with $\underline{x}_i^T$

$$\underline{x}_i^T \cdot \underline{A} \cdot \underline{x}_i - \lambda_i \cdot \underline{x}_i^T \cdot \underline{B} \cdot \underline{x}_i = 0$$

and solving the equation for λ_i, one obtains

$$\lambda_i = \frac{\underline{x}_i^T \cdot \underline{A} \cdot \underline{x}_i}{\underline{x}_i^T \cdot \underline{B} \cdot \underline{x}_i}. \tag{1.18}$$

With the iteration vector $\underline{\hat{x}}_i^{(k)}$ of the kth iteration step, the eigenvalue is

$$\lambda_i^{(k)} = \frac{\left(\underline{\hat{x}}_i^{(k)}\right)^T \cdot \underline{A} \cdot \underline{\hat{x}}_i^{(k)}}{\left(\underline{\hat{x}}_i^{k}\right)^T \cdot \underline{B} \cdot \underline{\hat{x}}_i^{(k)}}. \tag{1.18a}$$

The basic algorithm of the inverse iteration can be summed up as (see [2]):

Inverse iteration

Start:
1. Gauss decomposition of $\underline{A}$ in order to determine $\underline{L}$ and $\underline{D}$.
2. Selection of a starting $\underline{\hat{x}}_i^{(1)} \neq \underline{0}$.
3. Computation of the starting vector $\underline{b} : \underline{b} = \underline{A} \cdot \underline{\hat{x}}_i^{(1)}$.

Iteration steps $(k = 1, 2, \ldots)$:
1. Rayleigh quotient:
   $$\underline{f} = \underline{B} \cdot \underline{\hat{x}}^{(k)}, \ N = \left(\underline{\hat{x}}^{(k)}\right)^T \cdot \underline{f}, \ Z = \left(\underline{\hat{x}}^{(k)}\right)^T \cdot \underline{b}, \lambda^{(k)} = Z/N;$$
2. Normalization: $\underline{x}^{(k)} = 1/\sqrt{N} \cdot \underline{\hat{x}}^{(k)}$;
3. Solution of the system of equations for $\underline{\hat{x}}^{(k+1)}$ with Gaussian elimination:
   Computation of the right side: $\underline{b} = 1/\sqrt{N} \cdot \underline{f}$.
   Forward substitution: $\underline{L} \cdot \underline{y} = \underline{b} \rightarrow \underline{y}$ and $\underline{D} \cdot \underline{c} = \underline{y} \rightarrow \underline{c}$.
   Back substitution: $\underline{L}^T \cdot \underline{\hat{x}}^{(k+1)} = \underline{c} \rightarrow \underline{\hat{x}}^{(k+1)}$.

Example 1.10 The smallest eigenvalue and the corresponding eigenvector are to be determined for the eigenvalue problem

$$
\left(
10^8 \cdot
\begin{bmatrix}
3.540 & -0.213 & -1.681 & 0 & 0 \\
-0.213 & 0.284 & 0 & -0.142 & 0.213 \\
-1.681 & 0 & 4.639 & -0.213 & -1.681 \\
0 & -0.142 & -0.213 & 0.142 & -0.213 \\
0 & -0.213 & -1.681 & -0.213 & 2.312
\end{bmatrix}
\right.
$$

$$
\left.
-\lambda \cdot
\begin{bmatrix}
1464 & 0 & 0 & 0 & 0 \\
0 & 318 & 0 & 0 & 0 \\
0 & 0 & 1606 & 0 & 0 \\
0 & 0 & 0 & 276 & 0 \\
0 & 0 & 0 & 0 & 1464
\end{bmatrix}
\right)
\cdot
\begin{bmatrix}
x_1 \\ x_2 \\ x_3 \\ x_4 \\ x_5
\end{bmatrix}
=
\begin{bmatrix}
0 \\ 0 \\ 0 \\ 0 \\ 0
\end{bmatrix}.
$$

To start with, the Gauss decomposition of the coefficient matrix $\underline{A}$ is done:

$$
\underline{L} =
\begin{bmatrix}
1 & 0 & 0 & 0 & 0 \\
-0.060 & 1 & 0 & 0 & 0 \\
-0.475 & -0.373 & 1 & 0 & 0 \\
0 & -0.524 & -0.070 & 1 & 0 \\
0 & 0.785 & -0.421 & -4.352 & 1
\end{bmatrix}
$$

$$
\underline{D} = 10^8 \cdot
\begin{bmatrix}
3.540 & 0 & 0 & 0 & 0 \\
0 & 0.271 & 0 & 0 & 0 \\
0 & 0 & 3.804 & 0 & 0 \\
0 & 0 & 0 & 0.049 & 0 \\
0 & 0 & 0 & 0 & 0.549
\end{bmatrix}
$$

The starting vector is chosen with all elements being ones. With it the vector $\underline{b}$ is determined as

$$
\underline{x}^{(1)} =
\begin{bmatrix}
1 \\ 1 \\ 1 \\ 1 \\ 1
\end{bmatrix},
\qquad
\underline{b} = \underline{A} \cdot \underline{x}^{(1)} = 10^8 \cdot
\begin{bmatrix}
1.646 \\ 0.142 \\ 1.065 \\ -0.426 \\ 0.639
\end{bmatrix}.
$$

The first iteration step will be shown in detail:

Rayleigh quotient:

$$\underline{f} = \underline{B} \cdot \underline{x}^{(1)} = \begin{bmatrix} 1464 \\ 318 \\ 1606 \\ 276 \\ 1464 \end{bmatrix}, \quad N = \left(\underline{x}^{(k)}\right)^{\mathrm{T}} \cdot \underline{f} = 5128,$$

$$Z = \left(\underline{x}^{(1)}\right)^{\mathrm{T}} \cdot \underline{b} = 3.066 \cdot 10^8, \quad \lambda^{(1)} = \frac{Z}{N} = 59790.$$

Normalization:

$$\underline{\hat{x}}^{(1)} = \frac{1}{\sqrt{N}} \cdot \underline{x}^{(1)} = \begin{bmatrix} 0.01396 \\ 0.01396 \\ 0.01396 \\ 0.01396 \\ 0.01396 \end{bmatrix}.$$

Solution of the system of equations:

$$\underline{b} = \underline{f} \cdot \frac{1}{\sqrt{N}} = \begin{bmatrix} 20.444 \\ 4.441 \\ 22.427 \\ 3.854 \\ 20.444 \end{bmatrix}.$$

$\underline{L} \cdot \underline{y} = \underline{b} \to \underline{y}$ and $\underline{D} \cdot \underline{c} = \underline{y} \to \underline{c}$ give

$$\underline{y} = \begin{bmatrix} 20.44 \\ 5.67 \\ 34.25 \\ 9.22 \\ 70.55 \end{bmatrix} \qquad \underline{c} = 10^{-7} \cdot \begin{bmatrix} 0.578 \\ 2.091 \\ 0.900 \\ 18.819 \\ 12.85 \end{bmatrix}.$$

The iteration vector $\underline{\hat{x}}^{(2)}$ is obtained (without normalization) from $\underline{L}^{\mathrm{T}} \cdot \underline{\hat{x}}^{(2)} = \underline{c}$ as

$$\underline{\hat{x}}^{(2)} = 10^{-6} \cdot \begin{bmatrix} 0.818 \\ 3.540 \\ 1.152 \\ 7.465 \\ 1.284 \end{bmatrix}.$$

The eigenvectors and eigenvalues in the following iteration steps are computed as

Eigenvalues and normalized eigenvectors ($\times 10^2$)	Iteration step k				
	1	2	3	4	5
$\lambda^{(k)}$	59790	4552	4343	4342	4342
$x_1^{(k)}$	1.396	0.518	0.467	0.465	0.465
$x_2^{(k)}$	1.396	2.244	2.323	2.330	2.331
$x_3^{(k)}$	1.396	0.730	0.669	0.666	0.666
$x_4^{(k)}$	1.396	4.732	4.834	4.837	4.837
$x_5^{(k)}$	1.396	0.814	0.738	0.733	0.733

Hence, the smallest eigenvalue and the corresponding eigenvector (rounded to 2 decimals) are

$$\lambda_1 = 4.342 \cdot 10^3, \quad \underline{x}_1 = 10^{-2} \cdot \begin{bmatrix} 0.47 \\ 2.33 \\ 0.67 \\ 4.84 \\ 0.73 \end{bmatrix}.$$

◀

It can be shown that the Rayleigh quotient converges to the eigenvalue with one order higher than the eigenvector. This means that the Rayleigh quotient gives the eigenvalue with about double (precise) decimals compared to the approximated eigenvector [2].

Inverse iteration also allows the computation of higher eigenvectors and eigenvalues. However, measures must be taken to prevent the convergence of the eigenvector toward the eigenvector of the smallest eigenvalue. This can be achieved by extracting the eigenvectors of the smaller eigenvalues from the iteration vector. For this, the starting vector for the iteration is chosen orthogonal to $\underline{B}$ according to (1.14). After the first iteration and in all further iteration steps, the iteration vectors are to be "purified" from components of the smaller eigenvectors which may arise from numerical rounding errors. The method is also denoted as *Gram–Schmidt orthogonalization*.

Assuming that the eigenvectors for the eigenvalues 1 to $(m - 1)$ are already determined and the iteration has to be done for the eigenvalue m. The extraction of the eigenvectors $\underline{x}_1, \underline{x}_2, \ldots, \underline{x}_{m-1}$ from the approximate vector $\tilde{\underline{x}}^{(k)}$ of the kth iteration step is obtained as ([2])

$$\hat{\underline{x}}^{(k)} = \tilde{\underline{x}}^{(k)} - \sum_{i=1}^{m-1} \left(\underline{x}_i^{\mathrm{T}} \cdot \underline{B} \cdot \tilde{\underline{x}}^{(k)} \right) \cdot \underline{x}_i. \tag{1.19}$$

The vector $\hat{\underline{x}}^{(k)}$ is $\underline{B}$-orthogonal to the $(m-1)$ eigenvectors already determined. Hence, the iteration converges to the eigenvector of the smallest eigenvalue whose eigenvector does not contain the eigenvectors $\underline{x}_1, \underline{x}_2, \ldots, \underline{x}_{m-1}$. This is the eigenvector of the eigenvalue m.

Example 1.11 For the eigenvalue problem given in Example 1.10, the second (smallest) eigenvalue and the corresponding eigenvector are to be determined by inverse iteration with *Gram–Schmidt orthogonalization*.

From the starting vector chosen as in Example 1.10, the component of the first eigenvector $\underline{x}_1$ is extracted as

$$\tilde{\underline{x}}^{(1)} = \begin{bmatrix} 1 \\ 1 \\ 1 \\ 1 \\ 1 \end{bmatrix}, \quad \hat{\underline{x}}^{(1)} = \tilde{\underline{x}}^{(1)} - \left(\underline{x}_1^{\mathrm{T}} \cdot \underline{B} \cdot \tilde{\underline{x}}^{(1)} \right) \cdot \underline{x}_1 = \begin{bmatrix} 0.772 \\ -0.142 \\ 0.674 \\ -1.370 \\ 0.641 \end{bmatrix}$$

$$\text{with } \underline{x}_1 = 10^{-2} \cdot \begin{bmatrix} 0.465 \\ 2.331 \\ 0.666 \\ 4.837 \\ 0.733 \end{bmatrix}.$$

With this starting vector, the vector $\underline{b}$ is obtained as $\underline{b} = \underline{A} \cdot \hat{\underline{x}}^{(1)} = 10^8 \cdot \begin{bmatrix} 1.631 \\ 0.126 \\ 1.043 \\ -0.454 \\ 0.616 \end{bmatrix}.$

The first iteration step is performed with this modified starting vector. Now, the Rayleigh quotient is computed as

$$\underline{f} = \underline{B} \cdot \hat{\underline{x}}^{(1)} = \begin{bmatrix} 1130 \\ -45 \\ 1082 \\ -378 \\ 938 \end{bmatrix}, \quad N = \left(\hat{\underline{x}}^{(1)} \right)^{\mathrm{T}} \cdot \underline{f} = 2727,$$

$$Z = \left(\hat{\underline{x}}^{(1)} \right)^{\mathrm{T}} \cdot \underline{b} = 2.961 \cdot 10^8, \quad \lambda^{(1)} = \frac{Z}{N} = 108569.$$

Therewith, the normalized eigenvector and the vector $\underline{b}$ are obtained as

$$\underline{x}^{(1)} = \frac{1}{\sqrt{N}} \cdot \hat{\underline{x}}^{(1)} = 10^{-2} \cdot \begin{bmatrix} 1.479 \\ -0.272 \\ 1.290 \\ -2.623 \\ 1.227 \end{bmatrix}, \quad \underline{b} = \frac{1}{\sqrt{N}} \cdot \underline{f} = \begin{bmatrix} 21.646 \\ -0.865 \\ 20.717 \\ -7.240 \\ 17.965 \end{bmatrix}.$$

The solution of the system of equations using the matrices $\underline{L}$ and $\underline{D}$ given in Example 1.10 gives the next approximation of the eigenvector:

$$\underline{L} \cdot \underline{y} = \underline{b} \rightarrow \underline{y} = \begin{bmatrix} 21.646 \\ 0.438 \\ 31.157 \\ -4.832 \\ 9.710 \end{bmatrix}, \quad \underline{D} \cdot \underline{c} = \underline{y} \rightarrow \underline{c} = 10^{-9} \cdot \begin{bmatrix} 0.612 \\ 0.161 \\ 0.819 \\ -9.851 \\ 1.767 \end{bmatrix}$$

$$\underline{L}^{\mathrm{T}} \cdot \hat{\underline{x}}^{(2)} = \underline{c} \rightarrow \hat{\underline{x}}^{(2)} = 10^{-6} \cdot \begin{bmatrix} 0.117 \\ -0.183 \\ 0.141 \\ -0.216 \\ 0.177 \end{bmatrix}.$$

For the next iteration step, this vector has to be orthogonized in relation to the eigenvector $\underline{x}_1$ as

$$\hat{\underline{x}}^{(1)} = \tilde{\underline{x}}^{(2)} - \left(\underline{x}_1^{\mathrm{T}} \cdot \underline{B} \cdot \tilde{\underline{x}}^{(2)} \right) \cdot \underline{x}_1 = 10^{-7} \cdot \begin{bmatrix} 1.773 \\ -0.182 \\ 1.415 \\ -2.142 \\ 1.770 \end{bmatrix}.$$

After normalization, the second eigenvalue and the corresponding eigenvector (rounded to two digits) are

$$\lambda_2 = 6.399 \cdot 10^4, \quad \underline{x}_2 = 10^{-2} \cdot \begin{bmatrix} 0.27 \\ -3.80 \\ 0.89 \\ -0.07 \\ 1.66 \end{bmatrix}.$$

In the next iteration steps, the following approximations of the eigenvalues and eigenvectors are obtained:

Eigenvalues and normalized eigenvectors $(\times 10^2)$	Iteration step k				
	1	2	3	4	5
$\lambda^{(k)}$	108569	85332	71014	65908	64483
$x_1^{(k)}$	1.479	1.065	0.712	0.494	0.379
$x_2^{(k)}$	−0.272	−1.654	−2.716	−3.281	−3.551
$x_3^{(k)}$	1.290	1.284	1.158	1.042	0.970
$x_4^{(k)}$	−2.623	−1.944	−1.173	−0.652	−0.365
$x_5^{(k)}$	1.227	1.607	1.729	1.700	1.700
	Iteration step k				
	6	7	8	9	10
$\lambda^{(k)}$	64115	64022	63998	63992	63991
$x_1^{(k)}$	0.321	0.292	0.278	0.270	0.267
$x_2^{(k)}$	−3.680	−3.743	−3.774	−3.790	−3.797
$x_3^{(k)}$	0.931	0.911	0.900	0.895	0.891
$x_4^{(k)}$	−0.217	−1.420	−0.105	−0.086	−0.071
$x_5^{(k)}$	1.681	1.670	1.664	1.661	1.658

◄

Besides the Gram–Schmidt orthogonalization, other methods to compute higher eigenvalues and eigenvectors are available. One is the so-called *shifting*. It denotes a shifting of the point of reference on the eigenvalue axis in the eigenvalue-determinant diagram (Fig. 1.1). With the relationship

$$\lambda = \lambda_0 + \bar{\lambda} \tag{1.20}$$

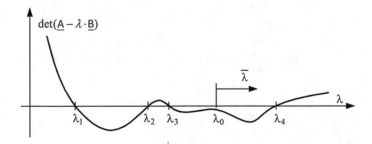

Fig. 1.1 Shifting of the reference point on the eigenvalue axis

a modified eigenvalue $\bar{\lambda}$ is introduced. It differs from the eigenvalue λ by a given value λ_0. By introducing (1.20) in the eigenvalue Eq. (1.12), the modified eigenvalue problem is obtained as

$$\left(\underline{A} - (\lambda_0 + \bar{\lambda}) \cdot \underline{B}\right) \cdot \underline{x} = \underline{0} \quad \text{or}$$

$$\left(\underline{\tilde{A}} - \bar{\lambda} \cdot \underline{B}\right) \cdot \underline{x} = \underline{0} \tag{1.20a}$$

with

$$\underline{\tilde{A}} = \underline{A} - \lambda_0 \cdot \underline{B}. \tag{1.20b}$$

This eigenvalue problem has the same eigenvectors as (1.12). However, its eigenvalues differ from the eigenvalues λ of the eigenvalue problem to be solved by the value λ_0. The matrix $\underline{\tilde{A}}$ is normally regular as it is obtained by the regular matrix $\underline{A}$ and the matrix $\underline{B}$. However, it is not necessarily positive definite as is the matrix $\underline{A}$.

Applying the method of inverse iteration on the modified eigenvalue problem, convergence is achieved toward the eigenvector with an eigenvalue closest to λ_0 for $\bar{\lambda}$ being positive. The convergence velocity is as higher as closer λ_0 is to the eigenvalue. The solution of the modified eigenvalue problem gives a modified eigenvalue $\bar{\lambda}$. Therewith, the eigenvalue λ is obtained by adding λ_0 according to (1.20).

Example 1.12 For the eigenvalue problem given in Example 1.10, all higher eigenvalues and eigenvectors are to be evaluated by inverse iteration with shifting.

The first shifting value is chosen to be

$$\lambda_0 = 50000.$$

The modified matrix $\underline{\tilde{A}}$ is obtained with (1.20b) and the matrices $\underline{A}$ and $\underline{B}$ given in Example 1.10 as

$$\underline{\tilde{A}} = 10^8 \cdot \begin{bmatrix} 2.808 & -0.213 & -1.681 & 0 & 0 \\ -0.213 & 0.125 & 0 & -0.142 & 0.213 \\ -1.681 & 0 & 3.836 & -0.213 & -1.681 \\ 0 & -0.142 & -0.213 & 0.004 & -0.213 \\ 0 & -0.213 & -1.681 & -0.213 & 1.587 \end{bmatrix}.$$

The Gauss decomposition of $\underline{\tilde{A}}$ gives

$$
\underline{L} = \begin{bmatrix}
1 & 0 & 0 & 0 & 0 \\
-0.076 & 1 & 0 & 0 & 0 \\
-0.599 & -1.171 & 1 & 0 & 0 \\
0 & -1.305 & -0.142 & 1 & 0 \\
0 & 1.957 & -0.534 & 0.586 & 1
\end{bmatrix}
$$

$$
\underline{D} = 10^8 \cdot \begin{bmatrix}
2.808 & 0 & 0 & 0 & 0 \\
0 & 0.109 & 0 & 0 & 0 \\
0 & 0 & 2.681 & 0 & 0 \\
0 & 0 & 0 & -0.235 & 0 \\
0 & 0 & 0 & 0 & 0.487
\end{bmatrix}.
$$

With a starting vector completely filled with ones, the inverse iteration (performed as shown in Example 1.10) gives after 9 iteration steps

$$
\bar{\lambda}_2 = 1.399 \cdot 10^4, \quad \lambda_2 = \bar{\lambda}_2 + \lambda_0 = 6.399 \cdot 10^4, \quad \underline{x}_2 = 10^{-2} \cdot \begin{bmatrix} 0.26 \\ -3.80 \\ 0.89 \\ -0.07 \\ 1.66 \end{bmatrix}.
$$

The iteration convergences toward the second eigenvalue since $\lambda_0 = 50000$ is closer to the second than to the first eigenvalue. If the shifting parameter is chosen to be $\lambda_0 = 20000$ which is closer to $\lambda_1 = 4342$ than to $\lambda_2 = 63990$, the iteration would have been toward the first eigenvalue.

The remaining three eigenvalues are determined by increasing the shifting parameter λ_0. One obtains,

with $\lambda_0 = 100000$ after 30 iteration steps

$$
\bar{\lambda}_3 = 2.7489 \cdot 10^5, \quad \lambda_3 = \bar{\lambda}_3 + \lambda_0 = 1.2749 \cdot 10^5, \quad \underline{x}_3 = 10^{-2} \cdot \begin{bmatrix} 1.29 \\ 2.75 \\ 0.94 \\ -3.41 \\ 0.59 \end{bmatrix}
$$

with $\lambda_0 = 200000$ after 10 iteration steps

$$\bar{\lambda}_4 = 2.454 \cdot 10^5, \quad \lambda_4 = \bar{\lambda}_4 + \lambda_0 = 2.245 \cdot 10^5, \quad \underline{x}_4 = 10^{-2} \cdot \begin{bmatrix} 1.74 \\ -1.99 \\ 0.51 \\ 1.06 \\ -1.56 \end{bmatrix}$$

and with $\lambda_0 = 400000$ after 4 iteration steps

$$\bar{\lambda}_s = 9.470 \cdot 10^3, \quad \lambda_5 = \bar{\lambda}_5 + \lambda_0 = 4.095 \cdot 10^5, \quad \underline{x}_5 = 10^{-2} \cdot \begin{bmatrix} 1.35 \\ -0.14 \\ 1.96 \\ 0.25 \\ 0.87 \end{bmatrix}.$$

Hence, all eigenvalues and eigenvectors are known since the matrices of the system of equation are of the fifth order.   ◀

The eigenvector toward the inverse iteration with shifting converges, depends on the choice of the shifting parameter λ_0. In order to check if all eigenvalues in a given interval have been found, the so-called *Sturm sequence check* is performed. For this, a Gauss decomposition of the matrix $\underline{\tilde{A}}$ according to (1.20b) is performed. With

$$\underline{\tilde{A}} = \underline{A} - \lambda_0 \cdot \underline{B} = \underline{L}^{\mathrm{T}} \cdot \underline{D} \cdot \underline{L}$$

the diagonal matrix $\underline{D}$ is obtained. It can be shown that the number of negative diagonal elements of $\underline{D}$ corresponds to the number of eigenvalues below λ_0. The test allows us to determine the number of eigenvalues in a given range in a rather inexpensive way.

The Sturm sequence check may also be used to localize eigenvalues by a repeated application of the check. In order to determine the number of eigenvalues in an interval, the check is applied at the interval limit points. The difference gives the number of eigenvalues in the interval.

Example 1.13 For the eigenvalue problem given in Example 1.10, the number of eigenvalues in the interval

$$50000 \leq \lambda < 300000$$

is to be determined with the Sturm sequence test.

For $\lambda_0 = 50000$, the Gauss decomposition of the matrix $\underline{\tilde{A}} = \underline{A} - \lambda_0 \cdot \underline{B}$ has been given in Example 1.12. It can be seen that the diagonal matrix $\underline{D}$ contains one negative value. This means that the eigenvalue problem possesses one eigenvalue below 50000. The Gauss decomposition of $\tilde{\underline{A}}$ for $\lambda_0 = 300000$ is obtained as

$$\tilde{\underline{A}} = 10^8 \cdot \begin{bmatrix} 0.853 & -0.213 & -1.681 & 0 & 0 \\ -0.213 & -0.670 & 0 & -0.142 & 0.213 \\ -1.681 & 0 & -0.179 & -0.213 & -1.681 \\ 0 & -0.142 & -0.213 & -0.686 & -0.213 \\ 0 & -0.213 & -1.681 & -0.213 & -2.072 \end{bmatrix}$$

$$\underline{L} = \begin{bmatrix} 1 & 0 & 0 & 0 & 0 \\ 0.250 & 1 & 0 & 0 & 0 \\ 1.971 & -0.681 & 1 & 0 & 0 \\ 0 & 0.230 & -0.091 & 1 & 0 \\ 0 & -0.345 & -0.449 & 0.589 & 1 \end{bmatrix}$$

$$\underline{D} = 10^8 \cdot \begin{bmatrix} -0.853 & 0 & 0 & 0 & 0 \\ 0 & -0.617 & 0 & 0 & 0 \\ 0 & 0 & 3.420 & 0 & 0 \\ 0 & 0 & 0 & -0.681 & 0 \\ 0 & 0 & 0 & 0 & -2.452 \end{bmatrix}.$$

The matrix $\underline{D}$ has four negative diagonal values. Hence, in the interval between 50000 and 300000, the eigenvalue problem possesses $(4 - 1) = 3$ eigenvalues. As can be seen from the solution in Example 1.12, these are the eigenvalues $\lambda_2 = 6.399 \cdot 10^4$, $\lambda_3 = 1.2749 \cdot 10^5$, and $\lambda_4 = 2.245 \cdot 10^5$. ◀

The inverse iteration is the basis of other numerical methods for the reliable and efficient solution of eigenvalue problems with large matrices which make use of special data storage techniques. One among them is the *simultaneous vector itera-tion* or *subspace iteration* [3, 5]. A subspace is understood as a set of a small number of eigenvectors which are iterated simultaneously observing the orthogonality condi-tions and the $\underline{B}$ normalization. The method is especially implemented in commercial finite element software for dynamic analyses.

Exercises

Problems

1.1

Consider the following system of equations

$$10 \cdot x_1 - 4 \cdot x_2 = y_1$$
$$-8 \cdot x_1 - x_2 = y_2$$

with

$$\begin{array}{ll}
10 \cdot y_1 + 2 \cdot y_2 = z_1 \\
6 \cdot y_1 + y_2 = z_2
\end{array}
\quad \text{and} \quad
\begin{array}{l}
3 \cdot z_1 + 4 \cdot z_2 = 20 \\
6 \cdot z_1 + 2 \cdot z_2 = 10.
\end{array}$$

(a) Establish the system of equations for the unknowns x_1, x_2 and the right side given by numbers in scalar notation.
(b) Use matrix notation for the 3 equations given and show that the matrix notation that leads to the same system of equations for the solution vector $\underline{x} = \begin{bmatrix} x_1 \\ x_2 \end{bmatrix}$ as the scalar notation according to (a).

1.2

Evaluate the matrix

$$\underline{C} = \underline{A}^{\mathrm{T}} \cdot c \cdot \underline{B}^{\mathrm{T}} \cdot \underline{B} \cdot \underline{A}$$

with

$$\underline{A} = \begin{bmatrix} 4 & 6 \\ 1 & 0 \\ 2 & 4 \end{bmatrix}, \quad \underline{B} = \begin{bmatrix} 2 & 0 & -6 \\ -2 & 2 & 0 \end{bmatrix}, \quad c = 1.5.$$

How can the matrix product be computed in the most efficient way, i.e. with the least number of multiplications?

1.3

Evaluate the inverses of

$$\underline{A} = \begin{bmatrix} 6 & -5 \\ -2 & 2 \end{bmatrix}, \quad \underline{B} = \begin{bmatrix} 10 & 0 & 2 \\ 0 & 0.5 & 0 \\ 5 & 0 & 2 \end{bmatrix}, \quad \underline{C} = \begin{bmatrix} 10 & 2 \\ 5 & 2 \end{bmatrix}.$$

Prove that for the matrices given, the following relations hold:

(a) $\underline{A} \cdot \underline{A}^{-1} = \underline{B} \cdot \underline{B}^{-1} = \underline{I}$

(b) $(\underline{A} \cdot \underline{C})^{-1} = \underline{C}^{-1} \cdot \underline{A}^{-1}$.

1.4

Evaluate the determinants of the matrices $\underline{A}$, $\underline{B}$, and $\underline{C}$ given in Exercise 1.3 as well as the determinants of their inverses.

1.5

Check the solvability of the following systems of equations:

(a) $6 \cdot x_1 - 5 \cdot x_2 = 4$
 $-2 \cdot x_1 + 2 \cdot x_2 = 1$

(b) $10 \cdot x_1 + 2 \cdot x_3 = 5$
 $0.5 \cdot x_2 = 2$
 $5 \cdot x_1 + 2 \cdot x_3 = 2.5$

(c) $10 \cdot x_1 + 2 \cdot x_3 = 5$
 $0.5 \cdot x_2 = 2$
 $-5 \cdot x_1 - x_3 = 2.5$

(d) $10 \cdot x_1 + 2 \cdot x_3 = 5$
 $0.5 \cdot x_2 = 2$
 $-5 \cdot x_1 - x_3 = -2.5$.

1.6

Solve the following system of equations by Gaussian elimination:

$$103850 \cdot x_1 + 5160 \cdot x_2 + 52000 \cdot x_3 = -150$$
$$5160 \cdot x_1 + 43290 \cdot x_2 + 24510 \cdot x_3 = 86.6$$
$$52000 \cdot x_1 + 24510 \cdot x_2 + 204370 \cdot x_3 = -100.$$

Evaluate the matrices $\underline{L}$ and $\underline{D}$.

1.7

Evaluate the Euclidean norm or the Frobenius norm, respectively, of

$$A = \begin{bmatrix} 6 & -5 \\ -2 & 2 \end{bmatrix}, \quad B = \begin{bmatrix} 10 & 0 & 2 \\ 0 & 0.5 & 0 \\ 5 & 0 & 2 \end{bmatrix}, \quad c = \begin{bmatrix} 2 \\ -1 \\ 5 \\ 2 \end{bmatrix}.$$

1.8

Evaluate the determinants and condition numbers of the coefficient matrices of the following systems of equations:

(a) $10 \cdot x_1 + 2 \cdot x_3 = 5$
$$0.5 \cdot x_2 = 2$$
$$-5 \cdot x_1 - x_3 = 2.5$$

(b) $10 \cdot x_1 + 2 \cdot x_3 = 5$
$$0.5 \cdot x_2 = 2$$
$$-5.01 \cdot x_1 - x_3 = 2.5$$

(c) $10 \cdot x_1 + 2 \cdot x_3 = 5$
$$0.5 \cdot x_2 = 2 \quad .$$
$$-5 \cdot x_1 - x_3 = 2.5$$

1.9

(a)   Consider the following eigenvalue problem:

$$\left(\begin{bmatrix} 36 & 6 \\ 6 & 12 \end{bmatrix} - \lambda \cdot \begin{bmatrix} 72 & 0 \\ 0 & 8 \end{bmatrix} \right) \cdot \begin{bmatrix} x_1 \\ x_2 \end{bmatrix} = \begin{bmatrix} 0 \\ 0 \end{bmatrix}.$$

Evaluate the characteristic polynomial and determine all eigenvalues.
(b)   Compute the corresponding eigenvectors normalized so that the greatest entry of each equals 1.
(c)   Prove that the eigenvectors fulfill the orthogonality conditions.
(d)   Normalize the eigenvectors with respect to the B matrix.

1.10

Evaluate the eigenvalue and the eigenvector of the eigenvalue problem given in Exercise 1.9 iteratively. Use the iteration rule of the inverse iteration with the starting vector $\begin{bmatrix} x_1 \\ x_2 \end{bmatrix} = \begin{bmatrix} 1 \\ 1 \end{bmatrix}$ and the Rayleigh quotient. Perform iteration steps and normalize the eigenvector so that its greatest entry equals 1.

Solutions

1.1

(a) $460 \cdot x_1 - 226 \cdot x_2 = 20$
    $608 \cdot x_1 - 302 \cdot x_2 = 10$

(b) $\underline{x} = \begin{bmatrix} x_1 \\ x_2 \end{bmatrix}, \quad \underline{y} = \begin{bmatrix} y_1 \\ y_2 \end{bmatrix}, \quad \underline{z} = \begin{bmatrix} z_1 \\ z_2 \end{bmatrix}, \quad \underline{b} = \begin{bmatrix} 20 \\ 10 \end{bmatrix}$

$\underline{A} = \begin{bmatrix} 10 & -4 \\ -8 & -1 \end{bmatrix}, \quad \underline{A} \cdot \underline{x} = \underline{y}$

$\underline{B} = \begin{bmatrix} 10 & 2 \\ 6 & 1 \end{bmatrix}, \quad \underline{B} \cdot \underline{y} = \underline{z}$

$\underline{C} = \begin{bmatrix} 3 & 4 \\ 6 & 2 \end{bmatrix}, \quad \underline{C} \cdot \underline{z} = \underline{b}$

$\underline{C} \cdot \underline{B} \cdot \underline{A} \cdot \underline{x} = \underline{b}$ with $\underline{C} \cdot \underline{B} \cdot \underline{A} = \begin{bmatrix} 460 & -226 \\ 608 & -302 \end{bmatrix}.$

1.2

$\underline{C} = \begin{bmatrix} 78 & 180 \\ 180 & 432 \end{bmatrix}.$

For minimization of the computational effort:
$\underline{D} = \underline{B} \cdot \underline{A} \rightarrow \underline{C}^* = \underline{D}^{\mathrm{T}} \cdot \underline{D} \rightarrow \underline{C} = c \cdot \underline{C}^*.$

1.3

(a) $\underline{A}^{-1} = \begin{bmatrix} 1 & 2.5 \\ 1 & 3 \end{bmatrix}, \quad \underline{B}^{-1} = \begin{bmatrix} 0.2 & 0 & -0.2 \\ 0 & 2 & 0 \\ -0.5 & 0 & 1 \end{bmatrix} \rightarrow \underline{A} \cdot \underline{A}^{-1} = \underline{I}$

(b) $\underline{C}^{-1} = \begin{bmatrix} 0.2 & -0.2 \\ 0.5 & 1 \end{bmatrix}, (\underline{A} \cdot \underline{C})^{-1} = \begin{bmatrix} 0 & -0.1 \\ 0.5 & 1.75 \end{bmatrix},$

$\underline{C}^{-1} \cdot \underline{A}^{-1} = \begin{bmatrix} 0 & -0.1 \\ 0.5 & 1.75 \end{bmatrix} \rightarrow (\underline{A} \cdot \underline{C})^{-1} = \underline{C}^{-1} \cdot \underline{A}^{-1}.$

1.4

$\det \underline{A} = 2, \qquad \det \underline{B} = 5, \qquad \det \underline{C} = 10$
$\det \underline{A}^{-1} = 0.5, \quad \det \underline{B}^{-1} = 0.2, \quad \det \underline{C} = 0.1.$

1.5

The systems of equations possess

(a) a single unique solution,
(b) a single unique solution,
(c) no solution,
(d) no unique solution (but infinitely many solutions: $x_2 = 4$,
$x_1 = 0.5 - 0.2 \cdot x_3$ for all x_3).

1.6

$$\underline{L} = \begin{bmatrix} 10.00000 & 0.00000 & 0.00000 \\ 0.04969 & 10.00000 & 0.00000 \\ 0.50072 & 0.50952 & 10.00000 \end{bmatrix},$$

$$\underline{D} = \begin{bmatrix} 10.038 & 0.000 & 0.000 \\ 0.000 & 0.430 & 0.000 \\ 0.000 & 0.000 & 10.672 \end{bmatrix} \cdot 10^5$$

$$x_1 = -13.46 \cdot 10^{-4}, \quad x_2 = 24.08 \cdot 10^{-4}, \quad x_3 = -4.35 \cdot 10^{-4}.$$

1.7

$$\|\underline{A}\| = 8.31, \quad \|\underline{B}\| = 11.54, \quad \|\underline{c}\| = 5.83.$$

1.8

(a) $\det(\underline{A}) = 5, \quad \kappa(\underline{A}) = 26.26$
(b) $\det(\underline{B}) = 0.01, \quad \kappa(\underline{B}) = 6511$
(c) $\det(\underline{C}) = 0, \quad \kappa(\underline{C})$ not defined because $\underline{C}$ is singular.

1.9

(a) $\lambda^2 - 2 \cdot \lambda + \dfrac{11}{16} = 0, \quad \lambda_1 = 0,440983, \quad \lambda_2 = 1,559017$

(b) $\underline{x}_1 = \begin{bmatrix} 1 \\ -0.708204 \end{bmatrix}, \quad \underline{x}_2 = \begin{bmatrix} 0.078689 \\ 1 \end{bmatrix}$

(c) $\underline{x}_1^{\mathrm{T}} \cdot \underline{B} \cdot x_1 = 76.012423, \qquad \underline{x}_1^{\mathrm{T}} \cdot \underline{B} \cdot x_2 = -2.4 \cdot 10^{-5} \approx 0,$

$\underline{x}_2^{\mathrm{T}} \cdot \underline{B} \cdot x_1 = -2.4 \cdot 10^{-5} \approx 0, \quad \underline{x}_2^{\mathrm{T}} \cdot \underline{B} \cdot x_2 = 8.445821$

(d) $\tilde{\underline{x}}_1 = \begin{bmatrix} 0.114698 \\ -0.08123 \end{bmatrix}, \quad \tilde{\underline{x}}_2 = \begin{bmatrix} 0.027077 \\ 0.344096 \end{bmatrix}.$

1.10

Starting vector: $\underline{x}^{(0)} = \begin{bmatrix} 1 \\ 1 \end{bmatrix}$, $\quad \underline{A}^{-1} \cdot \underline{B} = \begin{bmatrix} 2.181818 & -0.121212 \\ -1.090909 & 0.727273 \end{bmatrix}$.

Iteration 1: $\quad \underline{x}^{(1)} = \begin{bmatrix} 1 \\ -0.176471 \end{bmatrix}$, $\lambda^{(1)} = 0.474138$.

Iteration 2: $\quad \underline{x}^{(2)} = \begin{bmatrix} 1 \\ -0.553398 \end{bmatrix}$, $\lambda^{(2)} = 0.443710$.

Iteration 3: $\quad \underline{x}^{(3)} = \begin{bmatrix} 1 \\ -0.664050 \end{bmatrix}$, $\lambda^{(3)} = 0.441202$.

Iteration 4: $\quad \underline{x}^{(4)} = \begin{bmatrix} 1 \\ -0.695685 \end{bmatrix}$, $\lambda^{(4)} = 0.441001$.

References

1. Bronstein IN, Semendjajew KA, Musiol G, Mühlig H (2008) Taschenbuch der Mathematik, 7th edn. Verlag Harri Deutsch, Frankfurt am Main
2. Schwarz HR (1984) Methode der finiten Elemente. B.G. Teubner, Stuttgart
3. Bathe K-J (2014) Finite element procedures, 2nd edn., Watertown MA, USA. ISBN 978-0-9790049-5-7
4. Pfaffinger D (1989) Tragwerksdynamik. Springer, Wien
5. Bathe K-J (2013) The subspace iteration method–revisited. Comput Struct 126:177–183

Chapter 2
Basic Equations of the Theory of Elasticity

Abstract In this chapter the basic equations of the theory of elasticity are compiled as far as they are needed in the following chapters. It starts after the definition of state variables and the types of basic equations with the one- and two-dimensional states of stress. Especially the kinematic equations, the constitutive equations and, most important, the principle of virtual displacements are discussed. These are the basic equations from where the derivation of finite element equations starts in Chap. 4. The next section deals with the corresponding equations for shear flexible and shear rigid beams and plates in bending. In addition, special effects as the edge effect in plates are considered. The derivation of bending plate elements in Chap. 4 but also the discussion of modelling effects in bending plates will refer to this section. The last section deals with special aspects of three-dimensional structures. One topic is torsion including warping torsion in beams, another are the basic equations of three-dimensional solids. Chapter 3 for torsion of beams and Chap. 4 for solids will make reference to it.

2.1 Types of Structures

The finite element analysis for elastic structures is based on the theory of elasticity. Its basic equations are summed up in this chapter as far as they are required in the next chapters. For more details, standard books on the theory of elasticity such as [1–3] as well as the presentation in [19] are referred to.

The basic equations of the theory of elasticity can be formulated for different stress states or types of structures. The most basic is the one-dimensional state of stress which corresponds to a truss type of structure. The two-dimensional state of stress is characterized by two normal stresses and one shear stress. The corresponding type of structure is a plate in a *membrane state* as, for example, a deep beam. The most general type of structure is the three-dimensional state of stress with six stress components (Fig. 2.1).

Beams and plates in bending are types of structures with a special two- or three-dimensional state of stress which results from the presumption of the Bernoulli hypothesis. Therefore, beams can be considered as one-dimensional systems and plates as two-dimensional systems.

© Springer Nature Switzerland AG 2021
H. Werkle, *Finite Elements in Structural Analysis*, Springer Tracts
in Civil Engineering, https://doi.org/10.1007/978-3-030-49840-5_2

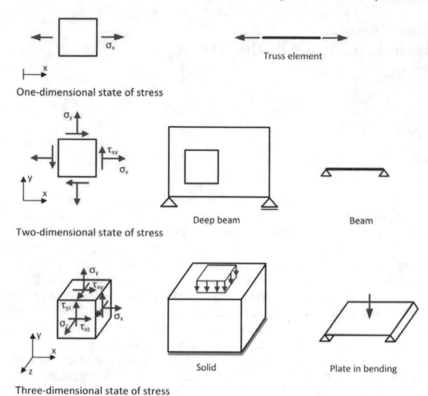

Fig. 2.1 States of stress and types of structures

The mechanical behavior of structures is described by state variables. These are

State variables

(a)   Displacements and their derivatives as strains and curvatures;
(b)   Stresses and stress resultants as forces and moments.

With the state variables, the following basic equations can be formulated:

Basic equations

(a)   The kinematic equations (compatibility of the strains and the displacements);
(b)   The constitutive equations (material law, e.g. Hooke's law);
(c)   The equations of equilibrium.

These equations have to be fulfilled at each point of the continuum. In addition, the boundary conditions have to be fulfilled, which may be related to the displacements (supports) or the forces (external loads) at the boundaries of the structure.

In the following, the basic equations for different types of structures are given. The equilibrium conditions, however, are omitted (except for some with an exemplary derivation for plane stress state). Instead of the equilibrium conditions, the principle of virtual displacements is formulated because of its fundamental importance for the finite element method.

2.2 One-Dimensional and Two-Dimensional States of Stress

A truss element has a one-dimensional or uniaxial state of stress whereas a plane stress element is exposed to a two-dimensional or plane state of stress. The deformation of the truss element can be described by the displacement $u(x)$ and the corresponding strain $\varepsilon_x = du/dx$ (Fig. 2.3). The only stress component is the normal stress σ_x.

The displacements for the *plane stress state* also denoted as *membrane state of stress* are described by two components $u(x, y)$ and $v(x, y)$ in the x- and y-directions, respectively. This leads to the strain variables ε_x, ε_y and γ_{xy}. They correspond to the stress components σ_x, σ_y and τ_{xy}.

The signs of the stress components in a uniaxial and in a plane state of stress are defined as follows (Fig. 2.2):

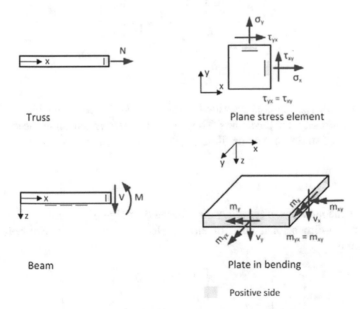

Fig. 2.2 Sign convention of stresses and section forces

Sign convention for stresses
Stresses are positive when they act in the positive coordinate direction at the positive side of an element. On a positive side, the outward normal vector is in the direction of the positive coordinate.

Stresses are often summed up to stress resultants or section forces. For the plane stress state, one obtains the section forces per length unit

$$n_x = \sigma_x \cdot t, \quad n_y = \sigma_y \cdot t, \quad , n_{xy} = \tau_{xy} \cdot t. \tag{2.1}$$

Herein, t is the plate thickness.

The strains are obtained as derivatives of the displacements (Fig. 2.3). In the uniaxial state of stress one obtains

$$\varepsilon_x = \frac{du}{dx} \tag{2.2a}$$

and in the state of plane stress

$$\begin{bmatrix} \varepsilon_x \\ \varepsilon_y \\ \gamma_{xy} \end{bmatrix} = \begin{bmatrix} \frac{\partial}{\partial x} & 0 \\ 0 & \frac{\partial}{\partial y} \\ \frac{\partial}{\partial y} & \frac{\partial}{\partial x} \end{bmatrix} \cdot \begin{bmatrix} u \\ v \end{bmatrix} \tag{2.2b}$$

$$\underline{\varepsilon} = \underline{L} \cdot \underline{u}. \tag{2.2c}$$

The constitutive equations described by the *material law* represent the relationship between the stresses and the strains. They allow the determination of the stresses for a given state of strain. In the one-dimensional state of stress, *Hooke's law* is

$$\sigma_x = E \cdot \varepsilon_x \tag{2.3a}$$

where E is the *modulus of elasticity* also denoted as *Young's modulus*.

In the two-dimensional state of stress, Hooke's law for an isotropic elastic continuum is

$$\begin{bmatrix} \sigma_x \\ \sigma_y \\ \tau_{xy} \end{bmatrix} = \frac{E}{1-\mu^2} \begin{bmatrix} 1 & \mu & 0 \\ \mu & 1 & 0 \\ 0 & 0 & \frac{1-\mu}{2} \end{bmatrix} \cdot \begin{bmatrix} \varepsilon_x \\ \varepsilon_y \\ \gamma_{xy} \end{bmatrix} \tag{2.3b}$$

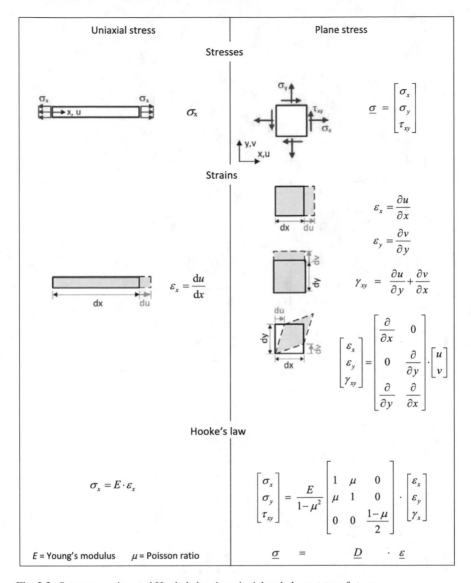

Fig. 2.3 Stresses, strains, and Hooke's law in uniaxial and plane states of stress

$$\sigma = \underline{D} \cdot \varepsilon. \qquad (2.3c)$$

The matrix $\underline{D}$ is denoted as the *elasticity matrix*. Both normal stress components are coupled due to the lateral strain effect.

Other material laws for the two-dimensional elastic material are given in Table 2.1 [4] (according to [5]), [18]. *Orthotropic material laws* are required for the analysis of wooden structures and special types of soil. Three-dimensional solids "infinitely

Table 2.1 Two-dimensional material laws

Isotropic material	Orthotropic material

Plane stress state

$$\begin{bmatrix} \sigma_x \\ \sigma_y \\ \tau_{xy} \end{bmatrix} = \frac{E}{1-\mu^2} \cdot \begin{bmatrix} 1 & \mu & 0 \\ \mu & 1 & 0 \\ 0 & 0 & \dfrac{1-\mu}{2} \end{bmatrix} \cdot \begin{bmatrix} \varepsilon_x \\ \varepsilon_y \\ \gamma_{xy} \end{bmatrix}$$

$$\begin{bmatrix} \sigma_x \\ \sigma_y \\ \tau_{xy} \end{bmatrix} = \frac{E_2}{(1-n\mu_2^2)} \begin{bmatrix} n & n\mu_2 & 0 \\ n\mu_2 & 1 & 0 \\ 0 & 0 & m(1-n\mu_2^2) \end{bmatrix} \cdot \begin{bmatrix} \varepsilon_x \\ \varepsilon_y \\ \gamma_{xy} \end{bmatrix}$$

Plane strain state

$$\begin{bmatrix} \sigma_x \\ \sigma_y \\ \tau_{xy} \end{bmatrix} = \frac{E(1-\mu)}{(1+\mu)(1-2\mu)} \cdot \begin{bmatrix} 1 & \dfrac{\mu}{1-\mu} & 0 \\ \dfrac{\mu}{1-\mu} & 1 & 0 \\ 0 & 0 & \dfrac{1-2\mu}{2(1-\mu)} \end{bmatrix} \cdot \begin{bmatrix} \varepsilon_x \\ \varepsilon_y \\ \gamma_{xy} \end{bmatrix}$$

$$\begin{bmatrix} \sigma_x \\ \sigma_y \\ \tau_{xy} \end{bmatrix} = \frac{E_2}{(1+\mu_1)\cdot(1-\mu_1-2n\mu_2^2)} \cdot$$

$$\cdot \begin{bmatrix} n(1-n\mu_2^2) & n\mu_2(1+\mu_1) & 0 \\ n\mu_2(1+\mu_1) & (1-\mu_1^2) & 0 \\ 0 & 0 & \ell \end{bmatrix} \cdot \begin{bmatrix} \varepsilon_x \\ \varepsilon_y \\ \gamma_{xy} \end{bmatrix}$$

$$\ell = m \cdot (1+\mu_1) \cdot (1-\mu_1-2n\mu_2^2)$$

(continued)

Table 2.1 (continued)

Isotropic material	Orthotropic material
Material constants	
E = Young's modulus μ = Poisson ratio G = Shear modulus with $G = \dfrac{E}{2 \cdot (1 + \mu)}$	Layers parallel to the x-z plane $E_1 = E_{\parallel}$ $ = E_x = E_z$ Modulus of elasticity for strains in the layer plane $E_2 = E_y = E_{\perp}$ Modulus of elasticity for strains normal to the layer plane $\mu_2 = \mu_{xy} = \mu_{yz}$ Poisson ratio for strains in the x- or z-direction due to σ_y $\mu_1 = \mu_{xz}$ Poisson ratio for strains in the x-direction due to σ_z $G_2 = G_{xy} = G_{yz}$ Shear modulus for shear deformations in the xy- or yz-plane $G_1 = G_{xz} = E_1/(2 \cdot (1 + \mu_1))$ $n = E_1/E_2$ $m = G_2/E_2$ $\mu_2 = G_2/E_2$

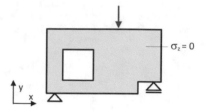

$\sigma_z = 0$

The stresses normal to the plane are zero

State of plane stress

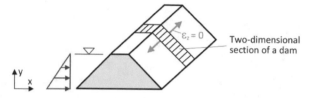

$\varepsilon_z = 0$

Two-dimensional
section of a dam

The strains normal to the plane are zero

State of plane strain

Fig. 2.4 States of plane stress and plane strain

long" in one direction can be described as two-dimensional systems if their cross section and loading in the direction regarded as arbitrarily long is uniform. Since the strains normal to the considered plane are zero, this state is denoted as *plane strain* whereas the stress state where the stresses normal to the plane are zero is denoted as *plane stress* (Fig. 2.4).

The equations of equilibrium in the uniaxial and the plane states of stress are formulated for an infinitesimally small element of the continuum. As an alternative to the equations of equilibrium, the principle of virtual displacements can be applied. It is the basis for the derivation of finite elements for plates and solids.

Principle of virtual displacements

If an arbitrary infinitesimally small virtual displacement is imposed upon a body which is in equilibrium, the total internal virtual work is equal to the external virtual work. The virtual displacements must fulfill the support conditions.

A virtual displacement is a small, fictitious displacement which is assumed in addition to the real displacements. The virtual external work is the work which would be done by the external forces if the virtual state of displacements would be imposed upon the system where the real loads are acting. The virtual internal work is the work which would have been done by the real section forces if the virtual states of displacements would be applied.

Hence, the principle of virtual displacement deals with work which would be done if a virtual displacement state would be imposed upon a structure being in equilibrium. In this case, the external work done by the real external forces with the virtual displacements must be equal to the internal work done by the internal section forces with the virtual deformations.

The function of the distribution of the assumed virtual displacements is arbitrary. However, it must be continuous, sufficiently differentiable, and compatible with the geometric boundary conditions.

The *external virtual work* is the product of the real loads and the virtual displacements in the direction of the corresponding loads. According to Fig. 2.6, the virtual external work for a single force is

$$\overline{W}_a = \bar{u} \cdot F, \tag{2.4a}$$

and for several forces

$$\overline{W}_a = F_1 \cdot \bar{u}_1 + F_2 \cdot \bar{u}_2 + F_3 \cdot \bar{u}_3 + \cdots$$

or

$$\overline{W}_a = [\bar{u}_1 \quad \bar{u}_2 \quad \bar{u}_3 \quad \cdots] \cdot \begin{bmatrix} F_1 \\ F_2 \\ F_3 \\ \vdots \end{bmatrix}$$

$$\overline{W}_a = \underline{\bar{u}}^{\mathrm{T}} \cdot \underline{F}. \tag{2.4b}$$

In order to indicate that the displacements and the external work are virtual quantities, they are marked with an overline.

The *internal virtual work* of an infinitesimal element of a truss with length $\mathrm{d}x$ is a product of the real normal force N and the virtual displacement $\bar{\varepsilon}_x \cdot \mathrm{d}x$ (Fig. 2.5). The virtual strain $\bar{\varepsilon}$ is determined for the virtual state of displacements of the truss element. For a truss element with length l, the internal virtual work is obtained by integration over its length by

$$\overline{W}_i = \int A \cdot \sigma_x \cdot \bar{\varepsilon}_x \, \mathrm{d}x$$

(omitting integration limits). This equation is identical to the well-known formula

$$\overline{W}_i = \int \frac{N\bar{N}}{EA} \, \mathrm{d}x$$

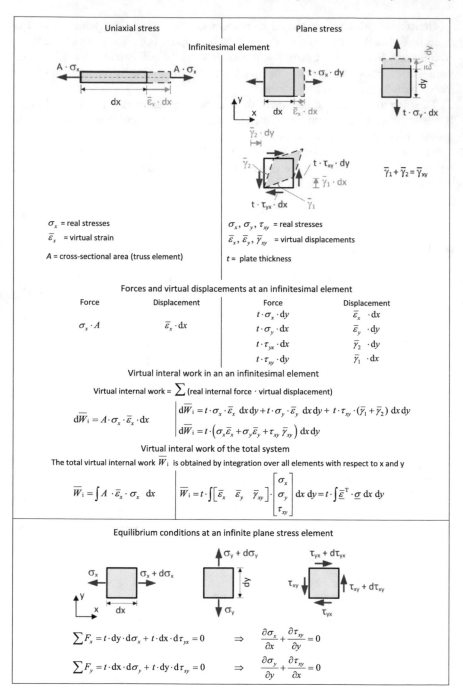

Fig. 2.5 Internal virtual work for uniaxial and plane stress states, and equilibrium conditions at an infinite plane stress element

which is applied in structural analysis to determine the displacements of truss systems and which is obtained by the principle of virtual forces. This can be shown easily by setting

$$A \cdot \sigma_x = N \qquad \bar{\varepsilon}_x = \frac{\bar{\sigma}_x}{E} = \frac{\bar{N}}{E\,A}$$

and therewith,

$$\int A \cdot \sigma_x \cdot \bar{\varepsilon}_x \; dx = \int \frac{N\bar{N}}{E\,A} \; dx.$$

In a plate in plane stress, the forces $\sigma_x \cdot dy \cdot t$, $\sigma_y \cdot dx \cdot t$ and $\tau_{xy} \cdot dx \cdot t$, $\tau_{xy} \cdot dy \cdot t$, which are obtained from the stress components, perform work with the associated virtual displacement components (Fig. 2.5).

Hence, the total virtual work is obtained for the truss as

$$\overline{W}_i = \int A \cdot \sigma_x \cdot \bar{\varepsilon}_x \; dx \tag{2.5a}$$

and for the plate in plane stress as

$$\overline{W}_i = t \cdot \int \begin{bmatrix} \bar{\varepsilon}_x & \bar{\varepsilon}_y & \bar{\gamma}_{xy} \end{bmatrix} \cdot \begin{bmatrix} \sigma_x \\ \sigma_y \\ \gamma_{xy} \end{bmatrix} dx \; dy$$

$$\overline{W}_i = t \cdot \int \underline{\bar{\varepsilon}}^\mathrm{T} \cdot \underline{\sigma} \; dx \; dy. \tag{2.5b}$$

According to the *principle of virtual displacements* the internal and external virtual work must be equal so that

$$\overline{W}_i = \overline{W}_a. \tag{2.5c}$$

The principle of virtual displacements corresponds to the equilibrium conditions and leads to identical results if all other fundamental equations are fulfilled exactly. It is well-known, for example, that the equations of equilibrium for rigid bodies can be solved applying the principle of virtual displacements [6].

However, the principle of virtual displacements has a characteristic which does not hold for the equations of equilibrium. It can be applied to stress distributions which do not fulfill the equations of equilibrium exactly in order to get an approximate solution. In this case, the equations of equilibrium are fulfilled "in mean" if the principle of virtual displacements is fulfilled. This is also called the "weak" form of

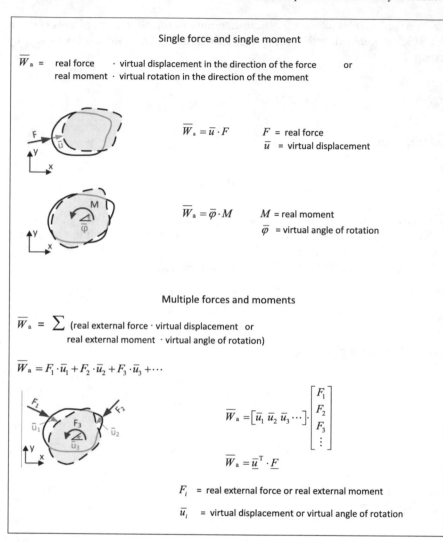

Fig. 2.6 External virtual work of single loads and moments

equilibrium conditions. The formulation of the finite element method makes use of this property of the principle of virtual displacements.

Another principle of work related to the principle of virtual displacements is the *principle of virtual forces*. With this principle, small virtual forces are applied as fictitious loading on the system. This principle can be formulated as:

Principle of virtual forces

If small virtual forces are applied on a body, the external virtual work done with real displacements equals the total internal work of the virtual section forces (caused by the virtual forces) done with real deformations.

The external virtual work is the work which the virtual forces would perform with the real displacements. The internal virtual work is the total work which the virtual section forces caused by the virtual forces would perform with the real displacements.

The principle of virtual forces can be derived by replacing the virtual displacements in Figs. 2.5 and 2.6 by the real displacements and the real forces by virtual forces. In structural analysis, the principle of virtual forces is applied to calculate deformations of beam structures.

2.3   Beams and Plates in Bending

Beams and plates in bending are considered to be one- or two-dimensional structures, respectively. Hence, the beam theory is not suited to describe local stress distributions, for example, at supports or in the vicinity of single loads. Similarly, the simplification of the three-dimensional continuum as a two-dimensional plate structure may be the cause of unrealistic section forces at certain points of the plate structure, e.g. at point supports or under single loads. Interpretation problems arising from this simplification are outlined in Sect. 4.11.6 (Modeling of Plates in Bending).

In beams and plates, stresses are summarized to shear forces and bending moments. In plates, in addition to the bending moments m_x and m_y, twisting moments m_{xy} and m_{yx} occur. The sign of the bending moments is defined with respect to their associated stresses, i.e. the moment m_x causes normal stresses σ_x in the x-direction, and the moment m_y normal stresses σ_y in the y-direction. For the twisting moments m_{xy} and m_{yx}, the first subscript denotes the direction of the normal to the side, and the second subscript the direction of the stress component (Fig. 2.2). The twisting moments m_{xy} and m_{yx} lead to shear stresses τ_{xy} and τ_{yx} (Fig. 2.7). The relationship for the shear stresses $\tau_{xy} = \tau_{yx}$, therefore, leads to the corresponding relationship $m_{xy} = m_{yx}$ for the twisting moments.

The sign convention of the section forces of the beam and of the shear forces for plates follows the same rule as for trusses and plates in plane stress:

Sign convention for section forces in beams and of shear forces in plates

Section forces in beams and shear forces in plates are positive when they act in the positive coordinate direction at the positive side of an element.

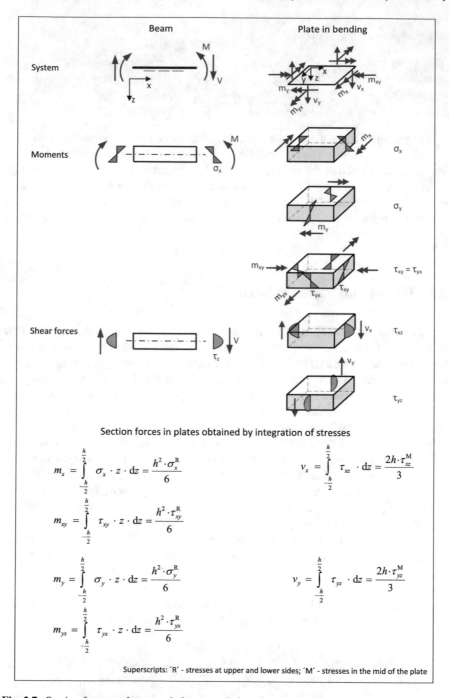

Fig. 2.7 Section forces and stresses in beams and plates in bending

The sign convention for bending moments in plates follows a different definition. It is based on the definition of the positive stresses caused by the moment:

Sign convention for moments in plate bending

Bending moments m_x and m_y are positive if they lead to positive normal stresses σ_x and σ_y, respectively, at the bottom side of the plate. The bottom side of the plate is at the positive side in the z-direction, i.e. on the positive z-side of the plate.

Twisting moments m_{xy} and m_{yx} are positive if they lead to shear stresses τ_{xy} and τ_{yx} in the positive x- and y-coordinates, respectively, at the bottom side of the plate.

The basic equations of the plate and the beam, taking shear deformations into account, are given in Fig. 2.8 [7, 8]. The equations of the shear rigid *Euler-Bernoulli beam* and shear rigid plate are obtained when the equations corresponding to the shear angle γ_{xz}, γ_{yz} are omitted. This corresponds to an "infinitely large" shear area A_s. The theory of the *shear rigid plate* is denoted as *Kirchhoff plate theory*.

While, in general, beam elements implemented in commercial finite element software consider the shear deformation, there are some finite element programs with plate elements for shear rigid plates on the basis of the Kirchhoff plate theory. From a practical point of view, for thin plates, the consideration of shear deformations is not necessary. However, shear flexible finite elements are often used, for reasons due to the finite element theory.

The theory of the *shear flexible plate* is also denoted as the *Reissner-Mindlin plate theory*. It is based on the assumption that points in a section normal to the undeformed midplane remain on a straight plane during deformation. However, it is no longer normal to the deformed midplane, as it is assumed in the Kirchhoff plate theory. This applies accordingly for the shear flexible beam which is also denoted as *Timoshenko beam*.

For the shear flexible beam, the total displacement $\mathrm{d}w$ of an infinitesimally small length $\mathrm{d}x$ is composed of the bending deformation $\varphi \cdot \mathrm{d}x$ and the shear deformation $\gamma \cdot \mathrm{d}x$ (Fig. 2.8):

$$\mathrm{d}w = -\varphi \cdot \mathrm{d}x + \gamma \cdot \mathrm{d}x. \tag{2.6}$$

The shear angle γ is caused by the shear force and is assumed to be constant in the cross section. The bending angle φ arises from the extension of the bottom fiber and the contraction of the top fiber due to a positive bending moment. The variation of the bending angle φ with respect to x is the curvature κ. It is obtained by differentiation as

$$\kappa = \frac{\mathrm{d}\varphi}{\mathrm{d}x} \tag{2.7a}$$

and with (2.6)

$$\gamma = \varphi + \frac{dw}{dx}. \tag{2.7b}$$

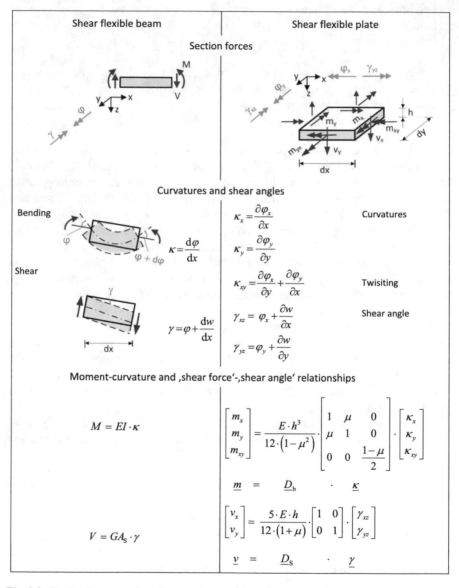

Fig. 2.8 Section forces, strain values, and material laws for beams and plates

The angles φ and γ are defined in such a way that they perform positive work with their associated section force at the positive side. This has as a consequence that the bending angle φ and the shear angle γ are defined as positive in opposite directions (Fig. 2.8).

The constitutive equations consist of the moment–curvature relationship and the "shear force"–"shear angle" relationship. They allow to determine the bending moment and the shear force from a given curvature and shear angle, respectively. With the curvature κ, the bending moment M for a constant bending stiffness $E \cdot I$ is obtained as

$$M = E \cdot I \cdot \kappa. \tag{2.8a}$$

The shear angle γ gives the shear force V and the corresponding shear stress $\tau = V/A_s$ (where A_s is the shear area), applying Hooke's law for shear deformations $\tau = G \cdot \gamma$ (where G is the shear modulus), as

$$V = \tau \cdot A_s$$

$$V = G \cdot A_s \cdot \gamma. \tag{2.8b}$$

For the plate, analogous relationships for the curvatures κ_x in the x-direction and κ_y in the y-direction as well as for the shear angles γ_{xz} and γ_{yz} are obtained. In addition, the twisting deformation κ_{xy} has to be considered. According to Fig. 2.8, the strain values of the shear flexible plate are obtained as

$$\begin{bmatrix} \kappa_x \\ \kappa_y \\ \kappa_{xy} \end{bmatrix} = \begin{bmatrix} 0 & \dfrac{\partial}{\partial x} & 0 \\ 0 & 0 & \dfrac{\partial}{\partial y} \\ 0 & \dfrac{\partial}{\partial y} & \dfrac{\partial}{\partial x} \end{bmatrix} \cdot \begin{bmatrix} w \\ \varphi_x \\ \varphi_y \end{bmatrix} \tag{2.9a}$$

$$\begin{bmatrix} \gamma_{xz} \\ \gamma_{yz} \end{bmatrix} = \begin{bmatrix} \dfrac{\partial}{\partial x} & 1 & 0 \\ \dfrac{\partial}{\partial y} & 0 & 1 \end{bmatrix} \cdot \begin{bmatrix} w \\ \varphi_x \\ \varphi_y \end{bmatrix}. \tag{2.9b}$$

The constitutive equations (or material laws) of the shear flexible plate, i.e. the moment–curvature relationship and the "shear force"–"shear angle" relationship are

$$\begin{bmatrix} m_x \\ m_y \\ m_{xy} \end{bmatrix} = \frac{E\,h^3}{12(1+\mu^2)} \begin{bmatrix} 1 & \mu & 0 \\ \mu & 1 & 0 \\ 0 & 0 & \dfrac{1-\mu}{2} \end{bmatrix} \cdot \begin{bmatrix} \kappa_x \\ \kappa_y \\ \kappa_{xy} \end{bmatrix} \tag{2.10a}$$

$$\underline{m} = \underline{D}_{\mathrm{b}} \cdot \underline{\kappa}$$

$$\begin{bmatrix} v_x \\ v_y \end{bmatrix} = \frac{5\,E \cdot h}{12(1+\mu)} \begin{bmatrix} 1 & 0 \\ 0 & 1 \end{bmatrix} \cdot \begin{bmatrix} \gamma_{xz} \\ \gamma_{yz} \end{bmatrix} \tag{2.10b}$$

$$\underline{v} = \underline{D}_{\mathrm{s}} \cdot \underline{\gamma}.$$

For orthotropic plates, the material laws can be generalized as

$$\begin{bmatrix} m_x \\ m_y \\ m_{xy} \end{bmatrix} = \begin{bmatrix} D_x & D_{xy} & 0 \\ D_{xy} & D_y & 0 \\ 0 & 0 & D_{\mathrm{D}} \end{bmatrix} \cdot \begin{bmatrix} \kappa_x \\ \kappa_y \\ \kappa_{xy} \end{bmatrix} \tag{2.10c}$$

$$\underline{m} = \underline{D}_{\mathrm{b}} \cdot \underline{\kappa}$$

$$\begin{bmatrix} v_x \\ v_y \end{bmatrix} = \begin{bmatrix} S_x & 0 \\ 0 & S_y \end{bmatrix} \cdot \begin{bmatrix} \gamma_{xz} \\ \gamma_{yz} \end{bmatrix} \tag{2.10d}$$

$$\underline{v} = \underline{D}_{\mathrm{s}} \cdot \underline{\gamma}.$$

Herein, D_x and D_y are the bending stiffnesses, D_{xy} the coupling stiffness, D_{D} the torsional stiffness, and S_x and S_y the shear stiffnesses. Formulae for the stiffness parameters for sandwich plates and plates with hollow sections are given in [9, 14]. For more details, see the extensive literature on plate theory in textbooks such as [10–14, 18].

The principle of virtual displacements for the shear flexible plate can be derived analogously to the plate in plane stress (Fig. 2.9). The imposed virtual plate bending corresponds to virtual curvatures and virtual shear deformations. It performs virtual internal work with the real bending moments and shear forces. The contributions of the bending and of the shear deformations are treated separately in the equation of the internal virtual energy. Here, the stress vector $\underline{\sigma}$ of the plate in plane stress corresponds to the section force vectors $\underline{m}$ or $\underline{q}$ of the plate in bending. The strain vector $\underline{\varepsilon}$ corresponds to the curvature vector $\underline{\kappa}$ or the strain vector $\underline{\gamma}$. With it, the virtual internal energy of the beam is obtained as

$$\overline{W}_{\mathrm{i}} = \int \bar{\kappa} \cdot M \,\mathrm{d}x + \int \bar{\gamma} \cdot V \mathrm{d}x. \tag{2.11a}$$

Shear flexible beam

Virtual internal work of a bending moment at an infinitesimal element

$$d\overline{W}_i = M \cdot d\overline{\varphi}$$

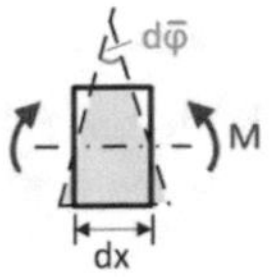

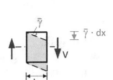

$$\overline{\kappa} = \frac{d\overline{\varphi}}{dx} \quad \rightarrow \quad d\overline{\varphi} = \overline{\kappa} \cdot dx$$

$$d\overline{W}_i = M \cdot \overline{\kappa} \cdot dx$$

  M = real moment
  $\overline{\kappa}$ = virtual curvature

Virtual internal work of a shear force at an infinitesimal element

$$d\overline{W}_i = V \cdot \overline{\gamma} \ dx$$

  V = real shear force
  $\overline{\gamma}$ = virtual shear angle

Total virtual internal work in a beam

$$\overline{W}_i = \int \overline{\kappa} \cdot M \, dx + \int \overline{\gamma} \cdot V \, dx$$

Shear flexible plate

Total internal work in a plate

$$\overline{W}_i = \int \begin{bmatrix} \overline{\kappa}_x & \overline{\kappa}_y & \overline{\kappa}_{xy} \end{bmatrix} \cdot \begin{bmatrix} m_x \\ m_y \\ m_{xy} \end{bmatrix} \, dx \, dy + \int \begin{bmatrix} \overline{\gamma}_{xz} & \overline{\gamma}_{yz} \end{bmatrix} \cdot \begin{bmatrix} v_x \\ v_y \end{bmatrix} \, dx \, dy$$

$$\overline{W}_i = \int \underline{\overline{\kappa}} \cdot \underline{m} \, dx \, dy + \int \underline{\overline{\gamma}} \cdot \underline{v} \, dx \, dy$$

Fig. 2.9 Principle of virtual displacements for the beam and the plate in bending

For the plate, it is given by

$$\overline{W}_i = \int \begin{bmatrix} \overline{\kappa}_x & \overline{\kappa}_y & \overline{\kappa}_{xy} \end{bmatrix} \cdot \begin{bmatrix} m_x \\ m_y \\ m_{xy} \end{bmatrix} \, dx \, dy + \int \begin{bmatrix} \overline{\gamma}_{xz} & \overline{\gamma}_{yz} \end{bmatrix} \cdot \begin{bmatrix} v_x \\ v_y \end{bmatrix} \, dx \, dy$$

$$\tag{2.11b}$$

$$\overline{W}_i = \int \underline{\overline{\kappa}}^T \cdot \underline{m} \, dx \, dy + \int \underline{\overline{\gamma}} \cdot \underline{q} \, dx \, dy. \tag{2.11c}$$

With

$$\bar{\kappa} = \frac{\bar{M}}{E \cdot I} \qquad \bar{\gamma} = \frac{\bar{V}}{G \cdot A_s}$$

the virtual internal energy of the beam according to (2.11a) yields the well-known expression of the work equation for beams in bending

$$\overline{W}_i = \int \frac{M \cdot \bar{M}}{E \cdot I} dx + \int \frac{V \cdot \bar{V}}{G \cdot A_s} dx.$$

In the *shear rigid plate* (Kirchhoff plate theory), the shear deformations are zero, i.e. $\gamma = 0$ or $\gamma_{xz} = 0$ and $\gamma_{yz} = 0$, and therefore, the angles of rotation are not independent state variables since they are obtained as derivatives of the displacement function. The deformations of a shear rigid plate are described by the curvatures κ_x and κ_y and the twisting deformation κ_{xy} only. They are obtained as second derivatives of the displacement function. The shear forces can no longer be obtained from the material law because the shear stiffness becomes "infinitely" large.

Furthermore, when solving the differential equation of the shear rigid plate, difficulties arise in fulfilling the boundary conditions. They can be solved approximately considering the so-called edge effect of plates in bending.

The *edge effect* occurs at simply supported and free edges of plates in bending (see Sect. 4.11.6). In the Kirchhoff plate theory, it allows an approximate fulfillment of the boundary conditions in the solution of the differential equation of the shear rigid plate. At each boundary, the state of three section forces (bending moment m_n, twisting moment m_{ns}, and shear force v_n to be zero or not) or of the corresponding displacements and rotations are to be defined (Fig. 2.10). For mathematical reasons, only the boundary conditions for two of the three existing section forces can be fulfilled for the shear rigid plate. In order to fulfill the boundary conditions approximately, the rate of change of the twisting moments and the shear forces at the plate boundary are expressed by *equivalent shear forces* as

$$v_n^* = v_n + \frac{dm_{ns}}{ds} \tag{2.12}$$

according to Fig. 2.11. They correspond to the force reactions of the support. In addition, point support forces act at the plate corners. At simply supported boundaries,

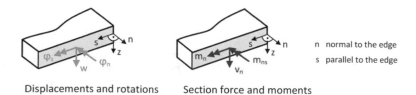

Displacements and rotations   Section force and moments

n normal to the edge
s parallel to the edge

Fig. 2.10 Section forces and displacements at a plate edge

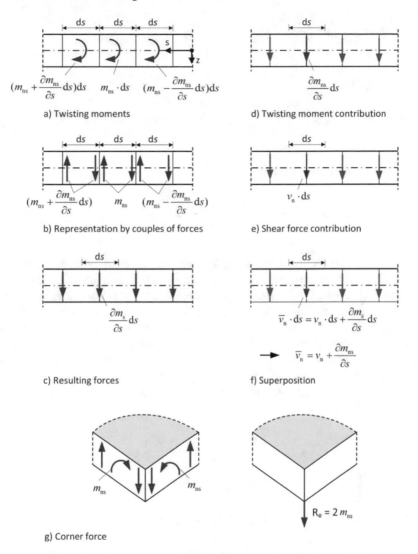

Fig. 2.11 Edge effect at the boundary of a plate in bending

the equivalent shear forces v_n^* according to (2.12) and corresponding to the support reaction differ by dm_{ns}/ds from the shear forces v_n at the plate edge. At unloaded free boundaries, according to the Kirchhoff plate theory, both twisting moments as well as shear forces exist. With $v_n^* = 0$ in (2.12), the shear forces at the boundary (which should be zero) are obtained as $v_n = -dm_{ns}/ds$, i.e. they correspond to the rate of change of the boundary twisting moments m_{ns}. At a clamped edge no twisting moments occur and hence, due to $dm_{ns}/ds = 0$, shear forces coincide with the support reactions.

At a discontinuity in plate thickness, the following transitional conditions apply

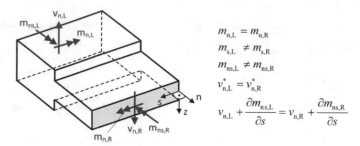

$$m_{n,L} = m_{n,R}$$
$$m_{s,L} \neq m_{s,R}$$
$$m_{ns,L} \neq m_{ns,R}$$
$$v_{n,L}^* = v_{n,R}^*$$
$$v_{n,L} + \frac{\partial m_{ns,L}}{\partial s} = v_{n,R} + \frac{\partial m_{ns,R}}{\partial s}$$

Fig. 2.12 Discontinuity in plate thickness

$$w_L = w_R, \; \varphi_{n,L} = \varphi_{n,R}, \; m_{n,L} = m_{n,R}, \; m_{s,L} \neq m_{s,R}, \; m_{ns,L} \neq m_{ns,R}, \; v_{n,L}^* = v_{n,R}^*. \tag{2.12a}$$

The bending moment and the twisting moment change abruptly since the curvature and twisting deformation are equal on both sides but the plate stiffness changes abruptly (Fig. 2.12). The equivalent shear forces are equal at both sides as well [14].

The theory of the *shear flexible plate* according to Reissner–Mindlin allows the fulfillment of all three boundary conditions. At free boundaries, the bending moments, twisting moments, and shear forces are set to zero. At simply supported boundaries, there are two possibilities for defining the boundary conditions. First, the displacements w, the bending moment m_n, and the angle of rotation φ_n may be set to be zero. In this case, the twisting moment m_{nt} becomes nonzero. For this so-called *hard support,* we have

$$w = 0 \quad\quad v_n \neq 0,$$

$$\varphi_n = 0 \quad\quad m_{ns} \neq 0,$$

$$\varphi_s \neq 0 \quad\quad m_n = 0. \tag{2.13}$$

In the so-called *soft support,* the twisting moment m_{nt} is set to zero, and therefore, the rotation φ_n becomes nonzero. The boundary conditions for the soft support are

$$w = 0 \quad\quad v_n \neq 0,$$

$$\varphi_n \neq 0 \quad\quad m_{ns} = 0,$$

$$\varphi_s \neq 0 \quad\quad m_n = 0. \tag{2.14}$$

At clamped edges, in addition to the deflections w and the bending angle φ_s, the twisting angle φ_n can be set to zero (hard support) or left free (soft support).

In analytical solutions of the shear flexible plate theory, the support reactions equal the shear forces at the plate boundary. Instead of point forces, peaks in the shear force diagram are obtained in the plate corners. For thin plates used in practice, the differences between both theories are significant only within a boundary strip of a width equal to the size of the plate thickness (approximately), and can thus normally be ignored.

2.4 Spatial Structures

2.4.1 Generals

The basic equations of spatial structures can be derived from those of plane structures. For the truss, they are identical and differ only with regard to the transformation into the spatial coordinate system. Shell structures can be understood as a superposition of the bearing behavior of plates in plane stress and plates in bending, i.e. of membrane stresses and bending stresses, respectively. The basic equations at the infinitesimal element are those for simultaneous membrane and bending action. Here again, the equations have to be transformed into the spatial coordinate system, leading to the equations of the shell theory. In the following, the torsion of the beam and the basic equations of a three-dimensional solid are examined as important cases of a three-dimensional state of stress.

2.4.2 Beams

In three-dimensional beams, in addition to bending on the two principal axes, torsion may occur. Normally, torsion is described by the St. Venant theory. However, for some types of sections, in addition, warping torsion may occur. This type of torsion will be considered in the following.

A torsional moment can be transferred in a cross section of a bar in two ways:

Torsion in beams

- *St. Venant torsion*: In the cross section, shear stresses and their associated shear angles arise. The stress resultant of the shear stresses is the torsional moment. The section remains plane.
- *Warping torsion*: Warping denotes a longitudinal displacement in the direction of the axis about which the torsional moment is applied. Parts of the section are in a state of bending. Herewith, normal stresses develop in the cross section. The corresponding stress resultant is a torsional moment. Due to the normal stresses, the section warps, i.e. it does not remain plane.

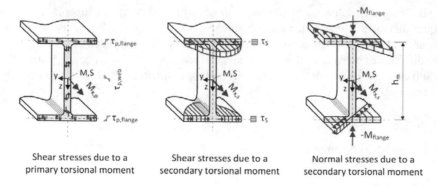

| Shear stresses due to a primary torsional moment | Shear stresses due to a secondary torsional moment | Normal stresses due to a secondary torsional moment |

Fig. 2.13 Warping torsion of an I-section

Non-warping cross sections such as circles or circular rings cannot sustain warping torsion, i.e. a torsional moment is transferred by St. Venant torsion only. Warping cross sections are able to transfer a part of a torsional moment by warping torsion, but only if the warping is constrained at a support. Warping cross sections are, for example, rectangular cross sections or thin open cross sections (excluding the case of a cross section consisting of two rectangles intersecting at one point).

Figure 2.13 shows the effect of warping torsion in the case of an I-beam constrained at the end. The cross section is doubly symmetric and its shear center M and center of gravity S are at the same point. The torsional moment is split into a part $M_{x,\mathrm{p}}$ of the *primary torsion* corresponding to the St. Venant torsion and a part $M_{x,\mathrm{s}}$ of *secondary torsion* due to warping, i.e.

$$M_x = M_{x,\mathrm{p}} + M_{x,\mathrm{s}}. \tag{2.15}$$

The primary torsional moment gives shear stresses which can, in general, be determined as

$$\tau_{\mathrm{p,max}} = \frac{M_{x,\mathrm{p}}}{W_{\mathrm{T}}}, \tag{2.16}$$

or for thin-walled open sections like the I-section

$$\tau_{\mathrm{p,web}} = \frac{M_{x,\mathrm{p}}}{I_{\mathrm{T}}} \cdot s, \quad \tau_{\mathrm{p,flange}} = \frac{M_{x,\mathrm{p}}}{I_{\mathrm{T}}} \cdot t, \tag{2.16a}$$

where W_{T} is the section modulus of torsion, I_{T} the torsional constant, and s, t are the thicknesses of the web and the flange, respectively.

For the I-section, the secondary torsional moment is obtained by the bending of the upper and lower flange. It corresponds to a force in the upper flange and an opposite force of the same magnitude in the lower flange representing a pair of forces with a moment arm h_{m}. This pair of forces represents the torsional moment $M_{x,\mathrm{s}}$.

The bending moments in the upper and lower flange are expressed by the *warping bi-moment* M_ω. For the I-section, one gets

$$M_\omega = M_{\text{flange}} \cdot h_{\text{m}}. \tag{2.17}$$

The normal stresses due to the warping bi-moment are

$$\sigma_\omega(y, z) = \frac{M_\omega}{I_\omega} \cdot \omega(y, z) \tag{2.18}$$

with the maximum value

$$\sigma_{\omega,\text{max}} = \frac{M_\omega}{I_\omega} \cdot w_{\text{m}}. \tag{2.18a}$$

The *unit warping function* $\omega(y, z)$, the value w_{m}, and the warping constant I_ω are characteristic parameters of a cross section. The diagrams of the shear stresses τ_{s} due to warping torsion in the upper and lower flange are parabolas. Their maximum value in the middle of the flange is

$$\tau_{\text{s,max}} = \frac{M_{x,\text{s}} \cdot w_{\text{m}}}{I_\omega} \cdot \frac{b}{4}, \tag{2.19}$$

where b is the width of the flange.

In the general case of a thin-walled open cross section, the relationships are much more complex and the warping bi-moment can no longer be interpreted physically [15–17].

The displacement value associated with the primary torsional moment is the angle of rotation φ_x of the x-axis of the cross section. It is also denoted as the angle of twist. The secondary torsional moment is related to the negative *rate of change of the angle of twist* (warping degree of freedom)

$$\psi = -\frac{d\varphi_x}{dx}, \tag{2.20}$$

which can be regarded as the "torsional curvature".

The shear stresses caused by the primary torsional moment result in a shear deformation of the section, whereas the warping bi-moment due to the secondary torsional moment causes a warping of the section, i.e. the section does not remain plane. The displacements due to warping in the longitudinal direction of the beam are described by

$$u(x, y, z) = \omega(y, z) \cdot \psi(x), \tag{2.21}$$

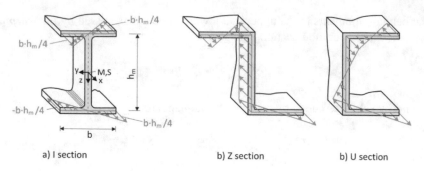

a) I section   b) Z section   b) U section

Fig. 2.14 Unit warping function $\omega(y, z)$

where $\omega(y, z)$ is the unit warping function. The unit warping function is a characteristic parameter of a cross section. For the I-section and two other sections, it is shown in Fig. 2.14.

The relationships between the primary and the secondary torsional moment and the angle of rotation φ_x are

$$M_{x,p} = G \cdot I_T \cdot \frac{d\varphi_x}{dx} = -G \cdot I_T \cdot \psi, \tag{2.22a}$$

$$M_{x,s} = - E \cdot I_\omega \cdot \frac{d^3\varphi_x}{dx^3}. \tag{2.22b}$$

The warping bi-moment is

$$M_\omega = -E \cdot I_\omega \cdot \frac{d^2\varphi_x}{dx^2} \tag{2.23}$$

and with it

$$M_{x,s} = \frac{dM_\omega}{dx}. \tag{2.23a}$$

With the distributed torsional moment $m_x = -dM_x/dx$ and (2.15) and (2.22a, 2.22b), the differential equation of a bar subjected to warping torsion is obtained. It is a linear fourth-order differential equation. Its solution with the appropriate boundary conditions is the basis of the derivation of the stiffness matrix of the beam element with warping torsion.

With the definition of the warping bi-moment according to (2.23, 2.23a) and Fig. 2.13, a positive secondary torsional moment $M_{x,s}$ according to (2.15) corresponds to a negative warping moment M_ω. It should also be pointed out that in the literature on torsional warping, occasionally, a deviating sign definition of the warping bi-moment and the unit warping function is found.

2.4.3 Solids

The displacements of a three-dimensional solid are described by the three components u, v, w in x-, y-, and z-directions, respectively. The strain components are defined as follows:

$$
\begin{bmatrix} \varepsilon_x \\ \varepsilon_y \\ \varepsilon_z \\ \gamma_{xy} \\ \gamma_{yz} \\ \gamma_{zx} \end{bmatrix}
=
\begin{bmatrix}
\dfrac{\partial}{\partial x} & 0 & 0 \\[2mm]
0 & \dfrac{\partial}{\partial y} & 0 \\[2mm]
0 & 0 & \dfrac{\partial}{\partial z} \\[2mm]
\dfrac{\partial}{\partial y} & \dfrac{\partial}{\partial x} & 0 \\[2mm]
0 & \dfrac{\partial}{\partial z} & \dfrac{\partial}{\partial y} \\[2mm]
\dfrac{\partial}{\partial z} & 0 & \dfrac{\partial}{\partial x}
\end{bmatrix}
\cdot
\begin{bmatrix} u \\ v \\ w \end{bmatrix}
\tag{2.24}
$$

$$
\underline{\varepsilon} = \underline{L} \cdot \underline{u}. \tag{2.24a}
$$

The corresponding stress components are

$$
\underline{\sigma}^{\mathrm{T}} = \begin{bmatrix} \sigma_x & \sigma_y & \sigma_z & \tau_{xy} & \tau_{yz} & \tau_{zx} \end{bmatrix}. \tag{2.25}
$$

Hooke's law of the three-dimensional state of stress can be written as

$$
\underline{\sigma} = \underline{D} \cdot \underline{\varepsilon} \tag{2.26}
$$

with the elasticity matrix $\underline{D}$ according to Fig. 2.15 for a homogeneous, isotropic material.

Furthermore, the inner work in a three-dimensional solid is given in Fig. 2.15 according to the principle of virtual displacements. The analogy to the two-dimensional stress state according to Table 2.1 and Fig. 2.5 is obvious.

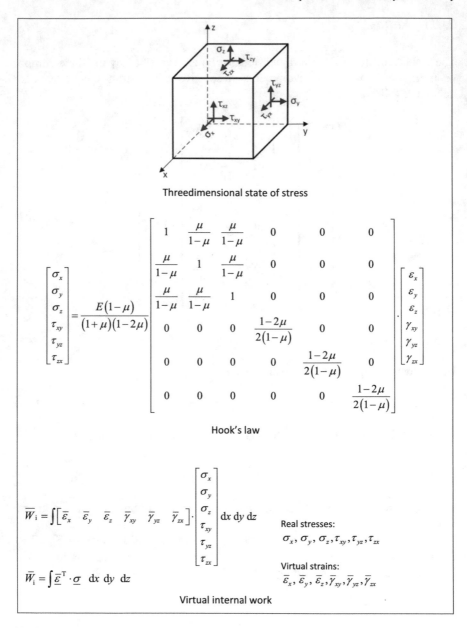

Threedimensional state of stress

$$\begin{bmatrix} \sigma_x \\ \sigma_y \\ \sigma_z \\ \tau_{xy} \\ \tau_{yz} \\ \tau_{zx} \end{bmatrix} = \frac{E(1-\mu)}{(1+\mu)(1-2\mu)} \begin{bmatrix} 1 & \dfrac{\mu}{1-\mu} & \dfrac{\mu}{1-\mu} & 0 & 0 & 0 \\[2mm] \dfrac{\mu}{1-\mu} & 1 & \dfrac{\mu}{1-\mu} & 0 & 0 & 0 \\[2mm] \dfrac{\mu}{1-\mu} & \dfrac{\mu}{1-\mu} & 1 & 0 & 0 & 0 \\[2mm] 0 & 0 & 0 & \dfrac{1-2\mu}{2(1-\mu)} & 0 & 0 \\[2mm] 0 & 0 & 0 & 0 & \dfrac{1-2\mu}{2(1-\mu)} & 0 \\[2mm] 0 & 0 & 0 & 0 & 0 & \dfrac{1-2\mu}{2(1-\mu)} \end{bmatrix} \cdot \begin{bmatrix} \varepsilon_x \\ \varepsilon_y \\ \varepsilon_z \\ \gamma_{xy} \\ \gamma_{yz} \\ \gamma_{zx} \end{bmatrix}$$

Hook's law

$$\overline{W}_i = \int \begin{bmatrix} \overline{\varepsilon}_x & \overline{\varepsilon}_y & \overline{\varepsilon}_z & \overline{\gamma}_{xy} & \overline{\gamma}_{yz} & \overline{\gamma}_{zx} \end{bmatrix} \cdot \begin{bmatrix} \sigma_x \\ \sigma_y \\ \sigma_z \\ \tau_{xy} \\ \tau_{yz} \\ \tau_{zx} \end{bmatrix} dx\ dy\ dz$$

Real stresses:
$\sigma_x, \sigma_y, \sigma_z, \tau_{xy}, \tau_{yz}, \tau_{zx}$

$$\overline{W}_i = \int \underline{\overline{\varepsilon}}^{\mathrm{T}} \cdot \underline{\sigma}\ \ dx\ dy\ dz$$

Virtual strains:
$\overline{\varepsilon}_x, \overline{\varepsilon}_y, \overline{\varepsilon}_z, \overline{\gamma}_{xy}, \overline{\gamma}_{yz}, \overline{\gamma}_{zx}$

Virtual internal work

Fig. 2.15 Material law and internal virtual work for a three-dimensional solid

References

1. Mang HA, Hofstätter G (2018) Festigkeitslehre, 5 edn. Springer Vieweg
2. Maceri A (2010) Theory of elasticity. Springer, Berlin
3. Gross D, Hauger W, Schröder J, Wall WA, Bonet J (2018) Engineering mechanics 2–mechanics of materials. Springer, Heidelberg
4. Zienkiewicz OC (1977) The finite element method, 3rd edn. McGraw-Hill, Maidenhead, England
5. Lekhnitskii SG (1963) Theory of elasticity of an anisotropic elastic body. Holden Day, San Francisco
6. Gross D, Hauger W, Schröder J, Wall WA, Rajapakse N (2013) Engineering mechanics 1–statics. Springer, Dordrecht
7. Reissner E (1947) On bending of elastic plates. Q Appl Math 5:55–68
8. Mindlin RD (1951) Influence of rotatory inertia and shear on flexural motions of isotropic, elastic plates. ASME J Appl Mech 18:31–38
9. Hinton E, Owen DRJ (1984) Finite element software for plates and shells. Pineridge Press Ltd., Swansea
10. Girkmann KG (1986) Flächenwerke. Springer, Berlin
11. Timoshenko S (1959) Theory of plates and shells. McGraw-Hill, New York
12. Gould PL (1988) Analysis of shells and plates. Springer, New York
13. Altenbach H, Altenbach J, Naumenko K (1998) Ebene Flächentragwerke. Springer, Berlin
14. Blaauewendraad J (2010) Plates and FEM. Springer, Dortrecht
15. Petersen C (2013) Stahlbau, 4th edn. Springer Vieweg, Wiesbaden
16. Francke W, Friemann H (2005) Schub und Torsion in geraden Stäben. Vieweg
17. Wunderlich W, Kiener G (2004) Statik der Stabtragwerke. Teubner Verlag/GWV Fachverlage, Wiesbaden
18. Neto MA, Amaro A, Roseiro L, Cirne J, Leal R (2015) Engineering computation of structures: the finite element method. Springer International, Cham, Switzerland
19. Bucalem ML, Bathe K-J (2011) The mechanics of solids and structures–hierarchical modeling and the finite element solution. Springer, Berlin

Chapter 3
Truss and Beam Structures

Abstract The presentation of the finite element method starts in this chapter with the explanation of the finite element method for truss and beam systems. The concept of stiffness matrices is shown first for a simple truss element. The procedure for setting up the equations for a finite element analysis is explained exemplary for a simple truss system. The procedure itself, however, is not limited to truss systems, it can be applied to other finite element systems as well, consisting of beam, plate and solid elements. The next sections deal with springs and with beam elements subjected to bending and their mechanical background. Two- and three-dimensional truss and beam systems are treated. It follows an extensive discussion of the modelling of beam structures. Symmetry conditions to be used for symmetrical structures are discussed as well. The chapter ends with some aspects of error sources including numerical errors and of the quality assurance of finite element analyses.

3.1 Introduction

3.1.1 The Finite Element Method in Structural Analysis

The analysis of statically indeterminate structural systems generally leads to an algebraic linear system of equations. Exceptions are investigations in which geometric or material nonlinearities are of importance. If the unknowns of this system of equations are forces and moments, then one speaks of the *force method*, if they are displacements and rotations, of the *displacement method*. Both the force method and the displacement method can be formulated in matrix notation and thus written in a form suitable for computer calculation. However, the displacement method is clearer and more easily schematizable than the force method and thus is better suitable for programming. Therefore, almost all program systems used in practice for structural calculations are based on the displacement method. The finite element method is based on the matrix displacement method. For beam and truss structures, the displacement method in matrix notation also is denoted as *direct stiffness method*.

Here, we will refer to the displacement method in matrix notation for beam and truss structures as the *finite element method* as well. This designation is adopted in

© Springer Nature Switzerland AG 2021
H. Werkle, *Finite Elements in Structural Analysis*, Springer Tracts
in Civil Engineering, https://doi.org/10.1007/978-3-030-49840-5_3

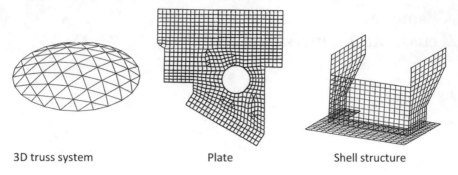

3D truss system   Plate   Shell structure

Fig. 3.1 Examples of finite element models

the following, since beam and truss structures are only a special case of the more general application of the finite element method to plate and shell structures or to three-dimensional continua. An overview of the method for beam and truss structures is given in [1–11, 54].

The basic idea of the finite element method consists of the subdivision of a complex structure into a huge number of elements with simple static properties (Fig. 3.1). These elements are assembled into the complete system considering the kinematic compatibility and the static equilibrium conditions. Since different structural elements such as beam, plate, and three-dimensional solid elements can be used in the same computational model, the method is extremely versatile and powerful.

3.1.2   Nodal Points, Degrees of Freedom, and Finite Elements

In the finite element analysis, the structure is discretized into so-called *finite elements*, which are connected at *nodal points*. Displacements and rotations as well as external loads and moments are defined at these nodal points. They refer to the *global coordinate system*, which is generally defined in Cartesian coordinates. Displacements or rotations of a node in global coordinates are denoted as *global degrees of freedom*.

Which degrees of freedom are assigned to a node depends on the type of the structure analyzed. In the general three-dimensional case, there are 6 degrees of freedom, i.e. the displacements in the x-, y-, and z-directions as well as the rotations about the x-, y-, and z-axes (Fig. 3.2). For plane systems and special types of structures, the number of degrees of freedom to be considered is reduced, e.g. a plate in bending in the xy-plane possesses three degrees of freedom per node: the displacement in the z-direction as well as the rotations about the x- and y-axes. The definition of the forces and moments for point loads, for example, corresponds to the sign definition of the corresponding degrees of freedom. Figure 3.3 presents an overview of typical element types and the corresponding degrees of freedom.

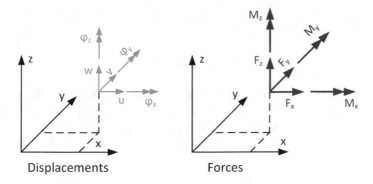

Fig. 3.2 Displacements and forces in global coordinates

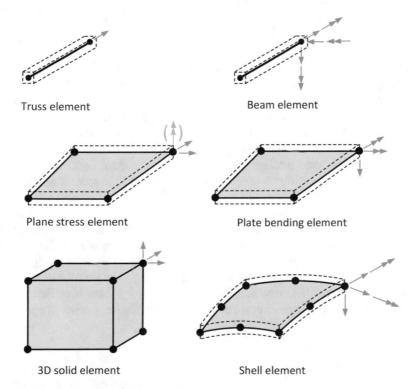

Fig. 3.3 Types of elements and degrees of freedom

3.1.3 Computational Method

Assuming a linear behavior of the system, the static analysis of a structure discretized into finite elements leads to a linear system of equations with the nodal point displacements and rotations as unknowns. Aggregating the forces and moments acting at the nodal points to the *load vector* $\underline{F}$ and the nodal point displacements to the *displacement vector* $\underline{u}$, the relationship can be written as

$$\underline{K} \cdot \underline{u} = \underline{F}. \tag{3.1}$$

For example, the plane framework in Fig. 3.4 is discretized into two beam elements and three nodal points. When the support conditions are taken into account, it has five degrees of freedom: u_1, φ_1, u_2, v_2 and φ_2. The corresponding external loads are F_{x1}, M_1, F_{x2}, F_{y2} and M_2. Therefore, the system of equations (3.1) has the form

$$\begin{bmatrix} k_{11} & k_{12} & k_{13} & k_{14} & k_{15} \\ k_{21} & k_{22} & k_{23} & k_{24} & k_{25} \\ k_{31} & k_{32} & k_{33} & k_{34} & k_{35} \\ k_{41} & k_{42} & k_{43} & k_{44} & k_{45} \\ k_{51} & k_{52} & k_{53} & k_{54} & k_{55} \end{bmatrix} \cdot \begin{bmatrix} u_1 \\ \varphi_1 \\ u_2 \\ v_2 \\ \varphi_2 \end{bmatrix} = \begin{bmatrix} F_{x1} \\ M_1 \\ F_{x2} \\ F_{y2} \\ M_2 \end{bmatrix}. \tag{3.1a}$$

The matrix $\underline{K}$ is called the *global stiffness matrix*. It is quadratic since $\underline{u}$ and $\underline{F}$ have the same size. The elements of the vectors $\underline{u}$ and $\underline{F}$ are ordered in such a way that a force F_i performs work with the displacement u_i (e.g. a force F_{y2} with v_2), i.e. the force F_i corresponds to a displacement in the direction u_i and a moment M_i corresponds to rotation φ_i about the rotational axis of the moment. The diagonal terms are (like spring constants) always positive. It can be shown that the stiffness matrix is symmetric and positive (semi-)definite under these conditions. In non-kinematic systems, the stiffness matrix is regular and thus the system of equations can be solved. In kinematic systems a singular stiffness matrix is obtained, i.e. the system of equations has no solution (see Sect. 1.4.2).

For each individual element, the relationship between the forces and the displacements at the element nodes can be expressed as a stiffness matrix similar to (3.1). This matrix is called the *element stiffness matrix*. The global stiffness matrix can be assembled from the element stiffness matrices following a simple scheme.

After the global stiffness matrix $\underline{K}$ and the load vector $\underline{F}$ have been determined, (3.1) is solved following a procedure for the solution of linear systems of equations,

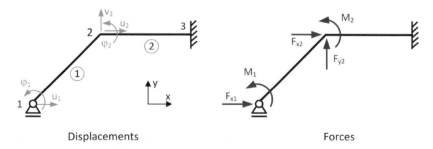

Displacements   Forces

Fig. 3.4 Degrees of freedom and nodal loads for a frame structure

e.g. the Gaussian elimination algorithm. Depending on the size of the problem, these can be systems of equations with several hundred, thousands, or hundred of thousands of unknowns. The greatest computational effort in a finite element analysis normally consists of the solution of this system of equations. After the solution of the system of equations with the displacements as unknowns, the support reactions are determined. The sectional forces are computed element by element using the nodal displacements. This leads to the following steps which are performed in each finite element analysis.

Computational steps in a finite element analysis
1. Computation of element stiffness matrices and evaluation of nodal point loads
2. Assembly of the global stiffness matrix using the element stiffness matrices and assembly of the load vector
3. Solution of the system of equations with displacements as unknowns
4. Determination of support reactions from the displacements
5. Determination of the element stresses (sectional forces) element by element from the displacements

3.2 Introductory Example: Plane Truss System

3.2.1 Structural System

The computational steps in a finite element analysis are explained using the example of a plane truss. Thereafter, spring and beam elements are introduced. The combination of different types of elements is shown with an example in Sect. 3.5.

A plane truss element has two degrees of freedom at each nodal point. For a truss in the xy-plane, these are the displacements u and v. Displacements in the z-direction as well as rotational degrees of freedom do not occur.

Example 3.1 The truss system in Fig. 3.5 is discretized into four nodal points and six truss elements. The elements numbered 5 and 6 are not connected at their crossing point. The truss system is loaded by the forces F_{x1} to F_{y4}. The type of the system of equations has to be determined.

Both the displacements of nodal point 4 as well as the vertical displacement of point 3 are fixed due to the support conditions. Hence, the system possesses 5 degrees of freedom. The system of equations with the displacements as unknowns has the form

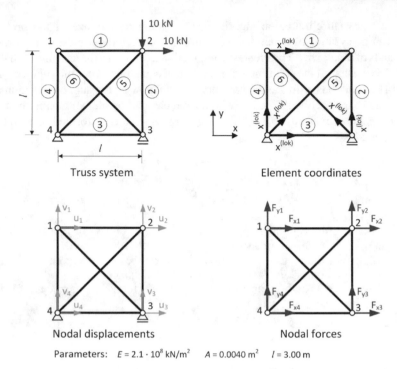

Fig. 3.5 Introductory example

$$
\begin{bmatrix}
k_{11} & k_{12} & k_{13} & k_{14} & k_{15} \\
k_{21} & k_{22} & k_{23} & k_{24} & k_{25} \\
k_{31} & k_{32} & k_{33} & k_{34} & k_{35} \\
k_{41} & k_{42} & k_{43} & k_{44} & k_{45} \\
k_{51} & k_{52} & k_{53} & k_{54} & k_{55}
\end{bmatrix}
\cdot
\begin{bmatrix}
u_1 \\ v_1 \\ u_2 \\ v_2 \\ u_3
\end{bmatrix}
=
\begin{bmatrix}
F_{x1} \\ F_{y1} \\ F_{x2} \\ F_{y2} \\ F_{x3}
\end{bmatrix} .
$$

◄

For the computation of the global stiffness matrix, the element stiffness matrices of the truss elements are needed. Therefore, the element stiffness matrix of a truss element is derived next.

3.2.2 Element Stiffness Matrix of a Truss Element

The truss element is a bar element which only admits the transfer of normal forces. Its element stiffness matrix is quite simple.

The *local coordinate* $x^{(\text{lok})}$ is introduced to describe element forces and displacements (Fig. 3.6). The element variables in the local coordinate system are indicated by the superscript "(lok)".

Fig. 3.6 Truss element in local coordinates

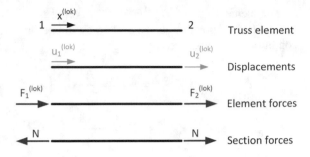

A bar with cross-sectional area A, Young's modulus E, and length l with a constant normal force N is considered. The elongation δ of the bar is

$$\delta = \frac{N \cdot l}{E \cdot A}.$$

Expressing the elongation of the bar as

$$\delta = u_2^{(lok)} - u_1^{(lok)}$$

with respect to the displacements $u_1^{(lok)}$ and $u_2^{(lok)}$ of the nodal points in the local coordinate system, the normal force N is obtained as

$$N = \frac{E \cdot A}{l} \cdot \delta = \frac{E \cdot A}{l} \cdot \left(-u_1^{(lok)} + u_2^{(lok)} \right)$$

or in matrix notation

$$N = \frac{E \cdot A}{l} \cdot \begin{bmatrix} -1 & 1 \end{bmatrix} \cdot \begin{bmatrix} u_1^{(lok)} \\ u_2^{(lok)} \end{bmatrix} \tag{3.2}$$

or as

$$N = \underline{S}_e^{(lok)} \cdot \underline{u}_e^{(lok)} \tag{3.2a}$$

where

$$\underline{S}_e^{(lok)} = \frac{E \cdot A}{l} \cdot \begin{bmatrix} -1 & 1 \end{bmatrix}. \tag{3.2b}$$

The matrix $\underline{S}_e^{(lok)}$ is called the *section force matrix*. Equation (3.2) is used to determine the normal force of a truss element by applying the nodal displacements. This can only be done after the solution of the global system of equations.

The element forces $F_1^{(lok)}$ and $F_2^{(lok)}$ act at the ends of the truss element. Both are defined as positive in the $x^{(lok)}$ coordinate direction. Due to the equilibrium conditions, they are given as

$$F_1^{(lok)} = -N = \frac{E \cdot A}{l} \cdot \left(u_1^{(lok)} - u_2^{(lok)} \right) \tag{3.2c}$$

$$F_2^{(lok)} = N = \frac{E \cdot A}{l} \cdot \left(-u_1^{(lok)} + u_2^{(lok)} \right). \tag{3.2d}$$

These equations define the relationship between the nodal displacements and the element forces. In matrix notation, they are

$$\frac{E \cdot A}{l} \cdot \begin{bmatrix} 1 & -1 \\ -1 & 1 \end{bmatrix} \cdot \begin{bmatrix} u_1^{(lok)} \\ u_2^{(lok)} \end{bmatrix} = \begin{bmatrix} F_1^{(lok)} \\ F_2^{(lok)} \end{bmatrix} \tag{3.3}$$

$$\underline{K}_e^{(lok)} \cdot \underline{u}_e^{(lok)} = \underline{F}_e^{(lok)} \tag{3.3a}$$

where the matrix

$$\underline{K}_e^{(lok)} = \frac{E \cdot A}{l} \cdot \begin{bmatrix} 1 & -1 \\ -1 & 1 \end{bmatrix} \tag{3.3b}$$

represents the *element stiffness matrix* of the truss element in local coordinates. It is, as all stiffness matrices, symmetric. The element stiffness matrix is singular, since the second row of the matrix can be obtained by multiplying the first row by -1. From a mechanical point of view, this means that the structural system—which here is a single truss element—is kinematic.

Example 3.2 The element stiffness matrices and the stress matrices in local coordinates have to be determined for all elements of the truss system in Fig. 3.5.

The stiffness matrices are obtained with $A = 0.004\,\text{m}^2$, $E = 2.1 \cdot 10^8\,\text{kN/m}^2$ according to (3.3) as:

Elements 1–4 (with $l = 3.00$ m):

$$\underline{K}_e^{(lok)} = 2.80 \cdot 10^5 \cdot \begin{bmatrix} 1 & -1 \\ -1 & 1 \end{bmatrix}$$

Elements 5 and 6 (with $l = 4.24$ m):

$$\underline{K}_e^{(lok)} = 1.98 \cdot 10^5 \cdot \begin{bmatrix} 1 & -1 \\ -1 & 1 \end{bmatrix}$$

The section force matrices are obtained with (3.2b) in local coordinates as:

Elements 1–4:

$$\underline{S}_e^{(lok)} = 2.80 \cdot 10^5 \cdot \begin{bmatrix} -1 & 1 \end{bmatrix}$$

Elements 5 and 6:

$$\underline{S}_e^{(lok)} = 1.98 \cdot 10^5 \cdot \begin{bmatrix} -1 & 1 \end{bmatrix} \qquad \blacktriangleleft$$

3.2.3 Coordinate Transformation

For truss systems with elements having an arbitrary position in space, the element's stiffness in local coordinates has to be transformed into global coordinates. In the following, this transformation is shown for an element of a plane truss system in the xy-plane.

The displacements can be expressed by the displacement components u, v in the global x- and y-directions, respectively, or by the local coordinates $u^{(lok)}$, $v^{(lok)}$. The relationship between the displacements $u^{(lok)}$, $v^{(lok)}$, and the displacements u, v are obtained by simple trigonometric relationships given in Fig. 3.7 as

$$\begin{bmatrix} u^{(lok)} \\ v^{(lok)} \end{bmatrix} = \begin{bmatrix} \cos\alpha & \sin\alpha \\ -\sin\alpha & \cos\alpha \end{bmatrix} \cdot \begin{bmatrix} u \\ v \end{bmatrix} \tag{3.4}$$

$$\underline{u}^{(lok)} = \underline{T}_u \cdot \underline{u}. \tag{3.4a}$$

Applying this coordinate transformation to the nodal point displacements $u_1^{(lok)}$, $u_2^{(lok)}$ of the truss element, the displacements $u_1^{(e)}$, $u_2^{(e)}$ in the x-direction and $v_1^{(e)}$, $v_2^{(e)}$ in the y-direction of the global coordinate system are

$$\begin{bmatrix} u_1^{(lok)} \\ u_2^{(lok)} \end{bmatrix} = \begin{bmatrix} \cos\alpha & \sin\alpha & 0 & 0 \\ 0 & 0 & \cos\alpha & \sin\alpha \end{bmatrix} \cdot \begin{bmatrix} u_1^{(e)} \\ v_1^{(e)} \\ u_2^{(e)} \\ v_2^{(e)} \end{bmatrix} \tag{3.5}$$

$$\underline{u}_e^{(lok)} = \underline{T} \cdot \underline{u}_e. \tag{3.5a}$$

The matrix $\underline{T}$ is the transformation matrix of the displacements of the truss element.

Coordinate systems

Coordinate transformation of the displacements

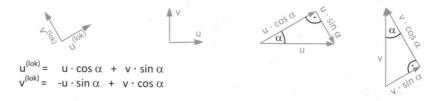

$$u^{(\text{lok})} = \quad u \cdot \cos \alpha \; + \; v \cdot \sin \alpha$$
$$v^{(\text{lok})} = \; -u \cdot \sin \alpha \; + \; v \cdot \cos \alpha$$

Coordinate transformation of the forces

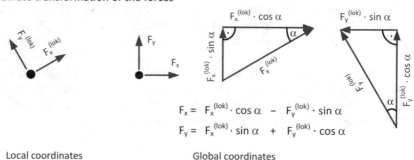

$$F_x = F_x^{(\text{lok})} \cdot \cos \alpha \; - \; F_y^{(\text{lok})} \cdot \sin \alpha$$
$$F_y = F_x^{(\text{lok})} \cdot \sin \alpha \; + \; F_y^{(\text{lok})} \cdot \cos \alpha$$

Local coordinates   Global coordinates

Fig. 3.7 Coordinate transformation of the nodal displacements and forces

The forces can be transformed similarly. For this, the forces $F_x^{(\text{lok})}$ and $F_y^{(\text{lok})}$ related to the local coordinate system are decomposed into global coordinates. The forces F_x and F_y in global coordinates are obtained according to Fig. 3.7 as

$$\begin{bmatrix} F_x \\ F_y \end{bmatrix} = \begin{bmatrix} \cos \alpha & -\sin \alpha \\ \sin \alpha & \cos \alpha \end{bmatrix} \cdot \begin{bmatrix} F_x^{(\text{lok})} \\ F_y^{(\text{lok})} \end{bmatrix}. \tag{3.6}$$

Applying this relationship to the element forces at both ends of the truss element gives the element forces and global coordinates as (Fig. 3.8)

Truss element

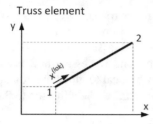
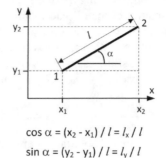

$$\cos \alpha = (x_2 - x_1) \,/\, l = l_x \,/\, l$$
$$\sin \alpha = (y_2 - y_1) \,/\, l = l_y \,/\, l$$

Nodal displacements

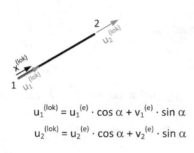

$$u_1^{(\text{lok})} = u_1^{(e)} \cdot \cos \alpha + v_1^{(e)} \cdot \sin \alpha$$
$$u_2^{(\text{lok})} = u_2^{(e)} \cdot \cos \alpha + v_2^{(e)} \cdot \sin \alpha$$

Nodal forces

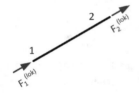
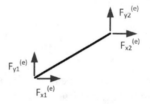

$$F_{x1}^{(e)} = F_1^{(\text{lok})} \cdot \cos \alpha \qquad F_{x2}^{(e)} = F_2^{(\text{lok})} \cdot \cos \alpha$$
$$F_{y1}^{(e)} = F_1^{(\text{lok})} \cdot \sin \alpha \qquad F_{y2}^{(e)} = F_2^{(\text{lok})} \cdot \sin \alpha$$

Local coordinates   Global coordinates

Fig. 3.8 Coordinate transformation of the truss element

$$
\begin{bmatrix} F_{x1}^{(e)} \\ F_{y1}^{(e)} \\ F_{x2}^{(e)} \\ F_{y2}^{(e)} \end{bmatrix}
=
\begin{bmatrix} \cos \alpha & 0 \\ \sin \alpha & 0 \\ 0 & \cos \alpha \\ 0 & \sin \alpha \end{bmatrix}
\cdot
\begin{bmatrix} F_1^{(\text{lok})} \\ F_2^{(\text{lok})} \end{bmatrix}
\tag{3.7}
$$

$$\underline{F}_e = \underline{T}^{\mathrm{T}} \cdot \underline{F}_e^{(\text{lok})}. \tag{3.7a}$$

The transformation matrix of the forces is, in general, the transpose of the transformation matrix of the displacements. It can be shown that this is always true under the conditions for the vectors $\underline{u}_e$ and $\underline{F}_e$ given in Sect. 3.1.3 (see Sect. 4.10.2).

The transformation matrices for the displacements and forces are now used to transform the stiffness matrix of the truss element from local to global coordinates. When the forces $\underline{F}_e^{(lok)}$ in (3.7a) are substituted by the stiffness relationship (3.3a)

$$\underline{F}_e^{(lok)} = \underline{K}_e^{(lok)} \cdot \underline{u}_e^{(lok)}$$

the forces in global coordinates are obtained as

$$\underline{F}_e = \underline{T}^{\mathrm{T}} \cdot \underline{K}_e^{(lok)} \cdot \underline{u}_e^{(lok)}.$$

Now the displacements $\underline{u}_e^{(lok)}$ are substituted using

$$\underline{u}_e^{(lok)} = \underline{T} \cdot \underline{u}_e$$

according to (3.5a) by the displacements $\underline{u}_e$ in global coordinates. This leads to the stiffness relationship in global coordinates

$$\underline{F}_e = \underline{K}_e \cdot \underline{u}_e \tag{3.8}$$

or

$$\underline{F}_e = \underline{T}^{\mathrm{T}} \cdot \underline{K}_e^{(lok)} \cdot \underline{T} \cdot \underline{u}_e. \tag{3.8a}$$

Hence, the element stiffness matrix $\underline{K}_e$ in global coordinates is obtained by the transformation of the stiffness matrix $\underline{K}_e^{(lok)}$ in local coordinates as

$$\underline{K}_e = \underline{T}^{\mathrm{T}} \cdot \underline{K}_e^{(lok)} \cdot \underline{T}. \tag{3.8b}$$

The stiffness relationship of the truss element in global coordinates according to Fig. 3.9 can thus be written as

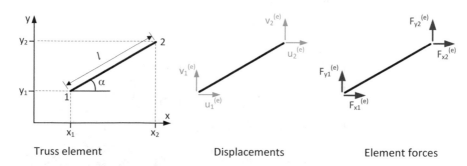

Truss element   Displacements   Element forces

Fig. 3.9 Truss element in global coordinates

$$\frac{E \cdot A}{l} \cdot \begin{bmatrix} c^2 & s \cdot c & -c^2 & -s \cdot c \\ s \cdot c & s^2 & -s \cdot c & -s^2 \\ -c^2 & -s \cdot c & c^2 & s \cdot c \\ -s \cdot c & -s^2 & s \cdot c & s^2 \end{bmatrix} \cdot \begin{bmatrix} u_1^{(e)} \\ v_1^{(e)} \\ u_2^{(e)} \\ v_2^{(e)} \end{bmatrix} = \begin{bmatrix} F_{x1}^{(e)} \\ F_{y1}^{(e)} \\ F_{x2}^{(e)} \\ F_{y2}^{(e)} \end{bmatrix} \tag{3.9}$$

$$\underline{K}_e \cdot \underline{u}_e = \underline{F}_e \tag{3.9a}$$

with

$$s = \sin\alpha \quad \text{and} \quad c = \cos\alpha$$

or alternatively as (see Fig. 3.8)

$$\frac{E \cdot A}{l^3} \cdot \begin{bmatrix} l_x^2 & l_x \cdot l_y & -l_x^2 & -l_x \cdot l_y \\ l_x \cdot l_y & l_y^2 & -l_x \cdot l_y & -l_y^2 \\ -l_x^2 & -l_x \cdot l_y & l_x^2 & l_x \cdot l_y \\ -l_x \cdot l_y & -l_y^2 & l_x \cdot l_y & l_y^2 \end{bmatrix} \cdot \begin{bmatrix} u_1^{(e)} \\ v_1^{(e)} \\ u_2^{(e)} \\ v_2^{(e)} \end{bmatrix} = \begin{bmatrix} F_{x1}^{(e)} \\ F_{y1}^{(e)} \\ F_{x2}^{(e)} \\ F_{y2}^{(e)} \end{bmatrix} \tag{3.9b}$$

with

$$l_x = x_2 - x_1, l_y = y_2 - y_1, l = \sqrt{l_x^2 + l_y^2}. \tag{3.9c}$$

In the same way, the section force matrix can be transformed from local to global coordinates. Substituting (3.5a)

$$\underline{u}_e^{(\text{lok})} = \underline{T} \cdot \underline{u}_e$$

in (3.2a)

$$N = \underline{S}_e^{(\text{lok})} \cdot \underline{u}_e^{(\text{lok})}$$

one obtains

$$N = \underline{S}_e \cdot \underline{u}_e \tag{3.10}$$

with

$$\underline{S}_e = \underline{S}_e^{(\text{lok})} \cdot \underline{T}$$

$$\underline{S}_e = \frac{E \cdot A}{l} \left[-\cos\alpha \quad -\sin\alpha \quad \cos\alpha \quad \sin\alpha \right] \tag{3.10a}$$

or

$$\underline{S}_e = \frac{E \cdot A}{l^2} \left[-l_x \quad -l_y \quad l_x \quad l_y \right]. \tag{3.10b}$$

Example 3.3 The stiffness and section force matrices of the elements of the truss system in Fig. 3.5 of Example 3.1 are to be determined in local coordinates.

The nodal displacements and element forces are denoted as given in Figs. 3.5 and 3.9, respectively. Both the element forces and the nodal displacements are positive in the global coordinate directions. The element forces are denoted with a superscript indicating the element number. The support conditions of the displacements are not yet taken into account.

Element 1 according to (3.9) with $\alpha = 0°$:

$$\begin{bmatrix} F_{x1}^{(1)} \\ F_{y1}^{(1)} \\ F_{x2}^{(1)} \\ F_{y2}^{(1)} \end{bmatrix} = 2.80 \cdot 10^5 \begin{bmatrix} 1 & 0 & -1 & 0 \\ 0 & 0 & 0 & 0 \\ -1 & 0 & 1 & 0 \\ 0 & 0 & 0 & 0 \end{bmatrix} \cdot \begin{bmatrix} u_1 \\ v_1 \\ u_2 \\ v_2 \end{bmatrix}$$

or with (3.3)

$$\begin{bmatrix} F_{x1}^{(1)} \\ F_{x2}^{(1)} \end{bmatrix} = 2.80 \cdot 10^5 \begin{bmatrix} 1 & -1 \\ -1 & 1 \end{bmatrix} \cdot \begin{bmatrix} u_1 \\ u_2 \end{bmatrix}$$

Element 2:

$$\begin{bmatrix} F_{y3}^{(2)} \\ F_{y2}^{(2)} \end{bmatrix} = 2.80 \cdot 10^5 \begin{bmatrix} 1 & -1 \\ -1 & 1 \end{bmatrix} \cdot \begin{bmatrix} v_3 \\ v_2 \end{bmatrix}$$

Element 3:

$$\begin{bmatrix} F_{x4}^{(3)} \\ F_{x3}^{(3)} \end{bmatrix} = 2.80 \cdot 10^5 \begin{bmatrix} 1 & -1 \\ -1 & 1 \end{bmatrix} \cdot \begin{bmatrix} u_4 \\ u_3 \end{bmatrix}$$

Element 4:

$$\begin{bmatrix} F_{y4}^{(4)} \\ F_{y1}^{(4)} \end{bmatrix} = 2.80 \cdot 10^5 \begin{bmatrix} 1 & -1 \\ -1 & 1 \end{bmatrix} \cdot \begin{bmatrix} v_4 \\ v_1 \end{bmatrix}$$

Element 5 with $\alpha = 45°$:

$$\begin{bmatrix} F_{x4}^{(5)} \\ F_{y4}^{(5)} \\ F_{x2}^{(5)} \\ F_{y2}^{(5)} \end{bmatrix} = 1.98 \cdot 10^5 \begin{bmatrix} 0.5 & 0.5 & -0.5 & -0.5 \\ 0.5 & 0.5 & 0.5 & -0.5 \\ -0.5 & -0.5 & 0.5 & 0.5 \\ -0.5 & -0.5 & 0.5 & 0.5 \end{bmatrix} \cdot \begin{bmatrix} u_4 \\ v_4 \\ u_2 \\ v_2 \end{bmatrix}$$

Element 6 with $\alpha = 135°$:

$$\begin{bmatrix} F_{x3}^{(6)} \\ F_{y3}^{(6)} \\ F_{x1}^{(6)} \\ F_{y1}^{(6)} \end{bmatrix} = 1.98 \cdot 10^5 \begin{bmatrix} 0.5 & -0.5 & -0.5 & 0.5 \\ -0.5 & 0.5 & 0.5 & -0.5 \\ -0.5 & 0.5 & 0.5 & -0.5 \\ 0.5 & -0.5 & -0.5 & 0.5 \end{bmatrix} \cdot \begin{bmatrix} u_3 \\ v_3 \\ u_1 \\ v_1 \end{bmatrix} \cdot$$

The section force matrices in global coordinates are:

Elements 1 and 3:

$$\underline{S}_e = 2.80 \cdot 10^5 \begin{bmatrix} -1 & 1 \end{bmatrix}$$

Elements 2 and 4:

$$\underline{S}_e = 2.80 \cdot 10^5 \begin{bmatrix} -1 & 1 \end{bmatrix}$$

Elements 5 with $\alpha = 45°$:

$$\underline{S}_e = 1.98 \cdot 10^5 \begin{bmatrix} -0.71 & -0.71 & 0.71 & 0.71 \end{bmatrix}$$

Elements 6 with $\alpha = 135°$:

$$\underline{S}_e = 1.98 \cdot 10^5 \begin{bmatrix} 0.71 & -0.71 & -0.71 & 0.71 \end{bmatrix}.$$

The stress matrices related to the same displacement vectors are the corresponding element stiffness matrices. ◀

3.2.4 Global Stiffness Matrix

With the stiffness matrices of all elements, the global stiffness matrix which is related
to all degrees of freedom of the system can easily be assembled. The procedure for
computing the global stiffness matrix is based on the compatibility equations at the
nodal points. In Fig. 3.10, a nodal point with several adjoining elements is shown.
The conditions for the connection of the elements with the nodal point are

– compatibility of the displacements of the elements with those of the nodal point,
– equilibrium conditions at the nodal point.

The compatibility conditions of the displacements are easy to satisfy. For this,
the global coordinate displacements only have to be introduced in place of the local
displacements of the elements. The displacements $u_2^{(a)}$, $u_1^{(b)}$ and $u_1^{(c)}$ of the elements
a, b and c in Fig. 3.10 are, for example, identical to the global displacement u_i
of the nodal point i and can, therefore, be substituted by u_i. The same applies to
$v_2^{(a)}$, $v_1^{(b)}$, $v_1^{(c)}$ and v_i.

In Example 3.3, this has been done already for the element stiffness matrices.
The relationship between the local and global degrees of freedom can be given as a
so-called *incidence matrix*. It defines for each element nodal point the corresponding
nodal point of the system (see Example 3.5).

Now the equilibrium conditions for the element forces and nodal loads at a nodal
point are considered. For point i in Fig. 3.10, they are

$$F_{x2}^{(a)} + F_{x1}^{(b)} + F_{x1}^{(c)} = F_{xi}$$

$$F_{y1}^{(b)} + F_{y1}^{(c)} = F_{yi}.$$

In these equations, the element forces $F_{x2}^{(a)}$, $F_{x1}^{(b)}$, $F_{y1}^{(b)}$, $F_{x1}^{(c)}$ and $F_{y1}^{(c)}$ are expressed by
the global nodal displacements using the stiffness matrices of the elements a, b and c.

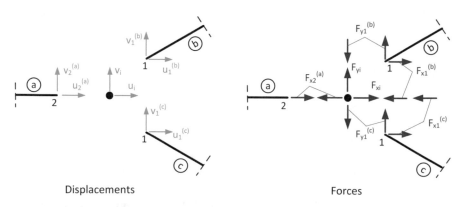

Displacements   Forces

Fig. 3.10 Displacements and forces at the nodal point i with truss elements a, b, c

For example, the element a may be connected with the nodal points h (initial point of the element) and i (end point of the element). The element force $F_{x2}^{(a)}$ can be expressed with the stiffness matrix of the element (3.3b) by the displacements u_h and u_i:

$$
\begin{bmatrix} F_{x1}^{(a)} \\ F_{x2}^{(a)} \end{bmatrix} = \begin{bmatrix} k_{11} & k_{12} \\ k_{21} & k_{22} \end{bmatrix} \cdot \begin{bmatrix} u_h \\ u_i \end{bmatrix}
$$

or

$$
F_{x1}^{(a)} = k_{11} \cdot u_h + k_{12} \cdot u_i
$$
$$
F_{x2}^{(a)} = k_{21} \cdot u_h + k_{22} \cdot u_i .
$$

Introducing the relationship for $F_{x2}^{(a)}$ in the equilibrium condition of the forces in the x-direction at nodal point i, one obtains

$$
k_{21} \cdot u_h + k_{22} \cdot u_i + \text{contributions of elements } b \text{ and } c = F_{xi}
$$

or

$$
k_{21} \cdot u_h + (k_{22} + \text{contributions of } b \text{ and } c \text{ to the degree of freedom } u_i) \cdot u_i
$$
$$
+ \text{contributions of } b \text{ and } c \text{ to other degrees of freedom} = F_{xi} .
$$

The contribution of element a to the equilibrium in the x-direction at nodal point i consists of adding the entries of the element stiffness matrix entrywise with respect to the corresponding degrees of freedom. By setting up the equations for the force components in all degrees of freedom, the stiffness relationship of the global system is obtained. Its coefficient matrix is the global stiffness matrix of the system. It is assembled by summing the entries of the element stiffness matrices entrywise in the corresponding degrees of freedom. The procedure is explained in Example 3.4.

Alternatively, the allocation of the element stiffness matrices to the global stiffness matrix can be described with so-called incidence matrices. An incidence matrix describes the assignment of the local degrees of freedom for a single element to the global degrees of freedom. It is obtained using an allocation table for the nodal points. It can be used to transform the stiffness matrix $\underline{K}_e^{(i)}$ of an element to global degrees of freedom, as shown in Example 3.5. Finally, the transformed element stiffness matrices are summed up to the global stiffness matrix. The stiffness matrix related to the global degrees of freedom of the system is as follows:

$$
\underline{K}_{e,f}^{(i)} \cdot \underline{u}_f = \underline{F}_{e,f}^{(i)} \tag{3.11}
$$

If the nodal forces $\underline{F}_{e,f}^{(i)}$ of all n elements are added, the external nodal point loads $\underline{F}_f$ are obtained, as shown in the example in Fig. 3.10, as

$$\sum_{i=1}^{n} \underline{F}_{e,f}^{(i)} = \underline{F}_f. \tag{3.12}$$

If the stiffness relationships of the elements according to (3.11) are used for the nodal forces, the result is as follows:

$$\sum_{i=1}^{n} \left(\underline{K}_{e,f}^{(i)} \cdot \underline{u}_f \right) = \underline{F}_f \tag{3.13}$$

or

$$\sum_{i=1}^{n} \left(\underline{K}_{e,f}^{(i)} \right) \cdot \underline{u}_f = \underline{F}_f \tag{3.14}$$

or

$$\underline{K}_f \cdot \underline{u}_f = \underline{F}_f \tag{3.15}$$

with

$$\underline{K}_f = \sum_{i=1}^{n} \underline{K}_{e,f}^{(i)}. \tag{3.15a}$$

The system stiffness matrix is thus the sum of the element stiffness matrices transformed to the global degrees of freedom.

In this way, the global stiffness matrix is obtained as

$$\underline{K}_f \cdot \underline{u}_f = \underline{F}_f$$

where

$\underline{K}_f$: global stiffness matrix (without support conditions);
$\underline{u}_f$: vector of nodal displacements;
$\underline{F}_f$: vector of nodal loads.

The support conditions are not yet considered in (3.15). The system can move freely in the xy-plane, which is indicated by the subscript f. Therefore, the matrix $\underline{K}_f$ is singular, just like the element stiffness matrices (3.3b) and (3.8).

Example 3.4 The global stiffness matrix of the truss system in Fig. 3.5 is to be assembled without yet considering the support conditions.

Fig. 3.11 Equilibrium at
nodal point 1

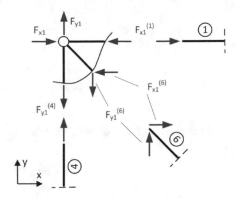

For nodal point 1, the equations for the stiffness matrix are given in detail. First, the equations for the equilibrium of the element forces and the external nodal loads at nodal point 1 are written with respect to the x- and y-directions (Fig. 3.11):

$$F_{x1}^{(1)} + F_{x1}^{(6)} = F_{x1}$$
$$F_{y1}^{(4)} + F_{y1}^{(6)} = F_{y1}$$

The element forces act as reacting forces on the nodal point in the negative x- or y-direction, respectively. The external load forces are defined as positive in the positive x- and y-directions.

The element forces given above are now expressed by the nodal point displacements using the element stiffness matrices. The support conditions are not yet considered. For example, the element force of element 1 is given by the first row of the element stiffness matrix

$$\begin{bmatrix} F_{x1}^{(1)} \\ F_{x2}^{(1)} \end{bmatrix} = 2.80 \cdot 10^5 \begin{bmatrix} 1 & -1 \\ -1 & 1 \end{bmatrix} \cdot \begin{bmatrix} u_1 \\ u_2 \end{bmatrix}$$

as

$$F_{x1}^{(1)} = 2.80 \cdot 10^5 \begin{bmatrix} 1 & -1 \end{bmatrix} \cdot \begin{bmatrix} u_1 \\ u_2 \end{bmatrix}.$$

In the same way, the element forces of element 6 at node 1 are obtained as (with the factor $2.8 \cdot 10^5$ instead of $1.98 \cdot 10^5$)

$$\begin{bmatrix} F_{x1}^{(6)} \\ F_{y1}^{(6)} \end{bmatrix} = 2.8 \cdot 10^5 \begin{bmatrix} -0.35 & 0.35 & 0.35 & -0.35 \\ 0.35 & -0.35 & -0.35 & 0.35 \end{bmatrix} \cdot \begin{bmatrix} u_3 \\ v_3 \\ u_1 \\ v_1 \end{bmatrix}$$

and the element force of element 4 is

$$F_{y1}^{(4)} = 2.80 \cdot 10^5 \begin{bmatrix} 1 & -1 \end{bmatrix} \cdot \begin{bmatrix} v_4 \\ v_1 \end{bmatrix}.$$

If these relationships are introduced into the equilibrium conditions given above for nodal point 1, the following relationship is obtained:

$$2.8 \cdot 10^5 \cdot \begin{bmatrix} 1.00 + 0.35 & -0.35 & -1.00 & 0 & -0.35 & 0.35 & 0 & 0 \\ -0.35 & 1.00 + 0.35 & 0 & 0 & 0.35 & -0.35 & 0 & -1.00 \end{bmatrix}$$

$$\cdot \begin{bmatrix} u_1 \\ v_1 \\ u_2 \\ v_2 \\ u_3 \\ v_3 \\ u_4 \\ v_4 \end{bmatrix} = \begin{bmatrix} F_{x1} \\ F_{y1} \end{bmatrix}.$$

These entries comprise already the first and the second row of the global stiffness matrix, The entries with factors ± 0.35 result from the element stiffness matrix of element 6, the entries with ± 1 from the element stiffness matrices of elements 1 and 4.

In order to achieve the complete global stiffness matrix, the equilibrium conditions at nodes 2, 3, and 4 have to be added.

Hence, the global stiffness matrix is assembled by accumulating the entries of the element stiffness matrices of all elements in the corresponding degrees of freedom. In the following, the assembly of the global stiffness matrix for the system in Fig. 3.5 is demonstrated.

The element stiffness matrices have already been evaluated in Example 3.3. They are added element-wise to the global stiffness matrix.

Stiffness of element 1:

$$2.8 \cdot 10^5$$

$$\cdot \begin{bmatrix} \mathbf{1.00} & 0 & \mathbf{-1.00} & 0 & 0 & 0 & 0 & 0 \\ 0 & 0 & 0 & 0 & 0 & 0 & 0 & 0 \\ \mathbf{-1.00} & 0 & \mathbf{1.00} & 0 & 0 & 0 & 0 & 0 \\ 0 & 0 & 0 & 0 & 0 & 0 & 0 & 0 \\ 0 & 0 & 0 & 0 & 0 & 0 & 0 & 0 \\ 0 & 0 & 0 & 0 & 0 & 0 & 0 & 0 \\ 0 & 0 & 0 & 0 & 0 & 0 & 0 & 0 \\ 0 & 0 & 0 & 0 & 0 & 0 & 0 & 0 \end{bmatrix} \cdot \begin{bmatrix} u_1 \\ v_1 \\ u_2 \\ v_2 \\ u_3 \\ v_3 \\ u_4 \\ v_4 \end{bmatrix} = \begin{bmatrix} F_{x1} \\ F_{y1} \\ F_{x2} \\ F_{y2} \\ F_{x3} \\ F_{y3} \\ F_{x4} \\ F_{y4} \end{bmatrix},$$

Addition of element 2:

$$2.8 \cdot 10^5$$

$$\cdot \begin{bmatrix} 1.00 & 0 & -1.00 & 0 & 0 & 0 & 0 & 0 \\ 0 & 0 & 0 & 0 & 0 & 0 & 0 & 0 \\ -1.00 & 0 & 1.00 & 0 & 0 & 0 & 0 & 0 \\ 0 & 0 & 0 & \mathbf{1.00} & 0 & \mathbf{-1.00} & 0 & 0 \\ 0 & 0 & 0 & 0 & 0 & 0 & 0 & 0 \\ 0 & 0 & 0 & \mathbf{-1.00} & 0 & \mathbf{1.00} & 0 & 0 \\ 0 & 0 & 0 & 0 & 0 & 0 & 0 & 0 \\ 0 & 0 & 0 & 0 & 0 & 0 & 0 & 0 \end{bmatrix} \cdot \begin{bmatrix} u_1 \\ v_1 \\ u_2 \\ v_2 \\ u_3 \\ v_3 \\ u_4 \\ v_4 \end{bmatrix} = \begin{bmatrix} F_{x1} \\ F_{y1} \\ F_{x2} \\ F_{y2} \\ F_{x3} \\ F_{y3} \\ F_{x4} \\ F_{y4} \end{bmatrix}$$

Addition of element 3:

$$2.8 \cdot 10^5$$

$$\cdot \begin{bmatrix} 1.00 & 0 & -1.00 & 0 & 0 & 0 & 0 & 0 \\ 0 & 0 & 0 & 0 & 0 & 0 & 0 & 0 \\ -1.00 & 0 & 1.00 & 0 & 0 & 0 & 0 & 0 \\ 0 & 0 & 0 & 1.00 & 0 & -1.00 & 0 & 0 \\ 0 & 0 & 0 & 0 & \mathbf{1.00} & 0 & \mathbf{-1.00} & 0 \\ 0 & 0 & 0 & -1.00 & 0 & 1.00 & 0 & 0 \\ 0 & 0 & 0 & 0 & \mathbf{-1.00} & 0 & \mathbf{1.00} & 0 \\ 0 & 0 & 0 & 0 & 0 & 0 & 0 & 0 \end{bmatrix} \cdot \begin{bmatrix} u_1 \\ v_1 \\ u_2 \\ v_2 \\ u_3 \\ v_3 \\ u_4 \\ v_4 \end{bmatrix} = \begin{bmatrix} F_{x1} \\ F_{y1} \\ F_{x2} \\ F_{y2} \\ F_{x3} \\ F_{y3} \\ F_{x4} \\ F_{y4} \end{bmatrix},$$

Addition of element 4:

$$2.8 \cdot 10^5$$

$$
\begin{bmatrix}
1.00 & 0 & -1.00 & 0 & 0 & 0 & 0 & 0 \\
0 & \mathbf{1.00} & 0 & 0 & 0 & 0 & 0 & \mathbf{-1.00} \\
-1.00 & 0 & 1.00 & 0 & 0 & 0 & 0 & 0 \\
0 & 0 & 0 & 1.00 & 0 & -1.00 & 0 & 0 \\
0 & 0 & 0 & 0 & 1.00 & 0 & -1.00 & 0 \\
0 & 0 & 0 & -1.00 & 0 & 1.00 & 0 & 0 \\
0 & 0 & 0 & 0 & -1.00 & 0 & 1.00 & 0 \\
0 & \mathbf{-1.00} & 0 & 0 & 0 & 0 & 0 & \mathbf{1.00}
\end{bmatrix}
\cdot
\begin{bmatrix}
u_1 \\ v_1 \\ u_2 \\ v_2 \\ u_3 \\ v_3 \\ u_4 \\ v_4
\end{bmatrix}
=
\begin{bmatrix}
F_{x1} \\ F_{y1} \\ F_{x2} \\ F_{y2} \\ F_{x3} \\ F_{y3} \\ F_{x4} \\ F_{y4}
\end{bmatrix}
$$

Addition of element 5:

$$2.8 \cdot 10^5$$

$$
\begin{bmatrix}
1.00 & 0 & -1.00 & 0 & 0 & 0 & 0 & 0 \\
0 & 1.00 & 0 & 0 & 0 & 0 & 0 & -1.00 \\
-1.00 & 0 & \mathbf{1.35} & \mathbf{0.35} & 0 & 0 & \mathbf{-0.35} & \mathbf{-0.35} \\
0 & 0 & \mathbf{0.35} & \mathbf{1.35} & 0 & -1.00 & \mathbf{-0.35} & \mathbf{-0.35} \\
0 & 0 & 0 & 0 & 1.00 & 0 & -1.00 & 0 \\
0 & 0 & 0 & -1.00 & 0 & 1.00 & 0 & 0 \\
0 & 0 & \mathbf{-0.35} & \mathbf{-0.35} & -1.00 & 0 & \mathbf{1.35} & \mathbf{0.35} \\
0 & -1.00 & \mathbf{-0.35} & \mathbf{-0.35} & 0 & 0 & \mathbf{0.35} & \mathbf{1.35}
\end{bmatrix}
\cdot
\begin{bmatrix}
u_1 \\ v_1 \\ u_2 \\ v_2 \\ u_3 \\ v_3 \\ u_4 \\ v_4
\end{bmatrix}
=
\begin{bmatrix}
F_{x1} \\ F_{y1} \\ F_{x2} \\ F_{y2} \\ F_{x3} \\ F_{y3} \\ F_{x4} \\ F_{y4}
\end{bmatrix}
$$

Addition of element 6:

$$2.8 \cdot 10^5$$

$$
\begin{bmatrix}
\mathbf{1.35} & \mathbf{-0.35} & -1.00 & 0 & \mathbf{-0.35} & \mathbf{0.35} & 0 & 0 \\
\mathbf{-0.35} & \mathbf{1.35} & 0 & 0 & \mathbf{0.35} & \mathbf{-0.35} & 0 & -1.00 \\
-1.00 & 0 & 1.35 & 0.35 & 0 & 0 & -0.35 & -0.35 \\
0 & 0 & 0.35 & 1.35 & 0 & -1.00 & -0.35 & -0.35 \\
\mathbf{-0.35} & \mathbf{0.35} & 0 & 0 & \mathbf{1.35} & \mathbf{-0.35} & -1.00 & 0 \\
\mathbf{0.35} & \mathbf{-0.35} & 0 & -1.00 & \mathbf{-0.35} & \mathbf{1.35} & 0 & 0 \\
0 & 0 & -0.35 & -0.35 & -1.00 & 0 & 1.35 & 0.35 \\
0 & -1.00 & -0.35 & -0.35 & 0 & 0 & 0.35 & 1.35
\end{bmatrix}
\cdot
\begin{bmatrix}
u_1 \\ v_1 \\ u_2 \\ v_2 \\ u_3 \\ v_3 \\ u_4 \\ v_4
\end{bmatrix}
=
\begin{bmatrix}
F_{x1} \\ F_{y1} \\ F_{x2} \\ F_{y2} \\ F_{x3} \\ F_{y3} \\ F_{x4} \\ F_{y4}
\end{bmatrix}
$$

Now the complete global stiffness matrix of the system has been obtained. During the assembly process, all those entries of the global system matrix inherit entries from the element stiffness matrices where they correspond to degrees of freedom connected by an element. The entries in the global stiffness matrix corresponding to degrees of freedom not connected by an element are zero. For example, the displacements u_1 and u_4 are not connected by an element (element 4 only has a stiffness in the y-direction). Hence, the 7th entry in the first row of the stiffness matrix as well as the first entry in the 7th row of the stiffness matrix are zero. ◀

Example 3.5 For the truss system in Example 3.1, the allocation table or incidence table of the nodal points is to be created. The transformation of the element stiffness matrix to global degrees of freedom is to be shown for a single element.

In the incidence table, the corresponding node numbers in the global system are given for each element. The incidence table for the system in Fig. 3.5 is given in Table 3.1.

The incidence table of the nodes allows the allocation of the global degrees of freedom to the local element degrees of freedom to be determined. For example, the displacements $u_1^{(5)}, v_1^{(5)}$ of the first nodal point of element 5 correspond to the global displacements u_4 and v_4 of the global nodal point 4. The displacements of the second nodal point of element 5 $u_2^{(5)}, v_2^{(5)}$ correspond to the global displacements u_2 and v_2 of nodal point 2 (see Figs. 3.5 and 3.9). This relationship can be described by the following incidence matrix consisting of only 0 and 1 entries.

$$
\begin{bmatrix} u_1^{(5)} \\ v_1^{(5)} \\ u_2^{(5)} \\ v_2^{(5)} \end{bmatrix} =
\begin{bmatrix}
0 & 0 & 0 & 0 & 0 & 0 & 1 & 0 \\
0 & 0 & 0 & 0 & 0 & 0 & 0 & 1 \\
0 & 0 & 1 & 0 & 0 & 0 & 0 & 0 \\
0 & 0 & 0 & 1 & 0 & 0 & 0 & 0
\end{bmatrix}
\cdot
\begin{bmatrix} u_1 \\ v_1 \\ u_2 \\ v_2 \\ u_3 \\ v_3 \\ u_4 \\ v_4 \end{bmatrix}
$$

$$\underline{u}_e^{(5)} = \underline{Z}^{(5)} \cdot \underline{u}_f.$$

For the forces one obtains similarly

$$\underline{F}_f^{(5)} = \underline{Z}^{(5)\mathrm{T}} \cdot \underline{F}_e^{(5)}.$$

Table 3.1 Incidence table of the truss system in Fig. 3.5

Element number	Element nodal point 1	Element nodal point 2
1	1	2
2	3	2
3	4	3
4	4	1
5	4	2
6	3	1

The stiffness relationship (3.9) of element 5

$$\underline{F}_e^{(5)} = \underline{K}_e^{(5)} \cdot \underline{u}_e^{(5)}$$

can be transformed with this equation as

$$\underline{F}_f^{(5)} = \underline{Z}^{(5)T} \cdot \underline{F}_e^{(5)} = \underline{Z}^{(5)T} \cdot \underline{K}_e^{(5)} \cdot \underline{u}_e^{(5)},$$

and therefore,

$$\underline{F}_f^{(5)} = \underline{Z}^{(5)T} \cdot \underline{K}_e^{(5)} \cdot \underline{Z}^{(5)} \cdot \underline{u}_f.$$

With the stiffness matrix $\underline{K}_e^{(5)}$ in Example 3.3, after performing the matrix multiplication, one obtains for element 5

$$
\begin{bmatrix}
F_{x1}^{(5)} \\
F_{y1}^{(5)} \\
F_{x2}^{(5)} \\
F_{y2}^{(5)} \\
F_{x3}^{(5)} \\
F_{y3}^{(5)} \\
F_{x4}^{(5)} \\
F_{y4}^{(5)}
\end{bmatrix}
= 2.8 \cdot 10^5 \cdot
\begin{bmatrix}
0 & 0 & 0 & 0 & 0 & 0 & 0 & 0 \\
0 & 0 & 0 & 0 & 0 & 0 & 0 & 0 \\
0 & 0 & 0.35 & 0.35 & 0 & 0 & -0.35 & -0.35 \\
0 & 0 & 0.35 & 0.35 & 0 & 0 & -0.35 & -0.35 \\
0 & 0 & 0 & 0 & 0 & 0 & 0 & 0 \\
0 & 0 & 0 & 0 & 0 & 0 & 0 & 0 \\
0 & 0 & -0.35 & -0.35 & 0 & 0 & 0.35 & 0.35 \\
0 & 0 & -0.35 & -0.35 & 0 & 0 & 0.35 & 0.35
\end{bmatrix}
\cdot
\begin{bmatrix}
u_1 \\
v_1 \\
u_2 \\
v_2 \\
u_3 \\
v_3 \\
u_4 \\
v_4
\end{bmatrix}
$$

$$\underline{F}_f^{(5)} = \underline{K}_{e,f}^{(5)} \cdot \underline{u}_f.$$

The stiffness matrices of all other elements transformed to the global degrees of freedom are obtained in the same way. Their sum is the global stiffness matrix $\underline{K}_f$.   ◄

3.2.5   Support Conditions

The vector $\underline{u}_f$ of the nodal point displacements contains all degrees of freedom of the system. However, not all displacements in the equation system (3.15) are unknowns; some degrees of freedom have a value which is known. These are the degrees of freedom which are fixed due to the support conditions of the system. Their value is zero for a simple support or nonzero in the case of a prescribed displacement of a support. Firstly, the case of a simple support is considered.

The matrix multiplication in (3.15) means that the columns of the matrix $\underline{K}_f$ corresponding to the fixed degrees of freedom are multiplied by zero. Hence, these columns can be omitted in (3.15). The rows in (3.15) corresponding to the fixed degrees of freedom cannot be used in the solution of the system of equations since they contain the support forces as unknown nodal forces on the right-hand side. The system of equations which is obtained after removing the columns of the fixed degrees of freedom is, therefore, divided into a system of equations containing the forces of the free degrees of freedom and a second system of equations describing the equilibrium at the fixed degrees of freedom.

The first system of equations is square since the same number of rows as columns of the original system of equations have been removed. It is the equation given already in (3.1):

$$\underline{K} \cdot \underline{u} = \underline{F}.$$

From this system of equations, the nodal point displacements are calculated. With the second system of equations, the support forces can be determined afterwards.

A prescribed displacement can be handled in a similar way. The corresponding column of the stiffness matrix is multiplied by the displacement and subtracted from both sides of the system of equations. The corresponding row is then omitted in the system of equations. It can be used to determine the reaction force at the nodal point with the prescribed displacement (see Example 3.10).

Example 3.6 For the truss system in Fig. 3.5, the support conditions have to be incorporated into the global stiffness matrix determined in Example 3.4. In addition, the equations for the computation of the support forces from the global displacements have to be determined.

The support conditions of the truss system in Fig. 3.5 are

$$v_3 = 0 \quad u_4 = 0 \quad v_4 = 0.$$

These relationships are introduced in the system of equations corresponding to (3.15) in Example 3.4. Thereafter, the equations are ordered according to the free and fixed degrees of freedom. Then the following systems of equations are obtained:

1. Equations for determining the displacements:

$$
2.80 \cdot 10^5
\begin{bmatrix}
1.35 & -0.35 & -1.00 & 0 & -0.35 \\
-0.35 & 1.35 & 0 & 0 & 0.35 \\
-1.00 & 0 & 1.35 & 0.35 & 0 \\
0 & 0 & 0.35 & 1.35 & 0 \\
-0.35 & 0.35 & 0 & 0 & 1.35
\end{bmatrix}
\cdot
\begin{bmatrix}
u_1 \\
v_1 \\
u_2 \\
v_2 \\
u_3
\end{bmatrix}
=
\begin{bmatrix}
F_{x1} \\
F_{y1} \\
F_{x2} \\
F_{y2} \\
F_{x3}
\end{bmatrix}.
$$

2. Equations for determining the support forces from the nodal displacements:

$$2.80 \cdot 10^5 \cdot \begin{bmatrix} 0.35 & -0.35 & 0 & -1.00 & -0.35 \\ 0 & 0 & -0.35 & -0.35 & -1.00 \\ 0 & -1.00 & -0.35 & -0.35 & 0 \end{bmatrix} \cdot \begin{bmatrix} u_1 \\ v_1 \\ u_2 \\ v_2 \\ u_3 \end{bmatrix} = \begin{bmatrix} F_{y3} \\ F_{x4} \\ F_{y4} \end{bmatrix}.$$

◀

Example 3.7 The stiffness matrix is to be interpreted by means of unit displacements in the individual degrees of freedom.

The procedure is shown for the horizontal displacement of nodal point 2. A horizontal unit displacement is applied at node 2 whereas all other degrees of freedom are fixed, i.e.

$$u_2 = 1$$
$$u_1 = v_1 = v_2 = u_3 = 0.$$

By introducing the displacements in both systems of equations, the reaction forces are now obtained as

$$\begin{bmatrix} F_{x1} \\ F_{y1} \\ F_{x2} \\ F_{y2} \\ F_{x3} \end{bmatrix} = 2.80 \cdot 10^5 \cdot \begin{bmatrix} -1.00 \\ 0 \\ 1.35 \\ 0.35 \\ 0 \end{bmatrix}$$

$$\begin{bmatrix} F_{y3} \\ F_{x4} \\ F_{y4} \end{bmatrix} = 2.80 \cdot 10^5 \cdot \begin{bmatrix} 0 \\ -0.35 \\ -0.35 \end{bmatrix}.$$

The vectors on the right-hand side correspond to the third column of the global stiffness matrix assigned to the degree of freedom u_2. In the same way, the other columns of the stiffness matrix can be obtained by applying unit displacements to the corresponding degrees of freedom and fixing all others. Hence, the columns of the stiffness matrix are the reaction forces caused by unit displacements in the corresponding degrees of freedom.   ◀

3.2.6   Solution of the System of Equations

The system of equations (3.1) with the nodal point displacements as unknowns is regular for stable (i.e. not kinematic) systems. For each load vector $\underline{F}$, a unique solution exists. In the case of a kinematic system, however, the global stiffness matrix becomes singular. In this case, the structural system has to be checked and corrected. The system of Eq. (3.1) can be solved numerically according to various methods, e.g. a Gauss–Cholesky elimination (see Sect. 1.4.3).

Example 3.8 For the truss system in Fig. 3.5, the nodal displacements are to be determined.

The global stiffness matrix has been established in Example 3.6. The loading consists of the forces

$$F_x = 10.0 \quad F_y = -10.0$$

at nodal point 2. Hence, the load vector is

$$\begin{bmatrix} F_{x1} \\ F_{y1} \\ F_{x2} \\ F_{y2} \\ F_{x3} \end{bmatrix} = \begin{bmatrix} 0 \\ 0 \\ 10 \\ -10 \\ 0 \end{bmatrix}.$$

The solution of the equation system with the stiffness matrix given in Example 3.6 was determined in the Examples 1.4–1.7 (without the prefactor $2.80 \cdot 10^5$ and with an accuracy of the stiffness matrix of two digits) by different methods. The nodal displacements are obtained as

$$\begin{bmatrix} u_1 \\ v_1 \\ u_2 \\ v_2 \\ u_3 \end{bmatrix} = \begin{bmatrix} 0.86 \\ 0.18 \\ 1.04 \\ -0.54 \\ 0.18 \end{bmatrix} \cdot 10^{-4}.$$

The unit of the displacements is m. ◀

3.2.7 Support Forces and Element Stresses

The support forces are obtained from the corresponding system of equations with the nodal displacements now known (see Sect. 3.2.5).

In order to obtain the element forces or element stresses, the section force matrices or the stress matrices, respectively, of the relevant element are multiplied by their vector of nodal point displacements.

Example 3.9 The support forces and the normal forces in the truss elements of the system in Fig. 3.5 are to be calculated using the nodal point displacements evaluated in Example 3.8.

The support forces are obtained by the system of equations given in Example 3.6, as

$$
2.80 \cdot 10^5 \cdot \begin{bmatrix} 0.35 & -0.35 & 0 & -1.00 & -0.35 \\ 0 & 0 & -0.35 & -0.35 & -1.00 \\ 0 & -1.00 & -0.35 & -0.35 & 0 \end{bmatrix} \cdot \begin{bmatrix} 0.86 \\ 0.18 \\ 1.04 \\ -0.54 \\ 0.18 \end{bmatrix} \cdot 10^{-4}
$$

$$
= \begin{bmatrix} 20.0 \\ -10.0 \\ -10.0 \end{bmatrix} = \begin{bmatrix} F_{y3} \\ F_{x4} \\ F_{y4} \end{bmatrix}.
$$

The support forces determined in this way are in equilibrium with the loads, i.e. the equilibrium conditions of the truss system

$$
\sum_{i=1}^{n} F_{xi} = 0, \qquad \sum_{i=1}^{n} F_{yi} = 0
$$

are fulfilled.

The element forces are obtained with the section force matrices evaluated in Example 3.3 as

Element 1:

$$
N_1 = 2.80 \cdot 10^5 \cdot \begin{bmatrix} -1.00 & 1.00 \end{bmatrix} \cdot \begin{bmatrix} 0.86 \\ 1.04 \end{bmatrix} \cdot 10^{-4} = 5.0
$$

Element 2:

$$
N_2 = 2.80 \cdot 10^5 \cdot \begin{bmatrix} -1.00 & 1.00 \end{bmatrix} \cdot \begin{bmatrix} 0.00 \\ -0.54 \end{bmatrix} \cdot 10^{-4} = -15.0
$$

Element 3:

$$
N_3 = 2.80 \cdot 10^5 \cdot \begin{bmatrix} -1.00 & 1.00 \end{bmatrix} \cdot \begin{bmatrix} 0.00 \\ 0.18 \end{bmatrix} \cdot 10^{-4} = 5.0
$$

Element 4:

$$
N_4 = 2.80 \cdot 10^5 \cdot \begin{bmatrix} -1.00 & 1.00 \end{bmatrix} \cdot \begin{bmatrix} 0.00 \\ 0.18 \end{bmatrix} \cdot 10^{-4} = 5.0
$$

Element 5:

$$
N_5 = 1.98 \cdot 10^5 \cdot \begin{bmatrix} -0.71 & -0.71 & 0.71 & 0.71 \end{bmatrix} \cdot \begin{bmatrix} 0.00 \\ 0.00 \\ 1.04 \\ -0.54 \end{bmatrix} \cdot 10^{-4} = 7.0
$$

Element 6:

$$N_6 = 1.98 \cdot 10^5 \cdot \begin{bmatrix} 0.71 & -0.71 & -0.71 & 0.71 \end{bmatrix} \cdot \begin{bmatrix} 0.18 \\ 0.00 \\ 0.86 \\ 0.18 \end{bmatrix} \cdot 10^{-4} = -7.0.$$

The element forces N_1 to N_6 are obtained in kN.

The complete results of the introductory example are

Nodal point	Displacements		Support forces		Element	Normal forces [kN]
	x [mm]	y [mm]	x [kN]	y [kN]		
1	0.086	0.018	–	–	1	5.0
2	0.104	−0.054	–	–	2	−15.0
3	0.018	–	–	20.0	3	5.0
4	–	–	−10.0	−10.0	4	5.0
					5	7.0
					6	−7.0

◀

Example 3.10 For the truss system in Fig. 3.5, the displacements caused by a prescribed displacement $u_1 = 1.0 \cdot 10^{-3}$ m of point 1 in the x-direction are to be determined. The system of equations is given in Example 3.6. Since no external loads are acting, it is

$$2.80 \cdot 10^5 \begin{bmatrix} 1.35 & -0.35 & -1.00 & 0 & -0.35 \\ -0.35 & 1.35 & 0 & 0 & 0.35 \\ -1.00 & 0 & 1.35 & 0.35 & 0 \\ 0 & 0 & 0.35 & 1.35 & 0 \\ -0.35 & 0.35 & 0 & 0 & 1.35 \end{bmatrix} \cdot \begin{bmatrix} u_1 \\ v_1 \\ u_2 \\ v_2 \\ u_3 \end{bmatrix} = \begin{bmatrix} F_{x1} \\ 0 \\ 0 \\ 0 \\ 0 \end{bmatrix}$$

or

$$2.80 \cdot 10^5 \cdot \left(\begin{bmatrix} 1.35 \\ -0.35 \\ -1.00 \\ 0 \\ -0.35 \end{bmatrix} \cdot u_1 + \begin{bmatrix} -0.35 & -1.00 & 0 & -0.35 \\ 1.35 & 0 & 0 & 0.35 \\ 0 & 1.35 & 0.35 & 0 \\ 0 & 0.35 & 1.35 & 0 \\ -0.35 & 0 & 0 & 1.35 \end{bmatrix} \cdot \begin{bmatrix} v_1 \\ u_2 \\ v_2 \\ u_3 \end{bmatrix} \right)$$

$$= \begin{bmatrix} F_{x1} \\ 0 \\ 0 \\ 0 \\ 0 \end{bmatrix}.$$

After rearranging the system of equations and omitting the first row, one obtains

$$
2.80 \cdot 10^5 \cdot
\begin{bmatrix}
1.35 & 0 & 0 & 0.35 \\
0 & 1.35 & 0.35 & 0 \\
0 & 0.35 & 1.35 & 0 \\
-0.35 & 0 & 0 & 1.35
\end{bmatrix}
\cdot
\begin{bmatrix}
v_1 \\ u_2 \\ v_2 \\ u_3
\end{bmatrix}
= 2.80 \cdot 10^5 \cdot
\begin{bmatrix}
0.35 \\ 1.00 \\ 0 \\ 0.35
\end{bmatrix}
\cdot u_1
$$

and with $u_1 = 1.0 \cdot 10^{-3}\,\mathrm{m}$

$$
2.80 \cdot 10^5 \cdot
\begin{bmatrix}
1.35 & 0 & 0 & 0.35 \\
0 & 1.35 & 0.35 & 0 \\
0 & 0.35 & 1.35 & 0 \\
-0.35 & 0 & 0 & 1.35
\end{bmatrix}
\cdot
\begin{bmatrix}
v_1 \\ u_2 \\ v_2 \\ u_3
\end{bmatrix}
= 2.80 \cdot 10^2 \cdot
\begin{bmatrix}
0.35 \\ 1.00 \\ 0 \\ 0.35
\end{bmatrix}
\cdot
$$

The solution of the system of equations is

$$
\begin{bmatrix}
v_1 \\ u_2 \\ v_2 \\ u_3
\end{bmatrix}
=
\begin{bmatrix}
2.07 \\ 7.93 \\ -2.07 \\ 2.07
\end{bmatrix}
\cdot 10^{-4}\,\mathrm{m}.
$$

The support force of node 1 in the x-direction is obtained from the first line of the original system of equations as

$$
F_{x1} = 2.80 \cdot 10^5 \cdot \left(1.35 \cdot u_1 + \begin{bmatrix} -0.35 & -1.00 & 0 & -0.35 \end{bmatrix} \cdot \begin{bmatrix} v_1 \\ u_2 \\ v_2 \\ u_3 \end{bmatrix} \right) = 116\,\mathrm{kN}.
$$

The normal forces of the elements can be determined using the element section force matrices and the nodal point displacements.   ◀

3.2.8 Flexibility Matrix

By multiplying the stiffness matrix with the displacement vector, the (external) nodal forces of the system are obtained. According to (3.1), the column j of the stiffness matrix corresponds to the vector of the restraining forces, which is obtained by the unit displacement of the system in the degree of freedom j when all other degrees of freedom are fixed. The inverse of the stiffness matrix is the *flexibility matrix*:

$$
\underline{H} = \underline{K}^{-1}. \tag{3.16a}
$$

With (3.1), one obtains

$$\underline{H} \cdot \underline{F} = \underline{u}. \tag{3.16b}$$

The column j of the flexibility matrix corresponds to the vector of displacements obtained when the system is loaded with a unit force in the degree of freedom j whereas all other loads are zero.

For some structural systems, it is easier to determine the flexibility matrix instead of the stiffness matrix. In such a case, the stiffness matrix can be computed by inversion of the flexibility matrix as

$$\underline{K} = \underline{H}^{-1}. \tag{3.16c}$$

The relationship can be used, for example, to determine the stiffness matrix condensed to a few degrees of freedom from a large finite element model.

3.3 Elastic Springs

3.3.1 Elastic Support of Nodal Points

In finite element models of structural systems, springs are used to represent elastic point supports and elastic restraints. Elastic springs exhibit a linear relationship between forces and displacements. For each degree of freedom, an individual spring constant is obtained.

Plane systems in the xy-plane may have three degrees of freedom at each node, i.e. the displacements in the x- and y-directions as well as the rotation about the z-axis. The relations between the forces and the displacements of an elastically supported nodal point i are (Fig. 3.12a):

$$F_x^{(e)} = k_x \cdot u_i \tag{3.17a}$$

$$F_y^{(e)} = k_y \cdot v_i \tag{3.17b}$$

$$M_z^{(e)} = k_{zz} \cdot \varphi_i \tag{3.17c}$$

with the spring constants
   k_x for the displacement in the x-direction (units, e.g. kN/m),
   k_y for the displacement in the y-direction (units, e.g. kN/m), and
   k_{zz} for the rotation about the z-axis (units, e.g. kNm).

Thus the elastic support of the point is defined by three individual springs which are not coupled. These relationships can also be written as a stiffness matrix for an elastic support as

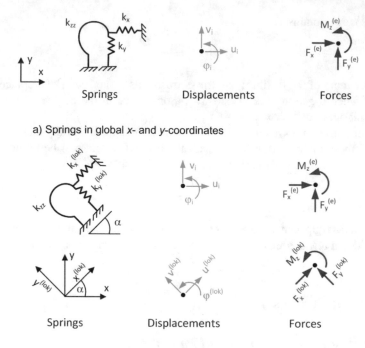

a) Springs in global x- and y-coordinates

b) Inclined springs in a local coordinate system

Fig. 3.12 Elastic support of a nodal point

$$
\begin{bmatrix} k_x & 0 & 0 \\ 0 & k_y & 0 \\ 0 & 0 & k_{zz} \end{bmatrix} \cdot \begin{bmatrix} u_i \\ v_i \\ \varphi_i \end{bmatrix} = \begin{bmatrix} F_x^{(e)} \\ F_y^{(e)} \\ M_z^{(e)} \end{bmatrix}
\tag{3.18}
$$

$$
\underline{K}_e \cdot \underline{u}_e = \underline{F}_e.
\tag{3.18a}
$$

In the stiffness matrix (3.18), only the diagonal terms are nonzero since the individual springs are not coupled.

After the solution of the global system of equations, the forces in the springs are obtained with Eqs. (3.17a–3.17c) or (3.18).

Example 3.11 In the truss shown in Fig. 3.5, two springs are added in the x-direction at nodes 1 and 3 (Fig. 3.13).

When constructing the global stiffness matrix, the two springs must be taken into account as additional elements. One obtains

Fig. 3.13 Truss system with
two springs

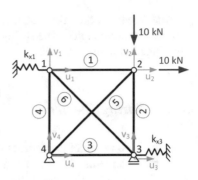

$$2.8 \cdot 10^5 \cdot \begin{bmatrix} 1.35 + \bar{k}_{x1} & -0.35 & -1.00 & 0 & -0.35 \\ -0.35 & 1.35 & 0 & 0 & 0.35 \\ -1.00 & 0 & 1.35 & 0.35 & 0 \\ 0 & 0 & 0.35 & 1.35 & 0 \\ -0.35 & 0.35 & 0 & 0 & 1.35 + \bar{k}_{x3} \end{bmatrix} \cdot \begin{bmatrix} u_1 \\ v_1 \\ u_2 \\ v_2 \\ u_3 \end{bmatrix} = \begin{bmatrix} 0 \\ 0 \\ 10 \\ -10 \\ 0 \end{bmatrix}$$

with $\bar{k}_{x1} = k_{x1}/2.8 \cdot 10^5$ and $\bar{k}_{x3} = k_{x3}/2.8 \cdot 10^5$. ◀

For displacement springs not aligned with the direction of the global x- and y-coordinates, the stiffness matrix has to be transformed similarly as for a truss element.

For springs $k_x^{(\text{lok})}$, $k_y^{(\text{lok})}$ inclined at an angle α toward the x-axis (Fig. 3.12b), the stiffness matrix in the local coordinate system with the displacements $u_i^{(\text{lok})}$, $v_i^{(\text{lok})}$ is

$$\begin{bmatrix} k_x^{(\text{lok})} & 0 \\ 0 & k_y^{(\text{lok})} \end{bmatrix} \cdot \begin{bmatrix} u_i^{(\text{lok})} \\ v_i^{(\text{lok})} \end{bmatrix} = \begin{bmatrix} F_x^{(\text{lok})} \\ F_y^{(\text{lok})} \end{bmatrix} \tag{3.19}$$

$$\underline{K}_e^{(\text{lok})} \cdot \underline{u}_e^{(\text{lok})} = \underline{F}_e^{(\text{lok})} \tag{3.19a}$$

with

$$\underline{K}_e = \underline{T}_u^{\text{T}} \cdot \underline{K}_e^{(\text{lok})} \cdot \underline{T}_u. \tag{3.20}$$

It can easily be transformed into the global x-y-coordinates system with the transformation matrix $\underline{T}_u$ according to (3.4). Including the rotational degree of freedom $\varphi_i = \varphi_i^{(\text{lok})}$, one obtains

$$\begin{bmatrix} k_x^{(\text{lok})} \cdot \cos^2 \alpha + k_y^{(\text{lok})} \cdot \sin^2 \alpha & (k_x^{(\text{lok})} - k_y^{(\text{lok})}) \cdot \sin \alpha \cdot \cos \alpha & 0 \\ (k_x^{(\text{lok})} - k_y^{(\text{lok})}) \cdot \sin \alpha \cdot \cos \alpha & k_x^{(\text{lok})} \cdot \sin^2 \alpha + k_y^{(\text{lok})} \cdot \cos^2 \alpha & 0 \\ 0 & 0 & k_{zz} \end{bmatrix} \cdot \begin{bmatrix} u_i \\ v_i \\ \varphi_i \end{bmatrix}$$

$$= \begin{bmatrix} F_x^{(e)} \\ F_y^{(e)} \\ M_z^{(e)} \end{bmatrix}. \tag{3.20a}$$

After the transformation, the displacements in the x- and y-directions of the stiffness matrix are coupled. The spring k_{zz} in the rotational degree of freedom remains unchanged in plane systems.

3.3.2 Spring Elements

The elastic coupling of the degrees of freedom of two nodal points can be described by spring elements. Figure 3.14a shows this for a displacement spring k_L in $x^{(lok)}$-direction (longitudinal spring). The spring force F and the element forces are obtained analogously to Fig. 3.6 for the truss element as

$$F_{x1}^{(lok)} = -F = k_L \cdot \left(u_1^{(lok)} - u_2^{(lok)} \right)$$
$$F_{x2}^{(lok)} = F = k_L \cdot \left(-u_1^{(lok)} + u_2^{(lok)} \right).$$

This results in the stiffness relationship

$$k_L \cdot \begin{bmatrix} 1 & -1 \\ -1 & 1 \end{bmatrix} \cdot \begin{bmatrix} u_1^{(lok)} \\ u_2^{(lok)} \end{bmatrix} = \begin{bmatrix} F_1^{(lok)} \\ F_2^{(lok)} \end{bmatrix} \tag{3.21}$$

$$\underline{K}_e^{(lok)} \cdot \underline{u}_e^{(lok)} = \underline{F}_e^{(lok)} \tag{3.21a}$$

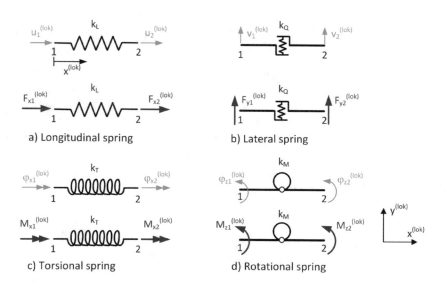

a) Longitudinal spring   b) Lateral spring

c) Torsional spring   d) Rotational spring

Fig. 3.14 Spring elements

wherein the matrix

$$\underline{K}_e^{(lok)} = k_L \cdot \begin{bmatrix} 1 & -1 \\ -1 & 1 \end{bmatrix}$$ (3.21b)

represents the stiffness matrix of the spring element in local coordinates. Similar to the element stiffness matrix of the truss element, it can be transformed into global coordinates with (3.8a, 3.8b).

Spring elements can be defined for all degrees of freedom. Analogous to the longitudinal spring, the following stiffness relationships are obtained for the springs shown in Fig. 3.14.

Lateral spring

$$k_Q \cdot \begin{bmatrix} 1 & -1 \\ -1 & 1 \end{bmatrix} \cdot \begin{bmatrix} v_1^{(lok)} \\ v_2^{(lok)} \end{bmatrix} = \begin{bmatrix} F_{y1}^{(lok)} \\ F_{y2}^{(lok)} \end{bmatrix}$$ (3.22a)

Torsional spring

$$k_T \cdot \begin{bmatrix} 1 & -1 \\ -1 & 1 \end{bmatrix} \cdot \begin{bmatrix} \varphi_{x1}^{(lok)} \\ \varphi_{x2}^{(lok)} \end{bmatrix} = \begin{bmatrix} M_{x1}^{(lok)} \\ M_{x2}^{(lok)} \end{bmatrix}$$ (3.22b)

Rotational spring

$$k_M \cdot \begin{bmatrix} 1 & -1 \\ -1 & 1 \end{bmatrix} \cdot \begin{bmatrix} \varphi_{z1}^{(lok)} \\ \varphi_{z2}^{(lok)} \end{bmatrix} = \begin{bmatrix} M_{z1}^{(lok)} \\ M_{z2}^{(lok)} \end{bmatrix}$$ (3.22c)

If the two spring nodal points 1 and 2 coincide geometrically, the spring merges into an elastic internal hinge in the respective degree of freedom.

3.4 Beams in Bending

3.4.1 Stiffness Matrix of the Beam Element

Beam elements are used to model continuous beams, frames, and beam grillages. Loads are transferred by bending moments and shear forces. Hence, the element forces of a beam element are a bending moment and a force perpendicular to the beam axis at each nodal point (Fig. 3.15). The corresponding degrees of freedom are a displacement perpendicular to the beam axis and a rotation. Hence, a beam element possesses four degrees of freedom in the local coordinate system. The displacement

Fig. 3.15 Beam element in
local coordinates

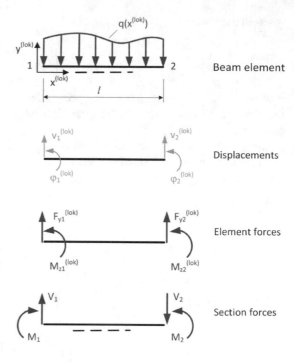

in the longitudinal direction corresponding to the normal forces is uncoupled with
those degrees of freedom and will be considered later (see Sect. 3.4.3).

The displacements $v^{(\text{lok})}(x^{(\text{lok})})$ can be obtained as a solution of the differential
equation of the shear rigid beam in bending

$$\frac{\mathrm{d}^4 v^{(\text{lok})}}{\mathrm{d}x^{(\text{lok})4}} = -\frac{q}{EI}. \tag{3.23}$$

with the boundary conditions $v^{(\text{lok})}(x^{(\text{lok})} = 0) = v_1^{(\text{lok})}$, $v^{(\text{lok})}(x^{(\text{lok})} = l) = v_2^{(\text{lok})}$,
$\varphi^{(\text{lok})}(x^{(\text{lok})} = 0) = \varphi_1^{(\text{lok})}$, and $\varphi^{(\text{lok})}(x^{(\text{lok})} = l) = \varphi_2^{(\text{lok})}$. By differentiation of the
displacements the angle of deflection and the bending moment as well as the shear
force can be determined. The relationship between the element forces and the corresponding displacements constitutes the stiffness matrix of the element. However,
since these solutions for the beam in bending are well known, here solutions given
in engineering handbooks are used to derive the stiffness matrix.

It has already been pointed out that the stiffness matrix can be interpreted with
the help of unit displacements in its individual degrees of freedom (cf. Example 3.7).
The columns of the stiffness matrix are the reaction forces obtained when applying a
unit displacement in its corresponding degrees of freedom, whereas all other degrees
of freedom are fixed. This procedure is used in the following to evaluate the stiffness
matrix of a beam element.

A single-span beam without shear deformation clamped at both ends is considered. In Fig. 3.16, the reaction forces and moments caused by unit displacements

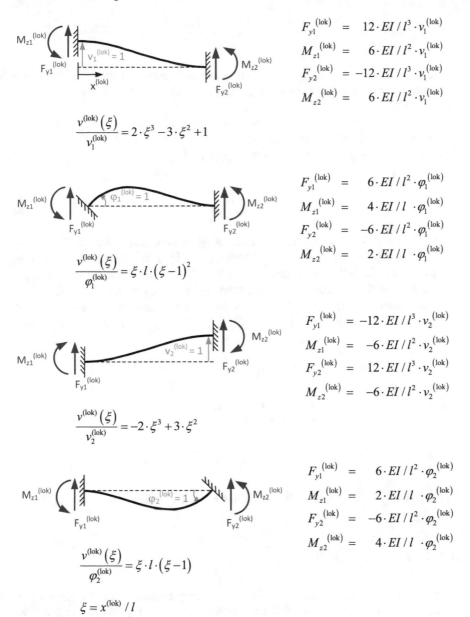

$$F_{y1}^{(lok)} = 12 \cdot EI / l^3 \cdot v_1^{(lok)}$$

$$M_{z1}^{(lok)} = 6 \cdot EI / l^2 \cdot v_1^{(lok)}$$

$$F_{y2}^{(lok)} = -12 \cdot EI / l^3 \cdot v_1^{(lok)}$$

$$M_{z2}^{(lok)} = 6 \cdot EI / l^2 \cdot v_1^{(lok)}$$

$$\frac{v^{(lok)}(\xi)}{v_1^{(lok)}} = 2 \cdot \xi^3 - 3 \cdot \xi^2 + 1$$

$$F_{y1}^{(lok)} = 6 \cdot EI / l^2 \cdot \varphi_1^{(lok)}$$

$$M_{z1}^{(lok)} = 4 \cdot EI / l \cdot \varphi_1^{(lok)}$$

$$F_{y2}^{(lok)} = -6 \cdot EI / l^2 \cdot \varphi_1^{(lok)}$$

$$M_{z2}^{(lok)} = 2 \cdot EI / l \cdot \varphi_1^{(lok)}$$

$$\frac{v^{(lok)}(\xi)}{\varphi_1^{(lok)}} = \xi \cdot l \cdot (\xi - 1)^2$$

$$F_{y1}^{(lok)} = -12 \cdot EI / l^3 \cdot v_2^{(lok)}$$

$$M_{z1}^{(lok)} = -6 \cdot EI / l^2 \cdot v_2^{(lok)}$$

$$F_{y2}^{(lok)} = 12 \cdot EI / l^3 \cdot v_2^{(lok)}$$

$$M_{z2}^{(lok)} = -6 \cdot EI / l^2 \cdot v_2^{(lok)}$$

$$\frac{v^{(lok)}(\xi)}{v_2^{(lok)}} = -2 \cdot \xi^3 + 3 \cdot \xi^2$$

$$F_{y1}^{(lok)} = 6 \cdot EI / l^2 \cdot \varphi_2^{(lok)}$$

$$M_{z1}^{(lok)} = 2 \cdot EI / l \cdot \varphi_2^{(lok)}$$

$$F_{y2}^{(lok)} = -6 \cdot EI / l^2 \cdot \varphi_2^{(lok)}$$

$$M_{z2}^{(lok)} = 4 \cdot EI / l \cdot \varphi_2^{(lok)}$$

$$\frac{v^{(lok)}(\xi)}{\varphi_2^{(lok)}} = \xi \cdot l \cdot (\xi - 1)$$

$$\xi = x^{(lok)} / l$$

Fig. 3.16 Unit displacements for a beam element

and rotations in the four degrees of freedom according to [12] are given. Applying the nodal point displacements and rotations in all four degrees of freedom simultaneously, the resulting forces and moments at the ends of the beam are obtained by superposition. For example, the force $F_{y1}^{(lok)}$ results as

$$F_{y1}^{(lok)} = \frac{12E \cdot I}{l^3} \cdot v_1^{(lok)} + \frac{6E \cdot I}{l^2} \cdot \varphi_1^{(lok)} - \frac{12E \cdot I}{l^3} \cdot v_2^{(lok)} + \frac{6E \cdot I}{l^2} \cdot \varphi_2^{(lok)}.$$

If the corresponding equations for all forces and moments at both ends of the beam element are determined and summarized in matrix notation, one obtains

$$\frac{E \cdot I}{l} \cdot \begin{bmatrix} 12/l^2 & 6/l & -12/l^2 & 6/l \\ 6/l & 4 & -6/l & 2 \\ -12/l^2 & -6/l & 12/l^2 & -6/l \\ 6/l & 2 & -6/l & 4 \end{bmatrix} \cdot \begin{bmatrix} v_1^{(lok)} \\ \varphi_1^{(lok)} \\ v_2^{(lok)} \\ \varphi_2^{(lok)} \end{bmatrix} = \begin{bmatrix} F_{y1}^{(lok)} \\ M_{z1}^{(lok)} \\ F_{y2}^{(lok)} \\ M_{z2}^{(lok)} \end{bmatrix} \tag{3.24}$$

$$\underline{K}_{be}^{(lok)} \cdot \underline{u}_{be}^{(lok)} = \underline{F}_{be}^{(lok)} \tag{3.24a}$$

where $E \cdot I$ is the bending stiffness and l the length of the beam element. The matrix $\underline{K}_{be}^{(lok)}$ is the stiffness matrix of the beam element in local coordinates. The equation does not yet include the element load q in Fig. 3.15, which will be considered later.

The section forces at the ends of the element can be easily derived from (3.24). For this, instead of the forces $\underline{F}_{be}^{(lok)}$, the section forces (according to Fig. 3.15) are introduced. With $V_1 = F_{y1}^{(lok)}$, $V_2 = -F_{y2}^{(lok)}$, $M_1 = -M_{z1}^{(lok)}$ and $M_2 = M_{z2}^{(lok)}$, one obtains

$$\begin{bmatrix} V_1 \\ M_1 \\ V_2 \\ M_2 \end{bmatrix} = \frac{E \cdot I}{l} \cdot \begin{bmatrix} 12/l^2 & 6/l & -12/l^2 & 6/l \\ -6/l & -4 & 6/l & -2 \\ 12/l^2 & 6/l & -12/l^2 & 6/l \\ 6/l & 2 & -6/l & 4 \end{bmatrix} \cdot \begin{bmatrix} v_1^{(lok)} \\ \varphi_1^{(lok)} \\ v_2^{(lok)} \\ \varphi_2^{(lok)} \end{bmatrix} \tag{3.25}$$

$$\underline{S} = \underline{S}_{be}^{(lok)} \cdot \underline{u}_{be}^{(lok)}. \tag{3.25a}$$

The matrix $\underline{S}_{be}^{(lok)}$ is the section forces matrix in local coordinates.

3.4.2 Element Loads

In the finite element method, external loads can only be included in the analysis by nodal forces in the global equation system. Nevertheless, it is still possible to take into account element loads as, for example, line loads on beam elements, exactly in the analysis.

In order to consider an arbitrary element load, first the reaction forces and moments of a beam clamped at both ends are determined. Now, in the superposition of the reaction forces and moments caused by the unit displacements and rotations of the beam supports, which leads to (3.24), the reaction forces and moments resulting from element loads are taken into account additionally. One obtains

$$\frac{E \cdot I}{l} \cdot \begin{bmatrix} 12/l^2 & 6/l & -12/l^2 & 6/l \\ 6/l & 4 & -6/l & 2 \\ -12/l^2 & -6/l & 12/l^2 & -6/l \\ 6/l & 2 & -6/l & 4 \end{bmatrix} \cdot \begin{bmatrix} v_1^{(lok)} \\ \varphi_1^{(lok)} \\ v_2^{(lok)} \\ \varphi_2^{(lok)} \end{bmatrix} = \begin{bmatrix} F_{y1}^{(lok)} \\ M_{z1}^{(lok)} \\ F_{y2}^{(lok)} \\ M_{z2}^{(lok)} \end{bmatrix} - \begin{bmatrix} F_{L1} \\ M_{L1} \\ F_{L2} \\ M_{L2} \end{bmatrix}$$

(3.26)

$$\underline{K}_{be}^{(lok)} \cdot \underline{u}_{be}^{(lok)} = \underline{F}_{be}^{(lok)} - \underline{F}_{bL}^{(lok)}.$$ (3.26a)

The vector $\underline{F}_{bL}^{(lok)}$ contains the reaction forces and moments of a clamped–clamped beam caused by the external load of the beam element with the sign convention in Fig. 3.15. They can be taken from engineering handbooks.

For some typical loadings, the reaction moments and forces are given in Tables 3.2 and 3.3 [8, 12]. These forces have to be entered in (3.26) with a negative sign on the right-hand side, i.e. the support reactions of the clamped–clamped beam are to be applied with opposite signs to the nodal loads on the global system.

The section forces at the ends of the beam are obtained as in (3.25) with $V_1 = F_{y1}^{(lok)}$, $V_2 = -F_{y2}^{(lok)}$, $M_1 = -M_{z1}^{(lok)}$ and $M_2 = M_{z2}^{(lok)}$ as

$$\begin{bmatrix} V_1 \\ M_1 \\ V_2 \\ M_2 \end{bmatrix} = \frac{E \cdot I}{l} \cdot \begin{bmatrix} 12/l^2 & 6/l & -12/l^2 & 6/l \\ -6/l & -4 & 6/l & -2 \\ 12/l^2 & 6/l & -12/l^2 & 6/l \\ 6/l & 2 & -6/l & 4 \end{bmatrix} \cdot \begin{bmatrix} v_1^{(lok)} \\ \varphi_1^{(lok)} \\ v_2^{(lok)} \\ \varphi_2^{(lok)} \end{bmatrix} + \begin{bmatrix} F_{L1} \\ -M_{L1} \\ -F_{L2} \\ M_{L2} \end{bmatrix}$$ (3.27)

$$\underline{S} = \underline{S}_{be}^{(lok)} \cdot \underline{u}_{be}^{(lok)} + \underline{F}_{LS}^{(lok)}.$$ (3.27a)

Table 3.2 Reaction forces of a truss element

$\alpha = a/l, \beta = b/l$		
Loading	F_{Lx1}	F_{Lx2}
n	$-n\dfrac{l}{2}$	$-n\dfrac{l}{2}$
H, a, b	$-H \cdot \beta$	$-H \cdot \alpha$
T, h Temperataure difference	$EA \cdot \alpha_T \cdot T$	$-EA \cdot \alpha_T \cdot T$

Table 3.3 Reaction moments and forces of a clamped–clamped beam

$\alpha = a/l,\ \beta = b/l$	F_{L1} ⎰ M_{L1}	F_{L2} ⎱ M_{L2}
Loading	F_{L1} M_{L1}	F_{L2} M_{L2}
	$q\dfrac{l}{2}$ $q\dfrac{l^2}{12}$	$q\dfrac{l}{2}$ $-q\dfrac{l^2}{12}$
	$\dfrac{l}{20}\left(7q_1+3q_2\right)$ $\dfrac{l^2}{60}\left(3q_1+2q_2\right)$	$\dfrac{l}{20}\left(3q_1+7q_2\right)$ $-\dfrac{l^2}{60}\left(2q_1+3q_2\right)$
	$q\dfrac{l}{20}\left(3+3\beta+3\beta^2-2\beta^3\right)$ $q\dfrac{l^2}{30}\left(1+\beta+\beta^2-\dfrac{3}{2}\beta^3\right)$	$q\dfrac{l}{20}\left(3+3\alpha+3\alpha^2-2\alpha^3\right)$ $-q\dfrac{l^2}{30}\left(1+\alpha+\alpha^2-\dfrac{3}{2}\alpha^3\right)$
  2nd order parabola	$q\dfrac{l}{3}$ $q\dfrac{l^2}{15}$	$q\dfrac{l}{3}$ $-q\dfrac{l^2}{15}$
  Sine	$q\dfrac{l}{\pi}$ $q\dfrac{2l^2}{\pi^3}$	$q\dfrac{l}{\pi}$ $-q\dfrac{2l^2}{\pi^3}$
	$F\beta^2\left(3-2\beta\right)$ $Fa\beta^2$	$F\alpha^2\left(3-2\alpha\right)$ $-Fb\alpha^2$
	$M\dfrac{6}{l}\alpha\beta$ $M\beta\left(3\alpha-1\right)$	$-M\dfrac{6}{l}\alpha\beta$ $M\alpha\left(3\beta-1\right)$
  $\Delta T = T_u - T_o$ Temperature difference	0 $EI\,\alpha_T\dfrac{\Delta T}{h}$	0 $-EI\,\alpha_T\dfrac{\Delta T}{h}$

In order to consider the element loads, the section forces $\underline{S}_{be}^{(lok)} \cdot \underline{u}_{be}^{(lok)}$ arising from the nodal displacements and rotations are superposed in (3.27) with the element forces $F_{LS}^{(lok)}$ of the clamped–clamped beam. $S_{be}^{(lok)}$ is the section force matrix.

Any type of element load can be taken into account in the same way. In principle, this also applies to other finite elements, e.g. truss elements. Hence, the procedure allows the user of a finite element program to deal "manually" with element load types which are not implemented in his program.

> **Method of equivalent loads of arbitrary element loads**
> 1. Determination of the reaction forces and moments of the clamped–clamped beam (Tables 3.2, 3.3)
> 2. Application of the reaction forces and moments with opposite signs as nodal loads to the global system
> 3. Computation of the global system with a finite element program
> 4. Superposition of the section forces obtained by the program with the section forces of the clamped–clamped beam

Reaction moments and forces of a clamped–clamped beam resulting from a jump or kink in the bending line of the beam are given in [5]. They can be used to determine the influence lines of shear forces or moments in beams.

Example 3.12 The truss system in Fig. 3.5 is to be investigated for a temperature load in element 1 of 30 °C.

The nodal loads correspond to the reaction forces of the truss element fixed at both ends (Fig. 3.17). They have to be applied with opposite signs. According to Table 3.2, for uniform temperature increase the following reaction force is obtained

$$F_{Lx1} = -F_{Lx2} = E \cdot A \cdot \alpha_T \cdot \Delta t$$

and with $\alpha_T = 1.2 \cdot 10^{-5} [1^0 C]$

$$F_{Lx1} = 2.1 \cdot 10^8 \cdot 0.004 \cdot 1.2 \cdot 10^{-5} \cdot 30 = 302.4 \,[kN], F_{Lx2} = -302.4 \,[kN].$$

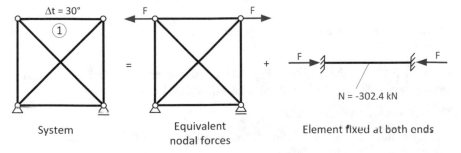

Fig. 3.17 Equivalent loads of a truss system with temperature load

Heating up the truss element fixed at both ends results in a compression force

$$N = -302.4\,[\text{kN}]$$

in the element. It has to be superposed with the element force obtained in the finite element solution with the nodal forces F_{Lx1} and F_{Lx2}.

For the truss system according to Example 3.6, the following system of equations is obtained:

$$2.80\cdot 10^5 \begin{bmatrix} 1.35 & -0.35 & -1.00 & 0 & -0.35 \\ -0.35 & 1.35 & 0 & 0 & 0.35 \\ -1.00 & 0 & 1.35 & 0.35 & 0 \\ 0 & 0 & 0.35 & 1.35 & 0 \\ -0.35 & 0.35 & 0 & 0 & 1.35 \end{bmatrix} \cdot \begin{bmatrix} u_1 \\ v_1 \\ u_2 \\ v_2 \\ u_3 \end{bmatrix} = \begin{bmatrix} -302 \\ 0 \\ 302 \\ 0 \\ 0 \end{bmatrix}.$$

Its solution gives the displacements

$$\begin{bmatrix} u_1 \\ v_1 \\ u_2 \\ v_2 \\ u_3 \end{bmatrix} = \begin{bmatrix} -0.540 \\ -0.112 \\ 0.428 \\ -0.112 \\ 0.112 \end{bmatrix} \cdot 10^{-3}\,\text{m}.$$

With these displacements, the following normal forces in the truss elements are obtained:

$$N_1 = 271.1\,\text{kN}, \quad N_2 = -31.3\,\text{kN}, \quad N_3 = 31.3\,\text{kN},$$
$$N_4 = -31.3\,\text{kN}, \quad N_5 = 44.3\,\text{kN}, \quad N_6 = 44.3\,\text{kN}.$$

The normal force in element 1 is now superposed with the normal force of the fixed element of -302.4 kN. The normal force in element 1 is thus finally obtained as

$$N_1 = 271.1 - 302.4 = -31.3\,\text{kN}.$$

All other element forces remain unchanged.   ◀

Example 3.13 For the system in Fig. 3.18 consisting of three beam elements, the section forces have to be determined using equivalent loads.

The solution is obtained in three steps:

(1) Determination of the reaction forces and moments of the clamped–clamped beam according to Table 3.3:

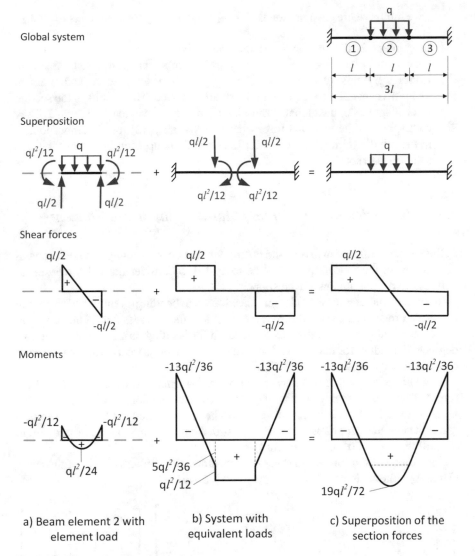

Fig. 3.18 Element load on a beam element

a) Beam element 2 with element load

b) System with equivalent loads

c) Superposition of the section forces

$$F_{L1} = F_{L2} = q \cdot l/2, \quad -M_{L1} = M_{L2} = -q \cdot l^2/12.$$

(2) The reaction forces F_{L1}, F_{L2} and moments M_{L1}, M_{L2} of the clamped–clamped beam are applied with opposite signs as equivalent loads to the structural system. In this way, the section force diagrams given in Fig. 3.18b are obtained. This step is performed by a finite element analysis. Here, however, instead of it a handbook solution given in [12] is used. The shear forces and the moments

have jumps where the point loads F_{L1}, F_{L2} and the imposed moments M_{L1}, M_{L2} are applied.

(3) In element 2, the section force diagrams given above are superposed with the section force diagrams of the clamped–clamped beam due to the element loading. By this superposition, the jumps in the shear force and moment diagrams caused by the equivalent loads are removed. This results in the section force diagrams of the global system loaded by uniform load on element 2. They correspond to the sectional force diagrams given in [12] for a member loaded in the middle third with a constant load q and clamped on both sides, i.e. the solution is exact.   ◀

3.4.3 Extension of the Stiffness Matrix of the Beam Element

In the element stiffness matrix of the beam, only the beam bending has been considered so far. In the following, it shall be extended to consider normal forces, shear stiffness, and a continuous elastic support.

In frame structures, a beam element, in addition to bending, is subjected to normal forces. In order to take normal forces into account, the stiffness matrix according to (3.24), containing only the degrees of freedom for bending, is expanded by the two degrees of freedom for the normal forces using the stiffness matrix (3.3) of the truss element already known.

In addition, beam elements may exhibit shear deformations when loaded by shear forces. These have already been included in the basic equations of the shear flexible beam in Sect. 2.3. However, they have not yet been considered in (3.24). In commercial finite element programs, normally, shear deformations are taken into account in the stiffness matrix of beam elements.

Including normal forces and shear deformations, the stiffness matrix of the beam element (without element loads) is given in [1, 5, 46] as

$$
\begin{bmatrix}
a_0 & 0 & 0 & -a_0 & 0 & 0 \\
0 & \dfrac{12 \cdot a_0}{l^2} & \dfrac{6 \cdot a_0}{l} & 0 & \dfrac{-12 \cdot a_0}{l^2} & \dfrac{6 \cdot a_0}{l} \\
0 & \dfrac{6 \cdot a_0}{l} & 4 \cdot a_2 & 0 & \dfrac{-6 \cdot a_0}{l} & 2 \cdot a_3 \\
-a_0 & 0 & 0 & a_0 & 0 & 0 \\
0 & \dfrac{-12 \cdot a_0}{l^2} & \dfrac{-6 \cdot a_0}{l} & 0 & \dfrac{12 \cdot a_0}{l^2} & \dfrac{-6 \cdot a_0}{l} \\
0 & \dfrac{6 \cdot a_0}{l} & 2 \cdot a_3 & 0 & \dfrac{-6 \cdot a_0}{l} & 4 \cdot a_2
\end{bmatrix}
\cdot
\begin{bmatrix}
u_1^{(lok)} \\
v_1^{(lok)} \\
\varphi_1^{(lok)} \\
u_2^{(lok)} \\
v_2^{(lok)} \\
\varphi_2^{(lok)}
\end{bmatrix}
=
\begin{bmatrix}
F_{x1}^{(lok)} \\
F_{y1}^{(lok)} \\
M_{z1}^{(lok)} \\
F_{x2}^{(lok)} \\
F_{y2}^{(lok)} \\
M_{z2}^{(lok)}
\end{bmatrix}
$$

$$(3.28)$$

or

$$K_e^{(lok)} \cdot \underline{u}_e^{(lok)} = \underline{F}_e^{(lok)} \tag{3.28a}$$

with

$$a_0 = \frac{E \cdot A}{l}, \qquad a_1 = \frac{E \cdot I}{l \cdot (1+m)}, \qquad a_2 = \frac{E \cdot I \cdot (4+m)}{4 \cdot l \cdot (1+m)}$$

$$a_3 = \frac{E \cdot I \cdot (2-m)}{2 \cdot l \cdot (1+m)}, \qquad m = \frac{12 \cdot E \cdot I}{G \cdot A_s \cdot l^2}. \tag{3.28b}$$

Here, G is the shear modulus and A_s the shear area of the beam element. For a beam without shear deformations, (3.28) corresponds with $A_s \to \infty$ and, therefore, $m = 0$ to the stiffness matrix (3.24) expanded by the normal force terms according to (3.3).

Beams may have *distributed elastic springs* along their axis, acting perpendicularly to the axis, which is referred to as a *continuous elastic support*. For beams with a continuous elastic support, exact analytical solutions of the stiffness matrix are available [5, 9]. In [13], an analytical solution in the form of series expansions is given, which avoids numerical difficulties with very small moduli of subgrade reaction and includes the case of a modulus of subgrade reaction equal to zero. Furthermore, numerical approximations of the stiffness matrix of a beam element with continuous elastic support were given, which have a discretization error, i.e. the accuracy of the solution depends on the element size [5, 7, 14–16]. They are obtained with shape functions similar to those for plane finite elements (see Sect. 4.3) and also apply to very small subgrade reaction moduli or a modulus being zero. Taking into account the analogy between the continuously supported beam and the beam under harmonic vibrations, the stiffness matrix of the continuously supported beam element is obtained as

$$\underline{K}_{Bed}^{(lok)} = \underline{K}_{be}^{(lok)} + \frac{\bar{k}_s}{\bar{m}} \cdot \underline{M}^{(e)} \tag{3.28c}$$

with the stiffness matrix $\underline{K}_{be}^{(lok)}$ of the beam element according to (3.24), $\underline{M}^{(e)}$ being the matrix according to (5.12b) obtained on the basis of shape functions, and $\bar{k}_s$ as spring constant per unit length. For a beam of width b with the modulus of subgrade reaction k_s, $\bar{k}_s = k_s \cdot b$ applies. In addition, there are other bedding models such as the *Pasternak-Kerr bedding*, in which a shear flexible layer is presumed to be between the beam and the spring [5].

The term elastic support is used only for continuous springs perpendicular to the beam axis. In principle, however, distributed springs, can occur in all degrees of freedom, i.e. also in rotational degrees of freedom. Stiffness matrices of beam elements with general distributed springs are given in [5, 14].

The stiffness matrix of a beam subdivided into sections can be determined using the *transfer matrix method*. A transfer matrix is set up for each section, which can have different properties (e.g. cross-section height) in each section. The product of the transfer matrices of all sections gives the transfer matrix of the entire beam element.

This can then be converted into a stiffness matrix. In detail, reference is made to [7, 11]. The method of transfer matrices is also very well suited for calculating the internal forces within a beam element divided into small sections, as they are required for the representation of bending, moment or shear force diagrams.

3.4.4 Coordinate Transformation

The stiffness matrix of the beam element in local coordinates can be transformed into global coordinates similarly as in the case of the truss element. The coordinate transformation of the displacements and the forces at the ends of the beam is shown in Fig. 3.19. We have

$$\underline{u}_e^{(\text{lok})} = \underline{T} \cdot \underline{u}_e \tag{3.29a}$$

$$\underline{F}_e = \underline{T}^{\text{T}} \cdot \underline{F}_e^{(\text{lok})} \tag{3.29b}$$

with

$$
\underline{T} = \begin{bmatrix}
\cos\alpha & \sin\alpha & 0 & 0 & 0 & 0 \\
-\sin\alpha & \cos\alpha & 0 & 0 & 0 & 0 \\
0 & 0 & 1 & 0 & 0 & 0 \\
0 & 0 & 0 & \cos\alpha & \sin\alpha & 0 \\
0 & 0 & 0 & -\sin\alpha & \cos\alpha & 0 \\
0 & 0 & 0 & 0 & 0 & 1
\end{bmatrix}
$$

$$
\underline{u}_e = \begin{bmatrix}
u_{e1} \\
v_{e1} \\
\varphi_{e1} \\
u_{e2} \\
v_{e2} \\
\varphi_{e2}
\end{bmatrix}
\qquad
\underline{F}_e = \begin{bmatrix}
F_{x1}^{(e)} \\
F_{y1}^{(e)} \\
F_{z1}^{(e)} \\
F_{x2}^{(e)} \\
F_{y2}^{(e)} \\
F_{z2}^{(e)}
\end{bmatrix}.
\tag{3.29c}
$$

The nodal forces in global coordinates are obtained with (3.28a), (3.29b), and (3.29a) as

$$\underline{F}_e = \underline{T}^{\text{T}} \cdot \underline{F}_e^{(\text{lok})} = \underline{T}^{\text{T}} \cdot \underline{K}_e^{(\text{lok})} \cdot \underline{u}_e^{(\text{lok})} = \underline{T}^{\text{T}} \cdot \underline{K}_e^{(\text{lok})} \cdot \underline{T} \cdot \underline{u}_e.$$

The transformed stiffness relationship is obtained as

$$\underline{K}_e \cdot \underline{u}_e = \underline{F}_e \tag{3.30}$$

Nodal displacements

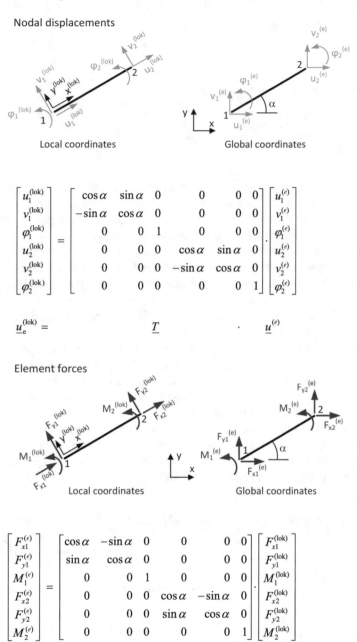

$$
\begin{bmatrix} u_1^{(lok)} \\ v_1^{(lok)} \\ \varphi_1^{(lok)} \\ u_2^{(lok)} \\ v_2^{(lok)} \\ \varphi_2^{(lok)} \end{bmatrix} = \begin{bmatrix} \cos\alpha & \sin\alpha & 0 & 0 & 0 & 0 \\ -\sin\alpha & \cos\alpha & 0 & 0 & 0 & 0 \\ 0 & 0 & 1 & 0 & 0 & 0 \\ 0 & 0 & 0 & \cos\alpha & \sin\alpha & 0 \\ 0 & 0 & 0 & -\sin\alpha & \cos\alpha & 0 \\ 0 & 0 & 0 & 0 & 0 & 1 \end{bmatrix} \cdot \begin{bmatrix} u_1^{(e)} \\ v_1^{(e)} \\ \varphi_1^{(e)} \\ u_2^{(e)} \\ v_2^{(e)} \\ \varphi_2^{(e)} \end{bmatrix}
$$

$$
\underline{u}_e^{(lok)} = \qquad\qquad \underline{T} \qquad\qquad \cdot \qquad \underline{u}^{(e)}
$$

Element forces

$$
\begin{bmatrix} F_{x1}^{(e)} \\ F_{y1}^{(e)} \\ M_1^{(e)} \\ F_{x2}^{(e)} \\ F_{y2}^{(e)} \\ M_2^{(e)} \end{bmatrix} = \begin{bmatrix} \cos\alpha & -\sin\alpha & 0 & 0 & 0 & 0 \\ \sin\alpha & \cos\alpha & 0 & 0 & 0 & 0 \\ 0 & 0 & 1 & 0 & 0 & 0 \\ 0 & 0 & 0 & \cos\alpha & -\sin\alpha & 0 \\ 0 & 0 & 0 & \sin\alpha & \cos\alpha & 0 \\ 0 & 0 & 0 & 0 & 0 & 1 \end{bmatrix} \cdot \begin{bmatrix} F_{x1}^{(lok)} \\ F_{y1}^{(lok)} \\ M_1^{(lok)} \\ F_{x2}^{(lok)} \\ F_{y2}^{(lok)} \\ M_2^{(lok)} \end{bmatrix}
$$

$$
\underline{F}^{(e)} = \qquad\qquad \underline{T}^{\mathrm{T}} \qquad\qquad \cdot \qquad \underline{F}_e^{(lok)}
$$

Fig. 3.19 Coordinate transformation of the nodal displacements and element forces of the beam element

where

$$\underline{K}_e = \underline{T}^{\mathrm{T}} \cdot \underline{K}_e^{(\mathrm{lok})} \cdot \underline{T} \tag{3.30a}$$

is the element stiffness matrix in global coordinates.

3.4.5 Hinges

Hinges for bending moments frequently occur in frameworks and other beam structures. In principle, however, a hinge effect can be defined for any degree of freedom,
i.e. shear hinges or normal force hinges are possible. These types of hinges are
required to model different effects, e.g. symmetry conditions, or to compute influence lines. A hinge removes a static link of an individual section force. This may be
a bending moment, a normal force, a shear force, or a torsional moment (Fig. 3.20).
In the hinge, the corresponding section force cannot be transferred, i.e. it is released.
It is also possible to add several releases simultaneously in one hinge.

When a link is removed at an end of a beam element, the beam element no longer
possesses a stiffness in the corresponding degree of freedom. Hence, in the stiffness
matrix the degree of freedom corresponding to the removed link no longer appears. In
order to consider the hinge effect, the stiffness matrix of the beam element according
to (3.24) or (3.28) has to be modified. As the degree of freedom has been removed,
a load in the corresponding degree of freedom cannot be transferred from the nodal
point to the element, i.e. it is not allowed.

The modification of the element stiffness matrix for hinges is shown in detail in
[4]. Here, a hinge for a bending moment (M-release) at the end of a shear fixed beam
element without element loads is considered (Fig. 3.21). According to (3.24), the
stiffness relationship of the beam element in local coordinates is

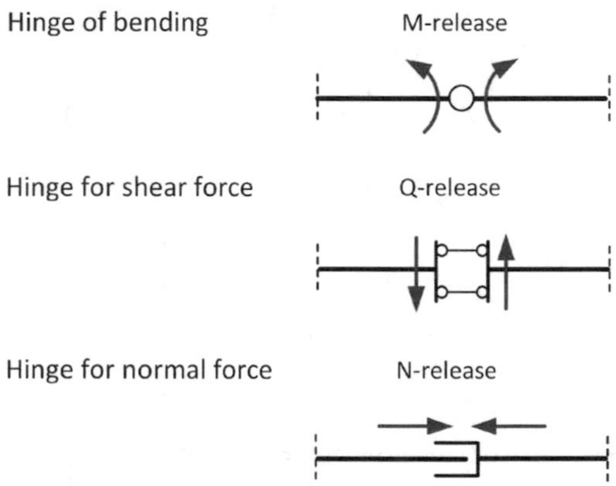

Fig. 3.20 Types of hinges

Hinge of bending	M-release

Hinge for shear force	Q-release

Hinge for normal force	N-release

Fig. 3.21 Beam element
with a hinge

$$\frac{E \cdot I}{l} \cdot \begin{bmatrix} 12/l^2 & 6/l & -12/l^2 & 6/l \\ 6/l & 4 & -6/l & 2 \\ -12/l^2 & -6/l & 12/l^2 & -6/l \\ 6/l & 2 & -6/l & 4 \end{bmatrix} \cdot \begin{bmatrix} v_1^{(lok)} \\ \varphi_1^{(lok)} \\ v_2^{(lok)} \\ \varphi_2^{(lok)} \end{bmatrix} = \begin{bmatrix} F_{y1}^{(lok)} \\ M_{z1}^{(lok)} \\ F_{y2}^{(lok)} \\ M_{z2}^{(lok)} \end{bmatrix}.$$

The removed link $M_{z2}^{(lok)} = 0$ corresponds to the elimination of the degree of freedom $\varphi_2^{(lok)}$. First the last equation of the stiffness relationship is solved for $\varphi_2^{(lok)}$ with $M_{z2}^{(lok)} = 0$. One obtains

$$EI/l \cdot \left(6/l \cdot v_1^{(lok)} + 2 \cdot \varphi_1^{(lok)} - 6/l \cdot v_2^{(lok)} + 4 \cdot \varphi_2^{(lok)}\right) = M_{z2}^{(lok)} = 0$$

and

$$\varphi_2^{(lok)} = -3/(2l) \cdot v_1^{(lok)} - 1/2 \cdot \varphi_1^{(lok)} + 3/(2l) \cdot v_2^{(lok)}.$$

If this expression for $\varphi_2^{(lok)}$ is introduced into the first, second, and third equations of (3.24), $\varphi_2^{(lok)}$ is eliminated from the system of equations. The stiffness matrix of the beam element with a hinge for a bending moment at node 2 is obtained as

$$\frac{E \cdot I}{l} \begin{bmatrix} 3/l^2 & 3/l & -3/l^2 \\ 3/l & 3 & -3/l \\ -3/l^2 & -3/l & 3/l^2 \end{bmatrix} \cdot \begin{bmatrix} v_1^{(lok)} \\ \varphi_1^{(lok)} \\ v_2^{(lok)} \end{bmatrix} = \begin{bmatrix} F_{y1}^{(lok)} \\ M_{z1}^{(lok)} \\ F_{y2}^{(lok)} \end{bmatrix}. \qquad (3.31)$$

The degree of freedom $\varphi_2^{(lok)}$ is now missing in the stiffness matrix. Therefore, the beam element does not resist a moment at node 2 of the element. A loading in the direction of degree of freedom $\varphi_2^{(lok)}$ is therefore not meaningful. This has to be considered at nodal points connecting several beam elements with hinges in order to avoid kinematic systems. Figure 3.22 shows a nodal point with a hinge connection of three beam elements. Hinges should be defined at the ends of two beam elements whereas for the remaining beam element no hinge definition should be provided (Fig. 3.22). If the nodal point is loaded by a moment, this moment is acting on the element without a hinge. Alternatively, a hinge can be defined at the ends of all

Fig. 3.22 Different hinge
definitions at a nodal point
with three beam elements

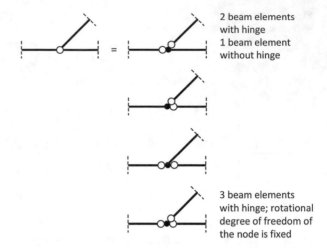

2 beam elements
with hinge
1 beam element
without hinge

3 beam elements
with hinge; rotational
degree of freedom of
the node is fixed

three elements and the rotational degree of freedom of the nodal point will be fixed.
Without fixing the degree of freedom, the system would be kinematic in the rotational
degree of freedom of this nodal point.

Example 3.14 A user of a finite element program has defined a hinge at the end of the
cantilever in Fig. 3.23 (erroneously). The consequences of this program input are to
be shown and possible remedies should be discussed. The horizontal displacements
have been fixed.

The system has two nodal points with the degrees of freedom v_1, φ_1, v_2, φ_2. Hence,
the global stiffness matrix has these four degrees of freedom. The element stiffness
matrix of the beam with the hinge at the right-hand side, according to (3.31), is added.
One obtains

$$
\frac{E \cdot I}{l} \cdot
\begin{bmatrix}
3/l^2 & 3/l & -3/l^2 & 0 \\
3/l & 3 & 3/l & 0 \\
-3/l^2 & -3/l & 3/l^2 & 0 \\
0 & 0 & 0 & 0
\end{bmatrix}
\cdot
\begin{bmatrix}
v_1 \\
\varphi_1 \\
v_2 \\
\varphi_2
\end{bmatrix}
=
\begin{bmatrix}
F_1 \\
M_1 \\
F_2 \\
M_2
\end{bmatrix}.
$$

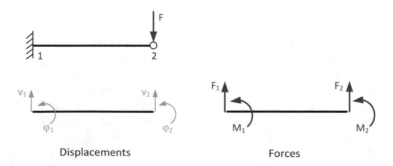

Displacements   Forces

Fig. 3.23 Cantilever with hinge

Considering the support conditions $v_1 = 0$ and $\varphi_1 = 0$ gives

$$\frac{EI}{l}\begin{bmatrix} 3/l^2 & 0 \\ 0 & 0 \end{bmatrix} \cdot \begin{bmatrix} v_2 \\ \varphi_2 \end{bmatrix} = \begin{bmatrix} F_2 \\ M_2 \end{bmatrix}.$$

The matrix of this system of equations is singular due to the second row and cannot be solved. The program stops the calculation unless special provisions have been taken within the program for this case. It can also be noted, that a loading by a moment M_2 is not meaningful.

There are the following remedies possible:

(a) Constraining the degree of freedom φ_2

In this case, the second equation is eliminated from the system of equations due to the support condition and one obtains with $F_2 = -F$:

$$3EI/l^3 \cdot v_2 = F_2 = -F, \quad -v_2 = \frac{Fl^3}{3EI}.$$

(b) Introduction of a rotational spring with a high stiffness at node 2

In this case, the system of equations is

$$\begin{bmatrix} 3EI/l^2 & 0 \\ 0 & k_\varphi \end{bmatrix} \cdot \begin{bmatrix} v_2 \\ \varphi_2 \end{bmatrix} = \begin{bmatrix} -F \\ M_2 \end{bmatrix}.$$

These are two uncoupled equations with the solution:

$$3EI/l^3 \cdot v_2 = -F \quad \rightarrow \quad v_2 = -Fl^3/3EI$$

$$k_\varphi \cdot \varphi_2 = M_2 \quad \rightarrow \quad \varphi_2 = M_2/k_\varphi.$$

A moment M_2 at nodal point 2 is carried by the spring since it cannot be transferred to the beam, due to the hinge. The angle φ_2 is the rotation of the spring. Many finite element programs have implemented this handling of zero diagonal terms in the stiffness matrix automatically with very high spring stiffnesses (e.g. 10^{20}).

(c) Substitution by a beam element without hinge

This is the normal case without any hinge definition. The system of equations with the stiffness matrix according to (3.24) is

$$\frac{E \cdot I}{l} \cdot \begin{bmatrix} 12/l^2 & 6/l & -12/l^2 & 6/l \\ 6/l & 4 & -6/l & 2 \\ -12/l^2 & -6/l & 12/l^2 & -6/l \\ 6/l & 2 & -6/l & 4 \end{bmatrix} \cdot \begin{bmatrix} v_1 \\ \varphi_1 \\ v_2 \\ \varphi_2 \end{bmatrix} = \begin{bmatrix} F_{y1} \\ M_{z1} \\ F_{y2} \\ M_{z2} \end{bmatrix}.$$

After the elimination of the rows and columns for v_1 and φ_1 in order to consider the support conditions, one obtains

$$\frac{EI}{l} \begin{bmatrix} 12/l^2 & -6/l \\ -6/l & 4 \end{bmatrix} \cdot \begin{bmatrix} v_2 \\ \varphi_2 \end{bmatrix} = \begin{bmatrix} F_{y2} \\ M_{z2} \end{bmatrix}.$$

The global stiffness matrix is regular. Herein the angle φ_2 is the rotational angle at the end of the cantilever.   ◀

3.5   Combined Beam and Truss Systems

A structural system of a plane framework may consist of truss and beam elements with some spring elements added. Nodal points connected with truss elements or displacement springs only possess the 2 degrees of freedom u and v, whereas nodal points connected with beam elements or rotational springs have the 3 degrees of freedom u, v and φ. All other degrees of freedom should be fixed (if not done by the program automatically) in order to avoid a singular global stiffness matrix.

Example 3.15 For the system shown in Fig. 3.24, the global stiffness matrix shall be established.

   The system has 3 degrees of freedom at the nodal points 1–3 connected with beam elements and 2 degrees of freedom at nodal point 4 connected with the truss element only. It is assumed that the degrees of freedom of the horizontal displacements are fixed, which has no influence on the result of the present system. Hence, the support conditions are

$$u_1 = u_2 = u_3 = u_4 = v_1 = v_4 = 0.$$

   First, the element stiffness matrices are formulated. For a shorter notation and better clarity, the support conditions are already taken into account in the element stiffness matrices.

   Element 1 is a beam element loaded by a uniform load q. Since the horizontal displacements u_1 and u_2 as well as the vertical displacements v_1 are zero, the corresponding rows and columns of the element stiffness matrix are omitted.

   Neglecting shear deformations, one obtains with (3.26) and the loading terms according to Table 3.3

System   Nodal points and elements

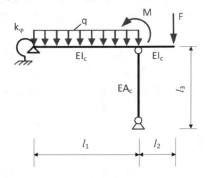

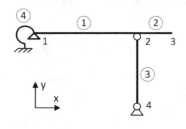

Element types   Degrees of freedom

Elements 1,2 :   Beam elements

Element 3 :   Truss element

Element 4 :   Spring element

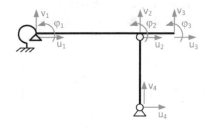

Fig. 3.24 Structural system

$$
\frac{E \cdot I_c}{l_1} \cdot
\begin{bmatrix}
4 & -\dfrac{6}{l_1} & 2 \\[2ex]
-\dfrac{6}{l_1} & \dfrac{12}{l_1^2} & -\dfrac{6}{l_1} \\[2ex]
2 & -\dfrac{6}{l_1} & 4
\end{bmatrix}
\cdot
\begin{bmatrix}
\varphi_1 \\[2ex] v_2 \\[2ex] \varphi_2
\end{bmatrix}
=
\begin{bmatrix}
M_{z1}^{(1)} \\[2ex] F_{y2}^{(1)} \\[2ex] M_{z2}^{(1)}
\end{bmatrix}
-
\begin{bmatrix}
\dfrac{q \cdot l_1^2}{12} \\[2ex] \dfrac{q \cdot l_1}{2} \\[2ex] -\dfrac{q \cdot l_1^2}{12}
\end{bmatrix}
$$

The stiffness relationship of element 2 is given by (3.24) as

$$
\frac{E \cdot I_c}{l_2} \cdot
\begin{bmatrix}
\dfrac{12}{l_2^2} & \dfrac{6}{l_2} & -\dfrac{12}{l_2^2} & \dfrac{6}{l_2} \\[2ex]
\dfrac{6}{l_2} & 4 & -\dfrac{6}{l_2} & 2 \\[2ex]
-\dfrac{12}{l_2^2} & -\dfrac{6}{l_2} & \dfrac{12}{l_2^2} & -\dfrac{6}{l_2} \\[2ex]
\dfrac{6}{l_2} & 2 & -\dfrac{6}{l_2} & 4
\end{bmatrix}
\cdot
\begin{bmatrix}
v_2 \\[2ex] \varphi_2 \\[2ex] v_3 \\[2ex] \varphi_3
\end{bmatrix}
=
\begin{bmatrix}
F_{y2}^{(2)} \\[2ex] M_{z2}^{(2)} \\[2ex] F_{y3}^{(2)} \\[2ex] M_{z3}^{(2)}
\end{bmatrix}
$$

Element 3 is a truss element. Considering the support condition $v_4 = 0$, the stiffness relationship according to (3.3) reduces to

$$\frac{E \cdot A_c}{l_3} \cdot v_2 = F_{y2}^{(3)}.$$

For the spring element with the spring constant k_φ (element 4), one obtains with (3.17c)

$$M_{z1}^{(4)} = k_\varphi \cdot \varphi_1.$$

The global stiffness matrix is composed of the element stiffness matrices element-by-element. The terms of the element stiffness matrices are added to the global stiffness matrices at those positions where the local element degrees of freedom correspond to the global degrees of freedom of the system.

In doing so, the stiffness matrix of the global system and the load terms are obtained as

$$
\begin{bmatrix}
k_\varphi + 4 \cdot c_1 & -6 \cdot \dfrac{c_1}{l_1} & 2 \cdot c_1 & 0 & 0 \\[2ex]
-6 \cdot \dfrac{c_1}{l_1} & 12 \cdot \dfrac{c_1}{l_1^2} + 12 \cdot \dfrac{c_2}{l_2^2} + \dfrac{E A_c}{l_3} & -6 \cdot \dfrac{c_1}{l_1} + 6 \cdot \dfrac{c_2}{l_2} & -12 \cdot \dfrac{c_2}{l_2^2} & 6 \cdot \dfrac{c_2}{l_2} \\[2ex]
2 \cdot c_1 & -6 \cdot \dfrac{c_1}{l_1} + 6 \cdot \dfrac{c_2}{l_2} & 4 \cdot c_1 + 4 \cdot c_2 & -6 \cdot \dfrac{c_2}{l_2} & 2 \cdot c_2 \\[2ex]
0 & -12 \cdot \dfrac{c_2}{l_2^2} & -6 \cdot \dfrac{c_2}{l_2} & 12 \cdot \dfrac{c_2}{l_2^2} & -6 \cdot \dfrac{c_2}{l_2} \\[2ex]
0 & 6 \cdot \dfrac{c_2}{l_2} & 2 \cdot c_2 & -6 \cdot \dfrac{c_2}{l_2} & 4 \cdot c_2
\end{bmatrix}
$$

$$
\cdot
\begin{bmatrix}
\varphi_1 \\[1.5ex]
v_2 \\[1.5ex]
\varphi_2 \\[1.5ex]
v_3 \\[1.5ex]
\varphi_3
\end{bmatrix}
=
\begin{bmatrix}
-q \cdot l_1^2/12 \\[1.5ex]
-q \cdot l_1/2 \\[1.5ex]
M + q \cdot l_1^2/12 \\[1.5ex]
-F \\[1.5ex]
0
\end{bmatrix}
$$

with $c_1 = E I_c / l_1$ and $c_2 = E I_c / l_2$.   ◄

3.6 Spatial Trusses and Beam Structures

3.6.1 3D Truss Element

Each nodal point of a spatial truss possesses three degrees of freedom for the displacements u, v, and w (Fig. 3.2). Hence, the stiffness matrix of a 3D truss element with two nodal points is a 6×6 matrix (Fig. 3.25). The 3D element stiffness matrix of a truss element is obtained with the element stiffness in local coordinates applying a transformation corresponding to (3.8b) for two-dimensional systems. Hence, the transformation matrix has to be formulated for the three-dimensional case.

The transformation of the displacements is obtained analogously to the two-dimensional case as [5, 15]

$$
\begin{bmatrix} u_1^{(\text{lok})} \\ u_2^{(\text{lok})} \end{bmatrix} = \frac{1}{l} \cdot \begin{bmatrix} l_x & l_y & l_z & 0 & 0 & 0 \\ 0 & 0 & 0 & l_x & l_y & l_z \end{bmatrix} \cdot \begin{bmatrix} u_1^{(e)} \\ v_1^{(e)} \\ w_1^{(e)} \\ u_2^{(e)} \\ v_2^{(e)} \\ w_2^{(e)} \end{bmatrix}
\tag{3.32}
$$

or

$$
\underline{u}_e^{(\text{lok})} = \underline{T}_{\text{3D,F}} \cdot \underline{u}_e
\tag{3.32a}
$$

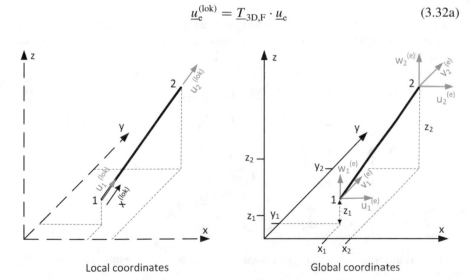

Local coordinates   Global coordinates

Fig. 3.25 Degrees of freedom of a three-dimensional truss element

with

$$l_x = x_2 - x_1, \quad l_y = y_2 - y_1, \quad l_z = z_2 - z_1, \quad l = \sqrt{l_x^2 + l_y^2 + l_z^2}. \qquad (3.32b)$$

The element forces are arranged in the same order as the degrees of freedom in $\underline{u}_e$. They are

$$\underline{F}_e = \begin{bmatrix} F_{x1}^{(e)} \\ F_{y1}^{(e)} \\ F_{z1}^{(e)} \\ F_{x2}^{(e)} \\ F_{y2}^{(e)} \\ F_{z2}^{(e)} \end{bmatrix}. \qquad (3.33)$$

Thus, the stiffness relationship in global coordinates is analogous to (3.8)

$$\underline{F}_e = \underline{K}_e \cdot \underline{u}_e \qquad (3.34)$$

with

$$\underline{K}_e = \underline{T}_{3D,F}^T \cdot \underline{K}_e^{(lok)} \cdot \underline{T}_{3D,F}. \qquad (3.34a)$$

The element stiffness matrix $\underline{K}_e$ of the truss element in global coordinates is obtained with the transformation matrix $\underline{T}_{3D,F}$ according to (3.32) and $\underline{K}_e^{(lok)}$ according to (3.3b) as

$$\underline{K}_e = \frac{E \cdot A}{l^3} \cdot \begin{bmatrix} l_x^2 & l_x \cdot l_y & l_x \cdot l_z & -l_x^2 & -l_x \cdot l_y & -l_x \cdot l_z \\ l_x \cdot l_y & l_y^2 & l_y \cdot l_z & -l_x \cdot l_y & -l_y^2 & -l_y \cdot l_z \\ l_x \cdot l_z & l_y \cdot l_z & l_z^2 & -l_x \cdot l_z & -l_y \cdot l_z & -l_z^2 \\ -l_x^2 & -l_x \cdot l_y & -l_x \cdot l_z & l_x^2 & l_x \cdot l_y & l_x \cdot l_z \\ -l_x \cdot l_y & -l_y^2 & -l_y \cdot l_z & l_x \cdot l_y & l_y^2 & l_y \cdot l_z \\ -l_x \cdot l_z & -l_y \cdot l_z & -l_z^2 & l_x \cdot l_z & l_y \cdot l_z & l_z^2 \end{bmatrix}. \qquad (3.35)$$

The normal force in the truss element can be determined analogously to (3.10) from

$$N = \underline{S}_e \cdot \underline{u}_e \qquad (3.36)$$

with

$$\underline{S}_e = \underline{S}_e^{(\mathrm{lok})} \cdot \underline{T}_{\mathrm{3D,F}}. \tag{3.36a}$$

After inserting the transformation matrix (3.32) and $\underline{S}_e^{(\mathrm{lok})}$ according to (3.2b) into this relationship, the section force matrix is obtained as

$$\underline{S}_e = \frac{E \cdot A}{l^2} \cdot \begin{bmatrix} -l_x & -l_y & -l_z & l_x & l_y & l_z \end{bmatrix}. \tag{3.37}$$

Example 3.16 For the spatial truss shown in Fig. 3.26a, consisting of 4 truss elements, the element forces are to be determined. The example has been analyzed in [12] by the force method. The nodal coordinates and the resulting member lengths as well as the cross-sectional areas of the members are given in Fig. 3.26a. Young's modulus is $E = 2.1 \cdot 10^8 \, \mathrm{kN/m^2}$.

The force $F = 200 \, \mathrm{kN}$ is acting at node 5. Its decomposition in x-, y-, and z-directions is given in Fig. 3.26b.

The system has only 3 degrees of freedom, which are freely movable, namely the displacements of the node 5 in x-, y-, and z-directions. The nodes 1–4 are fixed in all 3 degrees of freedom.

First, the element stiffness matrices for the 4 truss members are set up, whereby— due to the shorter notation—the support conditions are already taken into account. They are thus related to the displacement vector

$$\underline{u} = \begin{bmatrix} u_5 \\ v_5 \\ w_5 \end{bmatrix}$$

and the corresponding nodal forces.

For element 1 with

$$l_x = 8 \, \mathrm{m}, \quad l_y = -3 \, \mathrm{m}, \quad l_z = 9 \, \mathrm{m}, \quad l = 12, 41 \, \mathrm{m}$$

the element stiffness matrix is obtained according to (3.35) as

$$\underline{K}^{(1)} = \frac{2.1 \cdot 10^8 \cdot 0.006}{12.41^3} \begin{bmatrix} 8^2 & 8 \cdot (-3) & 8 \cdot 9 \\ 8 \cdot (-3) & (-3)^2 & (-3) \cdot 9 \\ 8 \cdot 9 & (-3) \cdot 9 & 9^2 \end{bmatrix} \frac{\mathrm{kN}}{\mathrm{m}}$$

$$= \begin{bmatrix} 42200 & -15820 & 47470 \\ -15820 & 5930 & -17800 \\ 47470 & -17800 & 53400 \end{bmatrix} \frac{\mathrm{kN}}{\mathrm{m}}.$$

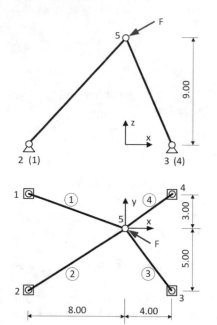

Nodal coordinates [m]

Nodal Point	x	y	z
1	-8	3	0
2	-8	-5	0
3	4	-5	0
4	4	3	0
5	0	0	9

Truss elements

Elem. no.	Node 1	Node 2	l [m]	A [m²]
1	1	5	12.41	0.0060
2	2	5	13.04	0.0070
3	3	5	11.05	0.0040
4	4	5	10.30	0.0030

a) Truss system

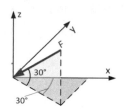

$$F_x = -F \cdot \cos(30°) \cdot \cos(30°) = -150 \text{ kN}$$
$$F_y = F \cdot \cos(30°) \cdot \sin(30°) = 86{,}6 \text{ kN}$$
$$F_z = -F \cdot \sin(30°) = -100 \text{ kN}$$

b) Decomposition of force F

Fig. 3.26 Spacial truss system

Accordingly, the element stiffness matrices for elements 2, 3, and 4 are as follows:

$$\underline{K}^{(2)} = \begin{bmatrix} 42440 & 26530 & 47750 \\ 26530 & 16580 & 29840 \\ 47750 & 29840 & 53720 \end{bmatrix} \frac{\text{kN}}{\text{m}}$$

$$\underline{K}^{(3)} = \begin{bmatrix} 9970 & -12470 & -22440 \\ -12470 & 15580 & 28050 \\ -22440 & 28050 & 50490 \end{bmatrix} \frac{\text{kN}}{\text{m}}$$

$$\underline{K}^{(4)} = \begin{bmatrix} 9240 & 6930 & -20780 \\ 6930 & 5200 & -15590 \\ -20780 & -15590 & 46760 \end{bmatrix} \frac{\text{kN}}{\text{m}}.$$

Hence, the global stiffness matrix is

$$\underline{K} = \underline{K}^{(1)} + \underline{K}^{(2)} + \underline{K}^{(3)} + \underline{K}^{(4)} = \begin{bmatrix} 103850 & 5160 & 52000 \\ 5160 & 43290 & 24510 \\ 52000 & 24510 & 204370 \end{bmatrix} \frac{kN}{m}$$

and the system of equations of the global system

$$\begin{bmatrix} 103850 & 5160 & 52000 \\ 5160 & 43290 & 24510 \\ 52000 & 24510 & 204370 \end{bmatrix} \cdot \begin{bmatrix} u_5 \\ v_5 \\ w_5 \end{bmatrix} = \begin{bmatrix} -150.0 \\ 86.6 \\ -100.0 \end{bmatrix}.$$

Its solution is

$$\begin{bmatrix} u_5 \\ v_5 \\ w_5 \end{bmatrix} = \begin{bmatrix} -13.460 \\ 24.075 \\ -4.355 \end{bmatrix} \cdot 10^{-4} \, m.$$

The normal forces in the truss elements are given by (3.36) with (3.37) as

$$\text{Element 1: } N^{(1)} = \frac{2.1 \cdot 10^8 \cdot 0.006}{12.41^3} \cdot \begin{bmatrix} 8 & -3 & 9 \end{bmatrix} \cdot \begin{bmatrix} -13.460 \\ 24.075 \\ -4.355 \end{bmatrix} \cdot 10^{-4} \, kN$$

$$= -179.3 \, kN$$

Elements 2, 3, and 4: $N^{(2)} = -22.9 \, kN$, $N^{(3)} = 93.0 \, kN$, $N^{(4)} = -34.2 \, kN$. ◄

3.6.2 3D Beam Element

Spatial frame structures with beam elements have, in addition, three rotational degrees of freedom at each nodal point (Fig. 3.2). Hence, the stiffness matrix of a 3D beam element with two nodal points is a 12 × 12 matrix (Fig. 3.27).

For beam elements, in addition to the 2D bending stiffness and normal stiffness given in (3.28), the bending stiffness with respect to the $z^{(lok)}$-axis and the torsional stiffness have to be considered before the transformation from local to global coordinates is done. The bending stiffness with respect to the $z^{(lok)}$-axis can easily be derived from the element stiffness matrix of the two-dimensional case where bending is related to the $y^{(lok)}$-axis. However, torsion occurs as an additional stress state.

At first only the St Venant torsion is considered. The stiffness matrix of an element subjected to torsion only can be evaluated easily. Its degrees of freedom are the torsional rotations at the nodal points of the element and the corresponding forces are the torsional moments (Fig. 3.28). The torsional angle caused by a moment T is

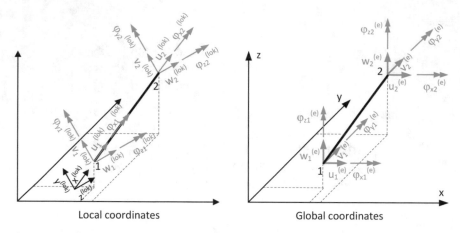

Local coordinates   Global coordinates

Fig. 3.27 Degrees of freedom of a three-dimensional beam element

$$\varphi = \left(\varphi_{x2}^{(\text{lok})} - \varphi_{x1}^{(\text{lok})}\right) = \frac{l}{GI_T} \cdot T. \tag{3.38}$$

Hence, the element moments are

$$M_{x1}^{(\text{lok})} = -T = \frac{GI_T}{l}\left(\varphi_{x1}^{(\text{lok})} - \varphi_{x2}^{(\text{lok})}\right) \tag{3.38a}$$

$$M_{x2}^{(\text{lok})} = T = \frac{GI_T}{l}\left(-\varphi_{x1}^{(\text{lok})} + \varphi_{x2}^{(\text{lok})}\right) \tag{3.38b}$$

or

$$\frac{GI_T}{l}\begin{bmatrix} 1 & -1 \\ -1 & 1 \end{bmatrix} \cdot \begin{bmatrix} \varphi_{x1}^{(\text{lok})} \\ \varphi_{x2}^{(\text{lok})} \end{bmatrix} = \begin{bmatrix} M_{x1}^{(\text{lok})} \\ M_{x2}^{(\text{lok})} \end{bmatrix}. \tag{3.39}$$

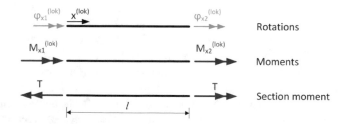

Fig. 3.28 Rotations and moments of a bar subjected to torsion

This stiffness matrix can be added to the element stiffness matrix (3.28) expanded by the degrees of freedom $\varphi_{x1}^{(lok)}$ and $\varphi_{x2}^{(lok)}$ as was shown in Sect. 3.4.3 for the longitudinal stiffness. The element stiffness matrix of the 3D-beam element is obtained by adding the degrees of freedom and the terms for bending with respect to the $z^{(lok)}$-axis. The stiffness matrix related to the coordinate system in Fig. 3.29 is

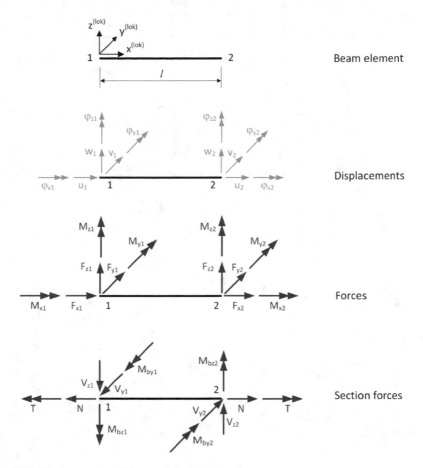

Fig. 3.29 Displacements and forces of the 3D beam element

$$
\left[
\begin{array}{cccccccccccc}
\dfrac{E\cdot A}{l} & 0 & 0 & 0 & 0 & 0 & -\dfrac{E\cdot A}{l} & 0 & 0 & 0 & 0 & 0 \\[2mm]
0 & \dfrac{12\cdot E\cdot I_z}{l^3} & 0 & 0 & 0 & \dfrac{6\cdot E\cdot I_z}{l^2} & 0 & -\dfrac{12\cdot E\cdot I_z}{l^3} & 0 & 0 & 0 & \dfrac{6\cdot E\cdot I_z}{l^2} \\[2mm]
0 & 0 & \dfrac{12\cdot E\cdot I_y}{l^3} & 0 & -\dfrac{6\cdot E\cdot I_y}{l^2} & 0 & 0 & 0 & -\dfrac{12\cdot E\cdot I_y}{l^3} & 0 & -\dfrac{6\cdot E\cdot I_y}{l^2} & 0 \\[2mm]
0 & 0 & 0 & \dfrac{G\cdot I_T}{l} & 0 & 0 & 0 & 0 & 0 & -\dfrac{G\cdot I_T}{l} & 0 & 0 \\[2mm]
0 & 0 & -\dfrac{6\cdot E\cdot I_y}{l^2} & 0 & \dfrac{4\cdot E\cdot I_y}{l} & 0 & 0 & 0 & \dfrac{6\cdot E\cdot I_y}{l^2} & 0 & \dfrac{2\cdot E\cdot I_y}{l} & 0 \\[2mm]
0 & \dfrac{6\cdot E\cdot I_z}{l^2} & 0 & 0 & 0 & \dfrac{4\cdot E\cdot I_z}{l} & 0 & -\dfrac{6\cdot E\cdot I_z}{l^2} & 0 & 0 & 0 & \dfrac{2\cdot E\cdot I_z}{l} \\[2mm]
-\dfrac{E\cdot A}{l} & 0 & 0 & 0 & 0 & 0 & \dfrac{E\cdot A}{l} & 0 & 0 & 0 & 0 & 0 \\[2mm]
0 & -\dfrac{12\cdot E\cdot I_z}{l^3} & 0 & 0 & 0 & -\dfrac{6\cdot E\cdot I_z}{l^2} & 0 & \dfrac{12\cdot E\cdot I_z}{l^3} & 0 & 0 & 0 & -\dfrac{6\cdot E\cdot I_z}{l^2} \\[2mm]
0 & 0 & -\dfrac{12\cdot E\cdot I_y}{l^3} & 0 & \dfrac{6\cdot E\cdot I_y}{l^2} & 0 & 0 & 0 & \dfrac{12\cdot E\cdot I_y}{l^3} & 0 & \dfrac{6\cdot E\cdot I_y}{l^2} & 0 \\[2mm]
0 & 0 & 0 & -\dfrac{G\cdot I_T}{l} & 0 & 0 & 0 & 0 & 0 & \dfrac{G\cdot I_T}{l} & 0 & 0 \\[2mm]
0 & 0 & -\dfrac{6\cdot E\cdot I_y}{l^2} & 0 & \dfrac{2\cdot E\cdot I_y}{l} & 0 & 0 & 0 & \dfrac{6\cdot E\cdot I_y}{l^2} & 0 & \dfrac{4\cdot E\cdot I_y}{l} & 0 \\[2mm]
0 & \dfrac{6\cdot E\cdot I_z}{l^2} & 0 & 0 & 0 & \dfrac{2\cdot E\cdot I_z}{l} & 0 & -\dfrac{6\cdot E\cdot I_z}{l^2} & 0 & 0 & 0 & \dfrac{4\cdot E\cdot I_z}{l}
\end{array}
\right]
\cdot
\left[
\begin{array}{c}
u_1 \\ v_1 \\ w_1 \\ \varphi_{x1} \\ \varphi_{y1} \\ \varphi_{z1} \\ u_2 \\ v_2 \\ w_2 \\ \varphi_{x2} \\ \varphi_{y2} \\ \varphi_{z2}
\end{array}
\right]
=
\left[
\begin{array}{c}
F_{x1} \\ F_{y1} \\ F_{z1} \\ M_{x1} \\ M_{y1} \\ M_{z1} \\ F_{x2} \\ F_{y2} \\ F_{z2} \\ M_{x2} \\ M_{y2} \\ M_{z2}
\end{array}
\right]
$$

$$\text{(3.40)}$$

In the stiffness matrix, the terms for longitudinal forces, torsion, bending about the $y^{(lok)}$-axis, and bending about the $z^{(lok)}$-axis are uncoupled. The section forces can easily be obtained, as they may differ from the element forces only with respect to their sign (Fig. 3.29).

The element stiffness matrix in global coordinates can be determined from the element stiffness matrix in the local coordinate system by a spatial transformation (Fig. 3.27). It is

$$\underline{K}_e = \underline{T}_{3D,B}^{T} \cdot \underline{K}_e^{(lok)} \cdot \underline{T}_{3D,B} \tag{3.41}$$

with $\underline{K}_e^{(lok)}$ according to (3.40). However, it should be noted that in order to set up the transformation matrix $\underline{T}_{3D,B}$, the spacial orientation of the principal axes of the beam element must be given. Usually, a point in the plane spanned by the $x^{(lok)}$- and $y^{(lok)}$-axis is defined for this purpose. The transformation matrix is given in [1, 2, 7, 15, 16].

3.6.3 Beam Element with Warping Torsion

A beam element with warping torsion possesses an additional degree of freedom. This degree of freedom is the warping ψ. The corresponding force is the warping bi-moment (Fig. 3.30). The stiffness matrix is therefore a 14×14 matrix.

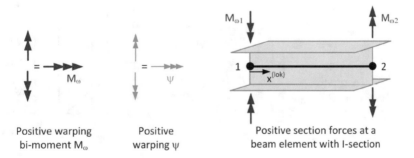

Positive warping bi-moment M_ω Positive warping ψ Positive section forces at a beam element with I-section

Fig. 3.30 Warping deformation and warping bi-moment

According to [7], one obtains (Fig. 3.31)

$$\begin{bmatrix}
\frac{E\cdot A}{l} & 0 & 0 & 0 & 0 & 0 & 0 & -\frac{E\cdot A}{l} & 0 & 0 & 0 & 0 & 0 & 0 \\[4pt]
0 & \frac{12\cdot E\cdot I_z}{l^3} & 0 & 0 & 0 & \frac{6\cdot E\cdot I_z}{l^2} & 0 & 0 & \frac{-12\cdot E\cdot I_z}{l^3} & 0 & 0 & 0 & \frac{6\cdot E\cdot I_z}{l^2} & 0 \\[4pt]
0 & 0 & \frac{12\cdot E\cdot I_y}{l^3} & 0 & \frac{-6\cdot E\cdot I_y}{l^2} & 0 & 0 & 0 & 0 & \frac{-12\cdot E\cdot I_y}{l^3} & 0 & \frac{-6\cdot E\cdot I_y}{l^2} & 0 & 0 \\[4pt]
0 & 0 & 0 & \frac{F\cdot E\cdot I_\omega}{l^3} & 0 & 0 & \frac{-H\cdot E\cdot I_\omega}{l^2} & 0 & 0 & 0 & \frac{-F\cdot E\cdot I_\omega}{l^3} & 0 & 0 & \frac{-H\cdot E\cdot I_\omega}{l^2} \\[4pt]
0 & 0 & \frac{-6\cdot E\cdot I_y}{l^2} & 0 & 4\frac{E\cdot I_y}{l} & 0 & 0 & 0 & 0 & \frac{6\cdot E\cdot I_y}{l^2} & 0 & 2\frac{E\cdot I_y}{l} & 0 & 0 \\[4pt]
0 & \frac{6\cdot E\cdot I_z}{l^2} & 0 & 0 & 0 & 4\frac{E\cdot I_z}{l} & 0 & 0 & \frac{-6\cdot E\cdot I_z}{l^2} & 0 & 0 & 0 & 2\frac{E\cdot I_z}{l} & 0 \\[4pt]
0 & 0 & 0 & \frac{-H\cdot E\cdot I_\omega}{l^2} & 0 & 0 & \frac{A'E\cdot I_\omega}{l} & 0 & 0 & 0 & \frac{H\cdot E\cdot I_\omega}{l^2} & 0 & 0 & B'\frac{E\cdot I_\omega}{l} \\[4pt]
-\frac{E\cdot A}{l} & 0 & 0 & 0 & 0 & 0 & 0 & \frac{E\cdot A}{l} & 0 & 0 & 0 & 0 & 0 & 0 \\[4pt]
0 & \frac{-12\cdot E\cdot I_z}{l^3} & 0 & 0 & 0 & \frac{-6\cdot E\cdot I_z}{l^2} & 0 & 0 & \frac{12\cdot E\cdot I_z}{l^3} & 0 & 0 & 0 & \frac{-6\cdot E\cdot I_z}{l^2} & 0 \\[4pt]
0 & 0 & \frac{-12\cdot E\cdot I_y}{l^3} & 0 & \frac{6\cdot E\cdot I_y}{l^2} & 0 & 0 & 0 & 0 & \frac{12\cdot E\cdot I_y}{l^3} & 0 & \frac{6\cdot E\cdot I_y}{l^2} & 0 & 0 \\[4pt]
0 & 0 & 0 & \frac{-F\cdot E\cdot I_\omega}{l^3} & 0 & 0 & \frac{H\cdot E\cdot I_\omega}{l^2} & 0 & 0 & 0 & \frac{F\cdot E\cdot I_\omega}{l^3} & 0 & 0 & \frac{H\cdot E\cdot I_\omega}{l^2} \\[4pt]
0 & 0 & \frac{-6\cdot E\cdot I_y}{l^2} & 0 & 2\frac{E\cdot I_y}{l} & 0 & 0 & 0 & 0 & \frac{6\cdot E\cdot I_y}{l^2} & 0 & 4\frac{EI_y}{l} & 0 & 0 \\[4pt]
0 & \frac{6\cdot E\cdot I_z}{l^2} & 0 & 0 & 0 & 2\frac{E\cdot I_z}{l} & 0 & 0 & \frac{-6\cdot E\cdot I_z}{l^2} & 0 & 0 & 0 & 4\frac{E\cdot I_z}{l} & 0 \\[4pt]
0 & 0 & 0 & \frac{-H\cdot E\cdot I_\omega}{l^2} & 0 & 0 & B'\frac{E\cdot I_\omega}{l} & 0 & 0 & 0 & \frac{H\cdot E\cdot I_\omega}{l^2} & 0 & 0 & A'\frac{E\cdot I_\omega}{l}
\end{bmatrix}
\cdot
\begin{bmatrix}
u_1 \\ v_1 \\ w_1 \\ \varphi_{x1} \\ \varphi_{y1} \\ \varphi_{z1} \\ \psi_1 \\ u_2 \\ v_2 \\ w_2 \\ \varphi_{x2} \\ \varphi_{y2} \\ \varphi_{z2} \\ \psi_2
\end{bmatrix}
=
\begin{bmatrix}
F_{x1} \\ F_{y1} \\ F_{z1} \\ M_{x1} \\ M_{y1} \\ M_{z1} \\ M_{\omega1} \\ F_{x2} \\ F_{y2} \\ F_{z2} \\ M_{x2} \\ M_{y2} \\ M_{z2} \\ M_{\omega2}
\end{bmatrix}$$

$$(3.42)$$

with

$$\varepsilon = l \cdot \sqrt{\frac{G \cdot I_\mathrm{T}}{E \cdot I_\omega}}$$

$$A' = \frac{\varepsilon \cdot (\sinh \varepsilon - \varepsilon \cdot \cosh \varepsilon)}{2 \cdot (\cosh \varepsilon - 1) - \varepsilon \cdot \sinh \varepsilon}, \quad B' = \frac{\varepsilon \cdot (\varepsilon - \sinh \varepsilon)}{2 \cdot (\cosh \varepsilon - 1) - \varepsilon \cdot \sinh \varepsilon},$$

$$D' = l^2 \cdot \frac{G \cdot I_\mathrm{T}}{E \cdot I_\omega},$$

$$H = A' + B', \quad F = 2 \cdot H + D'. \tag{3.42a-f}$$

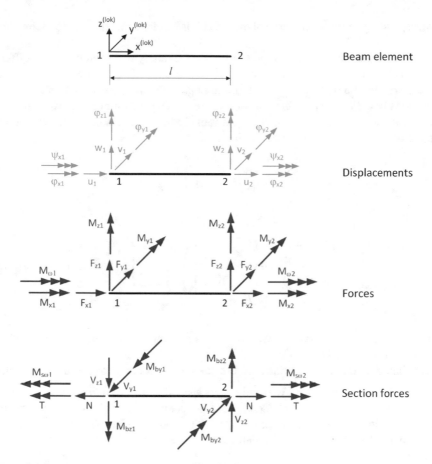

Fig. 3.31 Displacements and forces of the 3D beam element with warping torsion

The section forces can easily be obtained using the beam element forces. The separation of the torsional moment into primary and secondary torsion is obtained with (2.22a) and (2.20) as well as (2.15) as

$$M_{x,\mathrm{p}} = -G \cdot I_{\mathrm{T}} \cdot \psi \tag{3.43}$$

$$M_{x,\mathrm{s}} = M_x - M_{x,\mathrm{p}}. \tag{3.43a}$$

The element stiffness matrices can be added to the global stiffness matrix, as in the case of beams without warping torsion. However, the compatibility of the warping degrees of freedom at a nodal point is only given for elements with identical sections lying on a straight or curved line. In all other cases, it cannot be presumed that the warping degree of freedom at the end of the element is identical with the corresponding degree of freedom of the neighboring element. Hints for modeling are given in Sect. 3.7.3.

The stiffness matrix (3.42) represents the exact analytical solution of the differential equation for a beam with warping. Hence, results are independent of the discretization of a beam member into finite beam elements. However, there are also solutions based on shape functions, similar to the case of finite elements of plates. According to [5, 7, 14], the stiffness relationship for torsion is obtained with cubic shape functions for the torsional rotation as

$$
\left(\frac{G \cdot I_T}{30 \cdot l} \cdot \begin{bmatrix} 36 & -3 \cdot l & -36 & -3 \cdot l \\ -3 \cdot l & 4 \cdot l^2 & 3 \cdot l & -l^2 \\ -36 & 3 \cdot l & 36 & 3 \cdot l \\ -3 \cdot l & -l^2 & 3 \cdot l & 4 \cdot l^2 \end{bmatrix} + \frac{E \cdot I_\omega}{l^3} \right.
$$
$$
\left. \cdot \begin{bmatrix} 12 & -6 \cdot l & -12 & -6 \cdot l \\ -6 \cdot l & 4 \cdot l^2 & 6 \cdot l & 2 \cdot l^2 \\ -12 & 6 \cdot l & 12 & 6 \cdot l \\ -6 \cdot l & 2 \cdot l^2 & 6 \cdot l & 4 \cdot l^2 \end{bmatrix} \right) \cdot \begin{bmatrix} \varphi_{x1} \\ \psi_1 \\ \varphi_{x2} \\ \psi_2 \end{bmatrix} = \begin{bmatrix} M_{x1} \\ M_{\omega1} \\ M_{x2} \\ M_{\omega2} \end{bmatrix}. \qquad (3.43b)
$$

In this case a discretization error occurs, requiring a sufficiently fine element discretization of the beam.

It should be noted that because of $\lim\limits_{I_\omega \to 0} \varepsilon \to \infty$ according to (3.42a-f), the exact solution is not defined for $I_\omega = 0$ and for very small values I_ω numerical difficulties are to be expected. When implementing the element in a finite element code, appropriate measures must be taken for this case. The approximate solution according to (3.43b), on the other hand, is also defined for $I_\omega = 0$.

Example 3.17 The beam in Fig. 3.32 loaded by a torsional moment has to be analyzed, considering warping torsion and using the exact, analytical stiffness matrix. At the nodal points 1 and 3 the torsional degree of freedom is fixed, whereas the warping degrees of freedom are free at all nodal points. The cross-sectional parameters are (steel profile IPE 200):

$$ E = 21000\,\mathrm{kN/cm^2}, \quad G = 8100\,\mathrm{kN/cm^2}, \quad I_\omega = 12990\,\mathrm{cm^6}, \quad I_T = 6,98\,\mathrm{cm^4} $$

The example has been adopted from [7] where it is explained in detail. The calculation is carried out in the units kN and cm.

Fig. 3.32 Beam with torsional moment

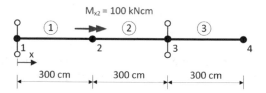

The degrees of freedom of the system are the torsional rotations φ_x at all nodal points and the warping degrees of freedom ψ at nodes 2 and 4. All other degrees of freedom are fixed.

The element stiffness matrices are obtained with (3.42) after elimination of the fixed degrees of freedom as

$$\underline{K}^{(1)} = \underline{K}^{(2)} = \underline{K}^{(3)} = \begin{bmatrix} 343 & -23214 & -343 & -23214 \\ -23214 & 5498721 & 23214 & 1465523 \\ -343 & 23214 & 343 & 23214 \\ -23214 & 1465524 & 23214 & 5498721 \end{bmatrix}.$$

The superposition of the element matrices and the observance of the support conditions give

$$\begin{bmatrix} 5498721 & 23214 & 1465524 & 0 & 0 & 0 \\ 23214 & 686 & 0 & -23214 & 0 & 0 \\ 1465524 & 0 & 10997443 & 1465524 & 0 & 0 \\ 0 & -23214 & 1465524 & 10997443 & 23214 & 1465524 \\ 0 & 0 & 0 & 23214 & 343 & 23214 \\ 0 & 0 & 0 & 1465524 & 23214 & 5498721 \end{bmatrix} \cdot \begin{bmatrix} \psi_1 \\ \varphi_{x2} \\ \psi_2 \\ \psi_3 \\ \varphi_{x4} \\ \psi_4 \end{bmatrix} = \begin{bmatrix} 0 \\ 100 \\ 0 \\ 0 \\ 0 \\ 0 \end{bmatrix}.$$

The solution of the global system of equations is

$$\begin{bmatrix} \psi_1 \\ \varphi_{x2} \\ \psi_2 \\ \psi_3 \\ \varphi_{x4} \\ \psi_4 \end{bmatrix} = \begin{bmatrix} -0.808 \\ 188.460 \\ 0.047 \\ 0.457 \\ -31.728 \\ 0.012 \end{bmatrix} \cdot 10^{-3}.$$

The element forces are given by the multiplication of the element stiffness matrix with the nodal displacements of the element. The section forces are obtained with (3.43), (3.43a), and Fig. 3.31 as

$$T_1 = -M_{x1}, \quad M_{x,p,1} = -G \cdot I_T \cdot \psi_1, \quad M_{x,s,1} = -M_{x,1} - M_{x,p,1}, \quad M_{S\omega 1} = -M_{\omega 1},$$
$$T_2 = M_{x2}, \quad M_{x,p,2} = -G \cdot I_T \cdot \psi_2, \quad M_{x,s,2} = M_{x,1} - M_{x,p,2}, \quad M_{S\omega 2} = M_{\omega 2}.$$

One obtains (units are kN and cm):

	Element 1	Element 2	Element 3
Primary torsional moment			
$M_{x,p,1}$	45.69	−2.65	−25.83
$M_{x,p,2}$	−2.65	−25.83	−0.69
Secondary torsional moment			
$M_{x,s,1}$	1.32	−50.34	25.83
$M_{x,s,2}$	49.66	−27.16	0.69
Total torsional moment			
T_1	47.01	−52.99	0
T_2	47.01	−52.99	0
Warping bi-moment			
$M_{Sω1}$	0	3448	−1794
$M_{Sω2}$	3448	−1794	0

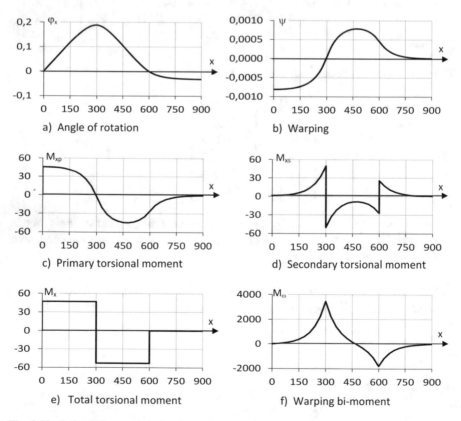

a) Angle of rotation

b) Warping

c) Primary torsional moment

d) Secondary torsional moment

e) Total torsional moment

f) Warping bi-moment

Fig. 3.33 Deformations and section forces (units are kN and cm)

Without considering warping torsion, the torsional moment would be $T = 50\,$kNcm in element 1 and $T = -50\,$kNcm in element 2, due to the symmetry of the system. However, the consideration of the warping torsion cancels the symmetry, since warping is restrained at point 3 whereas it is free at point 1. Due to the warping bi-moment transferred to element 3, secondary and primary torsional moments occur in element 3 which are, however, zero in their sum. The section force and deformation diagrams obtained with a discretization with 30 elements (each with $l = 30\,$cm) are given in Fig. 3.33. They agree exactly with the rotations, warping values and sectional forces of the model with 3 elements only (with $l = 300\,$cm) given above as the analytical solution of the stiffness matrix has been used in both models. ◄

Warping torsion is included in general finite element software packages like [SOF 1, SOF 2]. Beam elements with warping torsion are also implemented in [SOF 6]. The element in [SOF 6] is based on Hermite shape functions according to [53] and enables a description with inclined reference axes for one-sided haunches. As a standard, however, elements with torsional warping are not included in commercial programs for structural analysis. Often special programs are used for this purpose.

3.7 Modeling of Beam Structures

3.7.1 Supports

Realistic support conditions are of decisive importance for an adequate assessment of the load-bearing behavior of structural systems. In finite element programs, support conditions are defined by specifying the degrees of freedom fixed at a node (Fig. 3.34). These degrees of freedom are then eliminated from the system of equations according to a procedure described in Sect. 3.2.5.

Inclined supports can be analyzed exactly by transforming the degrees of freedom of the corresponding nodal point. For finite element programs that do not provide for this, inclined supports must be represented by extremely stiff spring or truss elements. The stiffness of the spring or the truss element, respectively, must be chosen such that their flexibility can be practically neglected in the total system (Fig. 3.35). Even for large values of the spring constant, numerical difficulties are not to be expected (see Sect. 3.8.1).

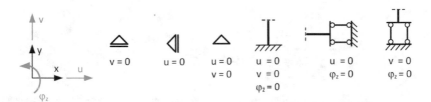

Fig. 3.34 Definition of supports for two-dimensional structures

Fig. 3.35 Definition of
inclined supports

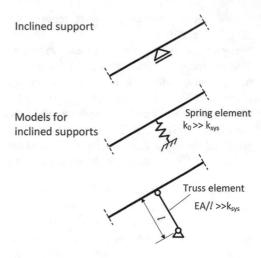

Inclined support

Models for
inclined supports

Spring element
$k_0 \gg k_{sys}$

Truss element
$EA/l \gg k_{sys}$

Prescribed support displacements or rotations can be handled easily as shown in Sect. 3.2.5. The displacements or rotations given are introduced in the global system of equations and result in a load vector on the right-hand side. If this procedure is not implemented in the finite element program, a spring with very high stiffness k_0 (e.g.10^{15}kN/m) has to be defined in the degree of freedom where the support displacement is prescribed. The spring is now loaded with a force

$$F_0 = k_0 \cdot u_0 \tag{3.44a}$$

or a moment

$$M_0 = k_0 \cdot \varphi_0, \tag{3.44b}$$

where u_0 or φ_0 are the prescribed displacements or rotations, respectively (Fig. 3.36). Since the spring is extremely stiff compared to the structural system, it takes the force

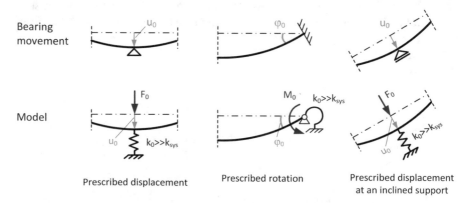

Bearing
movement

Model

Prescribed displacement   Prescribed rotation   Prescribed displacement
                                                              at an inclined support

Fig. 3.36 Modeling of prescribed displacement and rotation of supports

F_0 almost completely. If this is not so, the spring constant k_0 has to be augmented. This procedure is also known as the penalty method. Instead of a spring, a truss element can be used, where $k_0 = E \cdot A / l$, according to (3.3).

Three-dimensional structures have 6 or 3 degrees of freedom per node (Fig. 3.2). For beams with warping torsion, warping is added as a supplementary degree of freedom. Since each degree of freedom can be fixed individually, there is a large number of possible support types. In general, $(2^n - 1)$ combinations or support types can be defined with n degrees of freedom at the bearing point. With 6 degrees of freedom, these are $(6^n - 1) = 63$ possible support definitions. Some typical examples are shown in Fig. 3.37. Nevertheless, the large number of possibilities indicates that this is a possible source of errors in program input with possibly serious consequences. Special attention should be paid to the modeling of three-dimensional supports in order to realistically represent the structural behavior on the one hand and to avoid kinematic systems on the other hand.

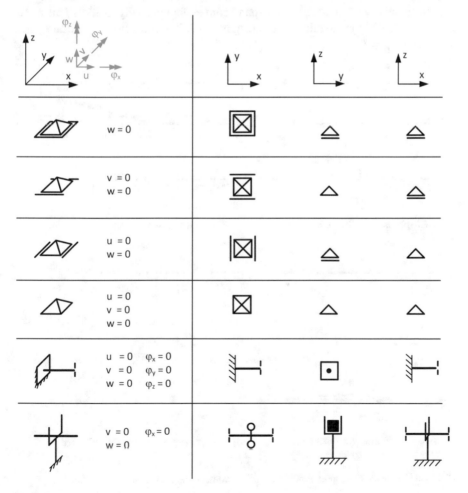

Fig. 3.37 Definition of supports for three-dimensional structures

Example 3.18 The two-span beam in Fig. 3.38 is analyzed with a finite element program as a three-dimensional beam structure. At the three supports, all degrees of freedom for displacements are fixed, while the rotational degrees of freedom are free. The example was taken from [17].

The moment diagrams calculated by a finite element program are given in Fig. 3.38. In Fig. 3.38a, the three supports are situated in a straight line and the moment diagram of a two-span beam is obtained as expected. The slight change of the geometry in Fig. 3.38b, i.e. a rotation of the right-hand span in the plan view by an extremely small kink angle φ, results in a totally different moment diagram. This effect can easily be explained by checking the moment equilibrium at the middle support. At this point, the bending moment M_{li} of the left-hand span is in equilibrium with the bending moment M_{re} and the torsional moment T of the right-hand span. It should be noted that the right-hand support is not capable of carrying a torsional moment. Hence, with $T = 0$ the vectorial equilibrium condition gives $M_{re} = 0$ (Fig. 3.38d). When torsion is restrained at the right-hand support, a moment diagram similar to the case of a straight two-span beam is obtained (Fig. 3.38c). This effect is independent of the magnitude of the angle φ, i.e. it occurs also for arbitrary small kink angles $\varphi \neq 0$.   ◀

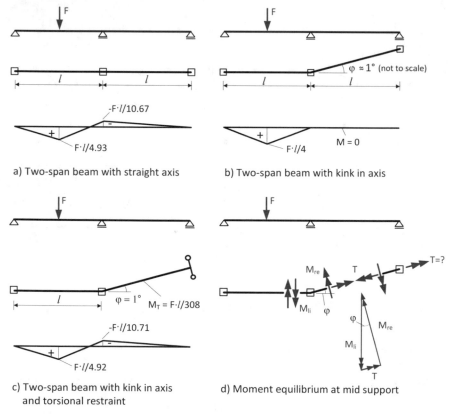

a) Two-span beam with straight axis

b) Two-span beam with kink in axis

c) Two-span beam with kink in axis and torsional restraint

d) Moment equilibrium at mid support

Fig. 3.38 Influence of support conditions on the moment diagrams of a two-span beam

3.7.2 Springs

Springs allow the modeling of any elastic flexibility in the structural model. Some examples for the modeling of adjacent beams by equivalent spring elements are given in Table 3.4. The flexibility of rigid foundations on the subsoil may also be represented by springs (see Sect. 5.8.5).

Table 3.4 Modeling of adjacent beams by equivalent spring elements

Beam system		Spring constant
![fixed-free beam with force F, length l, displacement δ]	$F = k_v \cdot \delta$	$k_v = \dfrac{EA}{l}$
![pinned-pinned beam with moment M and rotation φ]	$M = k_\varphi \cdot \varphi$	$k_\varphi = \dfrac{3 \cdot EI}{l}$
![pinned-fixed beam with moment M and rotation φ]	$M = k_\varphi \cdot \varphi$	$k_\varphi = \dfrac{4 \cdot EI}{l}$
![beam with moment M, rotation φ and rotational spring k₀]	$M = k_\varphi \cdot \varphi$	$k_\varphi = \dfrac{4 \cdot l \cdot k_0 + 12 \cdot EI}{4 \cdot l + \dfrac{l^2 \cdot k_0}{EI}}$
EA = Longitudinal stiffness		l = Length of a beam
EI = Bending stiffness		k_o = Rotational spring constant

Example 3.19 For modeling purposes, a simply supported beam has to be replaced by an equivalent spring in the global structural system (Fig. 3.39). The beam is connected in its midpoint at the structural system.

The displacement in the middle of a beam due to the force F is

$$v = \frac{Fl^3}{48EI}.$$

Hence, the relationship between the force and the displacement, i.e. the spring constant, is obtained as

Fig. 3.39 Modeling of a simply supported beam as an equivalent spring

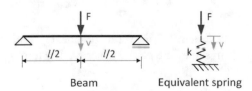

Beam   Equivalent spring

$$F = \frac{48EI}{l^3} \cdot v, \qquad F = k \cdot v, \qquad k = \frac{48EI}{l^3}.$$ ◄

Multiple springs can be replaced by an equivalent individual spring. They can be placed in series or in parallel. The equivalent springs are given in Table 3.5.

Table 3.5 Equivalent spring constants for springs in parallel and in series

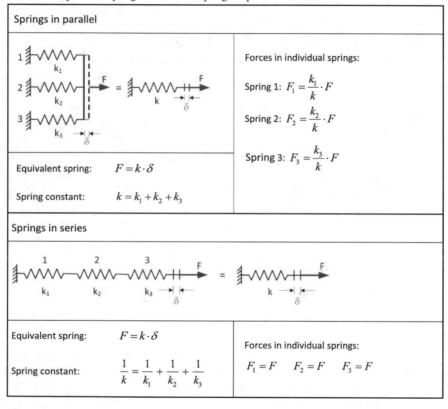

Example 3.20 The beams in Fig. 3.40 are to be replaced by springs. The spring constants have to be determined.

System "a" represents the elastic restraint of a horizontal beam by two vertical beams. The vertical beams can be regarded as rotational springs in parallel. The total spring constant is obtained as

$$k_{\varphi 1} = \frac{3EI_1}{l_1}, \qquad k_{\varphi 2} = \frac{4EI_2}{l_2}, \qquad k_{\varphi} = \frac{3EI_1}{l_1} + \frac{4EI_2}{l_2}.$$

In system "b", the vertical elements are aligned. Considering them as displacement springs, they are placed in series. The total spring constant is obtained as

$$k_1 = \frac{EA_1}{l_1}, \qquad k_2 = \frac{EA_2}{l_2}$$

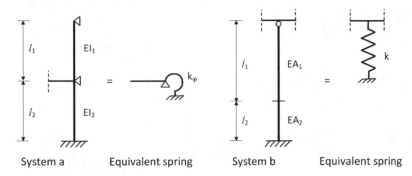

Fig. 3.40 Substitution of beams by springs

$$\frac{1}{k} = \frac{1}{k_1} + \frac{1}{k_2} = \frac{l_1}{EA_1} + \frac{l_2}{EA_2}, \qquad k = \frac{1}{\dfrac{l_1}{EA_1} + \dfrac{l_2}{EA_2}} = \frac{E \cdot A_1 \cdot A_2}{A_2 \cdot l_1 + A_1 \cdot l_2}. \qquad \blacktriangleleft$$

3.7.3 Beam Structures

Position of the beam axis

Frames and other beam structures are modeled using beam elements. It must be ensured that the beam axes correspond to the axes of the structural system. In the structural system, the axis is defined as the line connecting the centers of gravity of the cross sections at the beginning and the end of a beam.

For beams with an eccentric axis, the axes of the structural model and of the beam do not coincide (Fig. 3.41). There are several ways to model them.

Modeling of eccentric beam axes
- Multi-Point Constraint—MPC
- Beam element with an eccentric axis
- Equivalent model with beam elements with very high stiffness

Rigid kinematic coupling conditions can be used to model rigid connections of nodal points. They are also called *Multi-Point Constraints* (MPC). With correct implementation, these formulations are numerically unproblematic (see Sect. 4.10.3, RDT model). Another modeling option is using beam elements with eccentric axes. They allow to define the deviation of the beam axis from the system axis directly (Fig. 3.42). This corresponds to an MPC coupling. Rigid couplings can also be modeled approximately using equivalent beam elements with large cross-sectional values. In order to avoid numerical problems, it should be ensured that the artificially

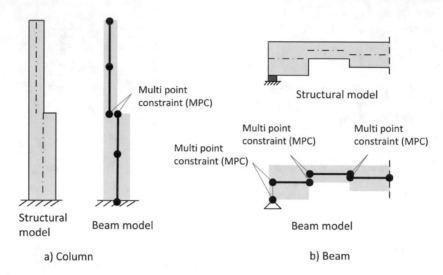

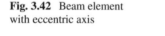

a) Column

b) Beam

Fig. 3.41 System with eccentric beam axes

Fig. 3.42 Beam element
with eccentric axis

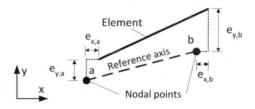

chosen stiffness does not differ by many orders of magnitude from the system stiffness
(see Sect. 3.8.1).

In the case of beams with an eccentric beam axis, as shown in Fig. 3.41b, for
example, horizontally restrained supports at both ends of a one-span beam will cause
normal forces in the beam. Therefore, horizontal support conditions (fixed or free)
have a significant influence on the internal forces (arching effect) and must be carefully modeled. Normally one end of the beam will be supported without horizontal
constraint, i.e. horizontally freely movable, unless its "immovability" appears to be
ensured by the existing high rigidity of the supporting structure. For the similar
problem of the plate in plane stress, this effect is explained in Example 4.24.

Eccentricities of the beam axis may be neglected for beams with normal forces
being zero, i.e. the beam can be modeled with an axis aligned but with piecewise
constant moments of inertia in the different sections. This approximation has only
little influence on bending moments and deflections. If, however, normal forces act
in the member (as in Fig. 3.41a), then the offset of the beam axis must necessarily
be taken into account in the model.

The center of gravity and shear center coincide in most cross sections of members used in reinforced concrete constructions. The local axes of the beam element correspond to the main axes of the cross section. Particularly in steel construction, cross sections are also used where the center of shear and the center of gravity are not identical. This case requires complex transformations. The application of the finite element method in steel construction is discussed in [14].

In principle, it is also possible to omit the reference to the main axes and to use any local coordinate system as the reference system for the cross section. In this case, it is no longer possible to consider the stresses from normal force, bending about the $y^{(lok)}$- and $z^{(lok)}$-axis, and torsion separately. With regard to the complex interrelations resulting from this and to the derivation of the stiffness matrix, reference is made to [18–21]. Questions concerning the modeling of beams with warping torsion are discussed below.

Haunches

Beams with haunches may be modeled with an inclined axis or approximated by a straight beam axis (Fig. 3.43). Even though solutions for beams with haunches have been derived, they are usually not implemented in commercial finite element programs [13, 48]. Modeling haunches with beam elements with piecewise constant sections causes a discretization error. In order to model the stiffness of haunches appropriately, a sufficiently fine discretization is required especially for statically undetermined systems. For beams with an inclined axis, it is important to model the horizontal degree of freedom appropriately, as in the case of beams with an eccentric axis. Offsets of the beam axis can be neglected approximately for beams without normal forces as shown, for example, in Fig. 3.43b, i.e. the beam can be modeled with a beam axis aligned but with piecewise different moments of inertia in the different sections. This approximation has only little influence on bending moments

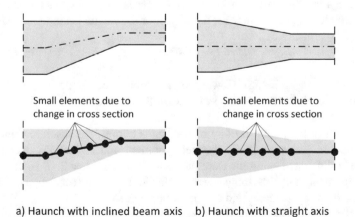

a) Haunch with inclined beam axis   b) Haunch with straight axis

Fig. 3.43 Beams with haunches

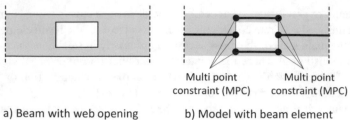

Multi point   Multi point
constraint (MPC)   constraint (MPC)

a) Beam with web opening   b) Model with beam element

Fig. 3.44 Modeling of a web opening in a beam

and deflections. If, however, normal forces act in the member, then the offset of the beam axis must necessarily be modeled.

Web openings

Large web openings in beams can be modeled with two beam elements, according to Fig. 3.44. As their axes differ from the system axis, elements with eccentric axes or multi-point constraint couplings have to be used. In reinforced concrete beams, a reduction of the longitudinal and bending stiffnesses of the element exposed to tension may be appropriate in order to model cracks in concrete. This can be achieved by a reduction of the Young's modulus in the elements concerned.

Frame corners

The mechanical behavior of the corner zone of frames influences the field and corner moments as well as the deflections of frames. Approximately frame corners can be assumed to be rigid. This is particularly justified for reinforced concrete frames. Such behavior can be modeled with rigid end connections, e.g. by bars with high stiffness or with a kinematic coupling in the area of the frame corner assumed to be rigid (Fig. 3.45a). Since the stiffness of the corner is overestimated by this model, it is recommended in [22] to increase the length of the elastic beam to

$$l^* = l - (1 - \eta)(r_i + r_j) \tag{3.45}$$

with $0 \leq \eta \leq 0{,}5$ (Fig. 3.45b) . However, the section forces for design are determined at the end of the elastic part of the beam, i.e. at the distances r_i and r_j from the nodes i or j, respectively.

As a result of the higher rigidity of the frame corner, the absolute values of the corner moments increase, while the field moment and deflections decrease. The influence is illustrated using the example of a frame with $l = 2 \cdot h$ in Table 3.6. In practice, for most frames the influence of an elastic calculation according to first-order theory is only a few percents and can often be neglected.

In steel constructions, the flexibility of bolted connections at frame corners can be taken into account by spring models. In most cases, these are hinges with rotational springs as shown in Fig. 3.14d, which are arranged close to the frame corner in the beam. The type and characteristic values of the springs depend on the design

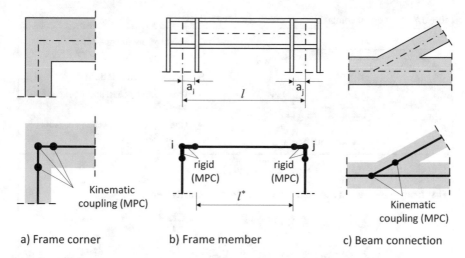

a) Frame corner   b) Frame member   c) Beam connection

Fig. 3.45 Modeling for rigid connections; beam–column joint models

Table 3.6 Modeling of frame corners as rigid connection

a/l	α_e	α_m	β_m
0	15.0	17.4	213
0.0125	14.4	18.0	229
0.0250	13.9	18.9	247
0.0500	13.0	20.9	290
0.1000	11.5	26.4	414

$$M_e = -\frac{ql^2}{\alpha_e}, \quad M_m = \frac{ql^2}{\alpha_m}, \quad v_m = \frac{ql^4}{\beta_m EI}$$

of the connection. Reference is made to the extensive literature on column–beam connections [14, 23–25].

For beam connections, the same applies as for frame corners (Fig. 3.45c). If beams are connected eccentrically in a node, the eccentricities must be taken into account in the modeling.

In timber construction, flexible connections such as dowel-type connections are also represented in the model by hinges with rotational springs according to Fig. 3.14d.

Bending shear, longitudinal and torsional deformations

Bending deformations are related to the principal axes of the cross section, which correspond to the local $y^{(lok)}$ and $z^{(lok)}$ axis of spatial beam elements. The center of gravity coordinates and moments of inertia can be determined numerically with sum formulas according to Table 3.7 for any polygon-like cross sections ([26] see also example in [27]).

Table 3.7 Numerical computation of cross-sectional values

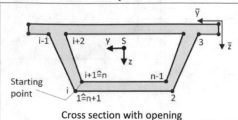

Coordinates of corner points:
$$y_i = \bar{y}_i - \bar{y}_S \qquad z_i = \bar{z}_i - \bar{z}_S$$
Polygonal line orientation:
• Cross section areas: positive, i.e. counter-clockwise
• Openings (areal reduction): negative, i.e. clockwise

Cross section with opening

Cross sectional area: $A = \dfrac{1}{2} \cdot \displaystyle\sum_{i=1}^{n} \left(\bar{y}_i \cdot \bar{z}_{i+1} - \bar{z}_i \cdot \bar{y}_{i+1} \right)$

First moment of area and coordinates of the center of gravity, related to $\bar{y}, \bar{z}$:

$$S_y = \frac{1}{6} \cdot \sum_{i=1}^{n} \left(\left(\bar{y}_i \cdot \bar{z}_{i+1} - \bar{z}_i \cdot \bar{y}_{i+1} \right) \cdot \left(\bar{z}_i + \bar{z}_{i+1} \right) \right) \qquad \bar{z}_s = \frac{S_y}{A}$$

$$S_z = \frac{1}{6} \cdot \sum_{i=1}^{n} \left(\left(\bar{y}_i \cdot \bar{z}_{i+1} - \bar{z}_i \cdot \bar{y}_{i+1} \right) \cdot \left(\bar{y}_i + \bar{y}_{i+1} \right) \right) \qquad \bar{y}_s = \frac{S_z}{A}$$

Second moment of area, related to y, z in the center of gravity:

$$I_y = \frac{1}{12} \cdot \sum_{i=1}^{n} \left(\left(y_i \cdot z_{i+1} - z_i \cdot y_{i+1} \right) \cdot \left(z_i^2 + z_i \cdot z_{i+1} + z_{i+1}^2 \right) \right)$$

$$I_z = \frac{1}{12} \cdot \sum_{i=1}^{n} \left(\left(y_i \cdot z_{i+1} - z_i \cdot y_{i+1} \right) \cdot \left(y_i^2 + y_i \cdot y_{i+1} + y_{i+1}^2 \right) \right)$$

$$I_{yz} = \frac{1}{24} \cdot \sum_{i=1}^{n} \left(\left(y_i \cdot z_{i+1} - y_{i+1} \cdot z_i \right) \cdot \left(2 \cdot y_i \cdot z_i + y_i \cdot z_{i+1} + z_i \cdot y_{i+1} + 2 \cdot y_{i+1} \cdot z_{i+1} \right) \right)$$

Normal force deformations are always taken into account when calculating frame structures using the finite element method, unless corresponding support conditions do prevent this. Thus, in frameworks the results can lead to slight deviations compared to classical frame formulas, which do not consider normal force deformations. If the normal force deformations are to be omitted in special cases, this can be done by entering a large cross-sectional area $\left(\text{e.g.} 1000\,\text{m}^2 \right)$. However, cross-sectional areas which are arbitrarily large by powers of 10 should not be used, since they can lead to numerical problems (cf. Example 3.24).

Shear deformations are usually included in the results of finite element programs, since the bending beam elements are normally implemented as shear-flexible beam elements. Shear areas of typical cross sections are listed in Table 3.8 [16, 28]. More information can be found in [29].

Shear deformations of slender beams are small and can be neglected [16]. For some systems, the ratio of the maximum shear deflection to the maximum bending deflection is given in Table 3.9. They apply to rectangular cross sections and isotropic materials with a Poisson ratio of $\mu = 0.2$. Thus, for a simply supported beam with a uniform load, the shear deformation is approximately 12% of the bending deformation. If necessary, shear deformations can be suppressed by defining an arbitrarily large shear area. In some finite element programs, it is possible to exclude

Table 3.8 Shear area A_S of cross sections

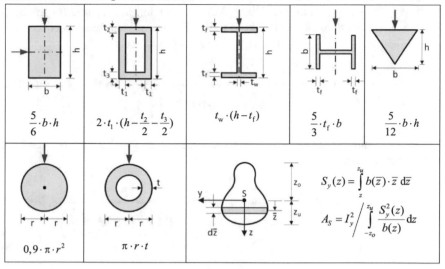

Table 3.9 Shear and bending deformations of beams

System	f_V / f_M	System	f_V / f_M
(cantilever, point load at m, length l)	$0.72 \cdot \left(\dfrac{h}{l}\right)^2$	*(simply supported, point load at m, l/2 + l/2)*	$2.88 \cdot \left(\dfrac{h}{l}\right)^2$
(cantilever, distributed load, length l)	$0.58 \cdot \left(\dfrac{h}{l}\right)^2$	*(simply supported, distributed load, l/2 + l/2)*	$2.30 \cdot \left(\dfrac{h}{l}\right)^2$
Cross section: *(rectangle of height h)*		**Displacements:** f_V : displacement at point m due to shear ($\mu = 0.2$) f_M : displacement at point m due to bending	

shear deformations by entering the shear area to zero. This is a program-specific input definition without any mechanical meaning.

For beams with semirigid connections, as they occur in timber structures, special models have been developed to describe their shear deformations. In the "lattice frame model", the beam is described by an equivalent truss system. The flexibility of the semirigid connection is represented by truss rods with adapted stiffnesses [30, 31]. In the "shear analogy method", the cross section is transformed into a three-part model cross section, whereby each partial cross section is assigned a different load-bearing behavior [32, 33].

Torsional deformations occur in spatial beam elements. The torsional moment of inertia which is required to describe St. Venant's torsion is specified for the most common cross sections in handbooks [55–57]. For rectangular cross sections with the side lengths h and $b \leq h$, the torsional moment of inertia is [12]

$$I_{\mathrm{T}} = \frac{h \cdot b^3}{\eta_{\mathrm{IT}}} \quad \text{with} \quad \eta_{\mathrm{IT}} \approx \frac{3}{1 - 0.63 \cdot \dfrac{b}{h} + 0.052 \cdot \left(\dfrac{b}{h}\right)^5} \quad \text{and} \quad b \leq h. \quad (3.46a)$$

The maximum shear stresses due to a torsional moment T are obtained for rectangular sections as

$$\tau = \frac{T}{W_{\mathrm{T}}} \quad \text{with} \quad W_{\mathrm{T}} = \frac{h \cdot b^2}{\eta_{\mathrm{WT}}} \quad \text{and} \quad \eta_{\mathrm{WT}} \approx \frac{3}{1 - 0.63\dfrac{b}{h} + 0.25 \cdot \left(\dfrac{b}{h}\right)^2}, b \leq h.$$

$$(3.46b)$$

For uniform cross sections the torsional moment of inertia, if the exact value is not known, can be estimated according to St. Venant as follows

$$I_{\mathrm{T}} = \frac{A^4}{4\pi^2 \left(I_y + I_z\right)} \quad (3.46c)$$

where A is the area and I_y, I_z are the moments of inertia of the cross section [12]. For elliptic cross sections (3.46c) is exact. A uniform cross section can also be replaced with (3.46c) approximately by an inscribed ellipse.

For the computation of torsional parameters of arbitrary cross sections, numerical methods including the finite element method are available [35, 36].

Beams on elastic foundation

Beams on elastic foundations can be modeled by special beam elements which exhibit a continuous elastic support. If such elements are not implemented in the finite element program used, the elastically supported beam can be modeled approximately as a series of beam elements and single springs attached at the nodes (Fig. 3.46). This, however, will result in a discretization error. Denoting the distance of the single springs by Δx, the beam width by b, and the subgrade modulus by k_s, their spring constant is obtained as $k_v = k_s \cdot b \cdot \Delta x$. With the characteristic length

$$L = \sqrt[4]{\frac{4 \cdot E \cdot I}{k_s \cdot b}} \quad (3.47)$$

the distance chosen for the single springs has to be restricted to

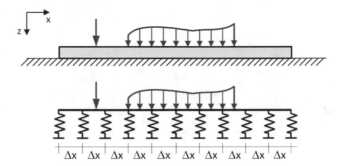

Fig. 3.46 Discretization of an elastically bedded beam

$$\Delta x = \frac{1}{4} L = \frac{1}{4} \cdot \sqrt[4]{\frac{4 \cdot E \cdot I}{k_s \cdot b}} \tag{3.48}$$

where EI is the bending stiffness of the beam. The internal forces obtained with this simplified model need an appropriate interpretation. In particular, due to the punctual application of forces at the individual springs, the moment diagram shows kinks at the nodal points and the shear force diagram shows jumps, which is not the case with beams with continuous support.

The deflections and internal forces of unilaterally and bilaterally infinitely long beams with a single load or moment are shown in Table 3.10 [37]. It can be seen that the diagrams decay almost to zero at a distance of approximately $x = 3L$ from the point of application and thus apply in good approximation also to beams with a length greater than $3L$. Table 3.10 can be useful when assessing the results of a finite element calculation as well as when selecting the distance Δx of the single springs.

Loads

The reference line of the beam element is the beam axis, which runs through the centers of gravity of the end cross sections of the element. Thus, referring to the beam axis, a single load can be represented as an equivalent line load (Fig. 3.47). Load propagation can be assumed to be 45°. This may lead to a noticeable reduction in the peak moment under the point load, e.g. in the case of a beam on the elastic subgrade.

Special loads which are not available in a finite element software can be treated as equivalent loads using the procedure specified in Sect. 3.4.2. Prestressing of beams, for example, can be analyzed with any program for the analysis of structures. The reaction moments of a clamped–clamped beam element for prestressing are given in [5, 6, 12]. Another option is to represent prestressing as temperature load. The compression and curvature of the beam element caused by prestress must correspond to the compression or curvature due to the temperature stress, respectively. Concrete shrinkage can be treated as a temperature load as well.

Table 3.10 Deformations and sectional forces of unilaterally and bilaterally infinitely long elastically supported beams

System	$w(\bar{x})$	$M(\bar{x})$	$V(\bar{x})$
	$\dfrac{F}{2 \cdot k_S \cdot L} \cdot f_1(\bar{x})$	$\dfrac{F \cdot L}{4} \cdot f_3(\bar{x})$	$-\dfrac{F}{2} \cdot f_2(\bar{x})$
	$\dfrac{M_0}{k_S \cdot L^2} \cdot f_2(\bar{x})$	$\dfrac{M_0}{2} \cdot f_4(\bar{x})$	$-\dfrac{M_0}{2 \cdot L} \cdot f_1(\bar{x})$
	$\dfrac{2 \cdot F}{k_S \cdot L} \cdot f_4(\bar{x})$	$-F \cdot L \cdot f_2(\bar{x})$	$-F \cdot f_3(\bar{x})$
	$-\dfrac{2 \cdot M_0}{L^2 \cdot k_S} f_3(\bar{x})$	$M_0 \cdot f_1(\bar{x})$	$-\dfrac{2 \cdot M_0}{L} \cdot f_2(\bar{x})$

$f_1(\bar{x}) = e^{-\bar{x}} \cdot (\cos(\bar{x}) + \sin(\bar{x}))$

$f_2(\bar{x}) = e^{-\bar{x}} \cdot \sin(\bar{x})$

$f_3(\bar{x}) = e^{-\bar{x}} \cdot (\cos(\bar{x}) - \sin(\bar{x}))$

$f_4(\bar{x}) = e^{-\bar{x}} \cdot \cos(\bar{x})$

$\bar{x} = x/L \quad L = \sqrt[4]{\dfrac{4 \cdot E \cdot I}{k_S \cdot b}}$

Soil pressure:
$p(\bar{x}) = k_S \cdot w(\bar{x})$

Functions f_1 to f_4

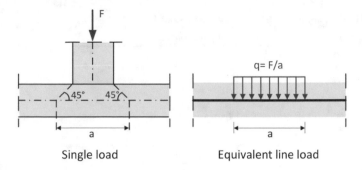

Fig. 3.47 Representation of a single load as a line load

Warping torsion

The sectional characteristics for torsion and warping of regular cross sections can be found in handbooks. For cross sections with arbitrary shape, numerical methods are available [34, 35, 41]. For some simple cross-sectional shapes, the warping constant is given in Table 3.11 [51, 52].

Special considerations are required for bars with warping torsion in order to model the restraints of the warping degree of freedom. The constructive design of the respective node is decisive for the question of whether warping deformations can occur freely or whether they are completely or partially restrained. Figure 3.48 shows the cross section at the end of an I-beam. If the cross section is free (Fig. 3.48a), it is allowed to warp freely and warping stresses do not develop. If an end plate is arranged (Fig. 3.48b), it prevents warping and the bending moments in the flanges caused by warping torsion can balance out between the upper and lower flanges. Its stiffness can be described by a warping spring as

$$M_\omega = k_\omega \cdot \psi. \tag{3.49}$$

Stiffener plates have a similar effect (Fig. 3.48c). Beam extensions beyond a support restrain warping deformations too (cf. Example 3.17). Warping springs for various constructive designs of the end of I-type cross sections are given in [14, 38, 58].

The same applies to connections of beam elements with a warping degree of freedom. The warping bi-moment can only be transferred from one beam element to the next if the construction allows it. For this, the warping at the end of both elements must be compatible (Figs. 2.13, 2.14). This requires that both elements have the same cross section and their axes are in line. In girder grids with identical cross sections of all beams as, for example, in Fig. 3.49a, the warping of the I-section can be transferred to the nodes. However, in most other cases as, for example, in the frame corner in Fig. 3.43b, the warping degree of freedom is not identical to the corresponding degree of freedom in the neighboring element. Hence, it must be modeled like a hinge for the warping bi-moment. The arrangement of an inclined plate in the frame corner (Fig. 3.49c) is similar to a end plate (Fig. 3.48b) and can be represented by a warping spring [14]. Further models are described in [39, 40].

Table 3.11 Warping constants of some cross sections [51]

Cross section	Warping constant I_ω
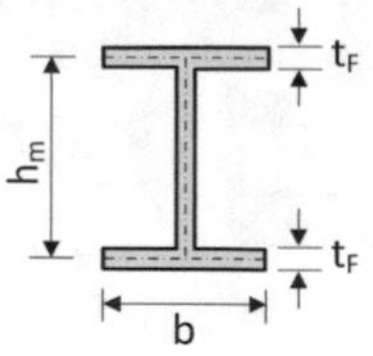	$I_\omega = I_F \cdot \dfrac{h_m^2}{2} = \dfrac{1}{24} \cdot A_F \cdot h_m^2 \cdot b^2$ $A_F = b \cdot t_F$ $I_F = \dfrac{t_F \cdot b^3}{12}$
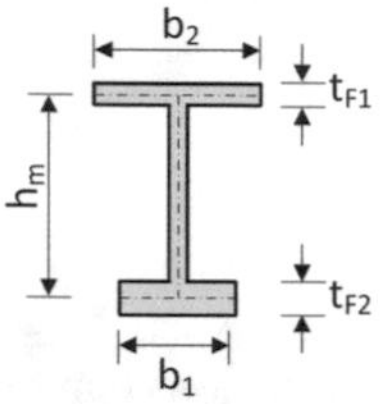	$I_\omega = \dfrac{I_{F1} \cdot I_{F2}}{I_{F1} + I_{F2}} \cdot h_m^2,$ $I_{F1} = \dfrac{t_{F1} \cdot b_1^3}{12},$ $I_{F2} = \dfrac{t_{F2} \cdot b_2^3}{12}$
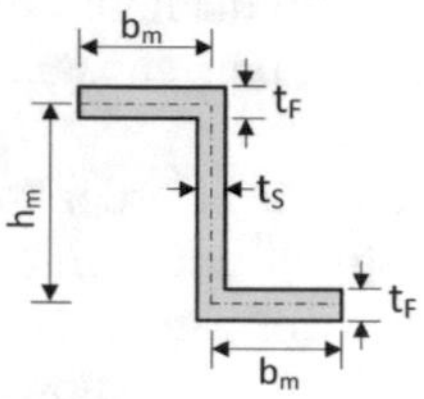	$I_\omega = A_F \cdot h_m^2 \cdot b_m^2 \cdot \left(\dfrac{1}{6} - \dfrac{A_F}{8 \cdot A_F + 4 \cdot A_S} \right)$ $A_F = b_m \cdot t_F,$ $A_S = h_m \cdot t_S$
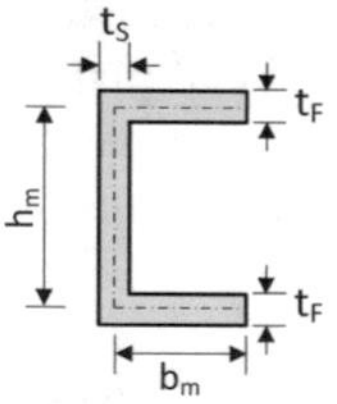	$I_\omega = A_F \cdot h_m^2 \cdot b_m^2 \cdot \left(\dfrac{1}{6} - \dfrac{A_F}{8 \cdot A_F + \dfrac{4}{3} \cdot A_S} \right)$ $A_F = b_m \cdot t_F,$ $A_S = h_m \cdot t_S$

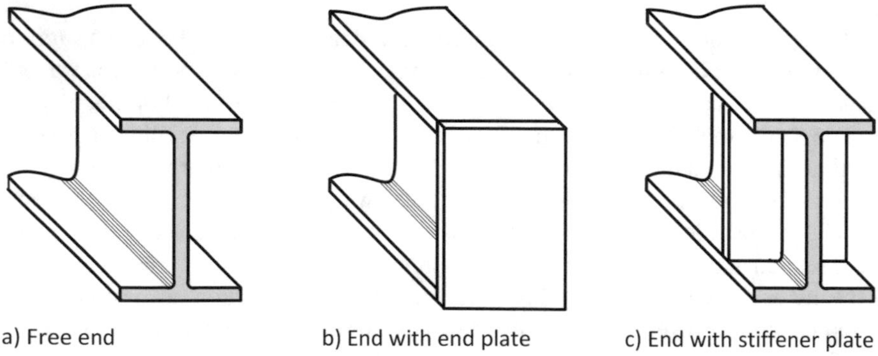

a) Free end   b) End with end plate   c) End with stiffener plate

Fig. 3.48 End cross section of a I-beam

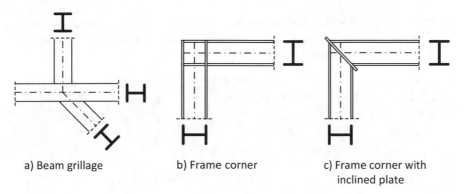

a) Beam grillage   b) Frame corner   c) Frame corner with inclined plate

Fig. 3.49 Connection of beams with a warping degree of freedom

Modeling of beams as special shell structures

In special cases, beams can also be modeled as shell structures. Examples of the classical theory of the strength of materials can be found in [50]. The assumptions of the technical bending theory like the Bernoulli hypothesis are dropped. This can be useful in order to determine the stresses as realistically as possible in regions of a beam structure in which the technical bending theory no longer applies, as in frame corners made of thin-walled profiles.

3.7.4 Symmetrical Structures

The displacements of a symmetrical structural system are symmetric for a symmetric loading and antisymmetric for antisymmetric loading. The symmetry properties of the sectional forces and stresses of symmetrical systems are given in Table 3.12. For systems with several axes of symmetry, these conditions hold at any axis of symmetry.

Table 3.12 Stresses and section forces in symmetrical systems

System	Section force/ Stress component	Loading	
		symmetric	antisymmetric
Beam structures	Bending moments Normal forces Torsional moments Shear forces	symmetric antisymmetric	antisymmetric symmetric
Plates in plane stress	Normal stresses Shear stresses	symmetric antisymmetric	antisymmetric symmetric
Plates in bending	Bending moments Shear forces Twisting moments	symmetric antisymmetric	antisymmetric symmetric

Symmetry conditions can be employed to reduce the size of a symmetric system to be analyzed. At the symmetry axes, appropriate support conditions resulting from the symmetry conditions have to be defined.

Symmetry and antisymmetry restraints

- For a symmetrical loading, the displacement vector component perpendicular to the plane of symmetry is zero, and the rotational vector components parallel to the plane of symmetry are zero
- For an antisymmetric loading, the displacements in the plane of symmetry are zero and the rotations normal to the plane of symmetry are zero

For plane systems with a symmetry axis in the x- or y-axis direction, the degrees of freedom to be restrained and which are free to move at the axes due to symmetry conditions are listed in Table 3.13.

Any asymmetric loading of a symmetrical system can be superimposed by symmetric and antisymmetric parts of the loading (Fig. 3.50). The sectional forces and displacements are to be computed with two finite element systems having different restraint conditions. The results of both systems have to be superposed. The advantage of the symmetry of a system is that it allows reducing the computer run time and memory required for the analysis, since only a portion of the actual structure needs to be modeled. However, it should be noted that not all finite element programs allow the automatic superposition of the results of different finite element systems, i.e. with different support conditions. Therefore, normally in case of arbitrary loading, the analysis of the complete system will be preferred even for symmetrical systems. On the other hand, in the case of a very large plate or shell structures with a symmetric (or antisymmetric) load, the option of modeling a symmetrical

Table 3.13 Symmetry conditions of plane symmetrical systems

System	Loading		
	symmetric		antisymmetric
Frame / plate in plane stress	$u = 0$ $v = 0$ $\varphi_z = 0$ $\varphi_z = 0$		$v = 0$ $u = 0$
Girder grid / plate in bending	$\varphi_y = 0$ $\varphi_x = 0$		$w = 0$ $w = 0$ $\varphi_x = 0$ $\varphi_y = 0$

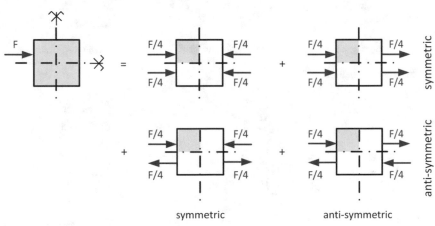

a) System with one axis of symmetry

b) System with two axes of symmetry

Fig. 3.50 Representation of a single load by symmetric and antisymmetric loads

subsystem should be checked in order to simplify the data input and, possibly, to reduce the computer time for solving the problem.

Example 3.21 For the symmetrical frame shown in Fig. 3.51, the equivalent systems for symmetric and antisymmetric loading have to be determined.

The decomposition of the loading into a symmetric and antisymmetric load case is shown in Fig. 3.51a. The equivalent systems are represented by one half of the actual system, with the support conditions for symmetric and antisymmetric loads according to Table 3.13. Beam elements on the axis of symmetry have to be defined with one half of their moment of inertia and cross-sectional area. Point loads on the axis of symmetry are represented with half of their values in the equivalent system. Both equivalent systems are shown in Fig. 3.51b. ◄

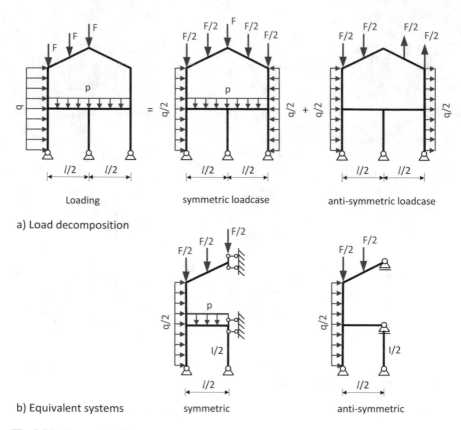

a) Load decomposition

b) Equivalent systems   symmetric   anti-symmetric

Fig. 3.51 Symmetrical frame

Example 3.22 An equivalent system for the floor slab in Fig. 3.52 with several axes of symmetry and a uniform distributed load have to be determined.

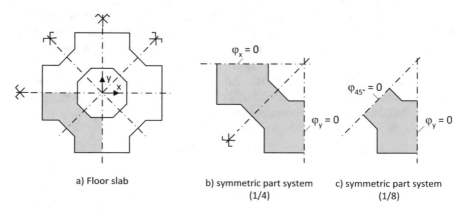

a) Floor slab   b) symmetric part system   c) symmetric part system
                              (1/4)   (1/8)

Fig. 3.52 Floor slab with several axes of symmetry and symmetric loading

An equivalent system of a quarter of the slab is given in Fig. 3.52. The support conditions on the symmetry axes are as given in Table 3.13. The system can be further simplified to one-eighth of the total system. However, in this case inclined supports have to be defined. ◄

Rotationally symmetric or axisymmetric structures can be represented by two-dimensional models. This considerably reduces computational effort, especially in the case of rotationally symmetric loading. Applications can be found for continua as well as for shell structures (see Sects. 4.8.3 and 4.9.3).

Structures may have a cyclic (rotational) symmetry. In this case, the total structure is composed of a number of identical partial structures which are arranged in a circle and repeat cyclically. The load must also be cyclic. Cyclic symmetry occurs mainly in three-dimensional components in mechanical engineering, such as a turbine wheel, for example, in which a blade with its connection to the shaft represents a cyclically symmetrical partial structure. For cyclic symmetric structures, the calculation can be reduced to a partial structure as well [46, 47].

3.8 Quality Assurance and Documentation

3.8.1 Sources of Error

Any result of a finite element analysis, as with any other computation, requires careful examination. For psychological reasons, the impressive performance of the software and the appealing graphical representation of the results on a computer screen may tempt the user to trust the results. However, a naive acceptance of computer results is always inappropriate. Each engineer should develop a strategy to check the results of a computer analysis before accepting them as correct.

Possible sources of errors and controls are described below.

Possible types of errors in the analysis of beam structures
- Insufficient knowledge of the software applied
- Error in the structural model
- Input data error
- Discretization error
- Numerical error
- Software error

Insufficient software knowledge

Commercial software for structural analysis is usually very powerful but also extremely complex. It requires the user to have basic knowledge of the underlying theory and thus of the software application area. In addition to program handling, the software manufacturer must also document the theoretical basics in the manual and provide verification examples. An uncritical application is prone to errors and can, under certain circumstances, lead to serious mistakes in the structural design. Before using the software in structural design, the user should familiarize himself with the software by studying the program extensively and by performing comparative calculations (with known results).

Error in the structural model

This type of error arises when the structure to be analyzed is not properly represented by the model to be analyzed. This includes the modeling of the beam or truss members, of hinges and supports. An error in the structural model can also occur in structural analyses with classical methods. However, finite element computations are more prone to this, as they allow the computation of any arbitrary structural system. However, with sufficient professional experience, an engineer should have the ability to model the structural system properly.

Input data error

Incorrect program input may be caused by inattention or by faulty or misunderstood information in the program manual. Input errors are by far the most usual type of errors. Careful control of the program input is necessary in order to avoid them. In case of ambiguities concerning the program input or output, the software company may provide helpful support.

Discretization error

Discretization errors describe the deviation of the finite element solution from the exact mathematical solution of the mechanical model. They are typical for shell structures. For beam elements, analytical solutions are usually available that are mathematically exact. This is the case, for example, for the stiffness matrices of the truss element according to (3.3), the beam element according to (3.24) or (3.28), and the beam element with torsional warping according to (3.42). For these elements, the element size does not influence the accuracy of the calculation, i.e. discretizing a beam element into many small elements leads to the same results as discretizing into a few large elements. Nevertheless, discretization errors can also occur in the analysis of beam structures, if approximate solutions are implemented in the program instead of the exact solution. Examples are the element stiffness matrices of continuously supported beams (Sect. 3.4.3) or for members with torsional warping according to (3.43b), but also for beams with haunches (Fig. 3.43). In these cases, the accuracy of the results depends on the element size, as it is the case with finite elements for plates, shells, and solid structures. The element size must be chosen so small that the error caused by the discretization is negligible.

Numerical error

Numerical errors may be caused by the limited computational accuracy of the computer. Numbers are represented in the computer by their mantissa and their exponents. Both values are limited, due to the storage capacity of the computer

Table 3.14 Representation of floating point numbers in the computer

Programing language	Type of number	Digits of the Mantissa	Max. Exponent
Fortran	real*4	~7	~37
	real*8	~16	~307
C/C++	float	~7	~37
Java	double	~16	~307

(Table 3.14). The number $\pi = 3.1415927\ldots$, for example, is represented in single precision (6 digits of the mantissa) as:

$$\pi = \underset{Mantissa}{\underline{0.314159}} \cdot \underset{Exponent}{\underline{10^1}}$$

All other digits are omitted when the number is stored in the computer.

The results of arithmetic operations are rounded, due to the limited length of the mantissa. In the case of the difference between two very large numbers, this may cause the loss of all significant digits. This may occur in the superposition of element stiffness matrices to the global stiffness matrix when the stiffness properties of neighboring elements are huge. Extreme differences in the stiffness properties are therefore to be avoided. They can be caused by extreme changes in the bending or longitudinal stiffness EI or EA, respectively, or by extremely short beam or truss lengths. Beam elements representing a rigid coupling between two nodal points should be defined with a moderately large stiffness (e.g. a cross section of a beam element of $10\,m^2$ but not $10^{10}\,m^2$).

The numerical accuracy of the solution of a linear system of equations can be assessed with a condition number. Denoting the number of digits considered by the computer with t, the number s of accurate digits of the numerical solution is given as [42]

$$s \geq t - \log_{10}(\kappa(\underline{K})). \tag{3.50}$$

In finite element computations, $\kappa(\underline{K})$ is the condition number (1.11) of the global stiffness matrix. The differences between the results obtained with the norms according to (1.10b) and (1.10d) are small.

Other numerical errors can be caused by the algorithm used in the program, e.g. termination criteria of the iterative solution of a system of equations. Normally the user of the program has no influence on a termination error.

Example 3.23 The numerical accuracy of a pocket calculator is determined. It can be checked by entering the difference between two huge numbers, for example:

$$1000 + 1 - 1000 = 1$$
$$10^9 + 1 - 10^9 = 1$$

$$10^{10} + 1 - 10^{10} = 0$$
$$10^{20} + 1 - 10^{20} = 0.$$

When calculating $(10^{10} + 1)$, the storage capacity of the mantissa is exceeded, i.e. the addition of 1 in the 11th digit is no longer possible and the 1 is truncated by rounding. Therefore, in the computer, the value 10^{10} is stored, instead of $(10^{10} + 1)$. The subtraction of 10^{10} consequently gives the (incorrect) value 0. This means that the pocket calculator used has a numerical accuracy of about 10 digits.   ◄

Example 3.24 In the truss system in Fig. 3.53 consisting of two truss elements, the nodal points 1 and 2 are connected by an element with an extremely large stiffness (case a). It is to be investigated whether numerical errors caused by rounding may occur when the program is written in C with single-precision numbers. Would there be any difference when double-precision numbers are used in the program?

The element stiffness matrices are according to (3.3):

$$\text{Element 1: } 10^{20} \cdot \begin{bmatrix} 1 & -1 \\ -1 & 1 \end{bmatrix}, \qquad \text{Element 2: } 10^{6} \cdot \begin{bmatrix} 1 & -1 \\ -1 & 1 \end{bmatrix}.$$

The global stiffness matrix is obtained as

$$\begin{bmatrix} 10^{20} & -10^{20} & 0 \\ -10^{20} & (10^{20} + 10^{6}) & -10^{6} \\ 0 & -10^{6} & 10^{6} \end{bmatrix} \cdot \begin{bmatrix} u_1 \\ u_2 \\ u_3 \end{bmatrix} = \begin{bmatrix} F \\ 0 \\ F_{x,3} \end{bmatrix}$$

and after consideration of the support condition $u_3 = 0$ as

$$\begin{bmatrix} 10^{20} & -10^{20} \\ -10^{20} & (10^{20} + 10^{6}) \end{bmatrix} \cdot \begin{bmatrix} u_1 \\ u_2 \end{bmatrix} = \begin{bmatrix} F \\ 0 \end{bmatrix}.$$

When numbers are represented with less than 14 digits, the sum $(10^{20} + 10^{6})$ is rounded to 10^{20}. If the program uses single-precision number representation (i.e. 6 digits accuracy), the existence of element 2 is lost, since it is represented by the stiffness value of 10^{6} in the analysis. After rounding, the system of equations obtained with single-precision numbers is

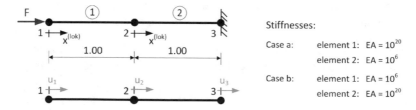

Stiffnesses:

Case a: element 1: EA = 10^{20}
            element 2: EA = 10^{6}

Case b: element 1: EA = 10^{6}
            element 2: EA = 10^{20}

Fig. 3.53 System with two truss elements

$$\begin{bmatrix} 10^{20} & -10^{20} \\ -10^{20} & 10^{20} \end{bmatrix} \cdot \begin{bmatrix} u_1 \\ u_2 \end{bmatrix} = \begin{bmatrix} F \\ 0 \end{bmatrix}.$$

The second row of the matrix is equal to multiplication of the first row by -1. Hence, the global stiffness matrix is singular. The system of equations cannot be solved and the analysis is terminated with a program error message. In terms of structural analysis, the singularity of the stiffness matrix means that element 2 is not considered in the computation (due to the rounding error).

In order to compute the sum correctly, a number precision of at least 15 digits is required. The possible accuracy of the solution can be checked with the condition number.

With (1.11) and the Frobenius norm (1.10b), the condition number is obtained as

$$\kappa(\underline{K}) = \|\underline{K}\| \cdot \|\underline{K}^{-1}\| = 4 \cdot 10^{14} \quad \text{with} \quad \underline{K} = \begin{bmatrix} 10^{20} & -10^{20} \\ -10^{20} & (10^{20} + 10^6) \end{bmatrix}.$$

If the number representation of the computer has $t = 16$ digits, the accuracy of the solution to be expected according to (3.49) is

$$s \geq t - \log_{10}(\kappa(\underline{K})) = 1.4,$$

i.e. an accuracy of 1–2 digits.

If the difference in the stiffness is less, the magnitude of the condition number decreases considerably. For two beams with the same stiffness $E \cdot A = 10^{20}$,

$$\text{with} \quad \underline{K} = \begin{bmatrix} 10^{20} & -10^{20} \\ -10^{20} & 2 \cdot 10^{20} \end{bmatrix}$$

the condition number is

$$\kappa(\underline{K}) = 7 \quad \text{and} \quad s = 15.1$$

i.e. an accuracy of 15 digits is obtained.

The magnitude of the condition number obtained with extremely different element stiffnesses indicates the loss of reliable digits. In the threshold range, badly conditioned systems of equations can result in totally erroneous solutions. In practice this situation is very rare; a termination of the program is more likely. However, extremely large magnitudes for element stiffnesses are to be avoided in practical calculations.

For comparison, now it will be assumed that the stiffness of element 2 is huge compared to element 1 (Fig. 3.53, case b). The stiffness matrix is obtained as

$$\underline{K} = \begin{bmatrix} 10^6 & -10^6 \\ -10^6 & (10^6 + 10^{20}) \end{bmatrix}$$

and after rounding with 14 digits precision as

$$\underline{K} = \begin{bmatrix} 10^6 & -10^6 \\ -10^6 & 10^{20} \end{bmatrix}.$$

Here, no significant digits are lost. This shows that very high stiffness values on the diagonal of the stiffness matrix are nonhazardous. Only off-diagonal terms are critical. This means that even extremely large spring constants used to model supports are uncritical from a numeric point of view.   ◀

Software error

Software errors are errors in the program code. They are very rare, but can never be excluded, even in the code of renowned software providers. They may occur when an uncommon combination of input data is present and the program executes an unusual program path. New program versions coming in updates may introduce program errors in the code, even in parts of the software that previously ran correctly. Theoretically, errors may occur even in tested program paths. Prevention of program errors requires carefully designed software testing, which should be part of the quality control process of the software producer [49].

3.8.2   Checking of Beam Structures Computations

Every finite element analysis requires a check for correctness. In practice, of course, it is not a matter of recalculating the results in detail. The aim is to avoid the sources of error mentioned above. Most important is a critical review of the program input and output.

For the practical implementation of the control of a comprehensive finite element calculation, a procedure in two steps is recommended. The first rough check, the preliminary check, provides a quick overview of the correctness of essential input values. After successful completion of the preliminary check, the more time-consuming final check is carried out with a detailed check of the input and output of the program. Checking the correctness of the software is only necessary in exceptional cases.

Reference should also be made to the guidelines for a software-supported structural analysis issued by engineering associations in [43–45].

Plausibility check

In a preliminary check, a check is carried out for coarse, easily identifiable errors.

Plausibility check of a finite element analysis of beam structures
- Check of the graphical display of the structural system
- Equilibrium control (sum of the loads by load cases or total sum)
- Check of the graphical display of the displacements
- Check of the graphical display of the sectional forces

The graphical display of the structural system, its deformations, and sectional forces give an overview of the structural behavior. It allows the detection of gross modeling errors such as missing supports or wrong signs of the loading. The curvature of beams is an illustrative measure of the size as well as the sign of bending moments. The sum of the loads given by the program can easily be checked by hand calculation and may give useful hints in the case of gross loading input errors. The plausibility of force and moment diagrams can be verified by simple hand calculations on simplified partial models. The equilibrium conditions of the support forces and the total loading can be checked easily.

In case the program stops before the calculation of the displacements is completed, input data or a kinematic system may be the reason. The error message of the program should provide further information.

In the case of incomplete input data, the following checks must be carried out:

Checks of input data
- Connectivity check: check whether for all nodes specified in the element data and the corresponding nodal data have also been defined
- Check whether cross-sectional values have been defined for all cross-section numbers defined with the element data
- Check whether material values have been defined for all material numbers defined in the element or cross-section data
- Further consistency checks (depends on the problem considered)

In the case of kinematic systems, the computation stops with the message such as "Kinematic system found", "Stiffness matrix singular", "Determinant is zero", or "Rigid body modes found". A program crash without any error message displayed may even occur. To detect the reason for a singular stiffness matrix, the following checks should be performed.

Checks in the case of a singular stiffness matrix
- Check of the completeness and the correctness of the support conditions
- Check for kinematic systems caused by hinges or by the combination of several hinge conditions
- Check for free nodal points with degrees of freedom not connected with any element
- Check if the longitudinal, bending, and torsional stiffness is unequal to zero for all elements
- Check for extreme jumps in the stiffness of neighboring elements, causing numerical errors

While the verification of the support conditions can normally be done easily, it is often difficult to detect kinematics caused by the combination of (e.g. three-dimensional) hinges. In general, all degrees of freedom not connected with an element should be restrained. This applies to free nodal points, which may be required for the definition of local coordinate systems for three-dimensional beams. However, in most finite element programs, this task is performed automatically by the software.

Final check

If the preliminary check was successful, a detailed final check must be carried out. All relevant input data must be reviewed and significant result values should be checked.

Final check

Detailed verification of input data
- nodal coordinates at significant locations
- connectivity check (elements with nodes)
- assignment of cross-sectional, material, and design values to elements
- cross-sectional, material, and design values
- spring constants, subgrade reaction modules, etc
- coupling conditions of nodal points (e.g. MPC)
- support and hinge definitions, enforced support displacements, and rotations
- location, size, and sign of actions (loads)
- coefficients for action combinations

Check of significant results
- sum of loads (equilibrium checks, load case checks)
- sectional forces and design results or stresses (approximate calculation with simplified structural models)
- deformations (plausibility checks)

The majority of errors in software-supported structural analyses are due to incorrect data input. Before a finite element analysis can be regarded as correct, all inputs must, therefore, be carefully checked in the program output file or display. The check of the node coordinates can be limited to significant locations, since an overall check has already been done when checking the graphical display of the structural system.

On the other hand, the cross-sectional, material, and design characteristic values and their assignment to members as well as loads must be checked in detail, since incorrect entries of these values can hardly be detected on the basis of the calculation results. The support and hinge definitions should also be checked again.

The final check also includes checking the plausibility of important calculation results. For internal forces, the check can be carried out approximately on simplified systems.

Special controls

The verification of the structural model and the checking on program errors are special controls that are to be considered if there are doubts about the correctness of the results and these cannot be eliminated by the preliminary and final checks.

Complete control of the calculation results does not require necessarily a recalculation of the system. In principle, it is sufficient to verify that the compatibility and the equilibrium conditions at the nodes and in the beam members are fulfilled. The computational effort, however, of such a comprehensive investigation is very high and hardly justified in practice.

In the case of very complex structural models, inconsistencies in the model may only be discovered after a detailed control of the analysis. The results of the analysis may reveal that the theoretical assumptions inherent in the model are violated. In these cases, the analysis must be redone with a new appropriate model.

Software errors are extremely seldom. If the above checks do not result in a diagnosis of the error, the possibility of a software error should also be considered. However, a quick diagnosis "software error" is not appropriate. Instead, one should try to clearly diagnose and document the suspected software error using simple examples. Where a software error is suspected, a stepwise simplification of the structural system such that the remaining partial system retains the supposed error is a good strategy. In the end, a simple system should remain where the error is obvious. It is the responsibility of the software producer to correct the program.

If there remain doubts about the correctness of the results after a careful examination of the program input and results, a new independent analysis with a different finite element program (by a different engineer) can be done as a basis for comparison.

3.8.3 Documentation of Finite Element Analyses

Input files of finite element programs are not well suited for documentation and long-term archiving of a structural analysis, as subsequent program versions are often unable to process them. Therefore, the program output describing the input data as well as the results of the calculation is the only documentation of the finite element analysis that remains over the long term. Their information must be complete so that the calculations carried out are comprehensible and reproducible. For readability, the most important results should also be presented graphically. Reference is made in detail to Sect. 4.12.3 and there in particular to Table 4.46.

Exercises

Problems

3.1

Calculate the nodal point displacements at nodal point 1 and the normal forces in truss elements 1 and 2 using FEM.

Check the normal forces (equilibrium at nodal point 1) and the y-displacement at node 1 by a simple hand calculation

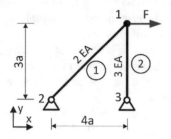

3.2

Determine by the FEM the nodal point displacements and the normal forces in all elements.

Remark: node restraints can be included at the element stiffness matrix level

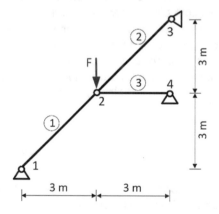

$E = 2.1 \cdot 10^8$ kN/m^2, $A = 10$ cm^2, $F = 10$ kN

3.3

Which are the degrees of freedom of the truss system? The corresponding vector of the nodal displacements of the global system has to be given. Note the structure of the stiffness matrix, label the degrees of freedom, and indicate for each entry of the stiffness matrix if it is equal to zero ("0") or unequal to zero ("x").

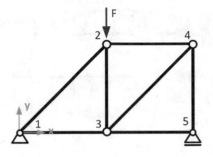

3.4

Determine the support reactions at nodal point 1, the normal force in the truss element, and the spring force with the FEM

Truss: $A = 10.3$ cm^2, $E = 2.1 \cdot 10^8$ kN/m^2

Spring: $k = 1.0 \cdot 10^5$ kN/m

Force: $F = 100$ kN

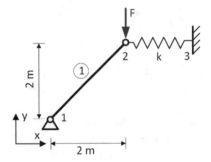

3.5

Derive the stiffness matrix of a truss element applying unit displacements.

Remark: A truss element exposed to a tension force F is elongated as $\delta = F \cdot L / (E \cdot A)$ where E is the Young's modulus, A the cross-sectional area, and L the element length.

3.6

Determine for the given element loads the equivalent nodal loads at nodal points 1 and 2 and sketch them.

$E = 2.1 \cdot 10^8$ kN/m^2, $A = 20$ cm^2, $\alpha_T = 1 \cdot 10^{-5}$

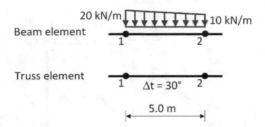

3.7

Analyze the system with two beam elements with the finite element method. Determine the bending moments at the nodal points and in the middle of the second span. Sketch the moment diagram.

Remark: Shear deformations are to be neglected

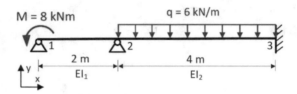

$EI_1 = 1.0$ kN m^2, $EI_2 = 2.0$ kN m^2

3.8

Analyze the frame structure using the finite element method. Normal force deformations are to be neglected, i.e. $A = \infty$

(a)   Calculate the rotational angles of all nodal points

(b)   Determine the sectional forces and draw the shear force and bending moment diagrams

(c)   How many degrees of freedom does the system have if normal forces are taken into account?

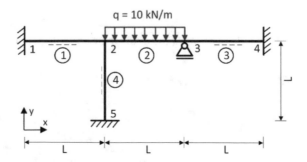

For all elements: $L = 4$ m; $EI = 1.0$ kN m^2

3.9

Determine the nodal displacements, the normal forces in the bar members, and the force in the spring when truss element 2 is cooled down by 40 °C.

Spring: $k = 0.25 \cdot 10^5$ kN/m

Truss elements: $EA = 3.15 \cdot 10^5$ kN m^2/m^2

$\alpha_T = 1 \cdot 10^{-5}$ K^{-1}

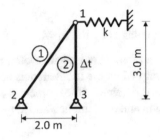

3.10

The system given has to be analyzed by the FEM.

(a) Determine all displacements
(b) Determine the normal forces in the bar members
(c) Check the normal forces in the bars by a hand calculation

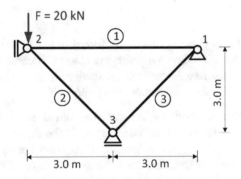

$EA = 1$ kNm^2/m^2

3.11

Which are the degrees of freedom of the plane frame system? The corresponding vector of the nodal displacements of the global system has to be given. Note the structure of the stiffness matrix, label the degrees of freedom, and indicate for each term of the stiffness matrix if it is equal to zero ("0") or unequal to zero ("x")

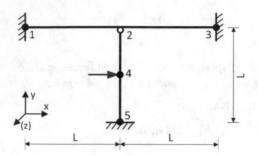

3.12

The system given has to be analyzed by the FEM.

(a) Element 2 is loaded by a moment at node 2. Determine the moment diagram.
(b) Give the moment diagram when the moment acts at the right end of element 1
(c) Which moment diagram is obtained if the moment is entered at node 2 and a hinge
     is defined both at the right end of element 1 and at the left end of element 2?

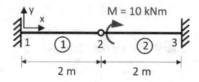

Elements 1 and 2: $EI = 1$ kNm^2, $A = \propto$

3.13

The girder grid in the xy-plane is loaded by forces in the z-direction as well as by
moments in the xy-plane. The beam members consist of I-beams with the web in the
z-direction. All members have the same cross section. In the nodal points 1, 2, 5, 11,
14, 20, 23, and 24, all degrees of freedom are fixed

(a) How many (not restrained) degrees of freedom does the system possess if
     warping torsion is taken into account? Specify the degrees of freedom at the
     nodal points 3, 6, and 7, e.g. $w_3 =$ displacement of nodal point 3 in the z-direction
(b) How many degrees of freedom does the system have if the beam elements
     between points 5–6–7–8–9–10–11 and 14–15–16–17–18–19–20 consist of an
     I-beam with 40 cm web height and the beam elements between points 1–3–7–
     12–16–21–23 and 2–4–9–13–18–22–24 consist of an I-beam with 25 cm web
     height?

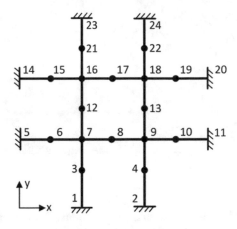

3.14

The three-dimensional beam system shown has 5 nodes. Which system of equations is set up for the system? The vector of the displacements of the global system of equations is to be given taking restraints into account. Two cases have to be distinguished

(a) beam without warping
(b) beam with warping

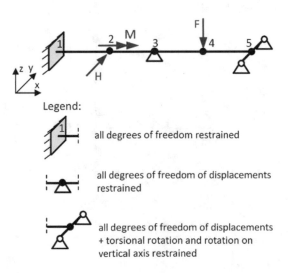

Legend:

all degrees of freedom restrained

all degrees of freedom of displacements restrained

all degrees of freedom of displacements + torsional rotation and rotation on vertical axis restrained

3.15

The beam system consisting of four beam elements with a cross section without warping is loaded with a torsional moment T. Computation is done using a finite element program written in C in which variables are defined in simple accuracy. Set up the global equations and investigate the program behavior for the following cases

(a)   Elements 1 and 3: $G \cdot I_T = 10^9$
       Element 2: $G \cdot I_T = 1$
(b)   Elements 1 and 3: $G \cdot I_T = 1$
       Element 2: $G \cdot I_T = 10^9$

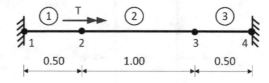

3.16

A beam system consists of a beam and two columns. Replace the two columns by rotational springs and determine the rotational deformations at nodal points 1 and 2 as well as the moment diagram of the system by a finite element analysis

Loading: $q = 10$ kN/m

Stiffnesses:

$E = 2.1 \cdot 10^8$ kN/m^2, $A = \infty$

beams 1-2: $I = 1000$ cm^4

beams 3-2, 2-4: $I = 500$ cm^4

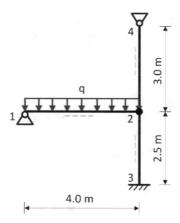

3.17

A 3D frame is symmetric in the xz-plane. Give all support conditions for the symmetric partial system at the plane of symmetry

(a)   for a symmetric loading
(b)   for an antisymmetric loading
       How do the beams on the plane of symmetry have to be modeled?

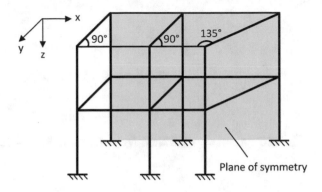

3.18

The foundation slab shown has the two symmetry axes A–A and B–B. For reasons of computing time, it should be calculated as a symmetrical partial system with the finite element method. Which degrees of freedom on the axes A–A and B–B are to be restrained, which ones are to be left free in the case of

(a) uniformly distributed load on the whole foundation slab?
(b) a positive distributed load on the left side of B–B and a negative distributed load of the same size on the right side?
(c) a positive distributed load on the partial areas I and III and a negative distributed load of the same size on the partial areas II and IV?

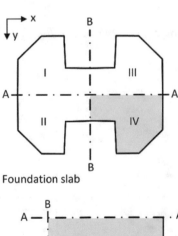

Foundation slab

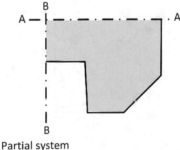

Partial system

Solutions

3.1

$$\begin{bmatrix} u_1 \\ v_1 \end{bmatrix} = \begin{bmatrix} 4.47 \\ -0.75 \end{bmatrix} \cdot \frac{F \cdot a}{E \cdot A}, \; N_1 = 1.25 \cdot F, \; N_2 = -0.75 \cdot F.$$

3.2

$$\begin{bmatrix} u_2 \\ v_2 \end{bmatrix} = \begin{bmatrix} 0.14 \\ -0.35 \end{bmatrix} \cdot 10^{-3} \mathrm{m}, \; N_1 = -7.07 \, \mathrm{kN}, \; N_2 = 7.07 \, \mathrm{kN}, \; N_3 = -10 \, \mathrm{kN}.$$

3.3

	u_2	v_2	u_3	v_3	u_4	v_4	u_5
u_2	x	x	0	0	x	0	0
v_2	x	x	0	x	0	0	0
u_3	0	0	x	x	x	x	x
v_3	0	x	x	x	x	x	0
u_4	x	0	x	x	x	x	0
v_4	0	0	x	x	x	x	0
u_5	0	0	x	0	0	0	x

3.4

$$\begin{bmatrix} u_2 \\ v_2 \end{bmatrix} = \begin{bmatrix} 1.00 \\ -3.62 \end{bmatrix} \cdot 10^{-3} \mathrm{m}, \; N_1 = -141.42 \, \mathrm{kN}, \; F_{\mathrm{Spring}} = 100.00 \, \mathrm{kN}.$$

3.5

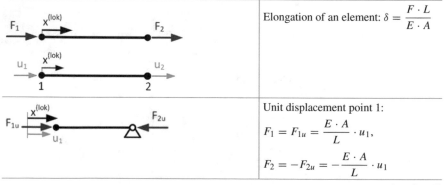

	Elongation of an element: $\delta = \dfrac{F \cdot L}{E \cdot A}$
	Unit displacement point 1: $F_1 = F_{1u} = \dfrac{E \cdot A}{L} \cdot u_1,$ $F_2 = -F_{2u} = -\dfrac{E \cdot A}{L} \cdot u_1$

(continued)

(continued)

	Unit displacement point 2:
	$F_1 = -F_{1u} = -\dfrac{E \cdot A}{L} \cdot u_2,$ $F_2 = \quad F_{2u} = \quad \dfrac{E \cdot A}{L} \cdot u_2$
Superposition:	Stiffness matrix:
$F_1 = \dfrac{E \cdot A}{L} \cdot u_1 - \dfrac{E \cdot A}{L} \cdot u_2$ $F_2 = -\dfrac{E \cdot A}{L} \cdot u_1 + \dfrac{E \cdot A}{L} \cdot u_2$	$k \cdot \begin{bmatrix} 1 & -1 \\ -1 & 1 \end{bmatrix} \cdot \begin{bmatrix} u_1 \\ u_2 \end{bmatrix} = \begin{bmatrix} F_1 \\ F_2 \end{bmatrix}$

3.6

3.7

$M_1 = -8.00\,\text{kNm}, M_2 = -1.14\,\text{kNm}, M_3 = -11.43\,\text{kNm},$
$M_{\text{mid-2-3}} = 5.71\,\text{kNm}.$

3.8

(a) $\varphi_2 = -5.80, \varphi_3 = 8.12$

(b) $M_1 = 2.90\,\text{kNm}, M_{2,\text{left}} = -5.80\,\text{kNm}, M_{2,\text{right}} = -11.60\,\text{kNm},$
     $M_{2,\text{column}} = 5.80\,\text{kNm}, M_3 = -8.12\,\text{kNm}, M_4 = 4.06\,\text{kNm},$
     $M_5 = -2.90\,\text{kNm}, M_{\text{mid-2-3}} = 10.14\,\text{kNm}, V_1 = -2.17\,\text{kN},$
     $V_{2,\text{left}} = -2.17\,\text{kN}, V_{2,\text{right}} = 20.87\,\text{kN}, V_{2,\text{column}} = -2.17\,\text{kN},$
     $V_{3,\text{left}} = -19.13\,\text{kNm}, V_{3,\text{right}} = 3.04\,\text{kN}, V_4 = 3.04\,\text{kN}, V_5 = -2.17\,\text{kN}$

(c) 5 DOFs.

3.9

$N_1 = -32.90\,\text{kN}, N_2 = 27.38\,\text{kN}, F_{\text{spring}} = -18.25\,\text{kN}.$

3.10
(a) $v_2 = -339.4\,\text{m}, u_3 = 169.7\,\text{m}$ (related to EA $= 1\,\text{kNm}^2/\text{m}^2$)
(b) $N_1 = 0, N_2 = -28.28\,\text{kN}, N_3 = -28.28\,\text{kN}.$

(c) $N_1 = 0$ since $u_1 = u_2 = 0$; N_2 and N_3 are obtained by the method of joints as given above.

3.11

	u_2	v_2	φ_2	u_4	v_4	φ_4
u_2	X	0	0	X	0	X
v_2	0	X	X	0	X	0
φ_2	0	X	X	0	0	0
u_4	X	0	0	X	0	X
v_4	0	X	0	0	X	0
φ_4	X	0	0	X	0	X

3.12

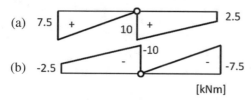

(a) 7.5, 10, 2.5

(b) -2.5, -10, -7.5

[kNm]

(c) The moment acts on the rotational degree of freedom $\varphi_{z,2}$ of node 2. Since a hinge was defined for both elements at node 2, this node has no rotational stiffness, i.e. the equation in the global stiffness matrix for this degree of freedom is $k \cdot \varphi_{z,2} = M$ with $k = 0$. This equation has no solution for $M \neq 0$. This means that the program gives an error message such as "kinematic system" as output or automatically inserts a spring with a large spring constant, e.g. $k = 10^{20}$, as a "repair measure" in this case. In the latter case, the program calculates the rotation of node 2 as $\varphi_{z,2} = M/k = 10^{-20} \cdot M$ and the moment diagram in elements 1 and 2 as zero.

3.13

(a) 16 nodes with 4 DOFs each, i.e. $16 \cdot 4 = 64$ DOFs
   Nodal point 3: $w_3, \varphi_{x,3}, \varphi_{y,3}, \psi_3$

(b) 4 additional degrees of freedom from warping at nodal points 7, 9, 16, 18, i.e. a total of 68 degrees of freedom.

3.14

(a) $\left[u_2 \ v_2 \ w_2 \ \varphi_{x,2} \ \varphi_{y,2} \ \varphi_{z,2} \ \varphi_{x,3} \ \varphi_{y,2} \ \varphi_{z,3} \ u_4 \ v_4 \ w_4 \ \varphi_{x,4} \ \varphi_{y,4} \ \varphi_{z,2} \ \varphi_{y,5} \right]^T$.

(b) $\begin{bmatrix} u_2 & v_2 & w_2 & \varphi_{x,2} & \varphi_{y,2} & \varphi_{z,2} & \psi_2 & \varphi_{x,3} & \varphi_{y,2} & \cdots\cdots \\ \cdots & \varphi_{z,3} & \psi_3 & u_4 & v_4 & w_4 & \varphi_{x,4} & \varphi_{y,4} & \varphi_{z,2} & \psi_4 & \varphi_{y,5} & \psi_5 \end{bmatrix}^{\mathrm{T}}.$

3.15

(a) Determinant of the global stiffness matrix: $\det(\underline{K}) = 4.10^9$;
Program gives the solution: $\varphi_{x,2} = 0.5 \cdot 10^{-9} \cdot T, \varphi_{x,3} = 0$

(b) Determinant of the global stiffness matrix: $\det(\underline{K}) = 0$; (with 6 digits precision)
Program fails to obtain a solution.

3.16

$k_\varphi = 2.73 \cdot 10^7$ kNm, $\varphi_1 = -8.67 \cdot 10^{-7}$ $\varphi_2 = 4.65 \cdot 10^{-7}$
$M_{2,\mathrm{left}} = -12.68$ kNm, $M_{2,\mathrm{upper}} = 4.88$ kNm, $M_{2,\mathrm{lower}} = -7.80$ kNm
$M_3 = 3.90$ kNm, $M_{\mathrm{mid}\text{-}1\text{-}2} = 13.66$ kNm.

3.17

Dof's	(a) Symmetric loading	(b) Asymmetric loading
Displacements in x-direction	0	x
Displacements in y-direction	x	0
Displacements in z-direction	0	x
Rotations about x-axis	x	0
Rotations about y-axis	0	x
Rotations about z-axis	x	0
	x = fixed 0 = free	

3.18

Dof's	(a)		(b)		(c)	
	A–A	B–B	A–A	B–B	A–A	B–B
Displacements in x-direction	0	x	0	0	x	0
Displacements in y-direction	x	0	x	x	0	x
Displacements in z-direction	0	0	0	x	x	x
Rotations about x-axis	x	0	x	x	0	x
Rotations about y-axis	0	x	0	0	x	0
Rotations about z-axis	x	x	x	0	0	0
	x = fixed 0 = free					

References

1. Przemieniecki JS (1968) Theory of matrix structural analysis. McGraw-Hill, New York
2. Szilard R (1982) Finite Berechnungsverfahren der Strukturmechanik, Band 1 Stabwerke. Ernst & Sohn, Berlin
3. Meißner U, Menzel A (1989) Die Methode der finiten Elemente. Springer, Berlin
4. Ahlert H (1992) Finite Elemente in der Stabstatik. Werner-Verlag, Düsseldorf
5. Ramm E, Hofmann T (1996) Stabtragwerke. In: Mehlhorn G (Hrsg) Der Ingenieurbau: Grundwissen, Bd. 5: Baustatik Baudynamik. Ernst & Sohn, Berlin
6. Duddeck H, Ahrens H (1998) Statik der Stabtragwerke, Beton-Kalender 1998. Ernst & Sohn, Berlin
7. Wunderlich W, Kiener G (2004) Statik der Stabtragwerke. Teubner/GWV Fachverlage, Wiesbaden
8. Krätzig WB, Harte R, Meskouris K, Wittek U (2005) Tragwerke 2, 4th edn. Springer, Berlin
9. Graf W, Vassilev T (2006) Einführung in die computerorientierten Methoden der Baustatik. Ernst & Sohn, Berlin
10. Könke C, Krätzig WB, Meskouris K, Petryna YS (2012) Theorie der Tragwerke. In: Zilch K, Diederichs CJ, Katzenbach R, Beckmann KJ (Hrsg.) Handbuch für Bauingenieure, 2nd edn. Springer, Heidelberg
11. Wagenknecht G (2018) Baustatik – Weggrößenverfahren. Beuth-Verlag, Berlin
12. Hirschfeld K (1969) Baustatik. Springer, Berlin
13. Rubin H (1993) Baustatik ebener Stabwerke. Stahlbau-Verlagsgesellschaft, Köln, Stahlbau-Handbuch
14. Kindmann R, Kraus M (2011) Steel design using FEM. Ernst & Sohn, Berlin
15. Steinke P (2015) Finite-Element-Methode, 5th edn. Springer Vieweg, Berlin
16. Merkel M, Öchsner A (2014) Eindimensionale Finite Elemente, 2nd edn. Springer Vieweg, Berlin
17. Katz C (2003) Warum Finite Elemente "falsch" rechnen, Fides Fachgespräch "Finite Elemente", Mainz, Oktober 2003. Fides DV-Partner Systemhaus Rhein-Main GmbH, Mainz
18. Kindmann R (1981) Traglastermittlung ebener Stabwerke mit räumlicher Beanspruchung. Dissertation, Ruhr-Universität Bochum, Technisch-wissenschaftliche Mitteilungen 81-3, Bochum
19. Roik K, Kindmann R (1982) Berechnung stabilitätsgefährdeter Stabwerke mit Berücksichtigung von Entlastungsbereichen. Der Stahlbau 51, Ernst & Sohn, Berlin, 310–318
20. Wagner W, Gruttmann F (2002) A displacement method for the analysis of flexural shear stresses in thin-walled isotropic composite beams, Universität Karlsruhe, Mitteilungen des Instituts für Baustatik 3, Karlsruhe
21. Cywinski Z (2009) Zur Entwicklung der Steifigkeitsmatrizen für dünnwandige Stäbe, Stahlbau 78, H. 2. Ernst & Sohn, Berlin
22. Wilson EL, Habibullah A (1989) SAP90 users manual. CSI, Computers and Structures, Berkeley, California
23. Ungermann D, Weynand K, Jaspart J-P, Schmidt B (2005) Momententragfähige Anschlüsse mit und ohne Steifen. In: Kuhlmann U (Hrsg.) Stahlbau-Kalender 2005. Ernst & Sohn, Berlin
24. Kuhlmann U, Rybinski M, Rölle L (2008) Anschlüsse im Stahl- und Verbundbau. Der Prüfingenieur, H. 32:36–49
25. Ungermann D, Schneider S (2015) Stahlbaunormen; DIN EN 1993-1-8: Bemessung von Anschlüssen. In: Kuhlmann U (Hrsg.) Stahlbaukalender 2015. Ernst & Sohn, Berlin
26. Fleßner H (1962) Ein Beitrag zur Ermittlung von Querschnittskennwerten mit Hilfe elektronischer Rechenanlagen, Bauingenieur 37, H. 4. Springer, Berlin, 146–149
27. Werkle H (2012) Mathcad in der Tragwerksplanung, 2nd edn. Vieweg-Teubner, Wiesbaden
28. Balke H (2014) Einführung in die Technische Mechanik - Festigkeitslehre. Springer Vieweg, Berlin
29. Petersen C (2013) Stahlbau, 4th edn. Springer Vieweg, Wiesbaden

30. Kneidl R (1994) Träger mit nachgiebigem Verbund - eine Diskretisierung mit STAR2, Sofistik, 7. Anwender-Seminar, Nürnberg
31. Kneidl R, Hartmann H (1995) Träger mit nachgiebigem Verbund – Eine Berechnung mit Stabwerksprogrammen. In: Bauen mit Holz, 285–290
32. Hartmann H (1999) Die Berücksichtigung elastisch-plastischer Verformungseigenschaften mechanischer Verbindungsmittel bei Verbundkonstruktionen im Ingenieurholzbau, Dissertation; Technische Universität München
33. Scholz A (2004) Ein Beitrag zur Berechnung von Flächentragwerken aus Holz, Dissertation, Technische Universität München
34. Gruttmann F, Wagner W, Sauer R (1998) Zur Berechnung von Wölbfunktion und Torsionskennwerten beliebiger Stabquerschnitte mit der Methode der finiten Elemente. Bauingenieur 73(3):138–143
35. Wagner W, Sauer R, Gruttmann F (1999) Tafeln der Torsionskenngrößen von Walzprofilen unter Verwendung von FE-Diskretisierungen, Universität Karlsruhe, Mitteilungen des Instituts für Baustatik, 5, Karlsruhe
36. Gruttmann F, Wagner W (2001) Shear correction factors in Timoshenko's beam theory for arbitrary shaped cross–sections. Comput Mech 27:199–207
37. Bornscheuer FW (1975) Vorlesungen in Baustatik – Teil E: Elastisch gebettete Balken, Institut für Baustatik, Universität Stutgart, WS 1975/76, Stuttgart
38. Kraus M (2004) Zur Anwendung der Wölbkrafttorsion auf Systeme mit Wölbfedern, RUBSTAHL-Bericht 2-2004. Ruhr-Universität Bochum, Bochum
39. Salzgeber G (1999) Modellierung von Konstruktionsdetails in räumlichen Stabberechnungen mittels geometrischer Zwangsgleichungen. In Meskouris: Baustatik-Baupraxis 7. Balkema, Rotterdam, 103–112
40. Leihkauf S (2001) Geometrisch und physikalisch nichtlineare Statik räumlicher Stabtragwerke aus Stahl unter Berücksichtigung realer Anschluß- und Aussteifungsbedingungen, Dissertation, TU Dresden
41. Kraus M (2005) Computerorientierte Berechnungsmethoden für beliebige Stabquerschnitte des Stahlbaus, Dissertation. Ruhr-Universität Bochum, Bochum
42. Bathe K-J (2014) Finite Element procedures, 2nd edn. K-J Bathe, Watertown M.A.
43. Ri-EDV-AP-2001 (2001) Richtlinie für das Aufstellen und Prüfen EDV-unterstützter Standsicherheitsnachweise, Ausgabe April 2001, Bundesvereinigung der Prüfingenieure für Bautechnik e.V., Hamburg
44. VDI 6201 Blatt 1 (2015) Softwaregestützte Tragwerksberechnung - Grundlagen, Anforderungen, Modellbildung, Verein Deutscher Ingenieure e.V., Düsseldorf
45. VDI 6201 Blatt 2 (2017) Softwaregestützte Tragwerksberechnung - Verifikationsbeispiele, Verein Deutscher Ingenieure e.V., Düsseldorf
46. Knothe K, Wessels H (2017) Finite Elemente, 5th edn. Springer Vieweg, Berlin
47. Cook RD, Malkus DS, Plesha ME (1989) Concepts and applications of finite element analysis. Wiley, New York
48. Herwig T, Wagner W (2017) Zur Behandlung von Vouten im Rahmen der FEM. Der Bauingenieur 92(10):444–453
49. Heidkamp H, Kimmich S, Rustler W (2014) Fehlerfreie Software – Quadratur des Kreises? Qualitätssicherung in Softwarehäusern, Baustatik-Baupraxis 12, TU München, 319–326
50. Schier K (2011) Finite Elemente Modelle der Statik und Festigkeitslehre. Springer, Berlin Heidelberg
51. Wiedemann J (2007) Leichtbau – Elemente und Konstruktion, 3rd edn. Springer, Berlin
52. Klein B, Gänsicke T (2019) Leichtbau-Konstruktion, 11th edn. Springer Vieweg, Wiesbaden
53. Schroeter H (1980) Berechnung idealer Kipplasten von Trägern veränderlicher Höhe mit Hilfe Hermit'scher Polynome. Mitteilungen aus dem Institut für Bauingenieurwesen I, H. 5, Technische Universität München
54. Neto MA, Amaro A, Roseiro L, Cirne J, Leal R (2015) Engineering computation of structures: the finite element method. Springer International, Cham, Switzerland

55. Rubin H (2020) Baustatik. In: Schneider - Bautabellen für Ingenieure, 24th edn., Reguvis Verlag, Köln
56. Roark RJ, Young WC (1975) Formulas for stress and strain, 5th edn. McGraw-Hill Book Company
57. Cook RD, Young WC (1985) Advanced mechanics of materials. Macmillan Publishing Company
58. Francke W, Friemann H (2005) Schub und Torsion in geraden Stäben, Vieweg, Wiesbaden

Finite Element Software

59. SOF 1 (2012) ADINA, Theory and Modeling Guide, Volume I: ADINA, Report ARD 12-8, ADINA R&D Inc, Watertown USA, Dec 2012
60. SOF 2 (2013) ANSYS ANSYS mechanical APDL element reference, release 15.0, ANSYS Inc, Canonsburg, USA, Nov 2013
61. SOF 6 (2014) SOFiSTiK Finite-Element-Software, Version 2014-9 ASE – Allgemeine Statik Finiter Element Strukturen, ASE Manual, Version 2014-9, Software Version, Sofistik 2014, SOFiSTiK AG, Oberschleissheim, Germany, 2015 TAPLA – 2D Finite Elemente in der Geotechnik, TALPA Manual, Version 2014-9, Software Version SOFiSTiK 2014, SOFiSTiK AG, Oberschleissheim, Germany

Chapter 4
Plate, Shell, and Solid Structures

Abstract In this chapter the finite element method is considered as a general method
for the analysis of plate and solid structures. While the method for truss and beam
structures – as it has been treated so far – is exact, for plate and solid structures it can
no longer be formulated as an exact method, i.e. here only an approximate solution
is obtained by a finite element analysis. The topic of the first part of the chapter
is the basic finite element theory of plate, shell and solid structures. After a short
historical outline the chapter begins by explaining the approximate character of the
finite element method for the example of a tapered truss element. Many characteristic
properties of a finite element analysis even for surface and solid structures can be
understood studying this basic element. It is followed by the derivation of a rectangular plane stress element used in the analysis of plates in plane stress. Further formulations of finite elements for plates in plane stress are discussed. The next sections
deal with plates in bending. To explain the basic concept, a shear flexible rectangular plate element is derived. Advanced formulations for shear flexible and shear
stiff plate elements and their properties are discussed. Elements for shells and three-
dimensional solids including axisymmetric elements are treated in the following.
A particular section addresses the problem of the connection of different types of
elements as e.g. beam with plate elements. The second also very comprehensive part
of the chapter is dedicated to the modelling of structural elements and buildings with
finite elements. General aspects such as mesh generation, the required mesh size, the
influence of singularities, the edge effect in bending plates are discussed in detail and
illustrated by examples. In the same way the modelling of columns in deep beams and
slabs, of RC beams partly integrated in slabs and of mounting parts are treated. An
own section deals with the modelling of soil for foundation slabs. Continuum models
of layered soils as well as subgrade reaction and two-parameter models are surveyed
and compared concerning their practical application. The finite element analysis of
three-dimensional building models is the topic of another section. The influence of
taking into account construction stages is investigated in detail. The final section of
the chapter deals with quality assurance of finite element analyses. It includes some
aspects of estimations of the discretization error. The chapter concludes with some
remarks on the documentation of finite elements computations.

© Springer Nature Switzerland AG 2021   191
H. Werkle, *Finite Elements in Structural Analysis*, Springer Tracts
in Civil Engineering, https://doi.org/10.1007/978-3-030-49840-5_4

4.1   Historical Background

The tremendous potential of the *finite element method* (FEM) in structural engineering is particularly apparent in the analysis of two- and three-dimensional shell and solid structures. The name "finite element" was, therefore, originally employed for two- and three-dimensional solid structures only, i.e. for plates in plane stress and bending, shells, and three-dimensional solids. Unlike for beam and truss structures considered so far, here the method is no longer mathematically exact, but only approximate.

Precursors of the finite element method in mathematical respect are the methods of Ritz [1] and Galerkin [2], for solving differential equations of mechanics, which were developed at the beginning of the twentieth century. They lead to systems of linear equations with the supporting points of piecewise linear approximations (Ritz) or the coefficients of polynomial approximations (Galerkin) as unknowns. At that time, before the availability of electronic computers, their solution was only possible "by hand", and thus only available for very small systems. The actual development of the finite element method began in the 1950s, starting in the aircraft and aerospace industry. Based on the stiffness method for trusses and beams as presented by Argyris [3], Turner, Clough, Martin, and Topp developed a general method for plate structures in 1956 [4]. Clough introduced the name "finite element" in 1960 [5]. The 1960s saw a rapid development of the finite element method. Significant contributions came, e.g. from Clough in Berkeley (USA), Zienkiewicz in Swansea (Great Britain), and Argyris in Stuttgart (Germany) [6–8]. Such fast progress in the FEM was enabled by the development of increasingly powerful computers.

In the following, the method has been extended for structural dynamics and for systems with geometric and material nonlinearities [9]. In the 1970s the FEM, which had thus far been developed only in engineering science, became understood as a field of research in mathematics, as well. Here problems of convergence and accuracy of the method have been treated. In the 1980s, questions of automatic mesh generation and mesh adaptation, as well as the calculation of nonlinear systems with large deformations were dealt with. Since 2000, new element types (isogeometric analysis) and calculation methods (meshless methods) have been at the forefront of development [10, 11]. Further, computational concepts were developed for coupled systems such as fluid-structure interaction or interdisciplinary multi-physics problems (simultaneous coupling of structural calculations with physical field calculations such as temperature or magnetic fields).

The first finite element programs specialized in civil engineering have been developed in the 1970s for the mainframe computers of that time. These were programs for the structural analysis of frameworks, plates, and shells with special design modules according to building standards. The development of PCs and workstations in the 1980s led to the widespread use of finite element software and the rapid adoption of the finite element method as the standard method in structural analysis. For the development history of the finite element method, see [12–15], and for the development of structural analysis, in particular, see [16].

4.2   Basic Concepts

For an analysis with the finite element method, a plate structure has to be discretized into finite elements, i.e. elements of finite size (Fig. 4.1). The elements are interconnected at nodal points. Hence, at the nodal points the equilibrium conditions of the nodal forces and the compatibility conditions of the displacements are fulfilled (see Sect. 3.2.4). Whereas for beam structures an exact solution is obtained, plate, shell and solid structures have to meet more extensive requirements for an exact solution.

Requirements for an exact solution of plate, shell and solid structures

(a)   The displacements at the boundaries of adjacent elements must coincide.
(b)   The stresses at the boundaries of adjacent elements must fulfill the equilibrium conditions.
(c)   At fixed boundaries the support conditions have to be fulfilled (geometric boundary conditions)
(d)   At free boundaries the equilibrium conditions of boundary loads and stresses have to be fulfilled (stress boundary conditions).

This means that for an exact solution, the compatibility conditions of the displacements and the equilibrium conditions of the stresses have to be met not only at the nodal points but at all points of the plate structure. Whereas the corresponding fundamental equations can be formulated for an infinitesimal element, exact analytical solutions are only available for simple support conditions and loadings. The finite element method can deal with arbitrary boundary conditions, however, the compatibility conditions are not fulfilled exactly. The capabilities of the method are thus contrasted with its approximate character. This must always be kept in mind when performing a finite element analysis in order to avoid possibly serious errors.

Finite elements can be based on different approximate assumptions (see an overview of methods in [17]). In the following, only displacement-based elements, i.e. elements with shape functions for the displacements, are treated in-depth. Historically, they are the basis for all finite elements and are implemented in many finite

Fig. 4.1 Finite element discretization of a plate

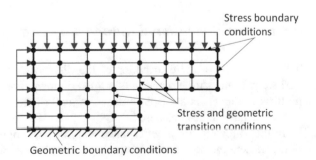

Stress boundary conditions

Stress and geometric transition conditions

Geometric boundary conditions

element programs. A short survey is also given on hybrid elements for plates and shells and for modern, advanced formulations.

Displacement-based elements are based on an assumption of the distribution of displacements into an element. For a triangular plate element in plane stress, e.g. a linear distribution of the displacements between two nodal points can be assumed. The displacements inside the element are obtained by interpolation. In order to be able to represent arbitrary distributions of displacements and stresses by this simplified assumption, small elements are required. They allow the representation of any displacement distribution with a close approximation. The requirements for the exact solutions are then fulfilled only partly.

Properties of the displacement-based finite element method

(a) The displacements coincide at the boundaries of adjacent elements (geometric transition conditions).
(b) The equilibrium conditions for the stresses at the boundaries of adjacent elements are not fulfilled. A jump in the stress diagram occurs which is not present in the exact solution, i.e. in reality (stress transition conditions).
(c) The support conditions (geometric boundary conditions) are fulfilled at fixed boundaries.
(d) At free boundaries, the equilibrium conditions between the loading and the stresses or the section forces (stress boundary conditions) are not fulfilled.

The approximate character of the finite element method means that the equilibrium conditions between the elements, as well as on boundaries are violated, i.e. they are only fulfilled approximately. On the other hand, the compatibility conditions for the displacements, as well as the support conditions are respected. An introduction is given by [18–26, 287], among others.

4.3 Approximation Character of the Finite Element Method

4.3.1 One-Dimensional Introductory Example

The assumption of the distribution of the displacements in an element allows the derivation of the stiffness matrix of a finite element. For two-dimensional elements, multiple components of displacements, as well as multidimensional states of stress and strain have to be considered.

In the following, for a better understanding, the approximation character of the finite element method is explained using a simple, one-dimensional example. It is a

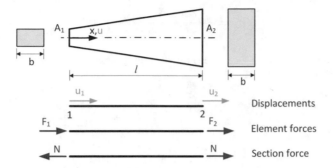

Fig. 4.2 Truss element with linearly varying cross section

truss element, with a linearly varying cross section (Fig. 4.2). The normal stress σ_x in the element is, therefore, also variable. The aim is to investigate the quality of the finite element approximation with simple displacement shape functions, as they are also used for "real" plane stress elements.

4.3.2   Analytical Solution

For shells and plates, an analytical solution is not available in general. For a truss element with linearly varying cross section, however, an analytical solution can be given. The exact solution will be derived in order to compare it with the approximate finite element solution and to assess their approximation character in this way.

The truss element has length l and its cross section A varies with x as

$$A = A_1 + \frac{x}{l} \cdot (A_2 - A_1).\tag{4.1}$$

The displacements at both ends of the element are u_1 and u_2 and the element forces are F_1 and F_2 (Fig. 4.2). The element has a constant normal force N. The normal stresses, strains, and displacements with respect to the local coordinate x are to be determined. A superscript "(lok)" of the coordinate x as in Sect. 3.2, is omitted here.

Generally, the stresses and displacements of a structural system for a given loading and for given support conditions are obtained by the solution of its differential equations with appropriate boundary conditions. In this case, the solution can be derived much more easily. The normal stress σ_x can be determined from a given normal force as

$$\sigma_x = \frac{N}{A} = \frac{N}{A_1 + x/l \cdot (A_2 - A_1)}$$

or

$$\sigma_x = \frac{N \cdot l}{A_1 \cdot l + x \cdot (A_2 - A_1)} . \tag{4.2}$$

With Hooke's law $\sigma_x = E \cdot \varepsilon_x$ (2.3a) the strains ε_x are obtained as

$$\varepsilon_x = \frac{\sigma_x}{E}$$

and with (4.2)

$$\varepsilon_x = \frac{N \cdot l}{E \cdot (A_1 \cdot l + x \cdot (A_2 - A_1))} . \tag{4.3}$$

From the strains, the displacements can be determined. Following the definition of the strains (2.2a) $\varepsilon_x = du/dx$ one obtains

$$u = \int_0^x \varepsilon_x \, d\bar{x} + u_1$$

and with (4.3)

$$u = \int_0^x \frac{N \cdot l}{E \cdot (A_1 \cdot l + x \cdot (A_2 - A_1))} \, d\bar{x} + u_1 .$$

After performing the integration, the displacements $u(x)$ are

$$u = \frac{N \cdot l}{E \, (A_2 - A_1)} \cdot \ln \left(\frac{l \cdot A_1 + x \, (A_2 - A_1)}{l \cdot A_1} \right) + u_1 . \tag{4.4}$$

Example 4.1 For the bar in Fig. 4.3, the stresses and the displacements in the middle of the bar and at $x = 0.0, 0.2l, 0.4l, \ldots, 1.0l$ are to be determined.

The displacements are obtained with (4.4), and with $u_1 = 0$ as

$$u = \frac{100 \cdot 500}{1000 \cdot (100 - 500)} \cdot \ln \left(\frac{500 \cdot 500 + x \cdot (100 - 500)}{500 \cdot 500} \right) + 0$$

$$u = -0.125 \cdot \ln (1 - 0.0016 \cdot x) \text{ with } u[\text{cm}] \text{ and } x[\text{cm}]$$

and the stresses with (4.2) as

$$\sigma_x = \frac{100 \cdot 500}{500 \cdot 500 + x \cdot (100 - 500)}$$

Table 4.1 Exact solution of the stresses and displacements

x[cm]	0	100	200	250	300	400	500
u[cm]	0	0.022	0.048	0.064	0.082	0.128	0.201
σ_x[kN/cm^2]	0.200	0.238	0.294	0.333	0.385	0.556	1.000

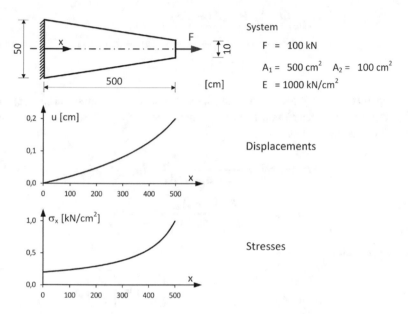

System

F = 100 kN

A$_1$ = 500 cm^2 A$_2$ = 100 cm^2

E = 1000 kN/cm^2

Displacements

Stresses

Fig. 4.3 Truss element with linearily varying cross section area

or

$$\sigma_x = \frac{100}{500 - 0.8 \cdot x} \quad \text{with } x\text{[cm] and } \sigma_x\left[\text{kN/cm}^2\right].$$

The results are given in Table 4.1, at different points and plotted in Fig. 4.3. ◀

The solution given above allows the derivation of the stiffness matrix of a truss element with linearly varying cross section. The displacement u_2 at the end of the element is obtained with (4.4) and $x = l$ as

$$u_2 = u_{(x=l)} = \frac{N \cdot l}{E\,(A_2 - A_1)} \cdot \ln\left(\frac{A_2}{A_1}\right) + u_1.$$

Solving this equation for the normal force N, one obtains

$$N = \frac{E\,(A_2 - A_1)}{l \cdot \ln\,(A_2/A_1)}\,(u_2 - u_1). \tag{4.5}$$

Substituting this relationship for N in (4.2) and (4.4), the stress σ_x and the displacement u are obtained as

$$\sigma_x = \frac{E \cdot (A_2 - A_1)}{(l \cdot A_1 + x \, (A_2 - A_1)) \cdot \ln \, (A_2/A_1)} \cdot (u_2 - u_1) \tag{4.6}$$

$$u = \frac{\ln \left(\dfrac{l \cdot A_1 + x \, (A_2 - A_1)}{l \cdot A_1} \right)}{\ln \left(\dfrac{A_2}{A_1} \right)} (u_2 - u_1) + u_1 \tag{4.7}$$

They depend on the displacements u_1 and u_2 at the ends of the element (Fig. 4.2). With (4.5) the element forces F_1 and F_2 are obtained from the equations of equilibrium at the ends of the element as

$$F_1 = -N = -\frac{E(A_2 - A_1)}{l \cdot \ln(A_2/A_1)} \cdot (u_2 - u_1)$$

$$F_2 = N = \frac{E(A_2 - A_1)}{l \cdot \ln(A_2/A_1)} \cdot (u_2 - u_1)$$

or in matrix notation

$$\frac{E(A_2 - A_1)}{l \cdot \ln(A_2/A_1)} \cdot \begin{bmatrix} 1 & -1 \\ -1 & 1 \end{bmatrix} \cdot \begin{bmatrix} u_1 \\ u_2 \end{bmatrix} = \begin{bmatrix} F_1 \\ F_2 \end{bmatrix} \tag{4.8}$$

$$\underline{K}_e \cdot \underline{u}_e = \underline{F}_e. \tag{4.8a}$$

The matrix $\underline{K}_e$ is the exact stiffness matrix of the tapered truss element with linearly varying cross section area. The stresses and displacements in the element can be calculated by (4.6) and (4.7), respectively.

4.3.3 FEM Approximate Solution with Linear Shape Functions

In general, the exact solution for structural mechanics problems, e.g. for plate structures, is not known. Therefore, the finite element solution is based on an assumption. It is assumed that the displacements between the nodal points and inside an element have a given distribution. With this assumption, the displacements at any point of the structure can be calculated when the nodal point displacements are known. The displacements are interpolated between the nodal points by so-called *shape functions*.

For the truss element with linearly varying cross section, a linear distribution of the displacements between the two nodal points can be assumed as

$$u = u_1 + \frac{x}{l} \cdot (u_2 - u_1). \tag{4.9}$$

It is apparent that this is an assumption since the exact displacements have been evaluated already in (4.7).

The strains corresponding to the linear distribution of the displacements in (4.9), are obtained by differentiating the displacements according to (2.2a) as

$$\varepsilon_x = \frac{du}{dx} = \frac{1}{l} \cdot (u_2 - u_1). \tag{4.10}$$

It can be seen that the strains based on the assumption of linear distribution of the displacements do not depend on x, i.e. they are constant in the element.

The stresses are obtained with Hooke's law (2.3a) as

$$\sigma_x = E \cdot \varepsilon_x = \frac{E}{l} \cdot (u_2 - u_1)$$

or

$$\sigma_x = \frac{E}{l} \cdot (-u_1 + u_2). \tag{4.11}$$

They are denoted as *element stresses*. Stresses and strains come out to be constant in each element, due to the assumption for the displacements. It is apparent that in this way the real stress distribution given in (4.6), cannot be represented. The equations of equilibrium $\sigma(x) \cdot A(x) = N$ at any section in the element and at the ends of the element cannot be fulfilled with a constant stress σ_x according to (4.11). Instead the stresses σ_x will be determined in such a way that the equilibrium conditions are fulfilled "in mean", i.e. the stress σ_x in (4.11), should be an averaged value of the real stress σ_x varying with x.

For this purpose, the principle of virtual displacements is applied. It can be used to fulfill the equations of equilibrium approximately, i.e. "in mean" or in a "weak" sense [17]. The stresses obtained by the principle of virtual displacements fulfill the equilibrium conditions exactly only if the assumption for the displacements contains the exact solution.

For the truss element, the principle of virtual displacements according to (2.5a) and (2.4a) is

$$\overline{W}_i = \overline{W}_a$$

with

$$\overline{W}_i = \int_0^l A \cdot \sigma_r \cdot \bar{\varepsilon}_r \, dx$$

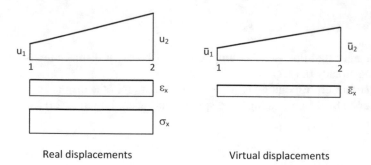

Real displacements   Virtual displacements

Fig. 4.4 Shape functions of the displacements and corresponding strains and stresses

This equation is valid for any virtual state of displacements which fulfills the boundary conditions, i.e. the conditions $\bar{u}_{(x=0)} = \bar{u}_1$ and $\bar{u}_{(x=l)} = \bar{u}_2$. From all the possible virtual states of displacements, now those are considered which have the same distribution as that assumed for the real displacements (Fig. 4.4). In the case of the truss element this is the linear assumption according to (4.9). It can be shown that with this assumption for the virtual displacements, a symmetric stiffness matrix is obtained. The virtual displacements then have the same distribution as the real displacements according to (4.9)

$$\bar{u} = \bar{u}_1 + \frac{x}{l} \cdot (\bar{u}_2 - \bar{u}_1)$$

(4.12)

where the virtual displacements $\bar{u}_1$ and $\bar{u}_2$ of the nodal points may assume any arbitrary value.

The strains in the virtual state of displacements are now obtained as

$$\bar{\varepsilon}_x = \frac{d\bar{u}}{dx} = \frac{1}{l} \cdot (\bar{u}_2 - \bar{u}_1).$$

(4.13)

The internal virtual work is therefore given by

$$\overline{W}_i = \int_0^l A_x \cdot \sigma_x \cdot \varepsilon_x \, dx$$

$$\overline{W}_i = \int_0^l \left(A_1 + \frac{x}{l} (A_2 - A_1) \right) \cdot \frac{E}{l} \cdot (u_2 - u_1) \cdot \frac{1}{l} \cdot (\bar{u}_2 - \bar{u}_1) \, dx.$$

(4.14)

The external virtual work is done by the real element forces F_1 and F_2 and the virtual displacements $\bar{u}_1$ and $\bar{u}_2$ at the ends of the element. It is given by

$$\overline{W}_a = F_1 \cdot \bar{u}_1 + F_2 \cdot \bar{u}_2.$$

(4.15)

Equating the internal and external virtual work, one obtains

$$\int\limits_0^l \left(A_1 + \frac{x}{l} \cdot (A_2 - A_1) \right) \cdot \frac{E}{l} \cdot (u_2 - u_1) \cdot \frac{1}{l} \cdot (\bar{u}_2 - \bar{u}_1) \, \mathrm{d}x = F_1 \cdot \bar{u}_1 + F_2 \cdot \bar{u}_2.$$

$$(4.16)$$

Since $\bar{u}_1$ and $\bar{u}_2$ are arbitrary virtual displacements, (4.16) is also valid for the displacement states

(a) $\bar{u}_1 = 1$; $\bar{u}_2 = 0$
(b) $\bar{u}_1 = 0$; $\bar{u}_2 = 1$.

By a weighted superposition of these two displacement states, any other displacement states can be obtained. Introducing the displacement state (a) in the work expression, one obtains

$$-\frac{E}{l^2} \cdot \int\limits_0^l \left(A_1 + \frac{x}{l} (A_2 - A_1) \right) \mathrm{d}x \cdot (u_2 - u_1) = F_1$$

$$-\frac{E}{l^2} \cdot \left[A_1 \cdot x + \frac{1}{2} \frac{x^2}{l} (A_2 - A_1) \right]_0^l \cdot (u_2 - u_1) = F_1$$

$$-\frac{E}{l^2} \cdot \left[A_1 \cdot l + \frac{1}{2} \frac{l^2}{l} (A_2 - A_1) \right] \cdot (u_2 - u_1) = F_1$$

or

$$\frac{E}{l} \cdot \frac{A_1 + A_2}{2} \cdot (u_1 - u_2) = F_1. \qquad (4.17a)$$

In the same way the virtual displacement state b) gives

$$\frac{E}{l} \cdot \frac{A_1 + A_2}{2} \cdot (-u_1 + u_2) = F_2. \qquad (4.17b)$$

The Eqs. (4.17a, 4.17b), are the stiffness relationship of the truss element. In matrix notation it is

$$\frac{E}{l} \cdot \frac{(A_1 + A_2)}{2} \cdot \begin{bmatrix} 1 & -1 \\ -1 & 1 \end{bmatrix} \cdot \begin{bmatrix} u_1 \\ u_2 \end{bmatrix} = \begin{bmatrix} F_1 \\ F_2 \end{bmatrix} \qquad (4.18)$$

$$\underline{K}_e \cdot \underline{u}_e = \underline{F}_e. \qquad (4.18a)$$

The matrix $\underline{K}_e$ is the stiffness matrix of the truss element with linearly varying cross section based on the assumption of a linear shape function. For a truss element with constant cross section, i.e. with $A_1 = A_2$, (4.18) corresponds to the exact

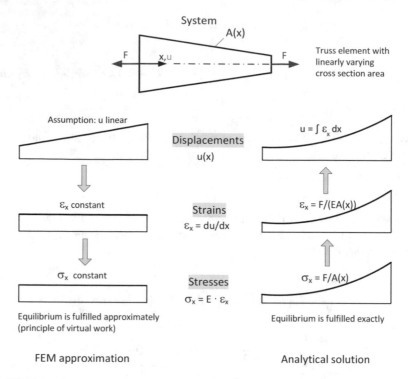

Fig. 4.5 Finite element solution with linear shape functions and exact solution

solution (3.3) since in this case, the displacement distribution (4.9), represents the exact solution.

The two derivations of the element stiffness matrix—with a linear shape function and as the exact—solution are compared in Fig. 4.5.

Example 4.2 For the truss element in Example 4.1, the solution with finite truss elements with linearly varying cross section has to be given. The bar is discretized into one or two elements, respectively.

When discretizing the bar into a single element, the stiffness relationship is obtained with (4.18), as

$$\frac{E}{l} \cdot \frac{(A_1 + A_2)}{2} \cdot \begin{bmatrix} 1 & -1 \\ -1 & 1 \end{bmatrix} \cdot \begin{bmatrix} u_1 \\ u_2 \end{bmatrix} = \begin{bmatrix} F_1^{(1)} \\ F \end{bmatrix}.$$

Considering the support condition $u_1 = 0$ and with $A_1 = 500\,\text{cm}^2$, $A_2 = 100\,\text{cm}^2$, $E = 1000\,\text{kN/cm}^2$, $l = 500\,\text{cm}$, and $F = 100\,\text{kN}$ one obtains (Fig. 4.6a):

$$\frac{E}{l} \cdot \frac{(A_1 + A_2)}{2} \cdot u_2 = F$$

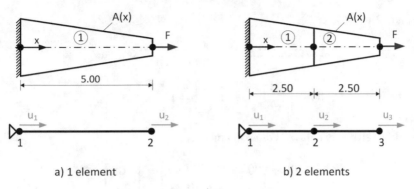

a) 1 element   b) 2 elements

Fig. 4.6   Discretization of a bar into one or two elements

$$\frac{1000}{500} \cdot \frac{500 + 100}{2} \cdot u_2 = 100$$

and therefore

$$u_2 = \frac{100}{600} = 0.167\,\text{cm}.$$

The element stress can be calculated with (4.11), as

$$\sigma_x^{(1)} = \frac{E}{l} \cdot (-u_1 + u_2) = \frac{1000}{500} \cdot (0 + 0.167) = 0.333\,\text{kN/cm}^2.$$

In order to discretize the bar into two finite elements, the cross section at $x = 250\,\text{cm}$ is determined with (4.1), and $A_1 = A(x = 0) = 500\,\text{cm}^2$, $A_2 = A(x = 500\,\text{cm}) = 100\,\text{cm}^2$ as

$$A = A_1 + \frac{x}{l} \cdot (A_2 - A_1) = 500 + \frac{250}{500} \cdot (100 - 500) = 300\,\text{cm}^2.$$

Therewith the element stiffness matrices according to (4.18), can be computed. The support condition $u_1 = 0$ is already considered when setting up the element stiffness matrix for element 1, i.e. the row and column for u_1 are eliminated.

Element 1:

$$\frac{E}{l} \cdot \frac{(A_1 + A_2)}{2} \cdot u_2 = F_2^{(1)}$$

$$\frac{1000}{250} \cdot \frac{500 + 300}{2} \cdot u_2 = F_2^{(1)}$$

$$1600 \cdot u_2 = F_2^{(1)}$$

Element 2:

$$\frac{E}{l} \cdot \frac{(A_1 + A_2)}{2} = \frac{1000}{250} \cdot \frac{300 + 100}{2} = 800$$

$$800 \cdot \begin{bmatrix} 1 & -1 \\ -1 & 1 \end{bmatrix} \cdot \begin{bmatrix} u_2 \\ u_3 \end{bmatrix} = \begin{bmatrix} F_2^{(2)} \\ F_3^{(2)} \end{bmatrix}$$

The global stiffness matrix is thus obtained as

$$\begin{bmatrix} 1600 + 800 & -800 \\ -800 & 800 \end{bmatrix} \cdot \begin{bmatrix} u_2 \\ u_3 \end{bmatrix} = \begin{bmatrix} 0 \\ 100 \end{bmatrix}.$$

The solution of the system of equations is

$$u_2 = 0.063 \,\text{cm} \quad u_3 = 0.188 \,\text{cm}.$$

The stresses in elements 1 and 2 are obtained with (4.11), as

$$\sigma_x^{(1)} = 0.250 \,\text{kN/cm}^2 \quad \sigma_x^{(2)} = 0.500 \,\text{kN/cm}^2.$$

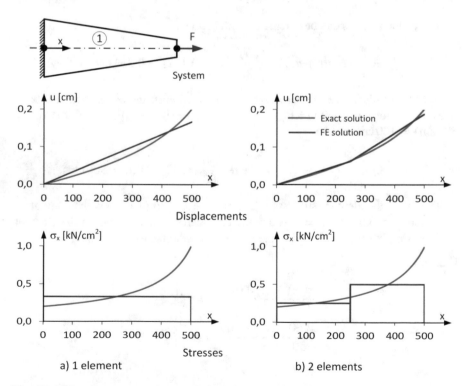

Fig. 4.7 Finite element approximate solution with one and two finite elements with linear shape functions

They are represented in Fig. 4.7, together with the exact solution (cf. Example 4.1). In the truss element the displacements are linear and the stress is constant. It can be seen that the displacements are approximated very well with only very few elements, whereas this is not the case for the stresses.

In addition, the so-called *nodal stresses*, i.e. the stresses at the nodal points, will be determined. The stress at a nodal point is obtained as average value of the element stresses of all elements connected to the nodal point. In nodal point 2, the stresses from element 1 and element 2 are to be averaged. Points 1 and 3 are each connected to only one element, i.e. the respective element stress is also the nodal point stress. One obtains

$$\sigma_{x,1} = 0.250 \, \text{kN/cm}^2,$$

$$\sigma_{x,2} = (0.500 + 0.250)/2 = 0.375 \, \text{kN/cm}^2,$$

$$\sigma_{x,3} = 0.500 \, \text{kN/cm}^2. \qquad \blacktriangleleft$$

4.3.4 FEM Approximate Solution with Quadratic Shape Functions

The quality of the solution can be improved with higher order shape functions. In the following, the stiffness matrix of a truss element with quadratic shape functions is derived. However, parabolas of a third or higher order may also be used as shape functions to derive higher order finite elements. For this element, the dimensionless coordinate r is introduced. It varies between -1 at the left and $+1$ at the right boundary of the element. In the middle of the element, r equals 0. The coordinate x in Fig. 4.8, can be transformed to r by

$$r = \frac{2 \cdot x}{l} - 1 \qquad (4.19a)$$

A quadratic parabola is defined by three sampling points. Therefore, in the middle of the element, a third nodal point is to be introduced. Hence, the element possesses three degrees of freedom, i.e. the displacements u_1, u_2 and u_3 (Fig. 4.8).

The cross section of the truss element varies linearly and is described by the areas A_1 and A_3 at the beginning and the end of the element by

$$A = \frac{1}{2}(1 - r)A_1 + \frac{1}{2}(1 + r)A_3. \qquad (4.19b)$$

The function for the displacements $u(r)$ is a parabola with the three sampling points u_1, u_2 and u_3:

Fig. 4.8 Finite element with
quadratic shape functions
(3-node element)

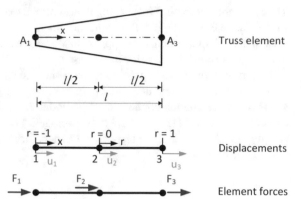

Truss element

Displacements

Element forces

$$u = \left[\frac{1}{2}(1-r) - \frac{1}{2}(1-r^2)\right] \cdot u_1 + (1-r^2) \cdot u_2 + \left[\frac{1}{2}(1+r) - \frac{1}{2}(1-r^2)\right] \cdot u_3$$

$$(4.20)$$

The correctness of the formula, which apparently is a quadratic parabola in r, can
easily be checked by introducing $r = -1$, $r = 0$ and $r = 1$ into the expression.

The strains are obtained by differentiation of the displacements with respect to x
as

$$\varepsilon_x = \frac{du}{dx} = \frac{du}{dr} \cdot \frac{dr}{dx}$$

and with (4.20)

$$\varepsilon_x = \frac{1}{l}(-1+2r) \cdot u_1 - \frac{4}{l}r \cdot u_2 + \frac{1}{l}(1+2r) \cdot u_3. \qquad (4.21)$$

It can be seen that the strains exhibit a linear distribution within the element.

The stresses are obtained by Hooke's law as

$$\sigma_x = E \cdot \varepsilon_x$$

or

$$\sigma_x = \frac{E}{l}(-1+2r) \cdot u_1 - \frac{4E}{l}r \cdot u_2 + \frac{E}{l}(1+2r) \cdot u_3. \qquad (4.22)$$

Like the strains, the stresses vary linearly in the element. At the nodal points, the
stresses have the values

$$\sigma_{x,1}^{(e)} = E \cdot (-3 \cdot u_1 + 4 \cdot u_2 - u_3)/l \qquad (4.22a)$$

$$\sigma_{x,2}^{(e)} = E \cdot (-u_1 + u_3)/l \tag{4.22b}$$

$$\sigma_{x,3}^{(e)} = E \cdot (u_1 - 4 \cdot u_2 + 3 \cdot u_3)/l. \tag{4.22c}$$

The virtual displacements are again described with the same shape functions as the real displacements. With (4.20), they are

$$\bar{u} = \left[\frac{1}{2}(1-r) - \frac{1}{2}(1-r^2)\right] \cdot \bar{u}_1 + (1-r^2) \cdot \bar{u}_2 + \left[\frac{1}{2}(1+r) - \frac{1}{2}(1-r^2)\right] \cdot \bar{u}_3. \tag{4.23}$$

The virtual strains thus become

$$\bar{\varepsilon}_x = \frac{1}{l}(-1+2r) \cdot \bar{u}_1 - \frac{4}{l}r \cdot \bar{u}_2 + \frac{1}{l}(1+2r) \cdot \bar{u}_3. \tag{4.24}$$

The internal virtual work is done by the real stresses (4.22), and the virtual strains, according to (4.24). The external virtual work is done by the forces F_1, F_2 and F_3 and the corresponding virtual displacements of the nodal points. Equating the internal and external virtual work, one obtains

$$\int_0^l \bar{\varepsilon}_x \cdot \sigma_x \cdot A \cdot dx = F_1 \cdot \bar{u}_1 + F_2 \cdot \bar{u}_2 + F_3 \cdot \bar{u}_3$$

or with (4.24) and (4.22)

$$\int_0^l \left[\frac{1}{l}(-1+2r) \cdot \bar{u}_1 - \frac{4}{l}r \cdot \bar{u}_2 + \frac{1}{l}(1+2r) \cdot \bar{u}_3\right]$$

$$\cdot \left[\frac{E}{l}(-1+2r) \cdot u_1 - \frac{4E}{l}r \cdot u_2 + \frac{E}{l}(1+2r) \cdot u_3\right]$$

$$\cdot (A_1 + \frac{x}{l}(A_3 - A_1)) \quad dx = F_1 \cdot \bar{u}_1 + F_2 \cdot \bar{u}_2 + F_3 \cdot \bar{u}_3$$

or

$$\frac{EA_1}{l}\left[\left[\left(\frac{7}{3} + \frac{\alpha}{2}\right) \cdot u_1 + \left(-\frac{8}{3} - \frac{2}{3}\alpha\right) \cdot u_2 + \left(\frac{1}{3} + \frac{\alpha}{6}\right) \cdot u_3\right] \cdot \bar{u}_1 \right.$$

$$+ \left[\left(-\frac{8}{3} - \frac{2}{3}\alpha\right) \cdot u_1 + \left(\frac{16}{3} + \frac{8}{3}\alpha\right) \cdot u_2 + \left(-\frac{8}{3} - 2\alpha\right) \cdot u_3\right] \cdot \bar{u}_2$$

$$+ \left[\left(\frac{1}{3} + \frac{\alpha}{6}\right) \cdot u_1 + \left(-\frac{8}{3} - 2\alpha\right) \cdot u_2 + \left(\frac{7}{3} + \frac{11}{6}\alpha\right) \cdot u_3\right] \cdot \bar{u}_3 \right]$$

$$= F_1 \cdot \bar{u}_1 + F_2 \cdot \bar{u}_2 + F_3 \cdot \bar{u}_3$$

with $\alpha = \dfrac{A_3 - A_1}{A_1}$.

As $\bar{u}_1$, $\bar{u}_2$, and $\bar{u}_3$ may assume any arbitrary value, three equations for the forces F_1 (with $\bar{u}_1 = 1$, $\bar{u}_2 = \bar{u}_3 = 0$), F_2 (with $\bar{u}_2 = 1$, $\bar{u}_1 = \bar{u}_3 = 0$) and F_3 (with $\bar{u}_3 = 1$, $\bar{u}_1 = \bar{u}_2 = 0$) are obtained. They can be written in matrix notation as

$$\frac{EA_1}{l} \cdot \begin{bmatrix} \dfrac{7}{3} + \dfrac{\alpha}{2} & -\dfrac{8}{3} - \dfrac{2}{3}\alpha & \dfrac{1}{3} + \dfrac{\alpha}{6} \\[2mm] -\dfrac{8}{3} - \dfrac{2}{3}\alpha & \dfrac{16}{3} + \dfrac{8}{3}\alpha & -\dfrac{8}{3} - 2\alpha \\[2mm] \dfrac{1}{3} + \dfrac{\alpha}{6} & -\dfrac{8}{3} - 2\alpha & \dfrac{7}{3} + \dfrac{11}{6}\alpha \end{bmatrix} \cdot \begin{bmatrix} u_1 \\[2mm] u_2 \\[2mm] u_3 \end{bmatrix} = \begin{bmatrix} F_1 \\[2mm] F_2 \\[2mm] F_3 \end{bmatrix} \tag{4.25}$$

$$\underline{K}_e \cdot \underline{u}_e = \underline{F}_e. \tag{4.25a}$$

This is the stiffness relationship of the element. The matrix $\underline{K}_e$ is the stiffness matrix of the truss element with linearely varying cross section and quadratic shape functions.

Example 4.3 The truss element with linearly varying cross section in Example 4.1, is investigated with one and two elements with quadratic shape functions.

When the bar is represented by a single element (Fig. 4.9a), the stiffness relationship (4.25), with the support condition $u_1 = 0$ is given by

$$\frac{EA_1}{l} \cdot \begin{bmatrix} \dfrac{16}{3} + \dfrac{8}{3}\alpha & -\dfrac{8}{3} - 2\alpha \\[2mm] -\dfrac{8}{3} - 2\alpha & \dfrac{7}{3} + \dfrac{11}{6}\alpha \end{bmatrix} \cdot \begin{bmatrix} u_2 \\[2mm] u_3 \end{bmatrix} = \begin{bmatrix} 0 \\[2mm] F \end{bmatrix}$$

With $A_1 = 500\,\text{cm}^2$, $A_3 = 100\,\text{cm}^2$, $E = 1000\,\text{kN/cm}^2$, $l = 500\,\text{cm}$, and $F = 100\,\text{kN}$ (Fig. 4.9) one obtains

$$1000 \cdot \begin{bmatrix} 3.200 & -1.067 \\ -1.067 & 0.867 \end{bmatrix} \cdot \begin{bmatrix} u_2 \\ u_3 \end{bmatrix} = \begin{bmatrix} 0 \\ 100 \end{bmatrix}.$$

The solution of the system of equations is

$$u_2 = 0.065\,\text{cm}, \quad u_3 = 0.196\,\text{cm}.$$

With (4.22a–4.22c) the stresses at the nodal points are

$$\sigma_{x,1}^{(1)} = E \cdot (-3 \cdot u_1 + 4 \cdot u_2 - u_3)/l$$
$$= 1000 \cdot (4 \cdot 0.065 - 0.196)/500 = 0.130\,\text{kN/cm}^2$$

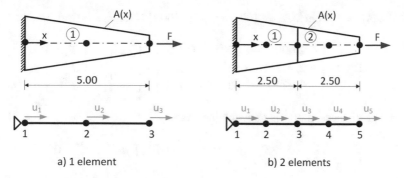

a) 1 element   b) 2 elements

Fig. 4.9 Discretization of a bar in elements with quadratic shape functions

$$\sigma_{x,2}^{(1)} = E \cdot (-u_1 + u_3)/l$$
$$= 1000 \cdot (0 + 0.196)/500 = 0.391\,\text{kN/cm}^2$$
$$\sigma_{x,3}^{(1)} = E \cdot (u_1 - 4 \cdot u_2 + 3 \cdot u_3)/l$$
$$= 1000 \cdot (-4 \cdot 0.065 + 3 \cdot 0.196)/500 = 0.650\,\text{kN/cm}^2.$$

When the bar is discretized into two finite elements, first the element stiffness matrices according to (4.25), have to be determined. The support condition $u_1 = 0$ will be already considered in the element stiffness matrix of element 1, i.e. the row and the column for u_1 are eliminated.

Element 1:

$$\frac{EA_1}{l} \cdot \begin{bmatrix} \dfrac{16}{3} + \dfrac{8}{3}\alpha & -\dfrac{8}{3} - 2\alpha \\[2mm] -\dfrac{8}{3} - 2\alpha & \dfrac{7}{3} + \dfrac{11}{6}\alpha \end{bmatrix} \cdot \begin{bmatrix} u_2 \\[2mm] u_3 \end{bmatrix} = \begin{bmatrix} F_2^{(1)} \\[2mm] F_3^{(1)} \end{bmatrix}$$

and with $A_1 = 500\,\text{cm}^2$, $A_3 = 300\,\text{cm}^2$, $E = 1000\,\text{kN/cm}^2$, $l = 250\,\text{cm}$:

$$2000 \cdot \begin{bmatrix} 4.267 & -1.867 \\ -1.867 & 1.600 \end{bmatrix} \cdot \begin{bmatrix} u_2 \\ u_3 \end{bmatrix} = \begin{bmatrix} F_2^{(1)} \\ F_3^{(1)} \end{bmatrix}$$

Element 2:

$$\frac{EA_1}{l} \cdot \begin{bmatrix} \dfrac{7}{3} + \dfrac{\alpha}{2} & -\dfrac{8}{3} - \dfrac{2}{3}\alpha & \dfrac{1}{3} + \dfrac{\alpha}{6} \\[2mm] -\dfrac{8}{3} - \dfrac{2}{3}\alpha & \dfrac{16}{3} + \dfrac{8}{3}\alpha & -\dfrac{8}{3} - 2\alpha \\[2mm] \dfrac{1}{3} + \dfrac{\alpha}{6} & -\dfrac{8}{3} - 2\alpha & \dfrac{7}{3} + \dfrac{11}{6}\alpha \end{bmatrix} \cdot \begin{bmatrix} u_3 \\[2mm] u_4 \\[2mm] u_5 \end{bmatrix} = \begin{bmatrix} F_3^{(2)} \\[2mm] F_4^{(2)} \\[2mm] F_5^{(2)} \end{bmatrix}$$

and with $A_1 = 300\,\text{cm}^2$, $A_3 = 100\,\text{cm}^2$, $E = 1000\,\text{kN/cm}^2$, $l = 250\,\text{cm}$:

$$1200 \cdot \begin{bmatrix} 2.000 & -2.222 & 0.222 \\ -2.222 & 3.556 & -1.333 \\ 0.222 & -1.333 & 1.111 \end{bmatrix} \cdot \begin{bmatrix} u_3 \\ u_4 \\ u_5 \end{bmatrix} = \begin{bmatrix} F_3^{(2)} \\ F_4^{(2)} \\ F_5^{(2)} \end{bmatrix}$$

The global stiffness matrix is now obtained as:

$$\begin{bmatrix} 8533 & -3733 & 0 & 0 \\ -3733 & 5600 & -2667 & 267 \\ 0 & -2667 & 4267 & -1600 \\ 0 & 267 & -1600 & 1333 \end{bmatrix} \cdot \begin{bmatrix} u_2 \\ u_3 \\ u_4 \\ u_5 \end{bmatrix} = \begin{bmatrix} 0 \\ 0 \\ 0 \\ 100 \end{bmatrix}$$

The solution of the system of equations is

$$u_2 = 0.028\,\text{cm} \quad u_3 = 0.064\,\text{cm}$$
$$u_4 = 0.115\,\text{cm} \quad u_5 = 0.200\,\text{cm}.$$

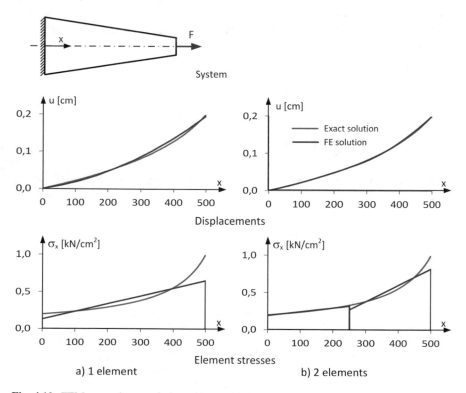

Fig. 4.10 FEM approximate solution with 1 and 2 finite elements with quadratic shape function

With Eqs. (4.22a–4.22c), the stresses in element 1 are

$$\sigma_{x,1}^{(1)} = 0.191 \, \text{kN/cm}^2 \quad \sigma_{x,2}^{(1)} = 0.255 \, \text{kN/cm}^2 \quad \sigma_{x,3}^{(1)} = 0.319 \, \text{kN/cm}^2$$

and in element 2

$$\sigma_{x,1}^{(2)} = 0.273 \, \text{kN/cm}^2 \quad \sigma_{x,2}^{(2)} = 0.545 \, \text{kN/cm}^2 \quad \sigma_{x,3}^{(2)} = 0.818 \, \text{kN/cm}^2.$$

The results are shown, together with the exact solution (according to Example 4.1), in Fig. 4.10. The displacements in the element are quadratic and the stresses are linear. The approximation of the displacements and stresses is significantly improved, compared to the element with a linear shape function in Example 4.2.

In addition, the nodal stresses are determined for the discretization in 2 elements as

$$\sigma_{x,1} = 0.191 \, \text{kN/cm}^2,$$
$$\sigma_{x,2} = (0.319 + 0.273)/2 = 0.296 \, \text{kN/cm}^2,$$
$$\sigma_{x,3} = 0.818 \, \text{kN/cm}^2.$$
◀

Example 4.4 The system given in Example 4.1, is examined for various finite element discretizations and the results are compared with the exact analytical solution. The absolute errors of the nodal displacements and element stresses are evaluated. A discretization of the bar into 4, 8, 16, and 32 elements with a linear shape function and into 4 and 8 elements with a quadratic shape function is investigated, i.e. the element size is halved in each mesh refinement step.

The displacements at four positions in the bar are given in Table 4.2, and shown in Fig. 4.11, for a subdivision into 4 elements. As the number of elements increases, they quickly converge toward the exact solution. The displacements obtained with

Table 4.2 Nodal displacements [cm]

Shape function	Number of elements	x[cm]				
		0	125	250	375	500
Linear	1	0	-	-	-	0.167
	2	0	-	0.063	-	0.188
	4	0	0.028	0.063	0.113	0.197
	8	0	0.028	0.063	0.114	0.200
	16	0	0.028	0.064	0.115	0.201
Quadratic	1	0	-	0.065	-	0.196
	2	0	0.028	0.064	0.115	0.200
	4	0	0.028	0.064	0.115	0.201
Exact	-	0	0.028	0.064	0.115	0.201

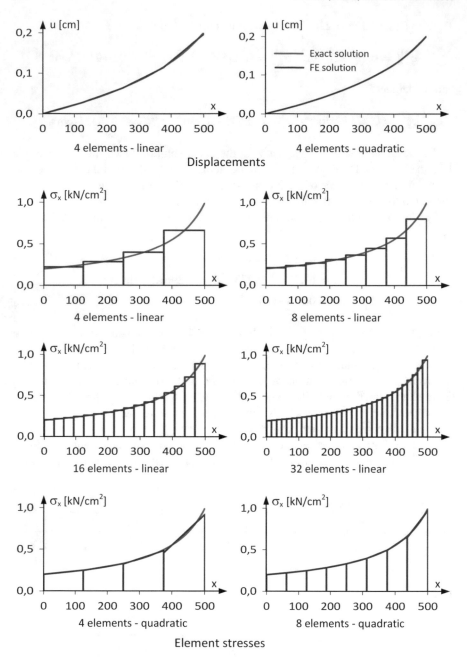

Fig. 4.11 Displacements and element stresses

Table 4.3 Error in the nodal displacements [cm]

Shape function	Number of elements	x[cm]				
		0	125	250	375	500
Linear	1	0	-	-	-	0.034
	2	0	-	0.001	-	0.013
	4	0	0.000	0.001	0.002	0.004
	8	0	0.000	0.001	0.001	0.001
	16	0	0.000	0.000	0.000	0.000
Quadratic	1	0	-	0.001	-	0.005
	2	0	0.000	0.000	0.000	0.001
	4	0	0.000	0.000	0.000	0.000

the finite element method are slightly lower than the exact displacements. The absolute error of the finite element solution, i.e. the absolute value of the difference of the exact values and the approximate finite element values is given in Table 4.3. The maximum displacement of 0.201 cm is calculated with 2 elements with quadratic shape function or 4 elements with linear shape function with a numerical error of less than 2%. For practical purposes, this means a high accuracy. With 4 elements with a quadratic shape function or 16 elements with a linear shape function, the displacements agree with the exact solution up to three decimals.

Figure 4.11 also shows the element stresses. They can be easily determined with (4.11) and (4.22a–4.22c), respectively, from the nodal displacements. For example, for a discretization in 4 elements with linear shape functions the stresses of 0.222, 0.286, 0.400, 0.667 kN/m^2 are obtained. The element stresses also converge against the exact solution with increasing mesh refinement. They are more accurate in the center of the element than at the element boundaries. Stress jumps occur at the nodal points. These become smaller with increasing mesh refinement and are, therefore, a measure of the accuracy of the analysis.

Nodal stresses are obtained by averaging the element stresses at the nodal points. They are given in Table 4.4, and shown for various discretizations in Fig. 4.12. They approximate the exact analytical solution particularly well in the middle region of the bar. The errors of the nodal stresses are listed in Table 4.5. The maximum error in the range 125 cm $\leq x \leq$ 375 cm is 0.008/0.500 = 1.6% with 8 elements with a linear shape function and 0.006/0.500 = 1.2% with 4 elements with a quadratic shape function. As expected, the discretization errors are higher at the right half of the bar with a high stress gradient than at the left half, where the nodal stresses converge more quickly. The accuracy of the nodal stresses at the boundaries of the bar is significantly lower than in the middle of the bar. This applies, in particular, to the right edge (x = 500 cm), where the stress gradient is high. In the case of 8 elements with a linear shape function, the error there is 0.200/1.000 = 20%, and in the case of 4 elements with a quadratic shape function, it is 0.077/1.000 = 7.7%. At the edge, the error of 20% is considerably higher than in the middle region, where it is only 1.6%. This is

Table 4.4 Nodal stresses [kN/cm^2]

Shape function	Number of elements	x[cm]				
		0	125	250	375	500
Linear	1	0.333	-	0.333	-	0.333
	2	0.250	0.250	0.375	0.500	0.500
	4	0.222	0.254	0.343	0.533	0.667
	8	0.211	0.251	0.336	0.508	0.800
	16	0.205	0.250	0.334	0.502	0.889
	32	0.203	0.250	0.333	0.500	0.941
Quadratic	1	0.130	-	0.391	-	0.652
	2	0.191	0.255	0.296	0.545	0.818
	4	0.198	0.248	0.327	0.474	0.923
	8	0.200	0.249	0.332	0.495	0.973
	16	0.200	0.250	0.333	0.499	0.992
Exact	-	0.200	0.250	0.333	0.500	1.000

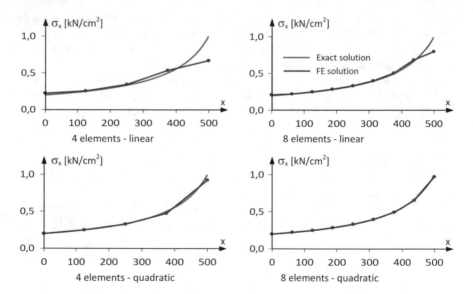

Fig. 4.12 Nodal stresses

due to the fact that it is not possible to average the stresses at the boundary node in the same way as at the nodal points in the center of the bar. The element with a quadratic shape function is able to better represent the stress diagram at the element boundaries and thus shows a significantly better convergence behavior. With 4 elements, the error of 7.7% at the right edge is relatively high compared to 1.2% in the middle region.

Table 4.5 Error in the nodal stresses [kN/cm^2]

Shape function	Number of elements	x[cm]				
		0	125	250	375	500
Linear	1	0.133	-	0	-	0.667
	2	0.050	0	0.042	0	0.500
	4	0.022	0.040	0.010	0.033	0.333
	8	0.011	0.001	0.003	0.008	0.200
	16	0.005	0.000	0.001	0.002	0.111
	32	0.003	0.000	0.000	0.000	0.059
Quadratic	1	0.070	-	0.058	-	0.348
	2	0.009	0.005	0.007	0.045	0.182
	4	0.002	0.002	0.006	0.006	0.077
	8	0.000	0.001	0.001	0.005	0.027
	16	0.000	0.000	0.000	0.001	0.008

However, with the same number of degrees of system freedom (8), it is considerably lower than the error with the element with a linear shape function (20%).

For a given discretization, e.g. in 8 elements with linear shape functions, the size of the stress jumps depends on the stress gradient, i.e. the slope of the stress diagram. At the right edge, where the stresses rise rapidly the stress gradient is

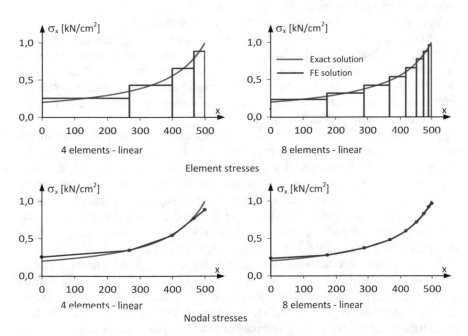

Fig. 4.13 Adaption of the element size to the stress gradient

high and the stress jumps are greater than at the left edge, where the stress gradient is low. This influence can be countered by an element size adapted to the stress gradient. With a uniform subdivision of the bar into 4 or 8 elements, these are each 125 cm or 67.5 cm, respectively, in size. On the other hand, Fig. 4.13, shows the element stresses of a discretization into 4 elements and 8 elements adapted approximately to the stress gradient. The element sizes are 267, 133, 67, and 33 [cm] for the discretization into 4 elements and 173, 116, 77, 51, 35, 23, 15, 10 [cm] for a discretization into 8 elements. The stresses at $x = 500$ cm are 0.883 kN/cm^2 for 4 elements and 0.962 kN/cm^2 for 8 elements with linear shape function. Hence, they possess a higher accuracy than the stresses obtained with a uniform discretization of the bar presuming the same number of elements.   ◀

4.3.5   Properties of the Finite Element Approximate Solution

Some important properties of the finite element method become apparent with the simple example of a truss element with linearly varying cross section. The approximation error of a finite element solution decreases when the number of elements increases or the element size decreases. Elements with higher order shape functions have a higher accuracy than elements with simple shape functions. Hence, the accuracy of the finite element solution can be increased by augmenting the number of elements or by applying elements with higher order shape functions.

The displacements are significantly more accurate than the stresses. The stresses are calculated in (2.3a), by multiplying the strains with the Young's modulus, whereas the strains are obtained in (4.10), by differentiation of the shape functions. In general, the error in an erroneous function is increased when the function is differentiated. With the integration, however, the error decreases. Since stresses in finite elements with shape functions for the displacements are obtained by differentiation, they are always less accurate than the displacements.

Stresses in the middle of an element are more accurate than at the element boundaries. Between two adjacent elements, jumps of the stresses occur due to the approximation of the displacements. The smaller the jump of the stresses is, the more accurate the result of a finite element calculation. It is, therefore, a measure for the accuracy of a finite element analysis. In a region with high stress gradient (e.g. at the right end of the bar in the example) the magnitude of the stress jump increases and the accuracy of the results decreases. In order to achieve a uniform accuracy, finite elements should be densified in regions of high stress gradient.

Nodal stresses are calculated by averaging the element stresses of the elements connected to the respective node. They are more accurate than the respective element stresses at the node. However, this does not apply to the boundary of the system. There, no averaging is possible and the nodal stresses can have comparatively high discretization errors, especially in combination with high stress gradients. It should also be noted that the information about the quality of the result, which is expressed in the size of the jumps of the element stresses at a node, is lost through the formation of nodal stresses.

It can be shown that for elements with displacement shape functions, the displacements obtained are on average too small. The system behaves too stiffly due to the displacement diagram prescribed by the shape functions in each element. In order to avoid artificial stiffness jumps which may cause erroneous stress redistributions in statically indeterminate systems, adjacent finite elements should not have size differences that are too large.

Some important results are summarized:

Properties of a finite element approximate solution

(a) The finite element solution approximates the exact solution. Its accuracy is increased by an augmentation of the number of elements or a reduction of the element size.

(b) For elements based on displacement shape functions only, the approximated nodal point displacements are on average too small, i.e. the system behaves too stiffly.

(c) Elements with higher order shape functions exhibit greater accuracy than elements with low-order shape functions.

(d) The finite element approximation is better in regions with low stress gradient compared to regions with high stress gradient if the element size is uniform.

(e) The element stresses in the middle of the element have higher accuracy than those at the element boundaries.

(f) The jump of the element stresses between two adjacent elements is a measure for the accuracy of the analysis at this point.

(g) Nodal stresses have higher accuracy than element stresses. However, this does not apply to the boundaries of the system. At the boundary, the nodal stresses are significantly less accurate than in the other areas.

These properties of the finite element method are valid in general for all plate and solid structures and should be considered in finite element modeling.

4.4 Rectangular Elements for Plates in Plane Stress

4.4.1 Shape Functions

The derivation of the stiffness matrix of a plane stress element is shown for a simple rectangular element. Its derivation is similar to the truss element with linearly varying cross section in the previous section. However, the equations are more complicated,

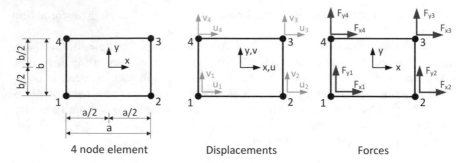

Fig. 4.14 Rectangular plate element for plane stress

since two displacement components and three strain and stress components have to be considered, instead of only one displacement and one strain and stress component.

At the beginning, the shape functions have to be chosen. The element has two degrees of freedom for the displacements at each nodal point (Fig. 4.14). Linear interpolation of the displacements $u(x, y)$ and $v(x, y)$ between the nodal points gives a bilinear function for the displacements

$$
\begin{aligned}
u &= \alpha_1 + \alpha_2 \cdot x + \alpha_3 \cdot y + \alpha_4 \cdot x \cdot y \\
v &= \beta_1 + \beta_2 \cdot x + \beta_3 \cdot y + \beta_4 \cdot x \cdot y
\end{aligned}
\tag{4.26}
$$

or

$$
\begin{bmatrix} u \\ v \end{bmatrix} =
\begin{bmatrix}
1 & x & y & xy & 0 & 0 & 0 & 0 \\
0 & 0 & 0 & 0 & 1 & x & y & xy
\end{bmatrix} \cdot
\begin{bmatrix}
\alpha_1 \\ \alpha_2 \\ \alpha_3 \\ \alpha_4 \\ \beta_1 \\ \beta_2 \\ \beta_3 \\ \beta_4
\end{bmatrix}
\tag{4.26a}
$$

$$
\underline{u} = \underline{N}_{\mathrm{a}} \cdot \underline{a}.
\tag{4.26b}
$$

If these functions are plotted over the element area, a curved surface is obtained which, however, runs linearly in sections parallel to the x- or y-axis (Figs. 4.15, 4.16).

The values $\alpha_1 - \alpha_4$ and $\beta_1 - \beta_4$ are free parameters which can be expressed by the nodal displacements u_1, v_1 to u_4, v_4. For this purpose the nodal point coordinates are introduced (4.26). One obtains the displacements

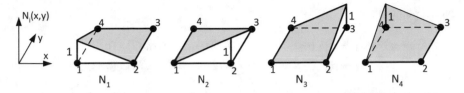

Fig. 4.15 Shape functions N_1–N_4

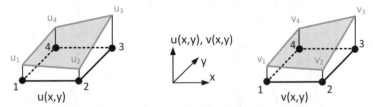

Element displacements based on shape functions

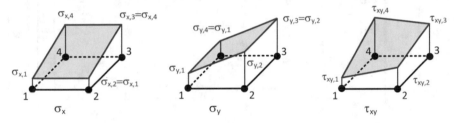

Stresses derived from element displacements ($\mu=0$)

Fig. 4.16 Displacements and element stresses of the rectangular finite element

Nodal point 1:

$$u_1 = \alpha_1 + \alpha_2 \cdot (-a/2) + \alpha_3 \cdot (-b/2) + \alpha_4 \cdot (-a/2) \cdot (-b/2)$$
$$v_1 = \beta_1 + \beta_2 \cdot (-a/2) + \beta_3 \cdot (-b/2) + \beta_4 \cdot (-a/2) \cdot (-b/2)$$

Nodal point 2:

$$u_2 = \alpha_1 + \alpha_2 \cdot (a/2) + \alpha_3 \cdot (-b/2) + \alpha_4 \cdot (a/2) \cdot (-b/2)$$
$$v_2 = \beta_1 + \beta_2 \cdot (a/2) + \beta_3 \cdot (-b/2) + \beta_4 \cdot (a/2) \cdot (-b/2)$$

Nodal point 3:

$$u_3 = \alpha_1 + \alpha_2 \cdot (a/2) + \alpha_3 \cdot (b/2) + \alpha_4 \cdot (a/2) \cdot (b/2)$$
$$v_3 = \beta_1 + \beta_2 \cdot (a/2) + \beta_3 \cdot (b/2) + \beta_4 \cdot (a/2) \; (b/2)$$

Nodal point 4:

$$u_4 = \alpha_1 + \alpha_2 \cdot (-a/2) + \alpha_3 \cdot (b/2) + \alpha_4 \cdot (-a/2) \cdot (b/2)$$
$$v_4 = \beta_1 + \beta_2 \cdot (-a/2) + \beta_3 \cdot (b/2) + \beta_4 \cdot (-a/2) \cdot (b/2).$$

These equations can be solved for the 8 parameters $\alpha_1 - \alpha_4$ and $\beta_1 - \beta_4$. One obtains

$$
\begin{bmatrix} \alpha_1 \\ \alpha_2 \\ \alpha_3 \\ \alpha_4 \\ \beta_1 \\ \beta_2 \\ \beta_3 \\ \beta_4 \end{bmatrix}
= \frac{1}{2}
\begin{bmatrix}
\frac{1}{2} & 0 & \frac{1}{2} & 0 & \frac{1}{2} & 0 & \frac{1}{2} & 0 \\
-\frac{1}{a} & 0 & \frac{1}{a} & 0 & \frac{1}{a} & 0 & -\frac{1}{a} & 0 \\
-\frac{1}{b} & 0 & -\frac{1}{b} & 0 & \frac{1}{b} & 0 & \frac{1}{b} & 0 \\
\frac{2}{a \cdot b} & 0 & -\frac{2}{a \cdot b} & 0 & \frac{2}{a \cdot b} & 0 & -\frac{2}{a \cdot b} & 0 \\
0 & \frac{1}{2} & 0 & \frac{1}{2} & 0 & \frac{1}{2} & 0 & \frac{1}{2} \\
0 & -\frac{1}{a} & 0 & \frac{1}{a} & 0 & \frac{1}{a} & 0 & -\frac{1}{a} \\
0 & -\frac{1}{b} & 0 & -\frac{1}{b} & 0 & \frac{1}{b} & 0 & \frac{1}{b} \\
0 & \frac{2}{a \cdot b} & 0 & -\frac{2}{a \cdot b} & 0 & \frac{2}{a \cdot b} & 0 & -\frac{2}{a \cdot b}
\end{bmatrix}
\begin{bmatrix} u_1 \\ v_1 \\ u_2 \\ v_2 \\ u_3 \\ v_3 \\ u_4 \\ v_4 \end{bmatrix}.
$$

$$\underline{a} = \underline{A} \cdot \underline{u}_e.$$

The functions $u(x, y)$ and $v(x, y)$ of the displacements can thus be expressed by the nodal point displacements. Substituting the equations for $\alpha_1 - \alpha_4$ and $\beta_1 - \beta_4$ in (4.26a), gives

$$\underline{u} = \underline{N}_a \cdot \underline{A} \cdot \underline{u}_e$$

and after performing the matrix multiplication

$$
\begin{bmatrix} u \\ v \end{bmatrix} =
\begin{bmatrix}
N_1 & 0 & N_2 & 0 & N_3 & 0 & N_4 & 0 \\
0 & N_1 & 0 & N_2 & 0 & N_3 & 0 & N_4
\end{bmatrix}
\begin{bmatrix} u_1 \\ v_1 \\ u_2 \\ v_2 \\ u_3 \\ v_3 \\ u_4 \\ v_4 \end{bmatrix}
\tag{4.27}
$$

$$\underline{u} = \underline{N} \cdot \underline{u}_e \tag{4.27a}$$

with

$$N_1 = \frac{1}{4} - \frac{1}{2a}x - \frac{1}{2b}y + \frac{1}{ab}xy \qquad (4.27b)$$

$$N_2 = \frac{1}{4} + \frac{1}{2a}x - \frac{1}{2b}y - \frac{1}{ab}xy \qquad (4.27c)$$

$$N_3 = \frac{1}{4} + \frac{1}{2a}x + \frac{1}{2b}y + \frac{1}{ab}xy \qquad (4.27d)$$

$$N_4 = \frac{1}{4} - \frac{1}{2a}x + \frac{1}{2b}y - \frac{1}{ab}xy. \qquad (4.27e)$$

The functions $N_1 - N_4$ are denoted as *shape functions*. They possess the value 1 at one nodal point and the value 0 at all other nodal points (Fig. 4.15).

4.4.2 Strains and Stresses

With shape functions of the displacements $u(x, y)$ and $v(x, y)$, the strains ε_x, ε_y and the shear angle γ_{xy} can be determined. They are obtained according to (2.2b), by differentiation of the shape functions (4.27), as

$$
\begin{bmatrix} \varepsilon_x \\ \varepsilon_y \\ \gamma_{xy} \end{bmatrix} = \begin{bmatrix} \dfrac{\partial u}{\partial x} \\ \dfrac{\partial v}{\partial y} \\ \dfrac{\partial u}{\partial y} + \dfrac{\partial v}{\partial x} \end{bmatrix}
$$

$$
= \begin{bmatrix} \dfrac{\partial N_1}{\partial x} & 0 & \dfrac{\partial N_2}{\partial x} & 0 & \dfrac{\partial N_3}{\partial x} & 0 & \dfrac{\partial N_4}{\partial x} & 0 \\ 0 & \dfrac{\partial N_1}{\partial y} & 0 & \dfrac{\partial N_2}{\partial y} & 0 & \dfrac{\partial N_3}{\partial y} & 0 & \dfrac{\partial N_4}{\partial y} \\ \dfrac{\partial N_1}{\partial y} & \dfrac{\partial N_1}{\partial x} & \dfrac{\partial N_2}{\partial y} & \dfrac{\partial N_2}{\partial x} & \dfrac{\partial N_3}{\partial y} & \dfrac{\partial N_3}{\partial x} & \dfrac{\partial N_4}{\partial y} & \dfrac{\partial N_4}{\partial x} \end{bmatrix} \cdot \begin{bmatrix} u_1 \\ v_1 \\ u_2 \\ v_2 \\ u_3 \\ v_3 \\ u_4 \\ v_4 \end{bmatrix}
$$

$$(4.28)$$

with

$$\frac{\partial N_1}{\partial x} = -\frac{1}{2a} + \frac{1}{ab}y, \qquad \frac{\partial N_1}{\partial y} = -\frac{1}{2b} + \frac{1}{ab}x$$

$$\frac{\partial N_2}{\partial x} = \frac{1}{2a} - \frac{1}{ab}y, \qquad \frac{\partial N_2}{\partial y} = -\frac{1}{2b} - \frac{1}{ab}x$$

$$\frac{\partial N_3}{\partial x} = \frac{1}{2a} + \frac{1}{ab}y, \qquad \frac{\partial N_3}{\partial y} = \frac{1}{2b} + \frac{1}{ab}x$$

$$\frac{\partial N_4}{\partial x} = -\frac{1}{2a} - \frac{1}{ab}y, \qquad \frac{\partial N_4}{\partial y} = \frac{1}{2b} - \frac{1}{ab}x \qquad\qquad \text{(4.28a-h)}$$

and therefore

$$
\begin{bmatrix} \varepsilon_x \\ \varepsilon_y \\ \gamma_{xy} \end{bmatrix} = \frac{1}{2ab}
\begin{bmatrix}
2y-b & 0 & -2y+b & 0 \\
0 & 2x-a & 0 & -2x-a \\
2x-a & 2y-b & -2x-a & -2y+b
\end{bmatrix} \dots\dots
$$

$$
\dots
\begin{bmatrix}
2y+b & 0 & -2y-b & 0 \\
0 & 2x+a & 0 & -2x+a \\
2x+a & 2y+b & -2x+a & -2y-b
\end{bmatrix} \cdot
\begin{bmatrix} u_1 \\ v_1 \\ u_2 \\ v_2 \\ u_3 \\ v_3 \\ u_4 \\ v_4 \end{bmatrix} \qquad \text{(4.29)}
$$

$$\underline{\varepsilon} = \underline{B} \cdot \underline{u}_e. \qquad\qquad\qquad \text{(4.29a)}$$

According to (4.29), the strain ε_x is constant in the x-direction and linearly varying in the y-direction within an element. The strain ε_y is constant in the y-direction and linearly varying in the x-direction. The shear deformation γ_{yx} is linearly varying in the x-direction, as well as in the y-direction.

The element stresses are obtained from the strains with Hooke's law (2.3b), as

$$
\begin{bmatrix} \sigma_x \\ \sigma_y \\ \tau_{xy} \end{bmatrix} = \frac{E}{1-\mu^2}
\begin{bmatrix}
1 & \mu & 0 \\
\mu & 1 & 0 \\
0 & 0 & \frac{1-\mu}{2}
\end{bmatrix} \cdot
\begin{bmatrix} \varepsilon_x \\ \varepsilon_y \\ \gamma_{xy} \end{bmatrix}
$$

or

$$\underline{\sigma} = \underline{D} \cdot \underline{\varepsilon}$$

and with the strains according to (4.29a), as

$$\underline{\sigma} = \underline{D} \cdot \underline{B} \cdot \underline{u}_e. \tag{4.30}$$

They are also denoted as *membrane stresses*. Since the stresses are obtained by multiplication of the strains with the matrix $\underline{D}$ they have, for Poisson ratio $\mu = 0$, the same distribution in the element as the strains (Fig. 4.16). In scalar notation, (4.30) may be written as

$$
\begin{aligned}
\sigma_x &= \frac{E}{(1 - \mu^2) \cdot 2 \cdot a \cdot b} \cdot [(2 \cdot y - b) \cdot (u_1 - u_2) + (2 \cdot y + b) \cdot (u_3 - u_4) \\
&\quad + \mu \cdot ((2 \cdot x - a) \cdot (v_1 - v_4) + (2 \cdot x + a) \cdot (-v_2 + v_3))] \\
\sigma_y &= \frac{E}{(1 - \mu^2) \cdot 2 \cdot a \cdot b} \cdot [\mu \cdot ((2 \cdot y - b) \cdot (u_1 - u_2) + (2 \cdot y + b) \cdot (u_3 - u_4)) \\
&\quad + (2 \cdot x - a) \cdot (v_1 - v_4) + (2 \cdot x + a) \cdot (-v_2 + v_3)] \\
\tau_{xy} &= \frac{E}{4 \cdot (1 + \mu) \cdot a \cdot b} \cdot [(2 \cdot y - b) \cdot (v_1 - v_2) + (2 \cdot y + b) \cdot (v_3 - v_4) \\
&\quad + (2 \cdot x - a) \cdot (u_1 - u_4) + (2 \cdot x + a) \cdot (u_3 - u_2)].
\end{aligned} \tag{4.30a}
$$

4.4.3   Stiffness Matrix

Based on the stresses (4.30), which were obtained with the shape functions for the displacements, the corresponding nodal forces are determined by the principle of virtual displacements. The distribution of the virtual displacements in the element is chosen similarly to the real displacements with its shape functions, i.e. analogously to (4.27a), as

$$\underline{\bar{u}} = \underline{N} \cdot \underline{\bar{u}}_e \tag{4.31}$$

where $\underline{\bar{u}}$ are the virtual displacements in the element and $\underline{\bar{u}}_e$ are the virtual nodal point displacements. The strains $\underline{\bar{\varepsilon}}$ corresponding to the virtual displacements are obtained with (4.29a) as

$$\underline{\bar{\varepsilon}} = \underline{B} \cdot \underline{\bar{u}}_e \tag{4.32}$$

or

$$\underline{\bar{\varepsilon}}^{\mathrm{T}} = \underline{\bar{u}}_e^{\mathrm{T}} \cdot \underline{B}^{\mathrm{T}}. \tag{4.32a}$$

The principle of virtual displacements states that the internal work of the real internal forces done with the corresponding virtual displacements equals the external work of the real external forces done with the corresponding virtual nodal displacements.

The virtual internal work is obtained according to (2.5b), as

$$\overline{W}_i = t \cdot \int \underline{\bar{\varepsilon}}^T \cdot \underline{\sigma} \, dx \, dy.$$

The integration is performed over the area of the element. Now the virtual strains $\underline{\bar{\varepsilon}}$ are expressed with (4.32a), by the virtual displacements of the nodal points. For the real stresses $\underline{\sigma}$, (4.30) is introduced, which contains the real nodal point displacements as unknowns. Therefore, the internal virtual work is

$$\overline{W}_i = t \cdot \int \underline{\bar{u}}_e^T \cdot \underline{B}^T \cdot \underline{D} \cdot \underline{B} \cdot \underline{u}_e \, dx \, dy$$

or, as the real nodal displacements $\underline{u}_e$, as well as the virtual displacements $\underline{\bar{u}}_e$, are independent of x and y

$$\overline{W}_i = \underline{\bar{u}}_e^T \cdot \int t \cdot \underline{B}^T \cdot \underline{D} \cdot \underline{B} \, dx \, dy \cdot \underline{u}_e. \qquad (4.33)$$

The external virtual work is the work done by the real forces with the virtual displacements. The nodal forces F_{x1}, F_{y1}, F_{x2} to F_{y4} perform external virtual work with the corresponding virtual nodal displacements. In addition, the virtual external work of surface loads caused, e.g. by mass forces or line loads at the element boundaries, have also to be considered. With (2.4a) and Fig. 2.6, the virtual work of the nodal forces are obtained as

$$\overline{W}_a^F = \begin{bmatrix} \bar{u}_1 & \bar{v}_1 & \bar{u}_2 & \bar{v}_2 & \bar{u}_3 & \bar{v}_3 & \bar{u}_4 & \bar{v}_4 \end{bmatrix} \cdot \begin{bmatrix} F_{x1} \\ F_{y1} \\ F_{x2} \\ F_{y2} \\ F_{x3} \\ F_{y3} \\ F_{x4} \\ F_{y4} \end{bmatrix} \qquad (4.34)$$

$$\overline{W}_a^F = \underline{\bar{u}}_e^T \cdot \underline{F}_e. \qquad (4.34a)$$

The forces p_x and p_y caused by the the surface loads $p_x \cdot dx \cdot dy$ at the infinitesimal element $p_x \cdot dx \cdot dy$ perform virtual external work with the virtual displacements $\bar{u}$ and $\bar{v}$ according to (4.31). One obtains

$$\overline{W}_a^P = \int (\bar{u} \cdot p_x + \bar{v} \cdot p_y) \, dx \, dy$$

$$= \int [\bar{u} \, \bar{v}] \cdot \begin{bmatrix} p_x \\ p_y \end{bmatrix} \, dx \, dy$$

or

$$\overline{W}_{\mathrm{a}}^{\mathrm{p}} = \int \underline{\bar{u}}^{\mathrm{T}} \cdot \underline{p} \; dx \; dy$$

and with $\bar{u}$ and $\bar{v}$ according to (4.31)

$$\overline{W}_{\mathrm{a}}^{\mathrm{p}} = \underline{\bar{u}}_{\mathrm{e}}^{\mathrm{T}} \cdot \int \underline{N}^{\mathrm{T}} \cdot \underline{p} \; dx \; dy.$$

The contribution of line loads at the element boundaries will be considered later. The total external virtual work is, therefore

$$\overline{W}_{\mathrm{a}} = \overline{W}_{\mathrm{a}}^{\mathrm{F}} + \overline{W}_{\mathrm{a}}^{\mathrm{p}} = \underline{\bar{u}}_{\mathrm{e}}^{\mathrm{T}} \cdot \underline{F}_{\mathrm{e}} + \underline{\bar{u}}_{\mathrm{e}}^{\mathrm{T}} \cdot \underline{F}_{\mathrm{L}} \qquad (4.34\mathrm{b})$$

where the vector

$$\underline{F}_{\mathrm{L}} = \int \underline{N}^{\mathrm{T}} \cdot \underline{p} \; dx \; dy \qquad (4.34\mathrm{c})$$

contains the nodal loads corresponding to the element loading.

Equating the internal and external virtual work according to (4.33) and (4.34b), one obtains

$$\underline{\bar{u}}_{\mathrm{e}}^{\mathrm{T}} \int t \cdot \underline{B}^{\mathrm{T}} \cdot \underline{D} \cdot \underline{B} \; dx \; dy \cdot \underline{u}_{\mathrm{e}} = \underline{\bar{u}}_{\mathrm{e}}^{\mathrm{T}} \cdot \underline{F}_{\mathrm{e}} + \underline{\bar{u}}_{\mathrm{e}}^{\mathrm{T}} \cdot \underline{F}_{\mathrm{L}}.$$

Since this equation is valid for any arbitrary virtual nodal displacements $\underline{\bar{u}}_{\mathrm{e}}^{\mathrm{T}}$, from this it follows

$$\int t \cdot (\underline{B}^{\mathrm{T}} \cdot \underline{D} \cdot \underline{B}) \; dx \; dy \cdot \underline{u}_{\mathrm{e}} = \underline{F}_{\mathrm{e}} + \underline{F}_{\mathrm{L}}$$

or without element loads

$$\underline{K}_{\mathrm{e}} \cdot \underline{u}_{\mathrm{e}} = \underline{F}_{\mathrm{e}} \qquad (4.35)$$

with

$$\underline{K}_{\mathrm{e}} = \int t \cdot \underline{B}^{\mathrm{T}} \cdot \underline{D} \cdot \underline{B} \; dx \; dy \qquad (4.35\mathrm{a})$$

The matrix $\underline{K}_{\mathrm{e}}$ is the stiffness matrix of the rectangular plane stress element with bilinear shape functions. After performing the matrix multiplication and integration for a constant plate thickness t the stiffness relationship is obtained according to [18, 27–30], as

$$\frac{E \cdot t}{12(1 - \mu^2)} \begin{bmatrix} k_{11} & k_{12} & k_{13} & k_{14} & k_{15} & k_{16} & k_{17} & k_{18} \\ k_{21} & k_{22} & k_{23} & k_{24} & k_{25} & k_{26} & k_{27} & k_{28} \\ k_{31} & k_{32} & k_{33} & k_{34} & k_{35} & k_{36} & k_{37} & k_{38} \\ k_{41} & k_{42} & k_{43} & k_{44} & k_{45} & k_{46} & k_{47} & k_{48} \\ k_{51} & k_{52} & k_{53} & k_{54} & k_{55} & k_{56} & k_{57} & k_{58} \\ k_{61} & k_{62} & k_{63} & k_{64} & k_{65} & k_{66} & k_{67} & k_{68} \\ k_{71} & k_{72} & k_{73} & k_{74} & k_{75} & k_{76} & k_{77} & k_{78} \\ k_{81} & k_{82} & k_{83} & k_{84} & k_{85} & k_{86} & k_{87} & k_{88} \end{bmatrix} \cdot \begin{bmatrix} u_1 \\ v_1 \\ u_2 \\ v_2 \\ u_3 \\ v_3 \\ u_4 \\ v_4 \end{bmatrix} = \begin{bmatrix} F_{x1} \\ F_{y1} \\ F_{x2} \\ F_{y2} \\ F_{x3} \\ F_{y3} \\ F_{x4} \\ F_{y4} \end{bmatrix}$$

$$\underline{K}_e \cdot \underline{u}_e = \underline{F}_e \tag{4.36}$$

with

$$k_{11} = k_{33} = k_{55} = k_{77} = 4b/a + 2(1 - \mu)a/b$$

$$k_{22} = k_{44} = k_{66} = k_{88} = 4a/b + 2(1 - \mu)b/a$$

$$k_{12} = k_{47} = k_{38} = k_{56} = 3/2(1 + \mu)$$

$$k_{13} = k_{57} = -4b/a + (1 - \mu)a/b$$

$$k_{14} = k_{27} = k_{58} = k_{36} = -3/2(1 - 3\mu)$$

$$k_{15} = k_{37} = -2b/a - (1 - \mu)a/b$$

$$k_{16} = k_{25} = k_{78} = k_{34} = -3/2(1 + \mu)$$

$$k_{17} = k_{35} = 2b/a - 2(1 - \mu)a/b$$

$$k_{18} = k_{23} = k_{67} = k_{45} = 3/2(1 - 3\mu)$$

$$k_{24} = k_{68} = 2a/b - 2(1 - \mu)b/a$$

$$k_{26} = k_{48} = -2a/b - (1 - \mu)b/a$$

$$k_{28} = k_{46} = -4a/b + (1 - \mu)b/a. \tag{4.36b}$$

In [19], a formulation of the stiffness matrix is given where the material law (2.3b), is not yet inserted explicitly. Thus, the stiffness matrix can be formulated with the material laws given in Table 2.1, also for the plane strain state or for orthotropic materials.

4.4.4 Element Loads

A rectangular plate element shall be loaded by a constant surface load and by line loads at its boundaries (Fig. 4.17). The equivalent nodal forces corresponding to this element loading shall be determined. According to (4.34c), the following applies for the equivalent nodal forces:

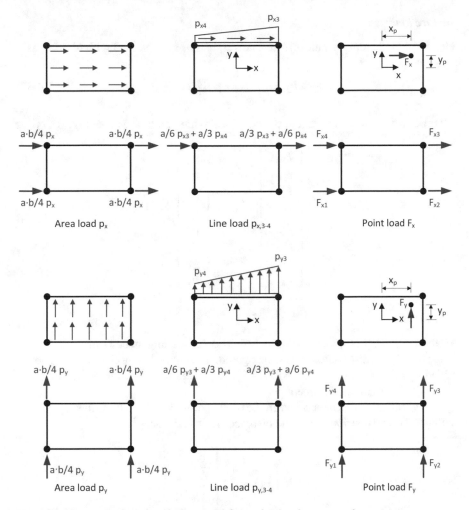

Fig. 4.17 Element loads and equivalent nodal forces for the plane stress element

Equivalent nodal forces of element loads

Equivalent nodal forces perform the same external work with the virtual nodal point displacements as the element loads with their associated virtual displacements.

The equivalent nodal forces will now be determined for different element loads. Similarly, the equivalent nodal point forces can be determined for other element loads.

Surface loads

The nodal loads for constant surface loads p_x and p_y are obtained with (4.34c), as

$$\underline{F}_{\mathrm{L}} = \int \underline{N}^{\mathrm{T}} \cdot \underline{p} \, \mathrm{d}x \, \mathrm{d}y. \tag{4.37}$$

After performing the integration, the nodal loads are obtained as (see Fig. 4.17)

$$\begin{bmatrix} F_{\mathrm{L}x1} \\ F_{\mathrm{L}y1} \\ F_{\mathrm{L}x2} \\ F_{\mathrm{L}y2} \\ F_{\mathrm{L}x3} \\ F_{\mathrm{L}y3} \\ F_{\mathrm{L}x4} \\ F_{\mathrm{L}y4} \end{bmatrix} = \frac{a \cdot b}{4} \begin{bmatrix} p_x \\ p_y \\ p_x \\ p_y \\ p_x \\ p_y \\ p_x \\ p_y \end{bmatrix}. \tag{4.37a}$$

Line loads

As an example of a line load, the linearly distributed loading in the x- and y-directions of the upper element boundary is investigated. The equivalent nodal loads are determined by equating the virtual work of the nodal loads and of the boundary loads caused by a virtual displacement.

The virtual displacement between the nodal points 3 and 4 is described by (4.31), with (4.27). The distribution of the displacements is linear, i.e.

$$\bar{v}_{3-4} = \begin{bmatrix} \bar{v}_3 & \bar{v}_4 \end{bmatrix} \cdot \begin{bmatrix} \frac{1}{2} + \frac{x}{a} \\ \frac{1}{2} - \frac{x}{a} \end{bmatrix}.$$

Similarly the linear distributed line load $p_{y,3-4}$ can be written

$$p_{y,3-4} = \begin{bmatrix} \left(\frac{1}{2} + \frac{x}{a} \right) & \left(\frac{1}{2} - \frac{x}{a} \right) \end{bmatrix} \cdot \begin{bmatrix} p_{y3} \\ p_{y4} \end{bmatrix}.$$

The virtual external work performed at the infinitesimal length $\mathrm{d}x$ by the line load $p_{y,3-4}$, i.e. by the force $p_{y,3-4} \cdot \mathrm{d}x$ and the virtual displacement $\bar{v}_{3-4}$ is

$$\overline{W}_{\mathrm{aL}} = \begin{bmatrix} \bar{v}_3 & \bar{v}_4 \end{bmatrix} \cdot \int_{-a/2}^{a/2} \begin{bmatrix} \frac{1}{2} + \frac{x}{a} \\ \frac{1}{2} - \frac{x}{a} \end{bmatrix} \cdot \begin{bmatrix} \left(\frac{1}{2} + \frac{x}{a} \right) & \left(\frac{1}{2} - \frac{x}{a} \right) \end{bmatrix} \mathrm{d}x \cdot \begin{bmatrix} p_{y3} \\ p_{y4} \end{bmatrix}.$$

Equating this work with the external virtual work of the equivalent nodal forces

$$\overline{W}_{aK} = \begin{bmatrix} \bar{v}_3 & \bar{v}_4 \end{bmatrix} \cdot \begin{bmatrix} F_{Ly3} \\ F_{Ly4} \end{bmatrix}$$

gives

$$\begin{bmatrix} \bar{v}_3 & \bar{v}_4 \end{bmatrix} \begin{bmatrix} F_{Ly3} \\ F_{Ly4} \end{bmatrix} = \begin{bmatrix} \bar{v}_3 & \bar{v}_4 \end{bmatrix} \int_{-a/2}^{a/2} \begin{bmatrix} \dfrac{1}{2} + \dfrac{x}{a} \\ \dfrac{1}{2} - \dfrac{x}{a} \end{bmatrix} \cdot \begin{bmatrix} \dfrac{1}{2} + \dfrac{x}{a} & \dfrac{1}{2} - \dfrac{x}{a} \end{bmatrix} dx \cdot \begin{bmatrix} p_{y3} \\ p_{y4} \end{bmatrix}.$$

Since this relationship holds for any virtual displacements $\bar{v}_3$ and $\bar{v}_4$ the nodal forces are obtained as

$$\begin{bmatrix} F_{Ly3} \\ F_{Ly4} \end{bmatrix} = \int_{-a/2}^{a/2} \begin{bmatrix} \dfrac{1}{2} + \dfrac{x}{a} \\ \dfrac{1}{2} - \dfrac{x}{a} \end{bmatrix} \cdot \begin{bmatrix} \dfrac{1}{2} + \dfrac{x}{a} & \dfrac{1}{2} - \dfrac{x}{a} \end{bmatrix} dx \cdot \begin{bmatrix} p_{y3} \\ p_{y4} \end{bmatrix}$$

and after performing the integration

$$\begin{bmatrix} F_{Ly3} \\ F_{Ly4} \end{bmatrix} = a \cdot \begin{bmatrix} \dfrac{1}{3} p_{y3} + \dfrac{1}{6} p_{y4} \\ \dfrac{1}{6} p_{y3} + \dfrac{1}{3} p_{y4} \end{bmatrix} \tag{4.38}$$

or

$$\begin{bmatrix} F_{Ly3} \\ F_{Ly4} \end{bmatrix} = \frac{a}{6} \cdot \begin{bmatrix} 2 & 1 \\ 1 & 2 \end{bmatrix} \cdot \begin{bmatrix} p_{y3} \\ p_{y4} \end{bmatrix} \tag{4.38a}$$

$$\underline{F}_{y,3-4} = \underline{A}_y \cdot \underline{P}_{y,3-4}. \tag{4.38b}$$

The equivalent nodal forces for $p_{x,3-4}$ can be derived similarly with the nodal displacements u_3 and u_4 (Fig. 4.17). In the same way other boundary loads can be handled.

Point loads

At a point inside the element with the element coordinates x_p, y_p two point forces F_x and F_y are considered (Fig. 4.17). With

$$\underline{F}_P = \begin{bmatrix} F_x \\ F_y \end{bmatrix} = \int \begin{bmatrix} p_x \\ p_y \end{bmatrix} dx\, dy = \int \underline{p}\, dx\, dy$$

one obtains according to (4.34c)

$$\underline{F}_{\mathrm{L,P}} = \int \underline{N}^{\mathrm{T}} \cdot \underline{p} \, \mathrm{d}x \, \mathrm{d}y$$

or

$$\underline{F}_{\mathrm{L,P}} = \underline{N}(x_{\mathrm{p}}, y_{\mathrm{p}})^{\mathrm{T}} \cdot \underline{F}_{\mathrm{P}} \tag{4.39}$$

and

$$\begin{bmatrix} F_{\mathrm{L}x1} \\ F_{\mathrm{L}y1} \\ F_{\mathrm{L}x2} \\ F_{\mathrm{L}y2} \\ F_{\mathrm{L}x3} \\ F_{\mathrm{L}y3} \\ F_{\mathrm{L}x4} \\ F_{\mathrm{L}y4} \end{bmatrix} = \begin{bmatrix} N_1(x_{\mathrm{p}}, y_{\mathrm{p}}) & 0 \\ 0 & N_1(x_{\mathrm{p}}, y_{\mathrm{p}}) \\ N_2(x_{\mathrm{p}}, y_{\mathrm{p}}) & 0 \\ 0 & N_2(x_{\mathrm{p}}, y_{\mathrm{p}}) \\ N_3(x_{\mathrm{p}}, y_{\mathrm{p}}) & 0 \\ 0 & N_3(x_{\mathrm{p}}, y_{\mathrm{p}}) \\ N_4(x_{\mathrm{p}}, y_{\mathrm{p}}) & 0 \\ 0 & N_4(x_{\mathrm{p}}, y_{\mathrm{p}}) \end{bmatrix} \cdot \begin{bmatrix} F_x \\ F_y \end{bmatrix}. \tag{4.39a}$$

4.4.5   Examples

Example 4.5 In order to verify the correct implementation of the rectangular element for plane stress with bilinear shape functions in a finite element program, a single element is investigated. In one degree of freedom a load of 1000 kN is applied, whereas all other degrees of freedom are fixed (Fig. 4.18). The displacement in the loaded degree of freedom and the restraining forces in all other degrees of freedom have to be verified.

Since the system consists of a single element, the global stiffness matrix is identical with the element stiffness matrix. All degrees of freedom except the displacement u_3 are fixed. Therefore, all rows and columns except those for the displacement u_3 have to be eliminated in the global stiffness matrix. One obtains with (4.36)

$$k_{55} \cdot u_3 = F_{x3}$$

with

$$\begin{aligned} k_{55} &= \frac{E \cdot t}{12 \cdot (1 - \mu^2)} \left(4 \cdot \frac{b}{a} + 2 \cdot (1 - \mu) \frac{a}{b} \right) \\ &= \frac{3 \cdot 10^7 \cdot 0.2}{12 \cdot (1 - 0.2^2)} \left(4 \cdot \frac{0.5}{1.0} + 2 \cdot (1 - 0.2) \frac{1.0}{0.5} \right) \\ &= 2.708 \cdot 10^6 \, \mathrm{kN/m} \end{aligned}$$

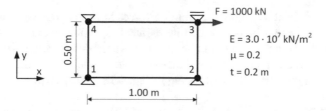

Fig. 4.18 Rectangular plate element with a single moveable degree of freedom

The displacement u_3 is obtained as

$$u_3 = F_{x3}/k_{55} = 1000/2.708 \cdot 10^6 = 3.69 \cdot 10^{-4} \text{m}$$

The restraining forces are obtained from the equations which have been eliminated for the support conditions. Since all displacements except u_3 are zero, only the stiffness values of a single column of the matrix have to be determined:

$$F_{x1} = k_{15} \cdot u_3 = 5.21 \cdot 10^5 \left(-2 \cdot \frac{0.5}{1.0} - (1 - 0.2) \cdot \frac{1.0}{0.5} \right) \cdot 3.69 \cdot 10^{-4}$$
$$= -1.354 \cdot 10^6 \cdot 3.69 \cdot 10^{-4} = -500 \,\text{kN}$$

$$F_{y1} = k_{25} \cdot u_3 = 5.21 \cdot 10^5 \left(-\frac{3}{2} \cdot (1 + 0.2) \right) \cdot 3.69 \cdot 10^{-4}$$
$$= -0.938 \cdot 10^6 \cdot 3.69 \cdot 10^{-4} = -346 \,\text{kN}$$

$$F_{x2} = k_{35} \cdot u_3 = 5.21 \cdot 10^5 \left(2 \cdot \frac{0.5}{1.0} - 2 \cdot (1 - 0.2) \cdot \frac{1.0}{0.5} \right) \cdot 3.69 \cdot 10^{-4}$$
$$= -1.146 \cdot 10^6 \cdot 3.69 \cdot 10^{-4} = -423 \,\text{kN}$$

$$F_{y2} = k_{45} \cdot u_3 = 5.21 \cdot 10^5 \left(\frac{3}{2} \cdot (1 - 3 \cdot 0.2) \right) \cdot 3.69 \cdot 10^{-4}$$
$$= 0.312 \cdot 10^6 \cdot 3.69 \cdot 10^{-4} = 115 \,\text{kN}$$

$$F_{y3} = k_{65} \cdot u_3 = 5.21 \cdot 10^5 \left(\frac{3}{2} \cdot (1 + 0.2) \right) \cdot 3.69 \cdot 10^{-4}$$
$$= 0.938 \cdot 10^6 \cdot 3.69 \cdot 10^{-4} = 346 \,\text{kN}$$

$$F_{x4} = k_{75} \cdot u_3 = 5.21 \cdot 10^5 \left(-4 \cdot \frac{0.5}{1.0} + (1 - 0.2) \cdot \frac{1.0}{0.5} \right) \cdot 3.69 \cdot 10^{-4}$$
$$= -0.208 \cdot 10^6 \cdot 3.69 \cdot 10^{-4} = -77 \,\text{kN}$$

$$F_{y4} = k_{85} \cdot u_3 = 5.21 \cdot 10^5 \left(-\frac{3}{2} \cdot (1 - 3 \cdot 0.2) \right) \cdot 3.69 \cdot 10^{-4}$$
$$= -0.312 \cdot 10^6 \cdot 3.69 \cdot 10^{-4} = -115 \,\text{kN}$$

The displacement u_3 and the restraining forces have to agree exactly with those of the program output if the same finite element is implemented. For a complete verification, the computation has to be repeated with all other degrees of freedom.

In addition, the complete stiffness matrix of the element (required later) will be specified. With the parameters according to Fig. 4.18, (4.36) gives

$$
\underline{K}_e =
\begin{bmatrix}
2.708 & 0.938 & -0.208 & -0.312 & -1.354 & -0.938 & -1.146 & 0.312 \\
0.938 & 4.583 & 0.312 & 1.667 & -0.938 & -2.292 & -0.312 & -3.958 \\
-0.208 & 0.312 & 2.708 & -0.938 & -1.146 & -0.312 & -1.354 & 0.938 \\
-0.312 & 1.667 & -0.938 & 4.583 & 0.312 & -3.958 & 0.938 & -2.292 \\
-1.354 & -0.938 & -1.146 & 0.312 & 2.708 & 0.938 & -0.208 & -0.312 \\
-0.938 & -2.292 & -0.312 & -3.958 & 0.938 & 4.583 & 0.312 & 1.667 \\
-1.146 & -0.312 & -1.354 & 0.938 & -0.208 & 0.312 & 2.708 & -0.938 \\
0.312 & -3.958 & 0.938 & -2.292 & -0.312 & 1.667 & -0.938 & 4.583
\end{bmatrix}
$$
$$\cdot\, 10^6 \ \mathrm{kN/m}$$

The 5th column contains the coefficients which were required to determine the nodal displacement and the support forces.   ◀

Example 4.6 For the plate in plane stress shown in Fig. 4.19a, the nodal displacements and element stresses are to be determined with a discretization into 3 finite elements.

The finite element model is shown in Fig. 4.19b. It has 8 nodal points and 12 degrees of freedom when the support conditions are taken into account.

All elements have the same parameters: $a = 5\,\mathrm{m}$, $b = 4\,\mathrm{m}$, $t = 0.4\,\mathrm{m}$, $E = 3\cdot10^7\,\mathrm{kN/m^2}$, $\mu = 0.2$. Therefore, the element stiffness matrices of the 3 elements are identical. With (4.36), they are obtained as

$$\underline{K}^{(1)} = \underline{K}^{(2)} = \underline{K}^{(3)} =$$

$$
1.042 \cdot 10^6 \cdot
\begin{bmatrix}
50.20 & 10.80 & -20.20 & -0.60 & -20.60 & -10.80 & -0.40 & 0.60 \\
10.80 & 60.28 & 0.60 & 10.22 & -10.80 & -30.14 & -0.60 & -40.36 \\
-20.20 & 0.60 & 50.20 & -10.80 & -0.40 & -0.60 & -20.60 & 10.80 \\
-0.60 & 10.22 & -10.80 & 60.28 & 0.60 & -40.36 & 10.80 & -30.14 \\
-20.60 & -10.80 & -0.40 & 0.60 & 50.20 & 10.80 & -20.20 & -0.60 \\
-10.80 & -30.14 & -0.60 & -40.36 & 10.80 & 60.28 & 0.60 & 10.22 \\
-0.40 & -0.60 & -20.60 & 10.80 & -20.20 & 0.60 & 50.20 & -10.80 \\
0.60 & -40.36 & 10.80 & -30.14 & -0.60 & 10.22 & -10.80 & 60.28
\end{bmatrix}
\frac{\mathrm{kN}}{\mathrm{m}}
$$

The global stiffness matrix can be set up in the same way as described for truss and beam structures in Sect. 3.2.

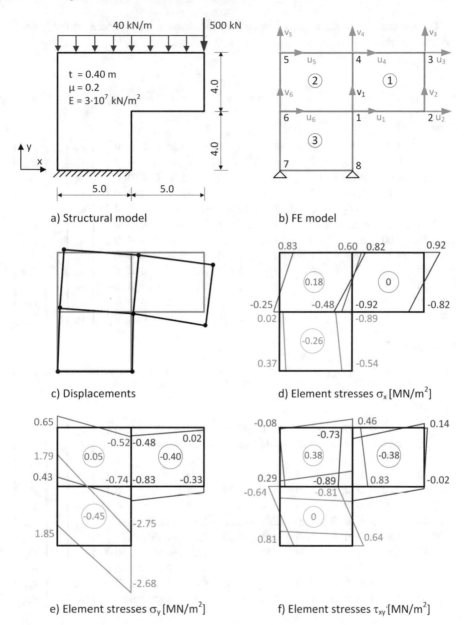

a) Structural model

b) FE model

c) Displacements

d) Element stresses σ_x [MN/m^2]

e) Element stresses σ_y [MN/m^2]

f) Element stresses τ_{xy} [MN/m^2]

Fig. 4.19   Plate in plane stress

The element stiffness matrices related to the degrees of freedom of the global system are obtained as

Element 1:

$$u_1^{(1)} \mathrel{\hat{=}} u_1 \quad v_1^{(1)} \mathrel{\hat{=}} v_1 \quad u_2^{(1)} \mathrel{\hat{=}} u_2 \quad v_2^{(1)} \mathrel{\hat{=}} v_2 \quad u_3^{(1)} \mathrel{\hat{=}} u_3 \quad v_3^{(1)} \mathrel{\hat{=}} v_3 \quad u_4^{(1)} \mathrel{\hat{=}} u_4 \quad v_4^{(1)} \mathrel{\hat{=}} v_4$$

$1.042 \cdot 10^6 \cdot$

$$
\begin{bmatrix}
5.20 & 1.80 & -2.20 & -0.60 & -2.60 & -1.80 & -0.40 & 0.60 & 0 & 0 & 0 & 0 \\
1.80 & 6.28 & 0.60 & 1.22 & -1.80 & -3.14 & -0.60 & -4.36 & 0 & 0 & 0 & 0 \\
-2.20 & 0.60 & 5.20 & -1.80 & -0.40 & -0.60 & -2.60 & 1.80 & 0 & 0 & 0 & 0 \\
-0.60 & 1.22 & -1.80 & 6.28 & 0.60 & -4.36 & 1.80 & -3.14 & 0 & 0 & 0 & 0 \\
-2.60 & -1.80 & -0.40 & 0.60 & 5.20 & 1.80 & -2.20 & -0.60 & 0 & 0 & 0 & 0 \\
-1.80 & -3.14 & -0.60 & -4.36 & 1.80 & 6.28 & 0.60 & 1.22 & 0 & 0 & 0 & 0 \\
-0.40 & -0.60 & -2.60 & 1.80 & -2.20 & 0.60 & 5.20 & -1.80 & 0 & 0 & 0 & 0 \\
0.60 & -4.36 & 1.80 & -3.14 & -0.60 & 1.22 & -1.80 & 6.28 & 0 & 0 & 0 & 0 \\
0 & 0 & 0 & 0 & 0 & 0 & 0 & 0 & 0 & 0 & 0 & 0 \\
0 & 0 & 0 & 0 & 0 & 0 & 0 & 0 & 0 & 0 & 0 & 0 \\
0 & 0 & 0 & 0 & 0 & 0 & 0 & 0 & 0 & 0 & 0 & 0 \\
0 & 0 & 0 & 0 & 0 & 0 & 0 & 0 & 0 & 0 & 0 & 0
\end{bmatrix}
\begin{bmatrix}
u_1 \\ v_1 \\ u_2 \\ v_2 \\ u_3 \\ v_3 \\ u_4 \\ v_4 \\ u_5 \\ v_5 \\ u_6 \\ v_6
\end{bmatrix}
=
\begin{bmatrix}
F_{x1} \\ F_{y1} \\ F_{x2} \\ F_{y2} \\ F_{x3} \\ F_{y3} \\ F_{x4} \\ F_{y4} \\ F_{x5} \\ F_{y5} \\ F_{x6} \\ F_{y6}
\end{bmatrix}
$$

Element 2:

$$u_1^{(2)} \mathrel{\hat{=}} u_6 \quad v_1^{(2)} \mathrel{\hat{=}} v_6 \quad u_2^{(2)} \mathrel{\hat{=}} u_1 \quad v_2^{(2)} \mathrel{\hat{=}} v_1 \quad u_3^{(2)} \mathrel{\hat{=}} u_4 \quad v_3^{(2)} \mathrel{\hat{=}} v_4 \quad u_4^{(2)} \mathrel{\hat{=}} u_5 \quad v_4^{(2)} \mathrel{\hat{=}} v_5$$

$1.042 \cdot 10^6 \cdot$

$$
\begin{bmatrix}
5.20 & -1.80 & 0 & 0 & 0 & 0 & -0.40 & -0.60 & -2.60 & 1.80 & -2.20 & 0.60 \\
-1.80 & 6.28 & 0 & 0 & 0 & 0 & 0.60 & -4.36 & 1.80 & -3.14 & -0.60 & 1.22 \\
0 & 0 & 0 & 0 & 0 & 0 & 0 & 0 & 0 & 0 & 0 & 0 \\
0 & 0 & 0 & 0 & 0 & 0 & 0 & 0 & 0 & 0 & 0 & 0 \\
0 & 0 & 0 & 0 & 0 & 0 & 0 & 0 & 0 & 0 & 0 & 0 \\
0 & 0 & 0 & 0 & 0 & 0 & 0 & 0 & 0 & 0 & 0 & 0 \\
-0.40 & 0.60 & 0 & 0 & 0 & 0 & 5.20 & 1.80 & -2.20 & -0.60 & -2.60 & -1.80 \\
-0.60 & -4.36 & 0 & 0 & 0 & 0 & 1.80 & 6.28 & 0.60 & 1.22 & -1.80 & -3.14 \\
-2.60 & 1.80 & 0 & 0 & 0 & 0 & -2.20 & 0.60 & 5.20 & -1.80 & -0.40 & -0.60 \\
1.80 & -3.14 & 0 & 0 & 0 & 0 & -0.60 & 1.22 & -1.80 & 6.28 & 0.60 & -4.36 \\
-2.20 & -0.60 & 0 & 0 & 0 & 0 & -2.60 & -1.80 & -0.40 & 0.60 & 5.20 & 1.80 \\
0.60 & 1.22 & 0 & 0 & 0 & 0 & -1.80 & -3.14 & -0.60 & -4.36 & 1.80 & 6.28
\end{bmatrix}
\begin{bmatrix}
u_1 \\ v_1 \\ u_2 \\ v_2 \\ u_3 \\ v_3 \\ u_4 \\ v_4 \\ u_5 \\ v_5 \\ u_6 \\ v_6
\end{bmatrix}
=
\begin{bmatrix}
F_{x1} \\ F_{y1} \\ F_{x2} \\ F_{y2} \\ F_{x3} \\ F_{y3} \\ F_{x4} \\ F_{y4} \\ F_{x5} \\ F_{y5} \\ F_{x6} \\ F_{y6}
\end{bmatrix}
$$

Element 3:

$$u_1^{(3)} = 0 \quad v_1^{(3)} = 0 \quad u_2^{(3)} = 0 \quad v_2^{(3)} = 0 \quad u_3^{(3)} \hat{=} u_1 \quad v_3^{(3)} \hat{=} v_1 \quad u_4^{(3)} \hat{=} u_6 \quad v_4^{(3)} \hat{=} v_6$$

$1.042 \cdot 10^6 \cdot$

$$
\begin{bmatrix}
5.20 & 1.80 & 0 & 0 & 0 & 0 & 0 & 0 & 0 & 0 & -2.20 & -0.60 \\
1.80 & 6.28 & 0 & 0 & 0 & 0 & 0 & 0 & 0 & 0 & 0.60 & 1.22 \\
0 & 0 & 0 & 0 & 0 & 0 & 0 & 0 & 0 & 0 & 0 & 0 \\
0 & 0 & 0 & 0 & 0 & 0 & 0 & 0 & 0 & 0 & 0 & 0 \\
0 & 0 & 0 & 0 & 0 & 0 & 0 & 0 & 0 & 0 & 0 & 0 \\
0 & 0 & 0 & 0 & 0 & 0 & 0 & 0 & 0 & 0 & 0 & 0 \\
0 & 0 & 0 & 0 & 0 & 0 & 0 & 0 & 0 & 0 & 0 & 0 \\
0 & 0 & 0 & 0 & 0 & 0 & 0 & 0 & 0 & 0 & 0 & 0 \\
0 & 0 & 0 & 0 & 0 & 0 & 0 & 0 & 0 & 0 & 0 & 0 \\
0 & 0 & 0 & 0 & 0 & 0 & 0 & 0 & 0 & 0 & 0 & 0 \\
-2.20 & 0.60 & 0 & 0 & 0 & 0 & 0 & 0 & 0 & 0 & 5.20 & -1.80 \\
-0.60 & 1.22 & 0 & 0 & 0 & 0 & 0 & 0 & 0 & 0 & -1.80 & 6.28
\end{bmatrix}
\cdot
\begin{bmatrix}
u_1 \\ v_1 \\ u_2 \\ v_2 \\ u_3 \\ v_3 \\ u_4 \\ v_4 \\ u_5 \\ v_5 \\ u_6 \\ v_6
\end{bmatrix}
=
\begin{bmatrix}
F_{x1} \\ F_{y1} \\ F_{x2} \\ F_{y2} \\ F_{x3} \\ F_{y3} \\ F_{x4} \\ F_{y4} \\ F_{x5} \\ F_{y5} \\ F_{x6} \\ F_{y6}
\end{bmatrix}
$$

The sum of the three element matrices gives the global stiffness matrix given below.

Alternatively, the terms of the element stiffness matrices can be added directly to the global stiffness matrix at the entries corresponding to the global degrees of freedom. This will be shown for the term $k_{2,8}$ of the global stiffness matrix, which represents the coupling of the second degree of freedom (v_1) with the eighth degree of freedom (v_4) of the global stiffness matrix. The global degree of freedom v_1 corresponds to $v_1^{(1)}$ of element 1, i.e. its second local degree of freedom, and to $v_4^{(2)}$ in element 2, i.e. the fourth local degree of freedom (Fig. 4.14).

The global degree of freedom v_4 corresponds to $v_4^{(1)}$ of element 1, i.e. its eighth local degree of freedom, and to $v_3^{(2)}$ of element 2, i.e. its sixth local degree of freedom. Element 3 does not connect the two degrees of freedom v_1 and v_4, and thus it is not included in $k_{2,8}$. The entry $k_{2,8}$, therefore, consists of the contribution of elements 1 and 2 exclusively as

$$k_{2,8} = k_{2,8}^{(1)} + k_{4,6}^{(2)} = 1.042 \cdot 10^6 \cdot (-4.36 + (-4.36)) = 1.042 \cdot 10^6 \cdot (-8.72) \; \frac{\text{kN}}{\text{m}}.$$

Accordingly, one gets the diagonal term $k_{2,2}$ associated to v_2 as

$$k_{2,2} = k_{2,2}^{(1)} + k_{4,4}^{(2)} + k_{6,6}^{(3)} = 1.042 \cdot 10^6 \cdot (6.28 + 6.28 + 6.28)$$

$$= 1.042 \cdot 10^6 \cdot 18.84 \; \frac{\text{kN}}{\text{m}}.$$

where the global degree of freedom v_2 is connected to all three elements.

The load vector is constructed from the element loads and the nodal loads. One obtains degree of freedom $v_3 : F_{y3} = -\dfrac{q \cdot l}{2} - F = -\dfrac{40 \cdot 5}{2} - 500 = -600\,\text{kN}$

degree of freedom $v_4 : F_{y4} = -\dfrac{q \cdot l}{2} - \dfrac{q \cdot l}{2} = -\dfrac{40 \cdot 5}{2} - \dfrac{40 \cdot 5}{2} = -200\,\text{kN}$

degree of freedom $v_5 : F_{y5} = -\dfrac{q \cdot l}{2} = -\dfrac{40 \cdot 5}{2} = -100\,\text{kN}.$

The global system of equations is thus in the units m and kN:

$1.042 \cdot 10^6 \cdot$

$$
\begin{bmatrix}
15.60 & 1.80 & -2.20 & -0.60 & -2.60 & -1.80 & -0.80 & 0 & -2.60 & 1.80 & -4.40 & 0 \\
1.80 & 18.84 & 0.60 & 1.22 & -1.80 & -3.14 & 0 & -8.72 & 1.80 & -3.14 & 0 & 2.44 \\
-2.20 & 0.60 & 5.20 & -1.80 & -0.40 & -0.60 & -2.60 & 1.80 & 0 & 0 & 0 & 0 \\
-0.60 & 1.22 & -1.80 & 6.28 & 0.60 & -4.36 & 1.80 & -3.14 & 0 & 0 & 0 & 0 \\
-2.60 & -1.80 & -0.40 & 0.60 & 5.20 & 1.80 & -2.20 & -0.60 & 0 & 0 & 0 & 0 \\
-1.80 & -3.14 & -0.60 & -4.36 & 1.80 & 6.28 & 0.60 & 1.22 & 0 & 0 & 0 & 0 \\
-0.80 & 0 & -2.60 & 1.80 & -2.20 & 0.60 & 10.40 & 0 & -2.20 & -0.60 & -2.60 & -1.80 \\
0 & -8.72 & 1.80 & -3.14 & -0.60 & 1.22 & 0 & 12.56 & 0.60 & 1.22 & -1.80 & -3.14 \\
-2.60 & 1.80 & 0 & 0 & 0 & 0 & -2.20 & 0.60 & 5.20 & -1.80 & -0.40 & -0.60 \\
1.80 & -3.14 & 0 & 0 & 0 & 0 & -0.60 & 1.22 & -1.80 & 6.28 & 0.60 & -4.36 \\
-4.40 & 0 & 0 & 0 & 0 & 0 & -2.60 & -1.80 & -0.40 & 0.60 & 10.40 & 0 \\
0 & 2.44 & 0 & 0 & 0 & 0 & -1.80 & -3.14 & -0.60 & -4.36 & 0 & 12.56
\end{bmatrix}
\begin{bmatrix}
u_1 \\ v_1 \\ u_2 \\ v_2 \\ u_3 \\ v_3 \\ u_4 \\ v_4 \\ u_5 \\ v_5 \\ u_6 \\ v_6
\end{bmatrix}
=
\begin{bmatrix}
0 \\ 0 \\ 0 \\ 0 \\ 0 \\ -600 \\ 0 \\ -200 \\ 0 \\ -100 \\ 0 \\ 0
\end{bmatrix}
$$

Its solution is the displacement vector of the system:

$$
u =
\begin{bmatrix}
u_1 \\ v_1 \\ u_2 \\ v_2 \\ u_3 \\ v_3 \\ u_4 \\ v_4 \\ u_5 \\ v_5 \\ u_6 \\ v_6
\end{bmatrix}
=
\begin{bmatrix}
0.204 \\ -0.344 \\ 0.08 \\ -1.613 \\ 1.088 \\ -1.635 \\ 0.936 \\ -0.429 \\ 0.818 \\ 0.302 \\ 0.26 \\ 0.237
\end{bmatrix}
\cdot 10^{-3}\,\text{m}
$$

The displacements are shown in Fig. 4.19c.

The stresses are determined element by element. For example, applying (4.30), for element 1 at the local point 3 with the local element coordinates $x = 2.5\,\text{m}$, $y = 2.0\,\text{m}$ (Fig. 4.14) the following stresses are obtained:

$$
\underline{\sigma}_3^{(1)} = \underline{D} \cdot \underline{B} \cdot \underline{u}^{(1)} \rightarrow \underline{\sigma}_3^{(1)} =
\begin{bmatrix}
\sigma_x \\ \sigma_y \\ \tau_{xy}
\end{bmatrix}
=
\begin{bmatrix}
915.3 \\ 18.1 \\ 137.5
\end{bmatrix}
\frac{\text{kN}}{\text{m}^2}.
$$

Herein with (4.29) and (2.3b) it is

$$
\underline{B} = \begin{bmatrix} 0 & 0 & 0 & 0 & 0.20 & 0 & -0.20 & 0 \\ 0 & 0 & 0 & -0.25 & 0 & 0.25 & 0 & 0 \\ 0 & 0 & -0.25 & 0 & 0.25 & 0.20 & 0 & -0.20 \end{bmatrix}
$$

$$
\underline{D} = 3.125 \cdot 10^7 \cdot \begin{bmatrix} 1.0 & 0.2 & 0 \\ 0.2 & 1.0 & 0 \\ 0 & 0 & 0.4 \end{bmatrix} \frac{\text{kN}}{\text{m}^2}, \qquad \underline{u}^{(1)} = \begin{bmatrix} 0.204 \\ -0.344 \\ 0.08 \\ -1.613 \\ 1.088 \\ -1.635 \\ 0.936 \\ -0.429 \end{bmatrix} \cdot 10^{-3}\,\text{m}.
$$

The vector $\underline{\sigma}_3^{(1)}$ gives the element stresses of element 1 in nodal point 3 of the finite element model (Fig. 4.19d–f).

Alternatively, the stresses can be determined with (4.30a). For example, the stress σ_y in the midpoint of element 3 is

$$
\sigma_{y,m}^{(3)} = \frac{30\,000\,000 \cdot 10^{-3}}{(1 - 0.2^2) \cdot 2 \cdot 5 \cdot 4} \cdot [0.2 \cdot 4 \cdot (0.204 - 0.206)
$$
$$
+ (-5) \cdot (-0.237) + 5 \cdot (-0.344)] = -450\,\text{kN/cm}^2.
$$

The element stresses at the element edges and at the element centers are shown in Fig. 4.19d–f. They have jumps at the element boundaries. The nodal stresses are calculated from the element stresses by averaging. For example, at node 4 one gets σ_x as

$$
\sigma_{x,4}^{(1)} = \frac{\sigma_{x,4}^{(1)} + \sigma_{x,4}^{(2)}}{2} = \frac{599 + 816}{2} = 708\,\frac{\text{kN}}{\text{m}^2}.
$$

Normally, finite element programs output only the nodal stresses and use them for graphical display of the stresses. The stress jump of $816/599 = 1.37$, i.e. of 37% indicates that the discretization of the plate is not sufficient. For practical purposes, much finer finite element meshes are required (cf. Sect. 4.11.3 and Example 4.19).

◀

Example 4.7 The deep beam in Fig. 4.20, is analyzed with a finite element program, in which the rectangular element with bilinear shape functions according to Sects. 4.4.1 to 4.4.4 is implemented. The analysis is performed with three finite element discretizations, i.e. with 2×2, 4×4, and 8×8 elements.

The element stresses in the sections A–A, B–B, and C–C are given for three finite element discretizations, Figs. 4.21, 4.22, 4.23, and Table 4.6. The stresses have

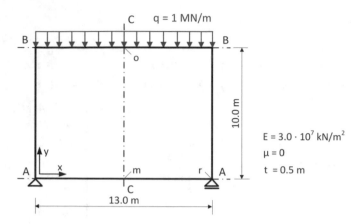

Fig. 4.20 Deep beam

pronounced jumps at the element boundaries. Inside the element they exhibit the distribution shown in Fig. 4.16. In section C–C, jumps of the stresses cannot occure due to the symmetry of the system and the loading.

For the design of a reinforced concrete deep beam, the longitudinal stresses σ_x in section C–C are considered. The stresses σ_x of the 2×2 discretization at the upper and lower edge of the deep beam are -0.431 MN/m^2 and 1.661 MN/m^2, respectively. They show, compared to the 8×8 discretization with -1.610 MN/m^2 and 4.216 MN/m^2, a large discretization error, and thus are useless for practical purposes. The stresses of the 4×4 and the 8×8 discretization at the upper edge of the deep beam differ considerably with

$$\frac{-1.610 - (-1.080)}{-1.610} = 0.33 \hat{=} 33 \%$$

whereas at the lower edge with

$$\frac{4.323 - 4.216}{4.216} = 0.025 \ \hat{=} 2.5 \%$$

a satisfyingly close agreement can be observed. If for the reinforced concrete design only the tension stresses at the lower edge of the deep beam are sought, the 8×8 discretization can be considered as sufficient for practical purposes. The accuracy of the stresses can be verified by further mesh refinement with a 16×16 discretization (Table 4.6) or by comparison with an analytical solution. For an engineering assessment of the results, the compression stresses, as well as the tension stresses, can be integrated separately in section C–C in order to obtain a compression and a tension force. The internal moment obtained by these forces should correspond to the total internal moment of the system in the mid of the deep beam.

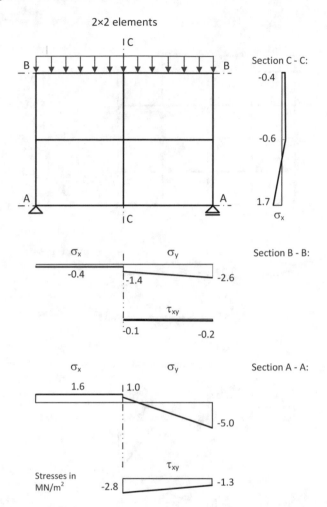

Fig. 4.21 Stresses for the 2×2 discretization

Some questions arise, however, in a more detailed analysis of the stress diagrams of the 8×8 discretization. Due to the vertical line load of $1.0\,\text{MN/m}$, the vertical compression stress at the upper edge of the deep beam, for example, should be $1.0\,\text{MN/m} / 0.5\,\text{m} = 2.0\,\text{MN/m}^2$ in all elements. The variation between $-1.9\,\text{MN/m}^2$ and $-2.1\,\text{MN/m}^2$, i.e. a deviation of ~5% from the exact value, is an indicator for the accuracy of the computed stresses. At the upper and lower edge, the shear stresses have a remarkably large value compared to the exact value of zero. The results can be improved by interpolation strategies (see Sect. 4.11.10). The averaging of the stresses of the nodal points which is implemented in finite element programs improve already the accuracy significantly. On the symmetry axis, however, the errors remain, e.g. shear stresses of -0.24 and $-0.48\,\text{MN/m}^2$ at the upper and lower edge

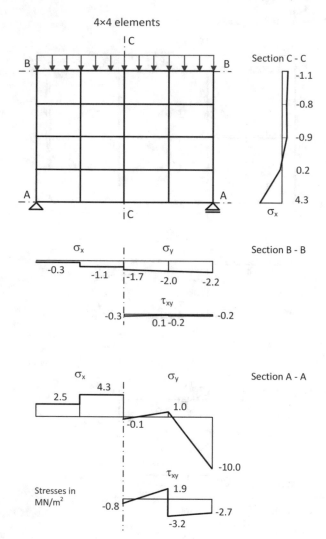

Fig. 4.22 Stresses for the 4 × 4 discretization

of the deep beam, respectively. Only by a finer discretization, these discretization errors can be reduced. This applies also for the considerably large jumps of the longitudinal stresses σ_x at the upper and lower edge of the deep beam.

The region near the support requires special attention. The supports have been defined as fixed nodal points where the support forces are acting as point forces. However, a point force applied to a plate in plane stress leads to a singularity of the stresses at the point where the load is applied. In a finite element analysis, stresses

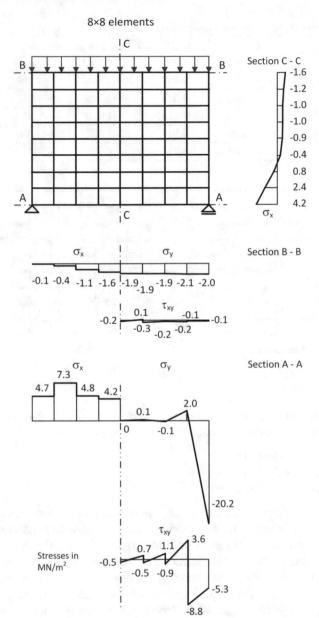

Fig. 4.23 Stresses for the 8×8 discretization

Table 4.6 Convergence of stresses and displacements

FE discretization	$\sigma_{x,o}$ [MN/m^2]	$\sigma_{x,m}$ [MN/m^2]	$\sigma_{y,o}$ [MN/m^2]	$\sigma_{y,r}$ [MN/m^2]	w_m [cm]
2×2	-0.431	1.661	-1.361	-4.954	-0.122
4×4	-1.080	4.323	-1.705	-9.991	-0.176
8×8	-1.610	4.216	-1.916	-20.170	-0.240
16×16	-1.795	4.247	-1.978	-40.382	-0.305
32×32	-1.844	4.256	-1.994	-80.760	-0.369

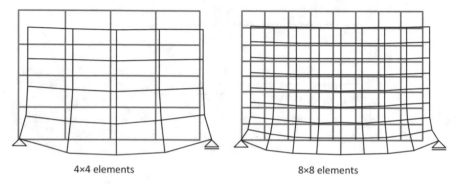

4×4 elements   8×8 elements

Fig. 4.24 Deformations of the finite element model

exhibit an extreme increase for each successive refinement of the finite element mesh. At the support point r, the following stresses σ_y, for example, are obtained

2×2-mesh: -4.954 MN/m^2
4×4-mesh: -9.991 MN/m^2
8×8-mesh: -20.170 MN/m^2.

A further mesh refinement at the support results in more and more increasing stresses, being meaningless for engineering purposes (Table 4.6). The displacements also have a singularity at the point support. These influence the displacements of the whole model, so that the displacements of the plate obtained are physically not meaningful. Table 4.6 shows the stresses at points m, o, and r and the displacements at point m according to Fig. 4.20, for various finite element discretizations. While the stresses in the middle region of the model converge, this is not the case with the stresses at the support point and the displacements of the whole plate. The singularities at the supports are also indicated by the strong deformations of the elements connected with the supports (Fig. 4.24). The remedies to these issues are concepts for a consistent modeling of structures and the appropriate interpretation of the results (see Sect. 4.11). ◄

4.5 Finite Elements for Plates in Plane Stress

4.5.1 Properties of Finite Elements

From 1960s till today, a large number of different finite elements have been developed. The aim is to obtain the best possible approximation of the stresses and displacements of the structural system with the largest and simplest possible elements (i.e. with the least possible computing effort).

Displacement-based finite elements differ by the choice of the shape functions. Hybrid elements are based on shape functions for element stresses in addition to displacement shape functions. Other finite elements, e.g. such with unknowns other than nodal point displacements have not been successful in practice.

All elements have to fulfill particular requirements.

Requirements for finite elements

(a) Rigid body displacements must not provoke nodal forces.
(b) Elements must be able to represent constant strain and, hence, constant stress states exactly.
(c) Elements must fulfill the condition of geometric isotropy.

A requirement arising from a mathematical point of view is the consistency of the displacement shape functions of adjacent elements. However, elements which do not fulfill this condition have been developed on a heuristic basis. These are called incompatible or nonconforming elements.

Rigid body displacements

If a finite element is displaced like a rigid body, nodal forces and or element stresses are not expected to occur in the element. Hence, for a plane stress element, it has to be postulated that rigid body displacements in the x- and y-directions, as well as rigid body rotations, do not lead to element forces and element stresses (Fig. 4.25). Any other type of a rigid body displacement can be composed of these displacements.

Example 4.8 It has to be shown that for a rigid body displacement in the x-direction of the rectangular plane stress element with bilinear shape functions, no nodal forces occur.

The rigid body displacement in the x-direction with a value of 1 according to Fig. 4.25, is described by

$$u_1 = u_2 = u_3 = u_4 = 1$$

$$v_1 = v_2 = v_3 - v_4 = 0.$$

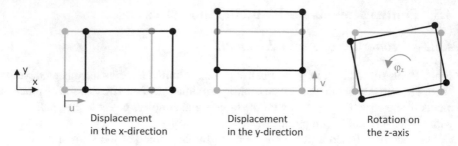

Fig. 4.25 Rigid body displacements of a plane stress element

The restraining forces are obtained by the multiplication of the displacement vector with the stiffness matrix according to (4.36), as

$$
\begin{bmatrix} F_1 \\ F_2 \\ F_3 \\ F_4 \\ F_5 \\ F_6 \\ F_7 \\ F_8 \end{bmatrix} = \frac{E \cdot t}{12(1 - \mu^2)} \begin{bmatrix} k_{11} & k_{12} & k_{13} & k_{14} & k_{15} & k_{16} & k_{17} & k_{18} \\ k_{21} & k_{22} & k_{23} & k_{24} & k_{25} & k_{26} & k_{27} & k_{28} \\ k_{31} & k_{32} & k_{33} & k_{34} & k_{35} & k_{36} & k_{37} & k_{38} \\ k_{41} & k_{42} & k_{43} & k_{44} & k_{45} & k_{46} & k_{47} & k_{48} \\ k_{51} & k_{52} & k_{53} & k_{54} & k_{55} & k_{56} & k_{57} & k_{58} \\ k_{61} & k_{62} & k_{63} & k_{64} & k_{65} & k_{66} & k_{67} & k_{68} \\ k_{71} & k_{72} & k_{73} & k_{74} & k_{75} & k_{76} & k_{77} & k_{78} \\ k_{81} & k_{82} & k_{83} & k_{84} & k_{85} & k_{86} & k_{87} & k_{88} \end{bmatrix} \cdot \begin{bmatrix} 1 \\ 0 \\ 1 \\ 0 \\ 1 \\ 0 \\ 1 \\ 0 \end{bmatrix}
$$

$$
= \frac{E \cdot t}{12(1 - \mu^2)} \begin{bmatrix} k_{11} + k_{13} + k_{15} + k_{17} \\ k_{21} + k_{23} + k_{25} + k_{27} \\ k_{31} + k_{33} + k_{35} + k_{37} \\ k_{41} + k_{43} + k_{45} + k_{47} \\ k_{51} + k_{53} + k_{55} + k_{57} \\ k_{61} + k_{63} + k_{65} + k_{67} \\ k_{71} + k_{73} + k_{75} + k_{77} \\ k_{81} + k_{83} + k_{85} + k_{87} \end{bmatrix} = \begin{bmatrix} 0 \\ 0 \\ 0 \\ 0 \\ 0 \\ 0 \\ 0 \\ 0 \end{bmatrix}
$$

with the terms k_{ij} according to (4.36b).

The result shows that for a rigid body displacement in the x-direction, the element forces are zero. In the same way, the verification for a rigid body displacement in the y-direction and a rotation on the z-axis can be performed.

Applying (4.30) or (4.30a), it can be shown easily that element stresses become zero when applying the displacement vector given above.   ◀

Constant strains

Any element must be able to represent a state of constant strain exactly. For the plane stress element these are the states where ε_x, ε_y and γ_{xy} assume constant values (Fig. 4.26). Since the stresses are obtained by the multiplication of the strains with

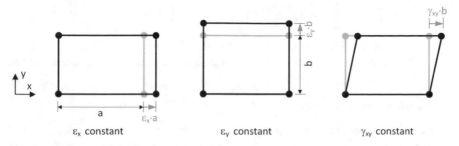

Fig. 4.26 Constant stress states for a plane stress element

the stress–strain material matrix (e.g. according to (2.3b)) this means that the element must be capable of representing constant stress states. This is of particular importance since the stresses in each element approximate a constant stress state when refining the mesh. If each finite element has the ability to represent a constant stress state exactly, it is expected that the numerical stress values converge to the exact solution which is constant in an infinitely small element.

Example 4.9 It has to be shown that the rectangular plane stress element with bilinear shape functions is able to represent constant stress states exactly. The element thickness is $t = 1$ m.

The proof can be done by hand with the stiffness Eqs. (4.36) and (4.30a), for the stresses or with a computer program for a numerical example. The last approach, which also checks the correctness of the implementation in the program, is followed here.

Figure 4.27 shows a single finite element which is supported such that the system is statically determinate, i.e. without any statically indeterminate constraint of the displacements. The loading consists of line loads at the element boundaries representing the stress states $\sigma_x = 1$, $\sigma_y = 1$ and $\tau_{xy} = 1$ in three different load cases. The analysis is done with a finite element program [SOF 6].

The results of the computation in the middle of the elements, as well as at the nodal points in the three load cases are

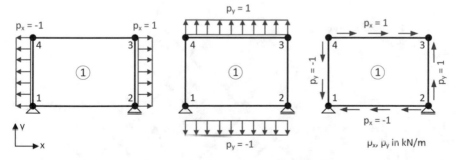

Fig. 4.27 Plane stress element with constant stress states

$$\sigma_x = 1, \quad \sigma_y = 1, \quad \tau_{xy} = 1 \ \left[\text{kN/m}^2\right],$$

respectively, whereas all other stress components of the corresponding load case are zero. Therefore, it is shown that the element is able to represent constant stress states exactly.   ◀

Geometric isotropy

The two-dimensional interpolation of the displacements should not prefer one of the two directions. This heuristic requirement is fulfilled when the shape functions contain all degrees of a polynomial, or at least the symmetrical terms such as $x^2 y$ and xy^2 (Fig. 4.28). In this case, the element is called geometrically isotropic. The plane stress element according to Sect. 4.4, for example, is isotropic since the shape functions in the x-direction, as well as in the y-direction are linear, i.e. they have the same polynomial degree. Elements implemented in programs used in practice are geometrically isotropic.

Continuity of the displacements

The requirement for the continuity of the displacements between elements has mathematical reasons. It means that the displacements of adjacent plane stress elements must not only coincide at the nodal points but also between them. Hence, the displacements of the elements may not lead to gaps between the elements (see Fig. 4.37). The element developed in Sect. 4.4, fulfills the continuity condition since the displacements between the nodal points are linear, and hence no gaps between the elements occur. Elements with continuous displacements are called compatible or conformal elements.

When the continuity conditions and the requirements according to (a) and (b) are fulfilled, the finite element method can be understood mathematically as a special type of the Ritz method in the calculus of variation [17, 20, 31, 32]. This means that all properties of the Ritz method also apply to the finite element method. The convergence of the method, and for a given element type also the convergence rate, can be proved mathematically. Especially, it can be shown that the convergence is monotonic, i.e. that for all geometrically possible displacement states (e.g. finite element shape functions) the solution minimizes the potential energy in the system.

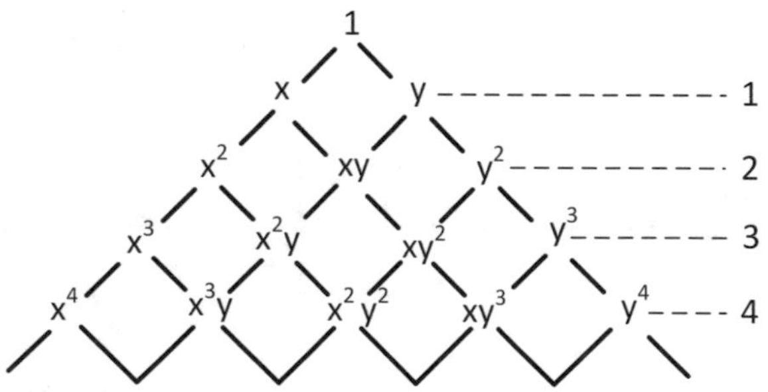

Fig. 4.28 Polynomial terms for complete polynomials (Pascal's triangle)

The absolute minimum of the potential energy would be achieved with the exact, analytical solution of the displacements.

Since the shape functions limit the possible deformations of the system, a finite element model (with continuous displacement shape functions) behaves too stiff. With increasing mesh refinement, the system becomes "softer", the displacements increase and approximate the exact solution. Since the displacements are always underestimated, they approximate the solution from one side, namely "from below", i.e. the convergence is monotonic. Furthermore, continuous displacement assumptions are the basis of a mathematically founded error estimation of the finite element method and the further development of automatic mesh refinement strategies adapted to the local error (see Sect. 4.12.2).

Since elements with continuous displacement assumptions behave too stiffly, attempts have been made to develop elements without this deficiency. Nonconforming elements, as well as hybrid elements, achieved some practical importance. Their derivation is based on a heuristic argumentation. With these elements the convergence of the solution is not monotonic, i.e. the deformations can be both over- and underestimated by the finite element solution.

Patch test

In order to show the convergence of elements with noncontinuous displacements for a mesh refinement, the so-called patch test according to Irons is performed [33]. This test requires that for an arbitrary configuration (patch) of incompatible elements under constant line loads corresponding to constant stresses, the element stresses of all elements reproduce the exact solution of the stresses (Fig. 4.29). However, it can be shown that the fulfillment of the patch test is not a sufficient criterion to prove the convergence of the finite element solution toward the exact solution. In detail, reference is made to the discussion in [19, 34].

Bending-type deformation of plane stress elements

In the following sections, different plane stress elements and their characteristics are discussed. First, the classical elements with continuous displacements are treated.

Fig. 4.29 Element configuration for a patch test

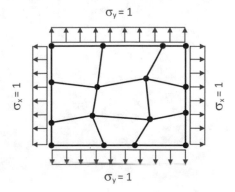

$\sigma_y = 1$

$\sigma_x = 1$ $\sigma_x = 1$

$\sigma_y = 1$

These elements overestimate the stiffness, which has a negative effect especially when subjected to bending, as it occurs, for example, in the web plate of an I-beam discretized in 4-node plane stress or shell elements. This phenomenon is called *shear locking*. There are various possibilities to improve the locking behavior in the plane stress state

---

Plane stress elements with improved locking behavior in bending state

- Conforme displacement-based finite elements with quadratic or higher order shape functions
- Non-conforme plane stress elements and EAS formulations
- Hybrid plane stress elements.

The properties of these elements are discussed below.

4.5.2 Displacement-Based Elements with Compatible Shape Functions

Classical finite elements are based on compatible displacement functions only. They are also known as conforming elements. Since the displacement functions are continuous, all mathematical properties of the Ritz method of the calculus of variations are valid. The element stiffness matrix of displacement-based elements is derived in the following steps.

Formulation of the element stiffness matrix for elements with displacement shape functions

1. Selection of the shape functions of the displacements; the unknowns of the displacement functions are the nodal point displacements. The order of the displacement function must be high enough so that all derivatives needed for the computation of the strains are not equal to zero.
2. Determination of the strains corresponding to the displacement functions

$$\underline{\varepsilon} = \underline{B} \cdot \underline{u}_{\mathrm{e}} \qquad (4.40\mathrm{a})$$

3. Formulation of the material law

$$\underline{\sigma} = \underline{D} \cdot \underline{\varepsilon} \qquad (4.40\mathrm{b})$$

4. The nodal forces corresponding to the chosen shape functions are obtained with the principal of virtual displacements as

$$\underline{K}_e \cdot \underline{u}_e = \underline{F}_e \tag{4.40c}$$

where

$$\underline{K}_e = \int \underline{B}^T \cdot \underline{D} \cdot \underline{B} \, dV \tag{4.40d}$$

represents is the stiffness matrix and dV is the infinitesimal volume element. Integration has to be done over the Volume V of the element. For a plane stress element with the thickness t the volume element is $dV = t \cdot dx \, dy$ and therewith

$$\underline{K}_e = \int\int t \cdot \underline{B}^T \cdot \underline{D} \cdot \underline{B} \, dx \, dy \tag{4.40e}$$

is the element stiffness matrix.

5. Determination of the nodal loads $\underline{F}_L$ equivalent to the element loads.

In this way, a great number of elements with different shape functions and numbers of nodal points can be derived. By appropriate selection of the material matrix $\underline{D}$ according to Table 2.1, all plane finite elements can be derived both for the plane strain and for the plane stress state with isotropic and orthotropic material law.

CST triangular element

The simple triangular element has three nodal points (Fig. 4.30). Between the nodal points and within the element, the displacements are linear with

$$u(x, y) = \alpha_1 + \alpha_2 \cdot x + \alpha_3 \cdot y, \tag{4.41a}$$

$$v(x, y) = \beta_1 + \beta_2 \cdot x + \beta_3 \cdot y. \tag{4.41b}$$

The strains which are obtained by differentiation of the displacement functions according to (2.1b), become constant, e.g. $\varepsilon_x = \partial u / \partial x = \alpha_2$. The strains $\varepsilon_x, \varepsilon_y$ and γ_{xy}, and therefore, also the stresses σ_x, σ_y and τ_{xy} are constant in the element. Therefore, the element is also known as a CST-element (**C**onstant **S**train **T**riangle). Its stiffness matrix can be determined explicitly according to (4.40e), and is given for the plane stress element with isotropic material in [21, 22, 27, 28, 32, 35].

The development of the finite element method began in 1956, with the derivation of the CST-element [4]. The element is implemented in many finite element programs. When used together with quadrilateral elements, however, stiffening effects are possible.

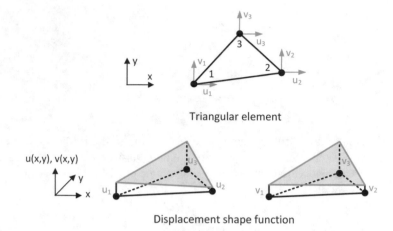

Triangular element

Displacement shape function

Fig. 4.30 Triangular element with linear shape function

Lagrange elements

Lagrangian elements are quadrilateral elements with inner nodes, where the Lagrangian polynomials are used to formulate the shape functions of the displacements. The inner nodes are not connected with other elements. They solely serve to increase the polynomial degree of the shape functions [7, 36, 37].

Isoparametric elements

A group of elements which may possess curved boundaries is known as isoparametric elements (Fig. 4.31) [7, 34]. To describe the geometry of the element boundaries, polynomials (straight lines, parabolas of second or third degree) are used. The displacements are interpolated with polynomials of the same degree as the polynomials describing the element geometry. For the 8-node element these are quadratic parabolas with three nodes at each side, i.e. two corner nodes and a (mid-)side node. The 4-node element is a general quadrilateral element and has linear shape functions between the nodal points. In the special form of a rectangular element, it is identical with the plane stress element derived in Sect. 4.4. The 3-node element has

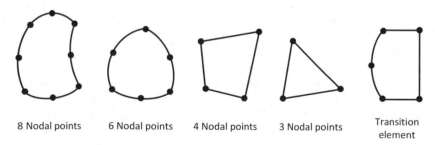

8 Nodal points   6 Nodal points   4 Nodal points   3 Nodal points   Transition element

Fig. 4.31 Isoparametric elements

linear shape functions between the nodal points and is identical with the CST-element already mentioned.

The shape functions of the isoparametric elements are described in curved local coordinates r and s (Fig. 4.32). They can be written as

$$u = \sum h_i \cdot u_i \tag{4.42a}$$

$$v = \sum h_i \cdot v_i \tag{4.42b}$$

where u_i and v_i are the displacements of the node i and h_i the interpolation functions in local coordinates r, s according to Fig. 4.32. The functions h_i are [34]

		$i = 5^*$	$i = 6^*$	$i = 7^*$	$i = 8^*$	$i = 9^*$
h_1	$= 1/4(1+r)(1+s)$	$-1/2\,h_5$	--	--	$-1/2\,h_8$	$-1/4\,h_9$
h_2	$= 1/4(1-r)(1+s)$	$-1/2\,h_5$	$-1/2\,h_6$	--	--	$-1/4\,h_9$
h_3	$= 1/4(1-r)(1-s)$	--	$-1/2\,h_6$	$-1/2\,h_7$	--	$-1/4\,h_9$
h_4	$= 1/4(1+r)(1-s)$	--	--	$-1/2\,h_7$	$-1/2\,h_8$	$-1/4\,h_9$
h_5	$= 1/2(1-r^2)(1+s)$	--	--	--	--	$-1/2\,h_9$
h_6	$= 1/2(1-s^2)(1-r)$	--	--	--	--	$-1/2\,h_9$
h_7	$= 1/2(1-r^2)(1-s)$	--	--	--	--	$-1/2\,h_9$
h_8	$= 1/2(1-s^2)(1+r)$	--	--	--	--	$-1/2\,h_9$
h_9	$= (1-r^2)(1-s^2)$	--	--	--	--	--

*if nodal point i is provided

$$\tag{4.43}$$

Fig. 4.32 Isoparametric element in local curved coordinates r, s

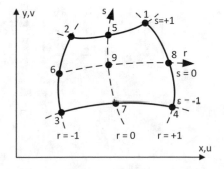

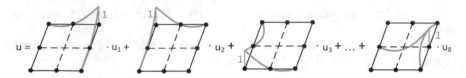

Fig. 4.33 Interpolation function as a weighted sum of shape functions

The number of side nodes is variable, i.e. the same formulation can be chosen for elements with different numbers of nodal points. The terms h_5, h_6, h_7, h_8 or h_9 in (4.43), are omitted if the corresponding nodal points do not exist. The functions h_i can be understood as shape functions. A shape function h_i has the value 1 for the degree of freedom i and is zero in all other degrees of freedom. Hence, the interpolation function is the sum of the shape functions multiplied by the corresponding nodal point displacements (Fig. 4.33).

The relationship between the global coordinates x, y and the element coordinates r, s are also described with the functions h_i as

$$x = \sum h_i \cdot x_i \tag{4.44a}$$

$$y = \sum h_i \cdot y_i \tag{4.44b}$$

where x_i and y_i are the coordinates of the nodal points. Since the same number of parameters is used to describe the geometry and the shape functions of the displacements, the elements are referred to as isoparametric elements ("isos" or "ἴσος", Greek "equal").

Between the global coordinates x, y and the element coordinates r, s a one-to-one relationship must exist. Especially, the so-called *Jacobian matrix* containing the first derivatives of x and y with respect to r and s must not be singular. Therefore, distorted element geometries, e.g. with interior angles greater than 180°, are not allowed (Fig. 4.34). The side nodes must always lie in the middle third of the element boundaries. Elements with a geometry "as rectangular as possible" and with side nodes lying in the middle between the corner nodes possess the highest accuracy.

In order to determine the strains and thus the matrix $\underline{B}$ according to (4.28), the derivatives of the displacements with respect to x and y are required. It is

$$\begin{bmatrix} \dfrac{\partial}{\partial r} \\[2ex] \dfrac{\partial}{\partial s} \end{bmatrix} = \begin{bmatrix} \dfrac{\partial x}{\partial r} & \dfrac{\partial y}{\partial r} \\[2ex] \dfrac{\partial x}{\partial s} & \dfrac{\partial y}{\partial s} \end{bmatrix} \cdot \begin{bmatrix} \dfrac{\partial}{\partial x} \\[2ex] \dfrac{\partial}{\partial y} \end{bmatrix} = \begin{bmatrix} j_{11} & j_{12} \\[2ex] j_{21} & j_{22} \end{bmatrix} \cdot \begin{bmatrix} \dfrac{\partial}{\partial x} \\[2ex] \dfrac{\partial}{\partial y} \end{bmatrix}, \tag{4.45}$$

$$\frac{\partial}{\partial r} = \underline{J} \cdot \frac{\partial}{\partial x}. \tag{4.45a}$$

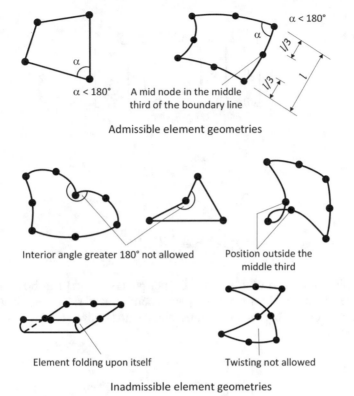

Admissible element geometries

Inadmissible element geometries

Fig. 4.34 Admissible and inadmissible geometries of isoparametric elements

Herein $\underline{J}$ is the so-called Jacobian matrix or Jacobian operator relating the element cooordinates r, s derivatives to the global coordinates x, y derivatives. The entries of the derivative matrix are functions of r, s and can be determined easily with (4.44a, 4.44b), as

$$j_{11} = \frac{\partial x}{\partial r} = \sum_i \frac{\partial h_i}{\partial r} \cdot x_i, \quad j_{12} = \frac{\partial y}{\partial r} = \sum_i \frac{\partial h_i}{\partial r} \cdot y_i$$

$$j_{21} = \frac{\partial x}{\partial s} = \sum_i \frac{\partial h_i}{\partial s} \cdot x_i, \quad j_{22} = \frac{\partial y}{\partial s} = \sum_i \frac{\partial h_i}{\partial s} \cdot y_i. \tag{4.46}$$

With the inverse of $\underline{J}$ one obtains

$$\frac{\partial}{\partial \underline{x}} = \underline{J}^{-1} \cdot \frac{\partial}{\partial \underline{r}} \tag{4.47}$$

or

$$\begin{bmatrix} \dfrac{\partial}{\partial x} \\[2ex] \dfrac{\partial}{\partial y} \end{bmatrix} = \begin{bmatrix} i_{11} & i_{12} \\ i_{21} & i_{22} \end{bmatrix} \cdot \begin{bmatrix} \dfrac{\partial}{\partial r} \\[2ex] \dfrac{\partial}{\partial s} \end{bmatrix} \tag{4.47a}$$

and

$$\frac{\partial}{\partial \underline{x}} = \underline{I}_J \cdot \frac{\partial}{\partial \underline{r}} \tag{4.47b}$$

with

$$\underline{I}_J = \underline{J}^{-1} = \begin{bmatrix} i_{11} & i_{12} \\ i_{21} & i_{22} \end{bmatrix}. \tag{4.47c}$$

An explicit inversion of $\underline{J}$ is generally not possible, but $\underline{J}$ can be numerically determined and inverted for the given point coordinates r, s. The matrix $\underline{B}$, which gives the relationship between the strains and the nodal displacements, is obtained according to (4.28), as

$$\begin{bmatrix} \varepsilon_x \\ \varepsilon_y \\ \gamma_{xy} \end{bmatrix} = \begin{bmatrix} \dfrac{\partial u}{\partial x} \\[1.5ex] \dfrac{\partial v}{\partial y} \\[1.5ex] \dfrac{\partial u}{\partial y} + \dfrac{\partial v}{\partial x} \end{bmatrix} = \begin{bmatrix} i_{11} \cdot \dfrac{\partial u}{\partial r} + i_{12} \dfrac{\partial u}{\partial s} \\[1.5ex] i_{21} \cdot \dfrac{\partial v}{\partial r} + i_{22} \dfrac{\partial v}{\partial s} \\[1.5ex] i_{11} \cdot \dfrac{\partial v}{\partial r} + i_{12} \dfrac{\partial v}{\partial s} + i_{21} \cdot \dfrac{\partial u}{\partial r} + i_{22} \dfrac{\partial u}{\partial s} \end{bmatrix}$$

$$= \begin{bmatrix} \sum\limits_i \left(i_{11} \cdot \dfrac{\partial h_i}{\partial r} + i_{12} \dfrac{\partial h_i}{\partial s} \right) \cdot u_i \\[2ex] \sum\limits_i \left(i_{21} \cdot \dfrac{\partial h_i}{\partial r} + i_{22} \dfrac{\partial h_i}{\partial s} \right) \cdot v_i \\[2ex] \sum\limits_i \left(\left(i_{21} \cdot \dfrac{\partial h_i}{\partial r} + i_{22} \dfrac{\partial h_i}{\partial s} \right) \cdot u_i + \left(i_{11} \cdot \dfrac{\partial h_i}{\partial r} + i_{12} \dfrac{\partial h_i}{\partial s} \right) \cdot v_i \right) \end{bmatrix}.$$

$$\tag{4.48}$$

Writing the nodal point displacements as vector $\underline{u}_e$, one finally obtains the matrix $\underline{B}$. The stiffness matrix can be written according to (4.40e)

$$\underline{K}_e = \int \int t \cdot \underline{B}^T \cdot \underline{D} \cdot \underline{B} \, dx \, dy,$$

where the integration has to be performed over the element area. For isoparametric elements, the integration is done over $-1 \le r \le 1$ and $-1 \le s \le 1$ in natural element coordinates. For it with

$$dx\, dy = \det(\underline{J}) \cdot dr\, ds \tag{4.49}$$

(see [38]) the equation is written as

$$\underline{K}_e = \int_{-1}^{1} \int_{-1}^{1} t \cdot \underline{B}^{\mathrm{T}} \cdot \underline{D} \cdot \underline{B} \cdot \det(\underline{J})\ dr\, ds. \tag{4.50}$$

The integral normally cannot be solved analytically. Therefore, the stiffness matrix of isoparametric elements has to be calculated by numerical integration. For this, the matrix $\underline{B}$, as well as the Jacobian matrix, are determined at the coordinates $r_j,\ s_k$ of the integration points j, k (from 1 to n) and one obtains

$$\underline{K}_e = t \cdot \sum_{j=1}^{n} \sum_{k=1}^{n} \alpha_j \cdot \alpha_k \cdot \underline{B}_{jk}^{\mathrm{T}} \cdot \underline{D} \cdot \underline{B}_{jk} \cdot \det \underline{J}_{jk}. \tag{4.51}$$

The parameters α_i for the numerical integration are given for a *Gaussian integration* in Table 4.7 [23]. The matrix $\underline{D}$ of the material law is given in Table 2.1, for the states of plane stress and plane strain. A more detailed derivation of stiffness matrix is given in [34].

For the general 4-node quadrilateral element in Fig. 4.35, the relationships required for the numerical determination of the stiffness matrix shall be given. The Jacobian matrix at the integration point $r = r_j$ and $s = s_k$ is

$$\underline{J}_{jk} = \begin{bmatrix} j_{11} & j_{12} \\ j_{21} & j_{22} \end{bmatrix}\ \text{with}$$

Fig. 4.35 4-node quadrilateral element

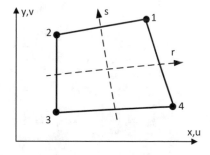

Table 4.7 Gaussian numerical integration of a function in the interval Δx

Integration order n	Formula $\int_{x_a}^{x_a+\Delta x} f(x)\, dx = \sum_i f(x_i) \cdot \alpha_i \cdot \frac{\Delta x}{2}$	Location of integration points $r = r_j$ and $s = s_k$ in plane finite elements
1-point integration $n = 1$	$\int_{x_a}^{x_a+\Delta x} f(x)\, dx = f(x_1) \cdot \alpha_1 \cdot \frac{\Delta x}{2}$ $\alpha_1 = 2$ Linear function is integrated exactly	
2-point integration $n = 2,\ a = \xi_1 \cdot \frac{\Delta x}{2}$	$\int_{x_a}^{x_a+\Delta x} f(x)\, dx =$ $(\alpha_1 f(x_1) + \alpha_2 f(x_2)) \cdot \frac{\Delta x}{2}$ $\alpha_1 = 1 \quad \alpha_2 = 1$ $\xi_1 = 1/\sqrt{3} \approx 0.577$ 3rd degree polynomial is integrated exactly	
3-point integration $n = 3,\ a = \xi_1 \cdot \frac{\Delta x}{2}$	$\int_{x_a}^{x_a+\Delta x} f(x)\, dx = (\alpha_1 \cdot f(x_1)$ $+ \alpha_2 \cdot f(x_2) + \alpha_3 \cdot f(x_3)) \cdot \frac{\Delta x}{2}$ $\alpha_1 = \alpha_3 = 5/9 \approx 0.556$ $\alpha_2 = 8/9 \approx 0.889$ $\xi_1 = \sqrt{3/5} \approx 0.775$ 5th degree polynomial is integrated exactly	

$$j_{11} = \left.\frac{\partial x}{\partial r}\right|_{r_j,s_k} = \frac{1}{4}\cdot(1+s_k)\cdot x_1 - \frac{1}{4}\cdot(1+s_k)\cdot x_2$$

$$- \frac{1}{4}\cdot(1-s_k)\cdot x_3 + \frac{1}{4}\cdot(1-s_k)\cdot x_4$$

$$j_{21} = \left.\frac{\partial x}{\partial s}\right|_{r_j,s_k} = \frac{1}{4}\cdot(1+r_j)\cdot x_1 + \frac{1}{4}\cdot(1-r_j)\cdot x_2$$

$$- \frac{1}{4}\cdot(1-r_j)\cdot x_3 - \frac{1}{4}\cdot(1+r_j)\cdot x_4$$

$$j_{12} = \left.\frac{\partial y}{\partial r}\right|_{r_j,s_k} = \frac{1}{4} \cdot (1 + s_k) \cdot y_1 - \frac{1}{4} \cdot (1 + s_k) \cdot y_2$$

$$- \frac{1}{4} \cdot (1 - s_k) \cdot y_3 + \frac{1}{4} \cdot (1 - s_k) \cdot y_4$$

$$j_{22} = \left.\frac{\partial y}{\partial s}\right|_{r_j,s_k} = \frac{1}{4} \cdot (1 + r_j) \cdot y_1 + \frac{1}{4} \cdot (1 - r_j) \cdot y_2$$

$$- \frac{1}{4} \cdot (1 - r_j) \cdot y_3 - \frac{1}{4} \cdot (1 + r_j) \cdot y_4. \tag{4.52}$$

The inversion of the Jacobian matrix at point $r = r_j$ and $s = s_k$ gives

$$\begin{bmatrix} i_{11} & i_{12} \\ i_{21} & i_{22} \end{bmatrix} = \begin{bmatrix} j_{11} & j_{12} \\ j_{21} & j_{22} \end{bmatrix}^{-1}.$$

Herewith the matrix $\underline{B}$ at $r = r_j$ and $s = s_k$ can be determined numerically. It is

$$\begin{bmatrix} \varepsilon_x \\ \varepsilon_y \\ \gamma_{xy} \end{bmatrix} = \begin{bmatrix} a_1 & 0 & a_2 & 0 & a_3 & 0 & a_4 & 0 \\ 0 & b_1 & 0 & b_2 & 0 & b_3 & 0 & b_4 \\ b_1 & a_1 & b_2 & a_2 & b_3 & a_3 & b_4 & a_4 \end{bmatrix} \cdot \begin{bmatrix} u_1 \\ v_1 \\ u_2 \\ v_2 \\ u_3 \\ v_3 \\ u_4 \\ v_4 \end{bmatrix} \tag{4.53}$$

$$\left.\underline{\varepsilon}\right|_{r_j,s_k} = \underline{B}_{jk} \cdot \underline{u}_e \tag{4.53a}$$

with

$$a_i = i_{11} \cdot \left.\frac{\partial h_i}{\partial r}\right|_{r_j,s_k} + i_{12} \left.\frac{\partial h_i}{\partial s}\right|_{r_j,s_k}, \quad b_i = i_{21} \cdot \left.\frac{\partial h_i}{\partial r}\right|_{r_j,s_k} + i_{22} \left.\frac{\partial h_i}{\partial s}\right|_{r_j,s_k} \tag{4.54a,b}$$

and

$$\left.\frac{\partial h_1}{\partial r}\right|_{r_j,s_k} = \frac{1}{4} \cdot (1 + s_k), \qquad \left.\frac{\partial h_2}{\partial r}\right|_{r_j,s_k} = -\frac{1}{4} \cdot (1 + s_k),$$

$$\left.\frac{\partial h_3}{\partial r}\right|_{r_j,s_k} = -\frac{1}{4} \cdot (1 - s_k), \qquad \left.\frac{\partial h_4}{\partial r}\right|_{r_j,s_k} = \frac{1}{4} \cdot (1 - s_k)$$

$$\left.\frac{\partial h_1}{\partial s}\right|_{r_j,s_k} = \frac{1}{4} \cdot (1 + r_j), \qquad \left.\frac{\partial h_2}{\partial s}\right|_{r_j,s_k} = \frac{1}{4} \cdot (1 - r_j),$$

$$\left.\frac{\partial h_3}{\partial s}\right|_{r_j,s_k} = -\frac{1}{4} \cdot (1 - r_j), \quad \left.\frac{\partial h_4}{\partial s}\right|_{r_j,s_k} = -\frac{1}{4} \cdot (1 + r_j). \qquad (4.55\text{a-h})$$

Herewith, all the relationships required for numerical determination of the stiffness matrix according to (4.40e) are provided. The stiffness matrix of isoparametric elements can be calculated numerically efficiently. Isoparametric elements for plane stress are implemented in many finite element programs and are popular in practical application.

In order to perform the numerical integration, Gaussian integration has been proven to be particularly efficient. The sampling points where the function which has to be integrated must be known, do not lie at the boundaries of the interval (as for the trapezoidal rule) but at certain points inside the interval, the so-called Gaussian points (Table 4.7). The position of the Gaussian points is determined such that an optimal accuracy of the numerical integration is achieved. In general, with n Gaussian sampling points a polynomial of order $(2n - 1)$ is exactly integrated. For example, the Gaussian 2-point integration integrates a third degree polynomial exactly, whereas the trapezoidal rule with two supporting points integrates exactly only a linear function. For plane finite elements this integration has to be done over both element coordinates. This is denoted as 2×2 or 3×3 Gausian integration, respectively.

Example 4.10 For the rectangular element shown in Fig. 4.36, the stiffness matrix shall be determined numerically using the equations of the general isoparametric quadrilateral element. The element parameters are identical to those in Example 4.5. The numerically determined stiffness matrix is to be compared with that in Example 4.5 which was determined with the analytical integration of the element matrix according to (4.35a, 4.36).

The nodal point coordinates are

$$x_1 = x_4 = 1.0, \quad x_2 = x_3 = 0, \quad y_1 = y_2 = 0.5, \quad y_3 = y_4 = 0 \text{ [m]}.$$

Matrix $\underline{B}$ is linear in x and y for the rectangular element according to (4.29). This results in a quadratic expression for the stiffness matrix according to (4.35a) or (4.40e). It is calculated exactly with the Gaussian 2-point integration used here (Table 4.7). The stiffness matrix computed with the 2-point integration formula must, therefore, exactly match that of the analytically integrated stiffness matrix of the rectangular element according to (4.36), in Example 4.5.

The integration points and the corresponding parameters α are (according to Table 4.7)

Fig. 4.36 4-node
rectangular element

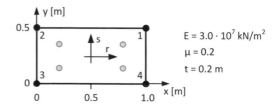

j	k	r_j	s_k	α_j	α_k
1	1	0.577	0.577	1.0	1.0
1	2	−0.577	0.577	1.0	1.0
2	1	−0.577	−0.577	1.0	1.0
2	2	0.577	−0.577	1.0	1.0

The derivatives of the shape functions at the integration points are obtained according to (4.55a-h), as

j	k	$\frac{\partial h_1}{\partial r}$	$\frac{\partial h_2}{\partial r}$	$\frac{\partial h_3}{\partial r}$	$\frac{\partial h_4}{\partial r}$	$\frac{\partial h_1}{\partial s}$	$\frac{\partial h_2}{\partial s}$	$\frac{\partial h_3}{\partial s}$	$\frac{\partial h_4}{\partial s}$
1	1	0.394	−0.394	−0.106	0.106	0.394	0.106	−0.106	−0.394
1	2	0.394	−0.394	−0.106	0.106	0.106	0.394	−0.394	−0.106
2	1	0.106	−0.106	−0.394	0.394	0.106	0.394	−0.394	−0.106
2	2	0.106	−0.106	−0.394	0.394	0.394	0.106	−0.106	−0.394

The Jacobian matrix and its determinant and inverse are now determined at the integration points with (4.52) and (4.47c). With (4.53) and 4.54a,b), the matrices $\underline{B}_{ij}$ are calculated.

Integration point $j = 1, k = 1$:

$$\underline{J}_{11} = \begin{bmatrix} 0.500 & 0 \\ 0 & 0.250 \end{bmatrix}, \quad \det \underline{J}_{11} = 0.125, \quad \underline{I}_{11} = \underline{J}_{11}^{-1} = \begin{bmatrix} 2.000 & 0 \\ 0 & 4.000 \end{bmatrix}$$

$$\underline{B}_{11} = \begin{bmatrix} 0.789 & 0 & -0.789 & 0 & -0.211 & 0 & 0.211 & 0 \\ 0 & 1.577 & 0 & 0.423 & 0 & -0.423 & 0 & -1.577 \\ 1.577 & 0.789 & 0.423 & -0.789 & -0.423 & -0.211 & -1.577 & 0.211 \end{bmatrix}$$

Integration point $j = 1, k = 2$:

$$\underline{J}_{12} = \begin{bmatrix} 0.500 & 0 \\ 0 & 0.250 \end{bmatrix}, \quad \det \underline{J}_{12} = 0.125, \quad \underline{I}_{12} = \underline{J}_{12}^{-1} = \begin{bmatrix} 2.000 & 0 \\ 0 & 4.000 \end{bmatrix}$$

$$\underline{B}_{12} = \begin{bmatrix} 0.789 & 0 & -0.789 & 0 & -0.211 & 0 & 0.211 & 0 \\ 0 & 0.423 & 0 & 1.577 & 0 & -1.577 & 0 & -0.423 \\ 0.423 & 0.789 & 1.577 & -0.789 & -1.577 & -0.211 & -0.423 & 0.211 \end{bmatrix}$$

Integration point $j = 2, k = 1$:

$$\underline{J}_{21} = \begin{bmatrix} 0.500 & 0 \\ 0 & 0.250 \end{bmatrix}, \quad \det \underline{J}_{21} = 0.125, \quad \underline{I}_{21} = \underline{J}_{21}^{-1} = \begin{bmatrix} 2.000 & 0 \\ 0 & 4.000 \end{bmatrix}$$

$$\underline{B}_{21} = \begin{bmatrix} 0.211 & 0 & -0.211 & 0 & -0.789 & 0 & 0.789 & 0 \\ 0 & 0.423 & 0 & 1.577 & 0 & -1.577 & 0 & -0.423 \\ 0.423 & 0.211 & 1.577 & -0.211 & -1.577 & -0.789 & -0.423 & 0.789 \end{bmatrix}$$

Integration point $j = 2, k = 2$

$$\underline{J}_{22} = \begin{bmatrix} 0.500 & 0 \\ 0 & 0.250 \end{bmatrix}, \quad \det \underline{J}_{22} = 0.125, \quad \underline{J}_{22}^{-1} = \underline{J}_{22}^{-1} = \begin{bmatrix} 2.000 & 0 \\ 0 & 4.000 \end{bmatrix}$$

$$\underline{B}_{22} = \begin{bmatrix} 0.211 & 0 & -0.211 & 0 & -0.789 & 0 & 0.789 & 0 \\ 0 & 1.577 & 0 & 0.423 & 0 & -0.423 & 0 & -1.577 \\ 1.577 & 0.211 & 0.423 & -0.211 & -0.423 & -0.789 & -1.577 & 0.789 \end{bmatrix}$$

The material law of a isotropic material in plane stress state is according to Table 2.1

$$\underline{D} = \frac{E}{1-\mu^2} \begin{bmatrix} 1 & \mu & 0 \\ \mu & 1 & 0 \\ 0 & 0 & \dfrac{1-\mu}{2} \end{bmatrix} = \begin{bmatrix} 3.125 & 0.625 & 0 \\ 0.625 & 3.125 & 0 \\ 0 & 0 & 1.250 \end{bmatrix} \cdot 10^7$$

The stiffness matrix is obtained with (4.51) or

$$\underline{K}_e = (\alpha_1 \cdot \alpha_1 \cdot \underline{B}_{11}^T \cdot \underline{D} \cdot \underline{B}_{11} \cdot \det \underline{J}_{11} + \alpha_1 \cdot \alpha_2 \cdot \underline{B}_{12}^T \cdot \underline{D} \cdot \underline{B}_{12} \cdot \det \underline{J}_{12}$$
$$+ \alpha_2 \cdot \alpha_1 \cdot \underline{B}_{21}^T \cdot \underline{D} \cdot \underline{B}_{21} \cdot \det \underline{J}_{21} + \alpha_2 \cdot \alpha_2 \cdot \underline{B}_{22}^T \cdot \underline{D} \cdot \underline{B}_{22} \cdot \det \underline{J}_{22}) \cdot t$$

as

$$\underline{K}_e = \begin{bmatrix} 2.708 & 0.938 & -0.208 & -0.312 & -1.354 & -0.938 & -1.146 & 0.312 \\ 0.938 & 4.583 & 0.312 & 1.667 & -0.938 & -2.292 & -0.312 & -3.958 \\ -0.208 & 0.312 & 2.708 & -0.938 & -1.146 & -0.312 & -1.354 & 0.938 \\ -0.312 & 1.667 & -0.938 & 4.583 & 0.312 & -3.958 & 0.938 & -2.292 \\ -1.354 & -0.938 & -1.146 & 0.312 & 2.708 & 0.938 & -0.208 & -0.312 \\ -0.938 & -2.292 & -0.312 & -3.958 & 0.938 & 4.583 & 0.312 & 1.667 \\ -1.146 & -0.312 & -1.354 & 0.938 & -0.208 & 0.312 & 2.708 & -0.938 \\ 0.312 & -3.958 & 0.938 & -2.292 & -0.312 & 1.667 & -0.938 & 4.583 \end{bmatrix}$$
$$\cdot 10^6 \text{ kN/m}.$$

As expected, it corresponds exactly with the matrix of the rectangular element with bilinear displacement shape functions calculated according to (4.36), in Example 4.5. (The different numbering of the nodes in Figs. 4.18 and 4.36 results in the same stiffness matrix after their cyclical interchanging or, alternatively, with $x_{\text{Figure 4.36}} = -x_{\text{Figure 4.18}}$ and $y_{\text{Figure 4.36}} = -y_{\text{Figure 4.18}}$).   ◀

Table 4.8 Integration order of isoparametric plane stress elements

Element type	Full integration	Element type	Full integration
4-node rectangular	2×2	8-node rectangular	3×3
4-node quadrilateral	3×3	8-node element	4×4

The required order of integration depends on the polynomial degree of the shape functions and of the element geometry. In a full integration, the stiffness matrix of rectangular elements as in Example 4.10, is computed exactly (Table 4.8). For elements whose geometrical shape deviates from the rectangle and for which an exact integration according to Gauss is not possible, the stiffness matrix is calculated with high accuracy with full integration order [34].

It has been suggested to integrate the stiffness matrix with lower order than for full integration in order to develop more precise and efficient elements, even if this seems mathematically inadmissible. This is called *reduced integration* or *underintegration*. *Selective integration* is defined as the integration of individual stress components with different integration orders. The aim of a reduced integration is to counteract the overestimation of elemental stiffness based on displacement shape functions. For this purpose, stiffness components are deliberately omitted in the summation of terms of the stiffness matrix in (4.51), and a "softer" element is obtained. To what extent the two influences compensate each other, however, cannot be proven exactly. Furthermore, the reduced integration is not always unproblematic. Some element types become kinematic due to the neglect of stiffness components, i.e. for certain deformation configurations (or loadings) the element has no stiffness and becomes unstable (e.g. the 4-node rectangle element with 1-point integration). The displacement patterns occurring are typically "zigzag-shaped" and are also referred to as "hourglass modes" or "zero energy modes". These modes have no deformation energy, and therefore, no influence on the stresses. Whether the disturbance of the displacements is only slight or considerably, depends on the system to be calculated and its loading. In order to avoid this effect, so-called *stabilization techniques* have been developed. Here, special stiffness contributions in the hourglass modes are added to the stiffness matrix [39, 40].

Even though in many cases it leads to better results, reduced integration must be viewed critically. Here, reference is made to the discussion in [14, 19, 24, 34]. In [34], Bathe asks the question whether the reliability of the elements is given in all practical situations. Furthermore, with reduced integration, the condition for monotonous convergence is no longer met due to the lack of stiffness components. The elements thus no longer possess the mathematically proven convergence properties of conformal elements. Bathe advises against a reduced integration in [34] (pages 469/472) and Knothe/Wessels in [19] (page 265).

4.5.3 Nonconforming Elements

The isoparametric 4-node quadrilateral element exhibits a stiff behavior when deformed in bending, which is also referred to as shear locking. To improve the bending behavior of the element, as early as the 1970s, an extension of the bilinear shape functions by quadratic terms has been suggested [41, 42]. For this, at each side of the element a displacement degree of freedom perpendicular to the element is introduced, which is not connected with other elements. When the element is deformed, these degrees of freedom show relative displacements to the adjacent elements, i.e. the displacement functions at the boundaries of two adjacent elements are not compatible between the nodal points (Fig. 4.37). The elements are denoted as *nonconforming elements* or *incompatible elements*, whereas elements with continuous displacements are called *conforming elements* or *compatible elements*. The additional degrees of freedom relating only to the corresponding element can be eliminated in advance by a static condensation. In this way a 4-node element with an 8×8 stiffness matrix is obtained [42].

With the additional quadratic shape functions the nonconforming element is able to reproduce pure bending deformations exactly. This is because the bending line of a beam under pure bending with the shear force $V = 0$ is a quadratic parabola.

The nonconforming element does not fulfill the continuity equation which is a precondition for monotonic convergence (see Sect. 4.5.1). However, rectangular and parallelogram-shaped elements fulfill the patch test as an indication for non-monotone convergence. A general proof of the convergence for general quadrilateral elements is not possible for the element according to [42]. Here, reference is made to the discussion in [19], but also to the further development of the element after [43], where the patch test is fulfilled. Many programs in which the isoparametric 4-node element is implemented also contain its extension as a nonconforming element.

An extension of the nonconforming elements is the *Enhanced Assumed Strain* (EAS) Method [44], which has been developed to avoid locking effects. Reference is made here to [45], and the newer development in [46–48] .

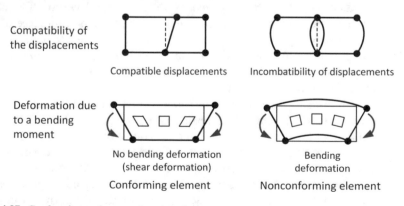

Fig. 4.37 Conforming and nonconforming elements

Example 4.11 The influence of incompatible displacement shape functions on the results of a finite element analysis will be shown by the example of a deep beam and a beam in bending modeled by plane stress finite elements.

The deep beam shown in Fig. 4.38a, has already been investigated in Example 4.7, with conforming elements and several finite element discretizations. It is now calculated with nonconforming 4-node elements according to [42] with [SOF 6]. In Table 4.9, the stresses $\sigma_{x,m}$ at the mid of the lower edge (point m) are given. While there are some differences between the results for the very coarse 2×2 and 4×4 meshes, this is not the case for practically relevant mesh sizes such as the 8×8 mesh. This means that in the case of the deep beam, the incompatible shape functions have practically no influence on the stresses.

If the height of the beam is reduced to 1.0 m and the load to 0.01 MN/m, the beam-like structure in Fig. 4.38b, is obtained with otherwise identical characteristic values. The bending stresses in the mid of the beam according to the elementary beam theory are regarded as the reference values. These are obtained with the bending moment

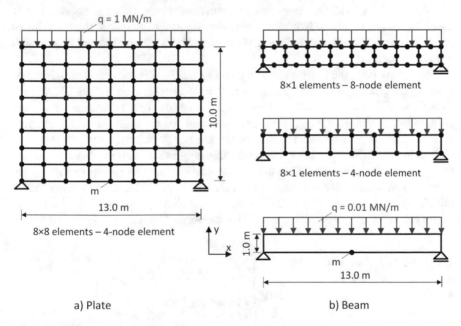

a) Plate   b) Beam

Fig. 4.38 Analysis of a deep beam and a beam in bending with conforming and nonconforming elements

Table 4.9 Convergence of the stresses $\sigma_{x,m}\left[\text{MN/m}^2\right]$ of a deep beam

FE mesh	Conforming elements	Nonconforming elements
2×2	1.66	2.28
4×4	4.32	4.72
8×8	4.22	4.22

M_m and the section modulus W

$$M_m = \frac{q \cdot l^2}{8} = \frac{0.01 \cdot 13^2}{8} = 0.211 \, \text{MNm}, \quad W = \frac{t \cdot h^2}{6} = 0.083 \, \text{m}^3$$

as

$$\sigma_{x,m} = \frac{M_m}{W} = \frac{0.211}{0.083} = 2.535 \, \text{MN/m}^2.$$

The beam in bending is discretized into a single element via its height (Fig. 4.38b). This corresponds to the Bernoulli hypothesis of the elementary beam theory as the bilinear displacement shape functions of the 4-node element are linear over the height of the element. The stresses $\sigma_{x,m}$ obtained with conforming and nonconforming 4-node elements are given in Table 4.10. It can be seen that the nonconforming elements show a much faster convergence than the conforming elements. With 8 nonconforming elements only, the reference value $\sigma_{x,m}$ is obtained with a deviation of 3%. To get $\sigma_{x,m}$ with the same accuracy with conforming elements approximately 64 elements are required. Nonconforming 4-node elements reflect the behavior of plates in plane stress subjected to bending much better than classical conforming 4-node elements.

For comparison, the beam in bending, in Fig. 4.38b is investigated with isoparametric (conforming) 8-node elements and a 2×2 integration with [SOF 2] (element PLANE82). The elements have quadratic shape functions of the displacements in the x-direction, similar to the additional displacements of the nonconforming element according to [42]. However, the side nodes are not "condensed out" so that they remain in the global equation system, which is larger than with the 4-node nonconforming element. The results for the 8-node element are also given in Table 4.10. The element is not fully integrated (see Table 4.8), so the element is softer than a fully integrated 8-node element, and the monotonic convergence of the displacements "from below" is not given. The stresses also converge "from above" against the limit

Table 4.10 Convergence of stresses $\sigma_{x,m} \left[\text{MN/m}^2 \right]$ of a beam in bending

FE mesh	4-node element		8-node conforming element 2×2 underintegrated
	Conforming Elements	Nonconforming Elements	
2×1	0.06	1.27	2.96
4×1	0.35	2.22	2.64
8×1	1.06	2.45	2.56
16×1	1.90	2.51	2.54
32×1	2.34	2.53	-
64×1	2.48	-	-
128×1	2.52	-	-

value. With 4 elements, the difference from the reference value is 4%, and with 8 elements, the reference value is only 1%. Convergence can, therefore, be considered to be very good.

The example shows that the conforming 4-node element is not well suited for the modeling of structural parts subjected to bending due to the shear locking effect. While the solution converges against the limit as it must be, it does it very slowly, requiring a large number of elements for sufficient accuracy. For this reason, nonconforming elements or conforming elements with quadratic or higher shape functions should be used for parts of plates in plane stress subjected to bending.   ◄

4.5.4 Hybrid Elements

Already in the early history of the finite element method, alternatives to displacement-based elements have been sought. As stresses are of main interest in engineering, shape functions for stresses are an evident choice. However, such approaches have not been successful since unknowns at the nodal points in addition to stresses no longer have a physical meaning, and the coupling of different types of structural elements with stresses as unknowns is no longer possible.

Exceptions are the *hybrid elements*. They are also denoted as hybrid stress models or *mixed finite element formulations*, while finite elements with displacement shape functions are also denoted as *displacement-based elements* or *displacement models*. Hybrid elements are based on additional stress shape functions. However, like the displacement models, the element matrices contain only displacements as nodal parameters. Hence, the resulting matrices are element stiffness matrices. This allows hybrid elements to be treated like elements with pure displacement shape functions within the framework of a finite element analysis. Therefore, the term "mixed finite element formulation" is more precise, since the formulation is based on shape functions for displacements, as well as for stresses.

The basic assumptions in the formulation of a finite element are shape functions for stresses inside the element and shape functions for displacements only at the element edges. The formulation is explained using the example of the simple rectangular plane stress element according to [49] (Fig. 4.39).

Linear shape functions for stresses are chosen according to [49], as (Fig. 4.40)

$$\sigma_x = \beta_1 + \beta_4 \cdot y$$

$$\sigma_y = \beta_2 + \beta_5 \cdot x$$

$$\tau_{xy} = \beta_3$$

or

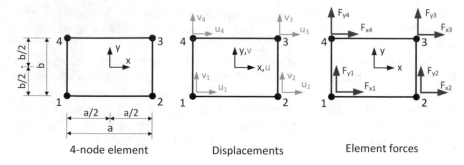

Fig. 4.39 Hybrid plane stress element

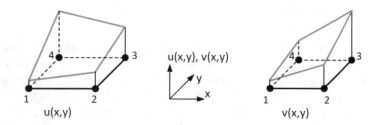

Shape functions for displacements at the element boundaries

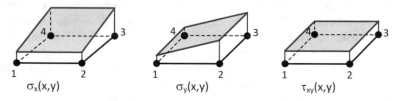

Shape functions for stresses inside the element

Fig. 4.40 Shape functions of a hybrid plane stress element

$$
\begin{bmatrix} \sigma_x \\ \sigma_y \\ \tau_{xy} \end{bmatrix} = \begin{bmatrix} 1 & 0 & 0 & y & 0 \\ 0 & 1 & 0 & 0 & x \\ 0 & 0 & 1 & 0 & 0 \end{bmatrix} \cdot \begin{bmatrix} \beta_1 \\ \beta_2 \\ \beta_3 \\ \beta_4 \\ \beta_5 \end{bmatrix} \tag{4.56}
$$

$$
\underline{\sigma} = \underline{P} \cdot \underline{\beta}. \tag{4.56a}
$$

Initially the parameters $\beta_1 - \beta_5$ are free parameters. The functions (4.56), fulfill the equilibrium conditions in the element as can be seen easily by introducing them into the equilibrium conditions according to Fig. 2.5

$$\frac{\partial \sigma_x}{\partial x} + \frac{\partial \tau_{xy}}{\partial y} = 0$$

$$\frac{\partial \sigma_y}{\partial y} + \frac{\partial \tau_{xy}}{\partial x} = 0.$$

If element loads are to be taken into account, the stress approach (4.56), must be extended so that the equilibrium conditions are also fulfilled by these. The stress functions correspond to boundary stresses σ_R at each edge of the finite element, which can be determined with (4.56), as

Edge 1–2:

$$\sigma_{y,1-2} = -\beta_2 - \beta_5 \cdot x$$
$$\tau_{xy,1-2} = -\beta_3$$

Edge 2–3:

$$\sigma_{x,2-3} = \beta_1 + \beta_4 \cdot y$$
$$\tau_{xy,2-3} = \beta_3$$

Edge 3–4:

$$\sigma_{y,3-4} = \beta_2 + \beta_5 \cdot x$$
$$\tau_{xy,3-4} = \beta_3$$

Edge 4–1:

$$\sigma_{x,4-1} = -\beta_1 - \beta_4 \cdot y$$
$$\tau_{xy,4-1} = -\beta_3$$

or

$$
\begin{bmatrix}
\sigma_{y,1-2} \\
\tau_{xy,1-2} \\
\sigma_{x,2-3} \\
\tau_{xy,2-3} \\
\sigma_{y,3-4} \\
\tau_{xy,3-4} \\
\sigma_{x,4-1} \\
\tau_{xy,4-1}
\end{bmatrix}
=
\begin{bmatrix}
0 & -1 & 0 & 0 & -x \\
0 & 0 & -1 & 0 & 0 \\
1 & 0 & 0 & y & 0 \\
0 & 0 & 1 & 0 & 0 \\
0 & 1 & 0 & 0 & x \\
0 & 0 & 1 & 0 & 0 \\
-1 & 0 & 0 & -y & 0 \\
0 & 0 & -1 & 0 & 0
\end{bmatrix}
\cdot
\begin{bmatrix}
\beta_1 \\
\beta_2 \\
\beta_3 \\
\beta_4 \\
\beta_5
\end{bmatrix}
\tag{4.57}
$$

$$\underline{\sigma}_R = \underline{P}_R \cdot \underline{\beta}. \tag{4.57a}$$

The signs are chosen in such a way that the stresses at the edges perform positive work with positive displacements u and v.

In addition to the approximation of stresses inside the element, hybrid elements also introduce an approximation for the displacements on the edges of the element. With the 4-node element, the displacements between the nodes are linear (Fig. 4.40). For the rectangular they are obtained as

$$
\begin{bmatrix} u_{1-2} \\ v_{1-2} \\ u_{2-3} \\ v_{2-3} \\ u_{3-4} \\ v_{3-4} \\ u_{4-1} \\ v_{4-1} \end{bmatrix}
=
\begin{bmatrix}
\frac{1}{2}-\frac{x}{a} & 0 & \frac{1}{2}+\frac{x}{a} & 0 & 0 & 0 & 0 & 0 \\
0 & \frac{1}{2}-\frac{x}{a} & 0 & \frac{1}{2}+\frac{x}{a} & 0 & 0 & 0 & 0 \\
0 & 0 & \frac{1}{2}-\frac{y}{b} & 0 & \frac{1}{2}+\frac{y}{b} & 0 & 0 & 0 \\
0 & 0 & 0 & \frac{1}{2}-\frac{y}{b} & 0 & \frac{1}{2}+\frac{y}{b} & 0 & 0 \\
0 & 0 & 0 & 0 & \frac{1}{2}+\frac{x}{a} & 0 & \frac{1}{2}-\frac{x}{a} & 0 \\
0 & 0 & 0 & 0 & 0 & \frac{1}{2}+\frac{x}{a} & 0 & \frac{1}{2}-\frac{x}{a} \\
\frac{1}{2}-\frac{y}{b} & 0 & 0 & 0 & 0 & 0 & \frac{1}{2}+\frac{y}{b} & 0 \\
0 & \frac{1}{2}-\frac{y}{b} & 0 & 0 & 0 & 0 & 0 & \frac{1}{2}+\frac{y}{b}
\end{bmatrix}
\cdot
\begin{bmatrix} u_1 \\ v_1 \\ u_2 \\ v_2 \\ u_3 \\ v_3 \\ u_4 \\ v_4 \end{bmatrix}
\tag{4.58}
$$

$$
\underline{u}_R = \underline{H}_R \cdot \underline{u}_e \tag{4.58a}
$$

The strains $\underline{\varepsilon}$ corresponding to the stresses $\underline{\sigma}$ are obtained by Hook's law (2.3c) and (4.56a), as

$$
\underline{\varepsilon} = \underline{D}^{-1} \cdot \underline{\sigma} = \underline{D}^{-1} \cdot \underline{P} \cdot \underline{\beta}. \tag{4.59}
$$

The strains must be kinematically compatible with each other and with the displacements $\underline{u}_R$ at the edges. Since the approximations for the displacements at the edges were chosen independently of the stress approximation inside the element, this is not the case at first. An approximate compatibility can be achieved by a suitable selection of the still free parameters $\underline{\beta}$. For this purpose, the principle of virtual stresses is applied, which corresponds to the principle of virtual forces in structural mechanics. Virtual stresses are introduced inside the element, for which the same shape functions are selected after (4.56a), as for the real stresses, i.e. in the element

$$\bar{\underline{\sigma}} = \underline{P} \cdot \bar{\underline{\beta}} \tag{4.60}$$

and on the edges

$$\bar{\underline{\sigma}}_R = \underline{P}_R \cdot \bar{\underline{\beta}}. \tag{4.61}$$

According to the principle of virtual stresses, the inner work done by the virtual stresses with the real displacements must be identical with the outer work that the virtual edge stresses perform with the real displacements $\underline{u}_R$ at the edges. This results in

$$\int_A t \cdot \bar{\underline{\sigma}}^{\mathrm{T}} \cdot \underline{\varepsilon} \, dx \, dy = \int_R t \cdot \bar{\underline{\sigma}}_R^{\mathrm{T}} \cdot \underline{u}_R \, ds \tag{4.62}$$

where A and R are the area and the boundary of the element, respectively, and s is the coordinate along the boundary. With the virtual stresses according to (4.60, 4.61), the real strains according to (4.59), and the real edge displacements according to (4.58a), one obtains

$$\bar{\underline{\beta}}^{\mathrm{T}} \cdot \int_A t \cdot \underline{P}^{\mathrm{T}} \cdot \underline{D}^{-1} \cdot \underline{P} \, dx \, dy \cdot \underline{\beta} = \bar{\underline{\beta}}^{\mathrm{T}} \int_R t \cdot \underline{P}_R^{\mathrm{T}} \cdot \underline{H}_R \, ds \cdot \underline{u}_e.$$

Since this relationship must be fulfilled for any values of the stress parameters $\bar{\underline{\beta}}$, it follows

$$\underline{E} \cdot \underline{\beta} = \underline{G} \cdot u_e \tag{4.63}$$

with

$$E = \int_A t \cdot \underline{P}^{\mathrm{T}} \cdot \underline{D}^{-1} \cdot \underline{P} \, dx \, dy \tag{4.63a}$$

$$\underline{G} = \int_R t \cdot \underline{P}_R^{\mathrm{T}} \cdot \underline{H}_R \cdot ds. \tag{4.63b}$$

The matrices $\underline{E}$ and $\underline{G}$ can be determined by numerical integration. Solving (4.63), for β, one obtains

$$\underline{\beta} = \underline{E}^{-1} \cdot \underline{G} \cdot \underline{u}_e. \tag{4.63c}$$

The stresses $\underline{\sigma}_R$ at the edges correspond to statically equivalent nodal forces $\underline{F}_e$

$$\underline{F}_e = \begin{bmatrix} F_{x1} \\ F_{y1} \\ F_{x2} \\ F_{y2} \\ F_{x3} \\ F_{y3} \\ F_{x4} \\ F_{y4} \end{bmatrix}. \tag{4.63d}$$

In order to achieve this, the principle of virtual displacements is applied, as with finite elements with displacement shape functions only, whereby the integration extends along the edge of the element. The same shape functions are selected for the virtual displacements at the edges as for the real displacements:

$$\bar{u}_R = \underline{H}_R \cdot \bar{u}_e \tag{4.64}$$

The work that the real stresses at the edges perform with the virtual edge displacements must be equal to the work that the real nodal forces perform with the virtual nodal displacements. After that it is

$$\int_R t \cdot \bar{u}_R^T \cdot \underline{\sigma}_R \, \mathrm{d}s = \bar{u}_e^T \cdot \underline{F}_e. \tag{4.65}$$

Introducing the expressions (4.64), for the virtual edge displacements $\bar{u}_R$, and (4.57a), for the real edge stresses $\underline{\sigma}_R$ gives

$$\bar{u}_e^T \cdot \int_R t \cdot \underline{H}_R^T \cdot \underline{P}_R \, \mathrm{d}s \cdot \underline{\beta} = \bar{u}_e^T \cdot \underline{F}_e$$

and with $\underline{G}$ according to (4.63b) and the parameters $\underline{\beta}$ according to (4.63c), one obtains

$$\bar{u}_e^T \cdot \underline{G}^T \cdot \underline{E}^{-1} \cdot \underline{G} \cdot u_e = \bar{u}_e^T \cdot \underline{F}_e.$$

Since the equation is valid for arbitrary virtual displacements $\underline{\bar{u}}_e^T$ it follows

$$\underline{G}^T \underline{E}^{-1} \underline{G} \cdot \underline{u}_e = \underline{F}_e \qquad (4.66)$$

or

$$\underline{K}_e \cdot \underline{u}_e = \underline{F}_e \qquad (4.66a)$$

with

$$\underline{K}_e = \underline{G}^T \cdot \underline{E}^{-1} \cdot \underline{G}. \qquad (4.66b)$$

The matrix $\underline{K}_e$ is the stiffness matrix of the hybrid element. Its computation is more time-consuming than for elements with displacement shape functions only since the numerical integration is to be accomplished for two matrices, and additionally, a matrix inversion is required.

The formulation of the stiffness matrix can be summarized as follows:

Formulation of the element stiffness matrix of a hybrid element

1. Determination of the shape functions of the edge displacements. The shape functions are polynomials with the nodal point displacements as unknowns:

$$\underline{u}_R = \underline{H}_R \cdot \underline{u}_e. \qquad (4.67a)$$

2. Determination of the shape functions for the stresses inside the element The stress shape functions must fulfill the equilibrium conditions. Their unknowns are the stress parameters β_i. A minimum of $(m - r)$ stress functions are required where m is the number of displacement degrees of freedom and r the number of rigid body modes (3 for the two-dimensional quadrilateral element):

$$\underline{\sigma} = \underline{P} \cdot \underline{\beta}. \qquad (4.67b)$$

The stresses at the edges are obtained as

$$\underline{\sigma}_R = \underline{P}_R \cdot \underline{\beta}. \qquad (4.67c)$$

3. Determination of the strains inside the element corresponding to the stress shape functions:

$$\underline{\varepsilon} = \underline{D}^{-1} \cdot \underline{\sigma} = \underline{D}^{-1} \cdot \underline{P} \cdot \underline{\beta}. \qquad (4.67d)$$

4. Determination of the stiffness matrix with the principles of virtual work:
   The displacements resulting from the strains inside the element are adopted
   approximately to the diplacement shape functions $\underline{u}_R$ at the edges applying
   the principle of virtual stresses. The stresses $\underline{\sigma}_R$ at the edges are transformed approximately to nodal forces applying the principle of virtual
   displacements. They are obtained as

$$\underline{K}_e \cdot \underline{u}_e = \underline{F}_e, \qquad (4.67e)$$

where

$$\underline{K}_e = \underline{G}^T \cdot \underline{E}^{-1} \cdot \underline{G} \qquad (4.67f)$$

is the element stiffness matrix with

$$\underline{E} = \int_G t \cdot \underline{P}^T \cdot \underline{D}^{-1} \cdot \underline{P} \, dx \, dy \qquad (4.67g)$$

$$\underline{G} = \int_R t \cdot \underline{P}_R^T \cdot \underline{H}_R \, ds. \qquad (4.67h)$$

By appropriate selection of the material matrix $\underline{D}$ according to Table 2.1 all plane
finite elements can be derived both for the plane strain state and for the plane stress
state with isotropic or orthotropic material law.

As in the case of elements with pure displacement shape functions, stress jumps
occur at the element boundaries in the case of hybrid elements, i.e. the equilibrium
conditions are violated at the element boundaries. On the other hand, the equilibrium
conditions in the element are fulfilled, which is not the case for elements with pure
displacement shape functions. However, hybrid elements show a further approximation: The strains caused by the stresses are, in general, neither compatible with each
other nor with the edge displacement shape functions. The quality of this approximation can be improved by increasing the order of the polynomials of the stress
shape functions, and thus also the number of stress parameters β_i. However, it can be
shown that this results in an increase of the stiffness of the element. In order to avoid
too stiff elements, the polynomial degree of the stress shape functions should not
be chosen too high. Here the hybrid elements show a parallel to the nonconforming
elements with pure displacement shape functions. In both cases, the elements are
made artificially "soft" by violating the continuity condition of the displacement
functions, thus attempting to partially compensate for the stiffness increase caused
by the displacement shape functions. Hybrid elements do not meet the requirements

for monotonic convergence. Instead, they are required to pass the patch test as a proof of non-monotonic convergence.

According to the above rules, a large number of elements can be derived with different geometric forms and shape functions. It has been shown that mixed finite element formulations have numerical advantages for incompressible or nearly incompressible materials with a Poisson ratio μ close to 0.5 (e.g. 0.499). Stiffening effects as they occur with finite elements based displacement shape functions can thus be avoided.

Hybrid plane stress elements with displacement degrees of freedom

The originally developed plane stress elements only had degrees of freedom from displacement at the nodal points. An example is the element derived above. A quadrilateral element on this basis is given in [50].

Hybrid plane stress elements with displacement and rotational degrees of freedom

Since for hybrid elements displacement shape functions are only defined on the element edges, it is possible to include the node rotations in the displacement shape functions (Fig. 4.41). At an edge without side node, the displacement function perpendicular to the edge is a third-order parabola similar to a beam in bending. On this basis, a series of elements can be derived [51].

For the quadrilateral element SV3KQ and the corresponding triangular element SD3KQ according to [51], for example, the following quadratic stress shape functions

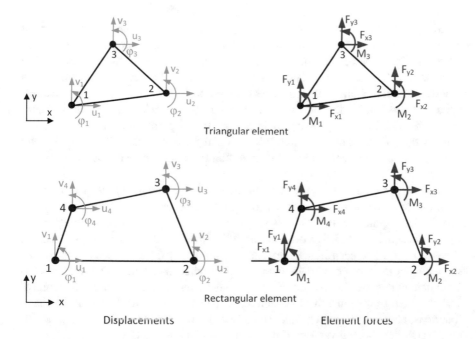

Triangular element

Rectangular element

Displacements   Element forces

Fig. 4.41 Plane stress elements with rotational degrees of freedom

with 12 stress parameters are chosen

$$
\begin{bmatrix} \sigma_x \\ \sigma_y \\ \tau_{xy} \end{bmatrix} = \begin{bmatrix} 1 & 0 & 0 & y & 0 & -x & 0 & y^2 & 0 & 2xy & 0 & -x^2 \\ 0 & 1 & 0 & 0 & x & 0 & -y & 0 & x^2 & 0 & 2xy & -y^2 \\ 0 & 0 & 1 & 0 & 0 & y & x & 0 & 0 & -y^2 & -x^2 & 2xy \end{bmatrix} \cdot \begin{bmatrix} \beta_1 \\ \beta_2 \\ \beta_3 \\ \beta_4 \\ \beta_5 \\ \beta_6 \\ \beta_7 \\ \beta_8 \\ \beta_9 \\ \beta_{10} \\ \beta_{11} \\ \beta_{12} \end{bmatrix}
$$

$$(4.68)$$

The displacements between two nodes are described by a linear function in the direction of the edge and perpendicular to it by a cubic displacement function.

4.5.5 Other Element Types

In addition to the elements mentioned above, a large number of other element types have been developed. For example, there are displacement-based plane stress elements with displacement shape functions that have degrees of freedom of rotation at the nodes. A 4-node quadrilateral element thus has 12 degrees of freedom and a 3-node triangular element has 9 degrees of freedom (Fig. 4.41). Such a 3-node triangular element can be derived from a triangular element with three additional mid nodes on the sides by expressing their displacements perpendicular to the element edge by three degrees of freedom of rotation at the corner nodes [52, 53]. A 4-node quadrilateral element with degrees of rotational freedom can be derived in a similar way [54, 55].

A special version of the finite element method is the so-called p-version. Starting from a coarse-meshed triangle mesh, the polynomial degree p of the shape functions of the displacements is successively increased until the desired accuracy of the solution is achieved [56]. The original mesh is kept while the polynomial degree can become $p = 10$ or even higher. Special polynomials [57], have been proposed as shape functions. It can be shown that the convergence is faster than with the (usual) h-version, where the element size h is successively reduced to increase the accuracy. The choice of the original mesh, which has to be adjusted to the degree of the shape functions, also has an influence. This leads to a combination of the p-version with the h-version, called hp-version of the finite element method [58].

4.5.6 Element Types in Finite Element Software for Structural Analysis

When developing a commercial finite element software package, the developer has to decide which element type or types should be implemented in the program. The choice varies depending on the individual preferences of the software developer [59]. Some finite element packages will be discussed.

As examples of general-purpose finite element software, ADINA® and ANSYS® are considered, which are also used in structural engineering for special problems (Table 4.11). In addition to static and dynamic problems of linear and nonlinear structural mechanics and fluid mechanics (CFD; **C**omputational **F**luid **D**ynamics), as well as fluid-structure interaction, they are suited to analyze temperature fields, electromagnetic fields, and coupled multi-physics problems. Programs developed especially for structural design, however, support the superposition of load cases and design calculations according to current building codes. Examples for this type of software are SAP2000®, Infograph, RFEM, SOFiSTiK®, and "iTWO structure fem/TRIMAS®" (Table 4.12) which are also available on the international market. In addition to static calculations, all programs allow dynamic calculations in the linear and often also in the nonlinear range for two- and three-dimensional structures.

In the following, finite elements for the linear analysis of plates in plane stress are discussed, which are implemented in the finite element software mentioned above.

Table 4.11 General purpose finite element software

Software	Development/software house
ADINA (Automatic Dynamic Incremental Nonlinear Analysis)	The beginnings of ADINA go back to the year 1974, when the program for nonlinear calculations of structural mechanics was created by Klaus-Jürgen Bathe [WEB 1]. Since 1975 K.J. Bathe is professor at the Massachusetts Institute of Technology (MIT), Boston, USA. He continues to lead the development of ADINA [279]. The program has been developed and distributed by ADINA R&D Inc., Watertown MA, USA since 1986
ANSYS (ANalysis SYStem)	The development of ANSYS goes back to John Swanson, who founded a software house in 1970. Since 1994 J. Swanson is no longer involved in the company [WEB 2]. After several changes of ownership, the software is developed and distributed today by ANSYS Inc. with headquarters in Canonsburg, PA, USA. An overview is given in [280]

Table 4.12 Finite element software for structural analysis

Software	Development/software house
InfoCAD	The development of InfoCad started in 1987 with the foundation of InfoGraph GmbH (Ingenieurgesellschaft für graphisch unterstützte Datenverarbeitung) [WEB 4], based on research work at RWTH Aachen University. The software is further developed and distributed by InfoGraph GmbH, Aachen, Germany
RFEM	Founded in 1987 by Georg Dlubal, the German-Czech software house Dlubal is cooperating closely with 'FEM consulting', Brno/Czech Republic, and there with Ivan Němec and Vladimír Kolář in the field of finite elements since 1993 [WEB 5]. Dlubal Software GmbH is located in Tiefenbach/Bavaria, Germany
SAP2000 (Structural Analysis Program)	The name SAP (Structural Analysis Program) traces back to a program of the same name, which has been developed in 1970 by Edward L. Wilson at the University of California, Berkeley, USA [281]. The SAPIV program, which was developed mainly in collaboration of Bathe and Peterson, became internationally known [279, 282]. In the 1970s, it could be obtained from the University of California, Berkeley, as Fortran source code and found worldwide distribution. Today's SAP2000 is a completely new development. Development and sales are carried out by Computers and Structures Inc. (CSi), Walnut Creek, CA, USA [WEB 3]
SOFiSTiK FEM	The beginnings of FEM at SOFiSTiK go back to a research project at the Technical University of Munich (TUM, Chair Prof. Werner) for the development of modular finite element software at the end of the 1970s [283]. Since 1982 Casimir Katz continued to develop the software concept up to a commercial software for structural design. This led in 1987 to the foundation of the software company SOFiSTiK [WEB 6], in which C. Katz played an important role in the further development of the software as well as in the management. SOFiSTiK AG is based in Oberschleißheim/Munich, Germany
iTWO structure fem/TRIMAS	In Germany, the development of programs for computer-oriented structural design began in 1961 with the founding of the "Recheninstitut im Bauwesen" (RIB) in Stuttgart [16, 284]. The first finite element programs for reinforced concrete constructions (i.e. with load case superposition and design) appeared in the 1970s. iTWO structures is a spatial finite element program which dates back to the Program TRIMAS newly developed in the 1990s. The RIB Software SE is located in Stuttgart, Germany [WEB 7]

ADINA

The basic plane stress element in ADINA is the isoparametric 3- to 9-node element according to Fig. 4.32, with the shape functions according to (4.43), [34]. It can be used as a triangular or quadrilateral element or as an element with curved edges (Fig. 4.31). The 8- and 9-node elements are recommended. The integration order for the numerical determination of the stiffness matrix according to (4.50) or (4.51), using Gaussian integration can be selected from 2×2 to 6×6. The default values are 2×2 integration for 4-node elements and 3×3 integration for all other elements.

The 4-node element can also be used with incompatible displacement shape functions [34, 41, 43]. It should be noted that in addition to the plane strain state and rotationally symmetric 2D models, a hybrid element is implemented in ADINA, which is recommended especially for materials with Poisson ratios close to 0.5.

As the formulation is not needed for such materials in plane stress conditions, the element is not available for the plane stress state [SOF 1].

ANSYS

ANSYS contains an element library with a number of elements suitable for plane stress analyses [SOF 2]. PLANE182 is an isoparametric 4-node element with bilinear shape functions. It can also be used as a triangle and then corresponds to the CST-element. The stiffness matrix of the general quadrilateral element has a reduced integration (see Table 4.8). Optionally a 2×2-point and a (stabilized) 1-point integration (standard setting) can be selected. PLANE183 is an isoparametric 8-node element with a 2×2-point integration. It can also be used as a 6-node triangle element with a 3-point integration (Fig. 4.31). The above-mentioned elements are also available with mixed finite element formulation in ANSYS, but only for the plane stress state and rotationally symmetric systems. The older (and not recommended by ANSYS) element PLANE42 is a 4-node element with additional incompatible Wilson/Taylor displacement shape functions.

Some shell elements can also be used as pure plate elements with a membrane state option. SHELL181 is an isoparametric 4-node membrane element similar to PLANE182. The SHELL281 element, preset as the standard element, is an isoparametric 8-node element with a 2×2-point integration and corresponds to PLANE183 in the membrane state.

The elements PLANE182 and PLANE183 can also be used for three-dimensional systems in the plain strain state and for axisymmetric systems.

InfoCad

In InfoCAD two different element types for the plane stress state are implemented [SOF 3]. The first is an isoparametric 6-node triangular element with quadratic shape functions. Furthermore, a hybrid 4-node quadrilateral element and a hybrid 3-node triangle element are available, which have rotational degrees of freedom in addition to the degrees of freedom of displacements (Fig. 4.41).

RFEM

In RFEM, a hybrid 4-node quadrilateral element and a hybrid 3-node triangle element are implemented as plane stress elements with rotational degrees of freedom (Fig. 4.41). These were formulated on the basis of quadratic shape functions for displacements and linear stress shape functions for stresses [52, SOF 4].

SAP2000

SAP2000 contains a shell element that is also used for calculations in the plane stress state [55]. It is implemented as a 4-node quadrilateral element with a 2×2 Gaussian integration and as a 3-node triangular element. The derivation is based on quadratic displacement shape functions, which results in a plane stress element with additional rotational degrees of freedom (Fig. 4.41) [SOF 5]. In addition, a 4-node plane stress element with incompatible displacements but without rotational degrees of freedom

("Plane") is implemented in SAP2000. This element can also be used for plane strain analyses.

For axisymmetric systems with axisymmetric loading, SAP2000 contains a two-dimensional, isoparametric 4-node element with a 2×2 integration ("Asolid"). Optionally, in addition, incompatible displacement shape functions can also be taken into account. The element is also available as a 3-node triangle element.

SOFiSTiK

Plate and shell structures, as well as solids, can be analyzed with the module ASE of the SOFiSTiK Software package. The plane stress element is an isoparametric 4-node quadrilateral element (Fig. 4.35), with additional noncompatible shape functions according to Wilson/Taylor [42], as standard setting (Sect. 4.5.3). It can also be used as a 3-node triangle element, but then without noncompatible shape functions, i.e. as a pure CST-element. The noncompatible shape functions can be "switched off" by the user so that the element then is a pure bilinear isoparametric quadrilateral element [SOF 6].

In addition, the TALPA program module also provides 2D elements for the plane strain state and for axisymmetric solids.

iTWO structure fem/TRIMAS

The shell elements implemented in "iTWO structure fem/ TRIMAS" are also applied to the analysis of structures in plane stress. They are available as 4- and 9-node quadrilateral elements and as 3- and 6-node triangle elements. These are isoparametric elements with additional incompatible displacements according to the EAS method for the membrane state [SOF 7].

4.6   Rectangular Element for Plates in Bending

4.6.1   Element Type

In the classical *Kirchhoff plate theory*, shear deformations are neglected. This is admissible for thin plates and simplifies the analytical solution of the differential equation of the plate. The more advanced plate theory including shear deformation is known as *Reissner–Mindlin plate theory*. This theory is generally preferred when developing finite plate elements.

A rectangular element on the basis of the plate theory with shear deformations is given in the following section. Section 4.7 gives an overview of other element types, including such according to the Kirchhoff theory.

4.6.2   Shape Functions

The deformations of a plate are described by the deflection w and the angles of rotation φ_x and φ_y at any location of the plate. Hence, the plate element has three degrees of freedom at each nodal point, i.e. the displacement w_i and the rotational angles φ_{xi} and φ_{yi}. The corresponding element forces are the force F_{zi} and the bending moment M_{xi} and M_{yi}. Figure 4.42 shows a 4-node element with the corresponding 12 degrees of freedom.

Whereas for the plate without shear deformations, the angles of rotation are given as the first derivative of the displacements, this is not the case for plates including shear deformations. Here the shear deformations result in an additional shear angle, and therefore, the angles of rotations and the displacements are independent functions. Therefore, in the derivation of the stiffness matrix for a plate element with shear deformations, independent shape functions are used for the displacements and the rotations.

The shape functions of a 4-node element are written as a bilinear interpolation of the corresponding nodal values. For bilinear interpolation the shape functions $N_1 - N_4$ have been given in the derivation of the plane stress element in (4.27b–4.27e). One obtains (Fig. 4.43)

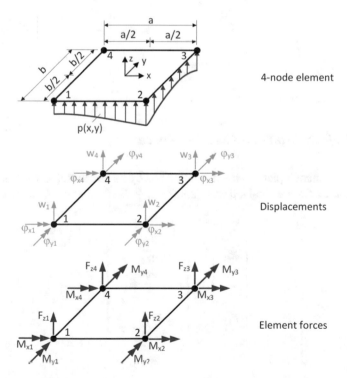

4-node element

Displacements

Element forces

Fig. 4.42 Rectangular element for plates in bending

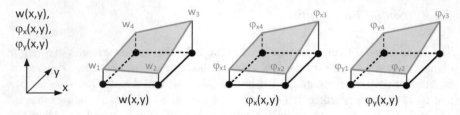

Fig. 4.43 Shape functions for a plate element with shear deformation

$$
\begin{bmatrix} w \\ \varphi_x \\ \varphi_y \end{bmatrix} = \begin{bmatrix} N_1 & 0 & 0 & N_2 & 0 & 0 & N_3 & 0 & 0 & N_4 & 0 & 0 \\ 0 & N_1 & 0 & 0 & N_2 & 0 & 0 & N_3 & 0 & 0 & N_4 & 0 \\ 0 & 0 & N_1 & 0 & 0 & N_2 & 0 & 0 & N_3 & 0 & 0 & N_4 \end{bmatrix} \begin{bmatrix} w_1 \\ \varphi_{x1} \\ \varphi_{y1} \\ w_2 \\ \varphi_{x2} \\ \varphi_{y2} \\ w_3 \\ \varphi_{x3} \\ \varphi_{y3} \\ w_4 \\ \varphi_{x4} \\ \varphi_{y4} \end{bmatrix} \tag{4.69}
$$

$$
\underline{u} = \underline{N} \cdot \underline{u}_e. \tag{4.69a}
$$

4.6.3 Deformations and Section Forces

The deformation of a plate element is described by its curvature, as well as its shear
deformations. They are obtained with (2.9a) and (2.9b) for the shape functions (4.69),
as

$$
\begin{bmatrix} \kappa_x \\ \kappa_y \\ \kappa_{xy} \end{bmatrix} = \begin{bmatrix} \dfrac{\partial \varphi_x}{\partial x} \\ \dfrac{\partial \varphi_y}{\partial y} \\ \dfrac{\partial \varphi_x}{\partial y} + \dfrac{\partial \varphi_y}{\partial x} \end{bmatrix} \tag{4.70a}
$$

or

$$
\begin{bmatrix} \kappa_x \\ \kappa_y \\ \kappa_{xy} \end{bmatrix} = \begin{bmatrix} 0 & \dfrac{\partial N_1}{\partial x} & 0 & 0 & \dfrac{\partial N_2}{\partial x} & 0 & 0 & \dfrac{\partial N_3}{\partial x} \\ 0 & 0 & \dfrac{\partial N_1}{\partial y} & 0 & 0 & \dfrac{\partial N_2}{\partial y} & 0 & 0 \\ 0 & \dfrac{\partial N_1}{\partial y} & \dfrac{\partial N_1}{\partial x} & 0 & \dfrac{\partial N_2}{\partial y} & \dfrac{\partial N_2}{\partial x} & 0 & \dfrac{\partial N_3}{\partial y} \end{bmatrix} \cdots
$$

$$
\begin{bmatrix} & 0 & 0 & \dfrac{\partial N_4}{\partial x} & 0 \\ \cdots\cdots & \dfrac{\partial N_3}{\partial y} & 0 & 0 & \dfrac{\partial N_4}{\partial y} \\ & \dfrac{\partial N_3}{\partial x} & 0 & \dfrac{\partial N_4}{\partial y} & \dfrac{\partial N_4}{\partial x} \end{bmatrix} \cdot \begin{bmatrix} w_1 \\ \varphi_{x1} \\ \varphi_{y1} \\ w_2 \\ \varphi_{x2} \\ \varphi_{y2} \\ w_3 \\ \varphi_{x3} \\ \varphi_{y3} \\ w_4 \\ \varphi_{x4} \\ \varphi_{y4} \end{bmatrix}
$$

$$
\underline{\kappa} = \underline{B}_b \cdot \underline{u}_e \tag{4.70b}
$$

and

$$
\begin{bmatrix} \gamma_{xz} \\ \gamma_{yz} \end{bmatrix} = \begin{bmatrix} \varphi_x + \dfrac{\partial w}{\partial x} \\ \varphi_y + \dfrac{\partial w}{\partial y} \end{bmatrix} \tag{4.70c}
$$

$$
\begin{bmatrix} \gamma_{xz} \\ \gamma_{yz} \end{bmatrix} = \begin{bmatrix} \dfrac{\partial N_1}{\partial x} & N_1 & 0 & \dfrac{\partial N_2}{\partial x} & N_2 & 0 & \dfrac{\partial N_3}{\partial x} & N_3 & 0 & \dfrac{\partial N_4}{\partial x} & N_4 & 0 \\ \dfrac{\partial N_1}{\partial y} & 0 & N_1 & \dfrac{\partial N_2}{\partial y} & 0 & N_2 & \dfrac{\partial N_3}{\partial y} & 0 & N_3 & \dfrac{\partial N_4}{\partial y} & 0 & N_4 \end{bmatrix} \cdot \begin{bmatrix} w_1 \\ \varphi_{x1} \\ \varphi_{y1} \\ w_2 \\ \varphi_{x2} \\ \varphi_{y2} \\ w_3 \\ \varphi_{x3} \\ \varphi_{y3} \\ w_4 \\ \varphi_{x4} \\ \varphi_{y4} \end{bmatrix}
$$

$$\underline{\gamma} = \underline{B}_s \cdot \underline{u}_e. \tag{4.70d}$$

The derivatives of the bilinear shape functions $N_1 - N_4$ have been given for the plane stress element in (4.28a-h). Hence, the element is capable of representing a constant curvature $\underline{\kappa}$ and a constant bending moment when fully integrated.

The section forces related to the curvature and shear angles can be determined by the moment/curvature relationship (2.10a) and the shear forces/shear angle relationship (2.10b)

$$\begin{bmatrix} m_x \\ m_y \\ m_{xy} \end{bmatrix} = \frac{E \cdot h^3}{12 \cdot (1 + \mu^2)} \begin{bmatrix} 1 & \mu & 0 \\ \mu & 1 & 0 \\ 0 & 0 & \dfrac{1 - \mu}{2} \end{bmatrix} \cdot \begin{bmatrix} \kappa_x \\ \kappa_y \\ \kappa_{xy} \end{bmatrix}$$

$$\underline{m} = \underline{D}_b \cdot \underline{\kappa}$$

$$\begin{bmatrix} v_x \\ v_y \end{bmatrix} = \frac{5 \cdot E \cdot h}{12 \cdot (1 + \mu)} \cdot \begin{bmatrix} 1 & 0 \\ 0 & 1 \end{bmatrix} \cdot \begin{bmatrix} \gamma_{xz} \\ \gamma_{yz} \end{bmatrix}$$

$$\underline{v} = \underline{D}_s \cdot \underline{\gamma}$$

as

$$\underline{m} = \underline{D}_b \cdot \underline{B}_b \cdot \underline{u}_e \tag{4.71a}$$

$$\underline{v} = \underline{D}_s \cdot \underline{B}_s \cdot \underline{u}_e. \tag{4.71b}$$

With these equations, the section forces in the element can be determined for given nodal point displacements and rotations.

4.6.4 Stiffness Matrix

The stiffness matrix represents the relationship between the nodal forces and moments and the nodal displacements and rotations. For its derivation, the element section forces (4.71a, 4.71b), are to be transformed to equivalent forces and moments. This is done with the principle of virtual displacements. For the virtual displacements, the same shape functions as for the real displacements are chosen (see (4.69)), i.e.

$$\underline{\bar{u}} = \underline{N} \cdot \underline{\bar{u}}_e. \tag{4.72}$$

The curvatures and shear angles for the virtual displacements are

$$\bar{\underline{\kappa}} = \underline{B}_b \cdot \bar{\underline{u}}_e \quad \text{or} \quad \bar{\underline{\kappa}}^T = \bar{\underline{u}}_e^T \cdot \underline{B}_b^T \tag{4.73a}$$

$$\bar{\underline{\gamma}} = \underline{B}_s \cdot \bar{\underline{u}}_e \quad \text{or} \quad \bar{\underline{\gamma}}^T = \bar{\underline{u}}_e^T \cdot \underline{B}_s^T. \tag{4.73b}$$

The internal virtual work is done by the real moments with the virtual curvatures and by the real shear forces with the virtual shear angles. It is, according to (2.11c)

$$\overline{W}_i = \int \bar{\underline{\kappa}}^T \cdot \underline{m} \cdot dx \, dy + \int \bar{\underline{\gamma}}^T \cdot \underline{v} \cdot dx \, dy.$$

Introducing the expressions for the virtual curvatures and shear angles (4.73a, b), as well as for the real moments and shear forces (4.71a, b), the virtual internal work is obtained as

$$\overline{W}_i = \int \bar{\underline{u}}_e^T \cdot \underline{B}_e^T \cdot \underline{D}_b \cdot \underline{B}_b \cdot \underline{u}_e \, dx \, dy + \int \bar{\underline{u}}_e^T \cdot \underline{B}_s^T \cdot \underline{D}_s \cdot \underline{B}_s \cdot \underline{u}_e \, dx \, dy$$

or, since $\underline{u}_e$ and $\underline{u}_e^T$ are independent on x and y

$$\overline{W}_i = \bar{\underline{u}}_e^T \cdot \left(\int \underline{B}_b^T \cdot \underline{D}_b \cdot \underline{B}_b \, dx \, dy + \int \underline{B}_s^T \cdot \underline{D}_s \cdot \underline{B}_s \, dx \, dy \right) \cdot \underline{u}_e. \tag{4.74}$$

The external virtual work is done by the real nodal forces with the virtual nodal displacements and by the real nodal moments with the virtual nodal rotations. The external work of element loads will be considered later. Hence, the external virtual work is

$$\overline{W}_a = \begin{bmatrix} \bar{w}_1 & \bar{\varphi}_{x1} & \bar{\varphi}_{y1} & \bar{w}_2 & \bar{\varphi}_{x2} & \bar{\varphi}_{y2} & \bar{w}_3 & \bar{\varphi}_{x3} & \bar{\varphi}_{y3} & \bar{w}_4 & \bar{\varphi}_{x4} & \bar{\varphi}_{y4} \end{bmatrix} \cdot \begin{bmatrix} F_{z1} \\ M_{x1} \\ M_{y1} \\ F_{z2} \\ M_{x2} \\ M_{y2} \\ F_{z3} \\ M_{x3} \\ M_{y3} \\ F_{z4} \\ M_{x4} \\ M_{y4} \end{bmatrix}$$

$$\overline{W}_a = \bar{\underline{u}}_e^T \cdot \underline{F}_e. \tag{4.75}$$

Equating the internal and external work gives

$$\underline{\bar{u}}_e^T \cdot \left(\int \underline{B}_b^T \cdot \underline{D}_b \cdot \underline{B}_b \, dx \, dy + \int \underline{B}_s^T \cdot \underline{D}_s \cdot \underline{B}_s \, dx \, dy \right) \cdot \underline{u}_e = \underline{\bar{u}}_e^T \cdot \underline{F}_e.$$

This equation is valid for any arbitrary virtual nodal point displacements $\underline{\bar{u}}_e^T$. This implies that

$$\left(\int \underline{B}_b^T \cdot \underline{D}_b \cdot \underline{B}_b \, dx \, dy + \int \underline{B}_s^T \cdot \underline{D}_s \cdot \underline{B}_s \, dx \, dy \right) \cdot \underline{u}_e = \underline{F}_e$$

or

$$\underline{K}^{(e)} \cdot \underline{u}_e = \underline{F}_e. \tag{4.76}$$

Here

$$\underline{K}^{(e)} = \int \underline{B}_b^T \cdot \underline{D}_b \cdot \underline{B}_b \, dx \, dy + \int \underline{B}_s^T \cdot \underline{D}_s \cdot \underline{B}_s \, dx \, dy \tag{4.76a}$$

$$\underset{Bending}{\big|} \qquad\qquad\qquad \underset{Shear}{\big|}$$

is the stiffness matrix of the plate element. It is composed of a term representing bending and another term for shear. It is a conforming element where C^0 continuity is given for both the displacements and the rotations.

If integration of the stiffness matrix according to (4.76a), is done exactly, which corresponds to a Gaussian 2×2 integration, the element behaves extremely stiffly for thin plates [19, 60]. This phenomenon is known as *shear locking*. The reason for this is that even in pure bending with shear forces equal to zero the shear deformations cannot become zero. As for elements in plane stress, a reduced numerical integration gives "softer" elements. Since a reduced 1-point Gaussian integration leads to kinematics, making the element unusable, the bending and shear contribution of the stiffness matrix are integrated with different order. This is called selective integration. Here, only the shear terms are integrated with the reduced order 1, whereas the bending terms are fully integrated (2×2 integration) [61]. For very thick plates where the shear deformations become relevant, a modified integration scheme is suggested in [61]. Other methods for avoiding shear locking are dealt with in Sect. 4.7.1.

4.6.5   Element Loads

Element loads are expressed by equivalent nodal loads. These perform the same external virtual work with the virtual displacements as the element loads. In the following, the case of a distributed load $p_z(x, y)$ is considered. Hence, the force $p_z \cdot dx \cdot dy$ acts on the infinitesimal area $dx \cdot dy$.

The virtual displacement $\bar{w}$ according to (4.72) and (4.69) is

$$\bar{w} = N_1 \cdot \bar{w}_1 + N_2 \cdot \bar{w}_2 + N_3 \cdot \bar{w}_3 + N_4 \cdot \bar{w}_4 = \sum_{i=1}^{4} N_i \cdot \bar{w}_i. \qquad (4.77)$$

The shape functions $N_1 - N_4$ are given in (4.27b–4.27e). With it the virtual work of the distributed load is obtained by integration over the area of the element as

$$\overline{W}_{a,p} = \int \bar{w} \cdot p_z \, dx \, dy = \int \left(\sum_{i=1}^{4} N_i \cdot \bar{w}_i \right) p_z \, dx \, dy. \qquad (4.78a)$$

The virtual work of the equivalent nodal loads with the virtual nodal displacements is

$$\overline{W}_{a,F} = \sum_{i=1}^{4} F_{zL,i} \cdot \bar{w}_i. \qquad (4.78b)$$

Equating the virtual work of the nodal loads and the virtual work of the distributed loads gives

$$\sum_{i=1}^{4} F_{zL,i} \cdot \bar{w}_i = \sum_{i=1}^{4} \int N_i \cdot p_z \, dx \, dy \cdot \bar{w}_i.$$

Since this relationship is valid for any arbitrary virtual displacements, the sums on both sides of the equation are equal. Hence, the equivalent nodal loads of an arbitrary distributed element load are

$$F_{zL,i} = \int N_i \cdot p_z \, dx \, dy \qquad (4.79)$$

or

$$\begin{bmatrix} F_{zL,1} \\ F_{zL,2} \\ F_{zL,3} \\ F_{zL,4} \end{bmatrix} = \int \begin{bmatrix} N_1 \\ N_2 \\ N_3 \\ N_4 \end{bmatrix} \cdot p_z \cdot dx \, dy. \qquad (4.79a)$$

The integration is to be done over the element area.

Constant distributed load

For a constant distributed load p_z the integration according to 4.79, gives (Fig. 4.44)

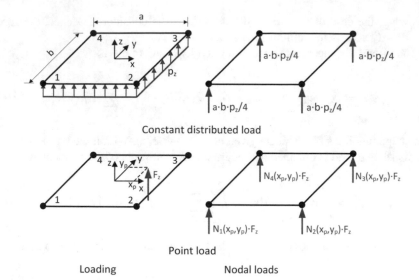

Fig. 4.44 Element loads and equivalent nodal loads of a plate element

$$F_{zL,i} = \frac{1}{4} a \cdot b \cdot p_z \qquad (4.80a)$$

or (Fig. 4.44)

$$\begin{bmatrix} F_{zL,1} \\ F_{zL,2} \\ F_{zL,3} \\ F_{zL,4} \end{bmatrix} = \frac{a \cdot b}{4} \cdot \begin{bmatrix} 1 \\ 1 \\ 1 \\ 1 \end{bmatrix} \cdot p_z. \qquad (4.80b)$$

Point load

For a point load F_z the equivalent nodal loads are obtained with (4.79), as (Fig. 4.44)

$$F_{zL,i} = N_i(x_p, y_p) \cdot F_z \qquad (4.81a)$$

or

$$\begin{bmatrix} F_{zL,1} \\ F_{zL,2} \\ F_{zL,3} \\ F_{zL,4} \end{bmatrix} = \begin{bmatrix} N_1(x_p, y_p) \\ N_2(x_p, y_p) \\ N_3(x_p, y_p) \\ N_4(x_p, y_p) \end{bmatrix} \cdot F_z. \qquad (4.81b)$$

The shape functions N_i according to (4.27b–4.27e), are taken at $x = x_p$ and $y = y_p$.

4.7 Finite Elements for Plates in Bending

4.7.1 Displacement-Based Shear Flexible Elements

Shear flexible plate elements have been developed in order to maintain the classical finite element theory of continuous shape functions for plates in bending consistently, and to consider shear deformations for thick plates in addition. They are also known as Reissner–Mindlin plate elements. The displacements and rotations of a shear flexible plate are independent unknowns for which only a C^0-continuity (continuity of the 0-th derivative, i.e. the function itself) is required at the boundaries of adjacent elements. Since the 1970s and 1980s, a huge number of shear flexible plate elements have been developed. The main problem with the derivation of shear flexible elements is to prevent unwanted shear locking. There are different approaches to this problem [62].

Shear flexible quadrilateral 4-node plate element

A simple rectangular element for a shear flexible plate has been derived in the last Section. This element can easily be extended to general quadrilaterals, as well as to triangles [19, 60], (Fig. 4.45).

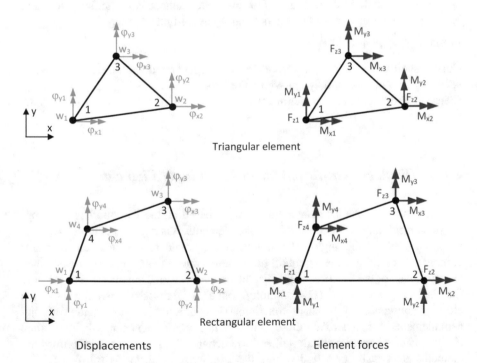

Triangular element

Rectangular element

Displacements   Element forces

Fig. 4.45 Quadrilateral and triangular plate elements

To avoid shear locking effects, the selective reduced integration, in which the plate shear components of the stiffness matrix are integrated one order lower than the bending components, has already been mentioned. However, this can lead to unfavorable behavior in geometrically distorted element shapes. Another possibility is the selection of additional shape functions for shear strains (**A**ssumed **N**atural **S**train - ANS) similar to mixed finite element formulations. From Hughes/Tezduyar an interpolation with bilinear shape functions has been proposed [63], which is related to the shear strains in the midpoints of the edges of the quadrilateral 4-node element. For the plate element of Bathe/Dvorkin, an independent linear distribution is assumed for each of the 4 plate shear strains in the element edges [64]. A generalization of these approaches is the **D**iscrete **S**hear **G**ap Method (DSG) by Bletzinger et al. [65]. There are also shear flexible elements which merge into the rigid DKT element (see below) in the case of thin plates [66]. With regard to details on this rather complex topic, reference is made to the extensive special literature.

Shear flexible isoparametric and Lagrange elements

For shear flexible plates, isoparametric and Lagrangian elements with higher shape functions can be derived in the same way as for plane stress elements. However, a parameter study shows that only the 16-node Lagrange element with full 4×4 integration is well suited as a general plate element, due to shear locking or the occurrence of undesired kinematics [19, 60]. The element is particularly suited to reproduce the shear forces of plates in bending with hight accuracy.

Plane shell elements

Curved shell elements can be formulated as "degenerated" solid elements. When used as plane elements, they are well suited for the analysis of plates in bending. Elements with more than 4 nodes are also available. Reference is made to Sect. 4.8.2 for details.

4.7.2 Displacement-Based Shear Rigid Plate Elements

The first formulations for plate elements were based on the Kirchhoff plate theory for shear rigid plates. For conforming plate elements, it is required that the displacements, as well as the angles of rotation, at the boundaries of adjacent elements coincide. According to the Kirchhoff plate theory, the angles of rotation are obtained as derivatives of the function of the displacement $w(x, y)$. This means that not only the displacements $w(x, y)$ (C^0-continuity), but also the first derivative normal to the element boundaries of the shape functions (C^1-continuity) must coincide for adjacent elements. When deriving finite elements, this requirement can only be fulfilled if significant disadvantages are taken into account [8]. Therefore, nonconforming elements are usually preferred, where the continuity condition is fulfilled for the displacements but not for the rotations between adjacent elements. An overview of the shear rigid plate elements developed according to the considerations above can be found in [18, 29].

Nonconforming 4-node rectangular element

A rectangular element with three degrees of freedom per node, i.e. one displacement and two rotations, can be derived based on the following interpolation function:

$$
w(x, y) = \alpha_1 + \alpha_2 \cdot x + \alpha_3 \cdot y + \alpha_4. \cdot x^2 + \alpha_5 \cdot xy + \alpha_6 \cdot y^2
$$
$$
+ \alpha_7 \cdot x^3 + \alpha_8 \cdot x^2 y + \alpha_9 \cdot xy^2 + \alpha_{10} \cdot y^3 + \alpha_{11} \cdot xy^3 + \alpha_{12} \cdot x^3 y
$$

$$(4.82)$$

This function is an incomplete polynomial (see Fig. 4.28). It fulfills the continuity condition for the displacements at the boundaries of adjacent elements. However, the continuity of the angles of rotation is no more given. Therefore, the element is a nonconforming plate element. The derivation of its stiffness matrix is given in [27] and [67], and for an orthotropic plate in bending together with the section forces matrix in [7]. An extension for general quadrilaterals is not possible. With the notations according to Fig. 4.42, the stiffness matrix can be given as

$$
\underline{K}^{(e)} = \frac{E \cdot t^3}{12 \cdot (1 - \mu^2) \cdot a \cdot b} \cdot
$$

$$
\begin{bmatrix}
k_{1,1} & k_{1,2} & k_{1,3} & k_{1,4} & k_{1,5} & k_{1,6} & k_{1,7} & k_{1,8} & k_{1,9} & k_{1,10} & k_{1,11} & k_{1,12} \\
k_{2,1} & k_{2,2} & k_{2,3} & k_{2,4} & k_{2,5} & k_{2,6} & k_{2,7} & k_{2,8} & k_{2,9} & k_{2,10} & k_{2,11} & k_{2,12} \\
k_{3,1} & k_{3,2} & k_{3,3} & k_{3,4} & k_{3,5} & k_{3,6} & k_{3,7} & k_{3,8} & k_{3,9} & k_{3,10} & k_{3,11} & k_{3,12} \\
k_{4,1} & k_{4,2} & k_{4,3} & k_{4,4} & k_{4,5} & k_{4,6} & k_{4,7} & k_{4,8} & k_{4,9} & k_{4,10} & k_{4,11} & k_{4,12} \\
k_{5,1} & k_{5,2} & k_{5,3} & k_{5,4} & k_{5,5} & k_{5,6} & k_{5,7} & k_{5,8} & k_{5,9} & k_{5,10} & k_{5,11} & k_{5,12} \\
k_{6,1} & k_{6,2} & k_{6,3} & k_{6,4} & k_{6,5} & k_{6,6} & k_{6,7} & k_{6,8} & k_{6,9} & k_{6,10} & k_{6,11} & k_{6,12} \\
k_{7,1} & k_{7,2} & k_{7,3} & k_{7,4} & k_{7,5} & k_{7,6} & k_{7,7} & k_{7,8} & k_{7,9} & k_{7,10} & k_{7,11} & k_{7,12} \\
k_{8,1} & k_{8,2} & k_{8,3} & k_{8,4} & k_{8,5} & k_{8,6} & k_{8,7} & k_{8,8} & k_{8,9} & k_{8,10} & k_{8,11} & k_{8,12} \\
k_{9,1} & k_{9,2} & k_{9,3} & k_{9,4} & k_{9,5} & k_{9,6} & k_{9,7} & k_{9,8} & k_{9,9} & k_{9,10} & k_{9,11} & k_{9,12} \\
k_{10,1} & k_{10,2} & k_{10,3} & k_{10,4} & k_{10,5} & k_{10,6} & k_{10,7} & k_{10,8} & k_{10,9} & k_{10,10} & k_{10,11} & k_{10,12} \\
k_{11,1} & k_{11,2} & k_{11,3} & k_{11,4} & k_{11,5} & k_{11,6} & k_{11,7} & k_{11,8} & k_{11,9} & k_{11,10} & k_{11,11} & k_{11,12} \\
k_{12,1} & k_{12,2} & k_{12,3} & k_{12,4} & k_{12,5} & k_{12,6} & k_{12,7} & k_{12,8} & k_{12,9} & k_{12,10} & k_{12,11} & k_{12,12}
\end{bmatrix}
\begin{bmatrix}
w_1 \\ \varphi_{x1} \\ \varphi_{y1} \\ w_2 \\ \varphi_{x2} \\ \varphi_{y2} \\ w_3 \\ \varphi_{x3} \\ \varphi_{y3} \\ w_4 \\ \varphi_{x4} \\ \varphi_{y4}
\end{bmatrix}
=
\begin{bmatrix}
F_{z1} \\ M_{x1} \\ M_{y1} \\ F_{z2} \\ M_{x2} \\ M_{y2} \\ F_{z3} \\ M_{x3} \\ M_{y3} \\ F_{z4} \\ M_{x4} \\ M_{y4}
\end{bmatrix}
$$

$$(4.83)$$

with $\beta = b/a$ and

$k_{1,1} = k_{4,4} = k_{7,7} = k_{10,10}$ $\qquad = 4 \cdot (\beta^2 + 1/\beta^2) + (14 - 4 \cdot \mu)/5$

$k_{1,2} = k_{2,1} = k_{4,5} = k_{5,4} =$

$-k_{7,8} = -k_{8,7} = -k_{10,11} = -k_{11,10}$ $\qquad = (2/\beta^2 + (1 + 4 \cdot \mu)/5) \cdot b$

$k_{2,2} = k_{5,5} = k_{8,8} = k_{11,11}$ $\qquad = (4/(3 \cdot \beta^2) + 4 \cdot (1 - \mu)/15) \cdot b^2$

$k_{1,3} = k_{3,1} = -k_{4,6} = -k_{6,4} =$

$-k_{7,9} = -k_{9,7} = k_{10,12} = k_{12,10}$ $\qquad = -(2 \cdot \beta^2 + (1 + 4 \cdot \mu)/5) \cdot a$

$k_{2,3} = k_{3,2} = -k_{5,6} = -k_{6,5} =$

$k_{8,9} = k_{9,8} = -k_{11,12} = -k_{12,11}$ $\qquad = -\mu \cdot a \cdot b$

$k_{3,3} = k_{6,6} = k_{9,9} = k_{12,12}$ $\qquad = 4 \cdot b^2/3 + 4 \cdot (1 - \mu) \cdot a^2/15$

$k_{1,4} = k_{4,1} = k_{7,10} = k_{10,7}$ $\qquad = -2 \cdot (2 \cdot \beta^2 - 1/\beta^2) - (14 - 4 \cdot \mu)/5$

$k_{2,4} = k_{4,2} = k_{1,5} = k_{5,1} =$

$-k_{7,11} = -k_{11,7} = -k_{8,10} = -k_{10,8}$ $\qquad = (1/\beta^2 - (1 + 4 \cdot \mu) \cdot 1/5) \cdot b$

$k_{3,4} = k_{4,3} = -k_{1,6} = -k_{6,1} =$

$k_{7,12} = k_{12,7} = -k_{9,10} = -k_{10,9}$ $\qquad = (2 \cdot \beta^2 + (1 - \mu)/5) \cdot a$

$k_{5,2} = k_{2,5} = k_{8,11} = k_{11,8}$ $\qquad = (2/(3 \cdot \beta^2) - 4 \cdot (1 - \mu)/15) \cdot b^2$

$k_{5,3} = k_{3,5} = k_{5,9} = k_{9,5} = k_{5,12} = k_{12,5} =$

$k_{6,2} = k_{2,6} = k_{6,8} = k_{8,6} = k_{6,11} = k_{11,6} =$

$k_{8,3} = k_{3,8} = k_{8,12} = k_{12,8} = k_{9,2} = k_{2,9} =$

$k_{9,11} = k_{11,9} = k_{11,3} = k_{3,11} = k_{12,2} = k_{2,12} = 0$

$k_{6,3} = k_{3,6} = k_{9,12} = k_{12,9}$ $\qquad = (2 \cdot \beta^2/3 - (1 - \mu)/15) \cdot a^2$

$k_{7,1} = k_{1,7} = k_{4,10} = k_{10,4}$ $\qquad = -2 \cdot (\beta^2 + 1/\beta^2) + (14 - 4 \cdot \mu)/5$

$k_{7,2} = k_{2,7} = -k_{4,11} = -k_{11,4} =$

$-k_{8,1} = -k_{1,8} = k_{10,5} = k_{5,10}$ $\qquad = (-1/\beta^2 + (1 - \mu)/5) \cdot b$

$k_{7,3} = k_{3,7} = -k_{9,1} = -k_{1,9} =$

$-k_{10,6} = -k_{6,10} = k_{12,4} = k_{4,12}$ $\qquad = (\beta^2 - (1 - \mu)/5) \cdot a$

$k_{7,4} = k_{4,7} = k_{10,1} = k_{1,10}$ $\qquad = 2 \cdot (\beta^2 - 2/\beta^2) - (14 - 4 \cdot \mu)/5$

$k_{7,5} = k_{5,7} = -k_{8,4} = -k_{4,8} =$

$k_{10,2} = k_{2,10} = -k_{11,1} = -k_{1,11}$ $\qquad = (-2/\beta^2 - (1 - \mu)/5) \cdot b$

$k_{7,6} = k_{6,7} = k_{9,4} = k_{4,9} =$

$-k_{10,3} = -k_{3,10} = -k_{12,1} = -k_{1,12}$ $\qquad = (\beta^2 - (1 + 4 \cdot \mu)/5) \cdot a$

$k_{8,2} = k_{2,8} = k_{11,5} = k_{5,11}$ $\qquad = (1/(3 \cdot \beta^2) + (1 - \mu)/15) \cdot b^2$

$k_{8,5} = k_{5,8} = k_{11,2} = k_{2,11}$ $\qquad = (2/(3 \cdot \beta^2) - (1 - \mu)/15) \cdot b^2$

$k_{9,3} = k_{3,9} = k_{12,6} = k_{6,12}$ $\qquad = (\beta^2/3 + (1 - \mu)/15) \cdot a^2$

$k_{9,6} = k_{6,9} = k_{12,3} = k_{3,12}$ $\qquad = (2 \cdot \beta^2/3 - 4 \cdot (1 - \mu)/15) \cdot a^2$

The stiffness matrix can be extended for a continuously elastic support of a plate. By means of the mathematical analogy between an elastic support and a harmonic oscillation, one obtains

$$\underline{K}_{\text{Pel}}^{(e)} = \underline{K}^{(e)} + \frac{k_S}{\bar{m}} \cdot \underline{M}^{(e)} \qquad (4.83a)$$

with the subgrade modulus k_S and $\underline{M}^{(e)}$ according to (5.12c).

Example 4.12 The simply supported quadratic plate in Fig. 4.46, is loaded in the middle by a point load F. The plate thickness is 0.20 m, the modulus of elasticity is $E = 3 \cdot 10^7$ kN/m^2, and the Poisson ratio is assumed to be $\mu = 0$. The displacement w_m in the middle of the plate is to be calculated with different regular $n \times n$ finite element meshes with $n = 2$ to 16.

For the 2×2 mesh, the plate can be computed by a single finite element, due to the symmetry of the system (Fig. 4.46). The global system of equations is obtained with the element stiffness matrix (4.83), and the nodal points according to Fig. 4.46, as

$$
10^4 \cdot
\begin{bmatrix}
3.200 & 0 & 0.800 & 0 & -0.320 \\
0 & 3.200 & 0 & 0.800 & 0.320 \\
0.800 & 0 & 3.200 & 0 & -0.880 \\
0 & 0.800 & 0 & 3.200 & 0.880 \\
-0.320 & 0.320 & -0.880 & 0.880 & 0.864
\end{bmatrix}
\cdot
\begin{bmatrix}
\varphi_{x1} \\
\varphi_{y1} \\
\varphi_{x2} \\
\varphi_{y3} \\
w_4
\end{bmatrix}
=
\begin{bmatrix}
0 \\
0 \\
0 \\
0 \\
25.0
\end{bmatrix}.
$$

Its solution is

$$
\begin{bmatrix}
\varphi_{x1} \\
\varphi_{y1} \\
\varphi_{x2} \\
\varphi_{y3} \\
w_4
\end{bmatrix}
=
\begin{bmatrix}
0.1116 \\
-0.1116 \\
0.8929 \\
-0.8929 \\
1.6741
\end{bmatrix}
\cdot 10^{-3}.
$$

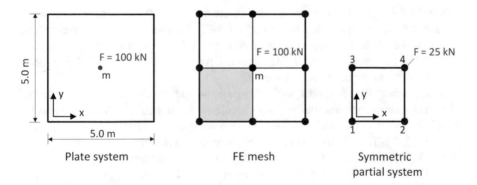

| Plate system | FE mesh | Symmetric partial system |

Fig. 4.46 Quadratic plate with a 2×2 mesh

A mesh refinement leads to the following deflections in the middle of the plate:

FE mesh	2×2	4×4	8×8	16×16
w_m [mm]	1.674	1.518	1.471	1.456

Compared to the 16×16 discretization, the error of the 4×4 discretization is 4.2%, and compared to the 8×8 mesh it is 1.0%. Due to the nonconformity of the shape functions, the convergence to the exact solution is not "from below". ◄

3-node triangle elements

Several shear rigid triangular plate elements have been developed, differing by the element type and the number of degrees of freedom. The basic problem in the derivation of shear rigid triangles is that for 3 degrees of freedom (the displacement and two rotations per node) an interpolation function with only 9 parameters can be chosen. Since the complete expression contains 10 parameters, one term has to be neglected or additional terms have to be introduced. A survey of the development of triangular elements in this sense is given in [18, 19].

DKT and DKQ elements

The discrete Kirchhoff theory assumes that the shear deformations of thin plates are very small and can be neglected. In DKT/DKQ elements the shear contribution in (4.76a), is omitted and the stiffness matrix is determined by

$$\underline{K}^{(e)} = \int \underline{B}_{\mathrm{DKT}}^{\mathrm{T}} \cdot \underline{D}_{\mathrm{b}} \cdot \underline{B}_{\mathrm{DKT}} \, \mathrm{d}x \, \mathrm{d}y . \tag{4.84}$$

In the matrix $\underline{B}_{\mathrm{DKT}}$, the rotations φ_x and φ_y are approximated by higher order shape functions than given in (4.69). For the additionally required equations, the condition is introduced so that the shear deformations at special (discrete) points in the element are set to be zero, as postulated by the Kirchhoff plate theory. DKT and DKQ elements are, therefore, shear rigid plate elements.

Finite triangular elements with quadratic shape functions and quadrilateral elements with incomplete cubic shape functions for the rotational angles and the condition that the shear deformations are zero at the element boundaries are given in [73, 285, 286]. The elements are also known as DKT (**D**iscrete **K**irchhoff **T**riangle) or DKQ (**D**iscrete **K**irchhoff **Q**uadrilateral) elements, respectively. They have proven to be particularly efficient in terms of computational effort. In [68], however, it is noted that the DKT element, with shells, may have convergence problems.

4.7.3 Hybrid Plate Elements

Due to the difficulties in the development of conforming Kirchhoff plate elements, hybrid elements for plates were developed as early as the 1960s. They are normally based on the Kirchhoff plate theory. However, there are also formulations for shear flexible hybrid plate elements [69, 70].

For shear rigid elements, the bending moments m_x, m_y and the twisting moment m_{xy} are approximated over the element area by shape functions. At the boundaries of the element, different shape functions for the displacement w and the angles of rotation φ_x and φ_y are chosen. Different combinations of shape function for the displacements and the bending moments are possible [51].

Hybrid shear rigid 4-node quadrilateral plate element

The element has 4 nodes with 3 degrees of freedom each (Fig. 4.45). For this classical hybrid element from the 1960s, the displacements at the boundaries are described by a parabola of third order [69]. With this, the angles of rotation about an axis normal to the element edge are parabolas of second order. This shape function corresponds to the bending line of a beam without distributed loads. The rotations on an axis in the direction of the element edge are assumed as a linear function. For the bending and twisting moments, quadratic shape functions with 17 parameters are chosen. An extension of the element for plates with continuous elastic support is given in [51].

Elements with mixed interpolation

The plate elements with additional shear strain shape functions, such as the Bathe/Dvorkin element in [64], are denoted as elements with mixed interpolation. In contrast to classical hybrid elements, they are not based on shape functions for stresses or internal forces but on an additional interpolation of shear angles. Reference is made to the further development in [46–48].

4.7.4   Other Element Types

The p-version of the finite element method is also available for plate elements (see Sect. 4.5.5). It appears attractive because it implicitly includes control of the global error in an energy norm. Reference is made to the literature [58, 71, 72].

4.7.5   Element Types in Finite Element Software for Structural Analysis

Among the great number of plate elements that have been developed in recent decades, some quite different elements have found their way into commercial software. The following plate elements are implemented in the finite element software according to Tables 4.11 and 4.12.

ADINA

In ADINA two element types for plates in bending have been implemented [SOF 1]. For thin plates a triangular element according to the Kirchhoff plate theory is available. It is a DKT element with 3 degrees of freedom per node (Fig. 4.45) [73, 74]. However, for analyses of plates, thin and moderately thick, the shell elements MITC3 + and MITC4 + based on Reissner-Mindlin plate theory are recommended [75]. These shell elements have 3 and 4 nodes, respectively, and are based on the formulation with mixed interpolation of tensorial components [46–48]. Higher order elements are also available [64, 76]. In this respect, reference is made to Sect. 4.8.4.

ANSYS

The two shell elements SHELL181 and SHELL282 (default setting) are available as shear flexible quadrilateral and triangular elements for a plate in bending [SOF 2]. Refer to Sect. 4.8.4, for details.

InfoCad

In InfoCad, 3-node DKT and 4-node DKQ elements are implemented and preset as standard elements [SOF 3]. Optionally, elements with mixed interpolation according to the Reissner–Mindlin plate theory for thick shells according to Bathe/Dvorkin are available [64, 76].

RFEM

For plates, the 3- or 4-node shell elements MITC3 and MITC4, respectively, of Bathe/Dvorkin are implemented. They are elements according to the Reissner–Mindlin plate theory with mixed interpolation based on the shell theory [64, 76, 77, SOF 4].

SAP2000

SAP2000 contains a 3-node triangle and a 4-node square plate element ("shell"). The elements are implemented both as formulations according to the Kirchhoff plate theory (as DKT/DKQ elements) and according to the Reissner–Mindlin plate theory [55, 78, SOF 5].

SOFiSTiK

In the program module ASE, a shear flexible 4-node quadrilateral plate element with incompatible shape functions for shear strain according to Hughes is implemented [63, 79, SOF 6]. The additional incompatible shape functions of the element are explained in [80].

iTWO structure fem/TRIMAS

The 4- and 9-node quadrilateral shell elements implemented in iTWO structure fem/ TRIMAS, as well as the 3- and 6-node triangular shell elements, are also available for the analysis of plates in bending [SOF 7]. Reference is made to Sect. 4.8.4.

4.8 Finite Elements for Shells

4.8.1 Plane Shell Elements as Superimposed Membrane and Bending Elements

Shells are structures carrying membrane and bending forces. Plane finite shell elements can, therefore, be composed of a plane stress element for membrane action and a plate element for bending action. Even curved shells may be approximated by flat elements appropriately (Fig. 4.47). Simply curved shells can be modeled with (flat) triangular and quadrilateral elements, whereas double curved shells with a complicated geometry can be modeled with triangular elements or non-flat quadrilateral elements. For non-flat quadrilateral elements that are (slightly) twisted, i.e. whose nodes do not lie exactly in a plane, the geometric deviations must be taken into account by an appropriate transformation.

A shell element composed of a plate and a plane stress element normally has five local degrees of freedom (Fig. 4.48). The sixth local degree of freedom corresponding

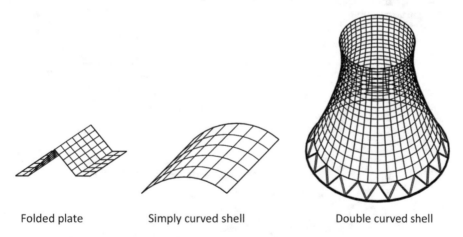

| Folded plate | Simply curved shell | Double curved shell |

Fig. 4.47 Modeling with flat shell elements

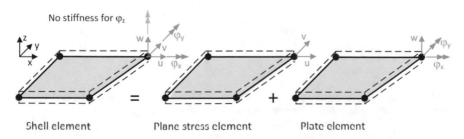

| Shell element | Plane stress element | Plate element |

Fig. 4.48 Composition of a flat shell element by a plate and a plane stress element

to the rotation on an axis normal to the element plane possesses no stiffness. Exceptions to this are plane stress elements with a rotational degree of freedom according to Fig. 4.41, as the hybrid element in Sect. 4.5.4, or the plane stress element of Ibrahimbegovic/Taylor/Wilson [54].

In general, six degrees of freedom per node are required for spatial shell structures. For shell elements with plane stress elements without rotational degree of freedom, special measures are, therefore, required to avoid kinematics (or a singularity of the global stiffness matrix) caused by the missing stiffness in the rotational degree of freedom. For this purpose, an artificial rotational spring about an axis perpendicular to the element plane is introduced at all nodes of the element (Fig. 4.49). Its spring constant must be chosen so small that it practically does not influence the stresses and displacements of the structure. For example, it can be selected to 1/10000 of the smallest diagonal term of the stiffness matrix of the shell element. However, the spring constant must also be large enough to prevent possible singularities of the global stiffness matrix and allow the correct solution of the system of equations.

Fig. 4.49 Artificial rotational springs for plane shell elements with five degrees of freedom per node

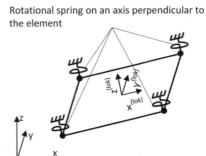

Rotational spring on an axis perpendicular to the element

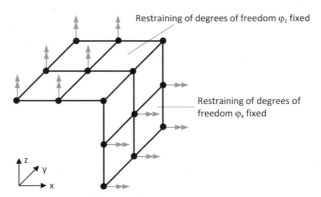

Fig. 4.50 Restraining of degrees of freedom of a folded plate with orthogonal partial surfaces and finite elements with plane stress elements without rotational degrees of freedom

A special case are folded plate structures whose partial surfaces lie parallel to the *xy*-, *yz*- or *xz*-plane. Here, all rotations about the axis perpendicular to the respective surfaces can be fixed if shell elements based on plane stress elements without rotational degrees of freedom are used (Fig. 4.50). The rotational degree of freedom not existing in the plane stress elements are thus eliminated from the global system of equations.

4.8.2 Curved Shell Elements as "Degenerated" Solid Elements

Shell elements can also be derived as special solid elements [34]. For this, the condition for plate deformation is introduced that planes perpendicular to the plate remain straight and perpendicular to the middle plane (Bernoulli hypothesis). In addition, the stress component normal to the plate has to be zero. With these conditions, the degrees of freedom of the nodal points can be expressed by the displacements and the rotations of the middle plane of the plate. These elements have the same properties as elements derived according to the theory of the shear flexible plate. This means that these shell elements must provide a strategy to deal with shear locking. With an isoparametric description of geometry, these elements may also be curved. Relating to their origin as (degenerated) solid elements, they are also known as *degenerated shell elements*. A typical example is the shear flexible shell element of Bathe/Dvorkin [76], which may have up to 16 nodes (Fig. 4.51). Reference is made to more recent developments with EAS formulations in [47, 48].

Shell models of higher order for very thick shells are dealt with in [81]. They possess an even closer approximation to the mechanical behavior of three-dimensional solids or solid elements.

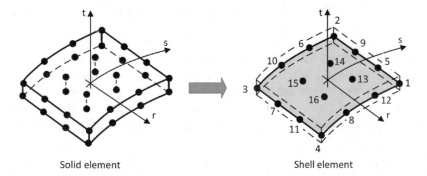

Solid element   Shell element

Fig. 4.51 Degenerated shell element

4.8.3   Axisymmetric Shell Elements

A special case are axisymmetric shells. Taking advantage of the symmetry, they can be analyzed with considerably less computational effort than for a fully three-dimensional finite element discretization of the shell.

If the loading is not axisymmetric, it must first be decomposed into a Fourier series. The radial, vertical, and tangential components of a distributed load are described by

$$p_r = \sum_n p_{r,n}^{\text{sym}} \cdot \cos(n \cdot \theta) + \sum_n p_{r,n}^{\text{ant}} \cdot \sin(n \cdot \theta) \qquad (4.85a)$$

$$p_z = \sum_n p_{z,n}^{\text{sym}} \cdot \cos(n \cdot \theta) + \sum_n p_{z,n}^{\text{ant}} \cdot \sin(n \cdot \theta) \qquad (4.85b)$$

$$p_\theta = -\sum_n p_{\theta,n}^{\text{sym}} \cdot \sin(n \cdot \theta) + \sum_n p_{\theta,n}^{\text{ant}} \cdot \cos(n \cdot \theta) \quad \text{with} \quad n = 0, 1, 2 \ldots . \qquad (4.85c)$$

The symmetric and antisymmetric terms of the load are obtained as Fourier decompositions of the load, e.g. for the load component p_r as

$$p_{r,0}^{\text{sym}} = \frac{1}{2\pi} \int_0^{2\pi} p_r \, d\theta \quad \text{for } n = 0 \qquad (4.86a)$$

$$p_{r,n}^{\text{sym}} = \frac{1}{\pi} \int_0^{2\pi} p_r \cdot \cos(n \cdot \theta) \, d\theta \quad \text{for} \quad n = 1, 2, \ldots . \qquad (4.86b)$$

$$p_{r,n}^{\text{ant}} = \frac{1}{\pi} \int_0^{2\pi} p_r \cdot \sin(n \cdot \theta) \, d\theta \quad \text{for} \quad n = 0, 1, \ldots \qquad (4.86c)$$

Figure 4.52 shows the lowest Fourier terms of a radial load. In the case of a symmetric loading, the load is completely represented by the term $n = 0$, all other Fourier terms are zero and can be dropped in the sum (4.85a–4.85c).

All stresses and displacements of the system are represented in cylindrical coordinates by a Fourier series. The displacements of a axisymmetric finite element are

$$u_r = \sum_n u_{r,n}^{\text{sym}} \cdot \cos(n \cdot \theta) + \sum_n u_{r,n}^{\text{ant}} \cdot \sin(n \cdot \theta) \qquad (4.87a)$$

$$u_z = \sum_n u_{z,n}^{\text{sym}} \cdot \cos(n \cdot \theta) + \sum_n u_{z,n}^{\text{ant}} \cdot \sin(n \cdot \theta) \qquad (4.87b)$$

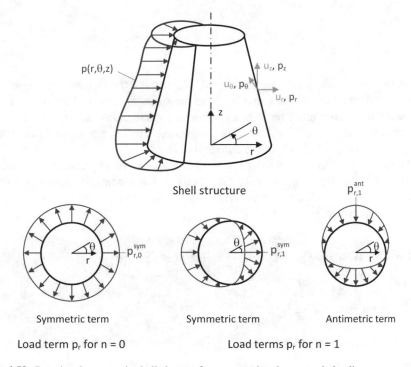

Shell structure

$p_{r,1}^{ant}$

Symmetric term   Symmetric term   Antimetric term

Load term p_r for $n = 0$ Load terms p_r for $n = 1$

Fig. 4.52 Rotational symmetric shell element for non-rotational symmetric loading

$$u_\theta = -\sum_n u_{\theta,n}^{sym} \cdot \sin(n \cdot \theta) + \sum_n u_{\theta,n}^{ant} \cdot \cos(n \cdot \theta) \quad \text{with} \quad n = 0, 1, 2 \ldots \quad (4.87c)$$

The rotations are decomposed in the Fourier series similarly.

The Fourier terms of the displacements $u_{r,n}^{sym}, u_{z,n}^{sym}, u_{\theta,n}^{sym}$ and $u_{r,n}^{ant}, u_{z,n}^{ant}, u_{\theta,n}^{ant}$, as well as the corresponding Fourier terms of the loads depend on the coordinates r and z only, i.e. they are independent of θ. The shape functions of axisymmetric elements are only for the coordinates r and z, whereas the θ-direction is omitted due to the Fourier decomposition. All other space dependent functions, like the strains and the stresses, are also represented as Fourier series. It can be shown that due to the orthogonality conditions of the trigonometric functions

$$\int_0^{2\pi} \sin(n \cdot \theta) \cdot \sin(m \cdot \theta) \, d\theta = 0 \quad \text{for} \quad n \neq m \qquad (4.88a)$$

$$\int_0^{2\pi} \cos(n \cdot \theta) \cdot \cos(m \cdot \theta) \, d\theta = 0 \quad \text{for} \quad n \neq m \qquad (4.88b)$$

$$\int\limits_0^{2\pi} \sin(n \cdot \theta) \cdot \cos(m \cdot \theta)\, d\theta = 0 \quad \text{for all } n,\, m \qquad (4.88c)$$

the individual Fourier terms are uncoupled. Introducing the Fourier decomposition of the displacements, stresses, and loads in the stiffness relationships of the finite elements, for each Fourier term n, a system of equations for the symmetrical and antisymmetrical state of displacements is obtained. The Fourier terms of the loads represent the right-hand side of the equations.

The solutions of these systems of equations are summarized by a Fourier synthesis, in order to obtain the complete solution. For the displacements, for example, this is done with (4.87a–4.87c). A Fourier synthesis also has to be performed for the stress and strain components.

Analysis with axisymmetrical finite elements and unsymmetrical loads
1. Decomposition of the loads into Fourier terms
2. Solution of the systems of equations and determination of the section forces for all Fourier terms
3. Composition of the displacements and section forces by their Fourier terms.

The finite shell elements are in general straight-lined elements in the rz-plane (Fig. 4.53). However, it is also possible to derive curved elements. There are formulations for shear flexible and shear rigid elements [82]. Since this is a two-dimensional problem, comparable to a beam in bending, analytical solutions for the stiffness matrix are also available [28, 83]. With regard to finite elements for axisymmetrical shells, reference is made to [7, 8, 20, 28, 34].

By the Fourier decomposition, the original 3D problem is replaced by several 2D problems. The number of 2D problems to be solved is proportional to the number of the Fourier terms required. The method is, therefore, especially efficient for axisymmetrical loading, since then only one Fourier term has to be considered. For this purpose special shell elements have been developed. The method is still efficient if

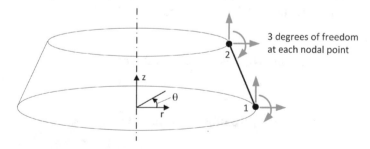

Fig. 4.53 Axisymmetric shell element for axisymmetric loading

the load can be represented by few Fourier terms, e.g. for a wind loading. However, it can become inefficient if many Fourier terms are required, for example, for a point load.

Modern computing power permits the analysis of huge general three-dimensional models. Therefore, the use of axisymmetric elements has lost its importance for practical applications. However, it is still attractive for comparison purposes of simplified axisymmetric systems, because of the low computational effort.

4.8.4   Element Types in Finite Element Software for Structural Analysis

The shell elements implemented in many commercial finite element software correspond to the membrane and bending plate elements. Either the shell elements are a superposition of the membrane and bending elements implemented in the program or a degenerated shell element derived as a modified solid element is implemented in the program which is also available in the analysis of plates in the plane stress and bending state. Many programs also contain formulations for layered cross sections or shells or plates with orthotropic material properties.

ADINA

Several elements are available. For analyses of shells, thin and moderately thick, the shell elements MITC3 + and MITC4 + based on Reissner-Mindlin plate theory are recommended [75]. These shell elements have 3 and 4 nodes, respectively, and are based on the formulation with mixed interpolation of tensorial components [46–48]. Higher order elements are also available [64, 76]. The 4-node element can also be used with incompatible displacement shape functions for the membrane state (cf. Sect. 4.5.3). The elements can be used automatically with 5 or 6 degrees of freedom at each node, i.e. they can be used without an artificial spring. Also, a 3D-shell element is available that is useful when through-the-thickness stresses and strains are important [75]. In addition, an axisymmetric shell element is implemented in ADINA for axisymmetric structures with axisymmetric loads [SOF 1].

ANSYS

ANSYS contains several shell elements in its element library [SOF 2]. SHELL181 is an isoparametric shear flexible 4-node element with additional shape functions for the plate shear according to Bathe/Dvorkin [76]. The default element SHELL281 is a curved, shear flexible, degenerated isoparametric 8-node element with a reduced 2×2-point integration and hourglass control. Both elements are also available as triangle elements with 3 or 6 nodes.

In addition, ANSYS has special shell elements for axisymmetric structures with axisymmetric loads. SHELL208 is an axisymmetric, shear flexible shell element with 2 nodes, SHELL209 a corresponding 3-node element. Both elements can also be used as pure membrane elements. Analyses of axisymmetric shells with non-axisymmetric loads, represented as Fourier series, can be done with the SHELL61 element.

InfoCad

The 3- and 4-node shell elements consist of the corresponding membrane and shell elements implemented in InfoCad (see Sects. 4.5.6 and 4.7.5). Since both, the membrane and the plate element, have 3 degrees of freedom per node each, elements with 6 degrees of freedom per node are obtained.

RFEM

The shell elements in RFEM consist of the membrane elements implemented in RFEM and the corresponding elements for plates in bending (see Sects. 4.5.6 and 4.7.5). Since both the membrane and the plate element have 3 degrees of freedom per node each, elements with 6 degrees of freedom per node are obtained.

SAP2000

The 3- and 4- node shell elements ("shell") consist of the corresponding membrane elements and the elements for plates in bending implemented in SAP2000 (see Sects. 4.5.6 and 4.7.5). Both the membrane element and the plate element each have 3 degrees of freedom per node, which results in a shell element with 6 degrees of freedom per node. The 4-node element needs not be planar, in which case the membrane and bending effects are coupled. In addition, a layered shell element is available. The layered shell considers transverse shearing deformation according to the Reissner–Mindlin plate theory.

SOFiSTiK

The shell element in the ASE program module consists of the 4-node plate element and the 4-node bending element (see Sects. 4.5.6 and 4.7.5). Perpendicular to the element plane, the shell element additionally has a rotational spring in each node with a very low stiffness. The shell element has additional terms for nonplanar (locally bilinear) geometries and can, therefore, also represent curved surfaces.

iTWO structure fem/TRIMAS

The curved shell element implemented in iTWO structure fem/ TRIMAS is a degenerated isoparametric shell element according to Reissner–Mindlin, in which the "Discrete Shear Gap Method" (DSG) is implemented to avoid shear locking and to improve the quality of shear forces [65]. Quadrilateral shell elements with 4 and 9 nodes, as well as triangular elements with 3 and 6 nodes, are available [SOF 7].

4.9 Solid Elements

4.9.1 Isoparametric Elements

Any mechanical structure can be considered as a three-dimensional solid. Three-dimensional solids are the most general way to model a structure. Assumptions such as the Bernoulli hypothesis for beams and plates in bending are not made. Solids can be modeled by three-dimensional finite solid elements.

The stiffness matrix of solid elements can be derived, based on the theory of elasticity for a three-dimensional continuum. Each nodal point has 3 degrees of freedom, i.e. the displacements u, v, w in the x-, y-, z-directions, respectively. Figure 4.54 shows some isoparametric solid elements. As for isoparametric plane stress elements, the number of nodal points is variable (see Sect. 4.5.2, isoparametric elements). A simple hexaeder element has 8 nodes (Fig. 4.54). The shape functions for the displacements

$$\underline{u} = \begin{bmatrix} u \\ v \\ w \end{bmatrix} \tag{4.89}$$

are polynomials in the local natural coordinates r, s, t similarly as in (4.42a, 4.42b), for the plane element. In this way, by differentiation, the strain components are obtained as

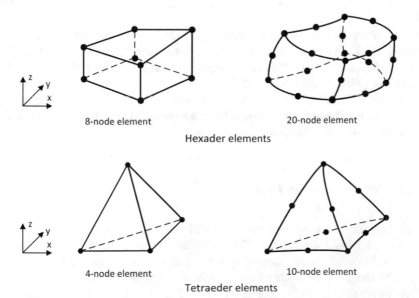

8-node element   20-node element

Hexader elements

4-node element   10-node element

Tetraeder elements

Fig. 4.54 Isoparametric solid elements

$$\underline{\varepsilon} = \underline{B} \cdot \underline{u} \tag{4.90}$$

with

$$\underline{\varepsilon}^{\mathrm{T}} = \begin{bmatrix} \varepsilon_x & \varepsilon_y & \varepsilon_z & \gamma_{xy} & \gamma_{yz} & \gamma_{zx} \end{bmatrix}. \tag{4.91}$$

With Hooke's law for the three-dimensional continuum

$$\underline{\sigma} = \underline{D} \cdot \underline{\varepsilon} \tag{4.92}$$

the stress components

$$\underline{\sigma}^{\mathrm{T}} = \begin{bmatrix} \sigma_x & \sigma_y & \sigma_z & \tau_{xy} & \tau_{yz} & \tau_{zx} \end{bmatrix} \tag{4.93}$$

can be determined. Applying the principle of virtual displacements, the stiffness matrix is obtained as

$$\underline{K}_{\mathrm{e}} = \int_V \underline{B}^{\mathrm{T}} \cdot \underline{D} \cdot \underline{B} \cdot \mathrm{d}V. \tag{4.94}$$

The integration is performed over the volume of the element, whereas the infinitesimal volume element

$$\mathrm{d}V = \det \mathbf{J} \cdot \mathrm{d}r \, \mathrm{d}s \, \mathrm{d}t \tag{4.94a}$$

is related to the local element coordinates by the Jacobian operator. The formulation is based on the basic equations of the three-dimensional continuum (cf. Fig. 2.13) and is analogous to that of the isoparametric plane stress element. For details of the formulation, see [34, Sect. 5.3].

Tetahedral elements, as well as pyramid-, cone-, wedge- and prism-shaped solid elements with and without nodes on the element sides can also be formulated in this way as isoparametric elements.

4.9.2 Other Element Types

Three-dimensional solid elements can exhibit the same locking effects as two-dimensional membrane elements. This concerns elements with linear or bilinear shape functions of the displacements such as the 8-node hexahedral element. As a remedy, the same methods as for membrane elements are used such as under-integration of the stiffness matrix with control of the hourglass modes, incompatible displacement approaches, enhanced assumed strain formulations, and mixed formulations. Solid elements with additional rotational degrees of freedom on x-, y-, and

z-axis, i.e. a total of 6 degrees of freedom per node, have been developed as well [84].

By merging nodes, more element types such as tetrahedra, pentahedra, prisms, wedges or with pyramid shapes can be developed from the hexahedral element. They are also known as "degenerated" element forms. However, very simple element forms such as the simple tetrahedron element with 4 nodes are only conditionally suitable for the representation of complex stress distributions. In principle, however, tetrahedral elements allow a very complex structural geometry, such as can be transferred from CAD systems to a finite element program, to be meshed very well. However, tetrahedral elements should always be used with higher shape functions, such as with 10-node elements with quadratic displacement shape functions which can represent linear stress distributions.

4.9.3 Axisymmetric Solid Elements

Axisymmetric systems with axisymmetric loads can be regarded as two-dimensional systems. This reduces the dimension of the model and thus the computational effort considerably. Solid elements for an axisymmetric continuum can be formulated in the same way as two-dimensional plane elements. The derivation is not given here, but the reference is made to the standard literature such as [7]. Figure 4.55 shows a 4-node element. With regard to possible locking effects, the same applies as explained for membrane elements. These can be countered by elements with higher shape functions.

For axisymmetric systems with non-axisymmetric loading, the loads can be represented as a Fourier series. As shown in Sect. 4.8.3, for shell structures, the analysis is performed separately for each Fourier term and then the results are superimposed. This requires special element formulations [8, 85, 86].

Fig. 4.55 4-node axisymmetric solid element

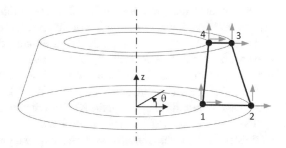

4.9.4   Element Types in Finite-Element Software
for Structural Analysis

Finite element meshes of three-dimensional solids easily lead to models with a very large number of degrees of freedom. The generation of three-dimensional meshes requires special strategies for irregular structures. The interpretation of the results is more complex than with two-dimensional structures. Therefore, in structural engineering, models with solid elements are limited to special applications. However, most finite element programs for structural design include the ability to analyze three-dimensional continua with solid elements.

ADINA

ADINA has implemented isoparametric hexaeder elements with 4 to 20 nodes at the corner points and element edges, as well as elements with 21 to 27 nodes, with additional internal nodes. These are elements with shape functions for the displacements [34, Sect. 5.3]. They can also be used in the form of tetrahedra, prisms, and pyramids.

In addition to the elements based on displacement shape functions, mixed finite element formulations are also available in which displacement and stress shape functions are combined. These are especially recommended for nearly incompressible materials whose Poisson's ratio is close to 0.5.

Axisymmetric structures with axisymmetric loading can be analyzed in ADINA with a two-dimensional, axisymmetric solid element. In addition, two-dimensional plane strain elements are available for the analysis of three-dimensional systems in the plane strain state (Sect. 4.5.6) [SOF 1].

ANSYS

ANSYS has implemented several solid elements. SOLID185 is an isoparametric hexahedron solid element with 8 nodes, SOLID186 a hexahedron element with 20 nodes. By merging nodes, the elements can also be used in the form of prisms, tetrahedra, and pyramids. The elements can be used as isoparametric elements with displacement shape functions and $2 \times 2 \times 2$ integration. The SOLID186 element can also be fully integrated (14 points). For the SOLID185 element, a 1-point integration with hourglass control or alternatively an enhanced assumed strain formulation to prevent shear locking effects are optionally available. Further, methods are available for nearly incompressible materials whose Poisson's ratio is close to 0.5.

SOLID187 is a tetrahedron element with 10 nodes (intermediate nodes on element edges) and quadratic shape functions of the displacements.

The two-dimensional elements PLANE182 and PLANE183 are used for the plane strain state and for axisymmetric systems with axisymmetric loading (see Sect. 4.5.6).

PLANE25 is an axisymmetric 4- or 3-node element with non-axisymmetric loading, PLANE83 is a corresponding element with quadratic shape functions and 8 nodes (quadrilateral) or 6 nodes (triangular). PLANE25 optionally provides incompatible displacement shape functions. The advanced axisymmetric elements

SOLID272 and SOLID273 facilitate the input of the non-axisymmetric loads [SOF 2].

InfoCad

The program contains two solid elements with different displacement shape functions. The hexahedral element is an isoparametric solid element with 8 nodes. More element geometries can be created by merging nodes.

A tetrahedral element with intermediate nodes on the element edges is also available. The element possesses 10 nodes. It has quadratic displacement shape functions and is able to represent linear stress distributions [SOF 3].

RFEM

A hexahedral element with 8 nodes is implemented in RFEM. In addition to the 3 degrees of freedom of displacements, the element also has three rotational degrees of freedom at each node [84]. Similar to the corresponding membrane elements with incompatible displacement shape functions, these are formulated on the basis of a 20 node hexahedral element by eliminating the intermediate nodes on the element edges [SOF 4].

SAP2000

SAP2000 contains an isoparametric hexahedral element ("solid") with 8 nodes and a $2 \times 2 \times 2$ integration [34]. Pyramid, tetrahedron, and wedge-shaped elements can be derived by merging nodes.

For axisymmetric systems, a two-dimensional 3- or 4-node element ("Asolid") is implemented in SAP2000. For analyses in the plane strain state, a 3- or 4-node element ("plane") is available (see Sect. 4.5.6) [SOF 5].

SOFiSTiK

The solid element implemented in the ASE software module is a hexahedral element with 8 nodes. In addition, it contains noncompatible shape functions in order to improve the element behavior in bending. The solid element is also available as a tetrahedron with internal quadratic shape functions.

Two-dimensional stress states can be investigated using the plane elements implemented in the TALPA program module for the plane strain state and for axisymmetric solids with axisymmetric loading (see Sect. 4.5.6) [SOF 6].

iTWO structure fem/TRIMAS

In iTWO structure fem/TRIMAS an isoparametric hexahedral element with 8 nodes and a $2 \times 2 \times 2$ integration is implemented [34]. Furthermore, a hexagonal isoparametric prism element with a triangular base is available [SOF 7].

4.10   Transition between Beam, Plate and Solid Elements

4.10.1   Generals

A major benefit of the finite element method compared to others, such as the classical finite difference method, is that different element types can be combined in the same structural model. A finite element model may include beam elements, as well as plate or solid elements. However, connecting different element types is not trivial for elements with different types of degrees of freedom resulting from different kinematic assumptions. Two-dimensional plane stress elements and three-dimensional solid elements are continuum elements. Their mechanical behavior is described by stresses, strains, and displacements. Beam, plate and shell elements, on the other hand, are elements subjected to bending where stresses are represented by stress resultants such as normal and shear forces, as well as bending and torsional moments. They are based on the kinematics of the Bernoulli assumption. The corresponding degrees of freedom are the displacements and rotations related to the member axis or centroid line. A consistent modeling of the transition between elements based on the Bernoulli assumption and continuum elements must take into account the different stress representations, as well as the different degrees of freedom of both element types.

In finite element models, the following transitions can occur between elements formulated for bending and solid elements:

- Two-dimensional plane models

  – Beam – plate in plane stress
  – Beam – plate in bending (e.g. columns in flat slabs)

- Three-dimensional models

  – Beam – shell
  – Beam – 3D solid
  – Shell – 3D solid (special case: plate in bending – 3D solid).

Examples are regions in plane stress models subjected to bending and modeled with beam elements, columns in flat slabs or the connection of foundation slabs with solid elements representing the soil.

In the following, the connection of a beam element with plane stress elements is considered where the connection area is perpendicular to the beam axis (Fig. 4.56). This case can be transferred to the above-mentioned transitions between different element types. When modeling the transition between beam and plate elements, it's normally assumed that the kinematics of the Euler–Bernoulli beam theory is valid at the connection. This means that cross sections after deformation remain plane and perpendicular to the deformed beam axis. If this kinematic assumption is transferred to the connection it means that its area remains plane and perpendicular to the deformed axis of the beam. Hence, it behaves like a rigid area, which performs

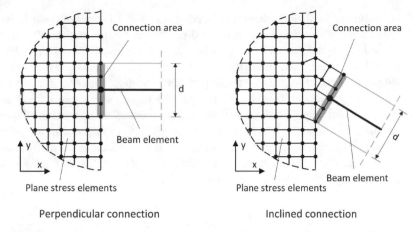

Fig. 4.56 Connection of a beam element with plane stress elements

displacements and rotations just like the cross section of the beam element according to the Bernoulli hypothesis.

However, this obviously does not correspond to the real behavior of a connection area, since it does not behave rigidly, but will deform. The Bernoulli hypothesis only applies to regular beam regions where bending strains are linear (bending- or B-regions) but not to discontinuities such as regions near supports or concentrated loads (discontinuity- or D-regions). An alternative to specifying the displacement distribution at the connection is the specification of a stress distribution. This means that the stresses in the beam element according to the Euler–Bernoulli beam theory are transferred to the plane stress elements and act on these as loading. The displacements of the plane stress elements thus have no constraint conditions as they would with the specification of a rigid connection area. However, they fulfill the compatibility condition with the displacements of the beam element "on average" in the sense of the energy principle. On the other hand, rigid inclusions in continuum models usually cause singularities of stresses and/or displacements. This is the case with the specification of a rigid connection area, but not with the model of stress transfer.

Both specifications, the one that is based on the assumption of the displacement distribution, as well as the one that is based on the stress distribution, are discussed in the following. The model with the assumption of a rigid connection area behaving like a rigid body will be denoted as the *RDT model* (**R**igid body **D**isplacement **T**ransformation model). The model of the stress transfer or *EST model* (**E**quivalent **S**tress **T**ransformation model) has been used so far to model columns as support of plates in bending for flat slabs [87–89]. Its application to the modeling of the connection of plates in plane stress with beams will be described in the following [90]. For three-dimensional beam connections with general cross sections, there is still a need for research.

The coupling of the displacements of different nodes and degrees of freedom by a condition described as a linear equation is called "**Multi-P**oint **C**onstraint" (MPC). Both the models, the RDT, as well as the EST model, can be described as a *Multi-Point Constraint* (MPC) condition. In the RDT model, the displacements and rotations of the nodal point of the beam (master node) prescribe the displacements of the nodes of the plane stress model (slave nodes). In the EST model, the displacements and rotation of the nodal point of the beam (slave node) are described as a weighted mean of the displacements of the plane stress model nodes (master nodes with weighting factors). Transition elements can be formulated for both models. In the RDT model, the transition elements are modified plane stress elements, in the EST model, the transition elements are modified beam elements.

4.10.2 Transformation of Element Matrices

First the transformation of the force and displacement vectors of one system into another system will be formulated. It can be applied to a coordinate transformation or to the connection area with two types of elements. First the system A with the force vector $\underline{F}_A$ and the displacement vector $\underline{u}_A$ is considered. The elements of both vectors are arranged in such a way that an entry of $\underline{F}_A$ performs work with the corresponding entry of $\underline{u}_A$. The system A is to be transformed into the system B with the force vector $\underline{F}_B$ and the displacement vector $\underline{u}_B$. System A is the original system, system B the target system.

The displacements of system A can be expressed by those of system B as

$$\underline{u}_A = \underline{T} \cdot \underline{u}_B. \tag{4.95a}$$

With (4.95a), the following relationship between the force vectors is valid:

$$\underline{F}_B = \underline{T}^T \cdot \underline{F}_A \tag{4.95b}$$

This can be shown by applying the principle of virtual displacements. The virtual external work is performed in system A by real forces $\underline{F}_A$ and the virtual displacements $\underline{\bar{u}}_A$ and in system B by the real forces $\underline{F}_B$ and the virtual displacements $\underline{\bar{u}}_B$. For the virtual displacements, the same relationship applies as for the real displacements, i.e.

$$\underline{\bar{u}}_A = \underline{T} \cdot \underline{\bar{u}}_B.$$

Equating the virtual work in both systems one obtains

$$\bar{u}_A^T \cdot F_A = \bar{u}_B^T \cdot F_B$$

and with (4.95a)

$$\bar{u}_B^T \cdot \underline{T}^T \cdot F_A = \bar{u}_B^T \cdot F_B,$$

from which the relationship (4.95b), follows for any arbitrary virtual displacements $\bar{u}_B^T$.

The stiffness matrix $\underline{K}_A$ which is defined in system A by

$$\underline{K}_A \cdot \underline{u}_A = \underline{F}_A \tag{4.96}$$

can now be transformed into system B. With (4.95a) and (4.95b), the equation for the transformation of the stiffness matrix is obtained as

$$\underline{T}^T \cdot \underline{K}_A \cdot \underline{T} \cdot u_B = \underline{F}_B$$

or

$$\underline{K}_B \cdot u_B = \underline{F}_B \tag{4.97}$$

with

$$\underline{K}_B = \underline{T}^T \cdot \underline{K}_A \cdot \underline{T}. \tag{4.97a}$$

If the stiffness matrix $\underline{K}_A$ is symmetric, it follows that $\underline{K}_B$ is symmetric too. This can easily be shown with $\underline{K}_A^T = \underline{K}_A$ and the relationships given in Table 1.1 as

$$\underline{K}_B^T = \left(\underline{T}^T \cdot \underline{K}_A \cdot \underline{T}\right)^T = \underline{T}^T \cdot \underline{K}_A^T \cdot \underline{T} = \underline{K}_B.$$

The equation has already been applied to the transformation of stiffness matrices from local to global coordinates, such as, e.g. to the spring element in (3.20). In general, the relations (4.97) to (4.97a), can be understood as equations for the transformation of element matrices.

4.10.3 Connections with Displacement Assumptions (RDT)

In the model based on a displacement assumption, the displacements in the connection area are specified on the basis of the Bernoulli assumption of the beam. The connection area is assumed to be a rigid area perpendicular to the beam axis. The nodes of the plane stress elements located in this connecting area are coupled by a rigid body condition (MPC). This is explained using the example of the connection of a beam element with plane stress elements.

RDT connection of a two-dimensional beam element with plane stress elements

Figure 4.57 shows a plate in plane stress state with the thickness t. It is discretized into isoparametric 4-node elements, which are to be connected to a beam element of height d and width t at n nodes corresponding to the beam height. The displacements of the beam are described by the two displacements u_B, v_B and a rotation φ_B. The

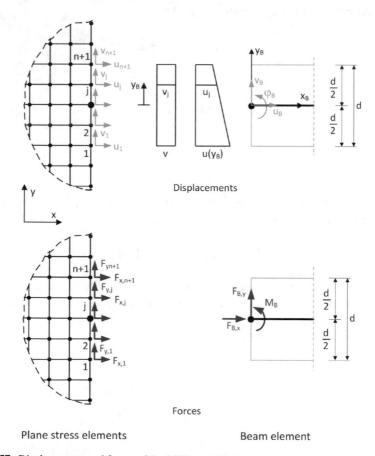

Fig. 4.57 Displacements and forces of the RDT model in the connection area

displacements of any node j of the plate in the connection area are prescribed as the rigid body motion as

$$u_j = u_B - y_{B,j} \cdot \varphi_B$$
$$v_j = v_B \tag{4.98}$$

where $y_{B,j}$ is the y_B-coordinate of nodal point j.

Summing up the displacements of all nodal points of the plane stress elements to a vector $\underline{u}_S$ and at the end of the beam to a vector $\underline{u}_B$, one obtains

$$
\begin{bmatrix}
u_1 \\
v_1 \\
\cdot \\
\cdot \\
u_j \\
v_j \\
\cdot \\
\cdot \\
u_{n+1} \\
v_{n+1}
\end{bmatrix}
=
\begin{bmatrix}
1 & 0 & -y_{B,1} \\
0 & 1 & 0 \\
\cdot & \cdot & \cdot \\
\cdot & \cdot & \cdot \\
1 & 0 & -y_{B,j} \\
0 & 1 & 0 \\
\cdot & \cdot & \cdot \\
\cdot & \cdot & \cdot \\
1 & 0 & -y_{B,n+1} \\
0 & 1 & 0
\end{bmatrix}
\cdot
\begin{bmatrix}
u_B \\
v_B \\
\varphi_B
\end{bmatrix}
. \tag{4.99}
$$

$$\underline{u}_S = \underline{T} \cdot \underline{u}_B \tag{4.99a}$$

The equation corresponds to (4.95a). Thus, with (4.95b), for the forces the relationship

$$\underline{F}_B = \underline{T}^T \cdot \underline{F}_S \tag{4.100}$$

is obtained with

$$
\underline{F}_S =
\begin{bmatrix}
F_{x,1} \\
F_{y,1} \\
\cdot \\
\cdot \\
F_{x,j} \\
F_{y,j} \\
\cdot \\
\cdot \\
F_{x,n+1} \\
F_{y,n+1}
\end{bmatrix}
, \quad
\underline{F}_B =
\begin{bmatrix}
F_{B,x} \\
F_{B,y} \\
M_B
\end{bmatrix}
. \tag{4.100a}
$$

The two Eqs. (4.99a, 4.100), for the transformation of the displacements and the forces are to be considered in the solution of the global system of equations. The degrees of freedom $\underline{u}_S$ depend on $\underline{u}_B$. Thus, they can be replaced by $\underline{u}_B$ in the global system of equations. However, for the implementation into a computer program this is a computationally complex task. Therefore, often other methods like *Penalty methods* or the method of the *Lagrange multipliers* are used to fulfill the MPC conditions (4.99a). However, it should be noted that the penalty method is an approximation method whose accuracy depends on the choice of a parameter α that corresponds to an artificial stiffness additionally introduced into the system. Refer to standard literature such as [34], for details.

The procedure can be transferred and generalized to three-dimensional connections of beam elements with shell and solid elements, as well as to the coupling of shell and shell elements [91, 92].

Example 4.13 A RDT connection of a beam element with two plane stress elements is considered. The matrix describing the MPC relationships has to be set up for the RDT connection of a beam with two plane stress elements (compare to Fig. 4.63b).

The matrix for the transformation of the displacements is obtained according to (4.99), as

$$
\underline{T} = \begin{bmatrix} 1 & 0 & d/2 \\ 0 & 1 & 0 \\ 1 & 0 & 0 \\ 0 & 1 & 0 \\ 1 & 0 & -d/2 \\ 0 & 1 & 0 \end{bmatrix} \text{ and } \underline{u}_S = \underline{T} \cdot \underline{u}_B \text{ with } \underline{u}_S = \begin{bmatrix} u_1 \\ v_1 \\ u_2 \\ v_2 \\ u_3 \\ v_3 \end{bmatrix} \quad \underline{u}_B = \begin{bmatrix} u_B \\ v_B \\ \varphi_B \end{bmatrix}.
$$

The transformation of the forces is given by (4.100), as

$$
\underline{F}_B = \underline{T}^T \cdot \underline{F}_S \text{ with } \underline{F}_S = \begin{bmatrix} F_{x,1} \\ F_{y,1} \\ F_{x,2} \\ F_{y,2} \\ F_{x,3} \\ F_{y,3} \end{bmatrix}.
$$
◀

Transition elements

If the assumption of a rigid connection area is not applied to the global equations but to the equations of an single finite element, so-called *transition elements* can be formulated [34]. These possess the degrees of freedom u_B, v_B and φ_B at one side of the element and thus can be connected with a beam element (Fig. 4.58). Transition elements are "degenerated" plane stress or solid elements (Fig. 4.59). The following simple example demonstrates the principle for a 4-node rectangular element.

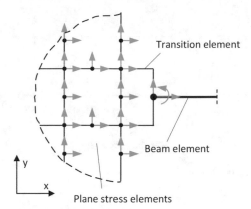

Fig. 4.58 Connection of a beam element with plane stress elements by a transition element

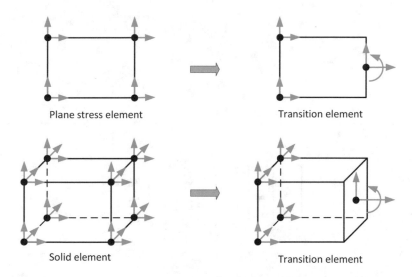

Fig. 4.59 Transition elements as „degenerated" continuum elements

Example 4.14 A two-dimensional transition element is considered. The stiffness matrix of the transition element given in Fig. 4.60, has to be formulated.

The element is a degenerated plane stress element. The stiffness matrix of the 4-node element shown in Fig. 4.14, has already been given in Sect. 4.4.3. For the

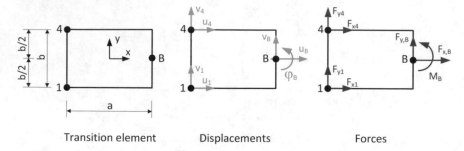

Transition element   Displacements   Forces

Fig. 4.60 3-node transition element

formulation as transition element the degrees of freedom of the nodes 2 and 3 are transformed into those of a beam. The corresponding relations are as follows:

$$u_2 = u_B + \frac{b}{2} \cdot \varphi_B$$

$$v_2 = v_B$$

$$u_3 = u_B - \frac{b}{2} \cdot \varphi_B$$

$$v_3 = v_B$$

or

$$\begin{bmatrix} u_2 \\ v_2 \\ u_3 \\ v_3 \end{bmatrix} = \begin{bmatrix} 1 & 0 & b/2 \\ 0 & 1 & 0 \\ 1 & 0 & -b/2 \\ 0 & 1 & 0 \end{bmatrix} \cdot \begin{bmatrix} u_B \\ v_B \\ \varphi_B \end{bmatrix}$$

$$\underline{u}_{2-3} = \underline{T} \cdot \underline{u}_B.$$

According to (4.100), the following applies to the forces

$$\underline{F}_B = \underline{T}^T \cdot \underline{F}_{2-3}.$$

With these relations the degrees of freedom u_2 to v_3 in the $\underline{B}$-matrix after (4.29) can be replaced by the degrees of freedom u_B, v_B and φ_B. Alternatively, instead of the shape functions of the displacement of the nodes 2 and 3, shape functions for the beam-type degrees of freedom u_B, v_B and φ_B can be added directly. Please refer to [34], for further information.

Here, the transformation relations are to be introduced directly into the stiffness matrix according to (4.36). For this purpose, (4.36) is subdivided into submatrices

$$
\begin{bmatrix}
\underline{K}_{11} & \underline{K}_{12} & \underline{K}_{13} \\
\underline{K}_{21} & \underline{K}_{22} & \underline{K}_{23} \\
\underline{K}_{31} & \underline{K}_{32} & \underline{K}_{33}
\end{bmatrix}
\cdot
\begin{bmatrix}
u_1 \\ v_1 \\ u_2 \\ v_2 \\ u_3 \\ v_3 \\ u_4 \\ v_4
\end{bmatrix}
=
\begin{bmatrix}
F_{x1} \\ F_{y1} \\ F_{x2} \\ F_{y2} \\ F_{x3} \\ F_{y3} \\ F_{x4} \\ F_{y4}
\end{bmatrix}
.
$$

With the transformation relations, one obtains

$$
\begin{bmatrix}
\underline{K}_{11} & \underline{K}_{12}\cdot\underline{T} & \underline{K}_{13} \\
\underline{T}^{\mathrm{T}}\cdot\underline{K}_{21} & \underline{T}^{\mathrm{T}}\cdot\underline{K}_{22}\cdot\underline{T} & \underline{T}^{\mathrm{T}}\cdot\underline{K}_{23} \\
\underline{K}_{31} & \underline{K}_{32}\cdot\underline{T} & \underline{K}_{33}
\end{bmatrix}
\cdot
\begin{bmatrix}
u_1 \\ v_1 \\ u_{\mathrm{B}} \\ v_{\mathrm{B}} \\ \varphi_{\mathrm{B}} \\ u_4 \\ v_4
\end{bmatrix}
=
\begin{bmatrix}
F_{x1} \\ F_{y1} \\ F_{x\mathrm{B}} \\ F_{y\mathrm{B}} \\ M_{\mathrm{B}} \\ F_{x4} \\ F_{y4}
\end{bmatrix}
.
$$

This is the stiffness matrix of the transition element. ◀

Transition elements have been formulated with higher shape functions for two- and three-dimensional systems [93–96]. Elements for the transition from axisymmetric shell elements to axisymmetric solid elements are given in [97].

4.10.4 Connections with Stress Assumptions (EST)

An alternative to the assumption of displacements as rigid body motion is the assumption of a stress distribution in the connection area. For this, the stresses of a beam in the linear elastic range according to the Euler–Bernoulli beam theory are applied. While the stress assumptions of both element types in the transition area are the same, the compatibility of the displacements is only fulfilled "on average" i.e. in the sense of the principle of virtual displacements. This avoids unwanted constraints in the transition area. Since the stresses between both element types are transformed in a consistent way, this concept is called *equivalent stress transformation* (EST). EST elements have been developed so far especially for modeling the support of membrane plates and bending plates on elastic supports (given by beam elements). Here, compared to the frequently used elastic (Winkler) springs, the stiffness matrices of EST elements represent interconnected springs. For bearings, the EST model is, therefore, also referred to as the *Coupling Spring Model*.

EST connection of beam elements with plane stress elements

A beam element with a rectangular cross section of height d and width t is considered. The stresses of the beam consist of normal and shear stresses.

First the normal stresses are considered. According to the Bernoulli hypothesis, with linear elastic material behavior, they are linearly distributed over the cross section height of the beam (Fig. 4.61). They result from the normal force N and the bending moment M. With the cross-sectional area A_B and the moment of inertia I_B they are given by

$$\sigma_x = \frac{N}{A_B} + \frac{M}{I_B} \cdot (-y_B). \tag{4.101}$$

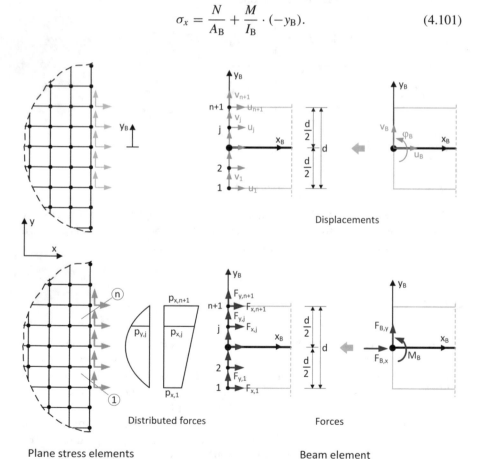

Fig. 4.61 Displacements and forces of the EST model in the connection area

The internal forces of the beam at $x_B = 0$ correspond to the element forces $F_{B,x} = -N$ and $M_B = -M$ (Figs. 3.6 and 3.15). If these are represented as line forces, one obtains

$$p_x = -\sigma_x \cdot t = \frac{-N}{A_B} \cdot t + \frac{-M}{I_B} \cdot t \cdot (-y_B) = \frac{F_{Bx}}{A_B} \cdot t + \frac{M_B}{I_B} \cdot t \cdot (-y_B)$$

or

$$p_x = \begin{bmatrix} 1 & -y_B \end{bmatrix} \cdot \begin{bmatrix} \dfrac{t}{A_B} & 0 \\ 0 & \dfrac{t}{I_B} \end{bmatrix} \cdot \begin{bmatrix} F_{B,x} \\ M_B \end{bmatrix} \qquad (4.102)$$

with t as plate thickness. The line forces are positive in the positive x-direction. Since positive (tensile) stresses at the left end of the bar point in the negative x-direction, the sign is reversed.

The beam element is connected to n plane stress elements. The ordinates of the line loads at the $n + 1$ nodes of the connection area are

$$\begin{bmatrix} p_{x,1} \\ \cdot \\ p_{x,j} \\ \cdot \\ p_{x,n+1} \end{bmatrix} = \begin{bmatrix} 1 & -y_{B,1} \\ \cdot & \cdot \\ 1 & -y_{B,j} \\ \cdot & \cdot \\ 1 & -y_{B,n+1} \end{bmatrix} \cdot \begin{bmatrix} \dfrac{t}{A_B} & 0 \\ 0 & \dfrac{t}{I_B} \end{bmatrix} \cdot \begin{bmatrix} F_{B,x} \\ M_B \end{bmatrix}, \qquad (4.103)$$

$$\underline{p}_{x,S} = \underline{X}_x \cdot \underline{I}_x \cdot \underline{F}_{x,B} \qquad (4.103a)$$

with $y_{B,j}$ as y_B - coordinate of point j.

Now, for the line loads acting on the elements, equivalent nodal forces are determined. According to (4.38a,b), these are obtained for a plane stress element with linear displacement shape functions at an edge and linear line loads according to Fig. 4.62 (or Fig. 4.17 with reference to the x-axis) to

$$\underline{F}_x^{(el)} = \underline{A}_x^{(el)} \cdot \underline{p}_x^{(el)} \qquad (4.104)$$

with

$$\underline{F}_x^{(j)} = \begin{bmatrix} F_{x,j} \\ F_{x,j+1} \end{bmatrix}, \quad \underline{p}_x^{(j)} = \begin{bmatrix} p_{x,j} \\ p_{x,j+1} \end{bmatrix}, \quad j = 1, 2, \ldots, n \qquad (4.104a)$$

Fig. 4.62 Line loads at a
4-node plane stress element

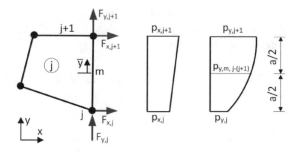

and

$$A_x^{(el)} = \frac{a_j}{6} \cdot \begin{bmatrix} 2 & 1 \\ 1 & 2 \end{bmatrix}, \quad a_j = y_{B,j+1} - y_{B,j} \quad \text{for element } j. \tag{4.104b}$$

The nodal forces of the individual elements are now superimposed to a force vector $\underline{F}_{x,S}$, which contains all forces in the x-direction in the connection area. For this purpose, a $2 \times (n + 1)$ incidence matrix consisting of only 0 and 1 entries is set up for each element. It assigns the forces of an element to those of the force vector in the connection area. It is

$$\underline{F}_{x,S}^{(el)} = \underline{Z}_x^{(el)T} \cdot \underline{F}_x^{(el)}. \tag{4.105a}$$

For the element j, after (4.104a), $\underline{Z}_{x,1,j}^{(j)} = 1$ and $\underline{Z}_{x,2,j+1}^{(j)} = 1$ applies, while all other entries of $\underline{Z}_x^{(j)}$ are zero. Correspondingly, the line loads at the element nodes of a single element are obtained as

$$\underline{p}_x^{(el)} = \underline{Z}_x^{(el)} \cdot \underline{p}_{x,S}. \tag{4.105b}$$

The vector $\underline{p}_{x,S}$ contains the ordinates of the line loads at the nodes. The nodal forces in x-direction are obtained from the sum of the contributions of all elements to

$$\underline{F}_{x,S} = \sum_{(el)} \underline{F}_{x,S}^{(el)} \tag{4.106}$$

with

$$\underline{F}_{x,S} = \begin{bmatrix} F_{x,1} \\ F_{x,2} \\ \cdot \\ F_{x,n+1} \end{bmatrix}. \tag{4.106a}$$

Introducing (4.105a), (4.104), and (4.105b) into (4.106), one obtains

$$\underline{F}_{x,S} = \sum_{(el)} \underline{F}_{x,S}^{(el)} = \sum_{(el)} \underline{Z}_x^{(el)T} \cdot \underline{A}_x^{(el)} \cdot \underline{Z}_x^{(el)} \cdot \underline{p}_{x,S},$$

and with (4.103a)

$$\underline{F}_{x,S} = \sum_{(el)} (\underline{Z}_x^{(el)T} \cdot \underline{A}_x^{(el)} \cdot \underline{Z}_x^{(el)}) \cdot \underline{X}_x \cdot \underline{I}_x \cdot \underline{F}_{B,x} \qquad (4.107)$$

or

$$\underline{F}_{x,S} = \underline{T}_x^T \cdot \underline{F}_{B,x} \qquad (4.108)$$

$$\underline{T}_x^T = \sum_{(el)} (\underline{Z}_x^{(el)T} \cdot \underline{A}_x^{(el)} \cdot \underline{Z}_x^{(el)}) \cdot \underline{X}_x \cdot \underline{I}_x. \qquad (4.108a)$$

Applying the matrix $\underline{T}_x^T$ the force $F_{B,x}$ and moment M_B of the beam are transformed into nodal forces related to the plane stress elements in the connection area.

For the displacements, according to Sect. 4.10.2, the following applies

$$\underline{u}_{B,x} = \underline{T}_x \cdot \underline{u}_S \qquad (4.109)$$

where

$$\underline{u}_S = \begin{bmatrix} u_1 \\ u_2 \\ \cdot \\ u_{n+1} \end{bmatrix} \qquad (4.110)$$

are the displacements related to the nodal points of finite elements and

$$\underline{u}_{B,x} = \begin{bmatrix} u_B \\ \varphi_B \end{bmatrix} \qquad (4.110a)$$

the displacements of the beam element.

Next, the shear stresses due to a shear force are investigated. For the rectangular beam they are described by a polynomial of second order as

$$\tau_{xy} = \frac{3}{2} \cdot \frac{1}{d \cdot t} \cdot \left(1 - 4 \cdot \frac{y_B^2}{d^2}\right) \cdot V. \qquad (4.111)$$

The shear force V at $x_B = 0$ corresponds to the element force $F_{B,y} = V$ (Fig. 3.15). If one converts the shear force into a line load, one obtains

$$p_y = \tau_{xy} \cdot t = \frac{3}{2} \cdot \frac{1}{d} \cdot \left(1 - 4 \cdot \frac{y_B^2}{d^2}\right) \cdot F_{B,y} \tag{4.112}$$

or

$$p_y = \begin{bmatrix} 1 & y_B^2 \end{bmatrix} \cdot \begin{bmatrix} \dfrac{3}{2 \cdot d} \\[2mm] \dfrac{-6}{d^3} \end{bmatrix} \cdot F_{B,y}. \tag{4.112a}$$

Three ordinates are required on a plane stress element to describe a quadratic line load. Therefore, besides the corner points, midpoints between the corner points are taken into account for the description of the line loads. The corresponding nodal values of the line loads are thus

$$
\begin{bmatrix} P_{y,1} \\ P_{y,m,1-2} \\ P_{y,2} \\ P_{y,m,2-3} \\ P_{y,3} \\ \cdot \\ P_{y,m,n-(n+1)} \\ P_{y,n+1} \end{bmatrix}
=
\begin{bmatrix}
1 & y_{B,1}^2 \\
1 & \frac{1}{4} \cdot (y_{B,1} + y_{B,2})^2 \\
1 & y_{B,2}^2 \\
1 & \frac{1}{4} \cdot (y_{B,2} + y_{B,3})^2 \\
1 & y_{B,3}^2 \\
\cdot & \cdot \\
\cdot & \cdot \\
1 & y_{B,j}^2 \\
1 & \frac{1}{4} \cdot (y_{B,j} + y_{B,j+1})^2 \\
1 & y_{B,j+1}^2 \\
\cdot & \cdot \\
1 & \frac{1}{4} \cdot (y_{B,n} + y_{B,n+1})^2 \\
1 & y_{B,n+1}^2
\end{bmatrix}
\cdot
\begin{bmatrix} \dfrac{3}{2 \cdot d} \\[2mm] \dfrac{-6}{d^3} \end{bmatrix}
\cdot F_{B,y} \tag{4.113}
$$

$$\underline{P}_{y,S} = \underline{X}_y \cdot \underline{I}_y \cdot F_{B,y}. \tag{4.113a}$$

The even-numbered entries of the vector refer to the midpoints.

To determine the nodal loads due to the line load p_y, a single element is considered (Fig. 4.62). The equivalent nodal forces in the y-direction are obtained as follows:

$$\begin{bmatrix} F_{y,j} \\ F_{y,j+1} \end{bmatrix} = \int \begin{bmatrix} N_{L,j} \\ N_{L,j+1} \end{bmatrix} \cdot p_y \cdot d\bar{y} \tag{4.114}$$

$$\underline{F}_y^{(el)} = \int \underline{N}_L \cdot p_y \cdot d\bar{y} \tag{4.114a}$$

with the shape functions of the virtual displacements

$$N_{L,j} = \frac{1}{2} - \frac{\bar{y}}{a}, \quad N_{L,j+1} = \frac{1}{2} + \frac{\bar{y}}{a}. \tag{4.114b}$$

The line loads p_y, are piecewise parabolas of second order, which are expressed by three ordinates to

$$p_y = \begin{bmatrix} N_{V,j} & N_{V,m} & N_{V,j+1} \end{bmatrix} \cdot \begin{bmatrix} p_{y,j} \\ p_{y,m} \\ p_{y,j+1} \end{bmatrix}, \quad j = 1, 2, \ldots, n \tag{4.115}$$

$$p_y = \underline{N}_V^T \cdot \underline{p}_y^{(el)} \tag{4.115a}$$

with

$$\underline{N}_V = \begin{bmatrix} -\dfrac{\bar{y}}{a} + 2 \cdot \left(\dfrac{\bar{y}}{a}\right)^2 \\[2mm] 1 - 4 \cdot \left(\dfrac{\bar{y}}{a}\right)^2 \\[2mm] \dfrac{\bar{y}}{a} + 2 \cdot \left(\dfrac{\bar{y}}{a}\right)^2 \end{bmatrix}. \tag{4.115b}$$

The nodal forces in y-direction are now obtained as

$$\underline{F}_y^{(el)} = \int \underline{N}_L \cdot \underline{N}_V^T \cdot d\bar{y} \cdot \underline{p}_y^{(el)} \tag{4.116}$$

or after integration from $\bar{y} = -a/2$ to $\bar{y} = a/2$

$$\underline{F}_y^{(el)} = \underline{A}_y^{(el)} \cdot \underline{p}_y^{(el)} \tag{4.116a}$$

with

$$\underline{A}_y^{(el)} = \frac{a}{6} \cdot \begin{bmatrix} 1 & 2 & 0 \\ 0 & 2 & 1 \end{bmatrix}. \tag{4.116b}$$

With the $3 \times (2n + 1)$ incidence matrix $\underline{Z}_{Ly}^{(el)}$ the line loads of a single element are related to the nodes of the plane stress elements in the connection area. For the y-direction, one obtains

$$\underline{p}_y^{(el)} = \underline{Z}_{Ly}^{(el)} \cdot \underline{p}_{y,S}. \tag{4.117}$$

In the incidence matrix $\underline{Z}_{Ly}^{(el)}$ of the element j, the entries are $\underline{Z}_{y,1,2j-1}^{(j)} = 1, \underline{Z}_{y,2,2j}^{(j)} = 1$ and $\underline{Z}_{y,3,2j+1}^{(j)} = 1$ whereas all other entries are zero. Correspondingly, the following applies to the nodal forces

$$\underline{F}_{y,S}^{(el)} = \underline{Z}_{Fy}^{(el)T} \cdot \underline{F}_y^{(el)} \tag{4.118}$$

with the $2 \times (n+1)$ incidence matrix

$$\underline{Z}_{Fy}^{(el)} = \underline{Z}_x^{(el)}. \tag{4.118a}$$

With (4.116a) to (4.118a), the nodal forces of a single element are

$$\underline{F}_{y,S}^{(el)} = \underline{Z}_x^{(el)T} \cdot \underline{A}_y^{(el)} \cdot \underline{Z}_{Ly}^{(el)} \cdot \underline{p}_{y,S} \tag{4.119}$$

and with (4.113a)

$$\underline{F}_{y,S}^{(el)} = \underline{Z}_x^{(el)T} \cdot \underline{A}_y^{(el)} \cdot \underline{Z}_{Ly}^{(el)} \cdot \underline{X}_y \cdot \underline{I}_y \cdot F_{B,y}. \tag{4.119a}$$

The sum of the forces of all elements is obtained to be

$$\underline{F}_{y,S} = \underline{T}_y^T \cdot F_{B,y} \tag{4.120}$$

with

$$\underline{T}_y^T = \sum_{(el)} (\underline{Z}_x^{(el)T} \cdot \underline{A}_y^{(el)} \cdot \underline{Z}_{Ly}^{(el)}) \cdot \underline{X}_y \cdot \underline{I}_y \tag{4.120a}$$

and

$$\underline{F}_{y,S} = \begin{bmatrix} F_{y,1} \\ F_{y,2} \\ \cdot \\ F_{y,n+1} \end{bmatrix}. \tag{4.120b}$$

The following applies accordingly to the transformation of the displacements

$$v_B = \underline{T}_y \cdot \underline{v}_{x,S} \tag{4.121}$$

with

$$\underline{v}_S = \begin{bmatrix} v_1 \\ v_2 \\ \cdot \\ v_{n+1} \end{bmatrix}. \tag{4.121a}$$

Until now, the degrees of freedom related to beam bending and shear have been treated separately. They are now merged. The displacements and forces of a finite element system are usually arranged according to degrees of freedom in the x- and y-direction at the nodal points. This requires a corresponding sorting in the transformation matrix $\underline{T}$, which summarizes the matrices $\underline{T}_x$ and $\underline{T}_y$. From (4.109) and (4.121), one obtains

$$\underline{u}_B = \underline{T} \cdot \underline{u}_S \tag{4.122a}$$

and from (4.108) and (4.120)

$$\underline{F}_S = \underline{T}^T \cdot \underline{F}_B \tag{4.122b}$$

with

$$\underline{u}_B = \begin{bmatrix} u_B \\ v_B \\ \varphi_B \end{bmatrix} \quad \underline{F}_B = \begin{bmatrix} F_{B,x} \\ F_{B,y} \\ M_B \end{bmatrix}, \quad \underline{u}_S = \begin{bmatrix} u_1 \\ v_1 \\ u_2 \\ v_2 \\ \cdot \\ \cdot \\ v_{n+1} \end{bmatrix}, \quad \underline{F}_S = \begin{bmatrix} F_{x1} \\ F_{y1} \\ F_{x2} \\ F_{y2} \\ \cdot \\ \cdot \\ F_{y,n+1} \end{bmatrix}. \tag{4.122c}$$

Equation (4.122a), represents the MPC condition of the EST model.

In the following, the transformation matrices for two cases are derived as examples. In the first case, the beam height corresponds to a single plane stress element, in the second case to two equally sized elements.

Example 4.15 An EST connection of a beam element to a single plane stress element is considered. The connection of the beam element with a single plane stress element is shown in Fig. 4.63. The transformation relationship for the force in the x-direction and the bending moment in the beam is obtained with

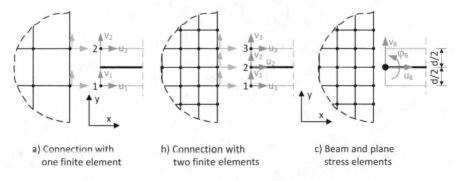

a) Connection with one finite element

b) Connection with two finite elements

c) Beam and plane stress elements

Fig. 4.63 EST model of the connection of a beam model with plane stress elements

$$\underline{Z}_x^{(1)} = \begin{bmatrix} 1 & 0 \\ 0 & 1 \end{bmatrix}, \quad \underline{A}_x^{(1)} = \frac{a}{6} \cdot \begin{bmatrix} 2 & 1 \\ 1 & 2 \end{bmatrix}, \quad \underline{X}_x = \begin{bmatrix} 1 & d/2 \\ 1 & -d/2 \end{bmatrix}, \quad \underline{I}_x = \begin{bmatrix} \dfrac{1}{d} & 0 \\ 0 & \dfrac{12}{d^3} \end{bmatrix}.$$

and $a = d$ according to (4.108a), as

$$\underline{T}_x^T = \underline{Z}_x^{(1)T} \cdot \underline{A}_x^{(1)} \cdot \underline{Z}_x^{(1)} \cdot \underline{X}_x \cdot \underline{I}_x$$

or

$$\underline{T}_x = \begin{bmatrix} 1/2 & 1/2 \\ 1/d & -1/d \end{bmatrix}. \tag{4.123}$$

Hence, the transformation relations of the forces and displacements are

$$\begin{bmatrix} u_B \\ \varphi_B \end{bmatrix} = \begin{bmatrix} 1/2 & 1/2 \\ 1/d & -1/d \end{bmatrix} \cdot \begin{bmatrix} u_1 \\ u_2 \end{bmatrix} \quad \text{and} \quad \begin{bmatrix} F_{x,1} \\ F_{x,2} \end{bmatrix} = \begin{bmatrix} 1/2 & 1/d \\ 1/2 & -1/d \end{bmatrix} \cdot \begin{bmatrix} F_{B,x} \\ M_B \end{bmatrix}. \tag{4.123a}$$

For the force and the displacements in y-direction with

$$\underline{Z}_{Ly}^{(1)} = \begin{bmatrix} 1 & 0 & 0 \\ 0 & 1 & 0 \\ 0 & 0 & 1 \end{bmatrix}, \quad \underline{A}_y^{(1)} = \frac{a}{6} \cdot \begin{bmatrix} 1 & 2 & 0 \\ 0 & 2 & 1 \end{bmatrix},$$

$$\underline{X}_y = \begin{bmatrix} 1 & \left(-\dfrac{d}{2}\right)^2 \\ 1 & 0 \\ 1 & \left(\dfrac{d}{2}\right)^2 \end{bmatrix}, \quad \underline{I}_y = \begin{bmatrix} \dfrac{-3}{2 \cdot d} \\ \dfrac{6}{d^3} \end{bmatrix}$$

the matrix is obtained with (4.120a), as

$$\underline{T}_y^T = \underline{Z}_x^{(1)T} \cdot \underline{A}_y^{(1)} \cdot \underline{Z}_{Ly}^{(1)} \cdot \underline{X}_y \cdot \underline{I}_y$$

or

$$\underline{T}_y = \begin{bmatrix} \dfrac{1}{2} & \dfrac{1}{2} \end{bmatrix}. \tag{4.124}$$

Hence, the transformation relations of the forces and displacements are

$$v_B = \begin{bmatrix} \dfrac{1}{2} & \dfrac{1}{2} \end{bmatrix} \cdot \begin{bmatrix} v_1 \\ v_2 \end{bmatrix} \quad \text{and} \quad \begin{bmatrix} F_{y1} \\ F_{y2} \end{bmatrix} = \begin{bmatrix} \dfrac{1}{2} \\ \dfrac{1}{2} \end{bmatrix} \cdot F_{B,y}. \qquad (4.124a)$$

Merging the matrices according to (4.123) and (4.124), into

$$\underline{T} = \begin{bmatrix} 0.5 & 0 & 0.5 & 0 \\ 0 & 0.5 & 0 & 0.5 \\ 1/d & 0 & -1/d & 0 \end{bmatrix} \qquad (4.125)$$

one gets

$$\begin{bmatrix} u_B \\ v_B \\ \varphi_B \end{bmatrix} = \begin{bmatrix} 0.5 & 0 & 0.5 & 0 \\ 0 & 0.5 & 0 & 0.5 \\ 1/d & 0 & -1/d & 0 \end{bmatrix} \cdot \begin{bmatrix} u_1 \\ v_1 \\ u_2 \\ v_2 \end{bmatrix}$$

and

$$\begin{bmatrix} F_{x1} \\ F_{y1} \\ F_{x2} \\ F_{y2} \end{bmatrix} = \begin{bmatrix} 0.5 & 0 & 1/d \\ 0 & 0.5 & 0 \\ 0.5 & 0 & -1/d \\ 0 & 0.5 & 0 \end{bmatrix} \cdot \begin{bmatrix} F_{B,x} \\ F_{B,y} \\ M_B \end{bmatrix}. \qquad (4.125a)$$

◀

Example 4.16 An EST connection of a beam element to two plane stress elements is considered. With two equally sized plane stress elements on the height d of the beam the following is obtained (Fig. 4.63b)

$$\underline{Z}_x^{(1)} = \begin{bmatrix} 1 & 0 & 0 \\ 0 & 1 & 0 \end{bmatrix}, \quad \underline{Z}_x^{(2)} = \begin{bmatrix} 0 & 1 & 0 \\ 0 & 0 & 1 \end{bmatrix}, \quad \underline{A}_x^{(1)} = \underline{A}_x^{(2)} = \frac{a}{6} \cdot \begin{bmatrix} 2 & 1 \\ 1 & 2 \end{bmatrix}$$

$$\underline{X}_x = \begin{bmatrix} 1 & d/2 \\ 1 & 0 \\ 1 & -d/2 \end{bmatrix}, \quad \underline{I}_x = \begin{bmatrix} \dfrac{1}{d} & 0 \\ 0 & \dfrac{12}{d^3} \end{bmatrix}$$

with $a = d/2$. The transformation matrix corresponding to the normal force and the bending moment is

$$\underline{T}_x^T = (\underline{Z}_x^{(1)T} \cdot \underline{A}_x^{(1)} \cdot \underline{Z}_x^{(1)} + \underline{Z}_x^{(2)T} \cdot \underline{A}_x^{(2)} \cdot \underline{Z}_x^{(2)}) \cdot \underline{X}_x \cdot \underline{I}_x$$

or

$$\underline{T}_x = \begin{bmatrix} 1/4 & 1/2 & 1/4 \\ 1/d & 0 & -1/d \end{bmatrix}. \qquad (4.126)$$

For the shear force one obtains with

$$\underline{Z}_{Ly}^{(1)} = \begin{bmatrix} 1 & 0 & 0 & 0 & 0 \\ 0 & 1 & 0 & 0 & 0 \\ 0 & 0 & 1 & 0 & 0 \end{bmatrix}, \quad \underline{Z}_{Ly}^{(2)} = \begin{bmatrix} 0 & 0 & 1 & 0 & 0 \\ 0 & 0 & 0 & 1 & 0 \\ 0 & 0 & 0 & 0 & 1 \end{bmatrix}$$

$$\underline{A}_y^{(1)} = \underline{A}_y^{(2)} = \frac{a}{6} \cdot \begin{bmatrix} 1 & 2 & 0 \\ 0 & 2 & 1 \end{bmatrix}, \quad \underline{X}_y = \begin{bmatrix} 1 & \left(-\dfrac{d}{2}\right)^2 \\ 1 & \left(-\dfrac{d}{4}\right)^2 \\ 1 & 0 \\ 1 & \left(\dfrac{d}{4}\right)^2 \\ 1 & \left(\dfrac{d}{2}\right)^2 \end{bmatrix} \quad \underline{I}_y = \begin{bmatrix} \dfrac{3}{2 \cdot d} \\ -\dfrac{6}{d^3} \end{bmatrix}$$

the transformation matrix

$$\underline{T}_y^T = (\underline{Z}_x^{(1)T} \cdot \underline{A}_y^{(1)} \cdot \underline{Z}_{Ly}^{(1)} + \underline{Z}_x^{(2)T} \cdot \underline{A}_y^{(2)} \cdot \underline{Z}_{Ly}^{(2)}) \cdot \underline{X}_y \cdot \underline{I}_y$$

or

$$\underline{T}_y = \begin{bmatrix} \dfrac{3}{16} & \dfrac{10}{16} & \dfrac{3}{16} \end{bmatrix}. \tag{4.127}$$

Combining both relationships, one gets the transformation matrix as

$$\underline{T} = \begin{bmatrix} 1/4 & 0 & 1/2 & 0 & 1/4 & 0 \\ 0 & 3/16 & 0 & 10/16 & 0 & 3/16 \\ 1/d & 0 & 0 & 0 & -1/d & 0 \end{bmatrix}. \tag{4.128}$$

Hence, displacements and forces are transformed as

$$\begin{bmatrix} u_B \\ v_B \\ \varphi_B \end{bmatrix} = \begin{bmatrix} 1/4 & 0 & 1/2 & 0 & 1/4 & 0 \\ 0 & 3/16 & 0 & 10/16 & 0 & 3/16 \\ 1/d & 0 & 0 & 0 & -1/d & 0 \end{bmatrix} \cdot \begin{bmatrix} u_1 \\ v_1 \\ u_2 \\ v_2 \\ u_3 \\ v_3 \end{bmatrix}, \tag{4.129a}$$

$$
\begin{bmatrix} F_{x1} \\ F_{y1} \\ F_{x2} \\ F_{y2} \\ F_{x3} \\ F_{y3} \end{bmatrix} = \begin{bmatrix} \dfrac{1}{4} & 0 & \dfrac{1}{d} \\ 0 & \dfrac{3}{16} & 0 \\ \dfrac{1}{2} & 0 & 0 \\ 0 & \dfrac{10}{16} & 0 \\ \dfrac{1}{4} & 0 & -\dfrac{1}{d} \\ 0 & \dfrac{3}{16} & 0 \end{bmatrix} \cdot \begin{bmatrix} F_{B,x} \\ F_{B,y} \\ M_B \end{bmatrix} . \tag{4.129b}
$$

Equations (4.129a, 4.129b) show that the displacements and the rotation of the beam element are evaluated as weighted means of the nodal displacements of the plane stress elements. The nodal forces of the plane stress elements are in equilibrium with the element forces of the beam.   ◄

For EST connections with finite elements of the same size, i.e. if

$$
a = a_j = \frac{d}{n} \text{ with } j = 1, 2, \ldots, n, \tag{4.130}
$$

the transformation matrices can be given explicitly. The entries of the matrices $\underline{T}_x$ and $\underline{T}_y$ are specified for this case in Table 4.13.

To apply the EST model to the connection of a bending beam with plane stress elements, the stiffness matrix of the beam according to (3.28), is transformed. The beam element with the nodes a and b (Fig. 4.64), is divided into submatrices, which refer to the two nodes

$$
\begin{bmatrix} \underline{K}_{aa} & \underline{K}_{ab} \\ \underline{K}_{ba} & \underline{K}_{bb} \end{bmatrix} \cdot \begin{bmatrix} u_a \\ u_b \end{bmatrix} = \begin{bmatrix} F_a \\ F_b \end{bmatrix} . \tag{4.131}
$$

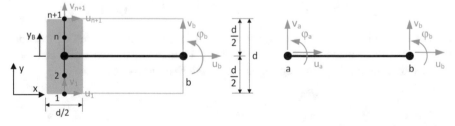

EST transition element   Beam element

Fig. 4.64 EST transition element

Table 4.13 Entries of the transformation matrices $\underline{T}_x$ and $\underline{T}_y$ for EST connections with equally sized plane stress elements

$\underline{T}_x$	Column j		
	$j = 1$	$2 \le j \le n$	$j = n+1$
Row 1	$\dfrac{1}{2 \cdot n}$	$\dfrac{1}{n}$	$\dfrac{1}{2 \cdot n}$
Row 2	$\dfrac{1}{d \cdot n} \cdot \left(3 - \dfrac{2}{n}\right)$	$\dfrac{6}{d \cdot n} \cdot \left(1 - 2 \cdot \dfrac{j-1}{n}\right)$	$-\dfrac{1}{d \cdot n} \cdot \left(3 - \dfrac{2}{n}\right)$

$\underline{T}_y$	Column j		
	$j = 1$	$2 \le j \le n$	$j = n+1$
Row 1	$\dfrac{1}{n^2} \cdot \left(1 - \dfrac{1}{2 \cdot n}\right)$	$\dfrac{1}{n^2} \cdot \left(6 \cdot j - 6 - \dfrac{1}{n} \cdot (6 \cdot j^2 - 12 \cdot j + 7)\right)$	$\dfrac{1}{n^2} \cdot \left(1 - \dfrac{1}{2 \cdot n}\right)$

Node a is connected to the plane stress elements. This means

$$\underline{u}_a = \underline{u}_B = \begin{bmatrix} u_B \\ v_B \\ \varphi_B \end{bmatrix}, \quad \underline{F}_a = \underline{F}_B = \begin{bmatrix} F_{x,B} \\ F_{y,B} \\ M_B \end{bmatrix}. \tag{4.131a}$$

The vectors are transformed into those which refer to the plane stress elements. With reference to point a (index a) one obtains with (4.122a, 4.122b)

$$\underline{u}_B = \underline{T}_a \cdot \underline{u}_{S,a}, \tag{4.132a}$$

$$\underline{F}_{S,a} = \underline{T}_a^T \cdot \underline{F}_B. \tag{4.132b}$$

By introducing both relationships into (4.131), one gets the stiffness matrix of the EST transition element.

$$\begin{bmatrix} \underline{T}_a^T \cdot \underline{K}_{aa} \cdot \underline{T}_a & \underline{T}_a^T \cdot \underline{K}_{ab} \\ \underline{K}_{ba} \cdot \underline{T}_a & \underline{K}_{bb} \end{bmatrix} \cdot \begin{bmatrix} u_{S,a} \\ u_b \end{bmatrix} = \begin{bmatrix} F_{S,a} \\ F_b \end{bmatrix}. \tag{4.133}$$

At the left end the degenerated beam element refers to the degrees of freedom corresponding to plane stress elements. When connecting to two finite elements (corresponding to the beam height), one gets (see Example 4.16)

$$\underline{T}_a = \begin{bmatrix} 1/4 & 0 & 1/2 & 0 & 1/4 & 0 \\ 0 & 3/16 & 0 & 10/16 & 0 & 3/16 \\ 1/d & 0 & 0 & 0 & -1/d & 0 \end{bmatrix}, \quad \underline{u}_{S,a} = \begin{bmatrix} u_1 \\ v_1 \\ u_2 \\ v_2 \\ u_3 \\ v_3 \end{bmatrix},$$

$$\underline{F}_{S,a} = \begin{bmatrix} F_{xa,1} \\ F_{ya,1} \\ F_{xa,2} \\ F_{ya,2} \\ F_{xa,3} \\ F_{ya,3} \end{bmatrix}. \quad (4.134)$$

The degrees of freedom of node b can also be transformed if necessary.

The EST model neglects the local increase in the stiffness of the plate in the region of the beam connection. This effect can be approximated with additional "artificial" beam elements. For example, a beam with the height $d/2$ and the thickness t may be inserted between the nodes 1-2-3 in Fig. 4.64 (highlighted in gray). The EST model with locally increased stiffness is called the ESTS model.

EST support element for plate in plane stress

An EST support of a plate represents a support on distributed springs, for example, of a RC deep beam on a column. As with connection to a beam, the displacement and rotational stiffness of the support are coupled. This is a significant advantage over a model as an uncoupled Winkler support, in which only either the displacement or the rotational stiffness of the column can be represented (Sect. 4.11.5).

A column with the length l, the cross-sectional area A and the bending stiffness EI is considered (Fig. 4.65). The relationship between the forces and displacements

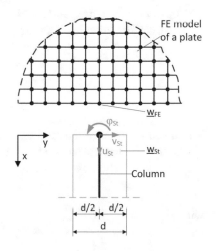

Fig. 4.65 EST support of a plate in plane stress

at the column head is given by

$$
\begin{bmatrix} F_{x,\mathrm{St}} \\ F_{y,\mathrm{St}} \\ M_{\mathrm{St}} \end{bmatrix} = \begin{bmatrix} k_{xx} & 0 & 0 \\ 0 & k_{yy} & k_{\varphi y} \\ 0 & k_{\varphi x} & k_{\varphi\varphi} \end{bmatrix} \cdot \begin{bmatrix} u_{\mathrm{St}} \\ v_{\mathrm{St}} \\ \varphi_{\mathrm{St}} \end{bmatrix}
\tag{4.135}
$$

or

$$
\underline{F}_{\mathrm{St}} = \underline{K}_{\mathrm{St}} \cdot \underline{w}_{\mathrm{St}}.
\tag{4.135a}
$$

The spring constants result from the stiffness of the column at the column head. They can be determined from (4.135). The column is considered as beam element according to (3.28), neglecting shear stiffness, i.e. with $A_{\mathrm{s}} \to \infty$ or $m = 0$.

For a column with all degrees of freedom being fixed at the lower end, one obtains with (3.28)

$$
k_{xx} = \frac{E \cdot A}{l}, \quad k_{\varphi\varphi} = 4 \cdot \frac{E \cdot I}{l}, \quad k_{\varphi y} = 6 \cdot \frac{E \cdot I}{l^2}, k_{yy} = 12 \cdot \frac{E \cdot I}{l^3}.
\tag{4.136a}
$$

For a column where only displacements are fixed but rotation is free at the lower end, the spring values are (see (3.31))

$$
k_{xx} = \frac{E \cdot A}{l}, \quad k_{\varphi\varphi} = 3 \cdot \frac{E \cdot I}{l}, \quad k_{\varphi y} = 3 \cdot \frac{E \cdot I}{l^2}, \quad k_{yy} = 3 \cdot \frac{E \cdot I}{l^3}.
\tag{4.136b}
$$

For a column head which is supposed to transfer the normal force of the column but no bending moment, one obtains

$$
k_{xx} = \frac{E \cdot A}{l}, \quad k_{\varphi\varphi} = k_{\varphi y} = k_{yy} = 0.
\tag{4.136c}
$$

The model is denoted as *fluid cushion model*.

The transformation relations are given by (4.122a, 4.122b), as

$$
\underline{w}_{\mathrm{St}} = \underline{T} \cdot \underline{w}_{\mathrm{FE}}
\tag{4.137a}
$$

$$
\underline{F}_{\mathrm{FE}} = \underline{T}^{\mathrm{T}} \cdot \underline{F}_{\mathrm{St}}.
\tag{4.137b}
$$

The transformation of the stiffness matrix to that with degrees of freedom of the plane stress finite elements (index "FE") is obtained as

$$
\underline{F}_{\mathrm{FE}} = \underline{K}_{\mathrm{FE}} \cdot \underline{w}_{\mathrm{FE}}
\tag{4.138}
$$

with

$$\underline{K}_{FE} = \underline{T}^{T} \cdot \underline{K}_{St} \cdot \underline{T}. \tag{4.138a}$$

After the solution of the global system of equations, the nodal displacements $\underline{w}_{FE}$ of the EST element are known. The sectional forces at the column head can be determined according to (4.133) and (4.137a), as

$$\underline{F}_{St} = \underline{K}_{St} \cdot \underline{T} \cdot \underline{w}_{FE}. \tag{4.139}$$

The application of the EST model on a deep beam support is demonstrated in Example 4.25.

EST support for a plate in bending

In the analysis of flat slabs, the supports of the slab with the columns have to be represented in the finite element model appropriately. The slab is modeled using plate bending elements, the beam theory applies to the columns. In the following, a support element based on the EST concept is given for this case.

The stiffness of the column is represented in the finite element model of the plate by an EST element (Fig. 4.66), [87–90, 97]. The column is subjected to biaxial bending and an axial force. With the sectional forces N, M_y and M_z (see Fig. 3.29 at nodal point 2) the normal stresses are given as

$$\sigma(x_B, y_B) = \frac{N}{A} + \frac{M_y}{I_y} \cdot z^{(lok)} - \frac{M_z}{I_z} \cdot y^{(lok)} = \frac{N}{A} - \frac{M_y}{I_y} \cdot x_B - \frac{M_z}{I_z} \cdot y_B \tag{4.140}$$

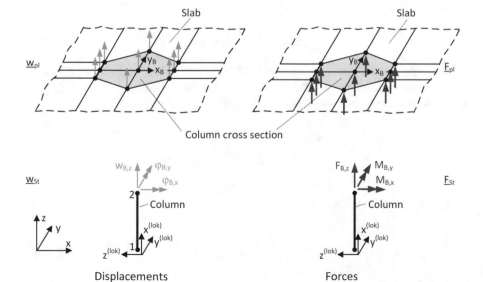

Fig. 4.66 EST element for a column of a flat slab

where A is the cross-sectional area and I_y, I_z are the second moments of area about the $y^{(\text{lok})}$- and $z^{(\text{lok})}$-axis, respectively. The principal axes of the column are $z^{(\text{lok})} = -x_B \,\hat{=}\, -x$ and $y^{(\text{lok})} = y_B \,\hat{=}\, y$ and hence $I_y = I_{By}$ and $I_z = I_{Bx}$. The sectional forces correspond (at $x^{(\text{lok})} = 0$) to the forces $F_{Bz} = N$, $M_{Bx} = -M_z$ and $M_{By} = M_y$. Considering the stresses as distributed forces p_z similar to loads gives

$$p_z(x_B, y_B) = \sigma(x_B, y_B) = \frac{F_{Bz}}{A} + \frac{M_{Bx}}{I_{Bx}} \cdot y_B - \frac{M_{By}}{I_{By}} \cdot x_B$$

or

$$p_z(x_B, y_B) = \begin{bmatrix} 1 & y_B & -x_B \end{bmatrix} \cdot \begin{bmatrix} 1/A & 0 & 0 \\ 0 & 1/I_{Bx} & 0 \\ 0 & 0 & 1/I_{By} \end{bmatrix} \cdot \begin{bmatrix} F_{Bz} \\ M_{Bx} \\ M_{By} \end{bmatrix}. \tag{4.141}$$

They are positive in the positive z-direction. The column cross section is connected with the plate at m nodal points whose location is described in the x_B-, y_B-coordinate system. The ordinates of the distributed forces at the nodal points are given by

$$\begin{bmatrix} p_{z,1} \\ p_{z,2} \\ \cdot \\ p_{z,j} \\ \cdot \\ p_{z,m} \end{bmatrix} = \begin{bmatrix} 1 & y_{B,1} & -x_{B,1} \\ 1 & y_{B,2} & -x_{B,2} \\ 1 & \cdot & \cdot \\ 1 & y_{B,j} & -x_{B,j} \\ \cdot & \cdot & \cdot \\ 1 & y_{B,m} & -x_{B,m} \end{bmatrix} \cdot \begin{bmatrix} 1/A_z & 0 & 0 \\ 0 & 1/I_{Bx} & 0 \\ 0 & 0 & 1/I_{By} \end{bmatrix} \cdot \begin{bmatrix} F_{Bz} \\ M_{Bx} \\ M_{By} \end{bmatrix} \tag{4.142}$$

$$\underline{P}_{z,\text{Pl}} = \underline{X}_{xy} \cdot \underline{I}_{yz} \cdot \underline{F}_{\text{St}}. \tag{4.142a}$$

Equivalent nodal forces are determined for the distributed forces acting similar to loads on the plate elements (cf. Sect. 4.6.5, Fig. 4.44). In the case of a 4-node plate element they can be evaluated with (4.79a), as

$$\begin{bmatrix} F_{z1} \\ F_{z2} \\ F_{z3} \\ F_{z4} \end{bmatrix} = \int\limits_{A^{(el)}} \begin{bmatrix} N_1 \\ N_2 \\ N_3 \\ N_4 \end{bmatrix} \cdot p_z \cdot dA \tag{4.143}$$

or

$$\underline{F}^{(el)} = \int\limits_{A^{(el)}} \underline{N} \cdot p_z \cdot dA . \tag{4.143a}$$

The forces F_{z1} to F_{z4} are the nodal forces at the 4 corner points of the element and p_z is the surface load that varies across the element surface. The integration has to be performed over the area $A^{(el)}$ of the element. Equation (4.143) also applies to a quadrilateral 4-node element.

The interpolation functions (4.42a, 4.42b), for the quadrilateral element are written as a function of the dimensionless variables r and s (Fig. 4.35), as

$$\underline{N}(r, s) = \frac{1}{4} \cdot \begin{bmatrix} (1+r) \cdot (1+s) \\ (1-r) \cdot (1+s) \\ (1-r) \cdot (1-s) \\ (1+r) \cdot (1-s) \end{bmatrix}. \tag{4.144}$$

For integration, the infinitesimal area $dA = dx \cdot dy$ is related to the coordinates r and s with (4.49), as

$$dA = \det(\underline{J}) \cdot dr \cdot ds \tag{4.145}$$

with the determinant $\det(\underline{J})$ of the Jacobian operator (see 4.45)

$$\underline{J} = \begin{bmatrix} \dfrac{\partial x}{\partial r} & \dfrac{\partial y}{\partial r} \\ \dfrac{\partial x}{\partial s} & \dfrac{\partial y}{\partial s} \end{bmatrix} \text{ with } x(r, s) = \underline{N}(r, s)^{\mathrm{T}} \cdot \begin{bmatrix} x_1 \\ x_2 \\ x_3 \\ x_4 \end{bmatrix} \text{ and } y(r, s) = \underline{N}(r, s)^{\mathrm{T}} \cdot \begin{bmatrix} y_1 \\ y_2 \\ y_3 \\ y_4 \end{bmatrix} \tag{4.146}$$

according to (4.44a, 4.44b). The coordinates x_1, x_2, x_3, x_4 and y_1, y_2, y_3, y_4 are the coordinates of the corner points of the respective finite element (Fig. 4.35).

The distributed force p_z is described by its nodal values and interpolated by the bilinear functions $\underline{N}(r, s)$ as

$$p_z(r, s) = \underline{N}(r, s)^{\mathrm{T}} \cdot \begin{bmatrix} p_1 \\ p_2 \\ p_3 \\ p_4 \end{bmatrix} \tag{4.147}$$

or

$$p_z(r, s) = \underline{N}(r, s)^{\mathrm{T}} \cdot \underline{p}^{(el)}. \tag{4.147a}$$

The nodal forces of an element are obtained with (4.143a), (4.145), and (4.147a), as

$$\underline{F}^{(el)} = \underline{A}^{(el)} \cdot \underline{p}^{(el)} \tag{4.148}$$

with

$$\underline{A}^{(el)} = \int_{-1}^{1} \int_{-1}^{1} \underline{N} \cdot \underline{N}^{\mathrm{T}} \cdot \det(\underline{J}) \ dr \ ds. \tag{4.148a}$$

For the quadrilateral element the integral can be solved analytically. The symmetrical matrix $\underline{A}^{(el)}$ is obtained as

$$\underline{A}^{(el)} = \begin{bmatrix} a_{1,1} & a_{1,2} & a_{1,3} & a_{1,4} \\ a_{2,1} & a_{2,2} & a_{2,3} & a_{2,4} \\ a_{3,1} & a_{3,2} & a_{3,3} & a_{3,4} \\ a_{4,1} & a_{4,2} & a_{4,3} & a_{4,4} \end{bmatrix} \tag{4.149}$$

with the entries

$$a_{1,1} = \frac{1}{72} \cdot \big[6 \cdot x_1 \cdot (y_2 - y_4) + 2 \cdot x_2 \cdot (-3 \cdot y_1 + y_3 + 2 \cdot y_4)$$
$$+ 2 \cdot x_3 \cdot (-y_2 + y_4) + 2 \cdot x_4 \cdot (3 \cdot y_1 - 2 \cdot y_2 - y_3) \big]$$

$$a_{1,2} = \frac{1}{72} \cdot \big[x_1 \cdot (3 \cdot y_2 - y_3 - 2 \cdot y_4) + x_2 \cdot (-3 \cdot y_1 + 2 \cdot y_3 + y_4)$$
$$+ x_3 \cdot (y_1 - 2 \cdot y_2 + y_4) + x_4 \cdot (2 \cdot y_1 - y_2 - y_3) \big]$$

$$a_{1,3} = \frac{1}{72} \cdot \big[x_1 \cdot (y_2 - y_4) + x_2 \cdot (-y_1 + y_3) + x_3 \cdot (-y_2 + y_4) + x_4 \cdot (y_1 - y_3) \big]$$

$$a_{1,4} = \frac{1}{72} \cdot \big[x_1 \cdot (2 \cdot y_2 + y_3 - 3 \cdot y_4) + x_2 \cdot (-2 \cdot y_1 + y_3 + y_4)$$
$$+ x_3 \cdot (-y_1 - y_2 + 2 \cdot y_4) + x_4 \cdot (3 \cdot y_1 - y_2 - 2 \cdot y_3) \big]$$

$$a_{2,2} = \frac{1}{72} \cdot \big[x_1 \cdot (6 \cdot y_2 - 4 \cdot y_3 - 2 \cdot y_4) + 6 \cdot x_2 \cdot (-y_1 + y_3)$$
$$+ x_3 \cdot (4 \cdot y_1 - 6 \cdot y_2 + 2 \cdot y_4) + 2 \cdot x_4 \cdot (y_1 - y_3) \big]$$

$$a_{2,3} = \frac{1}{72} \cdot \big[x_1 \cdot (2 \cdot y_2 - y_3 - y_4) + x_2 \cdot (-2 \cdot y_1 + 3 \cdot y_3 - y_4)$$
$$+ x_3 \cdot (y_1 - 3 \cdot y_2 + 2 \cdot y_4) + x_4 \cdot (y_1 + y_2 - 2 \cdot y_3) \big]$$

$$a_{2,4} = \frac{1}{72} \cdot \big[x_1 \cdot (y_2 - y_4) + x_2 \cdot (-y_1 + y_3) + x_3 \cdot (-y_2 + y_4)$$
$$+ x_4 \cdot (y_1 - y_3) \big] = a_{1,3}$$

$$a_{3,3} = \frac{1}{72} \cdot \left[2 \cdot x_1 \cdot (y_2 - y_4) + x_2 \cdot (-2 \cdot y_1 + 6 \cdot y_3 - 4 \cdot y_4) \right.$$
$$\left. + 6 \cdot x_3 \cdot (-y_2 + y_4) + x_4 \cdot (2 \cdot y_1 + 4 \cdot y_2 - 6 \cdot y_3) \right]$$

$$a_{3,4} = \frac{1}{72} \cdot \left[x_1 \cdot (y_2 + y_3 - 2 \cdot y_4) + x_2 \cdot (-y_1 + 2 \cdot y_3 - y_4) \right.$$
$$\left. + x_3 \cdot (-y_1 - 2 \cdot y_2 + 3 \cdot y_4) + x_4 \cdot (2 \cdot y_1 + y_2 - 3 \cdot y_3) \right]$$

$$a_{4,4} = \frac{1}{72} \cdot \left[x_1 \cdot (2 \cdot y_2 + 4 \cdot y_3 - 6 \cdot y_4) + 2 \cdot x_2 \cdot (-y_1 + y_3) \right.$$
$$\left. + x_3 \cdot (-4 \cdot y_1 - 2 \cdot y_2 + 6 \cdot y_4) + 6 \cdot x_4 \cdot (y_1 - y_3) \right] \qquad (4.149a\text{-}j)$$

All other entries result from the symmetry of $\underline{A}^{(el)}$. For a rectangular element with the side lengths a and b (Fig. 4.67) the matrix $\underline{A}^{(el)}$ is obtained as

$$\underline{A}^{(el)} = \frac{a \cdot b}{36} \begin{bmatrix} 4 & 2 & 1 & 2 \\ 2 & 4 & 2 & 1 \\ 1 & 2 & 4 & 2 \\ 2 & 1 & 2 & 4 \end{bmatrix} \qquad (4.150)$$

For each element, the element forces are related to the nodal points of the connection area (index "P") with an incidence matrix as

$$\underline{F}_{Pl}^{(el)} = \underline{Z}_{xy}^{(el)\,T} \cdot \underline{F}^{(el)}. \qquad (4.151)$$

Correspondingly, the following relationship applies between the distributed forces $\underline{p}^{(el)}$ at the nodes of a single element, and the distributed forces $\underline{p}_{z,Pl}$ at the nodal points of the connection area

$$\underline{p}^{(el)} = \underline{Z}_{xy}^{(el)} \cdot \underline{p}_{z,Pl}. \qquad (4.152)$$

With (4.151, 4.152) and (4.148), one gets for a single element

$$\underline{F}_{Pl}^{(el)} = \underline{Z}_{xy}^{(el)\,T} \cdot \underline{A}^{(el)} \cdot \underline{Z}_{xy}^{(el)} \cdot \underline{p}_{z,Pl}. \qquad (4.153)$$

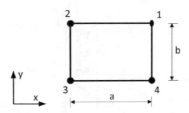

Fig. 4.67   4-node plate element

The nodal forces of the EST element are the sum of all element forces as

$$\underline{F}_{\mathrm{Pl}} = \sum_{(el)} \left(\underline{Z}_{xy}^{(el)\mathrm{T}} \cdot \underline{A}^{(el)} \cdot \underline{Z}_{xy}^{(el)} \right) \cdot \underline{p}_{z,\mathrm{Pl}} \tag{4.154}$$

and with (4.142a)

$$\underline{F}_{\mathrm{Pl}} = \sum_{(el)} \left(\underline{Z}_{xy}^{(el)\mathrm{T}} \cdot \underline{A}^{(el)} \cdot \underline{Z}_{xy}^{(el)} \right) \cdot \underline{X}_{xy} \cdot \underline{I}_{yz} \cdot \underline{F}_{\mathrm{St}} \tag{4.154a}$$

or

$$\underline{F}_{\mathrm{Pl}} = \underline{T}_{xy}^{\mathrm{T}} \cdot \underline{F}_{\mathrm{St}} \tag{4.155}$$

with

$$\underline{T}_{xy}^{\mathrm{T}} = \sum_{(el)} \left(\underline{Z}_{xy}^{(el)\mathrm{T}} \cdot \underline{A}^{(el)} \cdot \underline{Z}_{xy}^{(el)} \right) \cdot \underline{X}_{xy} \cdot \underline{I}_{yz} \tag{4.155a}$$

or

$$\underline{T}_{xy}^{\mathrm{T}} = \underline{A} \cdot \underline{X}_{xy} \cdot \underline{I}_{yz} \tag{4.155b}$$

and

$$\underline{A} = \sum_{(el)} \left(\underline{Z}_{xy}^{(el)\mathrm{T}} \cdot \underline{A}^{(el)} \cdot \underline{Z}_{xy}^{(el)} \right). \tag{4.155c}$$

With the matrix $\underline{T}_{x,y}^{\mathrm{T}}$ the column forces $\underline{F}_{\mathrm{St}}$ are transformed to nodal forces at the nodal points of the plate elements in the connection area.

In addition to the forces, the displacements must be transformed from the column displacements to the displacements of the plate elements. According to Sect. 4.10.2

$$\underline{w}_{\mathrm{St}} = \underline{T}_{xy} \cdot \underline{w}_{\mathrm{Pl}}, \tag{4.156}$$

applies where the vector

$$\underline{w}_{\mathrm{Pl}} = \begin{bmatrix} w_1 \\ w_2 \\ w_3 \\ \cdot \\ \cdot \\ w_n \end{bmatrix} \tag{4.156a}$$

denotes the displacements at the nodal points of the EST element at the plate and

$$\underline{w}_{St} = \begin{bmatrix} w_{Bz} \\ \varphi_{Bx} \\ \varphi_{By} \end{bmatrix} \tag{4.156b}$$

are the displacements at the column head. In (4.156), the displacements $\underline{w}_{St}$ are evaluated as weighted means of the nodal displacements $\underline{w}_{Pl}$ of the plate elements.

The stiffness of the column at the column head is written as

$$\begin{bmatrix} F_{Bz} \\ M_{Bx} \\ M_{By} \end{bmatrix} = \begin{bmatrix} k_z & 0 & 0 \\ 0 & k_{xx} & 0 \\ 0 & 0 & k_{yy} \end{bmatrix} \cdot \begin{bmatrix} w_{Bz} \\ \varphi_{Bx} \\ \varphi_{By} \end{bmatrix} \tag{4.157}$$

$$\underline{F}_{St} = \underline{K}_{St} \cdot \underline{w}_{St}. \tag{4.157a}$$

Here, k_z denotes the spring for vertical displacements, k_{xx} the rotational spring around the x_B-axis and k_{yy} the rotational spring around the y_B-axis of the column where x_B and y_B are the principal axes of the column cross section.

For a column with all degrees of freedom being fixed at the lower end, one obtains according to Table 3.4

$$k_z = \frac{E \cdot A}{h}, \quad k_{xx} = 4 \cdot \frac{E \cdot I_{xB}}{h}, \quad k_{yy} = 4 \cdot \frac{E \cdot I_{yB}}{h} \tag{4.157b}$$

and for a column end where only displacements are fixed but which is free to rotate

$$k_z = \frac{E \cdot A}{h}, \quad k_{xx} = 3 \cdot \frac{E \cdot I_{xB}}{h}, \quad k_{yy} = 3 \cdot \frac{E \cdot I_{yB}}{h}. \tag{4.157c}$$

Herein, E is the modulus of elasticity, h the length of the column, and I_{xB} and I_{yB} the moments of inertia around the x_B- and y_B-axis, respectively. The rotational spring stiffnesses of a column above the slab can be added, if necessary.

In the liquid cushion model, the column head is supposed to transfer the normal force of the column only, but no bending moments. The spring constants are

$$k_z = \frac{E \cdot A}{h}, \quad k_{xx} = 0, \quad k_{yy} = 0. \tag{4.157d}$$

The stiffness relationship (4.157a) of the column can easily be transformed to the nodes of the plate elements applying the relationships (4.155) for forces, and (4.156) for displacements. One obtains

$$\underline{F}_{Pl} = \underline{K}_{Pl} \cdot \underline{w}_{Pl}, \tag{4.158}$$

where

$$\underline{K}_{Pl} = \underline{T}_{xy}^T \cdot \underline{K}_{St} \cdot \underline{T}_{xy} \tag{4.158a}$$

is the stiffness matrix of the EST support element.

After solving the global system of equations, the element internal forces are calculated. For the EST element, the spring forces and moments are obtained as

$$\underline{F}_{St} = \underline{K}_{St} \cdot \underline{w}_{St} = \underline{K}_{St} \cdot \underline{T}_{xy} \cdot \underline{w}_{Pl}. \tag{4.159}$$

The column force calculated in this way and the moments at the column head can be used in the RC design of the column.

Example 4.17 An EST connection of a rectangular column with a slab is considered. The column has a rectangular cross section. The connection area of the slab is subdivided into 4 bending plate elements (Fig. 4.68). With the nodal point numbers given in Fig. 4.68, and their coordinates in the x_B-, y_B-coordinate system, one obtains with (4.142)

$$\underline{X}_{xy} = \begin{bmatrix} 1 & -b_S/2 & h_S/2 \\ 1 & -b_S/2 & 0 \\ 1 & -b_S/2 & -h_S/2 \\ 1 & 0 & h_S/2 \\ 1 & 0 & 0 \\ 1 & 0 & -h_S/2 \\ 1 & b_S/2 & h_S/2 \\ 1 & b_S/2 & 0 \\ 1 & b_S/2 & -h_S/2 \end{bmatrix}, \quad \underline{I}_{yz} = \begin{bmatrix} \dfrac{1}{h_S \cdot b_S} & 0 & 0 \\ 0 & \dfrac{12}{h_S \cdot b_S^3} & 0 \\ 0 & 0 & \dfrac{12}{h_S^3 \cdot b_S} \end{bmatrix}.$$

The assignment of the element degrees of freedom of the 4 elements to the 9 global degrees of freedom in the connection area is described for each element by the incidence matrices

$$\underline{Z}_{xy}^{(1)} = \begin{bmatrix} 0 & 0 & 0 & 0 & 1 & 0 & 0 & 0 & 0 \\ 0 & 0 & 0 & 1 & 0 & 0 & 0 & 0 & 0 \\ 1 & 0 & 0 & 0 & 0 & 0 & 0 & 0 & 0 \\ 0 & 1 & 0 & 0 & 0 & 0 & 0 & 0 & 0 \end{bmatrix}, \quad \underline{Z}_{xy}^{(2)} = \begin{bmatrix} 0 & 0 & 0 & 0 & 0 & 1 & 0 & 0 & 0 \\ 0 & 0 & 0 & 0 & 1 & 0 & 0 & 0 & 0 \\ 0 & 1 & 0 & 0 & 0 & 0 & 0 & 0 & 0 \\ 0 & 0 & 1 & 0 & 0 & 0 & 0 & 0 & 0 \end{bmatrix}$$

$$\underline{Z}_{xy}^{(3)} = \begin{bmatrix} 0 & 0 & 0 & 0 & 0 & 0 & 0 & 1 & 0 \\ 0 & 0 & 0 & 0 & 0 & 0 & 1 & 0 & 0 \\ 0 & 0 & 0 & 1 & 0 & 0 & 0 & 0 & 0 \\ 0 & 0 & 0 & 0 & 1 & 0 & 0 & 0 & 0 \end{bmatrix}, \quad \underline{Z}_{xy}^{(4)} = \begin{bmatrix} 0 & 0 & 0 & 0 & 0 & 0 & 0 & 0 & 1 \\ 0 & 0 & 0 & 0 & 0 & 0 & 0 & 1 & 0 \\ 0 & 0 & 0 & 0 & 1 & 0 & 0 & 0 & 0 \\ 0 & 0 & 0 & 0 & 0 & 1 & 0 & 0 & 0 \end{bmatrix}.$$

With

$$\underline{A}^{(1)} = \underline{A}^{(2)} = \underline{A}^{(3)} = \underline{A}^{(4)} = \frac{b_S \cdot h_S}{144} \begin{bmatrix} 4 & 2 & 1 & 2 \\ 2 & 4 & 2 & 1 \\ 1 & 2 & 4 & 2 \\ 2 & 1 & 2 & 4 \end{bmatrix}$$

according to (4.150), one obtains the matrix $\underline{A}$ with (4.155c), as

$$\underline{A} = \underline{Z}^{(1)\mathrm{T}} \cdot \underline{A}^{(1)} \cdot \underline{Z}^{(1)} + \underline{Z}^{(2)\mathrm{T}} \cdot \underline{A}^{(2)} \cdot \underline{Z}^{(2)} + \underline{Z}^{(3)\mathrm{T}} \cdot \underline{A}^{(3)} \cdot \underline{Z}^{(3)} + \underline{Z}^{(4)\mathrm{T}} \cdot \underline{A}^{(4)} \cdot \underline{Z}^{(4)}$$

$$\underline{A} = \frac{b_S \cdot h_S}{144} \cdot \begin{bmatrix} 4 & 2 & 0 & 2 & 1 & 0 & 0 & 0 & 0 \\ 2 & 8 & 2 & 1 & 4 & 1 & 0 & 0 & 0 \\ 0 & 2 & 4 & 0 & 1 & 2 & 0 & 0 & 0 \\ 2 & 1 & 0 & 8 & 4 & 0 & 2 & 1 & 0 \\ 1 & 4 & 1 & 4 & 16 & 4 & 1 & 4 & 1 \\ 0 & 1 & 2 & 0 & 4 & 8 & 0 & 1 & 2 \\ 0 & 0 & 0 & 2 & 1 & 0 & 4 & 2 & 0 \\ 0 & 0 & 0 & 1 & 4 & 1 & 2 & 8 & 2 \\ 0 & 0 & 0 & 0 & 1 & 2 & 0 & 2 & 4 \end{bmatrix}.$$

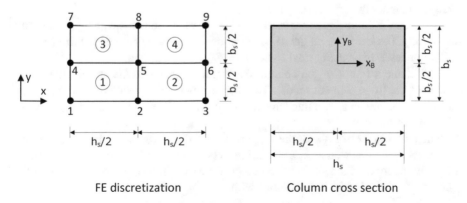

FE discretization   Column cross section

Fig. 4.68 Finite element discretization at a rectangular column

The transformation matrix $\underline{T}_{xy}$ is obtained with (4.155b), as

$$\underline{T}_{xy} = (\underline{A} \cdot \underline{X}_{xy} \cdot \underline{L}_{yz})^{\mathrm{T}}.$$

One obtains

$$
\underline{T}_{xy} =
\begin{bmatrix}
\dfrac{1}{16} & \dfrac{1}{8} & \dfrac{1}{16} & \dfrac{1}{8} & \dfrac{1}{4} \\[2mm]
-\dfrac{1}{4 \cdot b_S} & -\dfrac{1}{2 \cdot b_S} & -\dfrac{1}{4 \cdot b_S} & 0 & 0 \ \cdots \\[2mm]
\dfrac{1}{4 \cdot h_S} & 0 & -\dfrac{1}{4 \cdot h_S} & \dfrac{1}{2 \cdot h_S} & 0
\end{bmatrix}
$$

$$
\qquad\qquad
\begin{matrix}
\dfrac{1}{8} & \dfrac{1}{16} & \dfrac{1}{8} & \dfrac{1}{16} \\[2mm]
\cdots\cdots\cdots \quad 0 & \dfrac{1}{4 \cdot b_S} & \dfrac{1}{2 \cdot b_S} & \dfrac{1}{4 \cdot b_S} \\[2mm]
-\dfrac{1}{2 \cdot h_S} & \dfrac{1}{4 \cdot h_S} & 0 & -\dfrac{1}{4 \cdot h_S}
\end{matrix}
\Bigg].
\qquad (4.160)
$$

The transformation matrix illustrates the weighted averaging for the displacements according to (4.156). The weighting factors are shown in Fig. 4.69. The first row of the matrix (4.160), contains the weighting factors with which the node displacements of the plate are transformed into the displacement w_{Bz} of the column head. The highest weighting factor of 1/4 is assigned to node 5 in the center of the column. The nodes on the side lines have a weighting factor of 1/8, while the corner nodes are weighted with 1/16. The sum of all weighting factors is 1.

The rotational angles φ_{Bx} and φ_{By} are calculated with the displacement differences of the edge displacements, whereby here too the nodes on the side centers are weighted higher than the corner nodes.

Analogue, column forces are transferred to the plate elements with $\underline{T}_{xy}^{\mathrm{T}}$ according to (4.155). The center node gets 1/4, the nodes on the side centers get 1/8 and the corner nodes 1/16 of the normal force $N = F_{Bz}$ of the column.

The element forces of the column correspond exactly to the sum of the nodal forces of the finite element model. The displacements and rotations of the column head are determined as weighted average from the nodal displacements of the finite elements of the slab.

The stiffness matrix $\underline{K}_{St}$ of the support in (4.157), is determined with the spring constants k_z, k_{xx}, as well as k_{yy}, according to (4.157b–4.157d). The stiffness matrix of the EST support element is obtained according to (4.158a). Alternatively, (4.160), can be understood as MPC conditions with

$$
w_{Bz} = \frac{1}{16}w_1 + \frac{1}{8}w_2 + \frac{1}{16}w_3 + \frac{1}{8}w_4 + \frac{1}{4}w_5 + \frac{1}{8}w_6 + \frac{1}{16}w_7 + \frac{1}{8}w_8 + \frac{1}{16}w_9
$$

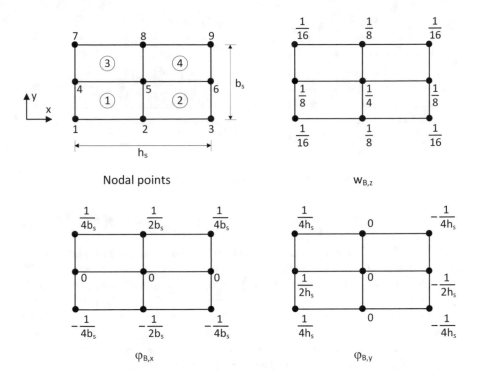

Fig. 4.69 Weighting factors of the finite element displacements of the slab

$$\varphi_{Bx} = -\frac{1}{4b_S}w_1 - \frac{1}{2b_S}w_2 - \frac{1}{4b_S}w_3 + 0 \cdot w_4 + 0 \cdot w_5 + 0 \cdot w_6$$
$$+ \frac{1}{4b_S}w_7 + \frac{1}{2b_S}w_8 + \frac{1}{4b_S}w_9$$

$$\varphi_{By} = \frac{1}{4h_S}w_1 + 0 \cdot w_2 - \frac{1}{4h_S}w_3 + \frac{1}{2h_S}w_4 + 0 \cdot w_5 - \frac{1}{2h_S}w_6$$
$$+ \frac{1}{4h_S}w_7 + \frac{1}{2h_S}w_8 + \frac{1}{4h_S}w_9$$

and entered as such in a finite element program.

The application of the EST model for flat slabs is shown in Examples 4.35 and 4.36.   ◀

For rectangular columns with a regular subdivision into finite elements of the same size, the transformation relationships can be specified explicitly. The cross section is subdivided into n_x elements in x-direction and n_y elements in x_B- and y_B-direction, respectively (Fig. 4.70). The position of the nodal points in the element grid is indicated by rows and columns—similar to the entries of a matrix. The weighting factors at the nodal points are specified for the degrees of freedom w_{Bz}, φ_{Bx} and φ_{By} separately. They are given in Table 4.14, as transformation matrices $\underline{T}_{wz}$, $\underline{T}_{\varphi x}$ and

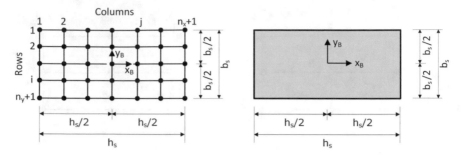

Fig. 4.70 Element numbering scheme of a rectangular cross section

$\underline{T}_{\varphi y}$ which correspond to the transformation matrix $\underline{T}_{xy}$ according to (4.156), and can easily be transformed into it.

The weighting factors in Table 4.14, can be derived from those of the plane system given in Table 4.13 by additionally weighing them perpendicular to the plane with the factors $1/n_s$ for the center nodes and $1/(2n_s)$ for the edge nodes, where n_s is the number of elements perpendicular to the plane. In this way, the transformation matrices for the element forces in the x- and y-directions, which correspond to the shear forces of the beam, can be specified with the matrix $\underline{T}_y$ in Table 4.13. With these transformation matrices, the three-dimensional beam to plate EST-connections can be formulated analogous to (4.133). The torsional degree of freedom can be neglected for columns of flat slabs. If necessary, however, the corresponding transformation matrix for the torsional moment can also be formulated, based on the shear stress distribution in a rectangular cross section.

Example 4.18 An EST transformation matrices for beams with a rectangular cross section and bending plate elements is considered. The transformation matrices for a 2×2-discretization (Fig. 4.68) and for a 4×6-discretization of a rectangular cross section have to be specified.

In the finite element mesh shown in Fig. 4.68, the number of elements are $n_x = n_y = 2$. The matrices $\underline{T}_{wz}$, $\underline{T}_{\varphi x}$ and $\underline{T}_{\varphi y}$ are thus obtained according to Table 4.14, as

$$\underline{T}_{wz} = \frac{1}{16} \cdot \begin{bmatrix} 1 & 2 & 1 \\ 2 & 4 & 2 \\ 1 & 2 & 1 \end{bmatrix}, \quad \underline{T}_{\varphi x} = \frac{1}{4 \cdot b_S} \cdot \begin{bmatrix} 1 & 2 & 1 \\ 0 & 0 & 0 \\ -1 & -2 & -1 \end{bmatrix},$$

$$\underline{T}_{\varphi y} = \frac{1}{4 \cdot h_S} \cdot \begin{bmatrix} 1 & 0 & -1 \\ 2 & 0 & -2 \\ 1 & 0 & -1 \end{bmatrix}.$$

They are related to the columns and rows of the nodel points according to Fig. 4.70. For example, nodal point 1 in Fig. 4.68, corresponds to the nodal point in row $i = 3$ and column $j = 1$ in Fig. 4.70. For this nodal point one obtains $\underline{T}_{wz,3,1} = 1/16$,

Table 4.14 Entries of the transformation matrices for beams with rectangular cross sections

$\underline{T}_{wz}$

Row i		Column j		
		$j = 1$	$2 \leq j \leq n_x$	$j = n_x + 1$
	$i = 1$	$\dfrac{1}{4 \cdot n_x \cdot n_y}$	$\dfrac{1}{2 \cdot n_x \cdot n_y}$	$\dfrac{1}{4 \cdot n_x \cdot n_y}$
	$2 \leq i \leq n_y$	$\dfrac{1}{2 \cdot n_x \cdot n_y}$	$\dfrac{1}{n_x \cdot n_y}$	$\dfrac{1}{2 \cdot n_x \cdot n_y}$
	$i = n_y + 1$	$\dfrac{1}{4 \cdot n_x \cdot n_y}$	$\dfrac{1}{2 \cdot n_x \cdot n_y}$	$\dfrac{1}{4 \cdot n_x \cdot n_y}$

$\underline{T}_{\varphi x}$

Row i		Column j		
		$j = 1$	$2 \leq j \leq n_x$	$j = n_x + 1$
	$i = 1$	$\dfrac{1}{2 \cdot b_S \cdot n_x \cdot n_y} \cdot \left(3 - \dfrac{2}{n_y}\right)$	$\dfrac{1}{b_S \cdot n_x \cdot n_y} \cdot \left(3 - \dfrac{2}{n_y}\right)$	$\dfrac{1}{2 \cdot b_S \cdot n_x \cdot n_y} \cdot \left(3 - \dfrac{2}{n_y}\right)$
	$2 \leq i \leq n_y$	$\dfrac{3}{b_S \cdot n_x \cdot n_y} \cdot \left(1 - 2\dfrac{i-1}{n_y}\right)$	$\dfrac{6}{b_S \cdot n_x \cdot n_y} \cdot \left(1 - 2\dfrac{i-1}{n_y}\right)$	$\dfrac{3}{b_S \cdot n_x \cdot n_y} \cdot \left(1 - 2\dfrac{i-1}{n_y}\right)$
	$i = n_y + 1$	$-\dfrac{1}{2 \cdot b_S \cdot n_x \cdot n_y} \cdot \left(3 - \dfrac{2}{n_y}\right)$	$-\dfrac{1}{b_S \cdot n_x \cdot n_y} \cdot \left(3 - \dfrac{2}{n_y}\right)$	$-\dfrac{1}{2 \cdot b_S \cdot n_x \cdot n_y} \cdot \left(3 - \dfrac{2}{n_y}\right)$

(continued)

Table 4.14 (continued)

$\underline{T}_{\phi y}$

Row i	Column j		
	$j = 1$	$2 \leq j \leq n_x$	$j = n_x + 1$
$i = 1$	$\dfrac{1}{2 \cdot h_S \cdot n_x \cdot n_y} \cdot \left(3 - \dfrac{2}{n_x}\right)$	$\dfrac{3}{h_S \cdot n_x \cdot n_y} \cdot \left(1 - 2\dfrac{j-1}{n_x}\right)$	$-\dfrac{1}{2 \cdot h_S \cdot n_x \cdot n_y} \cdot \left(3 - \dfrac{2}{n_x}\right)$
$2 \leq i \leq n_y$	$\dfrac{1}{h_S \cdot n_x \cdot n_y} \cdot \left(3 - \dfrac{2}{n_x}\right)$	$\dfrac{6}{h_S \cdot n_x \cdot n_y} \cdot \left(1 - 2\dfrac{j-1}{n_x}\right)$	$-\dfrac{1}{h_S \cdot n_x \cdot n_y} \cdot \left(3 - \dfrac{2}{n_x}\right)$
$i = n_y + 1$	$\dfrac{1}{2 \cdot h_S \cdot n_x \cdot n_y} \cdot \left(3 - \dfrac{2}{n_x}\right)$	$\dfrac{3}{h_S \cdot n_x \cdot n_y} \cdot \left(1 - 2\dfrac{j-1}{n_x}\right)$	$-\dfrac{1}{2 \cdot h_S \cdot n_x \cdot n_y} \cdot \left(3 - \dfrac{2}{n_x}\right)$

$\underline{T}_{\varphi x,3,1} = -1/(4 \cdot b_S)$ and $\underline{T}_{\varphi x,3,1} = 1/(4 \cdot h_S)$. This corresponds to the first column of the matrix $\underline{T}_{xy}$ in Example 4.17.

The weighting factors for a 6×4 subdivision of the column cross section are obtained with $n_x = 6$ and $n_y = 4$ as

$$\underline{T}_{wz} = \frac{1}{96} \cdot \begin{bmatrix} 1 & 2 & 2 & 2 & 2 & 2 & 1 \\ 2 & 4 & 4 & 4 & 4 & 4 & 2 \\ 2 & 4 & 4 & 4 & 4 & 4 & 2 \\ 2 & 4 & 4 & 4 & 4 & 4 & 2 \\ 1 & 2 & 2 & 2 & 2 & 2 & 1 \end{bmatrix}$$

$$\underline{T}_{\varphi x} = \frac{1}{288 \cdot b_S} \cdot \begin{bmatrix} 15 & 30 & 30 & 30 & 30 & 30 & 15 \\ 18 & 36 & 36 & 36 & 36 & 36 & 18 \\ 0 & 0 & 0 & 0 & 0 & 0 & 0 \\ -18 & -36 & -36 & -36 & -36 & -36 & -18 \\ -15 & -30 & -30 & -30 & -30 & -30 & -15 \end{bmatrix}$$

$$\underline{T}_{\varphi y} = \frac{1}{144 \cdot h_S} \cdot \begin{bmatrix} 8 & 12 & 6 & 0 & -6 & -12 & -8 \\ 16 & 24 & 12 & 0 & -12 & -24 & -16 \\ 16 & 24 & 12 & 0 & -12 & -24 & -16 \\ 16 & 24 & 12 & 0 & -12 & -24 & -16 \\ 8 & 12 & 6 & 0 & -6 & -24 & -8 \end{bmatrix}$$

In addition, the weighting factors of the displacements u_B and v_B in the direction of x_B and y_B, respectively, are to be specified. For the discretization in 2×2 elements according to Fig. 4.68, one obtains according to Table 4.13 or (4.127), with $n_x = n_y = 2$

$$u_B : \begin{bmatrix} \dfrac{3}{16} & \dfrac{10}{16} & \dfrac{3}{16} \end{bmatrix} \Rightarrow \underline{T}_{ux} = \begin{bmatrix} \dfrac{3}{64} & \dfrac{5}{32} & \dfrac{3}{64} \\ \dfrac{3}{32} & \dfrac{5}{16} & \dfrac{3}{32} \\ \dfrac{3}{64} & \dfrac{5}{32} & \dfrac{3}{64} \end{bmatrix},$$

$$v_B : \begin{bmatrix} \dfrac{3}{16} \\ \dfrac{10}{16} \\ \dfrac{3}{16} \end{bmatrix} \Rightarrow \underline{T}_{vy} = \begin{bmatrix} \dfrac{3}{64} & \dfrac{3}{32} & \dfrac{3}{64} \\ \dfrac{5}{32} & \dfrac{5}{16} & \dfrac{5}{32} \\ \dfrac{3}{64} & \dfrac{3}{32} & \dfrac{3}{64} \end{bmatrix}.$$

For a 6×4 subdivision of the column cross section one obtains with $n_x = 6$ and $n_y = 4$

$$u_B : \begin{bmatrix} \dfrac{11}{432} & \dfrac{29}{216} & \dfrac{47}{216} & \dfrac{53}{216} & \dfrac{47}{216} & \dfrac{29}{216} & \dfrac{11}{432} \end{bmatrix}$$

$$\Rightarrow \underline{T}_{ux} = \frac{1}{3456} \cdot \begin{bmatrix} 11 & 58 & 94 & 106 & 94 & 58 & 11 \\ 22 & 116 & 188 & 212 & 188 & 116 & 22 \\ 22 & 116 & 188 & 212 & 188 & 116 & 22 \\ 22 & 116 & 188 & 212 & 188 & 116 & 22 \\ 11 & 58 & 94 & 106 & 94 & 58 & 11 \end{bmatrix}$$

$$v_B : \begin{bmatrix} \dfrac{7}{128} \\ \dfrac{17}{64} \\ \dfrac{23}{64} \\ \dfrac{17}{64} \\ \dfrac{7}{128} \end{bmatrix} \Rightarrow \underline{T}_{vy} = \frac{1}{1536} \cdot \begin{bmatrix} 7 & 14 & 14 & 14 & 14 & 14 & 7 \\ 34 & 68 & 68 & 68 & 68 & 68 & 34 \\ 46 & 92 & 92 & 92 & 92 & 92 & 46 \\ 34 & 68 & 68 & 68 & 68 & 68 & 34 \\ 7 & 14 & 14 & 14 & 14 & 14 & 7 \end{bmatrix} \cdot$$

◄

4.10.5 Engineering Models

Engineering models are heuristic models. The aim is to represent an RDT connection using artificially introduced beam elements (Fig. 4.87). Please refer to Sect. 4.11.5, for more information.

4.10.6 Other Element Transitions

The concept of EST connections can also be applied to the transition from beam elements to shell elements (according to the EST support element given above). The formulation of transformation relationships between beam, plate, shell, and solid elements is possible.

Particular stress representations can be found in special finite elements for axisymmetric systems with non-axisymmetric displacements, such as the solid elements mentioned in Sect. 4.9.3. These elements can be combined with hexaeder and other isoparametric solid elements in a finite element model if the corresponding transformation relationships are available. In [98, 99], the transformation for an axisymmetric semifinite element for stratified soil, a so-called "transmitting boundary", is formulated for applications in soil dynamics. Figure 4.71 shows the three-dimensional cylindrical region to be discretized arbitrarily into isoparametric solid elements. The transmitting boundary element is an axisymmetric element with non-axisymmetric displacements. It is connected with the cylindrical boundary of the discretizised region. As an example, a single 20-node solid element in the discretizised region is shown which is connected to the transmitting boundary element at nodes 1–8. The transformation of the stiffness matrix of the transmitting boundary element the degrees of freedom of the discretizised three-dimensional region is obtained as

$$\underline{K}_{3D} = \frac{1}{2\pi} \cdot \left(\underline{S}_0^T \cdot \underline{K}_0 \cdot \underline{S}_0 + \underline{A}_0^T \cdot \underline{K}_0 \cdot \underline{A}_0 \right)$$
$$+ \frac{1}{\pi} \cdot \sum \left(\underline{S}_n^T \cdot \underline{K}_n \cdot \underline{S}_n + \underline{A}_n^T \cdot \underline{K}_n \cdot \underline{A}_n \right). \tag{4.161}$$

Herein $\underline{K}_n$ is the stiffness matrix of the n-th Fourier term of the transmitting boundary element and $\underline{S}_n$, $\underline{A}_n$ are transformation matrices of the symmetric and antimetric Fourier terms, respectively, specified in [98, 99]. $\underline{K}_{3D}$ is the stiffness matrix of the transmitting boundary transformed to the degrees of freedom of the region discretizised into isoparametric solid elements. It can be used for dynamic calculations of three-dimensional finite element models of the soil including the radiation

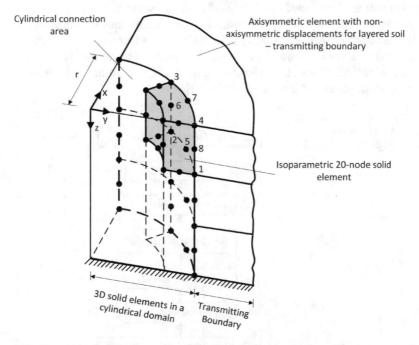

Fig. 4.71 Transition between a region discretized in isoparametric solid elements and an axisymmetric solid element (1/4 of the entire model)

damping at the cylindrical boundary of the model, but also for the static analysis of soil models.

The transformation (4.161), is generally valid and allows connecting a cylindrical region discretized in hexahedral or tetraeder solid elements to an axisymmetric solid element with non-axisymmetric displacements.

4.11   Modeling of Structural Elements and Buildings

4.11.1   Structural Models

In a static analysis, the idealization of a structure or its components is done in several steps (Fig. 4.72). In the first step, the real structure is idealized as a *structural model* that can be investigated using structural analysis methods. In the case of beam structures, the structural model is also denoted as structural system. In the second step, a *computational model* is established and analyzed using suitable mathematical methods. As a result, the state variables of the selected structural model are obtained. These represent the input values of a *design model*.

Typical models in structural design are thus the structural model or structural system, the mathematical model (e.g. finite element model), and the design model (e.g. in reinforced concrete beams: cracked cross section in failure state). In addition, there are further models for the description of the material behavior (linear, nonlinear) and the deformation behavior (theory I., II., III. order). With regard to the examination of the correctness of a model, a distinction is made between *verification* (within the framework of the model assumptions made) and *validation* (with regard to the purpose of the model) of the results [100–103].

Fig. 4.72 Models in structural design

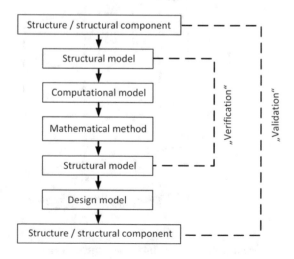

All model types can be simplified more or less detailed. This is referred to as the modeling depth of a model. On the one hand, there are simplifying engineering models, while on the other hand, there are rather physical models with high modeling depth. A beam, for example, can be modeled as a 3D solid model, as a plate in plane stress (2D model) and as a beam according to the Euler–Bernoulli beam theory (1D model). Each model uses certain state variables that cannot be easily used with other models. This complicates the transition between models of different modeling depths, such as, when connecting a beam element to plane stress or 3D solid elements (see Sect. 4.10). In practice, engineering models should be preferred due to their greater transparency. However, if used, they must be able to represent all significant mechanical aspects.

The different types of models are discussed using the example of the slab-column connection of a flat slab (Fig. 4.73). There is no unique idealization of a column and a slab as structural models accessible to an analysis. For example, for a linear analysis the following structural models may be applied:

- modeling of slab and column as three-dimensional solid,
- modeling of the slab as shear flexible or shear rigid plate with continuous elastic support,
- modeling of the slab as shear flexible or shear rigid plate with elastic or rigid point support.

The analysis of these structural models can be done with different computational methods requiring specific computational (or mathematical) models, as

- finite element method (FEM),
- finite difference method (FDM),
- boundary element method (BEM),
- analytical methods, based on series expansions (SE).

Of course, the results of different computational methods can only be expected to be identical if the underlying structural models are the same.

In the classical methods of structural analysis for hand calculation, the establishment of the structural model is usually obvious and limited by its calculability. By hand, the computation of surface structures is carried out using tables, which are based on specific structural models being often no longer questioned in practice. Problems arising in modeling and in the interpretation of results have already been considered by the author of the table in question. In tables for plates in plane stress, for example, support reactions are described by linearly distributed pressures. For plates in bending, the shear forces are given with and without edge effect. Computer-oriented computational methods such as the finite element method, on the other hand, are characterized by the fact that they make very complex structural models accessible to analysis. This means, however, that the responsibility for defining the structural model and interpreting the results is shifting to the user of the software. This applies, in particular, to three-dimensional models of entire buildings.

A special problem of structural models are singularities that may occur in the results. These are discussed in the next section. It is followed by a section on general rules for creating finite element models. The following sections deal with special

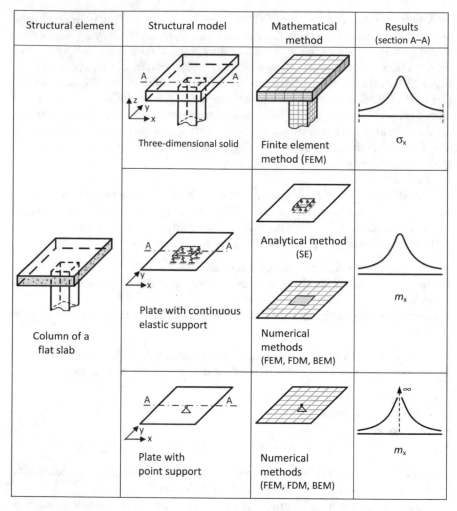

Structural element	Structural model	Mathematical method	Results (section A–A)
	Three-dimensional solid	Finite element method (FEM)	σ_x
Column of a flat slab	Plate with continuous elastic support	Analytical method (SE) / Numerical methods (FEM, FDM, BEM)	m_x
	Plate with point support	Numerical methods (FEM, FDM, BEM)	m_x

Fig. 4.73   Modeling the slab-column connection of a flat slab (see [167])

problems in structural models of surface structures and with the interpretation of their results. References to modeling in reinforced concrete constructions can be found in [104, 105], and for more general aspects in [57, 80, 103, 106, 107].

4.11.2   Singularities of Stresses and Displacements

When creating the structural model and interpreting the results of the calculation, special attention must be paid to points with singularities of internal forces or displacements. At a singularity point, an internal force or a displacement assumes an

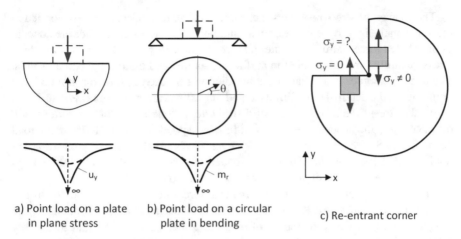

a) Point load on a plate
   in plane stress

b) Point load on a circular
   plate in bending

c) Re-entrant corner

Fig. 4.74 Points with singularities in structural models

unlimited value, i.e. in the limit one obtains ∞. Such a result is physically meaningless
and requires an engineering interpretation.

Singularities occur at locations with a physically inadequate structural modeling or
load representation. They are, therefore, a problem of modeling a structure, a support
or a load, and not a problem of the computational method applied. A simple example
is the case of a point load on a plate in plane stress (Fig. 4.74a). If, for example, the
load F introduced into the plate by a column is represented more realistically as a
line load $p = F/a$, the stress in plate under the load is obtained as $\sigma_y = F/a$. At
the transition to a point load with $a \to 0$, the stress under the load is obtained to be
$\sigma_y \to \infty$. A more detailed examination shows that not only the stress σ_y but also
the stresses σ_x and τ_{xy}, as well as the displacements under the point load, possess a
singularity [108, 109].

Singularities also occur in plates in bending. This is illustrated by the following
example. A circular plate is simply supported at the boundaries and loaded by the
point load F at its center (Fig. 4.74b). In an arbitrary section with the radius r
around the center, the shear force is obviously $v_r = F/(2 \cdot \pi \cdot r)$ for reasons of
symmetry. In the center of the circle, with $r \to 0$, the shear force v_r tends to ∞. A
more detailed investigation of the circular plate shows that not only the shear force
but also the bending moments at the point where the concentrated load is applied
possess a singularity [109]. The deflections under the point load, however, remain
finite for a shear rigid plate. On the other hand, at a shear flexible plate, the deflections
under the point load are also singular. If the point load is represented as a circularly
distributed constant load, finite values for the bending moments, the shear forces and
the displacements are obtained.

With regard to singularities in point and ring loads on a three-dimensional elastic
half-space, reference is made to [110, 111].

The concept of stress resultants, i.e. forces and moments, successfully applied in the static analysis of rigid bodies, truss, and beam structures contradict the concept of distributed forces (stresses or internal forces/moments per unit length) with which stress quantities are represented in surface structures and three-dimensional solids. The above examples show that this contradiction can lead to problems at individual points due to the depth of modeling assumed. If loadings are represented as distributed loads, the stresses in a surface or solid structure assume finite values. At a certain distance from the loaded area, the results correspond to those of the point load according to St. Venant's principle. However, the internal forces or stresses at the load center depend on the size of the loaded area and—unlike with the beam in bending—are highly sensitive to its arbitrary reduction.

Singularities of stresses and related internal forces occur at geometrically promi- nent points in the structure. Re-entrant corners in plates in plane stress or bending are a typical example. Here, the theory of plates and shells leads to singularities in the stresses at the corner point, which would not occur in a more realistic representation of the geometry. The problems in the structural model are illustrated in Fig. 4.74c. The stresses σ_y at the edge parallel to the x-axis are obviously zero, whereas the stresses σ_y at the edge perpendicular to the x-axis are unequal to zero. At the point in the corner "belonging" to both edges, there is a contradiction, i.e. the stress σ_y is not defined at this point. This results in a singularity of stresses. It can be eliminated by rounding off the corner.

Within the limits of finite element theory, it is not possible to obtain "infinite" stresses at a singularity point because the displacement shape functions and the finite stress values resulting thereof comprise only finite values. Therefore, finite stress values are obtained, which, however, do not converge to a final value with an increasing number of elements, but instead increase continuously (cf. stresses at the support in Example 4.7). Points of the structural model at which singularities of stresses and displacements occur can be identified with the help of the tables in Sects. 4.11.5 and 4.11.6.

Example 4.19 For the plate in plane stress shown in Fig. 4.75a, the singularity points of stresses and displacements in the structural model are to be identified. The implications on the results of a finite element analysis at these points are to be investigated.

Singularities can occur at corner points of the structural model and at points with point loads. Based on Sect. 4.11.5, Tables 4.17 and 4.21, the following findings are obtained for the points given in Fig. 4.75a:

- Points a, b: stress singularity, since $\alpha > 63°$
- Point c: stress singularity, since $\alpha > 180°$
- Point g: singularity of the stresses and displacements due to the point load

At the corner points e and d, and at all other points such as f, no singularities are to be expected, since with $\alpha < 180°$ they do not represent a re-entrant corner. At these points stresses and displacements are regular.

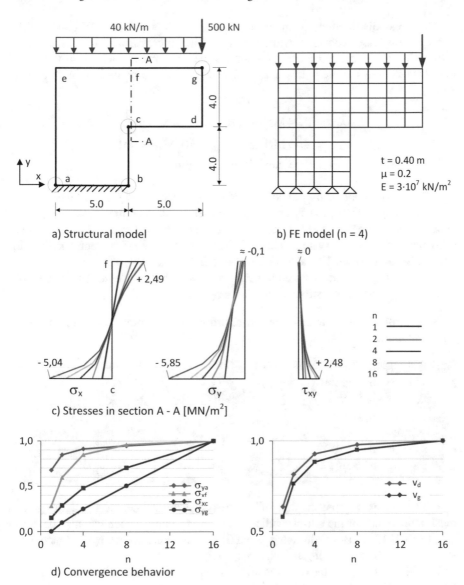

Fig. 4.75 Plate in plane stress with points of singularities

The analysis is performed using the 4-node rectangular element derived in Sect. 4.4. The plate is divided into three rectangular regions with a size of 4 × 5 m each. Each region is meshed with $n \times n$ elements, where $n = 1, 2, 4, 8, 16$ (Fig. 4.75b). The global stiffness matrix obtained with the corresponding element matrices and the solution of the system of equations is given in [112].

The stress distributions obtained in section A–A are shown in Fig. 4.75c. At point f, the stresses converge, while at point c, they increase with increasing mesh refinement, as expected. Figure 4.75d shows the convergence behavior of stresses and displacements at the points a, c, f, d, and g. They are related to the reference values

$$\sigma_{y,a} = 2.74 \, \text{MN/m}^2, \, \sigma_{x,c} = 5.04 \, \text{MN/m}^2,$$
$$\sigma_{x,f} = 2.49 \, \text{MN/m}^2, \, \sigma_{y,g} = 10.04 \, \text{MN/m}^2$$
$$v_d = 2.53 \, \text{mm}, \quad v_g = 2.83 \, \text{mm}.$$

The stress component $\sigma_{x,f}$ at point f and the displacement v_d at point d converge, where the displacement v_d has a higher convergence rate than the stress $\sigma_{x,f}$. Stress singularities can be observed at points c and g. The stresses $\sigma_{x,c}$ and $\sigma_{y,g}$ continue to increase even with very small element sizes during mesh refinement, and thus do not reach a final value. This also applies—to a lesser extent—to the stress $\sigma_{y,a}$. The displacement v_g also increases with mesh refinement, but to a lesser extent than the stresses. One speaks of a strong or a weak singularity.   ◄

Singularities can be avoided by an appropriate structural modeling and load representation.

Avoidance of singularities
- Represent point loads as loads distributed over the areas actually affected by the loading.
- Represent point supports as continuous elastic support with the size of the bearing area.
- Round out re-entrant corners.

For a more exact representation of the stress distribution near a singularity point, mesh refinements of the corresponding zones are required which can also be done adaptively [113, 114]. In addition, special finite elements have been developed with shape functions representing singularities [115].

In practice, however, the question arises to which extent a detailed knowledge of the stress distribution near points of singularities is required. Outside St. Venant's domain, the influence on the stress distribution is negligible. The accuracy with which internal forces or stresses are required in the singularity domain depends on the purpose of the analysis. More accurate results are required in steel constructions than in reinforced concrete constructions, where a certain stress equalization due to the formation of cracks and the ductility of the material can be assumed.

In steel construction in addition to a mesh refinement, further approaches based on fracture mechanics can be followed [116]. In reinforced concrete construction, a complex modeling to prevent singularities is usually avoided. Rather, singularities are accepted deliberately and practical solutions in the domain of singularities are obtained by an engineering interpretation of the results. However, one must understand that the stresses in the singularity domain obtained by a finite element analysis are inaccurate, and are subject to a certain hazard. Hints for the treatment of singularities in practice are given in Sects. 4.11.5 and 4.11.6.

4.11.3 Element Types and Meshing

Since the beginnings of the finite element method in the 1960s, a large number of different element types have been developed. Commercial finite element packages often have implemented different element types in the same software package. Research and development of new element types is still going on. The aim is to develop elements that are as robust and efficient as possible.

A comparison of different element types requires a sophisticated examination, so that a general assessment hardly seems possible. Some aspects of the selection of an element type, and of meshing will be given in the following.

Frequently, 4-node elements are used for the finite element modeling of surface structures. On the one hand, they allow a good adaptation to the geometry of a structure. On the other hand, their accuracy is higher than that of simple 3-node triangle elements, since they contain higher shape functions for the same number of degrees of freedom (two triangles correspond to a quadrilateral). All types of 4-node elements for plates implemented in commercial software can be expected to have comparable accuracy with respect to bending moments and deflections, provided they are used in a rectangular shape. However, in the case of distorted (not rectangular) elements, there may be significant differences with respect to the shear forces in plates in bending or the behavior near points of singularities.

In the case of 4-node quadrilateral elements for plates in plane stress, the classical conformal (isoparametric) element appears to be too stiff when subjected to bending. Alternatives are, e.g. nonconforming elements or hybrid elements (cf. Section 4.5.3 and Example 4.11). Elements with higher shape functions exhibit a considerably higher accuracy than 4-node elements. This applies to plates in plane stress but also for plates in bending (e.g. with the 16-node Lagrangian element [37, 60]). Special numerical techniques are required for shear flexible 4-node bending plate elements in order to obtain reliable shear forces, especially with thin plates.

For plates in plane stress, element types with and without rotational degrees of freedom are implemented in finite element software (Fig. 4.41). Rotational degrees of freedom do not occur in the theory of elasticity of plane stress. They are introduced for mathematical reasons only and do not have the same physical meaning as the degrees of freedom of rotation for beams and bending plates. Therefore, they must not be used for load application or for a connection with beam elements. The same applies to the degrees of freedom of rotation of shell elements about an axis perpendicular to the shell plane. When defining a fixed support, however, the rotational degrees of freedom of a plate must be fixed, as well as the translational degrees of freedom.

Table 4.15 Shape factors for finite elements

Geometrical characteristics	Shape factor	Optimal value
Rectangular through side midpoints Quadrilateral element Side midpoint	Aspect ratio $\dfrac{a}{b}$ with $a > b$	$\dfrac{a}{b} = 1$
Parallelogram	Skew α	$\alpha = 90°$
$A_1 \ldots A_4$: partial areas	Taper $$t_F = 4 \cdot \frac{\min(A_1, A_2, A_3, A_4)}{(A_1 + A_2 + A_3 + A_4)}$$	$t_F = 1$

The accuracy of a finite analysis is affected by the shape of the elements. The factors given in Table 4.15, have been proposed for the assessment of the quality of the shape of 4-node elements [117].

Shape factors of finite elements

(a) Aspect ratio a/b
(b) Skew defined as angle α
(c) Taper defined as ratio t_F.

The parameter a/b specifies the aspect ratio for rectangular elements. It is extended by the definition in Table 4.15, to the general quadrilateral shape. The optimum value is 1. Long, narrow elements with $a/b > 2$ should be avoided. Exceptions are cases in which the gradient of the stresses in the direction of the long side is very small.

The parameter α denotes the inside angle. A right angle is regarded as optimal. Discrepancies of more than 45°, i.e. inner angle with $\alpha < 45°$ or $\alpha > 135°$, are to be avoided.

The taper of a quadrilateral is a measure of its position between the rectangular and triangular shape. The optimum taper value $t_F = 1$ applies to rectangular elements. The taper value $t_F = 0$ stands for triangular elements.

The square is considered the best shape of a finite element. The second-best shape is the rectangle, the third-best the parallelogram, and then the general quadrilateral. Extremely distorted element shapes, such as quadrilaterals with re-entrant corners, are not permitted (Fig. 4.34). To which extent deviations of the form factors from those of the square influence the results of a finite element analysis quantitatively, depends on the element type and the result values considered (displacements, stresses, moments, shear forces). The properties of finite elements can be evaluated and compared with those of other element types by means of benchmark tests and sample problems, such as those specified in [118].

Example 4.20 For the cantilever plate with a uniform line load on an edge, shown in Fig. 4.76, the influence of mesh irregularities on sectional forces has to be investigated.

The square bending plate is computed using a regular and two irregular finite element meshes (Figs. 4.76 and 4.77). The regular mesh is optimal with regard to the shape factors given above. The irregular mesh 1 has an element topology mixed of triangular and quadrilateral elements, while the irregular mesh 2 consists exclusively of quadrilateral elements. The element shapes are distorted, which is indicated in the shape factors given in Fig. 4.77c. The irregular meshes have significantly more elements and degrees of freedom than the regular mesh and are typical for widening regions of finite element meshes.

For comparison the investigations are carried out with three element types:

(a) HSS: shear rigid hybrid plate element with quadratic stress and cubic displacement shape functions according to [51, SOF 8a],
(b) VSW: shear flexible 4-node plate element with displacement shape functions according to [63, 79, SOF 6].

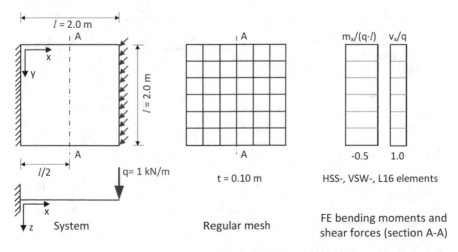

Fig. 4.76 Cantilever plate with regular finite element mesh

y[m]	HSS element		VSW element		L16 element	
	$m_x/(q \cdot l)$	v_x/q	$m_x/(q \cdot l)$	v_x/q	$m_x/(q \cdot l)$	v_x/q
0.00	-0.499	1.475	-0.505	1.144	-0.500	1.000
0.50	-0.501	1.094	-0.503	1.138	-0.500	1.000
1.03	-0.510	1.093	-0.501	1.046	-0.500	1.000
1.28	-0.506	1.496	-0.509	1.035	-0.500	1.000
1.48	-0.502	1.105	-0.511	1.137	-0.500	1.000
1.73	-0.510	1.020	-0.511	1.027	-0.500	1.000
2.00	-0.501	1.030	-0.501	0.946	-0.500	1.000
Max. absolute error	0.010	0.496	0.011	0.14	0	0
Max. relative error	2%	49.6%	2.2%	14%	0%	0%

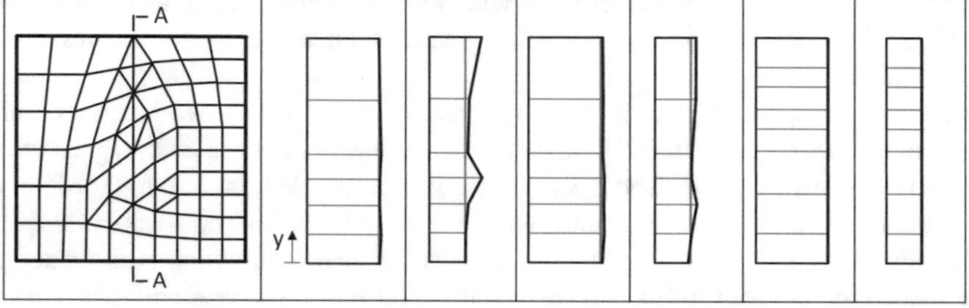

a) Irregular finite element mesh 1 with a mixed element topology (section A-A)

y[m]	HSS element		VSW element		L16 element	
	$m_x/(q \cdot l)$	v_x/q	$m_x/(q \cdot l)$	v_x/q	$m_x/(q \cdot l)$	v_x/q
0.00	-0.502	0.990	-0.500	1.007	-0.500	1.000
0.26	-0.501	1.000	-0.500	0.995	-0.500	1.000
0.43	-0.500	0.993	-0.506	0.946	-0.500	1.001
0.61	-0.502	0.899	-0.515	0.937	-0.500	0.999
0.80	-0.499	0.888	-0.492	0.945	-0.500	1.000
0.99	-0.500	1.328	-0.486	1.003	-0.500	1.000
1.35	-0.499	1.000	-0.498	1.055	-0.500	0.999
1.74	-0.500	1.000	-0.501	0.998	-0.500	1.000
2.00	-0.501	0.997	-0.499	0.983	-0.500	1.000
Max. absolute error	0.002	0.328	0.015	0.055	0	0.001
Max. relative error	0.4%	32.8%	3.0%	5.5%	0%	0.1%

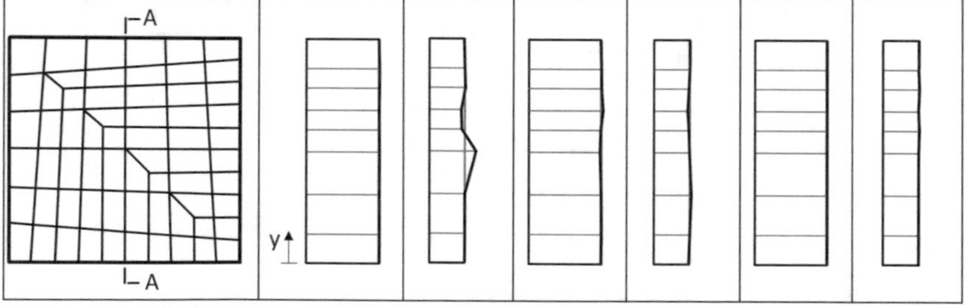

b) Irregular finite element mesh 2 with a quadrilateral elements solely (section A-A)

Fig. 4.77 Internal forces and moments in a cantilever plate

Finite element mesh	max/min shape factors		
	a/b	α	t_F
Regular FE mesh	1.0	90°	1.00
Irregular FE mesh 1	1.8	35°	0.00
Irregular FE mesh 2	1.9	59°	0.63

c) Shape factors

Fig. 4.77 (continued)

(c) L16: shear flexible 16-node plate element with Lagrangian shape functions and full 4×4 integration [37, 60, SOF 9].

With the regular 6×6 grid, the exact shear forces and moments $v_x = q$ and $m_x = -q \cdot l/2$, respectively, are obtained in section A–A in the middle of the cantilever slab with all element types (Fig. 4.76). However, the results for the two irregular meshes are different (Fig. 4.77a, b).

In the irregular mesh 1 with an element topology mixed of triangles and quadrilaterals, the HSS and VSW elements show an error of approx. 2% in the bending moments (Fig. 4.77a). The error in the shear forces is significantly higher. It is 14% for the shear flexible VSW 4-node element and 50% for the shear rigid hybrid HSS 4-node element. Despite the large number of elements, the shear forces are thus practically worthless. This also applies to other 4-node elements. For example, in the case of the DKT element, which depicts the moment diagram with mesh 1 with good accuracy, the maximum shear force in section A–A is 1.40, i.e. it possesses an error of 40%. With the shear-flexible 16-node Lagrange element, bending moments, and shear forces can be reproduced much better due to the higher shape functions. Despite the distorted element shapes, the results are nearly exact ([119, SOF 9]).

In the case of irregular mesh 2 according to Fig. 4.77b, in which solely quadrilateral elements are used for mesh refinement, the maximum errors of the shear forces of the HSS and VSW elements are lower. They are 33% for the hybrid element and 6% for the shear flexible element. The results show that the investigated 4-node elements in a distorted shape, i.e. with poor shape factors, are not able to reproduce the shear forces with sufficient accuracy. Since this applies to a constant value of the shear force, no increase in the accuracy of the shear forces can be expected even with decreasing the element size (cf. patch test). The shear flexible 16-node element gives practically exact internal forces also in the case of irregular mesh 1.

For the two shear flexible elements VSW and L16, the internal forces depend on the plate thickness, which is not the case for the shear rigid element (HSS). This applies, in particular, to the shear forces which are derived directly from the shear deformations (shear angles) in shear flexible elements. In Fig. 4.77, the ratio t/ℓ is $0.1/2 = 0.05$. In case of a variation of the plate thickness t, the deviations of the shear forces from the exact value $v_x/q = 1$ are given in Table 4.16. Thereafter, in the shear flexible VSW element, the accuracy of the shear forces significantly increases with an augmentation of the plate thickness. With the 16-node element, the shear forces are practically exact under the investigated ratios $t/l > 0.01$ due to the

Table 4.16 Shear force error with different FE meshes

Element type	t/l	v_x/q	
		FE mesh 1 (%)	FE mesh 2 (%)
HSS	-	1.50 (50 %)	1.33 (33 %)
VSW	0.01	1.92 (92 %)	1.56 (56 %)
	0.02	1.54 (54 %)	0.83 (17 %)
	0.05	1.14 (14 %)	1.06 (6 %)
	0.10	1.06 (6 %)	0.95 (5 %)
	0.30	1.03 (3 %)	0.97 (3 %)
L16	0.01	1.001 (0.1 %)	0.996 (0.6 %)
	0.02	1.000 (0.0 %)	0.997 (0.3 %)
	0.05	1.000 (0.0 %)	1.001 (0.1 %)
	0.10	1.000 (0.0 %)	1.000 (0.0 %)
	0.30	1.000 (0.0 %)	1.000 (0.0 %)

higher shape functions with which the derivatives required for the shear force can be evaluated more precisely. With smaller ratios t/l, however, the shear force errors also increase with the L16 element. Such thin plates can occur in steel construction. The shear forces obtained with the L16 element, however, are very accurate for the plate thicknesses typical in reinforced concrete construction, even with irregular element geometries.   ◀

The significance of the regularity of finite element meshes for elements with low shape functions (3- and 4-node elements) can be seen from the example given above. Especially with plates and shells, irregularities in the mesh can influence the internal forces. However, bending moments and normal stresses are less sensitive to meshing, while shear forces can be corrupted to the point of becoming useless in practice. The shape factors can be used to assess the regularity of a mesh. Elements with higher shape functions are significantly less sensitive to irregular meshing than elements with low shape functions.

For simple 3- or 4- node elements, meshes of quadrilateral elements should always be preferred to meshes of triangular elements. Exceptions are complicated biaxially curved shell structures, which cannot be represented by flat quadrilateral elements. In the case of regular meshes of triangular elements, the diagonals should alternate in order to avoid a directional influence of the diagonal inclination on the results (Fig. 4.78). Sharp needle-shaped elements are not permitted.

The finite element mesh should be refined in areas with high stress gradients unless a larger local discretization error is acceptable there. The accuracy is increased by reducing the size of the elements. This is also referred to as the h-method (h as element size). Figure 4.79 shows possibilities for mesh refinement with rectangular elements in structured meshes. Alternatively, the elements of the coarse and the finely discretized regions can be connected to each other at common boundaries by coupling conditions. Figure 4.80 shows an example. The two meshes are incompatible at

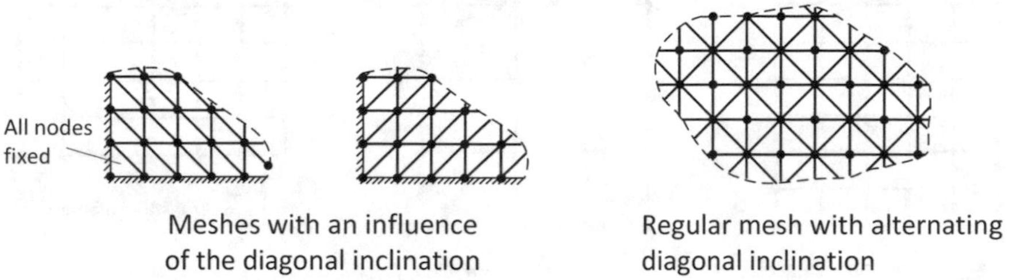

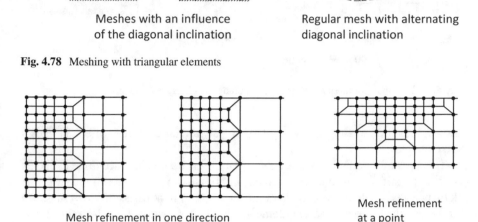

Meshes with an influence
of the diagonal inclination

Regular mesh with alternating
diagonal inclination

Fig. 4.78 Meshing with triangular elements

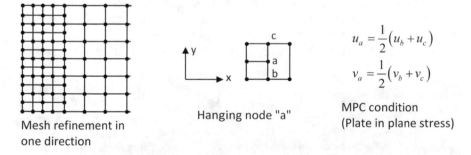

Mesh refinement in one direction

Mesh refinement
at a point

Fig. 4.79 Examples for mesh refinements with quadrilateral elements (*h*-version)

Mesh refinement in
one direction

Hanging node "a"

$$u_a = \frac{1}{2}\left(u_b + u_c\right)$$

$$v_a = \frac{1}{2}\left(v_b + v_c\right)$$

MPC condition
(Plate in plane stress)

Fig. 4.80 Mesh refinement with "hanging nodes" (*h*-method)

common edges. Those nodes of the finely discretized region that do not correspond
to a node in the coarsely discretized region are referred to as *hanging nodes*. Their
degrees of freedom must be transformed with MPC coupling conditions as "slave"
nodes into the degrees of freedom of the corresponding corner nodes ("master"
nodes). The procedure leads to the so-called *mortar methods* for the common surfaces
of two incompatible element meshes, which are used for contact problems.

As an alternative to reducing the size of the elements, their polynomial degree can
be increased (Fig. 4.81). This is called the *p* method (*p* for polynomial degree). For
isoparametric elements, the number of nodes is variable. This property can be used
to formulate elements that possess different shape functions at their edges and thus

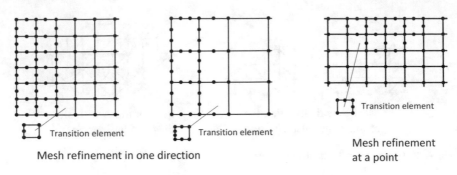

Fig. 4.81 Examples for mesh refinement with rectangular elements (*p*-method)

allow a transition to elements with a higher polynomial degree. Figure 4.81 shows two examples.

The sizes of adjacent elements should not differ too much. Since finite elements behave too stiff due to the displacement approximation, a large difference in the element sizes would correspond to an artificial jump in stiffness of the system. As a rule, the size ratio of adjacent elements should not exceed 1.5. The value 1.5 is to be understood as a recommended value dependent on the element type and the investigated system and its stress distribution.

Stiffness jumps entered by the program user should not be arbitrarily large, as they can lead to numerical difficulties (cf. Sect. 3.8.1, Example 3.24). This applies, in particular, to thickness changes of plates. As a guide, the ratio of 10:1 for the thicknesses of adjacent plate elements should not be exceeded.

In summary, the following rules apply to the generation of finite element meshes:

Rules for finite element meshing

(a) Finite element meshes should be regular, preferably with square or rectangular elements.

(b) For triangular and quadrilateral 3- or 4-node elements the following applies:

   – Quadrilateral elements are to be preferred to triangular elements
   – An element topology of quadrilateral elements is to be preferred to a mixed element topology of triangular and quadrilateral elements.

(c) Elements with higher shape functions (and additional nodal points) possess significantly higher accuracy than simple 3- or 4-node elements.

(d) Element meshes must be refined in areas with high stress gradients if constant accuracy is required.

(e) The element mesh refinement should be done uniformly in order to avoid artificial stiffness jumps.

(f) Stiffness jumps, e.g. as a result of thickness alterations in plates, must not
be arbitrarily large.

The rules also apply accordingly to three-dimensional meshes with solid elements.

Rules for the generation of finite element meshes are also referred to as "a priori
criteria". Although they do not permit a quantitative assessment of the quality of the
computational results, they should be observed in order to avoid possible sources of
error. More accurate error estimation and rules for mesh improvement by means of
adaptive meshing are provided by "a posteriori" estimates based on the computational
results (Sect. 4.12.2).

4.11.4 Mesh Generation

Mesh generation is an essential part of a finite element analysis. Finally, the quality
of the mesh is decisive for the accuracy of the computation. Today, powerful methods
are available for the automatic generation of two-dimensional finite element meshes.
There are procedures for a structured and an unstructured meshing. Mainly the
following methods are applied:

Methods for finite element mesh generation

- Structured meshing

  - Mesh forms a rectangular grid obtained by interpolation in macro-
    elements

- Unstructured meshing in regions with polygonal boundaries

  - Triangulisation
  - Advancing-front method
  - Successive domain subdivision.

Furthermore, there are methods which generate pure triangle meshes, mixed
meshes with triangle and quadrilateral elements or pure quadrilateral meshes. For
numerical reasons meshes with quadrilateral elements only, are to be preferred.

For a *structured meshing*, the mesh is composed of quadrilateral elements arranged
in a regular grid [120]. The domain to be analyzed is subdivided into triangular or
quadrilateral macro-elements (also called "cells"). The finite elements are generated by interpolation within the macro-elements. To this, a uniform rectangular
grid is described in local r-s-coordinates, which are defined similar to an isoparametric four-sided element, and transformed into x-y-coordinates using (4.44a, 4.44b)
(Fig. 4.82), [121]. Each inner node is connected with the same number of elements.

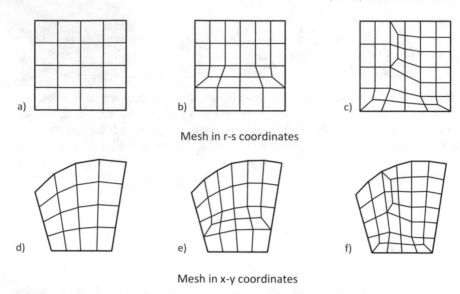

Mesh in r-s coordinates

Mesh in x-y coordinates

Fig. 4.82   Mesh generation in a macro-element

Mesh widening in one or two directions and continuously increasing element sizes is possible [122]. The method delivers numerically beneficial quadrilateral finite elements for surfaces that can be composed of quadrilaterals. However, for complex structures, the required subdivision into macro-elements is cumbersome. For structures, for which the method is suitable, however, structured meshing should be preferred. But, because of its limitations, it is not sufficient as the only generation method in a finite element program.

In an *unstructured meshing*, there are no restrictive requirements on the geometry of the generation domain ("free meshing"). Normally, however, more distorted elements are created which are not optimal from a numerical point of view. Even a rectangular area is divided into an irregular element mesh and not into rectangular elements. The main advantage of unstructured meshes is that they can be flexibly adapted to any geometry [123–125].

Triangulation is another way of meshing a domain with a polygonal boundary. The domain is subdivided into triangles. For this purpose, points are generated within the domain in a given distribution density, which are then connected to triangles [126]. Often the Delauney criterion is used, which maximizes the angles of the triangles and thus avoids triangles with very small, sharp angles. By joining two triangles together, mixed element meshes with triangle and quadrilateral elements can be generated. With advanced strategies, it is also possible to generate meshes that contain quadrilateral elements only.

With the *Advancing-Front method*, the mesh is generated step-by-step from the boundary of the domain to the interior. Nodes and elements are constructed simultaneously, whereby the quality of the element geometry is observed [127, 128].

Triangles can be combined to form quadrilaterals, resulting in pure quadrilateral meshes. The meshes can be further improved by smoothing methods.

Another way to generate finite element meshes is the *successive domain subdivision*. The domain with a polygonal boundary is first divided into convex subareas. These are subdivided successively by new dividing lines, which are selected according to a specific strategy, until only quadrilateral or triangular sub-domains of a given size are left, which then represent the finite elements [129]. This strategy can also be formulated in such a way that only quadrilateral elements are generated [130].

In addition to the generation methods mentioned above, there are a number of other methods that are less important in practice.

For mesh generation, complex systems are split into several regions that are connected at the edges. The location of supports and loads must be taken into account. The regions can possess different material properties and geometrical parameters (such as plate thickness) but also their own parameters for mesh generation (such as mean element size).

In adaptive mesh adjustments (Sect. 4.12.2), meshes are automatically refined at locations where the discretization error exceeds a specified limit. This can be done by successively subdividing the domain during mesh generation, as well as by introducing a density function that controls the local element size in the triangularization procedures and the advancing-front method [127, 128, 131]. Special attention is required when multiple load cases are to be investigated [132, 133].

Finite element meshes for shell structures can be generated using the same methods as for plane surface structures if the plane mesh is projected onto the curved surface [128, 134]. The basis for this is a geometry description of the structure that is independent of the finite element model. For this purpose, freeform surfaces resulting from a digital building model can be used for finite element meshing [135].

For the meshing of three-dimensional solid structures with solid elements, the same methods are available as for the meshing of two-dimensional structures. Especially for unstructured meshes, these methods are mathematically complex. Reference is made to [125, 136, 137].

Example 4.21 For the floor slab shown in Fig. 4.83a, with a circular opening, supporting walls and two columns, a structured mesh based on macro-elements and an unstructured mesh based on the advancing-front method are to be created.

Figure 4.83b shows the subdivision into quadrilateral macro-elements defined by the program user. The automatic mesh generation in the macro-elements can be performed with a mixed element topology (Fig. 4.83c), or with quadrilateral elements only (Fig. 4.83d) [122]. Two other meshes are shown in Fig. 4.83e, f. They were generated with the program FE MESH [138], according to the Advancing-Front Method on the basis of [127], with quadrilateral elements. The mesh T1 (Fig. 4.83e), has a uniform mesh density, while the mesh T2 (Fig. 4.83f), is locally refined at the re-entrant corner points. According to Tables 4.22 and 4.23, a singularity of the internal forces occurs at these points in the case of rigid support. A local increase of the internal forces can be expected for an elastic support as well.

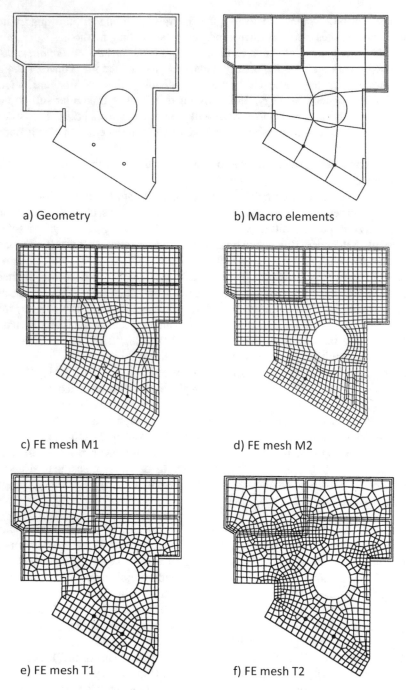

a) Geometry

b) Macro elements

c) FE mesh M1

d) FE mesh M2

e) FE mesh T1

f) FE mesh T2

Fig. 4.83 Generation of different finite element meshes of a floor slab (structural model with walls, columns and opening)

The element size for all meshes was selected in such a way that a sufficient accuracy of the bending moments in the plate with 3- and 4-node elements can be expected. The meshes in Fig. 4.83d–f, which contain only quadrilateral elements are to be preferred for numerical reasons. The meshes M1 and M2 generated by the quadrilateral macro-elements are more regular than the meshes T1 and T2 generated by the Advancing-Front method. On the other hand, the advancing-front method is characterized by a lower input effort.   ◀

4.11.5 Modeling of Plates in Plane Stress

A typical application of the finite element method is the analysis of stresses in concrete walls and deep beams. In steel and timber construction, the finite element method is mostly used in the context of general studies with a more specific character, e.g. for the investigation of zones with load applications with plane stress models or of corners of frames with shell models. In the following, rules to be observed when creating the structural model and the finite element model are given. They mainly refer to deep beams and reinforced concrete shear walls.

Panels of plates in plane stress

The term "plate panel" is used to define quadrilateral regions of plates without large openings, which are statically relevant. Plate panels should be discretized as uniform as possible in accordance with the rules specified in Sect. 4.11.3. The number of elements depends on the element type and loading. As a rule, at least eight to twelve 4-node elements should be provided between two supports (cf. Example 4.7). Examples of the accuracy, obtained for a regular plate panel with different finite element programs are given in [139].

Singularities of stresses can occur at corner points of the structural model. The critical angles above which the stresses according to [140], become singular are given in Table 4.17, for various types of edge support. According to this, singularities are to be expected at re-entrant corners ($\alpha > 180°$) for all types of supports. If one edge at the corner is fixed and the other edge is free, singularities already occur at angles $\alpha > 63°$. If the critical angles are only slightly exceeded, the singularities are weak, i.e. in a finite element analysis the stresses at the corner point increase only slowly with mesh refinement. At larger angles, such as at a corner with $\alpha = 270°$, the stresses increase rapidly with mesh refinement. In [141], critical angles are given for more cases with different restraint conditions of displacements perpendicular and parallel to the edge.

In reinforced concrete walls, in the vicinity of stress peaks, a stress redistribution may be assumed due to the ductility of the material. The reinforcement has to be designed based on the resulting tensile forces which are obtained by integrating the tensile stresses in the zone of the stress peak. The procedure is explained in Example 4.22. It can be noted that the tensile force changes much less than the peak value of the stress and, unlike the corner stress, converges to a constant value. As an integral

Table 4.17 Critical angles for singularities of stresses at corner points of plates in plane stress [140, 141]

Support condition	Stresses
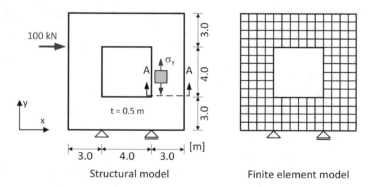	$\alpha > 180°$
	$\alpha > 180°$
	$\alpha > 63°(\mu = 0.3)$

Support: ---- free edge   ⁄⁄⁄⁄⁄ fixed edge

of an erroneous function (the finite element stresses), it has a higher accuracy than a single finite element stress value. In practice, it is, therefore, often sufficient to determine the tensile force and to provide reinforcement for it. For this purpose, a regular mesh according to the above-mentioned criteria without mesh refinement at the location of the singularity is often adequate.

Example 4.22 The reinforced concrete wall shown in Fig. 4.84, possesses stress singularities in the corner points of the opening. The magnitude and position of the tensile force used to determine the vertical reinforcement in section A–A at the corner has to be given for different finite element discretizations.

The computations are performed for comparison with the following 4-node rectangular elements:

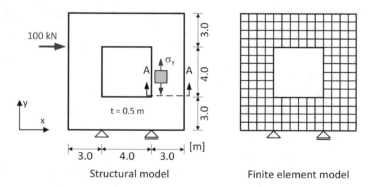

Structural model   Finite element model

Fig. 4.84 Reinforced concrete wall with an opening and finite element model

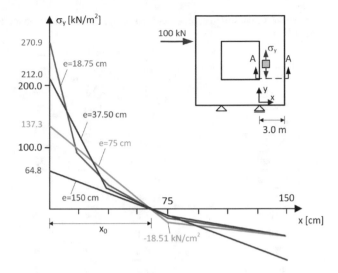

Fig. 4.85 Stresses σ_y in section A–A

(a) Hybrid elements according to [51, SOF 8a] with shape functions according to (4.68), averaged nodal stresses
(b) Isoparametric elements with displacement shape functions according to Sect. 4.4, averaged nodal stresses
(c) Isoparametric elements with displacement shape functions according to Sect. 4.4, element stresses at nodal points.

Figure 4.85 shows the nodal stresses determined with hybrid elements in model (a) with [SOF 8a] for the element sizes 150.00, 75.00, 37.50, and 18.75 cm.

As expected, the corner stresses increase with decreasing element size (cf. Table 4.17). For an element size of $e = 75$ cm, the tensile force Z to be taken into account in reinforcement design in the corner is obtained for a wall thickness of 0.50 m in the following way:

Edge distance of the zero-point of the stresses:

$$x_0 = 0.75 \cdot 137.3/(137.3 + 18.51) = 0.66 \, \text{m}$$

Resulting tensile force:

$$Z = 137.3 \cdot 0.66 \cdot 0.50 \cdot 0.5 = 22.6 \, \text{kN}$$

The corner stresses σ_y and the resulting tensile forces of the other finite element meshes and finite element models are shown in Table 4.18. The tensile stresses increase rapidly with increasing mesh refinement and indicate the presence of a stress singularity. The resulting stresses, i.e. the tensile forces, however, only increase slowly and apparently approach a final value. The resulting tensile force for the

Table 4.18 Peak values of stresses in the corner and at $x = e$, distances of the point of zero stress and resulting tensile forces

Model	FE size e [m]	Number of elements	Stress σ_y [kN/m^2]		Distance x_0 [m]	Resultant Z [kN]
			$x = 0$	$x = e$		
Hybrid elements, nodal stresses	1.5000	2	64.8	−53.9	0.82	13.3
	0.7500	4	137.3	−18.5	0.66	22.6
	0.3750	8	212.0	31.5	0.64	24.9
	0.1875	16	270.9	94.6	0.64	25.6
Isoparametric elements, nodal stresses	1.5000	2	27.4	−51.2	0.52	3.6
	0.7500	4	101.6	−22.2	0.62	15.6
	0.3750	8	180.6	27.2	0.62	21.2
	0.1875	16	277.7	87.4	0.62	25.0
Isoparametric elements, element stresses	1.5000	2	78.1	−43.6	0.96	18.8
	0.7500	4	156.4	−13.3	0.69	27.0
	0.3750	8	252.5	34.5	0.69	29.6
	0.1875	16	379.5	93.1	0.67	31.5

element size of 18.75 cm differs in model (a) only by $(25.6 - 24.9)/25.6 = 14\%$ from the tensile force determined above for the element size of 75 cm, while the stress peaks differ by a factor of 2. Similar results are obtained with the nodal stresses of the isoparametric elements in model (b) according to [SOF 6a]. Only with the very coarse element meshes with $e = 1.50$ m and $e = 0.75$ m the tensile forces differ clearly from the values given above, whereby the hybrid elements converge somewhat faster than the isoparametric elements.

In most finite element programs, the stress output consists of the averaged nodal stresses only. However, integration based on the element stresses makes more sense, since this avoids the error caused by averaging the stress jump in the corner (see Sect. 4.11.10). Table 4.18 shows the element stresses computed with [SOF 6a] at the nodes of the elements. They are significantly higher than the averaged nodal stresses in the corner. In a distance of one element length at $x = e$, however, the differences become smaller. To the left of the re-entrant corner, the stress should be $\sigma_y = 0$ because of the free edge. The low stress values of the left adjacent element are now included in the mean value of the stresses in the corner point and lead to a systematic error. Therefore, both calculations converge to a different resulting tensile force in the corner. While the force converges to $Z = 25$ kN when calculated on the basis of averaged nodal stresses in models (a) and (b), the value $Z = 32$ kN is obtained by using element stresses with model (c). Stress resultants should, therefore, better be calculated on the basis of the element stresses.

In the load case considered here, it is adequate to discretize the section A–A according to the rules for plane panels by eight finite elements and to design the reinforcement for the resulting tensile force determined on the basis of the element stresses. The reinforcement is arranged at a distance x_0 from the corner according to the center of gravity of the tensile stress distribution. Alternatively, the reinforcement determined from the nodal stresses of the elements can be provided. This leads to the same amount of reinforcement. When analyzing a finer finite element mesh, the amount of reinforcement remains practically constant. Only the distribution of the reinforcement may be influenced slightly by the mesh refinement.   ◄

Beam-like structural parts of plates in plane stress

Long, narrow regions of plates in plane stress are subjected to bending stresses similar to those of a beam in bending. Their representation with conforming 4-node quadrilateral elements requires a very fine subdivision in the direction of the "longitudinal axis" (of the "beam") in order to be able to represent stress changes in this direction. For CST triangle elements, a very fine discretization with approximately 10 element layers over the "height" is additionally required. With a less fine meshing, there is a risk that the stresses will be roughly underestimated. In the case of hybrid elements or conforming elements with quadratic displacement shape functions, considerably fewer elements are sufficient to determine the stresses with the same accuracy (cf. Example 4.11).

When modeling plate structures, it often is useful to model parts of them as beams with beam elements (Fig. 4.86). This not only reduces the computational effort, but also results in the internal forces required for beam design, without the need for integration of stresses in plate elements. However, in addition to the two degrees of freedom of displacements, beam elements possess a degree of freedom of rotation in addition. As this is not the case with plane stress elements, special models are needed to connect the degree of freedom of rotation with the plane stress elements. This also applies to plane stress elements with rotational degrees of freedom, since these do not contain any real stiffness from the plane stress formulation [106].

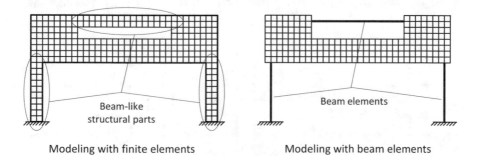

Beam-like
structural parts   Beam elements

Modeling with finite elements   Modeling with beam elements

Fig. 4.86 Plate in plane stress with beam-like structural parts

In addition to the models already discussed in Sect. 4.10, heuristic engineering models will also be considered here. In particular, there are the following models:

> **Models of connecting beam and plane stress elements**
>
> - Rigid surface model (RDT)
>
>   - Kinematic coupling
>   - Method of Lagrange multipliers
>   - Penalty method
>
> - Equivalent stress transformation (EST)
> - Transition elements
> - Heuristic engineering models

Rigid surface model (RDT model)

In the rigid surface model, it is assumed that the connecting surface with the beam element remains plane (Fig. 4.87a, see Sect. 4.10.3). This allows to formulate rigid surface conditions for the displacements of the nodes of the plane stress elements as a function of the displacements and rotations of the beam end. For example, for the connection of a beam element with plane stress elements shown in Fig. 4.87a, the following applies for the displacements of nodes 1, 2, and 3 in the x- and y-direction

$$u_1 = u_2 + \varphi_2 \cdot d/2, \quad u_3 = u_2 - \varphi_2 \cdot d/2, \quad v_1 = v_2, \quad v_3 = v_2.$$

The displacements of the "slave nodes" 1 and 3 are prescribed by the displacements and rotation of the "master node" 2 to which the beam element is connected. This method is also called the transformation method. Other methods to formulate rigid surface conditions are the Lagrange parameter method and the penalty method.

The formulation of kinematic rigid body conditions seems to make sense because of the Bernoulli hypothesis saying that a cross section of beams remains plane.

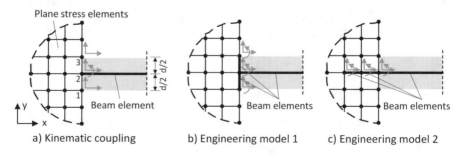

Fig. 4.87 Models for the coupling of beam and plane stress elements

Nevertheless, this model also exhibits some shortcomings. Rigid inclusions in elastic structures often lead to singularities of the stresses, also here. They do not correspond to physical reality either, since the Bernoulli hypothesis only applies to undisturbed beam regions, but not to the end zones of a beam. This is also confirmed by studies of the displacement distribution on plane stress models. Nevertheless, the model is consistent with the assumptions of the theory of beams and frequently used in practice.

Equivalent stress transformation (EST model)

An approach that does not presume a rigid connection surface is the equivalent stress transformation (EST model, cf. Sect. 4.10.4). Here it is assumed that the stresses transferred from the beam to the plane stress elements are linear in the connection surface. No assumption is made for the displacements at the connection of beam and plane stress elements. Compared to kinematic rigid body coupling, this is a soft connection that avoids the stress singularities that occur with rigid inclusions in elastic structures.

The EST model is a model that is consistent with the stress assumptions of the elastic beam theory. However, the linear stress distribution according to Bernoulli does not apply to the end sections of a beam, nor does the linear displacement distribution of the RDT model. A EST connection is "softer" than the rigid surface model. In combination with the finite elements, which on average are to "stiff", it leads to more accurate results than in the rigid surface model.

Engineering models

In many cases, simple engineering models are proposed to allow the beam bending moment to be introduced into the plane stress elements. In engineering model 1, also denoted as *transversal beam model* or *end plate model*, the beam elements are connected to the plane stress elements in the same way as by an end plate, modeled by artificial beam elements (Fig. 4.87b). In engineering model 2, the beam elements are simply extended into the finite element mesh of the plate (Fig. 4.87c). Such models should be handled with care, especially when the stiffness of the connection is important in statically indeterminate systems. The transverse beam model simulates the rigid surface model and is equivalent to it, provided that the stiffness of the artificially introduced beams is sufficiently high. On the other hand, it must not be too large to avoid numerical difficulties. The engineering model 2, in which the beam elements are extended one or two element rows into the finite element model of the plate, has no mechanical legitimation (Fig. 4.87c). The internal forces in the plate are disturbed in the connection region and the rigidity of the connection can no longer be interpreted mechanically. In statically indeterminate systems, the end moments in the beam, therefore, depend randomly on parameters such as the beam cross section and the embedding length. This model is, therefore, not recommended.

The EST model, the rigid surface model, and among the engineering models, the transverse beam model with sufficiently stiff artificial beam elements are the only consistent models.

Example 4.23 The wall shown in Fig. 4.88, with a large opening is analyzed in [142] with different models. The wall thickness is 0.5 m, the modulus of elasticity 30000 MN/m^2 and the Poisson ratio is assumed to be zero. In model 1, the wall is completely modeled with conforming isoparametric 4-node elements according to Sect. 4.4. In model 2, the structural parts above and below the openings subjected to bending are modeled with shear flexible beam elements according to (3.28), which are connected to the plane stress elements with an EST connection (see Sect. 4.10.4). The element size is varied from $e = 0.100$ m to $e = 0.125$ m.

The moment and the normal force in section A–A are examined. These are determined from element stresses, as well as from nodal stresses.

First, the normal force and the bending moment in the cut of the beam shall be calculated by integrating the element stresses. For this, the element stresses σ_x are multiplied by the plate thickness and replaced by equivalent forces according to Fig. 4.17. In the section A–A one obtains with the element stresses indicated in Fig. 4.89, for an element size $e = 0.5$ m:

$$F_a = (-16.32 \cdot 0.5/3 - 4.02 \cdot 0.5/6) \cdot 0.5 = -1.527 \, \text{MN}$$
$$F_b = (-4.02 \cdot 0.5/3 - 16.32 \cdot 0.5/6) \cdot 0.5$$
$$+ (-4.02 \cdot 0.5/3 + 13.8 \cdot 0.5/6) \cdot 0.5 = -0.775 \, \text{MN}$$

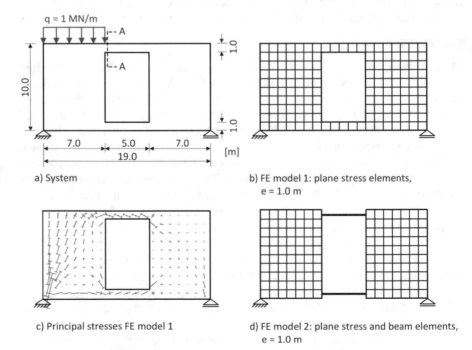

a) System

b) FE model 1: plane stress elements, e = 1.0 m

c) Principal stresses FE model 1

d) FE model 2: plane stress and beam elements, e = 1.0 m

Fig. 4.88 Wall with a large opening

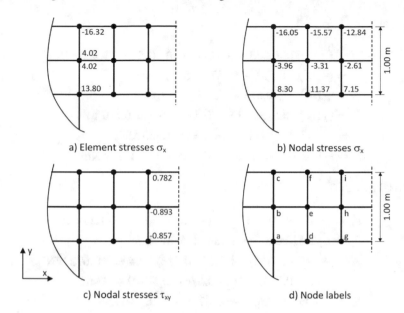

Fig. 4.89 FE stresses at section A–A [MN/m²], model 1 with $e = 0.5$ m

$$F_c = (13.8 \cdot 0.5/3 - 4.02 \cdot 0.5/6) \cdot 0.5 = 0.982 \, \text{MN}$$

$$N = F_a + F_b + F_c = -1.527 - 0.775 + 0.982 = -1.32 \, \text{MN}$$

$$M = F_a \cdot 0.5 - F_b \cdot 0.5 = -1.527 \cdot (-0.5) + 0.982 \cdot (0.5) = 1.254 \, \text{MNm}$$

Since according to Fig. 4.16, for $\mu = 0$ the element stresses σ_x are constant in the x-direction, the internal forces apply to the entire element and are most accurate in the middle of the element (and not in the section A–A).

In most finite element programs, the stress output consists of the averaged nodal stresses. In section A–A, however, the nodal stresses in the corner point are problematic due to the stress singularity occurring there. Not only the element stresses of the elements in the structural part subject to bending are included in the averaged stress value, but also those of the adjacent plate region (cf. Fig. 4.89a, b). Therefore, in the following, the internal forces in section A–A are extrapolated from the internal forces in the first and second nodal rows right of section A–A. The internal forces in the two cuts are obtained from the nodal stresses. The procedure is explained by means of the nodal stresses σ_x given in Fig. 4.89b, for an element size of $e = 0.5$ m. Similar to the element stresses one gets:

Nodal row 1:

$$F_d = (-15.57 \cdot 0.5/3 - 3.31 \cdot 0.5/6) \cdot 0.5 = -1.435\,\text{MN}$$
$$F_e = (-3.31 \cdot 0.5/3 - 15.57 \cdot 0.5/6) \cdot 0.5$$
$$\qquad + (-3.31 \cdot 0.5/3 + 11.27 \cdot 0.5/6) \cdot 0.5 = -0.731\,\text{MN}$$
$$F_f = (11.27 \cdot 0.5/3 - 3.31 \cdot 0.5/6) \cdot 0.5 = 0.801\,\text{MN}$$
$$N_1 = -1.435 - 0.731 + 0.801 = -1.365\,\text{MN},$$
$$M_1 = -1.435 \cdot (-0.5) + 0.801 \cdot (0.5) = 1.118\,\text{MNm}$$

Nodal row 2:

$$F_g = (-12.84 \cdot 0.5/3 - 2.61 \cdot 0.5/6) \cdot 0.5 = -1.179\,\text{MN}$$
$$F_h = (-2.61 \cdot 0.5/3 - 12.84 \cdot 0.5/6) \cdot 0.5$$
$$\qquad + (-2.61 \cdot 0.5/3 + 7.15 \cdot 0.5/6) \cdot 0.5 = -0.672\,\text{MN}$$
$$F_i = (7.15 \cdot 0.5/3 - 2.61 \cdot 0.5/6) \cdot 0.5 = 0.487\,\text{MN}$$
$$N_2 = -1.179 - 0.672 + 0.487 = -1.364\,\text{MN},$$
$$M_2 = -1.179 \cdot (-0.5) + 0.487 \cdot (0.5) = 0.833\,\text{MNm}$$

Section A–A (extrapolated):

$$N = -1.365 + (-1.365 - (-1.364)) = -1.366\,\text{MN},$$

$$M = 1.118 + (1.118 - 0.833) = 1.403\,\text{MNm}.$$

The internal forces in section A–A are listed for different element sizes in Table 4.19. It can be seen that the bending moments determined from the extrapo-

Table 4.19 Sectional forces in section A–A [142]

Model	Beam model	Sectional force	Element size e [m]			
			1.0	0.5	0.25	0.125
1	Plane stress elements σ-element	Normal force [MN]	−1.35	−1.32	−1.36	−1.35
		Bending moment [MNm]	0.83	1.25	1.45	1.54
	Plane stress elements σ−node	Normal force [MN]	−1.36	−1.37	−1.37	−1.37
		Bending moment [MNm]	1.05	1.40	1.53	1.57
		Shear force at $a = 1$ m [MN]	0.00	−0.43	−0.59	−0.63
2	Beam element & EST	Normal force [MN]	−1.36	−1.36	−1.36	−1.36
		Bending moment [MNm]	1.59	1.59	1.59	1.59
		Shear force [MN]	−0.64	−0.64	−0.64	−0.64

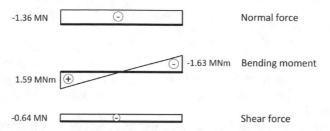

Fig. 4.90 Shear forces of the upper beam element, model 2

lated nodal stresses converge somewhat faster than those determined from the element stresses.

If one calculates directly with the nodal stresses in section A–A without extrapolating, one obtains with an element size of $e = 0.5$ m:

$$F_a = (-16.05 \cdot 0.5/3 - 3.98 \cdot 0.5/6) \cdot 0.5 = -1.503 \, \text{MN}$$
$$F_b = (-3.98 \cdot 0.5/3 - 16.05 \cdot 0.5/6) \cdot 0.5$$
$$+ (-3.98 \cdot 0.5/3 + 8.3 \cdot 0.5/6) \cdot 0.5 = -0.986 \, \text{MN}$$
$$F_c = (8.3 \cdot 0.5/3 - 3.98 \cdot 0.5/6) \cdot 0.5 = 0.526 \, \text{MN}$$
$$N_1 = -1.503 - 0.986 + 0.526 = -1.963 \, \text{MN},$$
$$M_1 = -1.503 \cdot (-0.5) + 0.526 \cdot (0.5) = 1.014 \, \text{MNm}.$$

The internal forces have a poorer quality than the values extrapolated from the nodal stresses of the first and second row of nodes due to the critical corner stress caused by averaging.

The shear forces in the structural part subjected to bending can be determined by integrating the shear stresses of the plane stress elements. They are determined at a distance of 1.0 m from section A–A in order to keep discrepancies from the beam theory small. For model 1 with $e = 0.5$ m and the shear stresses according to Fig. 4.89c, the shear force is obtained as

$$V = (-0.782 \cdot 0.25 - 0.893 \cdot 0.5 + 0.857 \cdot 0.25) \cdot 0.5 = -0.428 \, \text{MN}.$$

The shear forces for the element sizes of 0.25 m and 0.125 m are given in Table 4.19.

In model 2, in which the part of the structure subjected to bending is represented as a beam, the shear force in the beam is obtained directly. All internal forces of the upper beam are shown in Fig. 4.90. They correspond well with the integrated internal forces of model 1 with an element size of 0.125 m. A rather fine discretization is thus required in order to obtain internal forces comparable to those of the beam element. The example shows that it makes sense to model structural parts subjected to bending with beams elements and to connect them to the plane stress elements in a suitable manner.   ◀

Supports

Horizontal and vertical support of plates in plane stress

In the structural model, support conditions must be representing the physical reality. In the case of concrete walls or deep beams, special attention must also be paid to horizontal displacements, as they have a significant influence on the size of the arching effect.

In the case of multi-span deep beams, a realistic modeling of the elastic supports in the vertical direction is also important. Due to the high stiffness of deep beams, these react particularly sensitively to the flexibility of bearings or to imposed bearing displacements in the case of statically indeterminate support conditions.

In coupled shear walls, the stiffness of the coupling can noticeably influence the distribution of the load on the individual shear walls. In order to avoid uncertainties, lower/upper limit analyses are recommended (e.g. concrete bars in tension with cracked, as well as with uncracked sections). Examples of the influence of support stiffness of walls are given in [104, 106, 139].

Example 4.24   For the deep beam shown in Fig. 4.91, of Example 4.7 the influence of the horizontal fixing of the right support on the stresses shall be investigated.

The analysis is performed with a mesh of 8×8 conforming 4-node elements. Figure 4.91 shows the principal stresses and the stresses σ_x in the section C–C for two support conditions (cf. Example 4.7). In one case, the right-hand bearing is fixed in a vertical direction, whereas in the other case, it is fixed horizontally as well. The stress distributions show that the arching effect increases considerably as a result of the horizontal restraint of the right-hand bearing. In particular, the compressive stress extends much further into the lower plate region than is the case with a horizontally movable bearing. The large difference in the horizontal stresses at the lower edge of the plate shows clearly that a realistic representation of the horizontal displaceability of bearings in plates in plane stress is most important.   ◄

Columns as supports of plates in plane stress (deep beams)

Columns as supports of reinforced concrete deep beams which are modeled as plates in plane stress can be regarded as beam-like structural parts of a plate structure. In addition to the models already mentioned for connecting beam and plate elements, there are other models.

Models for columns as supports of plates in plane stress
- Finite element modeling of the column with plane stress (or solid) elements
- Modeling of the column with beam elements and a RDT/EST-connection with the plane stress elements of the plate
- Engineering model 1 (transversal beam elements)

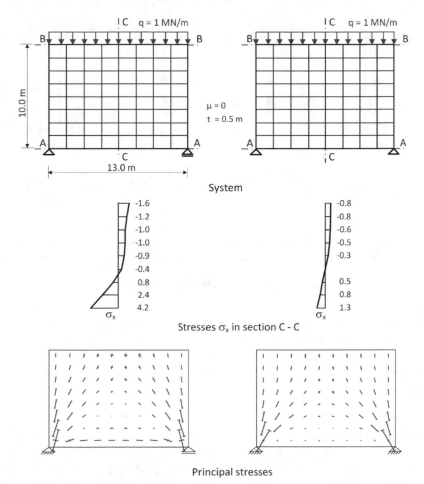

Fig. 4.91 Influence of the support conditions on the stresses in a deep beam

- Engineering model 2 (extended beam elements)
- Elastic springs connected with the rigid surface model (RDT) or with the equivalent stress model (EST)
- Column modeled as continuous elastic support
- Elastic point support
- Rigid point support

Finite element modeling of a column

In the analysis of plates in plane stress, columns can be represented as plate-like structural parts and modeled with finite elements, which is particularly useful for wide columns (Fig. 4.92a). For column modeling, plane stress elements suitable for

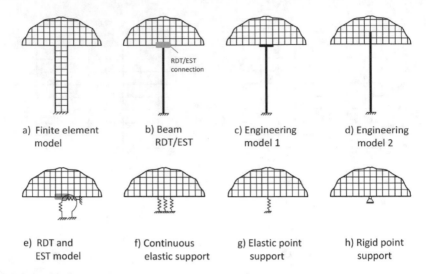

Fig. 4.92 Modeling of columns as supports of plates in plane stress

structural parts subjected to bending must be used (cf. Example 4.11). The internal
forces in the column that are required for the reinforced concrete design are obtained
by integrating the stresses. A stress singularity occurs in the re-entrant corner. For
the interpretation of averaged nodal stresses in the corner, cf. Examples 4.23 and
4.25 (see also [143]).

Modeling of the column with beam elements

If the column is modeled with beam elements, the column sectional forces required
for design are obtained directly (Fig. 4.92b–d). The connection with the plane stress
elements can be done with the rigid surface model (RDT) or the equivalent stress
model (EST), as given in Sect. 4.10. In addition, heuristic engineering models are
frequently used in practice. The transfer of the bending moment of the column into
the finite element model is achieved by artificially introduced beam elements that are
as stiff as possible. The engineering model 1 or transverse beam model (Fig. 4.92b),
corresponds to the rigid surface model (RDT model), provided that the artificially
inserted cross beams are sufficiently rigid. For numerical reasons, however, they
must not be too rigid (cf. Example 3.24). Engineering model 2 (Fig. 4.92c), in which
the beam elements of the column are extended into the plate element mesh, is not
recommended because it is not mechanically motivated. It leads to ambiguous stiffness conditions in the connection region, which results in disturbances of the stress
distributions in the plate and errors in the internal forces of the column.

Modeling of the column as rigid surface or by springs

A column clamped or simply supported at its base may be represented by springs as
given in (4.135, 4.136a–4.136c) (Fig. 4.92e). The connection with the plane stress
elements can be performed as a rigid surface with the RDT model or as a flexible

surface with the soft EST model or with the engineering model 1. The advantages and disadvantages of these models have already been dealt with in Sect. 4.10. With respect to modeling of supports for plates in plane stress, the EST model represents a further enhancement of the continuous elastic support. It eliminates the contradiction in determining the modulus of subgrade reaction by introducing coupling terms (i.e. terms outside the diagonals) in the stiffness matrix of the bearing element. It is, therefore, also called coupling spring model. For the RDT model reference is made to Sect. 4.10.3, for the EST model to Sect. 4.10.4, and for a comparative study to Example 4.25.

Continuous elastic support model

In the continuous elastic support model, the column is represented in the finite element model by an elastic line support or by a series of individual springs (Fig. 4.92f). However, the magnitude of the modulus $k_{St,Bett}$ of subgrade reaction of the column depends on the type of deformation at the top of the column. Reference is made to the derivation in Sect. 4.11.6 and the Eqs. (4.166–4.167c). The line spring related to the length can be obtained from

$$k_{St,Lin} = k_{St,Bett} \cdot t = \alpha \cdot \frac{E_{St}}{l_{St}} \cdot t \qquad (4.162)$$

with E_{St} as modulus of elasticity, I_{St} as moment of inertia and l_{St} as length of the column. Presuming a vertical displacement at the top of the column $\alpha = 1$ is obtained. Presuming a rotation $\alpha = 3$ for a simply supported, and $\alpha = 4$ for a clamped column base is obtained. The vertical force and the bending moment at the top of the column can be determined from the spring forces. The model possesses no horizontal stiffness.

Point support

A very simple modeling of the support of a plate in plane stress is a rigid or elastic point support (Fig. 4.92g, h). It can be easily modeled by fixing the displacements of a node of the finite element model or by defining an elastic spring. It gives a hinge-type support of the plate. At the same time, however, it also has decisive shortcomings. On the one hand, the size of the column is not considered in the model, which leads to inaccuracies with wide columns. Furthermore, the point support represents a point of a singularity of the stresses and displacements. As explained in Sect. 4.11.2, the peak values of the stresses at the support are not physically meaningful and require an engineering interpretation. For reinforced concrete plates, the design of the support zone can be carried out using truss and tie models [144]. Point supports do not give physically meaningful displacements in the whole plate due to the singularity of the support displacements (cf. Example 4.7).

However, it is suitable as a simple model for plate analyses in which only stresses at a sufficiently large distance from the point support are of interest and for slender columns. In general, however, other structural models that do not show any stress or displacement singularities at the supports should be chosen [145].

Recommended models

The bending behavior of the column and its connection to the plate are modeled consistently by the representation of the column by finite elements (Fig. 4.92a) or as beam (Fig. 4.92b), with an RDT or EST connection. As an alternative to the RDT model, engineering model 1 (Fig. 4.92c), can be selected, whereby the stiffness of the artificially inserted transverse beam elements must be carefully chosen. In models using beam elements, the sectional forces at the top of the column are determined as element forces without the need for integration of the stresses. They are consistent with the displacement or stress distribution in the column and give reliable results without any requirements on the column finite element discretization.

Example 4.25 The deep beam shown in Fig. 4.93, is connected with two columns with a width of 1.0 m. With $E = 30000 \, \text{MN/m}^2$, $\mu = 0$ and $t = 0.5$ m it has the same parameters as the deep beam investigated in Example 4.7 which, however, is supported by two point supports.

The following column models are examined:

- Finite element modeling of the column

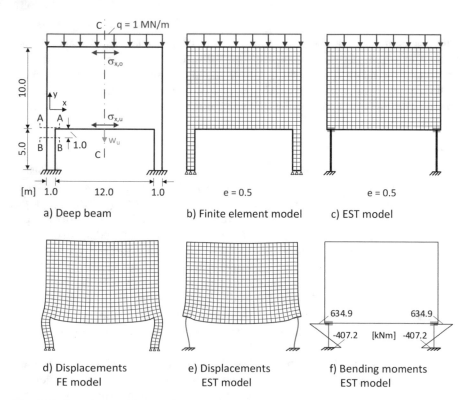

a) Deep beam b) Finite element model c) EST model

d) Displacements e) Displacements f) Bending moments
FE model EST model EST model

Fig. 4.93 Deep beam with two columns

- Column as beam with an RDT connection (rigid surface model)
- Column as beam with EST- and ESTS-connection (coupling spring model)
- Engineering models 1 and 2 ($e = 0.5$ m)

The finite element model with elements of size $e = 0.5$ m is shown in Fig. 4.93b. In addition, models with different element sizes are examined to assess the convergence behavior of the solution. The calculations are performed with the rectangular element with bilinear displacement shape functions according to Sect. 4.4.

Selected computational results are summarized in Table 4.20. The stresses $\sigma_{x,o}$ and $\sigma_{x,u}$ at the upper and lower edges of the plate in section C–C and the deflection w_u at the lower edge possess a satisfying accuracy for all models. Compared to the analysis with point supports in Example 4.7, the stresses σ_x at the lower edge are smaller. This is caused by the consideration of the columns width of 1.0 m in the current analysis, and the resulting smaller clear span width.

The deflections in the mid of the lower edge converge in all models against a value of $w_u \approx 4.0$ mm, while in Example 4.7, they diverge due to the point support. The deformed finite element meshes are shown for the FE model and the EST model of the column in Fig. 4.93d, e. It can be seen that the section A–A in the column section shows a slight upward curvature in both the FE model and the EST model, which is not the case in the RDT model due to the assumption of a rigid surface made there.

The sectional forces at the top of the column are required for the reinforced concrete design of the column. These are the normal force, the bending moment, and the shear force. In the EST and RDT models, they are obtained directly as a

Table 4.20 Computational results for different models of a deep beam [142]

Column model Element size e [m]	Finite elements—section C–C			Column—section A–A		
	$\sigma_{x,u}$ [MN/m^2]	$\sigma_{x,o}$ [N/m^2]	w_u [mm]	N [MN]	V^* [MN]	M [MNm]
Finite element model						
1.000	3.51	−1.80	3.80	−7.000	-	−0.500
0.500	3.60	−1.92	3.99	−7.000	−0.160	−0.600
0.067	3.66	−1.96	4.04	−7.000	−0.203	−0.613
EST model						
1.000	3.61	−1.84	3.94	−7.000	−0.222	−0.682
0.500	3.64	−1.93	4.01	−7.000	−0.208	−0.635
RDT model						
1.000	3.61	−1.84	3.94	−7.000	−0.222	−0.685
0.500	3.62	−1.92	3.99	−7.000	−0.225	−0.689
Engineering model 1						
0.500	3.67	−1.94	4.11	−7.000	−0.188	−0.560
Engineering model 2						
0.500	3.73	−1.99	4.00	−7.000	−0.130	−0.370

*In section B–B

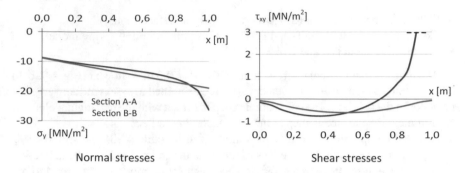

Normal stresses   Shear stresses

Fig. 4.94 Normal and shear stress distributions in the column, $e = 1/15$ m [142]

result of the analysis. If the column is modeled with finite elements, the sectional forces must be calculated by integrating the element stresses. Figure 4.94 shows the distribution of the normal and shear stresses in the column in sections A–A and B–B (nodal stresses). The section A–A depicts the stress singularity in the re-entrant corner, while section B–B shows the linear normal stress distribution and the parabolic shear stress distribution according to the beam theory.

In order to investigate the different behavior of the models more closely, a convergence study for the bending moment in section A–A is carried out. As reference model the FE model with a discretization of the column cross section height in 15 elements, i.e. with an element size of $e = 1/15$ m $= 0.067$ m is used. The moment for this extremely fine model is 0.613 MNm. The moments obtained with the FE model of the column, the EST model, and the RDT model are plotted over the number n of elements on the column height in Fig. 4.95. The FE model naturally converges against the reference value of 0.613 MNm. The RDT model converges against a higher value than the reference value, the EST model against a lower value. This means that the RDT model overestimates the stiffness of the connection while the EST model underestimates it. The ESTS model improves the EST model by introducing additional artificial beams in section A–A with column cross section height h and the width of the column, which takes into account the local increase of the plate

Fig. 4.95 Convergence of the bending moment of the column in section A–A

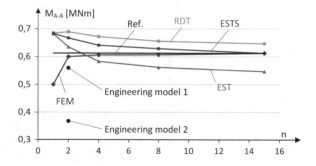

stiffness in the column zone. The EST model gives good results with a discretization in 2-4 elements over the column cross section height suitable for practice. The overestimation of the local stiffness by the shape functions of the isoparametric finite elements is compensated by the slight underestimation of the local stiffness in the EST model. In the RDT model, however, the stiffness of the connection is always overestimated due to the kinematic rigid surface coupling conditions. Therefore, the EST model proves to be best suited for connecting beam elements with plane stress elements in practical element sizes.

Table 4.20 and Fig. 4.95, also show the results of the two engineering models for an element size of $e = 0.5$ m. Engineering model 1 results in a bending moment of $M = 0.560$ MNm which is comparable to the EST or RDT model. The stiffness of the transverse beams was assumed to be the same as the column stiffness. If the stiffness of the transverse beams is increased, the moment approaches that of the RDT model. In engineering model 2, the artificial beam elements were defined with the same stiffness as the column and extended into the plate across 2 rows of elements. However, the moment of $M = 0.370$ MNm, obtained, is considerably too low. It depends on arbitrary parameters such as the number of rows of elements over which the beam elements are extended into the mesh of the plate and the stiffness of the artificial beams. While model 2 proves to be not suitable, engineering model 1 provides reliable column sectional forces if the stiffness of the transverse beams is selected appropriately. ◄

Loads

The types of loads which may act on a plate in plane stress are point loads, line loads, and area loads. These are described in the finite element model by equivalent nodal loads. The singularity points occurring at supports and points where loads are applied are listed in Table 4.21 [110]. According to this, singularities occur mainly

Table 4.21 Singularities at supports and loads of plates in plane stress

Support or point load	Singularity		Support or line load	Singularity	
	Stresses	Displacements		Stresses	Displacements
	Yes	Yes		No	No
	Yes	No		Yes (σ_x)	No
	Yes	Yes		No	No
	Yes	Yes		No	No

at points where point loads are applied and at point supports. Line loads and area loads usually show no singularity points. Singularities can occur at the boundaries of a line load applied parallel to the edge of a plate.

With point loads, singularities always occur unless they are represented more realistically as line loads. The situation here is similar to that with point supports. Either the single load is represented as a point load and the local results are interpreted using engineering methods, or a more realistic representation by a line load is chosen and with a sufficiently fine finite element mesh, reliable stresses in the loading zone are obtained.

In practice, however, the exact local stress distribution at points where point loads are applied is usually of less interest, since simplified engineering models as strut-and-tie models are available for design. Therefore, a detailed modeling of a point load as a line load and an adequate mesh refinement of the loading zone will often not be necessary.

4.11.6   Modeling of Plates in Bending

The most widespread application of the finite element method in structural engineering is the analysis of slabs in reinforced concrete construction. These are mostly floor slabs or foundation slabs of buildings with isotropic behavior. The rules for modeling given in the following mainly refer to this.

In the structural models of plates in bending, a distinction is made between thin plates, which can be described as shear rigid plates according to the Kirchhoff plate theory, and thick, shear flexible plates according to the Reissner–Mindlin theory. According to [146], plates with a plate thickness to plate span ratio of less than 1/10 are regarded as thin plates. Most finite elements are based on the Reissner–Mindlin theory of the shear flexible plate. However, there are also shear rigid finite elements for thin plates according to the Kirchhoff plate theory.

Panels of plates in bending

The term "plate panel" describes a quadrilateral region of a plate without large openings being statically relevant. Plate panels should be discretized as uniform as possible in accordance with the rules specified in Sect. 4.11.3. The number of elements depends on the element type and loading. As a rule, at least 6 to 10 4-node elements should be provided between two supports. With it, sufficiently accurate results can be expected for bending moments and deflections.

The accuracy of shear forces depends considerably on the element type and the shape of the elements (rectangle, parallelogram, general quadrilateral, triangle). The shear forces in plates calculated by the finite element method can be quite inaccurate (cf. Example 4.20). If the shear forces are to be determined reliably within a finite element analysis, considerably finer element meshes are required than for the determination of bending moments and deflections. If there are doubts on the accuracy of the shear forces of the program used, the investigation of simple structural systems and loading conditions, e.g. with constant or linearly increasing shear forces, and convergence studies with different meshes are recommended.

Singularities of the internal forces can occur at the corner points of the structural model. Typical, as in the case of the plate in plane stress, is the case of a re-entrant corner. A closer look shows that the critical angles for singularities are different for bending moments and shear forces. These are given in Table 4.22, for the shear rigid plate [147, 148], and in Tables 4.23 and 4.24, for the shear flexible plate [149], with different types of support.

Table 4.22 Critical angles for singularities of moments and shear forces at corner points of shear rigid plates in bending [147, 148]

Support	Moments	Shear forces	Support	Moments	Shear forces
(diagram)	$\alpha > 180°$	$\alpha > 77.8°$	(diagram)	$\alpha > 95.3°$	$\alpha > 52.1°$
(diagram)	$\alpha > 90°$	$\alpha > 51.1°$	(diagram)	$\alpha > 128.7°$	$\alpha > 90°$
(diagram)	$\alpha > 90°$	$\alpha > 60°$	(diagram)	$\alpha > 180°$	$\alpha > 126°$

Support: ---- free edge   —— simply supported edge   ⁄⁄⁄⁄ fixed edge

Table 4.23 Critical angles for singularities of moments and shear forces at corner points of shear flexible plates in bending with soft support (twisting dof free) [149]

Support	Moments	Shear forces	Support	Moments	Shear forces
(diagram)	$\alpha > 180°$	$\alpha > 180°$	(diagram)	$\alpha > 90°$	$\alpha > 90°$
(diagram)	$\alpha > 180°$	$\alpha > 90°$	(diagram)	$\alpha > 90°$	$\alpha > 180°$
(diagram)	$\alpha > 180°$	$\alpha > 180°$	(diagram)	$\alpha > 90°$	$\alpha > 180°$

Support: ---- free edge   —— simply supported edge (s=soft support)   ⁄⁄⁄⁄ fixed edge, (s=soft support)

Table 4.24 Critical angles for singularities of moments and shear forces at corner points of shear flexible plates in bending—special support conditions [149]

Support	Moments	Shear forces	Support	Moments	Shear forces
	$\alpha > 180°$	$\alpha > 180°$		$\alpha > 61.7°$ ($\mu = 0.29$)	$\alpha > 90°$
	$\alpha > 128.7°$	$\alpha > 90°$		$\alpha > 90°$	$\alpha > 180°$
	$\alpha > 90°$	$\alpha > 180°$		$\alpha > 180°$	$\alpha > 180°$
	$\alpha > 128.7°$	$\alpha > 180°$		$\alpha > 61.7°$ ($\mu = 0.29$)	$\alpha > 180°$
	$\alpha > 90°$	$\alpha > 180°$		$\alpha > 45°$	$\alpha > 90°$

Support: ---- free edge
       —— simply supported edge   s= soft support   h= hard support
       ⁄⁄⁄⁄ fixed edge   s= soft support   h= hard support

In the case of the shear flexible plate, a distinction must be made between the hard and soft support with regard to the twisting moment (cf. Sect. 2.3). With the hard support ("h"), the rotation φ_n perpendicular to the edge is fixed (Fig. 2.10), with the soft support ("s") it is free. The critical angles for more support conditions are given in [149].

Example 4.26 Possible points with singularities of the slab shown in Fig. 4.96, are to be identified. The slab is simply supported on the walls modeled as rigid line supports. The columns are modeled as point supports and the single load as point load.

Singularity points at point supports and loads are given in Table 4.35, and at the corner points of shear rigid (SSP) and shear flexible (SWP) plates in Tables 4.22 and 4.23, respectively. Thereafter, singularities result at the following points:

- Point a: Singularity of the bending moments and shear forces, since $\alpha > 180°$ (SSP and SWP).
- Point b: Singularity of the bending moments at SSP and the shear forces at SSP and SWP as well. For the bending moments of a SWP the point is regular, since $\alpha < 180°$.

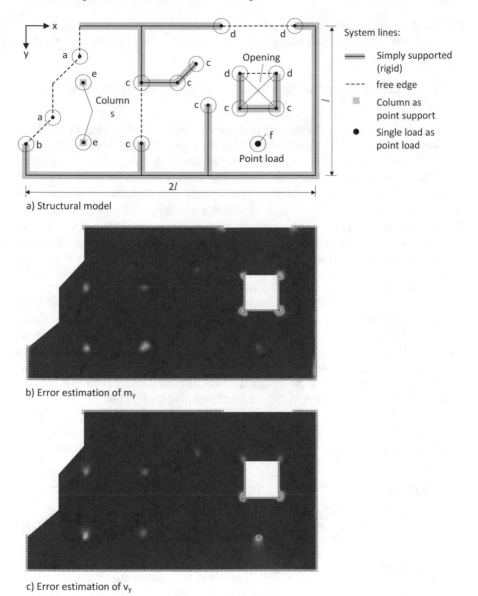

a) Structural model

b) Error estimation of m_y

c) Error estimation of v_y

Fig. 4.96 Points of singularities of a plate of bending

- Points c: Singularity of the bending moments and shear forces, since $\alpha >$ 180°(SSP and SWP).
- Points d: Singularity of the bending moments and shear forces, since $\alpha >$ 90°(SSP) and $\alpha > $ 180°(SWP), respectively.

- Points e: Singularity of the bending moments and shear forces at the column, since it is modeled as point support. For a SWP in addition the displacements are singular.
- Point f: Singularity of the bending moments and the shear forces, since the single load has been modeled as point load. For a SWP in addition the displacements are singular.

At singularity points, there is no convergence of forces, moments or displacements, respectively, with increasing mesh refinement. This becomes apparent by relatively high discretization errors. For comparison, the slab loaded by a uniform load and a single load is calculated as a shear flexible plate with [SOF 6] and the errors according to Zienkiewicz/Zhu are determined with an element size of $e = 1/64$ (cf. Section 4.12.2). Figure 4.96b,c show an error estimation of the bending moments m_y and the shear forces v_y. The high discretization errors at points c, d, e, and f can be noticed. The errors at points a and b are less pronounced. There are obviously weaker singularities, at least in the load case investigated.   ◀

The example shows that a multitude of singularity points can be expected for standard building construction slabs. However, most of them can be avoided by suitable modeling (see below). Since the internal forces singularities are local effects, the internal forces related to the unit of length can be integrated into a corner zone and introduced into the design as resultant forces or moments, as it has already been explained for plates in plane stress. Alternatively, the internal forces determined by the program can be used for the design, and the reinforcement can be redistributed constructively if required due to the existence of a singularity point.

At the boundaries of plates in bending, the so-called *edge effect* appears, which affects the shear forces and twisting moments in the edge zone. The relationships are summarized in Table 4.25 (cf. Fig. 2.10). For simply supported edges with soft

Table 4.25 Edge effect

Support condition at the edge	Equivalent shear force	Effects at the edge
Free	$\bar{v}_n = 0$ $v_n = -\dfrac{dm_{ns}}{ds}$	Both, calculated shear forces v_n and twisting moments m_{ns} occur.
Simply supported	$\bar{v}_n = v_n + \dfrac{dm_{ns}}{ds}$	The equivalent shear forces or support reactions $\bar{v}_n$ differ from the shear forces v_n by $\dfrac{dm_{ns}}{ds}$. Concentrated forces in the plate corners
Clamped	$\bar{v}_n = v_n$	No edge effect ($m_{ns} = 0$)

support conditions, the supporting forces are converted into plate shear forces and a part corresponding to the variation of the twisting moments at the edge. This redistribution takes place in a narrow edge strip whose width corresponds approximately to the plate thickness. The shear forces in the immediate vicinity of the edge, therefore, do not correspond to the support forces which are also denoted as *equivalent shear forces*. With free edges, the resulting "supporting" forces at the edge are zero. At a short distance from the edge, however, shear forces occur which correspond to the variation in the twisting moments at the edge. The edge effect does not occur with clamped edges.

Since the edge effect affects a narrow zone at the plate edge, it can only be detected in finite element analyses with very small 4-node elements. Alternatively, larger finite elements with higher shape functions [150], or three-dimensional solid models can be used. Three-dimensional solid models, however, are of interest for basic studies but hardly suitable for practical purposes. The edge effect is always part of the results obtained by a finite element analysis with the element sizes commonly used in practice. This leads to the fact that when performing an equilibrium check, the shear forces at the edge do not agree with the supporting forces (Example 4.27). At free plate edges, finite element computations show both shear forces and twisting moments due to the edge effect [151].

In the shear design of (directly) supported reinforced concrete slabs, the shear forces are not required directly at the slab edge according to reinforced concrete standards, but at a distance d from the edge corresponding to the slab thickness. Therefore, the shear forces at the edge of simply supported slabs containing the edge effect are used in the shear design of reinforced concrete slabs. At free edges of reinforced concrete slabs, the edge effect can also be explained with strut-and-tie models. In [152], the necessity of stirrup reinforcement at free edges is thus explained.

In finite element analyses of simply supported plate edges, the rotational degrees of freedom perpendicular to the edge (corresponding to the twisting moments) may be defined as freely movable or as fixed. This is referred to as soft or hard support. In general, bearings are defined as soft supports. In a hard support additional twisting moments (moment vector perpendicular to the edge) appear as support reactions. A consistent load transmission requires that these twisting moments are taken into account as loads of the structural components representing the support. This can be done by converting the nodal moments into additional nodal forces (Fig. 4.97a–c). These can be combined with the support forces at the nodes and represented as distributed forces (Fig. 4.97d–h). The resulting equivalent support forces in a hard support are composed of the support forces given as output of the finite element analysis and the additional contribution of the twisting moments at the support. With soft supports, the degree of freedom from rotation perpendicular to the edge is free, and therefore, no fixing moments occur. The redistribution described above takes place in a similar way in the edge strip of the plate, so that similar support forces are obtained.

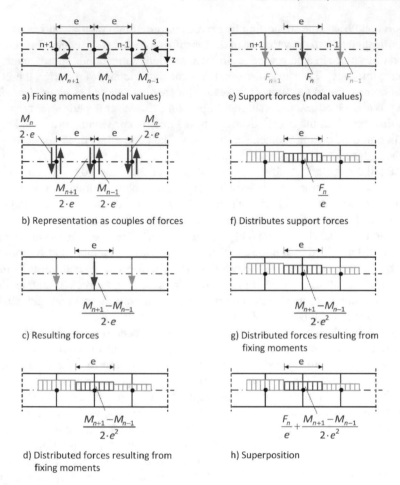

Fig. 4.97 Fixing moments at a hard support of simply supported edge of a plate in bending

Example 4.27 The square plate shown in Fig. 4.98a is simply supported on all edges and loaded with a uniformly distributed load q. It is a thin plate with a slenderness ratio of $t/l = 0.2/10 = 0.02$. The supports at the edges are defined as soft support. The plate is to be investigated with two types of plate elements and different discretizations.

The first plate element is a shear rigid hybrid 4-node element according to [51, SOF 8a]. The second element is a shear flexible 4-node element with displacement shape functions and additional shape functions for shear stresses according to [63, 79, SOF 6]. Two regular finite element discretizations are shown in Fig. 4.98b. Their shape factors according to Table 4.15, are optimal. According to Tables 4.22 and 4.23, singularity points of the bending moments do not exist.

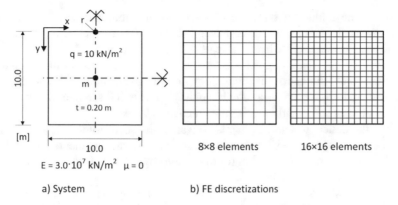

Fig. 4.98 Simply supported square plate with a uniformly distributed load q

Table 4.26 Convergence of internal forces and displacements of a simply supported square plate

Structural model	Solution method	FE mesh	e [m]	w_m [cm]	$m_{x,m}$ [kNm/m]	$v_{yr,m}$ [kN/m]
	Analytical solution	–	–	2.03	36.76	33.78
Shear rigid plate	FEM with 4-node element (hybrid element with quadratic bending moment shape functions and cubic displacement shape functions)	2×2	5.0000	1.43	36.70	33.86
		4×4	2.5000	1.82	34.33	29.33
		8×8	1.2500	1.93	36.39	31.07
		16×16	0.6250	2.02	36.69	32.30
		32×32	0.3125	2.03	36.80	33.01
Shear flexible plate, soft support	FEM with 4-node element (displacement shape functions and additional shape functions for shear)	2×2	5.0000	1.80	35.11	4.28
		4×4	2.5000	2.07	39.41	10.77
		8×8	1.2500	2.09	38.84	24.39
		16×16	0.6250	2.08	37.84	31.24
		32×32	0.3125	2.08	37.50	34.16
Shear flexible plate, hard support	FEM with 4-node element (displacement shape functions and additional shape functions for shear)	2×2	5.0000	1.68	27.83	1.13
		4×4	2.5000	2.07	39.79	10.08
		8×8	1.2500	2.07	38.50	25.02
		16×16	0.6250	2.04	37.31	30.50
		32×32	0.3125	2.04	36.96	32.21

The deflection w_m and the bending moment $m_{x,m}$ in the middle of the plate, as well as the shear force $v_{yr,m}$ at the midpoint of the edge, are investigated. For comparison, the analytical solution according to [153], is given.

The results are presented in Table 4.26. First, they are examined for a soft support condition. The deflections w_m and bending moments $m_{x,m}$ obtained with the hybrid element in the middle of the plate converge toward the analytical solution with increasing mesh refinement. The displacements w_m and moments $m_{x,m}$ of the shear flexible plate differ slightly from those of the shear rigid plate due to the additional shear components. Taking the 32×32 discretization as reference values, the 8×8 discretization determines the bending moments with an accuracy of 1% for the shear rigid plate and 4% for the shear flexible plate. The 8×8 discretization can thus be regarded here as sufficiently precise for practical purposes.

While the deflections and moments converge, the shear forces $v_{yr,m}$ at the plate edge continue to increase slightly with increasing mesh refinement, even with the 16×16 discretization. The reason for this is the edge effect. This will be examined in more detail in the following.

Figure 4.99 shows the shear forces v_y and twisting moments m_{xy} obtained at the nodes in the middle of the upper plate edge ($x = 5.00$ m) for the very fine 32×32 discretization with an element size of 31.25 cm. The right half of the symmetry is depicted. Due to the symmetry of the system, the shear forces are symmetrical to the symmetry axis and the twisting moments are antimetric.

First, the shear rigid plate is examined. According to (2.12), the support force in the midpoint of the edge is composed of the element shear force $v = -33.0$ kN/cm and a part due to the variation in the twisting moments. With this part, which is obtained from the element size of $10/32 = 0.3125$ m and the twisting moment values of 0 at $x = 0$ and -3.69 kNm/m at the adjacent node, the equivalent shear force, which corresponds to the distributed support force, results to

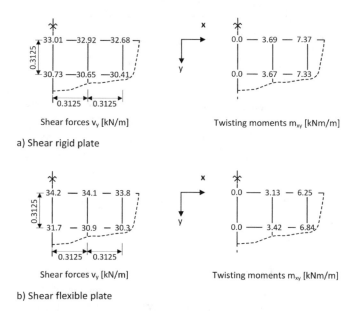

Fig. 4.99 Shear forces and twisting moments at the edge of the plate for soft support conditions (32×32 discretization)

$$\bar{v} = v_n^* = 33.0 + \frac{3.69 - 0.0}{0.3125} = 44.8 \, \text{kN/m}.$$

It corresponds well with the distributed support force determined from the nodal force 14.26 kN at the support point as

$$a_z = \frac{14.26}{0.3125} = 45.6 \text{kN/m}$$

as well as with the analytical solution of

$$\bar{v} = \frac{10.10}{2.19} = 45.7 \, \text{kN/m}$$

according to [153]. For the shear flexible plate, the equivalent shear force is obtained as

$$\bar{v} = v_n^* = 34.2 + \frac{3.13 - 0.0}{0.3125} = 44.2 \, \text{kN/m}$$

and with the nodal force of 14.29 kN the distributed support force a_z as

$$a_z = \frac{14.29}{0.3125} = 45.73 \, \text{kN/m}.$$

In both the shear rigid and the shear flexible plate, the shear forces at the edge reflect the edge effect and correspond approximately to the shear forces given in table books for shear rigid plates such as [153]. The equivalent shear forces of the finite element calculation correspond well with the distributed support forces.

Figure 4.100 shows the shear forces and twisting moments of the shear flexible plate for a 10 × 10 discretization corresponding to an element size of 1.00 m, which rather corresponds to element sizes used in practice. With this one obtains

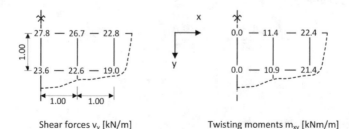

Shear forces v_y [kN/m]   Twisting moments m_{xy} [kNm/m]

Fig. 4.100 Shear forces and twisting moments at the edge of the plate for soft support conditions (10 × 10 discretization)

$$\bar{v} = v_n^* = 27.8 + \frac{11.4 - 0.0}{1.00} = 39.2 \text{ kN/m}$$

and with the nodal force of 45.63 kN

$$a_z = \frac{45.63}{1.00} = 45.63 \text{ kN/m.}$$

While the distributed support force a_z is calculated with good accuracy, both the shear force at the edge and the equivalent shear force differ significantly from the analytical solution. This demonstrates that the element size in the edge zone of the plate must not be too large if reliable shear forces are to be determined—taking into account the edge strip effect (cf. Example 4.29).   ◄

Example 4.28 The plate investigated in Example 4.27, with a soft support is to be analyzed with a hard support. All other parameters are retained (Fig. 4.98a).

The displacements w_m and moments $m_{x,m}$ in the middle of the plate, as well as the shear forces v_y at the edge of the plate, are given in Table 4.26. They differ only slightly from the values for a soft support.

Figure 4.101 shows the shear forces and twisting moments at the plate edge with hard supports for the 32×32 and 10×10 discretizations. In addition, the support reactions, i.e. the support forces F_z and fixing moments M_y at the nodal points are given. For load transfer to the line support, the fixing moments M_y should be converted into distributed support forces. This is always necessary if the fixing moments M_y cannot be transmitted directly by the line support, such as in the case of a wall modeled with plane stress elements without rotational degrees of freedom. According to Fig. 4.97, the distributed support forces (positive against the z-direction) are obtained for the 32×32 and the 10×10 discretization as

32×32 elements:

$$a_z = -\left(-\frac{10.6}{0.3125} + \frac{-(+1.15) - 1.15}{2 \cdot 0.3125^2}\right) = 45.7 \text{ kN/m.}$$

10×10 elements:

$$a_z = -\left(-\frac{34.8}{1.00} + \frac{-(+10.6) - 10.6}{2 \cdot 1.00^2}\right) = 45.4 \text{ kN/m.}$$

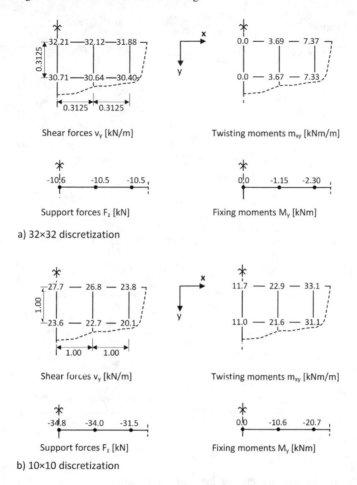

Fig. 4.101 Shear forces and twisting moments at the edge of the plate for hard support conditions

For both discretizations, these correspond to the distributed support forces of the soft support. The "redistribution" of the twisting moment variation into support forces takes place directly in the support, whereas it takes place in the edge strip in the case of a soft support. ◀

Example 4.29 For the simply supported square plate shown in Fig. 4.102, with a uniformly distributed load and soft support conditions on all sides, the edge strip effect has to be verified. The slenderness ratio is $t/l = 1/25$, the Poisson ratio is assumed to $\mu = 0$.

The plate is calculated with several element discretizations $n \times n$, where n varies from 4 to 256. If $n = 25$, the element size e corresponds to the plate thickness. The analysis is done with [SOF 6] and the implemented shear flexible 4 node element. The (unit-free) results are shown in Fig. 4.103. The bending moments and the deflection in the middle of the plate, as well as the support force in the midpoint of the edge,

Fig. 4.102 Simply
supported square plate with
uniformly distributed load

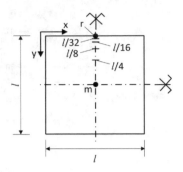

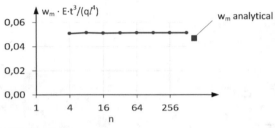

a) Deflection w_m

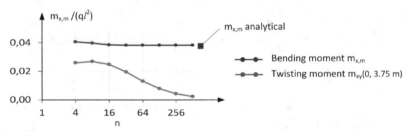

b) Bending moment $m_{x,m}$ and twisting moment m_{xy}

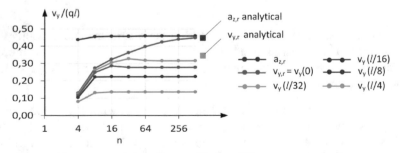

c) Support force $a_{z,r}$ and shear forces $v_y(y)$ at $x = l/2$

Fig. 4.103 Convergence of the forces, moments and displacements

converge rapidly (Fig. 4.103a,b). However, due to the shear flexibility of the plate, the final values differ slightly from the analytical values given for the shear rigid plate according to [153].

Now the shear forces at the edge at $y = 0$ and at the points $y = l/32$, $l/16$, $l/8$, $l/4$ are examined (Fig. 4.103c). It can be seen that the shear forces converge fast at $y = l/32$, $l/16$, $l/8$, $l/4$, whereby the further a point is from the edge, the better is the convergence. In a discretization with 16×16 elements, the final values are reached at practically all points. The shear force directly at the edge (point r) exhibits a different behavior. It converges extremely slowly against the support force, with which it must coincide for equilibrium reasons. Only with an extremely fine mesh of 512×512 elements, an approximate equality is achieved. For a plate with $l = 5.00$ m, this corresponds to an element size of approximately 1 cm. The same applies to the twisting moments at the edge, which should be zero (Fig. 4.103b). They decrease with an increasing number of elements, but only become nearly zero with extremely small element sizes.

The results illustrate the edge effect: At the edge, the support force is equal to the edge shear force. This value is also achieved with extremely small elements in the finite element method. The edge shear force is converted within a very small strip along the edge into a significantly lower shear force and a twisting moment. This shear force is applied to shear design in reinforced concrete plates. With usual element sizes, such as the 8×8 or 16×16 discretizations, this shear force value is obtained approximately at the edge of the finite element model as edge shear force, since the finite elements – due to their size – are not able to represent the redistribution process.

The edge strip effect can also be explained using strut-and-tie models. It is the reason why in reinforced concrete constructions the shear design at free and simply supported edges is based on the shear force at the distance of the slab thickness from the edge [152].

The shear force exhibits a special characteristic in another respect as well. Figure 4.104 shows the distribution of the shear force determined with [SOF 9] and a 16-node Lagrange rectangular plate element according to [37, 60]. Along the edge parallel to the y-axis, the shear force assumes extraordinarily high (positive and negative) values. To investigate this effect in more detail, the plate is modeled with

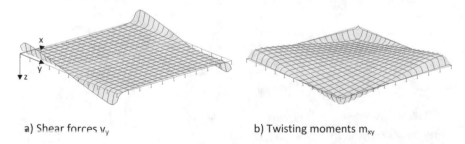

a) Shear forces v_y b) Twisting moments m_{xy}

Fig. 4.104 Shear force v_y and twisting moment m_{xy} of a square plate with $t/l = 1/25$ [119]

64-node solid elements with [SOF 9, 119]. The Lagrange element used has 3rd order parabolas as displacement shape functions and is, therefore, able to exactly represent quadratic stress distributions. The very fine discretization in 41 × 41 elements is shown in Fig. 4.105a. At the edge, the plate is continuously, elastically supported in a 10 cm wide strip to avoid stress singularities ($k_S = 1.2 \cdot 10^7$ kN/m^3). Figure 4.105b shows the shear stresses corresponding to the shear forces. Figure 4.105b shows the shear stresses τ_{yz} corresponding to the shear forces v_y. Like the shear forces,

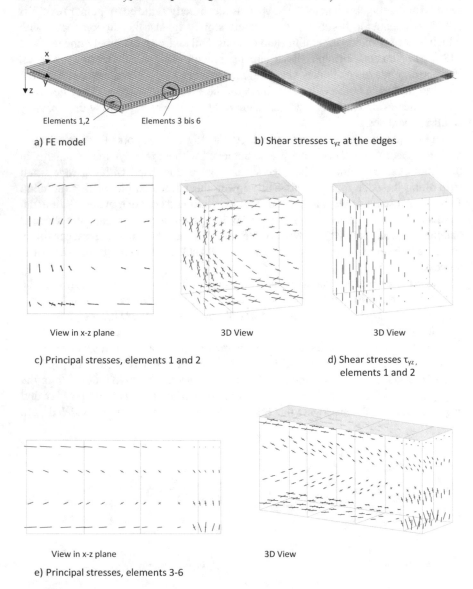

a) FE model

b) Shear stresses τ_{yz} at the edges

View in x-z plane

3D View

3D View

c) Principal stresses, elements 1 and 2

d) Shear stresses τ_{yz}, elements 1 and 2

View in x-z plane

3D View

e) Principal stresses, elements 3-6

Fig. 4.105 Finite element modeling of a plate with three-dimensional solid elements [119]

the stresses have peaks at the edges parallel to the y-direction. For a more detailed examination of the stress distribution, the two finite elements at the edges are examined. Element 1 is located in the 10 cm wide elastically supported strip, element 2 in the slab. Figure 4.105d shows the high shear stresses at the edge. However, if one examines the principal stresses in the two elements shown in Fig. 4.105c, one can see that these do not have any peaks. The stresses are similar to the normal stresses resulting from the twisting moments at the edge. The more detailed modeling thus demonstrates that the high shear force peaks of v_y at the edge parallel to the y-axis are practically uncritical and not relevant for the RC design.

In addition, the distribution of the principal stresses in elements 3-6 of the solid model is examined (Fig. 4.105e). Here the twisting moments are zero (Fig. 4.104b). Similar to the beam at an edge (D-region), the principal stresses are inclined and can no longer be understood as bending and shear stresses in the sense of bending beam theory. For the RC design, therefore, the stresses at a distance approximately equal to the plate thickness are relevant in which a beam-like stress state has developed.

From an engineering point of view, an exact determination of the shear force distribution shown in Fig. 4.104, is not required as it is not relevant in reinforced concrete design. Thus, a discretization at the edge with node-elements with the size of the slab thickness is normally sufficient to compute the shear forces (cf. Example 4.27). According to the Kirchhoff plate theory, shear rigid elements are well suited for thin plates, i.e. the more accurate Reissner–Mindlin theory does not provide any relevant information in additional. In [146], due to its shortcomings, it is disadvised to use shear flexible elements for thin plates. ◄

Example 4.30 The influence of the plate thickness t is investigated for the simply (soft) supported plate with a uniformly distributed load q shown in Fig. 4.102. The calculations are carried out for two Poisson ratios and with a shear flexible 4-node element according to [SOF 6].

The deflections and bending moments in the middle of the plate are presented in Table 4.27. For the shear rigid plate, the analytical solutions according to [153], are given for a Poisson's ratio $\mu = 0$. They are independent of the t/l ratio. The shear soft plate is examined with slenderness ratios t/l between 0.02 and 0.40. For thin

Table 4.27 Influence of plate thickness

Structural model	t/l	$w_m \cdot \dfrac{E \cdot h^3}{q \cdot l^4}$		$m_{x,m} \dfrac{1}{q \cdot l^2}$	
		$\mu = 0$	$\mu = 0.2$	$\mu = 0$	$\mu = 0.2$
Shear rigid plate	-	0.0487	-	0.0368	0.0442
Shear flexible plate, soft support	$0.02 = 1/50$	0.0499	0.0439	0.0375	0.0449
	$0.04 = 1/25$	0.0515	0.0490	0.0382	0.0455
	$0.10 = 1/10$	0.0568	0.0536	0.0404	0.0475
	$0.20 = 1/5$	0.0682	0.0642	0.0439	0.0505
	$0.40 = 1/2.5$	0.0986	0.0954	0.0489	0.0545

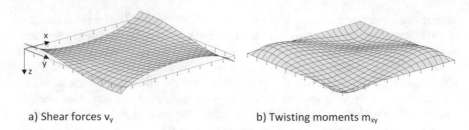

a) Shear forces v_y b) Twisting moments m_{xy}

Fig. 4.106 Shear force v_y and twisting moment m_{xy} of a square plate with $t/l = 1/2.5$ [119]

plates with a slenderness $t/l < 0.1$, the deflections and bending moments of the shear soft plate correspond well with those of the shear rigid plate. For $t/l = 0.1$ the differences in the deflections and moments are 16% and 10%, respectively, (with $0.0568/0.0487 = 1.16$ and $0.0404/0.0368 = 1.10$). With $t/l = 0.4$ the differences between the Kirchhoff and Reissner–Mindlin plate theory are considerably larger. The deflections of the shear flexible plate are about twice as large and the bending moments in the middle of the plate are about 33% larger than with the shear rigid plate.

The influence of the Poisson's ratio μ on the deflections and moments of the plate is also shown in Table 4.27. Assuming that Poisson's ratio does not influence the curvatures, a simplified formula can be derived with the moment to curvature relationship given in Fig. 2.8. Denoting the moments for $\mu = 0$ as $m_{x,0}$, $m_{y,0}$, $m_{xy,0}$ the moments for any value of μ can be written approximately as

$$
\begin{aligned}
m_{x,\mu} &\approx m_{x,0} + \mu \cdot m_{y,0} \\
m_{y,\mu} &\approx m_{y,0} + \mu \cdot m_{x,0} \\
m_{xy,\mu} &\approx (1 - \mu) \cdot m_{xy,0}.
\end{aligned}
\tag{4.163}
$$

In general, the Poisson's ratio for reinforced concrete is assumed to be $\mu = 0.2$. However, there are arguments too for an assuming $\mu = 0$ due to crack formation in concrete [152].

Figure 4.106 shows the shear forces v_y and twisting moments m_{xy} of a thick plate with $t/l = 1/2.5$. The sharp peaks of the shear forces at the edges of the thin plate in Fig. 4.104, do no longer appear. The increase of the twisting moments from the edges is much more flat than in the case of the thin plate. This behavior can only be described properly by the Reissner–Mindlin theory of the shear flexible plate. Thick plates must, therefore, be modeled with shear flexible plate elements.   ◀

Therefore, in addition to the general rules for mesh generation given in Sect. 4.11.3, the following aspects must be considered when modeling plate panels.

Modeling of plate panels

(a) Thin plates with a ratio $t/l < 0.1$ should be analyzed with the theory of shear rigid plates; thick plates must be computed according to the Reissner–Mindlin theory for shear flexible plates.

(b) The edge effect must be observed at free and simply supported edges. It influences the shear forces and twisting moments at the edge obtained by a finite element calculation. If shear forces at the edge of the plate are to be determined reliably, a finite element discretization at the edge with an element size similar to the plate thickness is required. However, smaller element sizes should not be used.

(c) Supports of plates in bending are usually to be defined as soft supports. In the case of rigid supports, the fixing twisting moments must be taken into account in the transmission of the support reactions to the supporting structural elements. This can be achieved by redistributing the variation of the twisting moments obtained as support reaction into additional support forces.

(d) Singularities of the internal forces can occur at corner points. Singularity locations of shear rigid and shear flexible plates in bending are given in Tables 4.22, 4.23 and 4.24.

Continuous elastic support of plates

Plates can be supported on points, lines or areas. The case of bedding on an elastic area, i.e. the continuous elastic support of plates, can easily be handled. The modeling can be done either with special finite plate elements, which consider the elastic bedding in the element formulation or with single springs at the nodes. Single springs have a spring constant

$$k_v = k_S \cdot A_k \tag{4.164}$$

where k_S is the modulus of subgrade reaction and A_k the reference area of the node, e.g. the element size in a regular rectangular mesh.

A continuous elastic support of plates appears mainly as elastic support of foundation slabs. Section 4.11.7 deals with special features of the modeling of the subsoil in the analysis of foundation slabs. The modeling of columns in flat slabs as continuously supported partial surface of the plate is covered below.

Line supports of plates

Walls as support of floor slabs are normally modeled as line supports. They can, however, be modeled also as continuous support of the slab distributed on an area taking the wall thickness b_W into account (c.f. Fig. 4.92f, for plates in plane stress). Often the influence of the wall thickness is neglected and walls are modeled as rigid or elastic line supports.

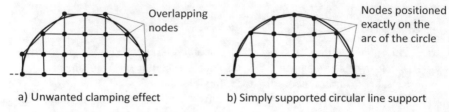

a) Unwanted clamping effect   b) Simply supported circular line support

Fig. 4.107 Simply supported circular plate

Rigid line supports are normally defined by fixing the vertical displacements, while the rotational degrees of freedom are left free. Concerning the twisting degree of freedom at the edges of the plate, this corresponds to a soft support.

In the case of simply supported edges of circular plates, care must be taken to ensure that the nodes of the line support lie exactly on the arc of the circle. If this is not the case, in soft supports an unwanted clamping effect will occur. Figure 4.107 shows, for example, a circular edge which is approximated by rectangular elements partly overlapping the edge, as well as a meshing, where the nodes are positioned exactly on the arc of the circle. The so-called "Babuska's paradox" has been detected for plate edges on a circular arc which is simply supported but in a hard condition concerning the twisting degree of freedom [154]. It states that with an increasing number of elements on the polygonal edge (or with decreasing element size) the solution deviates more and more from the analytically exact solution. Since the plates are usually supported in a soft condition, this phenomenon is of little practical importance. However, it points to another problem of fixing the twisting degree of freedom in hard supports.

Unwanted clamping effects can also occur with walls whose axis has a small offset in some sections. In this case, it should be checked whether the assumption of a simplified straight-line course of the wall without clamping effects does not correspond more closely to the desired load-bearing behavior.

At corner points of rigid line supports, singularities of the internal forces may occur (see Tables 4.22, 4.23). A particularly strong singularity occurs with discontinuous line supports. The handling of singularities has been discussed in Sect. 4.11.2. Either the singularity is accepted in the structural model and the design is based on integral internal forces only, or the structural model is modified in such a way that singularity points are avoided.

In order to avoid singularities, the flexibility of the line support can be taken into account, and an elastic line support can be introduced instead of a rigid one in the structural model. The spring constant of a wall (per m) can easily be determined from its stiffness when loaded by a normal force as

$$k_v = \frac{E \cdot b_W}{h} \left[\text{kN/m}^2\right]. \tag{4.165}$$

Herein h is the height of the wall under the slab, b_W its thickness and E the modulus of elasticity of the wall. In general, walls should be represented as elastic supports, since rigid line supports, e.g. in the case of an intermittent support, can strongly disturb the internal forces and support reactions.

According to [155], short intermittences of a line support, e.g. due to door and window openings, can be neglected in the structural model for ratios $l_{open}/t \leq 7$. Here, l_{open} is the length of the missing support and t the slab thickness. In this case, the line support is modeled as being continuous. Additional reinforcement at the support intermittence is to be provided based on structural detailing.

Example 4.31 Skew plates in bending which are simply supported on the four edges are used as benchmark tests for parallelogram-shaped finite elements. The plate in Fig. 4.108, is loaded by a uniformly distributed load. Support conditions are soft concerning the twisting degree of freedom. The plate is investigated with two element types: a rigid shear hybrid 4-node plate element with the quadratic moment and cubic displacement shape functions according to [51] and [SOF 8a], and with a shear flexible 4-node plate element with displacement shape functions according to [63, 79, SOF 6]. The results are compared with the analytical solution of the shear rigid plate in [156].

A mesh with 8×8 elements and the corresponding principal moments are shown in Fig. 4.108. The load is transmitted mainly in the direction of the "short span" e-f.

The results of the analysis are summarized in Table 4.28. In the case of the rigid supports, the principal moments in the mid of the slab converge rapidly. They correspond well with the analytical solution. An 8×8 discretization is required at least in order to get a satisfactory accuracy.

In the case of a rigid support and a shear rigid plate, a singularity of the bending moments occurs in point e, whereas no singularity is to be expected in the case of

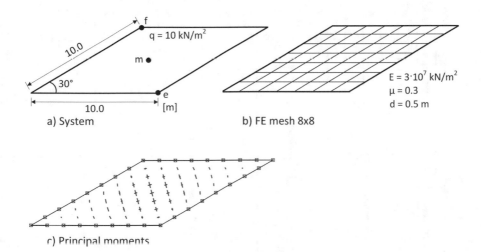

a) System   b) FE mesh 8x8

c) Principal moments

Fig. 4.108 Skew plate in bending, simply supported at 4 edges

Table 4.28 Displacements and principal moments of a skew plate

Structural model	Mathematical model (FE mesh)	w_m [mm]	$m_{I, m}$ [kNm/m]	$m_{II, m}$ [kNm/m]	$m_{I, e}$ [kNm/m]	$m_{II, e}$ [kNm/m]
	Analytical solution	0.12	19.1	10.8	-	∞
Shear rigid plate, rigid support	2×2	0.13	18.4	12.0	2.3	3.4
	4×4	0.13	20.9	13.3	0.3	7.7
	8×8	0.12	18.5	10.8	1.1	14.6
	16×16	0.12	19.9	11.5	1.6	25.9
	32×32	0.12	-	-	1.8	44.8
Shear rigid plate, elastic support	4×4	-	21.2	12.9	0.4	6.1
	8×8	-	20.0	11.8	0.1	6.6
	16×16	-	20.3	12.0	-0.3	5.6
	2×2	0.10	11.04	7.06	2.86	-0.95
	4×4	0.12	17.84	10.02	-2.45	2.30
	8×8	0.13	20.11	11.97	-7.37	6.23
Shear flexible plate, rigid support	16×16	0.13	20.18	12.03	-9.02	10.12
	32×32	0.13	20.08	11.85	-9.01	11.65
	64×64	0.13	20.11	11.84	-8.18	9.96
	128×128	0.13	20.13	11.85	-6.88	6.91
	256×256	0.13	20.14	11.86	-5.37	4.15
Shear flexible plate, elastic support	4×4	-	18.4	10.1	2.43	1.62
	8×8	-	20.9	12.1	1.47	3.83
	16×16	-	20.9	12.2	0.80	4.86
	32×32	-	20.7	12.1	0.42	4.67

the shear flexible plate (c.f. Tables 4.22, 4.23). This behavior of the solution is also reflected in the principal moments in point e given in Table 4.28, for a fixed support. Whereas the bending moments of the shear rigid plate increase continuously with the mesh refinement, this is not the case with the shear soft plate. Here the moments in the corner point ascend to a discretization with 32×32 elements, only to descend again. Similar to the edge effect, discussed above, local redistribution effects are the reason for this behavior. For practical purposes, the moments calculated in point e for the shear flexible plate with rigid support appear just as unsuitable as those for the shear rigid plate with rigid support.

The singularity at point e can be avoided by adopting an elastic support of the plate. As this assumption is closer to the physical reality than a fixed support it represents an improvement of the structural model. With a spring constant of $1 \cdot 10^6 \left[kN/m^2 \right]$, the values given in Table 4.28, are obtained. The very small increase in the field moments is due to the elastic support and the resulting lower "clamping effect" of the corners.

The same plate was examined in [51], with different finite elements. In [157], it is pointed out that for certain element types the element shape (rectangle or parallelogram) can have a significant influence on the internal forces of skew plates. Calculation results of a similar plate with a 4×4 finite element mesh obtained with various commercial finite element programs are given in [139].   ◄

Example 4.32 Intermittent supports of floor slabs are often encountered in practice. With rigid line supports, strong singularities of the plate internal forces occur at the free wall ends. The plate shown in Fig. 4.109a, is examined with rigid and with elastic line supports when loaded by a uniformly distributed load. In particular, the bending moments and bearing pressures at point m that are crucial for the design are to be determined. The calculations are carried out with [SOF 6] using a shear flexible 4-node plate element.

First, the case of rigid support is investigated. The principal stresses and deflections in Fig. 4.109b,c illustrate the loading of the plate. At the free end of the wall, high bending moments occur in the x- and y-directions. The moments m_x and m_y given in Table 4.29, for different element sizes e at point m at the free end of the wall show the singularity to be expected for rigid supports. This is also apparent from the error estimation of m_x according to Zienkiewicz/Zhu in Fig. 4.109d (cf. Sect. 4.12.2). The bearing pressures a_y also exhibit a strong singularity. In order to determine the total bending moment M_y, the moments m_y are integrated into section A–A between the two points of zero moments. While the bending moments m_y increase with decreasing element size, the total moment M_y converges toward a value of -49 kNm. The total moment and the resulting total reinforcement remain constant. Only the distribution of the reinforcement can change slightly with increasing mesh refinement.

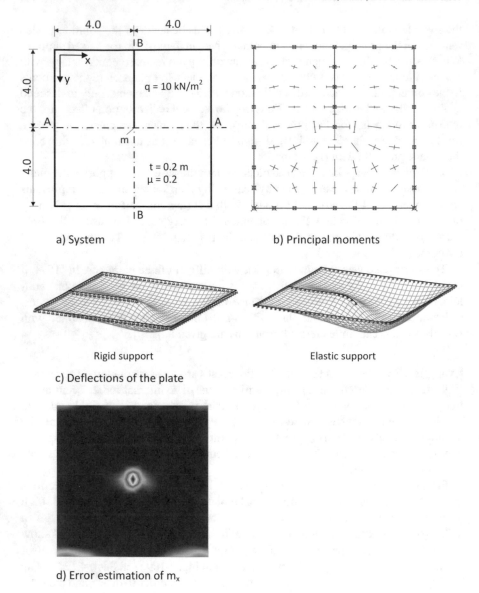

a) System   b) Principal moments

Rigid support   Elastic support

c) Deflections of the plate

d) Error estimation of m_x

Fig. 4.109 Floor slab with a protruding wall

An improved structural model takes into account the elastic stiffness of the support. For this purpose, a spring constant of $1 \cdot 10^6 \text{kN/m}^2$ is introduced as wall stiffness. In the case of elastic support, the plate moments at point m do not possess a singularity point. The total moment M_y of -24 kNm is significantly lower than with rigid supports, while the field moments are slightly higher. The extreme peak values of the bearing force at the free wall end with a rigid line support are significantly reduced

Table 4.29 Deflections and internal forces

Structural model	e [m]	Section A–A		Section B–B		
		$max\ m_x$ [kNm/m]	$m_{x,\,m}$ [kNm/m]	$max\ m_y$ [kNm/m]	$m_{y,\,m}$ [kNm/m]	$M_{y,\,m}$ [kNm]	$a_{y,\,m}$ [kN/m]
Shear flexible plate, rigid support	1.000	12.6	−27.8	15.3	−23.6	−43.4	347
	0.500	11.9	−42.5	13.8	−39.4	−48.0	725
	0.250	11.8	−56.3	14.1	−53.8	−48.8	1491
	0.125	11.8	−70.0	14.0	−66.4	−48.6	2797
Shear flexible plate, elastic support	1.000	12.7	−23.4	16.5	−15.5	−26.4	235
	0.500	12.2	−30.9	15.5	−19.7	−23.6	303
	0.250	12.2	−33.0	15.8	−19.9	−23.7	352
	0.125	12.2	−34.0	15.8	−20.2	−23.5	382

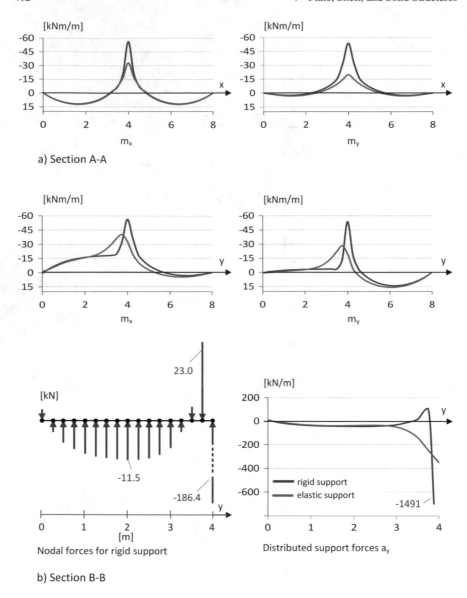

Fig. 4.110 Bending moments and support forces in sections A–A and B–B ($e = 0.25$ m)

by the elasticity of the support (Table 4.29 and Figs. 4.109c and 4.110). Therefore, elastic line supports should be preferred when defining the structural model.   ◄

Point supports (columns)

Various models have been proposed for modeling the support of slabs on columns. These will be discussed below. Furthermore, the finite element mesh in the zone

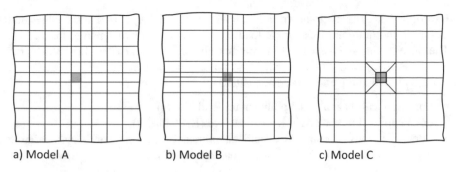

a) Model A   b) Model B   c) Model C

Fig. 4.111 Finite element discretization of the plate at a column

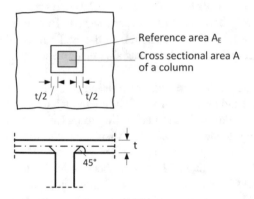

Fig. 4.112 Reference area of a column

of the column head must be adapted to the column geometry. At least four 4-node plate elements should be provided at the column cross section. This is the case with models B and C in Fig. 4.111. Model A is only suitable for plate elements with higher shape functions. If necessary, mesh refinements should be provided due to the high gradients of the internal forces in the support zone.

The reference area of the action is the mid surface of the plate. It is obtained according to Fig. 4.112, assuming a force propagation of 45°. In the case of thin slabs, due to the limited influence on the calculation results, the support surface is usually not enlarged, i.e. the column cross section is assumed to be the reference area.

A column can be represented by beam elements. However, a simple connection of a beam element with the plate elements of the slab leads to singularities in the internal forces of the plate (Table 4.35). Furthermore, the rotational degree of freedom possesses a singularity.

Example 4.33 The plate shown in Fig. 4.113, is simply supported at the edges and loaded in the midpoint by a single load and a moment in two separate load cases. The plate thickness is $t = 0.20$ m, the modulus of elasticity $E = 3.0 \cdot 10^7$ kN/m^2 and the Poisson ratio $\mu = 0$. The load is applied on a surface of 50×50 cm. Three models of load application are investigated:

- Load application at a nodal point (point load model)
- Load application via a rigid loaded surface (RDT model)
- Representation of the single load and the moment by a constant or linearly variable distributed load, respectively, on the load application surface (EST model).

The calculations are performed with shear flexible 4-node elements according to [SOF 6]. The mean element size e is varied between 50 cm and 6.25 cm. For a regular mesh with $n = 500/e$ [cm], this corresponds to a grid of 10×10 or 80×80 elements, respectively. Models with coarser meshes of up to 2×2 elements were also investigated for the point load model. In the EST model, the load application surface of the model with $e = 50$ cm was also discretized by 2×2 elements ($e = 25$ cm each).

The displacement w_m and the angle of rotation $\varphi_{y,m}$ in point m are examined for the point load model and the rigid surface loading model. In the EST model, the mean displacements and angles of rotation according to (4.156), are determined from the displacements of the loaded surface. The displacements for the point load model converge with the element implemented in [SOF 6] to a value of 1.54 mm, which is slightly larger than the value of 1.46 mm obtained in Example 4.12, for the shear rigid plate. In the RDT and EST models, the displacements converge rapidly, toward 1.25 mm in the RDT model and 1.49 mm in the EST model. However, the rotational degree of freedom shows no convergence in the point load model. If a moment is introduced in a single nodal point, the angle of rotation $\varphi_{y,m}$ increases continuously with increasing mesh refinement. This illustrates the singularity given in Table 4.35, for this case. The rotations of the RDT and EST models, on the other hand, converge rapidly against a value of 0.63 and 0.86, respectively.

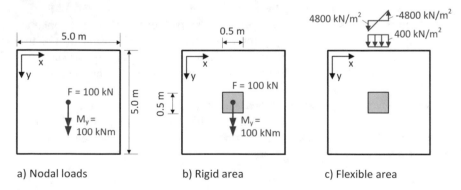

Fig. 4.113 Plate with load application areas (not in scale)

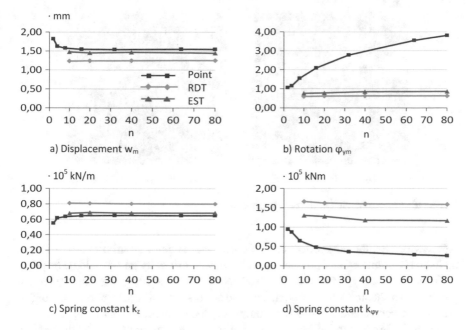

Fig. 4.114 Displacements, rotations and spring constants with different models of load application

From the displacements w_m and rotations $\varphi_{y,m}$ obtained with the force $F = 100\,\text{kN}$ and the moment $M = 100\,\text{kNm}$, respectively, the translational springs $k_z = F/w_m$ and the rotational springs $k_{\varphi y} = M/\varphi_{y,m}$ are determined. They are shown in Fig. 4.114. The translational spring constants converge rapidly toward the values $6.48 \cdot 10^4\,\text{kN/m}$ with the point load model, to $8.00 \cdot 10^4\,\text{kN/m}$ with the RDT model and $6.83 \cdot 10^4\,\text{kN/m}$ with the EST model. For the rotational spring constants one obtains $1.59 \cdot 10^5\,\text{kNm}$ with the RDT model and $1.17 \cdot 10^5\,\text{kNm}$ with the EST model. Due to the assumption of a rigid area in the RDT model, the spring constants are higher with the RDT model than with the EST model. With the point load model, the calculated rotational springs depend on the size of the finite elements and show no convergence. Due to this singular behavior, it is not feasible to connect beam elements to a single node of the plate elements. It is remarkable that with the RDT and EST models, a 2×2 discretion of the load application area ($n = 20$) gives satisfactory spring constants.

In the case of the RDT model, the elastic plate elements are clamped in the rigid load application area. Therefore, singularities of moments and shear forces occur at its corner points according to Table 4.24. Figure 4.115a shows as an example an error estimation of the twisting moments m_{xy} caused by the force F according to Zienkiewicz/Zhu for $e = 6.25$ cm. Similar errors occur with the bending moments m_x and m_y. They indicate a strong singularity. The bending moments and shear forces in the corner zones are considerably disrupted. In the EST model, however, the error estimation of the moments does not show any singularities (Fig. 4.115b).

a) Rigid load area (RDT model)   b) Flexible load area (EST model)

Fig. 4.115 Error estimation of the twisting moment m_{xy}

At the supports, the edge effect can be noted. Here again, it can be seen that rigid inclusions in finite element models must be viewed critically and should be avoided.

In the load case with the force $F = 100\,\text{kN}$ and the 20×20 discretization (element size 25 cm) a square section through the plate at a distance of 12.5 cm from the loaded area will be considered. For control purposes, the element shear forces v_x and v_y, respectively, are integrated into this section to give a resultant force. With the rigid load area model, a resulting force of 88.5 kN is obtained. This corresponds to an error of 12% compared to the applied load of 100 kN. With the flexible load area, the resulting force is 95.7 kN, i.e. the error is 4%. Even in the fine 20×20 mesh with square elements, which possess optimum shape factors, the rigid inclusion is apparent in the error of the shear forces. With distorted elements with poor shape factors, significantly higher errors of the shear forces are to be expected.   ◄

Different structural models have been developed to represent the stiffness of columns in the analysis of slabs (Fig. 4.116). They are similar to the models of columns in deep beams (Sect. 4.11.5).

Models for columns as supports of plates in bending

- Finite element modeling of the column with three-dimensional solid elements
- Modeling of the column with beam elements and of a RDT/EST-connection with the plate elements
- Engineering model (transversal beam elements)
- RDT/EST spring model at a point
- Continuous elastic support of the plate at the column head
- Fluid cushion model
- Rigid point support.

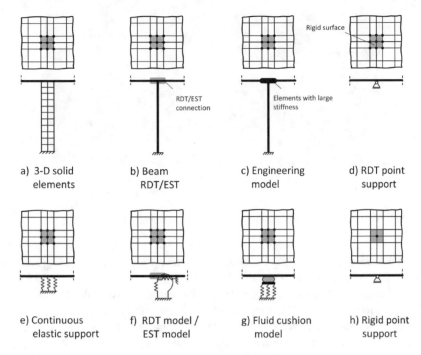

a) 3-D solid elements

b) Beam RDT/EST

c) Engineering model

d) RDT point support

e) Continuous elastic support

f) RDT model / EST model

g) Fluid cushion model

h) Rigid point support

Fig. 4.116 Modeling of columns as supports of plates in bending

The models in Fig. 4.116a–c, explicitly represent the column and are suited for three-dimensional finite element models of buildings. This is not the case for the models in Fig. 4.116d–h. They were developed for two-dimensional slab analyses. The models (a) to (c), (e), and (f) represent in addition the rotational stiffness of the column, which may influence the internal forces of a slab [158]. The other models assume a hinged connection.

The slab design is generally carried out based on the bending moments at the face of the columns. Classical design methods for hand calculation are based on total moments integrated into the column and middle strips perpendicular to the direction of reinforcement [155].

Column as a finite element model with three-dimensional solid elements

Columns can be represented as three-dimensional solid models, while the floor slab is represented with plate elements (Fig. 4.116a). The model includes the description of the local stiffness increase of the slab at the column head. Simple 8-node solid elements, as well as 4-node plate elements, show stiffening effects. In order to represent the bending of the column properly, a sufficiently fine vertical subdivision of the column is necessary or elements with higher shape functions must be used. The column internal forces are obtained by numerical integration of the stresses in the solid elements. The results of this model generally agree well with the EST model

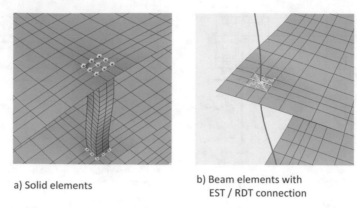

a) Solid elements

b) Beam elements with
EST / RDT connection

Fig. 4.117 Modeling of columns in a three-dimensional model of a building [159]

and other models [159, 160]. The model is particularly suited for modeling columns in three-dimensional building models (Fig. 4.117a).

In addition to the column, the plate can also be represented with three-dimensional solid elements [161, 162]. This very complex modeling allows the stress distribution in the zone of the column head to be determined. This may be valuable in research, but is too complex in practice. Furthermore, the question remains how the calculated three-dimensional stress components can be converted into stress resultants suited for design models used in reinforced concrete constructions.

Column as beam model with RDT/EST connection

In general, columns in three-dimensional building models are modeled with beam elements. A direct connection of the beam end node with a node of the plate is not allowed because of the singularity of the rotational freedom of the slab (cf. Example 4.33). However, the connection to the plate elements can be done as RDT or EST connection, taking into account the cross-sectional area of the column (Fig. 4.116b). With the RDT connection, a rigid connection surface is assumed. With the EST connection, the connection surface is flexible, whereas a linear stress distribution of the normal stresses and a parabolic distribution of the shear stresses of a (rectangular) column are assumed. For this purpose, reference is made to Sects. 4.10.3 and 4.10.4, and to the discussion of both models in the following sections.

Column with engineering models

In the engineering model, the beam end node of the column is connected with artificial beam or plate elements with high stiffness in the area of the column cross section (Fig. 4.116c). This simulates a rigid surface as in the RDT model. The same restrictions apply as for the RDT model. In addition, numerical difficulties may occur if the artificial stiffnesses are selected to be too large.

Column head as RDT model (rigid surface)

When calculating slabs as two-dimensional plate models, the plate is supported directly on the column head. In the RDT model, the column head is understood as a rigid surface which is supported in its center of gravity. The point support may be represented by a rigid point support or by a spring which represents the vertical column stiffness (Fig. 4.116d). The rigid surface can be described using kinematic coupling conditions (cf. Section 4.10.3). Numerical difficulties in the system of equations as with plate elements with artificially high stiffness do not occur. However, the plate internal forces in the vicinity of the column head can be considerably disturbed by the jump in stiffness. This is particularly the case when, in addition to the displacements of the slab, its rotational degrees of freedom are also coupled. The latter must, therefore, be left free, while the rigid surface conditions are only introduced for the degrees of freedom of the displacements. A rigid column head with a coupling including the rotational degrees of freedom doesn't have any curvature and bending moments in the corresponding plate elements. If the rotational degrees of freedom are left free, linear bending moment distributions can occur in the slab in the column area.

Column head as continuous elastic support of the plate

Columns are frequently represented as a continuous elastic support of the plate at the column head (Fig. 4.116e). This avoids singularities of the moments and shear forces of the plate in the vicinity of the column and gives a rounded moment distribution at the column head.

The modulus of subgrade reaction can be determined from the vertical stiffness of the column. With the column cross-sectional area A and column length h, the modulus of subgrade reaction is obtained with $k_z = E \cdot A/h$ (cf. Table 3.4) as

$$k_{S,z} = \frac{\sigma_z}{w_z} = \frac{F_z}{A_E \cdot w_z} = \frac{k_z}{A_E} = \frac{E}{h} \cdot \frac{A}{A_E} \qquad (4.166)$$

where σ_z is the pressure acting on the plate, w_z the displacement of the column head and A_E the reference area of the column. If the reference area is assumed to be equal to the cross-sectional area of the column, one gets

$$k_{S,z} = \frac{E}{h}. \qquad (4.166a)$$

The formula reproduces the stiffness of the column in the case of a vertical displacement of the column head. However, another modulus of subgrade reaction is obtained by assuming that the column head is subjected to a rotation. For a bending moment M_y at the column head, the stress distribution in the reference area is obtained as $\sigma_z = z \cdot M_y/I_{y,E}$, where $I_{y,E}$ is the moment of inertia of the reference area. For a column pinned at its base, the rotational spring is obtained as $k_{\varphi y} = M_y/\varphi_y = 3 \cdot E \cdot I_{y,E}/h$ (cf. Table 3.4). This leads to a modulus of subgrade reaction

$$k_{S,z} = \frac{\sigma_z}{w_z} = \frac{M_y \cdot z}{I_{y,E} \cdot \varphi_y \cdot z} = \frac{k_{\varphi z}}{I_{y,E}} = 3 \cdot \frac{E}{h} \cdot \frac{I_y}{I_{y,E}}. \tag{4.167}$$

If again the reference area is assumed to be equal to the cross-sectional area of the column, one obtains with $I_y = I_{y,E}$

$$k_{S,z} = 3 \cdot \frac{E}{h}. \tag{4.167a}$$

For a column clamped at its base the rotational spring is obtained as $k_{\varphi y} = M_y / \varphi_y = 4 \cdot E \cdot I_{y,E} / h$ (Table 3.4) and with it the modulus of subgrade reaction as

$$k_{S,z} = 4 \cdot \frac{E}{h} \cdot \frac{I_y}{I_{y,E}} \tag{4.167b}$$

and assuming $I_y = I_{y,E}$ one gets

$$k_{S,z} = 4 \cdot \frac{E}{h}. \tag{4.167c}$$

In the case of a column in the upper floor and a column in the lower floor, the subgrade moduli corresponding to the rotational springs of both columns are to be added up.

Different moduli of subgrade reaction are obtained for the translational and the rotational degree of freedom. The vertical stiffness and the rotational restraint of a column cannot be described without contradiction by a single subgrade modulus. It should also be noted that a model with a modulus of subgrade reaction E/h resulting from a constant stress distribution defines a clamping of the slab in the column as well.

The moment applied at the head of the column should be taken into account in the design of the column. In case of doubt, the modulus of subgrade reaction should be selected on the basis of the column's rotational stiffness in order to model the rotational stiffness of the column properly (cf. Example 4.36). With this assumption, the translational stiffness is estimated too large, but this has only a minor effect on the results.

Column head as EST model (coupling spring model)

The EST model represents an extension of the model of the continuous elastic support of the plate at the column head, but avoids its inconsistencies [88, 97]. The model represents both the longitudinal and the bending stiffness of the column. The elastic support is not represented by individual (distributed) springs as with a continuous elastic support, rather the "springs" are coupled to each other. This is why it is also referred to as the coupling spring model. This results in a stiffness matrix for the elastic support. For derivation, reference is made to Sect. 4.10.4, and in particular to Example 4.17.

Another approach also based on the internal forces of the columns can be found in [163]. According to [163], the stress distribution at the column head resulting from the bending moments and the normal force is introduced into the work equation of the finite element system via a mathematical constraint with Lagrange multipliers.

Column head as fluid cushion model

In the fluid cushion model it is assumed that only constant normal pressures are transferred from the column to the slab, i.e. clamping moments are avoided by the modeling (Fig. 4.116g). The assumption of a constant pressure has also been used in the classical analytical solution of the problem of the regular flat slab in [164, 165]. The mathematical formulation for finite element models is somewhat more elaborate.

In [51], the Hellinger-Reissner principle is used to develop a hybrid plate element in which a constant support pressure is applied as a mathematical constraint. The modeling of columns within the framework of a finite element method with large finite plate elements and shape functions with high polynomial degrees is described in [166]. The stresses introduced into the plate by the column are described as a constant pressure. Therefore, the stresses determined in the supported area of a large-size finite element are improved iteratively until a constant stress distribution is achieved. The formulation based on the EST approach is facile. The EST connection is simply written with a vertical spring only, so that $k_{xx} = k_{yy} = 0$ in (4.157). Since only constant vertical pressures are acting on the plate, this corresponds to the fluid cushion model.

The model possesses no horizontal stiffness. In practice, the fluid cushion model is used when, for constructional reasons, the column connection is to be regarded as a hinge. Cracks in concrete are to be expected and their width should be verified.

Column head as point support

With a point support, a nodal point of the finite element model is fixed in the vertical direction (Fig. 4.116h). Since the support force acts as a point load on the plate, the internal forces at the support are singular for both the shear rigid and the shear soft plate. The consideration of the clamping effect of the column by a rotational spring at the nodal point is not permitted, since the rotation due to a moment at a point is a singular quantity (cf. Table 4.35, Example 4.33). Despite the shortcomings mentioned, the model of the point support is used in practice. However, certain restrictions should be observed.

For regular flat slabs, sufficiently accurate moments at the column face are obtained if the ratio of column width d to span width l is $d/l \leq 0.1$ [167]. In the case of thicker columns, the point support model underestimates the moment at the column face. Point supports are not recommended for columns with a cross section other than square, as well as for edge and corner columns [168].

Another method is the design of a slab for total moments [143]. For this, the moment is integrated perpendicular to the reinforcement direction according to the subdivision of the slab into column and middle strips (Fig. 4.118) [155]. In [143], it is pointed out that the integration of the element moments results in more accurate

Fig. 4.118 Design for
resulting total moments
(column strip)

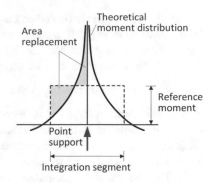

total moments than the integration of the averaged nodal moments. Another way
of correcting the assumption of the punctiform application of the column force is
described in [169]. There the superposition with an equilibrium group of area loads
at the column is suggested.

The point support can be modeled as rigid if other supports as, e.g. line supports
for walls are rigid too. If, however, walls are described as elastic line supports,
the columns should be represented by vertical springs with the spring constant (cf.
Table 3.4)

$$k_z = E \cdot A / h \tag{4.168}$$

in order to be consistent with the model of the line supports.

Example 4.34 For a regular infinite flat slab with a square column grid (interior panel), the bending moments m_x for a point support are to be determined
(Fig. 4.119a). For different conditions of d/l (column width d, span width l), the
moments at the column face are to be compared with the analytical solution according
to [164, 165], in which the column force was assumed as a constant pressure in the
area of the column cross section. Due to the symmetry of the system, the analysis is
performed on a quarter of a slab panel with several regular meshes (Fig. 4.119b).

Figure 4.119c shows the bending moment m_x in a section through a centerline of
columns for the different finite element meshes. At the point support, the structural
model has a singularity point. This results in an increase in the support moment
when refining the finite element mesh. In addition to the moment distributions, the
analytically determined moments at the column faces are plotted for the ratios $d/l =$
0.05, 0.10, 0.15, 0.20, 0.25, and 0.30. For slim columns with $d/l = 0.05$–0.10, these
are close to the moments obtained for a point support with the finite element analysis.
With larger ratios d/l, the moments at the column faces are underestimated with the
point support model. However, for ratios, $d/l \leq 0.1$, the point support model yields
quite reliable moments at the column faces.

It should be mentioned that the formula given in [146, p. 153] for the maximum
fixing moment in the mid of a circular slab with rotational restraints at the edge only

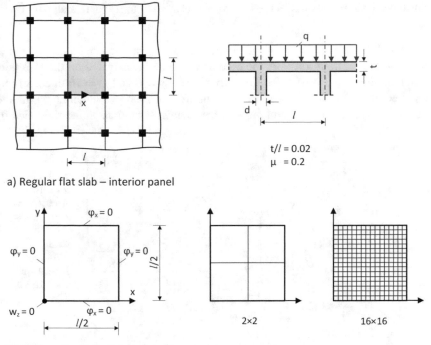

a) Regular flat slab – interior panel

b) FE discretization of a quarter of a panel

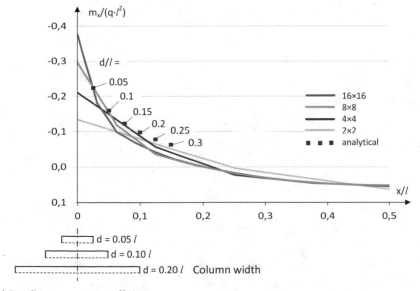

c) Bending moments coefficients

Fig. 4.119 Finite element analysis of the interior panel of a flat slab with the point support model

leads to very similar results as [164, 165] if it is applied to a square slab panel with equal area.   ◀

Example 4.35 The influence of the column modeling on the bending moments of a slab and the reaction forces in the columns is to be investigated on the slab with 4 columns shown in Fig. 4.120a. The two edges at $y = 0$ and $y = 16$ m are simply supported, the edges at $x = 0$ and $x = 24$ m are free. The two columns on the left side are represented by different models, the identical columns on the right side by

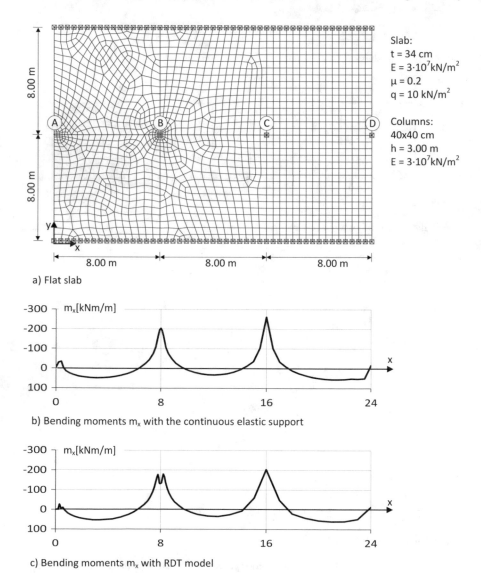

a) Flat slab

b) Bending moments m_x with the continuous elastic support

c) Bending moments m_x with RDT model

Fig. 4.120 Modeling of columns of a flat slab

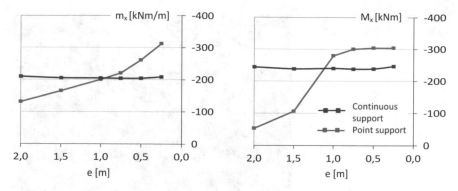

Fig. 4.121 Bending moments m_x and total moments M_x (integration segment: 1.6 m)

point supports. All columns are hinged at the base. The calculations are carried out with a shear flexible plate element with [SOF 6]. The mesh has a global mesh size of $e = 0.5$ m and is densified at the two load application areas of the left supports.

Some significant results are presented in Table 4.30. First, the two left columns are modeled as continuous elastic supports with the modulus of subgrade reaction $k_S = E/h$ resulting from the axial stiffness of the columns. The bending moments m_x in a section at $y = 8$ m through the slab are shown in Fig. 4.120b. The support moment of -204 kNm/m at the left column with continuous elastic support is significantly lower than the moment of -261 kNm/m on the right rigid point support. The support forces are also significantly higher at the rigid point supports (823 kN | 318 kN) than at the continuous elastic supports (775 kN | 346 kN). Even if the two rigid supports are replaced by individual springs, representing the axial stiffnesses of the columns a significant difference remains.

The influence of the finite element size on the two support moments is shown in Fig. 4.121a. Thereafter, with the point support m_x increases continuously with the reduction of the global element size from 2.00 m to 0.25 m. This expresses the singularity at point supports. In the case of continuous support, however, the support moment m_x remains constant even with large global element sizes. In order to get the total moment M_x in a column strip, the moments are integrated over a width of 1.6 m (Fig. 4.121b). With a mesh refinement, the total moment converges to a limit value also in the case of point support. However, the total moment for the point support differs by 20% from that for the continuous support, i.e. the total moment is overestimated by rigid point support. The error estimate of m_x with a global element size of 0.5 m shown in Fig. 4.122 also indicates the inaccuracy of the moments at the two point supports.

The slab is now examined with a modulus of subgrade reaction $k_S = 3 \cdot E/h$ corresponding to the rotational stiffness of the columns. The results are given in Table 4.30. The bending moment in the edge column and the slab moment at the edge column is significantly increased compared to the analysis with the axial stiffness of the columns. All other values remain practically unchanged.

Fig. 4.122 Error estimation
for m_x ($e_{min} = 0.50$ m)

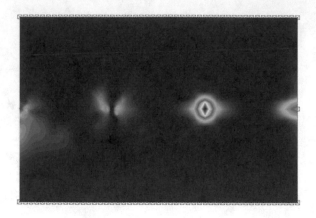

Table 4.30 Bending moments in the slab and sectional forces and moments in the columns

Model	Floor slab		Column		
	m_x [kNm/m]		F_z [kN]		M_y [kNm]
	Column A	Column B	Column A	Column A	Column A
Point support (columns C, D)	−17	ca. −260	318/321[a]	823/806[a]	–
Continuous el. support E/h	−34	−204	346	775	−32
Continuous el. support $3E/h$	−73	−203	352	778	−75
EST model	−66	−204	349	784	−79
Fluid cushion model	−11	−207	335	798	–
RDT model (rigid column head)[b]	−27	−178[c]	348	841	–

[a]With vertical spring instead of a rigid support
[b]MPC displacements only are coupled
[c]In the face of the column

    The EST model of the columns yields similar results as a continuous support that corresponds to the rotational stiffness. For example, the bending moment of the edge column for the EST model is 79 kNm and 75 kNm for a continuous support bedding with $k_S = 3 \cdot E/h$. A subgrade modulus, which corresponds to the rotational stiffness of the column, represents the rotational stiffness correctly. At the same time, however, the increase in the axial stiffness associated with the continuous elastic support has only a little influence on the support forces, i.e. the normal forces in the columns. Therefore, if one does not use the consistent EST model, one should determine the modulus of subgrade reaction in the continuous elastic support model from the rotational stiffness of the column and not from its axial stiffness.

    In the RDT model with a rigid point support, the MPC coupling conditions were formulated only for the displacements. At the face of the central column B, the bending moment of 178 kNm/m is similar to the moment of 166 kNm/m obtained with the continuous support model. The moments in the plate directly above the column are not zero—because of the rotational degrees of freedom allowed to move freely—but physically not meaningful. An additional coupling of the rotational degrees of freedom, however, would cause strong disturbances of the internal forces of the plate in the vicinity of the column. As with the rigid point support model, the support forces are significantly higher than with an elastic support. The mesh in the right slab area has an element size of 1.0 m (not shown here), resulting in a moment of −204 kNm at the point support of column C.

The example illustrates that the EST model with its coupled continuous support conditions shows significant advantages over a point support or the RDT model. Almost as good results are obtained with the model of a continuous elastic support if the modulus of subgrade reaction is determined from the rotational stiffness of the column. The fluid cushion model is well suited for columns to which no moments have to be assigned for design reasons.   ◄

Example 4.36 A flat slab with 3×3 panels is loaded by a constant distributed load q, its thickness is $t = l/30$, its Poisson's number $\mu = 0.2$. The square columns have a cross section of $d \times d$ with $d = l/20$, their height is $h = l/2$ and they are hinged at the base (Fig. 4.123), [159, 160].

Figure 4.124 shows a comparison of the bending moments of the EST model with those of the continuous support model with a modulus of subgrade reaction of $k_{S,z} = E/h$ and $k_{S,z} = 3E/h$, corresponding to the axial and bending stiffness of the column. The bending moments at the column faces are practically the same in all models.

For the maximum moments the model with the continuous support and a subgrade reaction modulus of $k_{S,z} = 3E/h$ gives similar results as the more sophisticated EST model. The internal forces required for the design of the column at its top are given in Table 4.31.

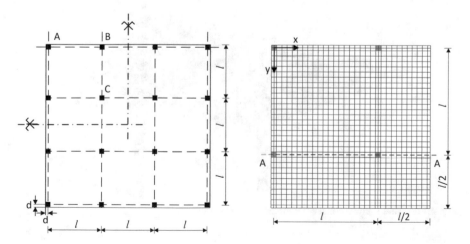

Fig. 4.123 Flat slab with 3×3 panels

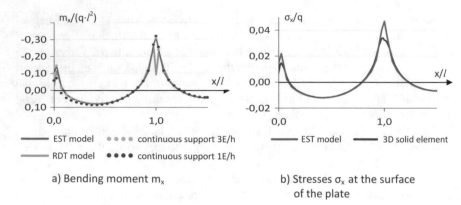

a) Bending moment m_x

b) Stresses σ_x at the surface of the plate

Fig. 4.124 Bending moments and stresses in section A–A

Table 4.31 Internal forces of the columns

Model	Column	$F_z/(q \cdot l^2)$	$M_x/(q \cdot l^3)$	$M_y/(q \cdot l^3)$
EST model	A	0.219	0.0134	−0.0134
	B	0.474	0.0218	0.0027
	C	1.157	−0.0040	−0.0040
Solid model	A	0.221	0.0136	−0.0136
	B	0.474	0.0212	0.0028
	C	1.159	−0.0040	−0.0040
RDT model	A	0.219	0.0141	−0.0141
	B	0.473	0.0218	0.0028
	C	1.159	−0.0041	−0.0041

Fig. 4.125 Three-dimensional solid model of a flat slab [160]

The stresses in the vicinity of the top of the column can be analyzed more realistically with a three-dimensional solid model than with the plate model. For comparison, the model shown in Fig. 4.125, has been examined. The column internal forces determined with the EST model and the axial stresses at the top of the slab correspond well with the solid model (Fig. 4.124b). However, the stresses in the slab at the column are lower than according to the EST model or the model of a continuous elastic support. This effect is due to the stiffening of the plate by the column and demonstrates that a design for the maximum plate moment at the column is not necessary. Rather, the design can be based on the moment at the face of the column. ◄

Flat slabs with drop panels

Drop panels at columns (or column head enlargements) are represented in the structural model as plate regions with increased thickness. A drop panel requires a finer discretization of the plate than would be the case without drop panel.

Figure 4.126 shows a regular flat slab with drop panels. At the edge of the drop panel a jump (or discontinuity) occurs in the diagram of bending moments with a

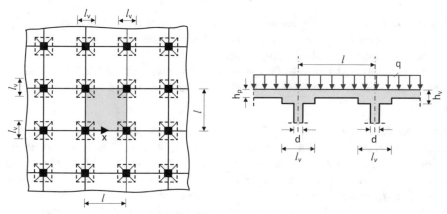

a) Regular flat slab – interior panel

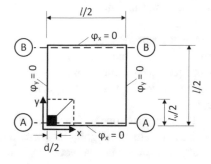

b) Model of one quarter of a slab panel

Fig. 4.126 Regular flat slab with drop panels

moment vector perpendicular to the edge. Its magnitude can be deduced from the relationship between moments and curvature. An edge of the drop panel, parallel to the y-axis is considered (Fig. 4.126b). Variables referring to the drop panel are denoted with the index "v", whereas variables for the plate without drop panel with "p". Due to equilibrium conditions, bending moments $m_{x,v}$ and $m_{x,p}$ at the edge are the same. This does not apply to the moments $m_{y,v}$ and $m_{y,p}$. Rather, the curvatures $\kappa_{y,v}$ and $\kappa_{y,p}$ to the left and right of the edge are identical. From $\kappa_{y,v} = \kappa_{y,p}$ using (2.10a), one obtains the ratio of the moments $m_{y,v}$ and $m_{y,p}$ at the drop panel edge as

$$\frac{m_{y,v}}{m_{y,p}} = \left(\frac{h_v}{h_p}\right)^3 \Bigg/ \left(1 + \mu \cdot \left(\left(\frac{h_v}{h_p}\right)^3 - 1\right) \cdot \frac{m_{x,v}}{m_{y,v}}\right) \tag{4.169}$$

and for the Poisson ratio $\mu = 0$

$$\frac{m_{y,v}}{m_{y,p}} = \left(\frac{h_v}{h_p}\right)^3. \tag{4.169a}$$

For $\mu = 0$ the jump in the moments depends on the jump in the plate thickness only. For $\mu \neq 0$ it depends in addition on the ratio of the moments $m_{x,v}$ and $m_{y,v}$ in x- and y-direction, respectively, in the drop panel.

In the reinforced concrete design of the slab two critical sections must be considered. One is at the face of the column, the other is at the face of the drop panel.

Example 4.37 The influence of drop panels in a flat slab is investigated for the interior panel of the regular flat slab shown in Fig. 4.126 [170]. The span of the column grid is $l = 6.0$ m, the slab thickness is $t = h_p = l/30 = 20$ cm, the square column cross section has a side length of $d = 0.05 \cdot l = 30$ cm. The drop panel has a length of $l_v = 0.30 \cdot l = 1.80$ m and a thickness of $h_v = 2.0 \cdot h_p = 40$ cm. The Poisson ratio is $\mu = 0.2$. The column has a height of $h_{St} = l/2$. It is represented by 4 continuously supported plate elements with a modulus of subgrade reaction of $3 \cdot E/h_{St}$. The analysis is performed taking into account the symmetry of the system for a quarter of an interior slab panel (cf. Fig. 4.126b).

The bending moments of the plate are shown for various regular finite element meshes in Fig. 4.127. The moments m_y have a jump at the edge of crop panel. With $m_{x,v}/m_{y,v} = 0.0656/0.2044 = 0.321$ in section A–A the ratio $m_{y,v}/m_{y,p} = 5.5$ is obtained after (4.169). This corresponds well to the value of $m_{y,v}/m_{y,p} = 0.2044/0.0656 = 5.7$ according to Fig. 4.127a.

The quarter of a plate panel is examined with discretizations in 5×5, 10×10 and 15×15 elements [170]. For the 5×5 mesh the results are slightly less accurate than for finer discretizations. For practical purposes, however, the accuracy with

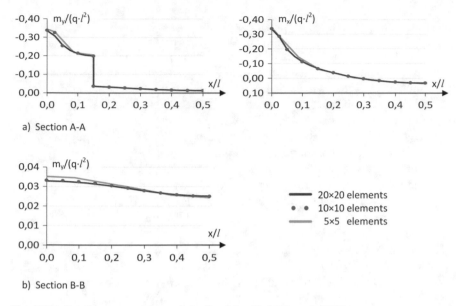

a) Section A-A

b) Section B-B

Fig. 4.127   Bending moments in a regular flat slab with drop panels [170]

5×5 elements, which corresponds to an discretization of the whole slab panel with 10×10 elements, is satisfying.   ◄

Conclusions

The model of a continuous elastic support is well suited to model columns in the analysis of flat slabs. Singularities are avoided and the reduction of the span, as well as the restraining effect of large column cross sections, are taken into account. However, the selection of the modulus of subgrade reaction is not straightforward. Its magnitude is particularly important for edge and corner columns, while it has less influence on the results for internal columns. The EST or coupling spring model is more consistent than the model of a continuous support, and therefore, is recommended. However, if the EST element is not implemented in the finite element program used, the continuous support model with a modulus of subgrade reaction, which is determined from the bending stiffness of the column, provides the best results (cf. Examples 4.35, 4.36). The moments resulting from the continuous support must be taken into account in addition to the normal forces when designing the columns. For columns with a hinged connection to the slab the fluid cushion model is a good choice.

The models can be used to determine the moments required for the reinforced concrete design at the face of the column, as well as the total moments in column and middle strips.

In summary, the following aspects apply to the modeling of columns:

Modeling of columns in flat slabs

- The EST model is best suited for columns clamped in a slab. Alternatively, a column can be represented as a continuous elastic support. The modulus of subgrade reaction should then be determined from the bending stiffness of the columns. If necessary, supports above the slab can be taken into account as well.
- The fluid cushion model is suited for columns connected to the slab in a hinged manner, i.e. for columns to which no moments are to be assigned for design reasons. This model can also be regarded as a special case of an EST model whose rotational stiffnesses are set to zero.
- The floor slab should be discretized into at least 2×2 4-node elements in the cross-sectional area of the column. In the vicinity of the column, the mesh should be densified if necessary.
- Flat slabs with drop panels require an even finer discretization in the region of the drop panels. In RC design, moments and shear forces must be checked at the column faces and additionally at the faces of the drop panel.
- Shear forces or shear stresses in critical sections around columns, which are required for punching shear checks, should be determined on the basis of the column forces. Shear stresses in the plate may only be applied in cases where a very fine finite element discretization and control of the resulting total force can ensure that the discretization error is small enough.

RC beams partly integrated into a slab

Reinforced concrete floor systems normally consist of slabs and beams underneath supporting the slab. The beams have an eccentricity to the axis of the slab and act together with the slab to resist loads. The T-shaped beams resulting thereof are composed of a part of the slab (the flanges) and of the beam below the slab (the web). They are denoted as T-beams. Other beams strengthening a slab may be L-shaped (inverted T-beam) or Z-shaped (Fig. 4.128). Since the neutral axis of the beam does not lie in the middle surface of the slab, normal forces occur in the slab. Their explicit consideration in the structural model requires the analysis of the system as a folded plate structure with finite shell elements instead of plate elements.

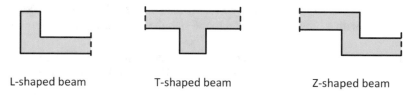

L-shaped beam   T-shaped beam   Z-shaped beam

Fig. 4.128 RC beams partly integrated in a slab

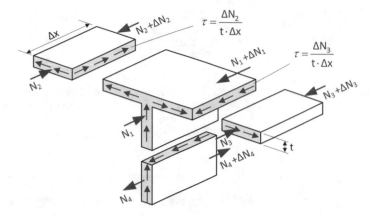

Fig. 4.129 Shear stresses in a T-beam cross section

In hand calculation methods the normal force in the slab is considered in the model of a T-beam with a specified effective flange width. The change in the normal forces in the plate results in shear stresses which are introduced into the web of the T-section (Fig. 4.129). This results in a load-bearing behavior as an overall T-shaped cross section. In the design of large reinforced concrete girders with a T-shaped cross section, the shear stresses are to be checked, whereas in the design of floor systems the check is usually omitted. Design models in RC are also based on a T-beam cross section, whereas a cracked tensile zone and nonlinear stress–strain behavior of the materials are taken into account.

Different structural models for finite element analysis of T-beams have been proposed [171–177]. There are two types of models (Fig. 4.130). Most general are the *folded plate models* where the cross section of the beam, as well as the slab, are modeled as a three-dimensional model including normal forces in the slab. Another type of models are the *plate in bending model* where the slab is modeled as a bending plate without normal forces and the stiffening effect of the beam is included approximately in the analysis.

Structural models for T-beams

- Folded plate models: Modeling of the slab with shell elements

    - Modeling of the web with 3D-shell elements
    - Modeling of the web with shell elements with an eccentric axis
    - Modeling of the web with beam elements with an eccentric axis

- Plate in bending models: Modeling of the slab with elements for bending plates

    - Modeling of the web with beam elements with an eccentric axis

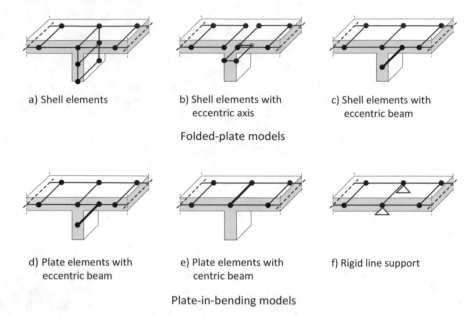

a) Shell elements   b) Shell elements with   c) Shell elements with
                              eccentric axis   eccentric beam

                           Folded-plate models

d) Plate elements with   e) Plate elements with   f) Rigid line support
   eccentric beam   centric beam

                           Plate-in-bending models

Fig. 4.130 Structural models for T-beam-like supports of slabs

> – Modeling of the web with beam elements with a centric axis
> – Rigid line support of the slab.

Folded-plate models

Folded plate models explicitly take into account the normal forces and the axial stiffness of the slab. It is, therefore, not required to determine an effective flange width.

The web may be modeled with shell elements (Fig. 4.130a). In this model, the shell elements overlap in the region of the transition from the web to the slab, which lead to a slight overestimation of the stiffness, especially in the case of compact cross sections. Furthermore, a sufficiently fine finite element meshing must be ensured in order to avoid stiffening effects (cf. Example 4.11).

The modeling of the web using shell elements with an eccentric reference axis in the slab is much easier (Fig. 4.130b). Due to Bernoulli's assumption for the plate elements, the cross section of the web is assumed to remain plane.

In the folded plate model in Fig. 4.130c, the web is modeled as an eccentric beam. In this model, inconsistencies exist between the beam and the shell elements, which lead to an additional discretization error and to jumps in the internal moment and internal force distributions. They have the same cause as in the model of the plate with an eccentric beam and are explained in this context (see below). Therefore, in

[174], it is recommended to use the model with eccentrically arranged shell elements if a beam is to be represented with a folded plate model.

Since each nodal point of a folded plate model has 6 degrees of freedom, the computational effort is considerably higher compared to plate in bending models with 3 degrees of freedom per node. This also applies to the modeling effort, since three-dimensional support conditions must be specified respecting the arching effect in the web. If applicable, a statically determined and thus constraint-free support in horizontal direction should be assumed.

Folded plate models with very high webs constitute the transition to large three-dimensional folded plate and shell structures. In practice, folded plate models are used when it is important to take into account the normal force distribution in the slab. This is the case, for example, with prestressed slabs and slab bridges [177].

Plate in bending models: centric beam model

Due to the easier modeling and the lower computational effort, plate models are usually preferred to folded plate models. First, the modeling of the web with centric beam elements according to Fig. 4.130e, is dealt with. The model is also denoted as a centerline beam model. In this model, the web is represented by a beam with a cross section of the T-beam. The normal force distribution in the slab is not determined explicitly. Instead, an effective flange width of a T-beam-section must be specified. The distribution of the normal forces n in the slab is replaced by a constant normal force n_{max} acting over the effective width b of the plate (Fig. 4.131). The approximation contained herein is justified, since the computational results depend only slightly on the magnitude of the effective flange width. In particular cases, the effective flange width can be specified as stepwise constant but varying over the beam length. With it the narrowing at single loads and supports can be taken into account [176]. For information on determining the effective flange width in accordance with [178], for single and multi-span T-beams, see Table 4.32. Reference is also made to [155].

The appropriate representation of the bending stiffness of the beam is very important as it has a significant influence on the internal forces especially the bending moments in the slab. The cross section of a T-beam is shown in Fig. 4.132. Its center

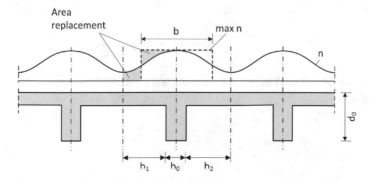

Fig. 4.131 Normal force distribution in the slab and effective flange width

Table 4.32 Effective flange width for continuous slabs according to DIN EN 1992 5.3.2.1 [178]

Single-span T-beam

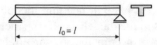

$l_0 = l$ l_0 : Distance of the points of zero moments

Multi-span T-beam   (with $\dfrac{2}{3} \leq \dfrac{l_i}{l_{i+1}} \leq \dfrac{3}{2}$)

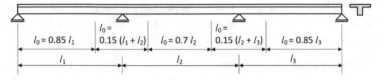

| $l_0 = 0.85 \, l_1$ | $l_0 = 0.15 \,(l_1 + l_2)$ | $l_0 = 0.7 \, l_2$ | $l_0 = 0.15 \,(l_2 + l_3)$ | $l_0 = 0.85 \, l_3$ |

l_1 l_2 l_3

Effective flange width: $b = b_0 + \overline{b}_1 + \overline{b}_2 \leq b_0 + b_1 + b_2$

$\overline{b}_1 = 0.2 \cdot b_1 + 0.1 \cdot l_0 \leq 0.2 \cdot l_0$ $\overline{b}_2 = 0.2 \cdot b_2 + 0.1 \cdot l_0 \leq 0.2 \cdot l_0$
$\quad\quad \leq b_1$ $\quad\quad \leq b_2$ b_0, b_1, b_2 according to Fig. 4.131

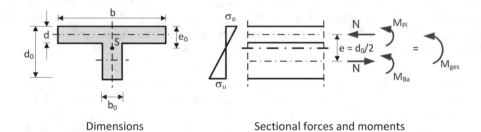

Dimensions   Sectional forces and moments

Fig. 4.132 T-beam

of gravity is given by

$$e_0 = \frac{d_0}{2} \cdot \frac{b_0 \,(d_0 - d)}{b \cdot d + b_0 \,(d_0 - d)} + \frac{d}{2}. \tag{4.170}$$

The center of gravity and hence the neutral axis of the T-beam lies inside the slab ($e_0 \leq d$), if

$$\frac{d_0}{d} \leq 1 + \sqrt{\frac{b}{b_0}}. \tag{4.170a}$$

The moment of inertia (i.e. the 2nd moment of area) of the T-beam is obtained as

$$I_{ges} = I_{Pl} + I_{Ba} + I_{St} \tag{4.171}$$

with

$$I_{Pl} = \frac{b \cdot d^3}{12} \tag{4.171a}$$

$$I_{Ba} = \frac{b_0 \cdot (d_0 - d)^3}{12} \tag{4.171b}$$

$$I_{St} = \frac{b \cdot d \cdot b_0 \cdot (d_0 - d)}{b \cdot d + b_0 \cdot (d_0 - d)} \cdot \frac{d_0^2}{4}. \tag{4.171c}$$

The total moment of inertia of the cross section of the T-beam is composed of the parts I_{Pl}, I_{Ba} and I_{St}, which correspond to the moments of inertia of the plate and the beam as well as to the Steiner's component, respectively. This moment of inertia must be reduced by the part of the plate as this part is already contained in the finite element model of the slab. Thus, the moment of inertia of the centric beam representing the contribution of the T-beam in the finite element model is obtained as

$$I_{FE} = I_{ges} - I_{Pl}$$

or

$$I_{FE} = I_{Ba} + I_{St}. \tag{4.172}$$

In the limit case of a hidden beam with $d = d_0$, the moment of inertia of the T-beam is obtained according to (4.172), with $I_{Ba} = I_{St} = 0$ as $I_{FE} = 0$.

The model contains the assumption of the effective flange width. It can be determined according to Table 4.32. If the effective flange width b approaches ∞ in the limit, the moment of inertia and the location of the center of gravity are obtained from (4.170) and (4.172), respectively, with $b \to \infty$ as

$$e_{0,\infty} = d/2 \tag{4.173}$$

$$I_{FE,\infty} = I_{Ba} + I_{St,\infty} \tag{4.174}$$

with

$$I_{St,\infty} = b_0 \cdot (d_0 - d) \cdot \frac{d_0^2}{4}. \tag{4.174a}$$

Similarly, a torsional constant can also be determined for the T-section. In practice, however, the torsional constant is mostly reduced or completely neglected to take possible crack formation into account.

The representation of a centric beam according to Fig. 4.130e, can also be done by a plate element with the width b_0 of the web and an equivalent height, which can be derived from the moment of inertia I_{FE} to

$$h_{ers} = \sqrt[3]{\frac{12 \cdot I_{FE}}{b_0}}. \tag{4.175}$$

It should be noted, however, that the Poisson ratio in the corresponding plate elements should be defined as $\mu = 0$ in order to avoid an influence of the "T-beam" plate elements on the transverse bending moments in the slab.

Plate in bending models: eccentric beam model

Another option the modeling of the web by beam elements with eccentric reference axis (Fig. 4.130d). The beam elements has the cross section of the web. According to (4.171b), its cross-sectional area and moment of inertia are

$$A_{Ba} = b_0 \cdot (d_0 - d) \tag{4.176a}$$

$$I_{Ba} = \frac{b_0 \cdot (d_0 - d)^3}{12}. \tag{4.176b}$$

The beam elements are rigidly coupled to the plate elements of the slab with an eccentricity $e = d_0/2$. If the moment of inertia is related to the mid surface of the plate, one obtains with (4.174), the contribution according to Steiner's Theorem as

$$A_{Ba} \cdot e^2 = b_0 \cdot (d_0 - d) \cdot \frac{d_0^2}{4}.$$

The representation of the web by a beam with an eccentric reference axis is, therefore, identical to the model of a centric beam with an infinite effective flange width. It is obvious that the bending stiffness of the T-beam is overestimated by the assumption of an infinitely large effective flange width, especially with huge web heights.

In addition, in the model of eccentrically connected beams, a discretization error occurs. It is caused by the axial stiffness of the beam and the related normal force which is constant. Hence, it can only describe an element-wise constant value in its contribution $N \cdot e_{0,\infty}$ to the total bending moment of the T-beam [179]. Therefore, in [173], the model of the centric beam with an infinite effective flange width or the eccentric beam model is recommended only up to a ratio of $d_0/d = 3$. In general, the model with a finite effective flange width should be preferred.

Rigid line support

Beams can only be represented as a rigid line support of a floor slab if their stiffness is considerably higher than that of the slab (Fig. 4.130f). The assumption of a rigid support is not appropriate with the usual dimensions of beams in reinforced concrete floor systems.

Design models

After the solution of the finite element system of equations, the internal forces required for the design of the T-beam are to be determined. In the folded plate model shown in Fig. 4.130a, normal forces and bending moments are obtained in the shell elements. With these sectional forces and moments, a RC design for a shell structure can be performed. However, this only makes sense for large folded plate structures. A beam in a concrete floor system should be designed as a T-beam cross section. In this aspect, the design model differs from the structural model. For design purposes, the resulting forces of the T-beam cross section must be determined by integrating the shell internal forces. It is taken into account that the slab is providing a pressure zone of the T-beam according to Fig. 4.131.

In general, the moment of the T-beam according to Fig. 4.132, is composed of three parts

$$M_{\text{ges}} = M_{\text{Pl}} + M_{\text{Ba}} + N \cdot e. \tag{4.177}$$

In the *folded plate model with shell elements* according to Fig. 4.130a, the resulting internal forces of an (assumed) T-beam cross section are obtained by integrating the internal forces of the shell elements. In this way the moments M_{ges} and M_{PlBa}^* (see below) can be determined. This also applies to the folded plate model with shell elements with an eccentric axis according to Fig. 4.130b.

With the *folded plate model with eccentric beams* according to Fig. 4.130c, the internal forces according to (4.177), are obtained explicitly. Due to the discretization, both the normal forces and the bending moments of the beam show jumps at the nodes. The design of the shell elements of the slab is carried out with their internal forces, i.e. their normal forces and bending moments. Therefore, for the RC design of the T-beam, it is sufficient to use the remaining part of the moment

$$M_{\text{PlBa}}^* = M_{\text{Ba}} + N \cdot e + \int_{b_0} m_{\text{Pl}} \cdot \mathrm{d}s. \tag{4.178}$$

Herein, m_{Pl} is the plate moment over the web with the width b_0. The latter portion in the integral is often small, but can be significant for beams with large web widths b_0.

In the *plate in bending model with eccentric beam* according to Fig. 4.130d, M_{Ba} and N are the beam internal moment and normal force, respectively, obtained from the finite element analysis. They also show jumps. The moment M_{Pl} is obtained by integrating the bending moments of the plate elements over the effective flange width. With it, the total Moment M_{ges} is obtained from (4.177) [180]. The distributed normal force in the plate elements of the slab can be approximated by

$$n = -N/b \qquad (4.179)$$

with a constant distribution over the effective flange width. It should be noted, however, that the magnitude of the normal force outside the web is overestimated (Fig. 4.131), and that its distribution should be assumed by engineering judgment if necessary. This allows the plate elements of the slab (flanges) to be designed taking the normal force into account. The plate moments parallel to the T-beam axis result from the curvature of the slab and must therefore not be neglected when designing the plate, even if the T-beam is designed for the moment M_{ges}. They are of particular importance in the case of short T-beams where the compression area lies in the slab only.

For the *plate in bending model with centric beam* according to Fig. 4.130e, the total moment M_{ges} can be obtained by multiplying the curvature κ by the bending stiffness according to (4.171) and (2.8a)

$$M_{\text{ges}} = \kappa \cdot E I_{\text{ges}} = \kappa \cdot E I_{\text{Pl}} + \kappa \cdot E I_{\text{Ba}} + \kappa \cdot E I_{\text{St}}. \qquad (4.180)$$

The moment parts according to (4.177)

$$M_{\text{ges}} = M_{\text{Pl}} + M_{\text{Ba}} + N \cdot e$$

can be assigned to the corresponding stiffness parts, i.e. it is $M_{\text{Pl}} = \kappa \cdot E I_{\text{Pl}}$, $M_{\text{Ba}} = \kappa \cdot E I_{\text{Ba}}$ and $N \cdot e = \kappa \cdot E I_{\text{St}}$ [172]. Since the T-beam acts as a homogeneous cross section, the curvature κ in the slab and the beam is the same, so that applies

$$\kappa = \frac{M_{\text{ges}}}{E I_{\text{ges}}} = \frac{M_{\text{Pl}}}{E I_{\text{Pl}}} = \frac{M_{\text{Ba}}}{E I_{\text{Ba}}} = \frac{N \cdot e}{E I_{\text{St}}}. \qquad (4.180\text{a})$$

The internal moment of the centric beam with the stiffness $I_{\text{FE}} = I_{\text{Ba}} + I_{\text{St}}$ according to (4.172), is thus

$$M_{\text{PlBa}} = M_{\text{Ba}} + N \cdot e. \qquad (4.181)$$

With (4.180a), the normal force N is obtained as

$$N = \frac{M_{\text{Ba}}}{e} \cdot \frac{E I_{\text{St}}}{E I_{\text{Ba}}} \qquad (4.181\text{a})$$

and the bending moment M_{PlBa} as

$$M_{\text{PlBa}} = M_{\text{Ba}} + M_{\text{Ba}} \frac{E I_{\text{St}}}{E I_{\text{Ba}}}. \qquad (4181\text{b})$$

With it the bending moment M_{Ba} is

$$M_{Ba} = \frac{M_{PlBa}}{1 + \dfrac{I_{St}}{I_{Ba}}}, \tag{4.182a}$$

and the normal force

$$N = \frac{M_{Ba}}{e} \cdot \frac{I_{St}}{I_{Ba}} = \frac{M_{PlBa}}{e} \cdot \frac{I_{St}}{I_{Ba} + I_{St}}. \tag{4.182b}$$

The total moment of the T-beam is

$$M_{ges} = M_{Ba} \cdot \frac{I_{ges}}{I_{Ba}} = M_{PlBa} \frac{I_{ges}}{I_{Ba} + I_{St}}. \tag{4.182c}$$

As in (4.179), with the plate model with eccentric beam, the distributed normal force in the plate elements can be approximated with

$$n = -N/b$$

as constant over the effective flange width and used in the design of the plate (flange) elements of the slab.

It should be noted that the moment M_{Pl} is contained already in the internal forces of the slab where it appears in moments calculated in the finite element analysis superimposed with plate moments from other sources. If the design of the T-beam cross section is done with (4.182c), the moment M_{Pl} is included twice in the design, once in the finite element moments of the plate elements and then again in the design moment of the T-beam. This can be seen, for example, in the case of a hidden beam with $d = d_0$, where with $M_{Ba} = N = 0$ (since $I_{FE} = 0$, see above) and one would get design moment for the (fictious) T-beam of $M_{ges} = M_{Pl}$ with (4.177), in addition to the bending moments in the slab. In [172], it is, therefore, recommended to design the slab with the moments calculated by the finite element program and to carry out the design of the entire T-beam cross section with the remaining moment

$$M_{PlBa} = M_{Ba} + N \cdot e. \tag{4.182d}$$

This moment is the beam moment of the centrically arranged beam with the stiffness according to (4.172). In addition, the plate moment over the web zone must also be taken into account. This results in the following design moment for the T-beam cross section according to (4.178):

$$M_{PlBa}^* = M_{Ba} + N \cdot e + \int_{b_0} m_{Pl} \cdot ds$$

The plate elements are designed for the bending moments calculated by the finite element analysis and the distributed normal forces with (4.179), possibly with an assumed normal force distribution.

The shear force of the T-beam cross section consists of a slab part and a beam part. It is

$$V_{ges} = V_{Ba} + \int_b v \cdot ds. \tag{4.183}$$

While in a hand calculation of a T-beam cross section the entire shear force V_{ges} is assigned to the web, this may not always be meaningful here, similar as in the case of the total moment, and require additional considerations for design.

Example 4.38 The slab with two panels and a partly integrated beam in the mid shown in Fig. 4.133, is subjected to a uniformly distributed load. The sides of the slab are simply supported. Two cross sections with different web heights are considered for the beam. Cross section Q1 has a web height of 80 cm, cross section Q2 of 40 cm. The cross section Q2 thus has a significantly lower bending stiffness than cross section Q1.

The effective flange width of the T-beam is determined according to Table 4.32 as

$$b = 0.3 + 2 \cdot (0.2 \cdot 3.0 + 0.1 \cdot 8.0) = 3.1 \text{ m}.$$

The cross section properties obtained are listed in Table 4.33.

The calculations were performed with shear flexible shell/plate and beam elements with [SOF 6]. The average element size of the 4-node elements is 0.5 m.
The following models were examined:

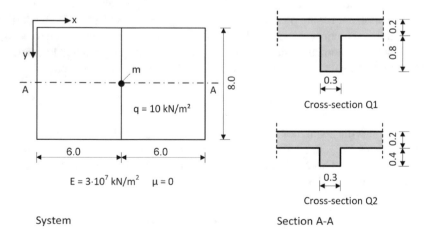

System   Section A-A

Fig. 4.133 Floor slab with a partly integrated beam

Table 4.33   Cross section properties

Cross section	e_0 [m]	A_{Ba} [m^2]	I_{Ba} [m^4]	I_{St} [m^4]	I_{Pl} [m^4]	I_{ges} [m^4]	I_{FE} [m^4]	$I_{FE,\infty}$ [m^4]
Q1	0.24	0.24	0.013	0.043	0.002	0.058	0.056	0.073
Q2	0.15	0.12	0.002	0.009	0.002	0.013	0.011	0.012

- A—Plate in bending model with centric beam and $b = 3.1$ m according to Fig. 4.130e
- B—Plate in bending model with centric beam and $b \to \infty$
- C—Plate in bending model with eccentric beam according to Fig. 4.130d
- C/FW—Folded plate model with eccentric beam according to Fig. 4.130c
- D—Plate with rigid line support according to Fig. 4.130f.

The bending moments m_x and m_y in the plate, the bending moments M_{ges}, M^*_{PlBa}, M_{PlBa} and M_{Ba} in the T-beam as well as the deflection f in the slab center are given in Table 4.34. The table also contains the forces N and the normal plate forces n_y. The fields highlighted in gray contain values that were not obtained directly as results of the finite element analysis, but were determined using the formulas given above. The distribution of internal forces in section A–A is shown in Fig. 4.135.

First, the cross section Q1 is considered. Figure 4.134 shows that the beam is noticeably deformed in comparison to the slab. The deflections in the middle of the beam are between 1.6 and 2.1 mm. In the middle of the panel, the maximum deflection of the slab is 3.8 mm for model A and 3.4 mm and 3.5 mm for models B and C, respectively. As expected, the lowest beam deflections result for model B with $b \to \infty$ where the bending stiffness of the beam is higher by a factor $I_{FE,\infty}/I_{FE} = 0.073/0.056 = 1.30$ than for model A. The difference in stiffness, on the other hand, has no significant effect on the bending moments in the plate and beam.

The agreement of the internal forces and deflections between the plate model A with centric beam and the folded plate model C/FW with the eccentric beam is remarkably good (Fig. 4.135). The "stiffer" models B and C with $b \to \infty$ show only slightly higher T-beam moments compared to model A. This means that the influence of the magnitude effective flange width b on the internal forces is small. The very small differences between models B and C are due to the discretization error mentioned above. The normal forces n_y in the plate above the beam are simulated well by the plate in bending models as well. While in the plate in bending models, they are constant over the effective flange width, in the folded plate model they drop to zero at a point that can be assumed approximately at a distance $\Delta x = \pm b/2$ from the beam's axis. When designing with plate in bending models, a linear interpolation of n_y from the axis of the beam to zero at $\Delta x = \pm b/2$ can be assumed.

Model D with a rigid support instead of a T-beam is presented for comparison purposes only. The deflection f and the beam moment M_{ges} were determined for a simply supported beam with the moment of inertia I_{ges} loaded by a trapezoidal line load. The line load was determined with the load distribution areas according to

Table 4.34 Bending moments and deflections in the center of the plate (point m)

Cross section Q1

Model	M_{ges} [kNm]	M^{*}_{PlBa} [kNm]	M_{PlBa} [kNm]	M_{Ba} [kNm]	N [kN]	n_y [kN/m]	m_y [kNm/m]	m_x [kNm/m]	f [mm]
A	496.3	480.4	478.7	109.3	738.7	−238.3	6.09	−30.6	2.0
B	507.2	493.5	492.1	86.5	811.0	−261.6	4.90	−32.0	1.6
C	506.9	493.1	491.6	90.5	802.2	−258.8	4.94	−31.9	1.6
C/FW	499.6	479.8	478.0	113.0	729.9	−234.3	6.10	−30.6	2.0
D	547.5	-	-	-	-	-	0	−38.1	2.1

Cross section Q2

Model	M_{ges} [kNm]	M^{*}_{PlBa} [kNm]	M_{PlBa} [kNm]	M_{Ba} [kNm]	N [kN]	n_y [kN/m]	m_y [kNm/m]	m_x [kNm/m]	f [mm]
A	384.5	328.0	322.0	48.4	912.0	−294.2	20.0	−14.3	6.6
B	392.2	338.9	333.2	43.0	967.3	−312.0	19.0	−15.4	6.3
C	394.1	341.5	335.8	45.1	969.2	−312.6	18.8	−15.6	6.2
C/FW	379.2	322.0	315.9	49.0	889.6	−283.0	20.4	−13.7	6.8
D	547.5	-	-	-	-	-	0	−38.1	9.4

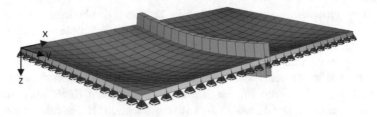

Fig. 4.134 Displacements of a slab with a beam (cross section Q1, model A)

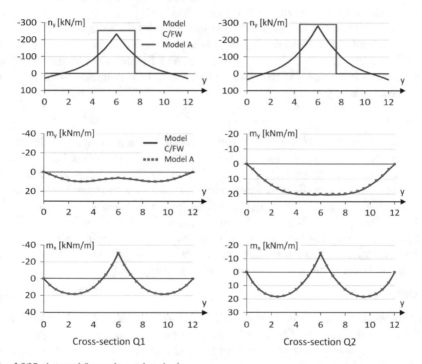

Cross-section Q1

Cross-section Q2

Fig. 4.135 Internal forces in section A–A

[155] and an angle of the load distribution of 60°. The beam moment is approximately 9% higher than on the more realistic models, the deflection matches the one of the C/FW folded plate model well. The support moment m_x of the plate above the web is overestimated by approx. 25% (absolute value), which leads to an underestimation of the span moment in the plate panel. For comparative calculations, the model appears to be suitable when the web height is large. In general, the models that represent a partly integrated beam as a T-beam should be preferred.

Further computational results for cross section Q1 obtained with different programs (but with $b = 3.7$ m) are given in [139]. They agree well with the results given above.

The T-beam with the cross section Q2 and the lower web height exhibits greater deflections and lower beam and slab moments (above the beam) than with cross section Q1 due to the lower stiffness (Table 4.34). The beam, therefore, participates to a minor extent in the load transfer. As with cross section Q1, the calculation results of the T-beam models A to C are in good agreement. This also applies to the folded plate model C/FW. Model D with the assumption of a rigid line support, however, gives results that differ greatly from the other models.

The example shows that all models examined, with the exception of model D, yield sufficiently accurate internal forces of the plate beam and the plate for practical use. The advantage of the C/FW folded plate model is that the distribution of the normal forces in the plate is determined, which can only be done approximately with the plate in bending models.   ◄

Conclusions

Plate in bending models with centric and eccentric beams, as well as the folded plate models are suited for modeling beams partly integrated into a slab. For plate in bending models, the centric beam model is recommended because it takes into account the size of the effective flange width and there are no jumps in the moment line as in the eccentric beam model. In RC design, the moments in the slab that the slab experiences due to the curvature in the axial direction of the beam should be considered. The normal forces in the slab should also be taken into account. In the folded plate models, these are calculated explicitly in the finite element analysis, for the plate in bending models, they can be determined approximately.

Mounting parts

Most mounting parts as mounting profiles and anchor plates have practically no influence on the static load-bearing behavior of plates. However, this is not always the case. One example are load-bearing thermal insulation elements that are used for the thermal separation of external structural components such as balconies or parapets. Reinforcement is passed through the insulation material and can transmit a negative bending moment and a shear force, but not a twisting moment. This load-bearing behavior must be represented in the structural model by an appropriate hinge. In addition, the stiffness of the thermal insulation element must be taken into account. The connection for the bending moment and the shear force is therefore represented as an elastic connection with a line hinge combined with a rotational spring and a translational spring for relative rotations and displacements, respectively (Fig. 4.136).

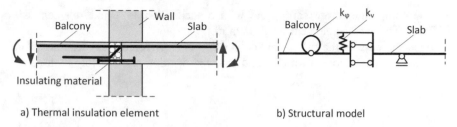

a) Thermal insulation element   b) Structural model

Fig. 4.136 Load-bearing thermal insulation element with insulation material and reinforcement

Example 4.39 The floor slab shown in Fig. 4.137, which is simply supported on all sides, is connected to a cantilever slab (balcony). The moments at the connection of the cantilever slab, as well as the bending moments of the slab panel, have to be investigated for the following two cases:

(a) the cantilever slab is monolithically connected to the floor slab
(b) the cantilever slab is connected to the floor slab with a load-bearing thermal insulation element.

The cantilever slab is connected to the floor slab with an elastic hinge for relative rotations and displacements. For the line springs, a spring constant of $k_\varphi = 10000\,\text{kNm/m}$ is assumed for relative rotations and of $k_v = 250000\,\text{kN/m}^2$ for transverse relative displacements [181].
The following cases shown in Fig. 4.137, are investigated:

- Load case 1: Loading of the floor slab
- Load case 2: Loading of the cantilever slab
- Load case combination K: Loading of the floor slab and the cantilever slab (here without partial safety factors).

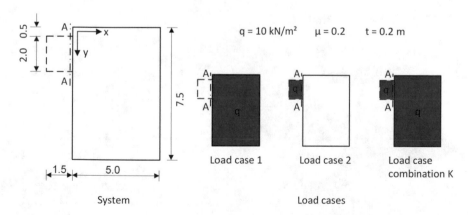

$q = 10\,\text{kN/m}^2$ $\mu = 0.2$ $t = 0.2\,\text{m}$

Load case 1   Load case 2   Load case combination K

System   Load cases

Fig. 4.137 Floor slab with a cantilever slab

The computations are carried out with a shear flexible 4-node element with [SOF 6]. The average element size is 0.25 m.

First, the case of the monolithic connection is considered. The deflections and principal moments are shown in Fig. 4.138, the bending and twisting moments in

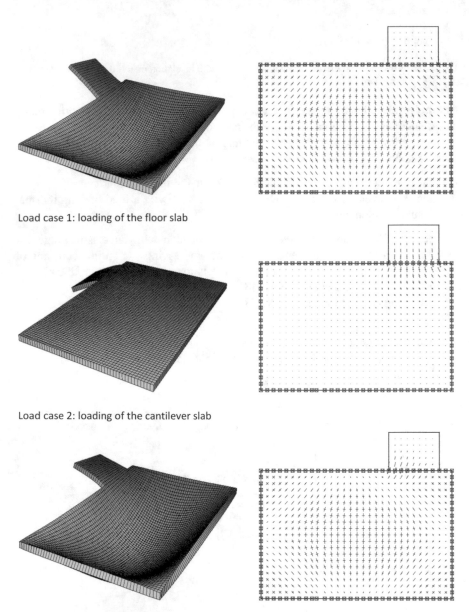

Load case 1: loading of the floor slab

Load case 2: loading of the cantilever slab

Load case combination K: loading of the floor slab and the cantilever slab

Fig. 4.138 Deflections and principal moments with a monolithic connection

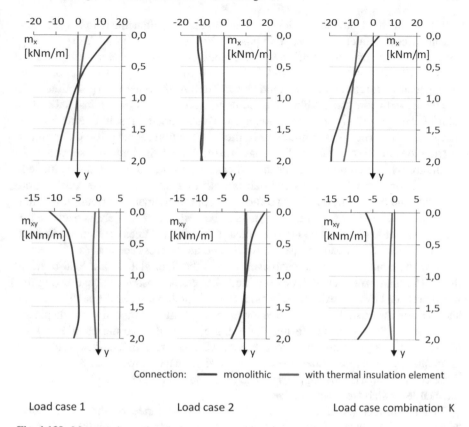

Connection: ▬▬ monolithic ▬▬ with thermal insulation element

Load case 1   Load case 2   Load case combination K

Fig. 4.139 Moments in section A–A

section A–A in Fig. 4.139. In load case 1, the rotational angles φ_y in section A–A increase from the corner of the panel to the center of the panel. This twisting of the slab corresponds to twisting moments which are transmitted in the cantilever slab and result in bending moments of considerable magnitude. A singularity of the moments occurs at the corner points of the cantilever slab in section A–A (Table 4.23). As a result, the moments increase rapidly with mesh refinement, especially at the point at $y = 2.5$ m. The maximum bending moment m_x obtained with $e = 0.25$ m is 14.9 kNm/m. The maximum moment m_x in the slab panel is obtained at 19.8 kNm/m. It is practically the same as the moment of 20.0 kNm/m obtained for a rectangular slab without a cantilever slab.

In load case 2, in which only the cantilever slab is loaded, the clamping of the cantilever slab into the floor slab becomes softer at the edge facing the slab center than at the edge facing the corner. This causes the slab to twist, which in turn corresponds to twisting moments. Again, the bending moments m_x are influenced by this, so that their distribution differs from the constant value $m_x = -q \cdot l_k^2/2 = 10 \cdot 1.5^2/2 = 11.25$ kNm/m. The integral over m_x in section A–A must of course again correspond to the total moment of 2 m $\cdot$ 11.25 kNm/m $= 22.5$ kNm.

If both, the field and the cantilever slab are loaded simultaneously in load combination K, both influences are superimposed. Whereas in section A–A the bending moments m_x in one end point become practically zero, they assume values of $m_x = -19 \, \text{kNm/m}$ at the other end. This means that at one point in section A–A no upper support reinforcement would be required. At the other end of section A–A one would get about the double reinforcement, which results from a manual calculation for $m_x = -q \cdot l_k^2/2 = 11.25 \, \text{kNm/m}$. This surprising result requires further discussion. In particular, the influence of the construction sequence on the moment distribution must be considered. The nonuniform moment distribution in section A–A results from imposed deformations in load case 1, in which the floor slab is loaded, e.g. by its own weight. If the floor slab is built in a first step of the construction sequence and the cantilever slab in a further step, this moment distribution will not occur. Load case 2, for which the system is to be designed separately, is decisive for the permanent loads. For live loads at most, the internal forces determined in load case 1 in section A–A can be relevant for an uncracked cross section.

Now the connection of the cantilever slab to the floor slab with a load-bearing thermal insulation element will be examined. The effects described above are significantly reduced by the flexible connection. In load combination K, the cantilever slab shows a downward deflection, whereas with the monolithic connection the displacements of the floor slab due to load case 1 are transmitted to the cantilever slab and result in an upward displacement (Fig. 4.140). Although in load case 1, bending moments appear in the cantilever slab, they are significantly lower than with a monolithic connection. This also applies to load combination K. The twisting moments in the cantilever slab are practically zero.

The above-mentioned considerations on the observance of the construction sequence also apply here. This means that a design for permanent loads in load case 2 must additionally be carried out here. This can also be achieved in approximate terms by checking the cantilever slab for the fixed support moment of $m_x = -q \cdot l_k^2/2$ [181].

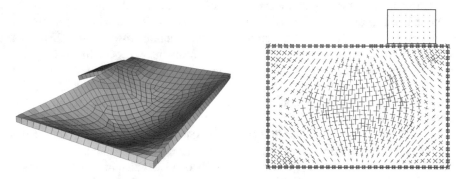

Fig. 4.140 Deflections and principal moments with a thermal insulation element in load combination K

The example shows that load-bearing thermal insulation elements can influence the local load-bearing behavior significantly. It also shows that influences of the construction method and sequence, of imposed deformations, of concrete creep and shrinkage and of indirect actions, which are usually not taken into account explicitly in finite element analyses, can be essential for the design and require special attention.   ◄

Corner lifting of slabs

In general, it is assumed that slabs are connected to their supports in a tension- and compression-proof manner. In the corners of plates with twisting stiffness, however, uplift forces occur, e.g. with a rectangular, simply supported and uniformly loaded plate. Computational tensile forces at a support are only allowed if they are exceeded by pressure forces (e.g. from walls on the slab) or anchored back accordingly. If this is not the case, the corners of plates lift up and the internal forces and stresses are affected [182]. These can be determined by a nonlinear analysis. A nonlinear analysis does not allow for load case superpositions, i.e. only load case combinations have to be investigated. Alternatively, in practice, RC plates are sometimes considered to be without twisting stiffness. Uplift forces are avoided by this simplifying model assumption.

Uplift forces are also to be expected at simply supported edges of plates with a partial offset in the support line (Fig. 4.141). They occur even with small offsets having the size of the wall thickness. Uplift forces can possibly be avoided by modeling the support as an elastic line support taking the vertical stiffness of the supporting walls into account. In the case of a small offset, the support line may be straightened and thus the modeling of the offset of the support can be omitted.

Loads

Plates in bending can be loaded by point, line or area loads. Singularities of the internal forces occur particularly with point forces and moments at the load application point (Table 4.35). It should be noted that the angle of rotation due to an applied single moment is singular. This means, among other aspects, that in a finite

Fig. 4.141 Uplift forces at a partial offset of a line support of a plate

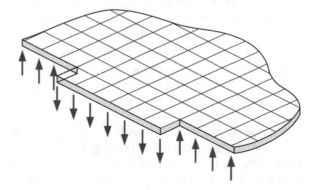

Table 4.35 Singularities in shear rigid plates at point A of load application

Load	Displacements	Sectional forces
F↓ A	No	Yes $(m_x,\ m_y,\ q_x,\ q_y)$
M A	Yes $(\varphi_y = \dfrac{dw}{dx})$	Yes $(m_{xy},\ q_y)$
p A A	No	Yes (q_x)

element model the pointwise connection of a beam with a plate element or with a shell element is not allowed.

For a more accurate representation, single loads on plates shall be presented as area loads. The area of load application refers to the center surface of the slab, i.e. the loaded area at the top of the slab can be increased accordingly in order to determine the reference area, as shown for beam structures in Fig. 3.47 and for supports of flat slabs in Fig. 4.112.

In order to determine the distribution of internal forces in the vicinity of single loads, a fine finite element meshing adapted to the stress gradient is required. In the case of a quadratic load area, a 2 × 2 discretization in 4-node elements is usually sufficient if bending moments are to be determined. For guidance, the finite element discretizations given in Fig. 4.111 for columns can be used. A finer meshing is required to determine the shear forces. The number of elements can be reduced by using elements with higher shape functions. This is of particular interest for the determination of shear forces.

Example 4.40 The simply supported slab shown in Fig. 4.142, is loaded in the midpoint by a single load of $F = 100$ kN. The area of load application at the top of the slab is 30 × 30 cm. Plate thickness of 20 cm results in a reference area of 50 × 50 cm at the mid surface of the plate. The single load is represented once as uniformly distributed load on the reference area (area load model) and for comparison as point or nodal load (point load model). The area of load application, i.e. the reference area, is discretized into $n \times n$ elements. In a convergence study, the parameter n is increased, starting with $n = 1$.

The computations were carried out with shear flexible 4-node elements according to [SOF 6]. Figure 4.143 shows the distribution of the bending moments m_x and the shear forces v_x in section A–A with an element size of $e = 12.5$ cm, i.e. for $n = 4$. The point load model shows the singularity of the shear forces and the bending

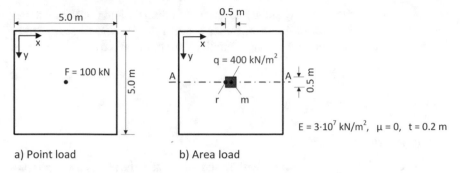

a) Point load   b) Area load

Fig. 4.142 Plate with single load

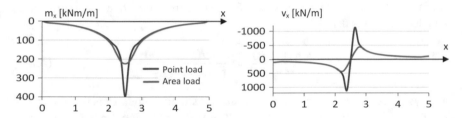

Fig. 4.143 Distribution of the internal forces in section A–A

moments to be expected according to Table 4.35, at the point of load application. The internal forces of the point load model and the area load model practically do not differ outside an area that corresponds to twice the dimensions of the load application area, i.e. of $1.0\,\text{m} \times 1.0\,\text{m}$.

To study the convergence of the internal forces at points m and r, the parameter n is varied from $n = 1$ to $n = 32$. This corresponds to an element size of 50.0 cm to 1.56 cm for a regular mesh with $e = 50/n$ [cm]. The element size is retained throughout the plate. Figure 4.144 shows the convergence of the bending moment m_x and the shear force v_x in the area load model. The internal forces are related to the reference values $m_{xm,\text{ref}} = 221.6\,\text{kNm/m}$, $m_{xr,\text{ref}} = 174.3\,\text{kNm/m}$ and $v_{xr,\text{ref}} = 536.8\,\text{kN/m}$. While the bending moments show good accuracy with a discretization of the loaded area in 2×2 elements only, this is not the case with the shear forces.

Fig. 4.144 Convergence of bending moment m_x and the shear force v_x in the points m and r

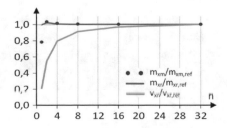

The shear forces converge only slowly. Even with a discretization of the load area into 88 elements, the error of the shear forces is still 10%. If shear forces are to be determined, elements with higher shape functions are more effective.   ◀

4.11.7   Foundation Slabs

Modeling of the soil

In the structural design of buildings with foundation slabs, the soil is usually represented as a linear, elastic continuum. Its material properties depend on the magnitude of stresses and on the load history. As a parameter to describe the elasticity of the soil, instead of the modulus of elasticity E, the constrained modulus M is often used in geotechnics. It is defined as

$$M = \frac{E \cdot (1 - \mu)}{1 - \mu - 2\mu^2} = \frac{E \cdot (1 - \mu)}{(1 + \mu) \cdot (1 - 2\mu)},   \tag{4.184}$$

where μ is the Poisson ratio of the soil. The constrained modulus is the gradient in the stress–strain diagram (or compression stress—compression strain diagram) with lateral strains inhibited on all sides. The increase of the constrained modulus is over-linear, i.e. the soil solidifies with increasing stress. Since the compression stresses increase with depth due to the superimposed load, the stiffness increases with depth too, even with an otherwise homogeneous, i.e. unstratified (cohesionless) soil. The stiffness is higher with de- and reloading than with initial loading, since the change in the pore space is irreversible. In general, an approximate constant constrained modulus is used, which takes the influences mentioned above into account.

The subsoil is often stratified, whereby in the case of inclined layer boundaries, there may be a different soil profile under each point of a foundation slab. The soil profile model is usually limited to the so-called limit depth z_{gr}, in which the additional vertical stress due to the structural load is equal to (or less than) 20% of the stress due to the dead load of the soil. Settlement components below the limit depth are neglected. In general, a value of $(1 \div 2) \times b$ with b as the reference length (shorter side) of the foundation is assumed for the limiting depth for rectangular foundations.

For the finite element analysis of foundation slabs different models for the representation of the load-bearing effect of the soil are available, which are called subsoil models.

Subsoil models
- Finite element model of the subsoil
- Continuum model of a layered soil (analytical solution)

- Subgrade reaction model
- Two-parameter elastic model (Pasternak model)

Finite element model of the subsoil

The soil below and next to the structure is represented by solid elements. Here, a sufficiently large soil region must be considered so that the ground surface settlements have decayed at the edge of the model. This can be done for soil models having an arbitrary geometry or with a geometry of a hemisphere (half-space) or a cylinder (layered soil with rigid lower boundary) (Fig. 4.145), [185, SOF 7]. Commercial finite element programs allow a mesh generation for spatially variable soil properties. These are described for horizontal or inclined layers by drilling profiles below the foundation slab. The vertical boundaries of the finite element model of the soil are assumed to be fixed in the horizontal direction and free to move in a vertical direction. Due to the size of the finite element model of the soil, extremely large systems of equations result and the computational effort for determining the ground reactions (and, in the case of 3D building models, also the building internal forces) is very high.

The soil model must be large enough to ensure that the influence of the boundaries of the finite element model on the stresses and displacements under the foundation slab is negligible. With special infinite elements, which are arranged at the boundaries and represent the stiffness of the soil, which is not discretized any further, the size of the soil model can be reduced and limited to the irregular soil region. In this respect, reference is made to [183, 184], and for horizontally stratified soils to [98, 99] (with the circular frequency $\Omega = 0$).

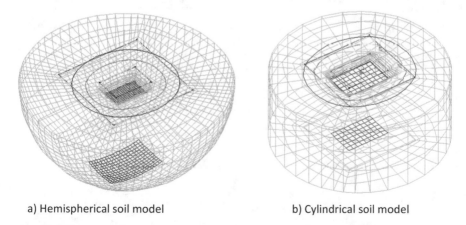

a) Hemispherical soil model   b) Cylindrical soil model

Fig. 4.145 Finite element models of a subsoil with a rectangular plate [SOF 7]

The finite element method is particularly suited for foundation slabs and subsoil conditions with complex geometry. The mutual influence of several buildings can also be investigated with finite element models. Stress superpositions and settlements induced by neighbor structures can be represented by the model [185].

Continuum model of a layered soil

In a continuum model, the soil is represented as an elastic half-space, which can be homogeneous or layered. It is also denoted as constrained modulus method. The constrained modulus M is a soil parameter. Typical values according to EAU [186], are given in Table 4.36.

In order to determine the stiffness matrix of the soil related to the nodal points of the foundation slab, first, the corresponding flexibility matrix is determined. For this, the displacements $f_{i,k}$ at the half-space surface at a point i due to a force F_k (or an equivalent surface load) at point k are required (Fig. 4.146). Closed solutions

Table 4.36 Constrained modulus

Soil type	M [MN/m^2]	Soil type	M [MN/m^2]
Sand, loose, round	20–40	Clay, stiff	2.5–5
Sand, loose, angular	40–80	Clay, soft	1–2.5
Sand, medium dense, round	50–100	Glacial till, firm	30–100
Sand, medium dense, angular	80–150	Silt, medium	5–20
Gravel without sand	100–200	Silt, soft	4–8
Clay, medium	5–10	Turf	0.4–1

Fig. 4.146 Determination of the flexibility matrix for a stratified half-space

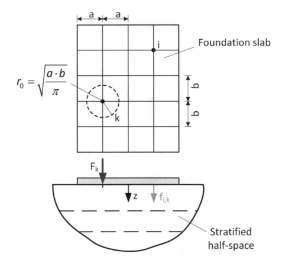

are only available for the homogeneous half-space. The vertical displacement w at distance r from a point force F is obtained as

$$w(r) = \frac{1 - \mu^2}{E} \cdot \frac{F}{\pi \cdot r}. \tag{4.185a}$$

Herein E is the modulus of elasticity and μ the Poisson ratio of the soil [187]. Under the point force, a singularity for displacements exists. Here the displacement can be determined by distributing the force on a circular area with the radius r_0. The radius is chosen so that the area $(\pi \cdot r_0^2)$ is equal to the influence area of the reference node (Fig. 4.146). With $p = F/(\pi \cdot r_0^2)$ the displacement in the center of the circular area is obtained as

$$w(0) = \frac{1 - \mu^2}{E} \cdot 2 \cdot r_0 \cdot p. \tag{4.185b}$$

There are also mechanically consistent solutions for soils with horizontal stratification, for example the semi-analytic method in [188] developed for dynamic problems. It can be applied to static problems setting the circular frequency as $\Omega = 0$. Nevertheless, most finite element programs are implemented with approximate solutions. In the case of a stratified soil, based on the assumption of a stress distribution $\sigma_z(z)$ of a homogeneous half-space, the displacements at the soil surface are determined by means of a settlement calculation. The settlement at point i of the stratified soil is thus obtained by integrating the strains $\varepsilon_z(z) = \sigma_z(z)/M(z)$ from $z = 0$ to the limiting depth z_{gr}. The inversion of the flexibility matrix results in the stiffness matrix of the soil, which is added to the stiffness matrix of the foundation or the building model [189–191]. In the case of large foundation slabs with many degrees of freedom, the inversion of the (fully occupied) flexibility matrix of the soil is very time-consuming.

Frequently, various soil layers or drilling profiles can be found at different points on a foundation slab. There are no mechanically consistent analytical solutions for this problem. One can remedy this by starting from the solution for horizontal layers. The diagonal term f_{kk} of the flexibility matrix is determined for the soil profile at point k and the non-diagonal term as $(f_{ik} + f_{ki})/2$, i.e. as a mean of the two different settlements f_{ik} (settlement at point i due to the force at k with the soil profile at k) and f_{ki} (settlement at point k due to the force at i with the soil profile at i). This approximation gives symmetrical flexibility and stiffness matrices.

Tables for circular slabs on an elastic half-space with point, line, and area loads are given in [192], an analytical method in [193]. They can be useful for simplified checking calculations of finite element computations. For example, one can obtain the deflections and bending moments of an infinitely extended elastic plate with a single load with the formulas given in Table 4.37. They can also be applied approximately to plates with a finite radius of at least $1.5 \cdot L$.

Table 4.37 Infinitely extended plate on an elastic half-space with a point load [192]

Elastic characteristic of the plate:

$$L = t \cdot \sqrt[3]{\frac{1}{6} \cdot \frac{1-\mu_{Bod}^2}{1-\mu_{Pla}^2} \cdot \frac{E_{Pla}}{E_{Bod}}} \qquad \rho = r/L \quad 1.5 \cdot L \le d$$

Deflection: $w = \bar{w} \cdot \dfrac{1-\mu_{Bod}^2}{E_{Bod} \cdot L} \cdot F$

Bending moments: $m_r = \bar{m}_r \cdot F \quad m_t = \bar{m}_t \cdot F$

$\rho = r/L$	0.00	0.05	0.10	0.20	0.30	0.40	0.60	0.80	1.00	1.20	1.40
$\bar{w}$	0.385	0.384	0.382	0.377	0.369	0.359	0.338	0.314	0.291	0.268	0.247
$\bar{m}_r$	∞	0.256	0.191	0.129	0.093	0.068	0.037	0.016	0.004	−0.005	−0.011
$\bar{m}_t$	∞	0.332	0.258	0.195	0.158	0.132	0.097	0.074	0.057	0.045	0.035

$\rho = r/L$	1.60	1.80	2.00	2.20	2.40	2.60	2.80	3.00	3.20	3.60	4.00
$\bar{w}$	0.226	0.207	0.189	0.173	0.159	0.146	0.135	0.124	0.115	0.099	0.087
$\bar{m}_r$	−0.014	−0.015	−0.016	−0.016	−0.016	−0.015	−0.014	−0.013	−0.012	−0.010	−0.007
$\bar{m}_t$	0.028	0.022	0.018	0.014	0.011	0.009	0.007	0.006	0.005	0.003	0.002

Subgrade reaction model

The subgrade reaction method also denoted as Winkler's method is a simple method to model the soil that has been established as standard in practice. The modulus of subgrade reaction is defined as

$$k_S = \frac{\sigma_k}{w_k} \qquad (4.186)$$

where σ_k is the compressive stress and w_k the displacement (settlement) at a point k of the foundation slab. Hence, the subgrade reaction module is a spring constant distributed over the unit area. Analytical methods and Tables for the analysis of continuously supported bending plates are given in handbooks as [194–196]. In finite element analyses, special plate elements have been formulated where the continuous elastic support is taken into account in the element formulation [51, 197, 198] (cf. (4.83a)). Alternatively, the elastic support can be discretized and represented by individual springs at the nodal points of the finite element mesh of the foundation slab. The spring constant at a nodal point can easily be determined by multiplying the modulus of subgrade reaction with the reference area of the point. The maximum element size according to (3.48), is normally observed due to the general requirements for the element size.

The advantage of the subgrade reaction model is its simplicity and ease of programming. It is therefore available in most finite element programs for structural analysis. For the modulus of subgrade reaction, tabular values such as for the constrained modulus cannot be specified, as it depends not only on the soil properties but also on the foundation geometry and load. This can be demonstrated by the example of a rigid circular foundation on an elastic soil layer of depth h or an elastic half-space (with $h \to \infty$) (Fig. 4.147). Its vertical spring constant is given in [199, 200] as

$$F = p \cdot A_F = \frac{4 \cdot G \cdot r}{1 - \mu} \cdot \left(1 + 1.28 \frac{r}{h}\right) \cdot w. \qquad (4.187a)$$

Herein, $G = 0.5 \cdot E/(1 + \mu)$ is the shear modulus of soil and $A_F = \pi \cdot r^2$ the area of the foundation slab. The compressive stress at the characteristic point (see below) is $\sigma_k = p = F/A_F$, the settlement is $w_k = w$. The modulus of subgrade reaction is thus obtained as

$$
\begin{aligned}
k_S &= \frac{\sigma_k}{w_k} = \frac{F}{A_F \cdot w_k} = \frac{4 \cdot G}{\sqrt{\pi} \cdot (1 - \mu)} \cdot \frac{1}{\sqrt{A_F}} \cdot \left(1 + 1.28 \frac{r}{h}\right) \\
&= \frac{2}{\sqrt{\pi}} \cdot \frac{1 - \mu - 2 \cdot \mu^2}{(1 + \mu) \cdot (1 - \mu)^2} \cdot \frac{M}{\sqrt{A_F}} \cdot \left(1 + 1.28 \frac{\sqrt{A_F}}{\sqrt{\pi \cdot h}}\right). \qquad (4.187b)
\end{aligned}
$$

The formula applies approximately to any foundation shape, which deviates from the circular shape. It can be seen that the modulus of subgrade reaction decreases with

Fig. 4.147 Rigid circular
foundation on an elastic soil
layer

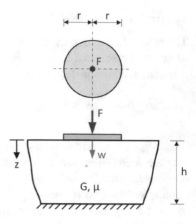

increasing foundation area, i.e. with the root of the foundation area A_F. If one doubles
the foundation area in the case of half-space ($h \to \infty$), the modulus of subgrade
reaction decreases by $1/\sqrt{2} = 0.71$ with otherwise identical soil properties, if one
halves the foundation area, it increases by $1/\sqrt{0.5} = 1.41$.

Normally, the subgrade reaction modulus is determined by a geotechnical consultant by means of a settlement analysis of the stratified soil. The Boussinesq equations provide the stresses in a homogeneous, isotropic half-space under a single load
distributed over a small rectangular area. With the elastic soil parameters, the vertical
strains are determined and, by integrating them from 0 to z_{gr}, the settlements on the
soil surface are obtained. In general, the settlement w_m is determined at the so-called
representative point of a foundation where the settlements are the same for a fully
flexible and a rigid foundation slab. The modulus of subgrade reaction is thus

$$k_{S,m} = \frac{\sigma_m}{w_m},   \tag{4.188}$$

where σ_m is the mean compression stress under the foundation slab. Further information and approaches for determining the modulus of subgrade reaction are given
in [201–204].

The subgrade reaction model does not take into account the mutual influence
of adjacent soil zones. It can therefore reproduce the mechanical behavior of the
half-space only roughly. This is demonstrated by two simple examples. Figure 4.148
shows the soil pressures under a constant uniformly distributed load on a totally
flexible foundation and under a rigid foundation slab. With the subgrade reaction
model, the bending moments are zero for a uniformly distributed load, regardless of
the bending stiffness of the foundation. In a flexible foundation with $E_F \cdot I_F \neq 0$
bending moments are obtained only with the continuum model. Another example
is given in Fig. 4.149. A one-sided uniformly distributed load is acting on one half
of the slab. In this case, the bending moments are antimetric by mistake, also if the

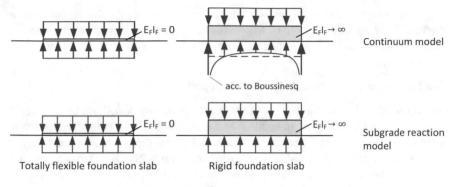

Soil stress distribution

Fig. 4.148 Soil pressures under a foundation slab with a uniformly distributed load

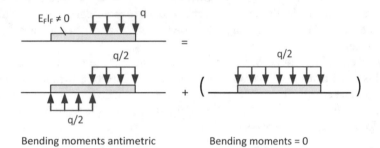

Bending moments antimetric   Bending moments = 0

Fig. 4.149 Subgrade reaction model of a flexible foundation slab with a one-sided uniformly distributed load

entire foundation is subjected to a uniformly distributed load in addition to the one-sided loading. Here, again an unrealistic result is obtained with the subgrade reaction model.

In order to achieve a better approximation to Boussinesq's stress distribution, the modulus of subgrade reaction in the edge zone of the foundation slab can be increased. Different distributions were proposed in literature. Typical is a doubling of the modulus of subgrade reaction in a strip with the width "$b = 0.1 \cdot slab\ width$" along the edge of the slab.

It is more consistent and accurate to adapt the distribution of the subgrade reaction moduli to the pressure-settlement distributions of the continuum model iteratively. This is only applicable for a single (the controlling) load combination. In the first step, the foundation slab is analyzed with a constant modulus of subgrade reaction. With the resulting soil pressure distribution, a settlement analysis with a continuum model is carried out. From the distribution of soil pressures and settlements over the foundation slab, using (4.186), the distribution of modulus of subgrade reaction is determined. With this distribution of the modulus of subgrade reaction, the foundation slab is analyzed again. The distribution of the moduli of subgrade reaction can be improved in further iteration steps until the internal forces of the foundation slab no longer change significantly [204]. Mathematically this corresponds to the iterative solution of the coupled system of equations of soil and structural model for a load combination.

Two-parameter elastic model (Pasternak model)

In order to improve the description of the elastic behavior of soil, multi-parameter-models have been developed. In addition to the modulus of subgrade reaction, multi-parameter models use other parameters to simulate the three-dimensional behavior of the soil (cf. [205]). A well-known two-parameter model is the Pasternak model. It is an extension of the subgrade reaction model with the aim of representing the mutual influence of adjacent soil zones in a simplified way. In addition to the elastic support, a "shear beam" is introduced. It can be understood as a near-surface soil layer of height h_{sb}, of which only its shear stiffness is taken into account. The model, discretized in individual springs, is shown in Fig. 4.150. Springs and shear beams extend beyond the foundation area in order to be able to represent a settlement trough. This zone should be chosen so large that the settlements at the edge have decayed almost to zero.

First, the equilibrium between the contact forces and the reaction forces of the ground is to be investigated for the two-dimensional case of a foundation beam of width Δy. The contact force $F_i = p_i \cdot \Delta x \cdot \Delta y$ acts at node i, where p is the surface pressure and $\Delta x \cdot \Delta y$ the reference area (Fig. 4.151). It is in equilibrium with the spring force $k \cdot w_i$ and the two shear forces $V_{zx,i,\mathrm{li}}$ and $V_{zx,i,\mathrm{re}}$. One obtains:

$$F_i = k \cdot w_i + V_{zx,i,\mathrm{li}} - V_{zx,i,\mathrm{re}} \tag{4.189}$$

The shear forces are determined with shear modulus G of the soil layer as

$$V_{zx,i,\mathrm{li}} = G \cdot h_{\mathrm{sb}} \cdot \Delta y \cdot \varphi_{y,\mathrm{li}}, \quad V_{zx,i,\mathrm{re}} = G \cdot h_{\mathrm{sb}} \cdot \Delta y \cdot \varphi_{y,\mathrm{re}}$$

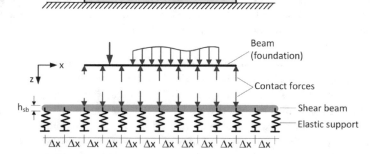

Fig. 4.150 Discretized Pasternak model

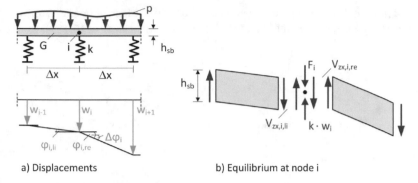

a) Displacements   b) Equilibrium at node i

Fig. 4.151 Equilibrium conditions

The spring constant is $k = k_0 \cdot \Delta x \cdot \Delta y$ with the modulus k_0 of subgrade reaction. Thus, the equilibrium conditions will be

$$p_i \cdot \Delta x \cdot \Delta y = k_0 \cdot \Delta x \cdot \Delta y \cdot w_i - G \cdot h_{sb} \cdot \Delta y \cdot \left(\varphi_{y,\text{re}} - \varphi_{y,\text{li}} \right)$$

or

$$p_i \cdot \Delta y = k_0 \cdot \Delta y \cdot w_i - G \cdot h_{sb} \cdot \Delta y \cdot \frac{\Delta \varphi_y}{\Delta x}$$

with $\Delta \varphi_y = \varphi_{y,\text{re}} - \varphi_{y,\text{li}}$. Taking the limit $\Delta x \to 0$, one obtains

$$p(x) \cdot \Delta y = k_0 \cdot \Delta y \cdot w(x) - G \cdot h_{sb} \cdot \Delta y \cdot \frac{\mathrm{d}\varphi_y}{\mathrm{d}x}$$

and with $\varphi_y = \mathrm{d}w/\mathrm{d}x$

$$p \cdot \Delta y = k_0 \cdot \Delta y \cdot w(x) - G \cdot h_{sb} \cdot \Delta y \cdot \frac{\mathrm{d}^2 w}{\mathrm{d}x^2}.$$

With the line load $\bar{p} = p \cdot \Delta y$ and the spring constant per unit length $\bar{k}_0 = k_0 \cdot \Delta y$ as well as with $\bar{k}_1 = G \cdot h_{sb} \cdot \Delta y$ the equation of the Pasternak support of a beam with the width Δy is

$$\bar{p}(x) = \bar{k}_0 \cdot w(x) - \bar{k}_1 \cdot \frac{\mathrm{d}^2 w}{\mathrm{d}x^2}. \tag{4.190}$$

For a foundation slab, the model must be extended. In addition to shear beams in the x-direction with a width Δy, shear beams in the y-direction with the width

Δx are provided. The model obtained is a "girder grid" of shear beams with vertical springs in the nodes. The equilibrium conditions result similar to (4.189), to

$$F_i = k \cdot w_i + V_{zx,i,\mathrm{li}} - V_{zx,i,\mathrm{re}} + V_{zy,i,\mathrm{li}} - V_{zy,i,\mathrm{re}}$$

with

$$V_{zy,i,\mathrm{li}} = G \cdot h_{\mathrm{sb}} \cdot \Delta x \cdot \varphi_{x,\mathrm{li}}, \quad V_{zy,i,\mathrm{re}} = G \cdot h_{\mathrm{sb}} \cdot \Delta x \cdot \varphi_{x,\mathrm{re}}.$$

Hence, it is

$$p_i \cdot \Delta x \cdot \Delta y = k_0 \cdot \Delta x \cdot \Delta y \cdot w_i - G \cdot h_{\mathrm{sb}} \cdot \Delta y \cdot \Delta\varphi_y - G \cdot h_{\mathrm{sb}} \cdot \Delta x \cdot \Delta\varphi_x$$

and

$$p_i = k_0 \cdot w_i - G \cdot h_{\mathrm{sb}} \cdot \left(\frac{\Delta\varphi_y}{\Delta x} + \frac{\Delta\varphi_x}{\Delta y} \right).$$

The equation of the Pasternak support of a foundation slab is obtained as

$$p(x) = k_0 \cdot w(x) - k_1 \cdot \left(\frac{\partial^2 w}{\partial x^2} + \frac{\partial^2 w}{\partial y^2} \right). \tag{4.191}$$

Here k_0 corresponds to the modulus of subgrade reaction and $k_1 = G \cdot h_{\mathrm{sb}}$ to the shear stiffness of the shear beam per length unit representing the shear stiffness of the soil layer.

In [107, 206] it is recommended, with reference to Pasternak [207], to select the parameters for a soil layer of height h above a rigid subsoil as follows:

$$k_0 = \frac{E}{h \cdot (1 - 2\mu^2)} \tag{4.192a}$$

$$k_1 = \frac{E \cdot h}{6 \cdot (1 + \mu)} = G \cdot \frac{h}{3}. \tag{4.192b}$$

Pasternak determined the parameter k_0 in [207] from the assumption that for the horizontal stresses in a three-dimensional soil $\sigma_x = \sigma_y = \mu \cdot \sigma_z$ model applies. With Hook's law of the three-dimensional continuum according to Fig. 2.15 one obtains

$$\varepsilon_z = \frac{1 - 2 \cdot \mu^2}{E} \cdot \sigma_z, \quad \varepsilon_x = \varepsilon_y = \frac{-\mu^2}{E} \cdot \sigma_z$$

and therefrom the displacement at the surface of a layer of height h (Fig. 4.147) to

$$w_s = \varepsilon_z \cdot h = \frac{1 - 2 \cdot \mu^2}{E} \cdot h \cdot \sigma_z.$$

From this, the parameter k_0 is determined to

$$k_0 = \frac{\sigma_z}{w_s} == \frac{E}{h \cdot (1 - 2 \cdot \mu^2)}$$

as given in (4.192a). The derivation of the parameter k_1 in [207] is based on a linear distribution of the horizontal displacements in the soil layer and thus on a constant shear angle.

With reference to Barwashov [208] the following parameters are given in [107]:

$$k_0 = \frac{E}{h \cdot (1 - \mu^2)}, \tag{4.193a}$$

$$k_1 = \frac{E \cdot h}{20 \cdot (1 - \mu^2)} = \frac{G}{(1 - \mu)} \cdot \frac{h}{10}. \tag{4.193b}$$

More detailed investigations on the parameters k_0 and k_1 for an elastic soil layer on a rigid subsoil were carried out in [209, 210]. Vlasov determines the two parameters as a function of the distribution of vertical displacements (settlements) with depth z [209]. It is approximately assumed that their distribution is independent on x, y. In addition, it is assumed that the horizontal displacements in soil are zero. Then, one obtains with the constrained modulus M

$$k_0 = \frac{M}{h} \cdot \varphi_k \tag{4.194a}$$

$$k_1 = G \cdot h \cdot \varphi_G \tag{4.194b}$$

with

$$\varphi_k = h \cdot \int_0^h \psi'(z)^2 \, dz \tag{4.195a}$$

$$\varphi_G = \frac{1}{h} \cdot \int_0^h \psi(z)^2 \, dz. \tag{4.195b}$$

The dimensionless function $\psi(z)$ specifies the (assumed) distribution of the vertical displacements in the layer, related to the displacement at the soil surface. It is $\psi(0) = 1$ and $\psi(h) = 0$. The function thus has the character of a shape function. With it, the coefficients k_0 and k_1 can be determined under the assumption of a restrained lateral deformation (horizontal strain $= 0$) for any vertical displacement distribution.

For an analytical description of the settlement distribution, the following functions are assumed in [209]:

$$\psi(z) = 1 - z/h \quad \text{for } \eta = 0 \tag{4.196a}$$

$$\psi(z) = \frac{\sinh(\eta \cdot (1 - z/h))}{\sinh(\eta)} \quad \text{for } \eta \neq 0 \tag{4.196b}$$

The parameter η determines the distribution of the settlements with depth (Fig. 4.152). The case $\eta = 0$ describes the linear distribution in a homogeneous soil layer of height h. The coefficients φ_k and φ_G are obtained as a function of η to (Fig. 4.153):

$$\varphi_k = \eta \cdot \frac{\sinh(\eta) \cdot \cosh(\eta) + \eta}{2 \cdot \sinh^2(\eta)} \tag{4.197a}$$

$$\varphi_G = \frac{1}{\eta} \cdot \frac{\sinh(\eta) \cdot \cosh(\eta) - \eta}{2 \cdot \sinh^2(\eta)} \tag{4.197b}$$

For a homogeneous soil layer with $\eta = 0$, the coefficients are obtained as $\varphi_k = 1$ and $\varphi_G = 1/3$. As expected, the coefficient k_0 which corresponds to the modulus of subgrade reaction of a soil layer in the case of restrained lateral displacements is

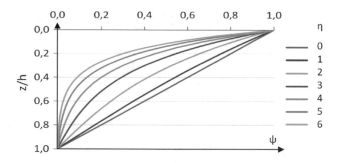

Fig. 4.152 Dimensionless parameter $\psi(z)$ for the distribution of the settlements with depth

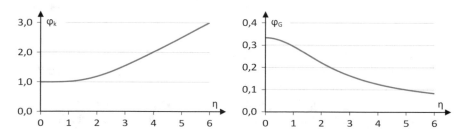

Fig. 4.153 Parameters φ_k and φ_G

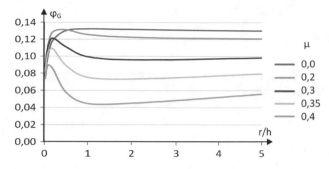

Fig. 4.154 Parameter φ_G for a rigid circular foundation on an elastic layer [213]

thus M/h. The coefficient k_1 of the homogeneous layer is $G \cdot h/3$, i.e. the height h_{sb} of the shear beam is one third of the height h of the soil layer.

The determination of the coefficients φ_k and φ_G for stratified soils under the above assumptions is discussed in [209, 211, 212].

A rigid circular foundation on an elastic layer is investigated by Tanahashi [213] with the Pasternak model (Fig. 4.147). The settlement through obtained were compared with those of a three-dimensional finite element model of the soil. In this way the parameters of the Pasternak model were adapted. The coefficients φ_G given in Fig. 4.154, were determined as a function of the Poisson's ratio and the ratio r/h under the assumption that $\varphi_k = 1$. They apply for $r/h \geq 0.2$ and $\mu \leq 0.3$. The values are significantly lower than the value of $1/3$ for a homogeneous soil layer with a surface load and constrained lateral strain.

For the Pasternak model, analytical solutions [214], as well as finite beam [215, 216] and plate elements [197, 198, 217, 218], have been derived. With the parameter $k_1 = 0$, they can also be applied to the subgrade reaction model.

Model comparison

The continuum model and the subgrade reaction model differ substantially in their mechanical model assumptions. The impact of this on the internal forces and settlements of a foundation slab is to be investigated in a parameter study on the example of a rectangular foundation. The slab is loaded separately by an area, line, and single load (Fig. 4.155) [219, 220]. The thickness of the foundation slab varies between 30 and 50 cm, the modulus of elasticity of the concrete is 31000 MN/m^2 and the Poisson ratio 0.2. The foundation slab is investigated with and without rigid stiffening walls at its edges (Fig. 4.156).

The soil properties and slab thicknesses of the parameter study are given in Table 4.38. The subgrade reaction moduli were determined according to (4.187b). A total of 4 cases with increasing system stiffness k are investigated. The calculations were carried out with a continuum model implemented in [SOF 6] (module HASE) and a discretization of the slab into 8×10 finite elements.

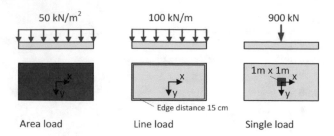

Area load   Line load   Single load

Fig. 4.155 Types of loading

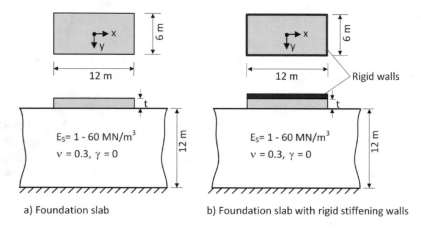

a) Foundation slab   b) Foundation slab with rigid stiffening walls

Fig. 4.156 Parameters of a rectangular foundation slab

Table 4.38 Values of the parameter study

Parameter	Case 1	Case 2	Case 3	Case 4
Soil type	Silty fine sand	Semi-stiff loam	Semi-stiff loam	Mushy clay
Constrained modulus M [MN/m^2]	60	13.5	13.5	1
Poisson ratio μ	0.3	0.3	0.3	0.3
Modulus of subgrade reaction k_S [MN/m^3]	9.87	2.22	2.22	0.16
Plate thickness t [m]	0.3	0.3	0.5	0.5
System stiffness k according to Meyerhof type of foundation slab	0.00067	0.003	0.014	0.19
	$k < 0.001$	$0.001 < k < 0.01$	$0.01 < k < 0.1$	$k > 0.1$
	$\rightarrow$ flexible	$\rightarrow$ semi-flexible	$\rightarrow$ semi-stiff	$\rightarrow$ stiff

Figures 4.157 and 4.158 show the bending moments and settlements of the foundation slab determined with the subgrade reaction model as well as with the continuum model under single and line loads. The diagrams show the distribution of the settlements and bending moments m_{xx} and m_{yy} in section 1–1 for case 2 ($t = 0.30$ m) and case 3 ($t = 0.50$ m) and semi-stiff loam. The maxima of the settlements and the

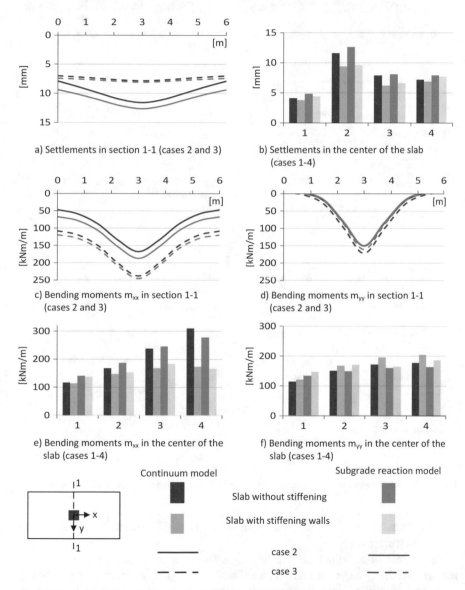

a) Settlements in section 1-1 (cases 2 and 3)

b) Settlements in the center of the slab (cases 1-4)

c) Bending moments m_{xx} in section 1-1 (cases 2 and 3)

d) Bending moments m_{yy} in section 1-1 (cases 2 and 3)

e) Bending moments m_{xx} in the center of the slab (cases 1-4)

f) Bending moments m_{yy} in the center of the slab (cases 1-4)

Continuum model   Subgrade reaction model

Slab without stiffening

Slab with stiffening walls

case 2

case 3

Fig. 4.157 Bending moments and settlements of a foundation slab with a single load

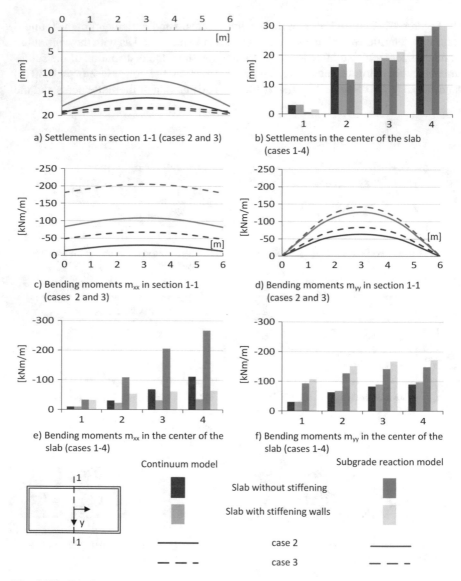

a) Settlements in section 1-1 (cases 2 and 3)

b) Settlements in the center of the slab (cases 1-4)

c) Bending moments m_{xx} in section 1-1 (cases 2 and 3)

d) Bending moments m_{yy} in section 1-1 (cases 2 and 3)

e) Bending moments m_{xx} in the center of the slab (cases 1-4)

f) Bending moments m_{yy} in the center of the slab (cases 1-4)

Continuum model   Subgrade reaction model

Slab without stiffening

Slab with stiffening walls

case 2

case 3

Fig. 4.158 Bending moments and settlements of a foundation slab with line loads at the edges

bending moments m_{xx} and m_{yy} are given in the center of the slab, with and without (rigid) stiffening walls.

In the case of a single load, the bending moments determined with the subgrade reaction model and the continuum model agree well. The stiffness of the stiffening walls has only little influence on the bending moments and settlements in the center of the slab. Only in case 4, the moment m_{xx} in the center of the slab shows distinct

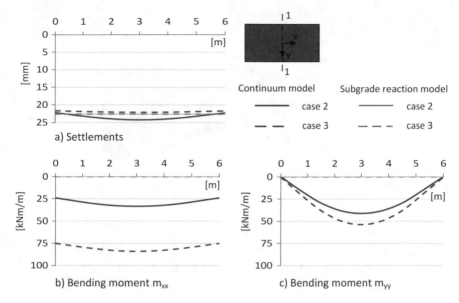

Fig. 4.159 Bending moments and settlements for a constant area load in section 1–1

differences for both models. Case 4, however, is practically irrelevant due to the low soil quality assumed.

The findings are different for the line load. The moments m_{xx} and m_{yy} determined in the center of the slab with the subgrade reaction model exceed those determined with the continuum model by a factor of 1.6–3.6. Increasing the modulus of subgrade reaction in an edge zone of the slab, as suggested by various authors, would be a possible remedy. Arranging stiffening walls on the edges reduces the moments m_{xx} significantly, while it increases the moments m_{yy}. If the wall stiffness is neglected, the internal forces of the foundation slab are normally on the safe side. The settlements determined with the subgrade reaction model and the continuum model differ only slightly—more with a line load than with a single load.

Foundation slabs may be subjected to area loads too. First, the case of a uniformly distributed load of 50 kN/m^2 according to Fig. 4.155, in cases 2 and 3 (semi-flexible and semi-stiff foundation) are considered. Figure 4.159 shows distributions of the settlements and bending moments m_{xx}. The settlements obtained with the subgrade reaction model and the continuum model agree well. In the subgrade reaction model, however—due to its model limitations—all bending moments are zero while there are positive moments m_{xx} (and m_{yy}) in the slab with the continuum model. The soil pressures according to the continuum model are—especially in the case of a rigid foundation—smaller in the middle and larger at the edge than the (constant) soil pressures according to the subgrade reaction model.

Very different moment distributions are obtained with the continuum and the subgrade reaction model in the case that one half of the slab is loaded with a uniform load q and the other half with a uniform load $2 \cdot q$ (Fig. 4.160). Figure 4.161

Fig. 4.160 Foundation slab
with unequal constant loads
on both halves

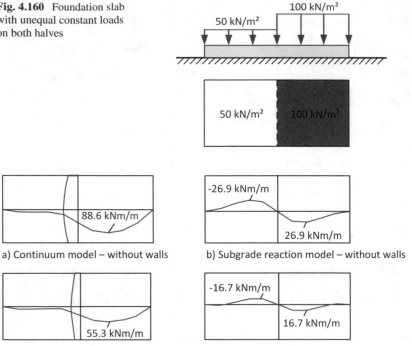

a) Continuum model – without walls   b) Subgrade reaction model – without walls

c) Continuum model – with stiffening walls   d) Subgrade reaction model – with stiffening walls

Fig. 4.161 Bending moments m_{xx} of a foundation slab with two unequal constant loads on both halves

shows the moments m_{xx} for a "semi-flexible foundation slab" (case 2) with and without stiffening by walls. Only with the continuum model, a mechanically appropriate moment distribution is obtained. With the subgrade reaction model, a positive bending moment is obtained in the right half of the slab and an equally large negative moment in the left half of the slab, which appears completely unrealistic. Here the subgrade reaction model fails. In addition, the moment obtained with the subgrade reaction model underestimates the magnitude of the moments, i.e. it is not on the safe side. The same applies to the moments m_{yy}. In practice, this may lead to a wrong distribution of reinforcement in the foundation slab. Stiffening walls significantly reduce the moments.

On the basis of the examples given above, it can be concluded that for determining the bending moments in foundation slabs the subgrade reaction model is well suited for single loads, conditionally suited for line loads and hardly suited for area loads. In practice, however, uniformly distributed area loads are only of minor importance for foundation slabs.

The distribution of the modulus of subgrade reaction under the foundation slab can be determined from the soil pressures and settlements. Figure 4.162 shows two

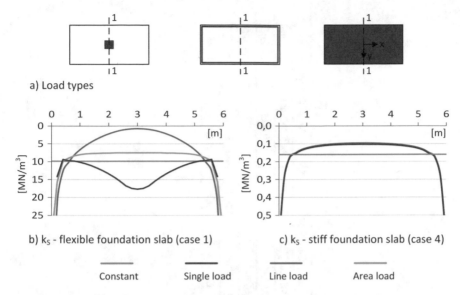

a) Load types

b) k_S - flexible foundation slab (case 1)   c) k_S - stiff foundation slab (case 4)

Constant   Single load   Line load   Area load

Fig. 4.162 Distributions of the modulus of subgrade reaction with different load types under a flexible and a stiff foundation slab

distributions in section 1–1 for cases 1 (flexible foundation slab) and 4 (stiff foundation slab). With the stiff foundation slab, the distribution of the modulus of subgrade reaction is the same for all load types. It is increased strongly at the edges with the continuum model due to the Boussinesq effect. In the case of the flexible foundation slab, the modulus of subgrade reaction in the continuum method is increased under the single load and the line loads, as to be expected. For computations using the subgrade reaction model, this can be taken into account approximately by increasing the modulus of subgrade reaction in an edge strip of the slab and, in the case of flexible slabs, also under single loads. At the same time, the modulus of subgrade reaction outside the edge strip should be reduced in order to maintain the overall stiffness. Increasing the modulus of subgrade reaction by a factor of 2 in an edge strip of $0.1 \cdot l$ is a common praxis with l being the smaller side length of a rectangular slab. The study shows, however, that higher factors can also occur (Fig. 4.162). This leads to an iterative improvement of the subgrade reaction moduli which is optimal but also time-consuming. The aim here is to improve the accuracy of the subgrade reaction model, especially for line loads but also for single loads.

For comparison, cases 1 to 4 are investigated with the two-parameter elastic model. With the soil parameters given in Table 4.38, the values k_0 and k_1 given in Table 4.39, are obtained. The results of the Pasternak model are compared with a reference model in which the soil is modeled with three-dimensional solid elements [221]. The hexaeder elements possess third-order polynomials as displacement shape functions [SOF 9]. The model has a size of $18 \times 15 \times 12$ m and comprises a quarter of the total system considering the symmetry conditions. As expected, its results agree very

Table 4.39 Parameters of the 2-parameter model according to Pasternak

	Case 1		Case 2		Case 3		Case 4	
	k_0 [MN/m^3]	k_1 [MN/m]	k_0 [MN/m^3]	k_1 [MN/m]	k_0 [MN/m^3]	k_1 [MN/m]	k_0 [MN/m^3]	k_1 [MN/m]
Pasternak[a]	4.530	68.571	1.019	15.429	1.019	15.429	0.075	1.143
Barwaschow[b]	4.082	29.388	0.918	6.612	0.918	6.612	0.068	0.490
Tanahash[c]	5.000	22.629	1.125	5.091	1.125	5.091	0.083	0.377
Vlasov $\eta = 0$[d]	5.000	68.571	1.125	15.429	1.125	15.429	0.083	1.143

[a] (4.192a, 4.192b)
[b] (4.193a, 4.193b)
[c] (4.194a, 4.194b) (Fig. 4.154)
[d] (4.194a, 4.194b)

well with those of the continuum model according to [SOF 6], discussed previously (Figs. 4.157, 4.158).

The bending moments in the middle of the plate obtained with the different models with [SOF 9] are shown in Fig. 4.163 [221]. For a single load, the moments obtained with the Tanahashi and Barwaschow parameters k_0 and k_1 correspond well with those of the continuum and the finite element model. The Pasternak formula, on the other hand, gives somewhat smaller bending moments. In the case of the line loads, all two-parameter models exhibit significantly lower moments than the continuum method. They are therefore on the unsafe side. The largest discrepancies are with the Pasternak parameters for k_0 and k_1. This load case also shows large discrepancies between the modulus of subgrade reaction model and the continuum model, but on the safe side (Fig. 4.158). For the area load, too, clear differences can be seen between the two-parameter models and the reference solution, albeit on the safe side. Of the two-parameter models examined, Tanahashi's approach yields the best results, followed by Barwaschow's approach. The Pasternak approach leads to larger

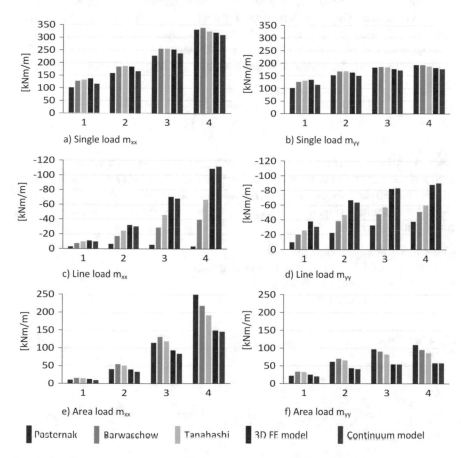

Fig. 4.163 Plate bending moments with the two-parameter elastic models [221]

differences compared to the reference solution and should therefore not be used in the light of the results presented here. This also applies to the settlements not shown here, which are determined too low with the Pasternak formulas. Probably the height of the shear beam is overestimated with (4.192b). Vlasow's approach with $\eta = 0$ gives similar parameters as Pasternak's approach and therefore leads to similar results. The models with $\eta = 0.33, \ 1.00, \ 3.00$ also examined in [221] showed no improvement compared to the approaches mentioned.

The results show that the two-parameter models still need to be investigated with regard to the methodology for determining the parameters k_0 and k_1 in order to achieve reliable results. This applies all the more to stratified soils with spatial variation in soil parameters.

Modeling of the foundation slab and the structural stiffness

The stiffness of rising reinforced concrete walls influences the internal forces in foundation slabs and must therefore not be neglected in the modeling process. Usually, walls are represented as L-shaped beams partly integrated into the slab and modeled by beam elements. The three-dimensional modeling of the lower stories as a folded structure with shell elements is more elaborate, but more realistic too.

In general, a structure can be flexible or stiff compared to the subsoil [201, 222]. In DIN 4018 [223], the system stiffness according to Meyerhof [224], is specified by the parameter

$$k = \frac{E_{\text{Pla}}}{12 \cdot M} \cdot \left(\frac{t}{l}\right)^3.$$
(4.198)

It describes the stiffness ratio of the foundation slab (or more precisely the superstructure) and the substructure. Here E_{Pla} is the modulus of elasticity of the concrete, M the constrained modulus of the soil, t the slab thickness and l the (largest) side length of a rectangular slab (Fig. 4.164). Depending on the system stiffness the following classification is made:

- $k < 0.001$ flexible foundation slab
- $0.001 \leq k < 0.01$ semi-flexible foundation slab
- $0.01 \leq k < 0.1$ semi-stiff foundation slab
- $0.1 < k$ stiff foundation slab

Fig. 4.164 Parameters of the system stiffness of a rectangular foundation slab

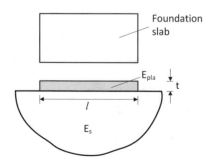

Fig. 4.165 Loads near the edges of foundation slabs

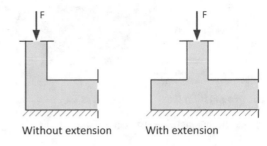

Without extension   With extension

A concentration of ground pressure appears under columns and walls. The more flexible the plate is compared to the subsoil, i.e. the smaller the parameter k, the higher is the concentration of soil pressures. The higher the concentration of soil pressures, the lower the bending moments in the foundation slab. The greatest bending moments occur with a stiff foundation slab.

Structural execution states may play a role with regard to structural stiffness. When concreting, the fresh concrete weight of a wall acts as a pure load. Only after the concrete has hardened the wall does possess its stiffness, which is effective for subsequent loads from the floors above and traffic loads. Shrinkage and creep of the concrete cause a further redistribution of the internal forces in the overall system.

The stiffening effect of walls is of particular importance when the load distribution on a foundation slab varies greatly. This is the case, for example, when two building units of different weights are located on a common foundation slab. Without modeling bracing walls connecting the two units, large settlements would occur under the heavy unit of the building, causing large curvatures in the transition area to the less loaded part of the foundation slab. If both units are connected by bracing walls, these curvatures and the corresponding bending moments do not occur in the foundation slab [225].

In the case of loads close to the edge of a foundation slab, their distance from the edge has considerable impact on the internal forces and bending moments in the foundation slab and—in the case of line loads from walls—of the rising walls. A slight extension of the foundation slab can noticeably reduce the bending moments (Fig. 4.165).

Example 4.41 The influence of the subsoil model on the internal forces is to be investigated for the foundation slab of a multi-storey building (Fig. 4.166) [219, 220]. The foundation slab has a size of 18×12 m, it is stiffened at the edges by rising basement walls. Two columns with a contact area of 0.40×0.40 m are acting on the slab. The foundation slab is loaded by the columns with two single loads of 1700 kN each, by the walls with a continuous line load of 138 kN/m and additionally by an area load of 10 kN/m² (Fig. 4.166). The resulting load is composed of the individual loads of $2 \cdot 1700 = 3400$ kN, the line loads of $(2 \cdot 18 + 2 \cdot 12) \cdot 138 = 8280$ kN and the area load of $18 \cdot 12 \cdot 10 = 2160$ kN.

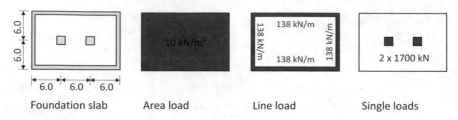

Fig. 4.166 Foundation slab with loading

Table 4.40 Soil parameters

Model	"Gravel"	"Clay"
Soil type	Gravel	Semi-stiff clay
Constrained modulus M [MN/m²]	100	15
Poisson ratio μ	0.3	0.45
Modulus of subgrade reaction k_S [MN/m³]	8.1	0.5
Plate thickness d [m]	0.3	0.5
System stiffness according to Meyerhof	$k = 0.001$ → flexible foundation slab	$k = 0.004$ → semi-flexible foundation slab

Two types of subsoil are being investigated: a gravel soil with a constrained modulus M of 100 MN/m² and a slab thickness of 0.3 m and a clay soil with a constrained modulus of 15 MN/m² at a slab thickness of 0.5 m (Table 4.40). The system stiffness k is 0.001 in the case of gravel soil and 0.004 in the case of clay soil, i.e. the foundation is flexible (gravel) or a semi-flexible slab (clay), respectively. The limiting depth z_{gr} was assumed to be 36 m.

For comparison, computations are carried out with three soil models:

- Continuum model
- Subgrade reaction model with a constant modulus of subgrade reaction
- Subgrade reaction model with a constant modulus of subgrade reaction with an augmentation of the subgrade modulus by a factor of 2 in an edge zone with a width of 1.20 m.

Figure 4.167 shows the bending moments m_{xx} for single, line, and area loads in a longitudinal section through a foundation slab without stiffening for both soil types. First, the case of the gravel soil is considered. For single loads, the moment distributions are similar to all models. The subgrade reaction model (411 kNm/m) yields somewhat higher values than the continuum model (355 kNm/m). For line loads, however, with the subgrade reaction model (-100 kNm/m), about 4 times the values of the continuum model (-26 kNm/m) are obtained. A significant improvement is achieved with the model where the modulus of subgrade reaction is augmented in an

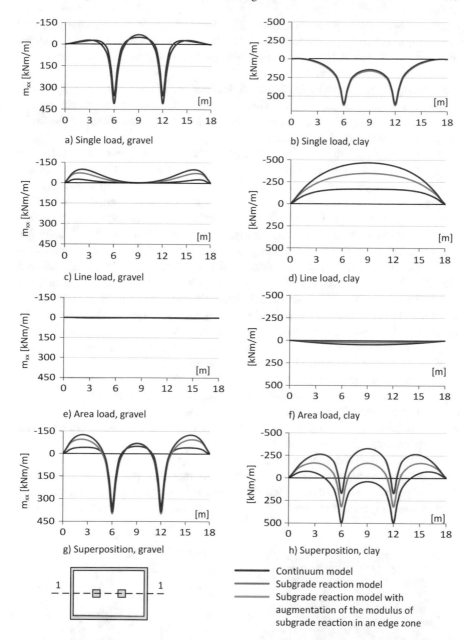

Fig. 4.167 Bending moments m_{xx} in section 1–1 with single, line and area loads for two types of soil

edge zone by a factor of 2 (−72 kNm/m). The bending moments due to the area load are small, in the subgrade reaction model with constant subgrade reaction modulus they are zero. Figure 4.167g shows the superposition of the bending moments from single, line, and area loads. For the gravel soil with a low slab thickness (30 cm), the bending moments due to line loads are low and the moments due to single loads under the column become decisive. These are well reproduced by all models, i.e. all models yield very similar maximum bending moments. The computations with the subgrade reaction model are on the safe side.

For the clay soil, the bending moments under columns are considered. With the single loads, the positive moments obtained with the subgrade reaction model (620/621 kNm/m) and the continuum model (622 kNm/m) agree well. However, with the line loads, the magnitude of the moments is considerably overestimated by the subgrade reaction model (−469 vs. −170 kNm/m). This also applies to the model with the augmented modulus of subgrade reaction at the edge zone (−348 kNm/m). The bending moments due to the area load are rather low. The superposition of the moments from single, line and area loads, reveals considerable differences between the maximum moments obtained with the subgrade reaction model (169 kNm/m or 316 kNm/m) and the continuum model (496 kNm/m). The bending moment under the columns is considerably underestimated by the subgrade reaction model. The reason can be found in the superposition of the negative moment due to line loads—the magnitude of which is overestimated by the subgrade reaction model—with the positive moment due to single loads, which is nearly correct—i.e. the same as in the continuum model. In principle, this also applies to the gravel soil. However, due to the larger stiffness of the gravel soil, the moments of the line loads are decayed at the column and contribute only little to the superposed total moment. Similar results are also obtained for the moments m_{yy}. The depicted effect can also occur on gravel soil with columns close to the edge.

The results with the subgrade reaction model can be improved significantly by an iterative adaptation of the distribution of the modulus of subgrade reaction to the "soil pressure/ settlement" distribution obtained with the continuum model. The conventional separation of responsibilities in structural design, according to which

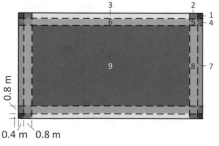

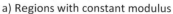

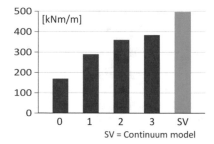

a) Regions with constant modulus   b) Convergence of m_{xx} under a column

Fig. 4.168 Iterative improvement of the distribution of the subgrade reaction modulus

Table 4.41   Distribution of the modulus of subgrade reaction [kN/m^3]

Region	Constant	Iteration 1	Iteration 2	Iteration 3
1	493	900	1350	1700
2	493	750	1000	1200
3	493	550	640	700
4	493	775	1050	1225
5	493	660	790	875
6	493	475	500	525
7	493	630	750	825
8	493	525	575	575
9	493	390	360	355

the structural engineer is responsible for the structure and the geotechnical consultant for the subsoil, is retained. In this example, an iteration is performed for the controlling load combination, i.e. for a superposition of the load cases single loads, line loads, and area loads. As the subgrade reaction moduli are in general not transferred as a general distribution from the geotechnical engineer to the structural engineer, they are divided into regions. For example, a subdivision into edge strips of 0.4 and 1.2 m width was selected on the basis of the settlement calculations (Fig. 4.168a). The subdivision chosen influences the computational results and the convergence velocity of the method. Starting with a constant modulus of subgrade reaction of 0.493 MN/m^3, the subgrade reaction moduli are adapted in different regions. Table 4.41 shows the modulus of subgrade reaction distribution in the 1st–3rd iteration step. The bending moments converge against those of the constrained modulus method. Figure 4.168b shows this for the moment m_{xx} under the column. After 3 iteration steps a suitable match is reached. For further improvement, the areas according to Fig. 4.168a, would have to be refined and the number of iterations increased.

Finally, the influence of the stiffening of the foundation slab by the surrounding basement walls is investigated. These have a height of 3.00 m and a thickness of 0.20 m. Two models are investigated. The basement walls are modeled as a folded

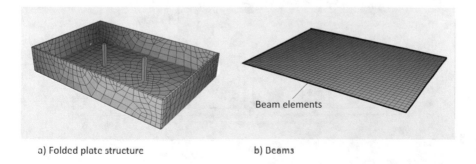

a) Folded plate structure   b) Beams

Fig. 4.169   Modeling of the basement walls

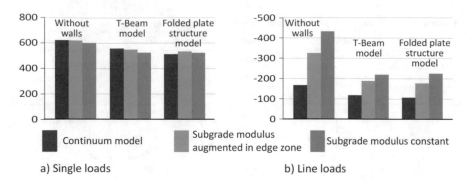

a) Single loads   b) Line loads

Fig. 4.170 Bending moment m_{xx} under the column with and without consideration of stiffening basement walls (clay soil)

structure with a hinge in the corner as well as with T-beams with an effective flange width of 2.0 m (Fig. 4.169). The folded structure model and the beam model provide similar results. In the case of clay soil (but also with gravel soil) and single loads, the moments under the column hardly differ with and without consideration of the stiffening walls. On the contrary, for line loads, the stiffening considerably influences the magnitude of the bending moments (Fig. 4.170). In this case, when designing the stiffening basement walls, the internal forces resulting from the analysis must also be taken into account.   ◄

Example 4.42 The statically indeterminate deep beam with two spans shown in Fig. 4.171 is positioned directly on a foundation slab. The soil is modeled with the subgrade reaction model. A soft soil (e.g. clay) with a subgrade modulus of 1800 kN/m^3 is assumed. The influence of the continuous elastic support of the foundation slab on the internal forces of the deep beam is to be investigated.

The foundation slab has a size of 16.00 m × 6.00 m, the edge distance of the deep beam is 1.0 m. A modulus of elasticity of $E = 3.0 \cdot 10^7$ kN/m^2 and a Poisson ratio

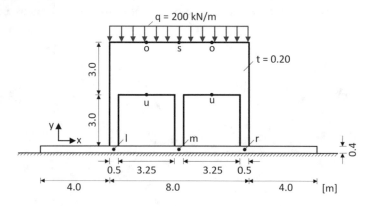

Fig. 4.171 Deep beam on a thin foundation slab

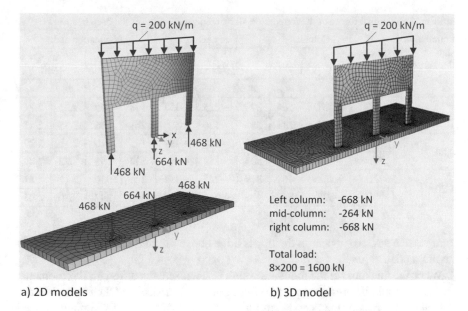

a) 2D models   b) 3D model

Fig. 4.172 Normal forces in the columns with the foundation slab with a thickness of 40 cm

of $\mu = 0.2$ are assumed for the foundation slab as well as for the deep beam. The plate thickness is assumed to be 40, 80, and 120 cm in a parameter study.

Two models are examined:

- 2D model: separate 2D models of the deep beam and the foundation slab
- 3D model: 3D model of the deep beam with the foundation slab

In the 2D model, the calculations for the deep beam and the foundation slab are carried out separately. The supporting forces are determined on the rigidly supported deep beam and applied to the foundation slab (Fig. 4.172a). With the 3D model, the calculation is carried out on the complete model, which consists of the deep beam and the foundation slab. First, the thin slab with a thickness of only 40 cm is examined. The 2D and the 3D models show a very different behavior in the column forces: The 2D model of the deep beam with rigid supports has the largest normal force in the mid-column. In the 3D model, the mid-column settles down due to the formation of the settlement trough of the foundation slab and the outer columns receive large normal forces (Fig. 4.172b). Due to the different column forces different moment distributions are obtained in the foundation slab. In the 2D model, the maximum moments m_x of the foundation slab are under the mid-column, where they are under the outer columns in the 3D model (Table 4.42). In the deep beam, the lower force in the mid-column in the 3D model reduces the continuous-beam effect and the bending moments in the field, and thus the stresses at the lower and upper edges of the deep beam increase. A pronounced compression arch is formed in the deep beam. Large

Table 4.42 Sectional forces of the foundation slab and normal forces of the deep beam

Foundation slab thickness [m]	Model	Columns N [kN]		Deep beam n_x [kN/m]			Foundation slab m_x [kN/m]		w_x [cm]	
		l/r	m	u	o	s	l/r	m	l/r	m
0.40	2D	−468	−664	379	−225	−191	148	229	2.31	2.71
	3D	−668	−264	553	−426	−593	261	56	2.35	2.47
0.80	2D	−468	−664	379	−225	−191	195	316	2.19	2.29
	3D	−595	−410	503	−359	−454	257	193	2.22	2.29
1.20	2D	−468	−664	379	−225	−191	206	336	2.17	2.21
	3D	−525	−550	436	−285	−311	242	287	2.20	2.23

compressive stresses appear at the upper edge above the mid-column, especially in the 3D model.

When the thickness of the foundation slab is increased, the differential settlements between the middle column and the outer columns decrease, and the column forces and internal forces of the deep beam approach those of the rigidly supported deep beam (Table 4.42). However, even with a 1.20 m thick foundation slab the column forces of a rigid support are not yet reached. The bending moments of the foundation slab in the 3D model are influenced by the restriction of the deformations by the deep beam, which is not represented by the 2D model.

Statically indeterminate deep beams are sensitive to the flexibility of the supports resulting in differential settlements. The settlement trough that develops can have a significant influence on the supporting forces and internal forces in the deep beam and the foundation slab. Boundary value analysis with min/max values must be performed in order to consider uncertainties in the modulus of subgrade reaction. In practice, the thickness of the foundation slab will be chosen as large as possible in order to reduce the influence of the soil–structure interaction on the internal forces of the deep beam.   ◄

Conclusions

The soil models discussed do not necessarily lead to the same results [206]. Reference models are the continuum model and the finite element modeling of the soil with three-dimensional solid elements. The subgrade reaction model which is generally used in practice for modeling the subsoil in the finite element analysis of foundation slabs normally leads to sufficiently accurate results. This is especially true for single loads from columns on foundation slabs. By contrast, the modulus of subgrade reaction model with a constant modulus of subgrade reaction is not suited for the area loads on a foundation slab, e.g. with a uniform load over large areas of the slab. In the case of line loads introduced by basement walls at the edge of a slab, the subgrade reaction model gives significantly larger negative bending moments than the continuum model. Especially with soft soils, this can lead to an underestimation of the resulting positive moments under columns when superimposing the positive moments due to column loads with the (too large) negative moments from line loads. This can be overcome by a more accurate modeling of the subsoil with the continuum model or an iterative adaptation of the subgrade reaction modulus distribution. Alternatively, the negative moments from line loads at the edges of a slab can be provided with a reduction factor when superimposing the positive moments under columns.

This leads to the following alternatives for selecting the subsoil model:

Recommended subsoil models
- Subgrade reaction model with constant subgrade modulus

  - Augmentation of the subgrade modulus at the edges of the foundation slab
  - Reduction factors when superposing plate bending moments of different signs

- Subgrade reaction model with the iterative adaption of the subgrade modulus distribution
- Continuum model (analytical solution)
- Three-dimensional finite element model of the soil with solid elements

In the case of the two-parameter elastic models, which are conceptually certainly of interest, more research on a reliable methodology for determining the parameters k_0 und k_1 is required.

Loss of contact between foundation slab and soil

Only compression stresses can be transmitted between a foundation and the ground. Therefore, after the finite element computation of a foundation slab, it must be checked on the basis of the soil pressures whether this is the case in all relevant load combinations. If tensile stresses occur in the calculation results, the calculation must be repeated with a model allowing for the formation of a contact loss between the foundation slab and the soil. This requires a nonlinear analysis. It can be done with all soil models mentioned above.

4.11.8 Modeling of Folded Plate and Shell Structures

Folded plate and shell structures are typical surface structures in structural engineering. The finite element method enables complex, three-dimensional structures to be realistically modeled and thus allows to take into account the interaction of the individual structural components. For particular structures, three-dimensional models achieve much more realistic results than the simplified, two-dimensional subsystems [139, 227 231].

The modeling of folded structures and shells does not differ substantially from that of plates in plane stress and bending. Some special features will be discussed in the following.

Panels

For the finite element discretization in panel regions, the rules specified for plates in plane stress and in bending apply. In addition, for some flat element types, the degrees of freedom of rotation about an axis perpendicular to the element surface must be fixed by inserting artificial rotational springs (see Sect. 4.8.1 and [227]).

For shells, bending moments at clamped edges possess peak values. At a short distance from the edge, however, the moments rapidly decrease, as the membrane action predominates there. An example of this are the clamping moments of a cylindrical shell under internal pressure, free at the upper edge and clamped at the lower edge. These have a characteristic distribution which is similar to the amplitude of a damped vibration [232]. The "period" of the solution, i.e. the distance between two (positive) maxima, is

$$L = \frac{2 \cdot \pi}{\sqrt[4]{3 \cdot (1 - \mu^2)}} \cdot \sqrt{t \cdot r}. \tag{4.199}$$

Herein, r is the radius of the cylindrical shell, t the wall thickness and μ the Poisson number. For $\mu = 0.2$

$$L = 4.82 \cdot \sqrt{t \cdot r}. \tag{4.199a}$$

Large cylindrical shells are defined by the condition $h > L$ where h is the height of the cylinder. At a small distance x from the clamped edge, the analytical solutions for the bending moments, shear forces, membrane forces and deflections are dominated of the term

$$e^{-\kappa \cdot x/r} \cdot \cos(\frac{\kappa \cdot x}{r}) \quad \text{with} \quad \kappa = 2 \cdot \pi \cdot \frac{r}{L}$$

[233]. The ratio of the solution at a distance $x = \sqrt{t \cdot r}$ to the solution at $x = 0$ can be estimated to be $e^{-\kappa \cdot \sqrt{t \cdot r}/r} \cdot \cos(\frac{\kappa \cdot \sqrt{t \cdot r}}{r})/1 = \sqrt[4]{3 \cdot (1 - \mu^2)}$. It is 7.2% for $\mu = 0.2$. This edge zone of the length

$$l_0 = \sqrt{t \cdot r} \tag{4.200}$$

where the moment peak occurs, must be discretized sufficiently fine [82]. Based on a study with 4-node DKQ elements, a subdivision into 2 to 4 elements is recommended if only the bending moments are to be determined. The reliable determination of shear forces requires an even finer discretization with at least 6 DKQ elements. Behind the edge zone, the solution decays rapidly. The ratio of two successive extreme values of an "damped vibration" in the distance $L/2$ is

$$e^{-\kappa \cdot (x+L/2)/r} \cdot \cos(\frac{\kappa \cdot (x+L/2)}{r}) \Big/ \left(e^{-\kappa \cdot x/r} \cdot \cos(\frac{\kappa \cdot x}{r})\right)$$

$$= -e^{-\kappa \cdot L/(2 \cdot r)} = -e^{-\pi} = 0.043 = 4.3\%$$

and in the distance L between two successive maxima

$$e^{-\kappa \cdot (x+L)/r} \cdot \cos(\frac{\kappa \cdot (x+L)}{r}) \Big/ \left(e^{-\kappa \cdot x/r} \cdot \cos(\frac{\kappa \cdot x}{r})\right)$$

$$= e^{-\kappa \cdot L/r} = e^{-2 \cdot \pi} = 0.0019 = 0.19\%.$$

Outside the edge zone of the length $l_0 = \sqrt{t \cdot r}$, larger element sizes can be chosen. These results apply to rigidly clamped edges. With elastic clamping, for example in a foundation slab, the gradient of the internal forces in the edge zone is lower than with rigid clamping, i.e. the element sizes selected for rigid clamping are on the safe side.

The solution behavior in the vicinity of clamped edges described above also applies to other rotationally symmetrical shells such as spherical shells (cf. Fig. 4.176). The edge zone can also be assumed here as $l_0 = \sqrt{t \cdot r}$ approximately. The discretization of a typical shell structure is discussed in [234]. Examples for the influence of the software used on the shell internal forces are given in [139], parameter studies with axisymmetric finite elements are presented in [20].

Example 4.43 The cylindrical shell shown in Fig. 4.173, is loaded by hydro-static pressure. The internal forces are to be determined for various finite element discretizations and compared with the analytical solution according to [233]. The parameters are chosen as in [233], as follows: $h = 5.5$ m, $r = 3.0$ m, $t = 0.28$ m, $E = 3 \cdot 10^7$ kN/m^2, $\mu = 0.2$ and $\gamma = 10$ kN/m^3.

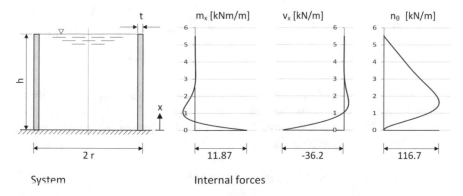

Fig. 4.173 Cylindrical shell with hydrostatic pressure

First, the analytical solution is considered. The internal pressure p_r increases linearly from the upper edge of the cylindrical shell downwards, so that with the specific weight of the liquid the following applies: $p_r = \gamma \cdot (h - x)$. The cylindrical shell is clamped at $x = 0$ (Fig. 4.173). Assuming that the influence of the upper edge on the internal forces at the lower edge is negligible (and vice versa), the internal forces according to [233] are obtained as:

Bending moment:

$$m_x(x) = \frac{E \cdot t^2}{2 \cdot r \cdot \sqrt{3 \cdot (1 - \mu^2)}} \cdot e^{\frac{-\kappa \cdot x}{r}} \cdot \left(C_1 \cdot \sin \frac{\kappa \cdot x}{r} - C_2 \cdot \cos \frac{\kappa \cdot x}{r} \right) \quad (4.201a)$$

Shear force:

$$v_x(x) = \frac{E \cdot t}{2 \cdot \kappa \cdot r} \cdot e^{\frac{-\kappa \cdot x}{r}} \cdot \left((-C_1 + C_2) \cdot \sin \frac{\kappa \cdot x}{r} + (C_1 + C_2) \cdot \cos \frac{\kappa \cdot x}{r} \right)$$
$$(4.201b)$$

Normal force:

$$n_\theta(x) = \frac{E \cdot t}{r} \cdot w(x) \quad (4.201c)$$

Displacement:

$$w(x) = e^{\frac{-\kappa \cdot x}{r}} \cdot \left(C_1 \cdot \cos \frac{\kappa \cdot x}{r} + C_2 \cdot \sin \frac{\kappa \cdot x}{r} \right) + \frac{\gamma \cdot r^2}{D \cdot (1 - \mu^2)} \cdot (h - x)$$
$$(4.201d)$$

with

$$D = \frac{E \cdot t}{1 - \mu^2}, \quad C_1 = \frac{-\gamma \cdot r^2}{D \cdot (1 - \mu^2)}, \quad C_2 = \frac{\gamma \cdot r^2}{D \cdot (1 - \mu^2)} \cdot \left(\frac{r}{\kappa} - h \right),$$
$$(4.201e)$$
$$\kappa = \sqrt[4]{3 \cdot (1 - \mu^2) \cdot \left(\frac{r}{t} \right)^2}$$

Herein E is the modulus of elasticity, μ the Poisson ratio number, t the wall thickness, r the radius and h the height of the cylindrical shell (Fig. 4.173). The solution given applies to high cylinders. The normal force in the upper part of the cylinder can be easily assessed by Barlow's formula as (Fig. 4.174)

$$n_\theta(x) = p_r(x) \cdot r = \gamma \cdot (h - x) \cdot r \quad (4.202)$$

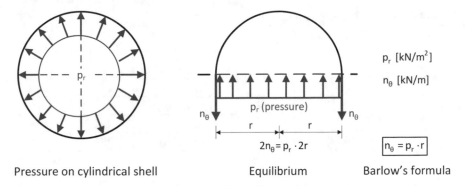

Pressure on cylindrical shell   Equilibrium   Barlow's formula

Fig. 4.174 Barlow's formula for a cylindrical shell with unconstrained deformations

Fig. 4.175 Cylindrical shell
- Part of the finite element
model with $e = 0.55$ m

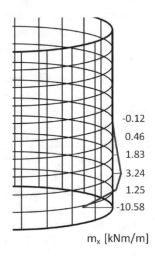

-0.12
0.46
1.83
3.24
1.25
-10.58

m_x [kNm/m]

With the parameters given above one obtains with (4.199) $L = 4.420$ m. Because of $h > L$ it represents a large cylindrical shell. The sectional forces are shown in Fig. 4.173.

The finite element analysis is performed with rectangular DKQ elements with [SOF 3a]. In the circumferential direction, the cylinder is discretized into 36 finite elements of a length of each $2 \cdot \pi \cdot r/36 = 0.524$ m. In the vertical direction, the element size is selected so that the edge zone of the length $l_0 = \sqrt{0.28 \cdot 3.00} = 0.917$ m is divided into 1, 1.7, 2 and 4 elements, respectively.

An example of a discretization and a typical moment distribution is shown in Fig. 4.175. The bending moment and the shear force at the clamped edge, as well as the maximum membrane force in the circumferential direction, are given in Table 4.43. The maximum membrane forces are obtained with good accuracy even with a discretization of l_0 in one element only. The error in the deflections is of the same magnitude, since the relationship (4.201c), between the deflections and the normal forces n_θ applies. In contrast, the bending moments possess a higher inaccuracy. With

Table 4.43 Internal forces of a cylindrical shell

	Element size [m]	$\dfrac{l_0}{e}$	m_x [kNm/m] (error)	q_x [kN/m]	n_θ [kN/m]
Analytical	-	-	11.8 (0%)	36.2	116.7
FEM	0.917	1	8.71 (26%)	13.9	123.0
	0.550	1.67	10.58 (10%)	21.5	119.9
	0.458	2	10.95 (7%)	23.8	116.9
	0.229	4	11.60 (2%)	29.8	117.0

an element size of $e = l_0/2$ the discretization error is $(11.80 - 10.95)/11.80 = 7.2\%$. The greatest inaccuracy occurs with the shear force. Here, with an element size of $e = l_0/4$, the error is $(36.2 - 29.8)/36.2 = 17.7\%$. The recommended discretization of the edge zone of the length l_0 into 2 to 4 elements leads to sufficiently accurate results for the bending moments. ◀

Supports and loads

In shell structures, the support conditions have a significant influence on the internal forces, especially on the bending moments in the vicinity of support. Since the magnitude of the bending moments reacts very sensitively to changes in the support conditions, in practice parameter studies should be performed. As an example, in [57] a spherical shell with a reinforcing ring at the outer edge is studied. The total opening angle φ_0 is $40°$. The internal forces between the shell and the reinforcing ring are given for three bearing conditions. The support condition of the outer ring is modeled as a liquid cushion support (vertical constant compressive stresses), a horizontally sliding, torsional stiff support (vertical displacements and torsional rotation of the edge beam fixed, horizontal displacements free), and a clamping (vertical and horizontal displacements, as well as torsional rotation of the edge beam fixed). In particular, in the bending moment in the shell at the connection of the shell and the ring, there are considerable differences between the three models. The greatest bending moments occur with the horizontal sliding support. The bending moments for clamping are lower. With liquid cushion support, the moments become very low. The cause of this load-bearing behavior is the arching effect, which is more pronounced with a horizontally fixed support (cf. Example 4.24).

A shell with support as a liquid cushion model is analytically investigated in [109], and denoted as "Girkmann's problem". For numerical analyses, in [235, 236], the following parameters are assumed: an opening angle of $\varphi_0 = 40°$, a radius of $r_0 = 15$ m, a shell thickness of $t = 0.06$ m, a modulus of elasticity $E = 2.059 \cdot 10^7$ kN/m^2, a Poisson ration of $\mu = 0$ and a dead load corresponding to a specific weight of the shell of 32.69 kN/m^3 (Fig. 4.176a). The dimensions of the edge beam are $a = 0.60$ m and $b = 0.50$ m. The distribution of the bending moment according to [237], is shown in Fig. 4.176b. A competition between computational engineers has been initiated, given this problem. The aim was to assess the verification of the computational results

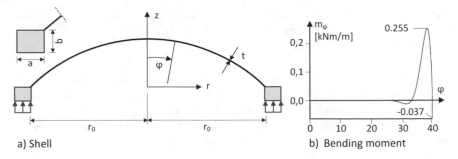

Fig. 4.176 Spherical shell with edge beam

with an accuracy of 5%. The findings of the competition, showing some deviations
in the results presented, are shortly discussed in [236] (cf. Example 4.48).

For modeling of single or area loads acting on surfaces of folded plates or shells,
the same applies as for plates in bending or plane stress. An example of the analysis of
a cylindrical shell loaded by a small, quadratic partial area load with different finite
element meshes is given in [104]. According to this, a subdivision of the loaded
square area into at least four 4-node elements is recommended for sufficiently high
accuracy in construction practice. The analysis of a prestressing force on a cylindrical
RC vessel is discussed in [234].

4.11.9   Three-Dimensional Building Models

Finite element analysis of buildings

In conventional structural analysis and design, buildings are subdivided into structural components such as floor slabs, columns, and walls. Forces are transferred
between the components. When buildings are analyzed as an integrated structural
model, however, all load-bearing components are taken into account in a single model
(Fig. 4.177). With integral models, for example, the influence of the flexibility of

Fig. 4.177 3D finite element
model of a building

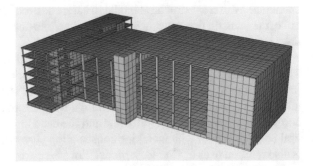

the foundation on the internal forces in the building—also in the upper floors—can be investigated. Even the influence of the change of the structural model within the construction sequence can be taken into account [226, 238–240]. In addition to vertical load transfer, horizontal loads from wind and earthquakes can be investigated with integral building models. Dynamic analyses of earthquake-excited vibrations, for example, can be done with an integral building model. Three-dimensional integral building models are comprehensive models for structural analysis that capture a large number of structural effects in the same model. There are complex structures with a spatial load-bearing behavior, for which the assessment of significant interactions in an integral model is mandatory.

The modeling effort for 3D structural models is not necessarily higher than for conventional 2D models with individual components, especially if a BIM model is available as part of the structural planning process. The definition of the load transfer between the components is omitted. Likewise, no additional definition of the support and restraint conditions resulting from subdividing the structure into individual components is required. This also eliminates sources of error. On the other hand, one drawback is that the structural model becomes more complex and hence the results can no longer be easily assessed. This can open up new sources of error. Changes at one component of the structural system require a recalculation and a new control of the entire structural system of the building. As a result, modifications of the building during the planning process become more time-consuming. Furthermore, integrated building models lead to extremely large finite element systems that require large computer capacities. For these reasons, some reservations are encountered in practice against a structural design with 3D integral structural models, as long as these are not indispensable for the assessment of complex three-dimensional structural interactions [241, 242].

Due to the complexity of the models, structural analysis with integral 3D building models requires special considerations, which are discussed below.

Construction sequence

If permanent loads and traffic loads are applied to an integral building model, all load-bearing components of the building participate in load transfer. In principle, this is correct for traffic loads applied after completion of the building. However, self-weight loads, which only become effective floor by floor during the construction process, encounter a different structural model in each construction phase. Therefore, in principle, the construction sequence must be taken into account in structural analysis for dead loads.

For integral building models, the effects of construction stages must be checked. This is a different approach to a conventional component-wise structural analysis, in which the load of the individual components is transferred exclusively to the underlying components. Only for buildings with less than 6–10 floors the effects of construction phases are generally low and the dead loads may be applied to the integral building model without taking construction phases into account. This modeling is also referred to as a "One-Cast Model". In the case of a typical building with a core and columns, without considering construction sequence, tensile stresses occur in the

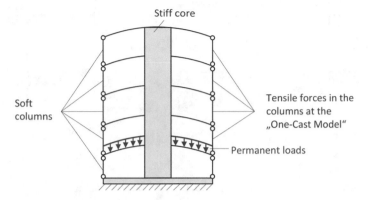

Fig. 4.178 Column loading without consideration of the construction sequence [238]

columns of the floors above, which cannot appear in reality. They are caused by the accumulation of the vertical deformations of the (soft) columns and the result in indirect actions on the floor slabs. As a result, if the construction sequence is neglected, the column forces are underestimated and the forces in the core are overestimated (Fig. 4.178). In principle, the computation with an integral building model thus gives inapplicable results for dead loads. With higher buildings and soft columns, these effects are significant so that construction phases must be taken into account.

The individual construction phases are taken into account by a system change in the structural model and the stresses of the individual systems are then superposed. Alternatively, one can do the analysis with the integral structural model and use differential deformations to ensure that the building sections that have not yet been constructed remain free of stress and deformation. This is referred to as "deformation compensation during construction phases" [243]. The shortening of columns, occasionally mentioned in the literature, by compensating for deformations in the leveling of floor slabs during the construction process is practical without any effect. Whether a column has a length of 3000 mm or 3005 mm (with 5 mm deformation), for example, will not noticeably influence the internal forces in the structural model.

The three methods of analysis are shown for a three-storey building in Fig. 4.179. Figure 4.179a shows the analysis of the integral model without taking the construction sequence into account. The column deformations at the connection to the floor slabs are depicted. They reflect the internal forces in the columns and the floor slabs. Even if the dead weight is applied on the first floor only, the upper floors participate in the load transfer and deformations $w_{2,1} \neq 0$ and $w_{3,1} \neq 0$ occur. The first index indicates the floor where the deformation is considered, the second the floor on which the load is applied. This continues accordingly with the load on the second floor. Only when applying the dead load on the third floor, the supporting structure, on which the deformations and internal forces are determined, also represents the structural system available at this construction stage. The summation of the three load cases results in the load case "total dead weight" with the loads g_1, g_2 and g_3 plus the specified imposed deformations.

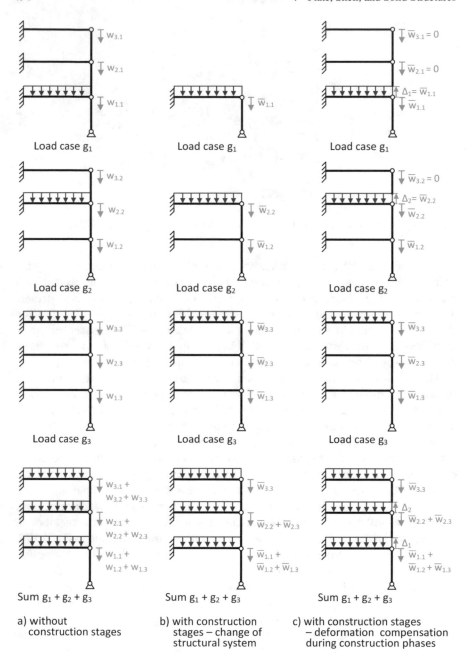

Fig. 4.179 Consideration of construction sequence in structural analysis of integral building models

Figure 4.179b shows the procedure to be followed when taking into account the structural stages caused by a system change [244]. The system corresponds to the construction progress in each load case. Already with the dead weight load g_1, the deformation is $\bar{w}_{1,1} \neq w_{1,1}$ since a "release" by tensile forces in the upper floors is not present. As construction progresses, the load g_2 is transferred to a supporting structure consisting of the two lower floors. On it, the deformations $\bar{w}_{2,2}$ and $\bar{w}_{1,2}$ are determined. Since g_3 acts on the top floor, the deformations here are identical to those in Fig. 4.179a, i.e. it applies $\bar{w}_{3,3} = w_{3,3}$, $\bar{w}_{2,3} = w_{2,3}$ and $\bar{w}_{1,3} = w_{1,3}$. The sum of the internal forces obtained in the three load cases correspond to the total loading of the system by dead loads with consideration of the construction sequence.

Figure 4.179c shows the procedure for considering the construction stages by means of "deformation compensation during construction phases". In all load cases, the analysis is performed on the integral building model. However, the actions are determined in such a way that sections of the structure that have not yet been constructed remain free of deformations and stresses in the corresponding construction phases. This is achieved by releasing bonds and introducing relative deformations between the already built and the not yet built sections. These correspond to the deformations of the already constructed building section. For example, with the load g_1 the relative imposed displacement $\Delta w_1 = \bar{w}_{1,1}$ is applied simultaneously. This ensures that the upper floors do not deform and are therefore free of internal forces. The same procedure is followed for g_2. The specification of relative displacements is not required for the top floor. The sum of the imposed deformations and the corresponding internal forces in the individual construction phases results in the total loading of the system by dead loads including the construction sequence. It is identical to that of the analysis with a change of the structural system according to Fig. 4.179b. However, all calculations can be performed on a single system, the integral building model, without changing the structural system. In addition to the loads g_1, g_2, g_3, as additional actions the imposed relative displacements Δw_1 and Δw_2 must be applied. However, these must be determined in advance using individual systems for each load case, e.g. by setting the stiffnesses of the not yet constructed parts of the structure to zero.

The deformations of reinforced concrete structures are influenced by shrinkage and creep. Since these rheological processes extend over long periods of time, they usually affect the integral building model. It is to be expected that the internal forces determined on the basis of the construction sequence will approximate those of the complete integral building model without consideration of the construction sequence.

Example 4.44 The influence of the construction process is investigated on the idealized reinforced concrete building shown in Fig. 4.180a, in [245]. The building is idealized as "infinitely long" in the y-direction. It has a width of 18.00 m and a floor height of 3.00 m. Concrete C25/30 is assumed as a building material. The walls of the building have a thickness of 0.30 m, the columns a cross section of 0.30 m $\times$ 0.30 m, the thickness of the floor slabs is 0.25 m. Due to their small cross section compared to the walls, the columns represent soft support of the floor slabs. The building has a stiff box-type basement, the subsoil is assumed to be rigid.

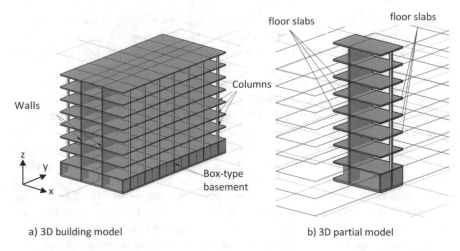

a) 3D building model   b) 3D partial model

Fig. 4.180 Three-dimensional structural model of a building [245]

The investigations are performed on a 5.0 m × 9.0 m large, three-dimensional partial model of the building considering the symmetry conditions (Fig. 4.180b). First, a 9-story building is examined. The number of floors is varied in a parameter study. The calculation was done with [SOF 6].

Various construction stages of the 9-story building are shown in Fig. 4.181. The building is examined once with and once without consideration of construction stages in the load case dead weight. For comparison, the column forces as well as the slab bending moments are examined in a section through the symmetry axis of the partial model (Fig. 4.182a, b).

For comparison, a conventional analysis of a single floor slab as a 5.0 m × 9.0 m large plate with appropriate supports and symmetry conditions is carried out. This model corresponds to the two-dimensional model of the slab within a conventional 2D structural analysis.

The normal forces of the columns and the bending moments of the floor slab in the field and at the wall clamping are shown in Fig. 4.183. Due to the compression of the lower soft columns, the slabs of the upper floors also participate in the load transfer of the lower floors, regardless of the construction progress. This is not the case if the construction sequence is taken into account. Thus, the normal force of the column on the 1st floor is 691.2 kN without and 712.8 kN with consideration of the construction progress. The vertical displacements of the columns are lower without consideration of the construction stages than with their consideration. The difference of the normal forces on the lowest floor of the 9-story building investigated here, however, is only $-(712.8 - 691.2)/- 691.2 = 3\%$ and can be neglected in this case (Fig. 4.183a, b). Compared to a conventional analysis with a 2D model in which the compression of the columns is not taken into account, there is a slightly greater difference. Here the normal force on the lower floor is -750.7 kN. It is thus $-(750 - 712.8)/- 712.8 = 5\%$ greater than the normal force taking into account the construction stages.

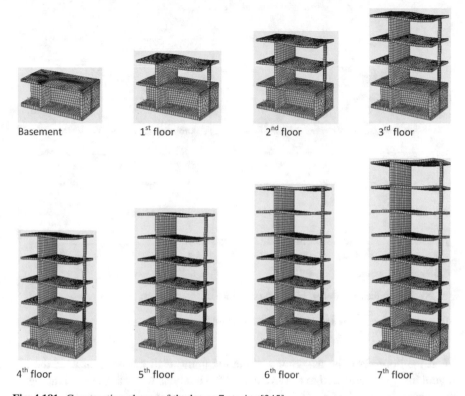

Fig. 4.181   Construction phases of the lower 7 stories [245]

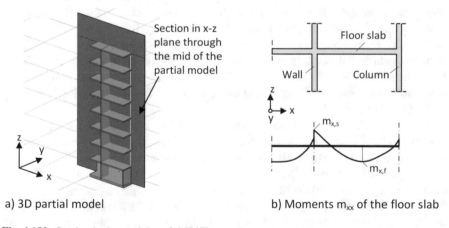

a) 3D partial model   b) Moments m_{xx} of the floor slab

Fig. 4.182   Section in the partial model [245]

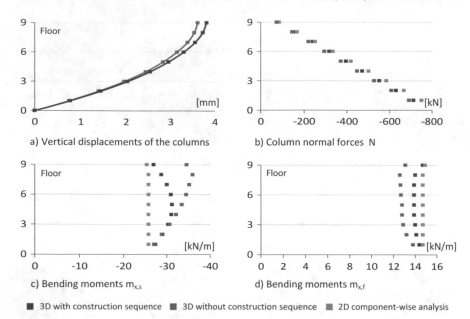

a) Vertical displacements of the columns

b) Column normal forces N

c) Bending moments $m_{x,s}$

d) Bending moments $m_{x,f}$

■ 3D with construction sequence   ■ 3D without construction sequence   ■ 2D component-wise analysis

Fig. 4.183 Deformations and internal forces of a building with 9 stories

In the floor slab, the magnitude of the clamping wall moments $m_{x,s}$ of the 3D model increase compared to a conventional 2D analysis whereas the field moments $m_{x,f}$ decrease (Fig. 4.183c, d). This is due to the vertical displacements of the column heads which support the floor slabs. Due to the larger cross section, the core has significantly smaller displacements than the columns. Without taking the construction sequence into account, the clamping moments $m_{x,s}$ increase with the number of floors in the building model. Taking the construction progress into account, they increase by up to half the height of the building and then decrease again. The column head displacements in the lower part of the structure are related to the column head displacements caused by the construction of the upper floors. These are decreasing in the upper floors.

In a parameter study in [245], it was examined to what extent the number of stories of a building influences the effect which the consideration of the construction sequence has on the results. For this purpose, partial models with 3, 6, 9, 12, and 15 stories were examined. The size of the columns has been maintained for better comparability. Figure 4.184a shows the column forces N_1 in the lowest floor as a function of the number n_S of stories of the building. The influence of the consideration of the construction sequence is small with buildings with 3 and 6 stories. For the higher buildings examined, in particular those with 12 and 15 stories, the differences which result from the construction stages are significant. The same applies to the conventional 2D analysis. For example, with a 15-story building, the normal force in the ground floor is obtained without consideration of the construction sequence to $N_{oB} = -1006\,\text{kN}$, with construction conditions $N_{mB} = -1097\,\text{kN}$ and from a 2D

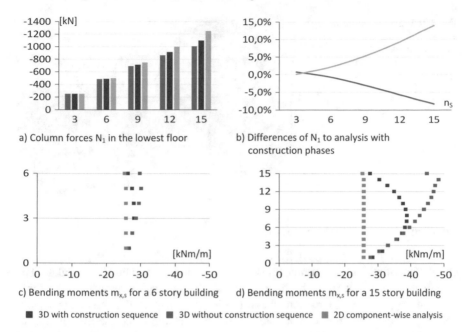

a) Column forces N_1 in the lowest floor

b) Differences of N_1 to analysis with construction phases

c) Bending moments $m_{x,s}$ for a 6 story building

d) Bending moments $m_{x,s}$ for a 15 story building

■ 3D with construction sequence   ■ 3D without construction sequence   ■ 2D component-wise analysis

Fig. 4.184 Internal forces for different numbers of stories of a building

analysis $N_{2D} = -1251$ kN. Compared to an analysis which takes construction stages into account, a difference of $\Delta N_{oB} = (N_{oB} - N_{mB})/N_{mB} = -8.3\%$ is obtained without taking them into account in an integral building model and of $\Delta N_{DB} = (N_{2D} - N_{mB})/N_{mB} = 14.0\%$ in comparison to a 2D analysis (Fig. 4.184b).

The influence of the consideration of the construction sequence on the bending moments in the floor slab is shown in Fig. 4.184c, d. It is small for low-rise buildings up to approximately 6 stories, but must not be neglected for high-rise buildings. For the 15-story building examined here, the clamping moment from a 2D analysis in a standard floor is -25.7 kNm/m, while an investigation of the integral building model with consideration of the construction stages results in -38.8 kNm/m in the middle floors. Without taking the construction stages into account, a value of -48.4 kNm/m is obtained on the second last floor. It should be noted, however, that the clamping moments partly are due to indirect actions which can be reduced by the formation of cracks in concrete.

The results also reported in [245] for larger column sizes (0.45 m × 0.45 m), which are required with an increase in the number of stories, as well as with additional consideration of a traffic load (3.0 kN/m²), have the same tendency. ◀

Coupling of different types of elements in the finite element model

Special attention should be paid to the coupling of elements of different modeling depths, such as beam elements with shell elements (Sect. 4.10). The same applies to the definition of (line) hinges.

Fig. 4.185 Deformations of
a three-dimensional
structural model (300 ×
exaggerated representation)
[159]

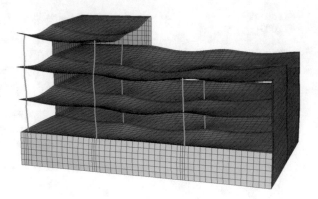

Interaction of structural components

In contrast to the conventional 2D analysis of structural components, a monolithic connection of all building components is assumed for integral building models. The internal forces result from the interaction of all building components taken into account (Fig. 4.185). Interactions that are otherwise neglected as "secondary load-bearing effects", such as compatibility torsion, may influence the distribution of internal forces significantly [246]. The disregarded crack formation can considerably overestimate the stiffness. If stiff components "unload" other components due to their high stiffness, boundary value analysis with min/max values should be carried out (e.g. by an additional analysis with stiffnesses taking cracking of concrete into account) in order to identify possible load redistributions. If required, clamping (such as floor slabs in walls) can be modeled as line hinges, presuming crack formation in the hinge. In this case, an appropriate check of allowable crack widths should be carried out.

Soil–structure interaction

The soil–structure interaction can be represented by various soil models (Sect. 4.11.7). For integral structural models, it affects the internal forces of the building. In addition, the construction sequence may be important. Without considering the construction sequence, the column forces are overestimated and the load transfer through the core is underestimated in a typical building with a core and columns (Fig. 4.186). If the construction sequence is taken into account, the column forces will be reduced.

The influence of soil–structure interaction depends on the ratio of structural stiffness to subsoil stiffness and can influence the internal forces in the building significantly, especially with soft soils. In such cases, the subsoil stiffness must be estimated realistically. Here too, boundary value analysis with min/max values are useful.

Fig. 4.186 Influence of
soil–structure interaction in
integral building models
[238]

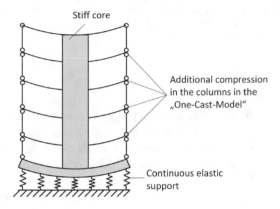

Stiff core

Additional compression
in the columns in the
„One-Cast-Model"

Continuous elastic
support

4.11.10 Interpretation of Results

After the solution of the global system of equations of a finite element model, the
nodal displacements and rotations are available. From these, the internal forces and
displacements can be determined at any point of an element, and thus at any point of
the finite element model. It has already been pointed out that the internal forces at the
center of the element are the most accurate. For elements with higher shape functions
whose stiffness matrices are calculated by numerical integration, the internal forces
at the integration points are the most accurate. On the other hand, the internal forces
at the element edges and nodes are relatively inaccurate and differing in adjacent
elements. There are, therefore, several possibilities for output and further processing
of the internal forces or stresses, respectively

Output points of stresses of a finite element analysis
- in the midpoints or the integration points of elements,
- in the nodal points by averaging the element stresses (at the nodal point) of
  all elements connected to the node,
- at any point of the finite element mesh by interpolation of the nodal values
  or the values at integration points.

All output options have advantages, but also disadvantages. The output at the integration points is very robust, i.e. it gives the best results even under critical conditions.
However, only the stresses in the center area of the elements, i.e. not at every point
of the finite element mesh are obtained. The values at the integration points of an
element can be used to calculate stresses at its nodal points by extrapolation. This is
particularly important on clamped or simply supported edges and under nodal loads,
where the maximum internal forces often occur.

Fig. 4.187 Determination of nodal stresses by averaging element stresses at nodal points

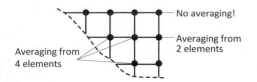

In most finite element programs, smoothed stresses or internal forces at the nodes are given as output. For this purpose, the program determines the element stresses or internal forces (at the node) of all elements connected to the respective node and calculates their mean value. This increases the accuracy of the relatively inaccurate nodal stresses in an individual element (cf. Example 4.4). However, this does not apply to nodes at the edge or in corners of the finite element mesh that have only one or no adjacent element. The nodal values of the stresses are therefore less precise there than at the inner nodes of the finite element mesh (Fig. 4.187).

However, jumps in the internal forces between two adjacent elements may also have static causes, e.g. the shear force jump in a plate under a line load (Fig. 4.188a). It should be noted that in many finite element programs such jumps are lost through averaging when the nodal stresses are output. However, jumps in the internal forces between two adjacent elements may also have static causes, e.g. the shear force jump in a plate under a line load (Fig. 4.188a). It should be noted that in most finite element programs such jumps are lost through averaging when nodal stresses are output. A similar problem occurs in the re-entrant corners of plates in plane stress or in bending. Here, too, the stresses are discontinuous. If the stresses of all elements adjacent to the corner point are averaged, the physically correct stress jump is lost (Fig. 4.188b). Instead the stresses in the individual elements should be used. Jumps in stress distributions also occur when the thickness of a plate changes sharply (Fig. 4.188c). In this case, for plates in bending the edge effect is also relevant (Sect. 2.3, Fig. 2.12). Difficulties with averaging can also occur at edges of folded plate structures or at the connection of planar elements of shell structures. In these cases, the averaging of the internal forces at the nodes must be avoided or suitable averaging methods must be used.

Stress values are often required at arbitrary points in the finite element model and thus at arbitrary points within a finite element. This applies, for example, to internal forces distributions along an arbitrary section through the finite element model and to isoline graphs of internal forces. If smoothed nodal stresses are available, consistent stress values within quadrilateral elements can be obtained by bilinear interpolation according to (4.40a, 4.40b). The extrapolation or interpolation of the more accurate stresses in the mid of the elements seems more appropriate. With a regular rectangular mesh with four common elements per node, this can be done by bilinear interpolation. The extrapolation of the element stresses to the edge often gives more accurate values [19].

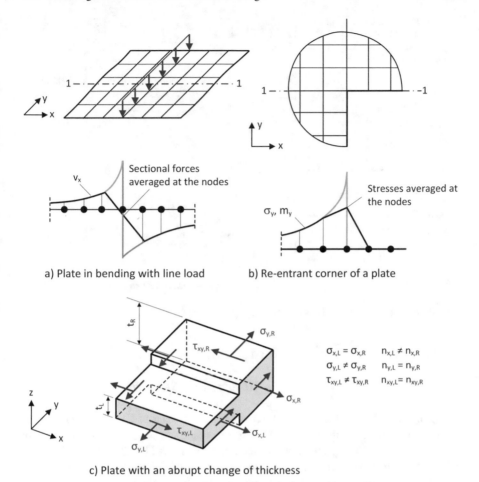

a) Plate in bending with line load   b) Re-entrant corner of a plate

$\sigma_{x,L} = \sigma_{x,R}$ $n_{x,L} \neq n_{x,R}$
$\sigma_{y,L} \neq \sigma_{y,R}$ $n_{y,L} = n_{y,R}$
$\tau_{xy,L} \neq \tau_{xy,R}$ $n_{xy,L} = n_{xy,R}$

c) Plate with an abrupt change of thickness

Fig. 4.188 Jumps in stresses and internal forces in plates

Example 4.45 The one-way spanning slab in Fig. 4.189a is clamped on both sides and loaded with a uniformly distributed load of $q = 10\,\text{kN/m}^2$. It is discretized into 5×5 shear flexible plate elements [SOF 6]. The Poisson's ratio is assumed to be $\mu = 0$. The plate thus exhibits the same behavior as a shear soft beam clamped on both sides.

With a span of $l = 6.00\,\text{m}$, the internal forces at the support are obtained as

$$v_{x,s} = q \cdot l/2 = 10 \cdot 6/2 = 30\,\text{kN/m} \quad \text{and}$$
$$m_{x,s} = q \cdot l^2/12 = 10 \cdot 6^2/12 = 30\,\text{kNm/m}$$

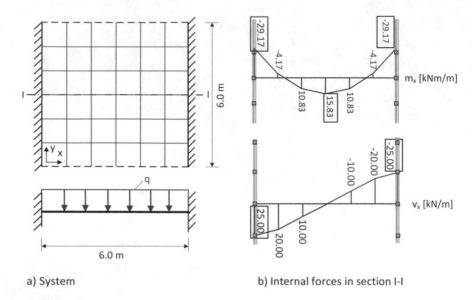

| a) System | b) Internal forces in section I-I |

Fig. 4.189 Finite element analysis of a one-way spanning slab

and the bending moment in the mid of the field is

$$m_{x,\mathrm{f}} = q \cdot l^2/24 = 10 \cdot 6^2/24 = 15\,\mathrm{kNm/m}.$$

The nodal values of the internal forces obtained in section I-I by a finite element analysis are shown in Fig. 4.189b. The clamping moment is computed with an error of $(30 - 29.17)/30 = 3\%$, the error of the field moment is $(15.83 - 15.00)/15.00 = 6\%$. The shear forces are correct in the mid zone of the plate. At the edge, the (unaveraged) shear force in the corner point of the element is given. It is less precise than the averaged node values inside the finite element mesh and shows an error of $(25.00 - 30.00)/30.00 = 17\%$. This shows that shear forces as derived quantities are numerically more sensitive than bending moments.

The support force is determined by the finite element program exactly with 30 kN/m. It is calculated using the equilibrium conditions upon which the global stiffness matrix of the system is based. Therefore it contains the finite element approximation only indirectly by the fact that the FEA only approximates the stiffness distribution in the system. The equilibrium conditions of the overall system, however, are always fulfilled by the support reactions.   ◀

4.12   Quality Assurance and Documentation

4.12.1   Types of Error

Setting up the model is an essential step in the finite element analysis of structures which may also be a potential source of error. Figure 4.190 shows the models formulated in Fig. 4.72 for the idealization of structures and the possible types of errors associated with them. These are a modeling error, a discretization error, and a solution error. In addition, there are general sources of error in computational structural analysis such as input and program errors. In this regard, reference is made to Sect. 3.8.1. A special aspect of surface and solid structures is the discretization error, which does not occur in beam structures except for a few special elements, since exact analytical solutions exist for these. This will be discussed in the following.

4.12.2   Error Estimation and Adaptive Meshing

The discretization error is the error that results from the approximation of the exact displacements or stresses by shape functions. With *a priori error estimators*, the expected maximum error is considered before and with *a posteriori error estimators* after a computation. Of particular practical interest is an a posteriori error estimate.

The discretization error of a finite element analysis can be specified in an error norm for the complete finite element system. Usually, it's based on the so-called *energy norm*. The error $\underline{e}$ of the displacements is defined as deviation of the displacements $\underline{u}^{(FE)}$ obtained with the finite element method from the exact solution $\underline{u}^{(exact)}$ as

$$\underline{e} = \underline{u}^{(FE)} - \underline{u}^{(exact)} \tag{4.203}$$

where the displacements $\underline{u}^{(FE)}$ and $\underline{u}^{(exact)}$ are to be understood as functions of the coordinates x and y. The error $\underline{e}$ is therefore also a function of the x- and y-coordinates.

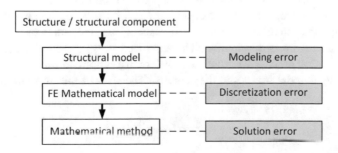

Fig. 4.190 Types of error in modeling

The energy norm is therewith

$$\|e\| = \sqrt{\int \left(\underline{L} \cdot \underline{e}\right)^{\mathrm{T}} \cdot \underline{D} \cdot \left(\underline{L} \cdot \underline{e}\right) \, \mathrm{d}V}. \tag{4.204}$$

For a plate in plane stress, $\underline{L}$ is the differential operator after (2.2c), $\underline{D}$ is the elasticity matrix after (2.3c), and the vectors $\underline{u}^{(\mathrm{FE})}$ and $\underline{u}^{(\mathrm{exact})}$ contain the displacement components u and v as in (2.2c). The integration is done over the volume of the finite element model. The energy norm is the root of the double strain energy of the error $\underline{e}$ of the displacements. In order to prove the convergence for a finite element, it must be proven that the energy norm converges toward zero with an increasing number of finite elements of equal size. Error norms can be applied to estimate the error of finite element calculation. *Error estimation* is understood as an upper limit value of an error, e.g. the error $\|e\|$, which is calculated on the basis of the results of a finite element analysis. Error estimates have been given for finite elements with displacement shape functions.

From a mathematical point of view, error norms are important for proving convergence and assessing the convergence speed of a finite element. From a practical point of view, however, the global error of a finite element analysis is less meaningful. There are also methods to determine the local errors in the energy and thus a distribution of the error in the finite element model [247, 248]. These are used for adaptive methods. Of particular interest is the local error of stresses and internal forces. For this purpose, there are error estimators on a mathematical basis, which provide a guaranteed bound of the error of a given static quantity, e.g. an internal force at an individual point of the finite element model. This is called a *goal-oriented error estimation*. However, the procedures required are too computationally intensive to be applied to all elements of a model and to determine the distribution of the error. For this purpose, other *local error estimators* have been developed, which provide computable but not mathematically proven error bounds with reasonable computational effort. In addition, there are heuristic error estimations based on the error estimator of Zienkiewicz/Zhu [249]. There are, however, currently no error estimation techniques that give a mathematically proven upper bound of the possible local error in the finite element mesh [250]. For error estimation in finite element analyses and the methods developed for this purpose, reference is made to the summary presentation in [251].

The heuristic error estimation of Zienkiewicz/Zhu is based on the difference of the stress distribution resulting from the displacement shape functions in the element and an enhanced stress distribution. The enhanced stress distribution results either from an averaging process according to [249], or in a simplified way, by interpolation of the averaged nodal stresses in the element. The error of the stress in the element is

$$e_\sigma = \sigma^* - \sigma^{(\mathrm{FE})} \tag{4.205}$$

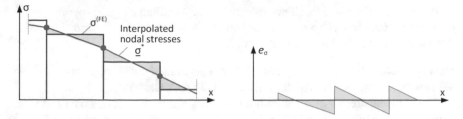

Fig. 4.191 Error estimation according to Zienkiewicz/Zhu

where $\sigma^{(FE)}$ are the element stresses and σ^* the enhanced stresses obtained by interpolation of the nodal stresses (Fig. 4.191). The same functions are used for interpolating the stresses in the element as for interpolating the displacements, i.e. the shape
functions. For the isoparametric elements, they are obtained according to (4.42a)

$$\sigma^* = \sum h_i \cdot \sigma_i^* \qquad (4.206)$$

where σ_i^* are the nodal values of the respective stress component. The estimated
error according to [249] is

$$\|e_\sigma\| = \alpha \cdot \sqrt{\frac{\int\limits_A \left(\sigma^* - \sigma^{(FE)}\right)^2 dA}{A}}. \qquad (4.207)$$

The integration extends over the element area. The factor α depends on the element
type. A factor of $\alpha = 1.1$ is given in [249] for quadrilateral plane stress elements with
a bilinear shape function. The technique can also be extended to plates in bending
[252] and is available in some commercial software [SOF 6], [253]. A similar error
estimate analysis is implemented in [SOF 2].

The error measure $\|e\|$ is used to estimate a mean error in the element and may
be interpreted as such, e.g. as an error of the stress in the center of the element.
However, it does not contain any information about the point-wise error such as at a
node. The point-wise error can become infinite, e.g. at a stress singularity, while the
error measure in the element concerned can be high, but always remains finite.

Discretization errors of a finite element analysis can also be interpreted as errors
in the applied loads as suggested by Hartmann et. al. [254–256]. The finite element
solution then represents the exact solution for a system with a modified loading, which
is composed of the actual loads and the error loads. One determines the distributed
loads acting by introducing the displacement shape functions of the finite element
into the analytical differential equation of equilibrium. In addition, line loads occur
at the edge of the element, which correspond to the edge stresses resulting from
the displacement shape functions. The magnitude of the error loads compared to
the actual loads is a measure for the quality of the finite element computation. In
practical cases, however, the results are not very comprehensible [80].

It should also be noted that if no error estimator is available in the program used, a comparative analysis with a very fine finite element mesh can be performed to assess the convergence behavior of the solution.

Example 4.46 The error of the stresses of all elements of the introductory Example 4.2 is to be determined for the discretization with four elements.

The element stresses given in Example 4.4 and the nodal stresses obtained from Table 4.4 are summarized in Table 4.44. The calculation of the error according to (4.207), is shown as an example for element 3 (from $x = 250$ to $x = 375$ cm). The stress σ^* has the values 0.343 and 0.533 at the nodes and a linear distribution between them according to the shape functions applied. The calculated finite element stress $\sigma^{(\text{FE})}$ has the value 0.400 and is constant in the element. The integration is carried out numerically according to Gauss with a 2-point scheme, which integrates the third-order parabola given here exactly. With

$$dA = h(x)\, dx = (300 - 0.8 \cdot x)\, dx \quad \text{with x [cm]}$$

and

$$A = \int h(x)\, dx = 3125 \ \text{cm}^2$$

the stress $\sigma^*(x)$ and the height $h(x)$ are obtained at the integration points (for a 2-point integration) as

$$x^{(el)} = 26.38 \ cm : \quad h = 27.89 \,\text{cm} \quad \sigma^* = 0.383 \ \text{kN/cm}^2$$

$$x^{(el)} = 98.63 \ \text{cm} : \quad h = 22.11 \,\text{cm} \quad \sigma^* = 0.494 \ \text{kN/cm}^2$$

The error $\|e_\sigma\|$ is obtained as

$$\|e_\sigma\| = 1.1 \cdot \sqrt{\frac{\left[(0.383 - 0.400)^2 \cdot 27.89 + (0.494 - 0.400)^2 \cdot 22.11\right] \cdot 125 \cdot 0.5}{3125}}$$

$$= 0.070 \,\text{kN/cm}^2.$$

Table 4.44 Element and nodal stresses and error measure of stresses

Stresses [kN/cm^2]				x [cm]				
	0		125		250		375	500
Element	-	0.222	-	0.286	-	0.400	-	0.667	-
Node	0.222	-	0.254	-	0.343	-	0.533	-	0.667
$\|e_\sigma\|$	-	0.014	-	0.025	-	0.070	-	0.114	-

The errors in all elements are given in Table 4.44. It can be seen that the distribution of the quality of the finite element approximation is well reproduced by the error $\|e_\sigma\|$. However, the comparison with Table 4.5 also clearly shows that the error estimation according to Zienkiewicz/Zhu cannot be expected to provide an upper limit of the error of the stresses in an element.　◄

Example 4.47 Error estimators are usually applied to locate zones of finite element meshes that require a mesh refinement. However, error estimators can also be used to locate errors caused by distorted element shapes.

In Example 4.20 a cantilevered slab with 3 element meshes was investigated. For meshes with distorted element shapes, considerable discretization errors of the shear force occur. Figure 4.192 shows the discretization errors according to Zienkiewicz/Zhu of the shear force v_x for the regular mesh with square elements and for the two irregular meshes according to [SOF 6]. The maximum error occurs in the elements marked in red. In the regular mesh, it is approximately 10^{-6} and is, therefore, negligible. The two irregular meshes show a maximum error of 0.956 kN/m (mesh 1) and 0.247 kN/m (mesh 2) with a shear force of 1 kN/m. These errors are substantial. However, they overestimate the actual errors according to Fig. 4.77, of 14% (mesh 1) and 5.5% (mesh 2). The distribution of the error in the finite element mesh is correctly reproduced in both irregular meshes.　◄

Local error measures allow to estimate the magnitude and distribution of the error of stresses and internal forces in shell structures. This can be used to automatically control the mesh refinement. For this process, which is called *adaptive mesh refinement*, a suitable local error measure is required. It provides a limit value of the error measure from where on the mesh is locally refined. In addition, a strategy for mesh refinement is required (Fig. 4.193). As an alternative to the local refinement of the mesh (*h*-adaptation), the order of the shape functions of the elements can be increased locally (*p*-adaptation). In the *hp*-adaptation, both methods are combined. For this purpose, elements with polynomials of a higher order are used as shape functions and additional abstract degrees of freedom are introduced. The polynomial order is determined adaptively on the basis of an error estimation [257]. Other methods have been proposed that switch from two-dimensional to three-dimensional models in critical structural zones, i.e. they adapt the structural model [258]. Adaptive methods can also be applied to nonlinear problems [259].

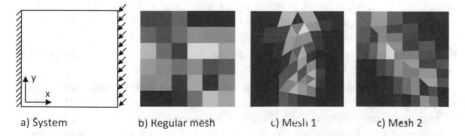

a) System　　　b) Regular mesh　　　c) Mesh 1　　　c) Mesh 2

Fig. 4.192 Error estimation of shear force v_x

Fig. 4.193 Example of an adaptive meshing according to Zienkiewicz/Zhu [249]

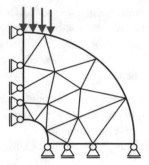
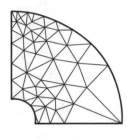

4.12.3 Checking of Surface and Spatial Structure Computations

Checks on program input and output of the results are of utmost importance in practice. The mistakes, occurring most frequently in the application of the finite element method, are in modeling and interpretation of results. This not only concerns the mechanically and mathematically correct computation, but also the consistency with the design model. The results of a finite element analysis, finally, must be implemented in the design of a reinforced concrete construction (cf. Table 4.45). Examples of this are described in the literature, for example in [144, 260–262]. A particularly spectacular case was the collapse of the Sleipner oil platform due to a serious design problem [104, 263]. The reason for this was an insufficient finite element discretization and an incorrect shear force resulting thereof, but also a roughly incorrect reinforcement layout. A checklist for standard building construction, which also includes design issues, is given in Table 4.46.

The guidelines of the German VPI [264] and the VDI guideline 6201 [102, 265] provide information on the checking of structural analyses with a special focus on finite element computations.

In the field of mechanical engineering, requirements for the verification of finite element calculations are also formulated (cf. [101]). Conceptual considerations with regard to structural engineering are presented in [266].

To check the results, the same strategy can be applied for shell structures as for beam structures (cf. Sect. 3.8.2). However, special aspects arise from the higher complexity of the structural systems to be examined and from the additionally required estimation of the discretization error [267, 268]. The verification of the programs used is even more demanding for surface and spatial structures than for beam structures. Verification manuals should document evaluation examples and benchmark tests [265, 269].

The verification of a structural analysis of an integral three-dimensional building model is particularly complex as it is less transparent than a conventional 2D analysis of individual components. The difficulties encountered have been pointed out in several papers [241, 242, 270–273]. It is therefore of utmost importance to carefully

Table 4.45 Design issues in RC constructions [225]

D-regions	Irregular regions of structural components as e.g. zones where loads are applied, supports, abrupt changes and kinks of cross sections require an additional constructural reinforcement which is not part of the results of a finite element analysis. It can be determined using strut-and-tie models [144].
Anchorage of reinforcement	Anchorage lengths and shift rules are to be taken into account in addition to the reinforcement evaluated by the finite element program. This applies not only for beams but also for deep beams and slabs.
Mounting parts	Load-bearing thermal insulation elements which are used for the thermal separation of balconies are able to transmit a negative bending moment and a shear force, but not a twisting moment. This load-bearing behavior must be represented in the structural model (cf. Sect. 4.11.6, Example 4.39).
T-beams	The bending reinforcement of T-beams consists of the reinforcement of the web and the slab reinforcement over the web. For beams with a large width of the web the latter exhibits a significant contribution (cf. Sect. 4.11.6).
Punching shear	The punching shear resistance is to be determined on the basis of the support reactions of a column (and not with shear stresses in a slab). This applies too on columns connected to beams partly integrated into a slab (T-beams). Here, the punching shear check can be done with the difference of the column force and the (shear) forces introduced by the beams.

Table 4.46 Checklist [225]

Modeling	Has the stiffness of all relevant components been adequately modeled, in particular the clamping stiffness of slabs in beams slabs and the building stiffness in foundation slabs?
	Have any subsequent modifications of the structure been made that affect the finite element model, e.g. introduction of thermal isolation elements?
	Are the output points for RC design correctly selected, e.g. at the edge of a bearing?
	Is the finite element mesh sufficiently fine in all regions, especially when shear forces are substantial?
Results	Have the input values been carefully checked again?
	Do the deformations and internal forces correspond to the expected load-bearing behavior? If necessary, perform simple checks of the force distribution and the resulting internal forces by hand.
	Do nontransferable tensile stresses or uplift forces occur?
	Have the design aspects according to Table 4.45 been considered?
	How was dimensioning at singular points? Is a redistribution of the reinforcement useful?
	Have the necessary checks for serviceability been carried out (permissible deflections, crack formation)?

examine the results, to make transparent the load transfer in the building, and to fully document the model data and results. Alternative models of load transfer can be detected and visualized in this way too.

Plausibility check

A plausibility check includes the check of the applied load by load cases (equilibrium checks), an approximate check of the load transfer as well as the check of the distribution of the deformations, relevant internal forces, and important design results.

Furthermore, the finite element model must be checked. Since the lack of a single finite element is not easy to detect when displaying a finite element mesh, the output of element numbers, "shrunk" element contours or a colored display of the elements can be useful. The graphical representation of deformations usually indicates gross errors in the support definition or a missing connection between elements.

The graphical representation of the internal forces in individual sections as principal stresses or principal moments allows an engineer to roughly assess the calculation. If there is any doubt about the correctness of the analysis, it may be helpful to carry out calculations using a simplified structural system.

Final check

After a successful plausibility check, all relevant input values have to be checked once again. A graphic display of important parameters (slab thicknesses, loads, load patterns, etc.) is helpful here.

If there is not sufficient experience, the magnitude of the discretization error should be accounted for. There are several ways of doing this.

Checks for estimating the discretization error

- Display of the element stresses/ internal forces and assessment of the jumps between the elements
- Convergence study by means of recalculation with substantially refined mesh
- Display of element-related error indicators
- Equilibrium checks on cut sections of the structural model.

Example 4.48 The numerical solution of the spherical shell with an edge beam shown in Fig. 4.176, the so-called Girkmann problem, with (commercial) finite element programs has been submitted in 2008 in a competition in [235]. It was required to determine the bending moment and the shear force of the shell at the connection to the edge beam, as well as the maximum bending moment in the shell and its position angle φ with a proven accuracy of 5%.

Of the 15 solutions received, 4 were based on the p-version and 11 on the h-version of the finite element method. The permitted 5% was achieved by all calculations of the p-version, but only by 2 of the 11 calculations with the h-version [237]. For the clamping moment of -0.03681 kNm (with 2% accuracy) values between -0.205 and 17.977 kNm were submitted. As a result, in [237], an increasing mesh refinement with subsequent extrapolation of the results is recommended as the best strategy for achieving the required accuracy.   ◄

Simple checks are also important for complex structures [274]. Equilibrium checks between the internal forces by cutting sections of the structure and the external loads and support reactions are very useful to determine the correctness of the integrated internal forces (but not their distribution). This is especially true for the shear forces of plates, which can easily be subject to significant errors. As shear forces relevant for the design often occur at the supports or at single loads, they should be determined from the (in sum exact) support forces or the loads. As a rough approximation, which, however, fulfills the equilibrium conditions, the load distribution with tributary areas or load distribution factors can be carried out according to [155]. Information on equilibrium checks on flat slabs are given in [275].

The output of element-related error indicators allows an estimation of the error of the element stresses. Obviously, the absolute error is decisive. The relative error given in % for small values may locally be higher than the usually allowed 5% of an internal force or stress.

The convergence behavior of internal forces during mesh refinement can provide information on how accurate internal forces are. If the order of accuracy is known, a more accurate value of the internal forces can be extrapolated from the knowledge of the internal forces of two finite element meshes (Richardson extrapolation, [19, 276]). With the computer power available today, structural components can be computed with very fine meshes, so that convergence studies with mesh refinements are feasible in practice. Still too time-consuming from the point of view of the computing capacity, however, are intensive convergence studies with three-dimensional integral building models.

When assessing the internal forces of a finite element analysis, it should be known at which points of the structural model singularities occur. At these particular zones, the distribution of the internal forces determined by the finite element program is not necessarily decisive for the RC design. Although the sum of the calculated reinforcement should be provided, the peak value of the reinforcement is to a certain extent arbitrary. If it seems to be useful for design reasons, the reinforcement calculated by the program can be redistributed "by hand" at singularity points (cf. Example 4.22). This can also be done with the support of special software [277].

Furthermore, the measures for the control and quality assurance of a finite element calculation specified for beam structures in Sect. 3.8.2 apply.

4.12.4   Documentation of Finite Element Analyses of Surface and Spatial Structures

The finite element analysis of shell structures should be documented under the same aspects as for beam structures. The specification of node and element numbers, but not the representation of the finite element mesh, can be omitted if the input data are presented completely in graphical form. The types of the finite elements and the software used should be specified in the documentation of the analysis. The calculation carried out must be documented completely and comprehensibly (Table 4.47). Recommendations can be found in [102, 264, 278].

Table 4.47 Documentation of finite element analyses of surface and spatial stuctures

General data	Finite element software and software version
	Type of structure, type of finite elements (e.g. shear flexible/stiff plate elements)
Description of the model	Representation and description of the structural model
	Geometrical data of the model
	Material properties and cross-sectional parameters and their assignment to regions of the structural model, imperfections
	Regions of the models with element types
	Support conditions (rigid, elastic), springs and hinges, coupling conditions
	Subsoil model with parameters
	Design parameters and position of reinforcement
	Finite element model
Loads and actions	Actions/ loads (type, location, size, direction)
	Combinations of actions/load cases with partial safety factors and combination coefficients
	Combinations, relevant for design (must be identifiable)
Results	All relevant internal forces in form of numerical values and graphical 3D representation e.g. with contour lines (single actions/loadcases as well as relevant combinations)
	Presentation of the principal stresses/moments in relevant load cases.
	Presentation of internal forces in relevant section cuts through the structure
	Support reactions for single actions/load cases and action combinations
	Graphical presentation of the deformations with maximum values (single actions/load cases as well as relevant combinations)
	Dimensioning or other static calculations (graphical representation in the form of numerical values and contour lines)

Exercises

Problems

4.1

A stiffness matrix based on a displacement shape function has been determined for a truss element with 4 nodal points. The element has a linearly variable cross section. With which accuracy are the displacements, the normal stresses and the strains in the element represented, if a polynomial distribution is assumed?

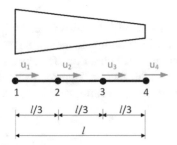

4.2

The cross-sectional area $A(x)$ of a truss element has an arbitrary distribution with x. Derive the stiffness matrix of a truss member with 2 nodes using the principle of virtual displacements

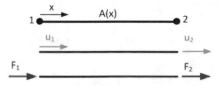

4.3

A truss member has a cross-sectional area with a quadratic distribution, which is described as $A(x) = A_1 + \dfrac{x}{l}(-3A_1 + 4A_m - A_2) + \dfrac{x^2}{l^2}(2A_1 - 4A_m + 2A_2)$ where A_1, A_2, A_m are the cross-sectional areas at points 1, 2 and at midpoint m of the element $(x = l/2)$, respectively. Determine the stiffness matrix of the element analytically by means of the result of problem 4.2. Check your result using the stiffness matrix determined in Example 4.2 for a member with the following parameters:

$l = 500$ cm, $E = 1000$ kN/cm^2, $A_1 = 500$ cm^2, $A_m = 300$ cm^2, $A_2 = 100$ cm^2

4.4

A truss member with quadratically variable cross-sectional area (cf. Problems 4.2, 4.3) has the parameters

$l = 500$ cm, $E = 1000$ kN/cm^2,
$A_1 = 900$ cm^2, $A_m = 400$ cm^2, $A_2 = 100$ cm^2

Determine the stiffness matrix of the truss element

(a)   analytically with the stiffness matrix according to Problem 4.3
(b)   numerically with a Gaussian-2-point integration

4.5

A bar subjected to tension is modeled with isoparametric plane stress elements. Draw in row 0 of the table the exact distribution of the normal stresses for the loadings A, B, and C and give their mathematical equation. The rows 1 to 4 show different finite element models. Indicate with "exact" or "approximate" if the FE models give the exact solution or not when loaded according to columns A, B, and C

		A 	B 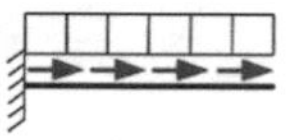	C 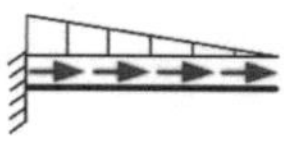
0	Exact solution of stress distribution			
1	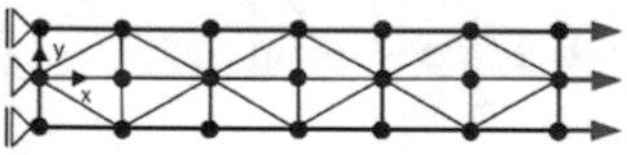			
2	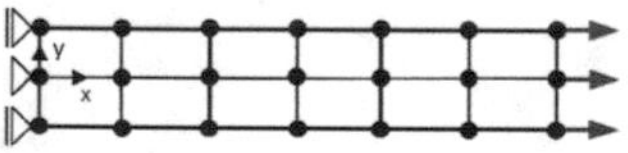			
3	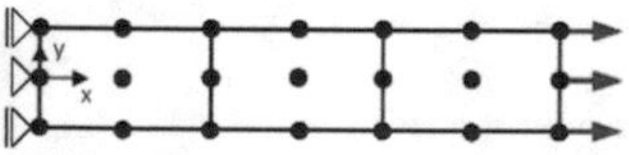			
4	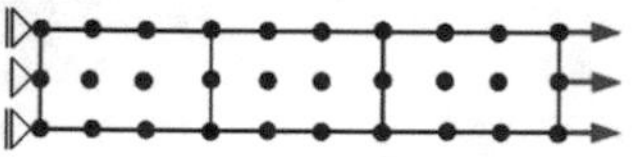			

4.6

The stiffness matrix of a FE model for plane stress is considered. Which are the degrees of freedom of the given FE model in plane stress? Show the structure of the stiffness matrix, label the degrees of freedom and indicate for each term of the stiffness matrix if it is equal to zero ("0") or unequal to zero ("x")

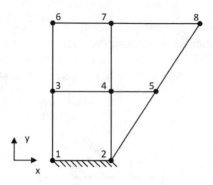

4.7

What is the order $n \times n$ of the stiffness matrix of the two solid elements shown?

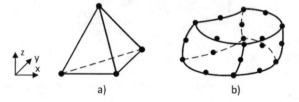

a)   b)

4.8

Solve the integral with numerical Gaussian 3-point integration:

$$\int_{0.5}^{1} \frac{\sqrt[3]{\sin(3 \cdot x)}}{x^2 + 30} \, dx$$

4.9

A force F_x is applied on a conforming, rectangular 4-nodal plane stress element. Determine the equivalent nodal forces

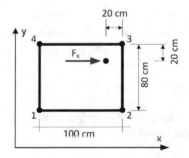

4.10

A finite element program has calculated the nodal displacements of a plane stress element given below. The element used is a conformal rectangular element with a bilinear shape function. How large are the displacements, stresses, and strains at point p?

Nodal displacements:

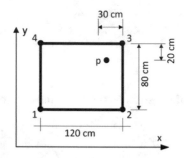

$E = 3 \cdot 10^7 \text{ kN/m}^2, \quad \mu = 0.2$

Node	u [mm]	v [mm]
1	0.1690	−0.2811
2	0.1596	−0.2808
3	0.1530	−0.2871
4	0.1693	−0.2875

4.11

Verify that the bar element subjected to torsion fulfills the rigid body motion condition required for finite elements

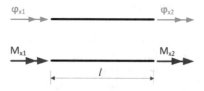

4.12

Verify that the truss member inclined at the angle α fulfills the rigid body displacement conditions required for finite elements

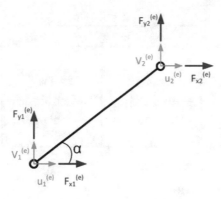

4.13

The finite element system shown consists of a plane stress element and a spring. Determine all displacements, the stresses in the middle of the element, and the force in the spring.

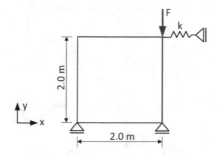

Parameters:

Plane stress element: $t = 0.20\,\text{m}$, $\quad E = 30000\,\text{MN/m}^2$

Spring: $k = 1000\,\text{MN/m}$

Load: $F = 0.500\,\text{MN}$

4.14

A plate is modeled with two plane stress elements. Determine the nodal displacements and the stresses (all components) of point 3 by the FEM. Assess the vertical displacement and the vertical stress in point 3 by a simple hand calculation. How do you assess the accuracy of the FEM calculation? Give the displacements in the midpoint of element 1. The thickness of the plate is $t = 0.10\,\text{m}$.

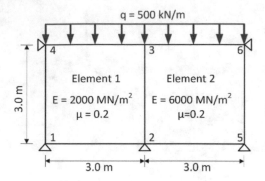

4.15

A system consisting of four plane stress elements is investigated. Determine the nodal displacements of point 5 and the stresses (all components) in the midpoint of element 1.

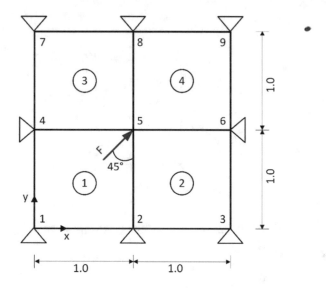

$E = \; 3 \cdot 10^7 \; \text{kN/m}^2, \quad \mu = 0.2, \quad t = 0.30 \; \text{m}, \quad F = 10000 \; \text{kN}$

4.16

A square 4-node element for plates in bending with bilinear displacement shape functions is loaded with a distributed load

$$p(x, y) = N_1(x, y) \cdot p_1 \text{ with } N_1 = \frac{1}{4} - \frac{1}{2a}x - \frac{1}{2b}y + \frac{1}{ab}xy$$

Determine the equivalent nodal forces using a numerical 2-point Gaussian integration.

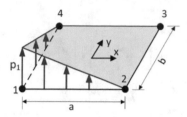

$a = 1.00 \text{ m}, \quad b = 1.00 \text{ m}, \quad p_1 = 1.0 \text{ kN/m}^2, \quad p_2 = p_3 = p_4 = 0$

4.17

A plate clamped on three sides with a free edge and a line load is discretized into two nonconforming, shear rigid plate elements. At the clamped edges, the translational, as well as all rotational degrees of freedom, are fixed. Determine the deflection and rotational angles in the middle of the free edge.

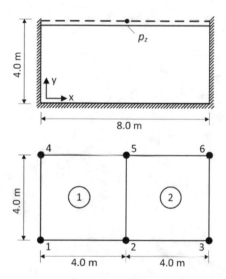

$E = 3.10^7 \text{ kN/m}^2, \quad \mu = 0.2, \quad t = 0.3 \text{ m}, \quad p_z = 10 \text{ kN/m}$

Solutions

4.1

Displacements are exactly represented up to 3rd order polynomials, stresses, and strains up to 2nd order polynomials.

4.2

$$u(x) = u_1 + \frac{x}{l} \cdot (u_2 - u_1) = \left[1 - \frac{x}{l} \quad \frac{x}{l}\right] \cdot \begin{bmatrix} u_1 \\ u_2 \end{bmatrix} \qquad \varepsilon = \frac{du}{dx} = \underline{B} \cdot \underline{u}$$

$$\underline{B} = \left[-\frac{1}{l} \quad \frac{1}{l}\right] \qquad \underline{u} = \begin{bmatrix} u_1 \\ u_2 \end{bmatrix} \qquad \underline{D} = [E]$$

$$\underline{K} = \int \underline{B}^{\mathrm{T}} \cdot \underline{D} \cdot \underline{B} \mathrm{d}V = \int_0^l \underline{B}^{\mathrm{T}} \cdot \underline{D} \cdot \underline{B} \cdot A(x) \ \mathrm{d}x = \frac{E}{l^2} \cdot \begin{bmatrix} 1 & -1 \\ -1 & 1 \end{bmatrix} \cdot \int_0^l A(x) \ \mathrm{d}x$$

4.3

General solution: $\underline{K} = \dfrac{E}{l} \cdot \begin{bmatrix} 1 & -1 \\ -1 & 1 \end{bmatrix} \cdot \left(\dfrac{1}{6} A_1 + \dfrac{2}{3} A_m + \dfrac{1}{6} A_2\right)$

Solution with the numerical values given: $\underline{K} = 600 \cdot \begin{bmatrix} 1 & -1 \\ -1 & 1 \end{bmatrix} \dfrac{\mathrm{kN}}{\mathrm{cm}}$

4.4

(a) and (b): $\underline{K} = 866.7 \cdot \begin{bmatrix} 1 & -1 \\ -1 & 1 \end{bmatrix} \dfrac{\mathrm{kN}}{\mathrm{cm}}$

4.5

		A	B	C
0	Exact solution for the bar	constant	linear	quadratic
1		Exact	Approximate	Approximate
2		Exact	Approximate	Approximate
3		Exact	Exact	Approximate
4		Exact	Exact	Exact

4.6

	u_3	v_3	u_4	v_4	u_5	v_5	u_6	v_6	u_7	v_7	u_8	v_8
u_3	X	X	X	X	0	0	X	X	X	X	0	0
v_3	X	X	X	X	0	0	X	X	X	X	0	0
u_4	X	X	X	X	X	X	X	X	X	X	X	X
v_4	X	X	X	X	X	X	X	X	X	X	X	X
u_5	0	0	X	X	X	X	0	0	X	X	X	X
v_5	0	0	X	X	X	X	0	0	X	X	X	X
u_6	X	X	X	X	0	0	X	X	X	X	0	0
v_6	X	X	X	X	0	0	X	X	X	X	0	0
u_7	X	X	X	X	X	X	X	X	X	X	X	X
v_7	X	X	X	X	X	X	X	X	X	X	X	X
u_8	0	0	X	X	X	X	0	0	X	X	X	X
v_8	0	0	X	X	X	X	0	0	X	X	X	X

4.7

(a) 12×12,

(b) 60×60.

4.8

$\int = 0.014$

4.9

$F_{Lx1} = 0.05\ F_x, \quad F_{Lx2} = 0.20\ F_x, \quad F_{Lx3} = 0.60\ F_x, \quad F_{Lx4} = 0.15\ F_x$

4.10

$u_p = 0.158$ mm, $\quad v_p = -0.286$ mm,

$\varepsilon_x = -1.215 \cdot 10^{-4}, \quad \varepsilon_y = -0.791 \cdot 10^{-4}, \quad \gamma_{xy} = -0.578 \cdot 10^{-4}$

$\sigma_x = -4290$ kN/m^2, $\quad \sigma_y = -3230$ kN/m^2, $\quad \tau_{xy} = -723$ kN/m^2

4.11

Nodal moments when both nodes are rotated with φ_{rigid}:

$$\frac{G \cdot I_T}{\ell} \cdot \begin{bmatrix} 1 & -1 \\ -1 & 1 \end{bmatrix} \cdot \begin{bmatrix} \varphi_{\text{rigid}} \\ \varphi_{\text{rigid}} \end{bmatrix} = \begin{bmatrix} 0 \\ 0 \end{bmatrix}$$

i.e. a rigid body rotation does not generate nodal moments.

4.12

Nodal forces when both nodes are subjected to the nodal displacements u_{rigid}, v_{rigid} ($s = \sin\alpha, \ c = \cos\alpha$):

$$\frac{E \cdot A}{l} \cdot \begin{bmatrix} c^2 & s \cdot c & -c^2 & -s \cdot c \\ s \cdot c & s^2 & -s \cdot c & -s^2 \\ -c^2 & -s \cdot c & c^2 & s \cdot c \\ -s \cdot c & -s^2 & s \cdot c & s^2 \end{bmatrix} \cdot \begin{bmatrix} u_{\text{rigid}} \\ v_{\text{rigid}} \\ u_{\text{rigid}} \\ v_{\text{rigid}} \end{bmatrix} = \begin{bmatrix} 0 \\ 0 \\ 0 \\ 0 \end{bmatrix}$$

i.e. a rigid body displacement does not generate nodal forces.

4.13

$u_1 = 0.044$ mm, $\quad v_1 = -0.186$ mm, $\quad F_{\text{spring}} = -44.5$ kN   (compression),

$\sigma_x = 57$ kN/m^2, $\quad \sigma_y = -1381$ kN/m^2, $\quad \tau_{xy} = -441$ kN/m^2

4.14

Nodal point 3:

$\quad u_3 = -0.636$ mm, $\quad v_3 = -3.959$ mm

Element 1:
$$\sigma_x^{(1)} = -992 \text{ kN/m}^2, \quad \sigma_y^{(1)} = -2838 \text{ kN/m}^2, \quad \tau_{xy}^{(1)} = -1277 \text{ kN/m}^2$$

Element 2:
$$\sigma_x^{(2)} = -992 \text{ kN/m}^2, \quad \sigma_y^{(2)} = -2838 \text{ kN/m}^2, \quad \tau_{xy}^{(2)} = -1277 \text{ kN/m}^2$$

Nodal point 3:
$$\sigma_x = -658 \text{ kN/m}^2, \quad \sigma_y = -5411 \text{ kN/m}^2, \quad \tau_{xy} = 746 \text{ kN/m}^2$$

Approximate solution:
$$v_{3,\text{approx}} = \frac{F \cdot l}{E \cdot A} = \frac{500 \cdot 3 \cdot 3}{0.5 \cdot (2 \cdot 10^6 + 6 \cdot 10^6) \cdot 0.1 \cdot 3}$$
$$= 3.75 \cdot 10^{-3}\text{m} = 3.75 \text{ mm} \quad \text{(downwards)}$$

$$\sigma_{y,\text{approx}} = -\frac{F}{A} = -\frac{500}{0.1} = -5000 \text{ kN/m}^2,$$
$$\sigma_{x,\text{approx}} = \mu \cdot \sigma_{y,\text{approx}} = -1000 \text{ kN/m}^2$$

Displacements in the midpoint of element 1:
$$u_m^{(1)} = -0.159 \text{ mm}, \quad v_m^{(1)} = -0.990 \text{ mm}$$

4.15

$$u_5 = 0.404 \text{ mm}, \quad v_5 = 0.404 \text{ mm},$$
$$\sigma_{x,m}^{(1)} = 7576 \text{ kN/m}^2, \quad \sigma_{y,m}^{(1)} = 7576 \text{ kN/m}^2, \quad \tau_{xy,m}^{(1)} = 5051 \text{ kN/m}^2$$

4.16

$$F_{z,1} = 0.111 \text{ kN}, \quad F_{z,2} = 0.056 \text{ kN}, \quad F_{z,3} = 0.028 \text{ kN}, \quad F_{z,4} = 0.056 \text{ kN, e.g.}$$

$$F_{z,2} = \frac{1}{4} \cdot (N_2(-0.289, -0.289) \cdot N_1(-0.289, -0.289)$$
$$+ N_2(0.289, -0.289 \cdot N_1(0.289, -0.289)+$$
$$+ N_2(0.289, -0.289 \cdot N_1(0.289, -0.289)$$
$$+ N_2(-0.289, -0.289 \cdot N_1(-0.289, -0.289))$$
$$= 0.056 \text{ kN}$$

4.17

$$u_5 = 0.647 \text{ mm}, \quad \varphi_{x,5} = 0.247 \cdot 10^{-3}, \quad \varphi_{y,5} = 0 \quad \text{(due to symmetry)}$$

References

1. Ritz W (1908) Über eine neue Methode zur Lösung gewisser Variationsprobleme der mathematischen Physik, Zeitschrift für Angewandte Mathematik und Mechanik, vol 135, H 1, 47–55
2. Galerkin BG (1915) Series occurring in various questions concerning the elastic equilibrium of rods and plates. Engineers Bulletin (Vestnik Inzhenerov) 19:897–908 (in Russian)
3. Argyris JH (1957) Die Matrizentheorie der Statik, Ingenieurarchiv 25. Springer, Berlin, 174–194
4. Turner MJ, Clough RW, Martin HC, Topp LJ (1956) Stiffness and deflection analysis of complex structures. J Aeronaut Sci 23:803–823, 853
5. Clough RW (1960) The finite element in plane stress analysis. In: Proceedings 2nd ASCE conference on electronic computation, Pittsburgh, Pa
6. Clough RW (1970) Areas of application of the finite element method, computers & structures 4. Pergamon Press, Oxford
7. Zienkiewicz OC (1977) The finite element method. McGraw-Hill Book Company Maidenhead Berkshire, England
8. Argyris J, Mlejnek H-P (1986) Die Methode der Finiten Element Band 1. Vieweg, Braunschweig
9. Bathe K-J, Wilson E (1976) Numerical methods in finite element analysis. Prentice Hall, Englewood Cliffs, NJ
10. Hughes TJR, Cottrell JA, Bazilevs Y (2005) Isogeometric Analysis: CAD, Finite Elements, Nurbs, Exact Geometry and Mesh Refinement. Comput Methods Appl Mech Eng 194 (39–41):4135–4195
11. Belytschko T, Krongauz Y, Organ D, Fleming M, Krysl P (1996) Meshless methods: an overview and recent developments. Computer methods in applied mechanics and engineering, vol 139, Issue 1–4. Elsevier, 3–47
12. Gupta KK, Meek JL (1996) A brief history of the beginning of the finite element method. International journal for numerical methods in Eng 39:3761–3774 (Wiley)
13. Clough RW (2004) Early history of the finite element method from the view point of a pioneer. Int J Numer Methods Eng 60:283–287 (Wiley)
14. Hahn M, Reck M (2018) Kompaktkurs Finite Elemente für Einsteiger. Springer, Wiesbaden
15. Rannacher R, Stein E, Ramm E, Schweizerhof K, Möller KH, Bayer U et al. (1997) Finite Elemente: Zur Geschichte der FEM, Spektrum der Wissenschaften, Heidelberg, März
16. Kurrer K-E (2018) The history of the theory of structures, 2nd edn. Ernst & Sohn, Berlin
17. Meißner U, Menzel A (1989) Die Methode der finiten Elemente. Springer, Berlin
18. Krätzig WB, Basar Y (1997) Tragwerke 3 – Theorie und Anwendung der Methode der Finiten Elemente. Springer, Berlin
19. Knothe K, Wessels H (2017) Finite elemente, 5th edn. Springer, Berlin
20. Kämmel G, Franeck H, Recke H-G (1990) Einführung in die Methode der Finiten Elemente. Carl Hanser Verlag, München
21. Klein B (2015) FEM - Grundlagen und Anwendungen der Finite-Element-Methode im Maschinen- und Fahrzeugbau, 10th edn. Springer Vieweg, Wiesbaden
22. Betten J (2003) Finite Elemente für Ingenieure 1, 2nd edn. Springer, Berlin
23. Steinbuch R (1998) Finite Elemente – ein Einstieg. Springer, Berlin
24. Wagner M (2017) Lineare und nichtlineare FEM. Springer Vieweg, Wiesbaden
25. Neto MA, Amaro A, Roseiro L, Cirne J, Leal R (2015) Engineering computation of structures: the finite element method. Springer International, Cham, Switzerland
26. Cook RD, Malkus DS, Plesha ME (1989) Concepts and applications of finite element analysis. Wiley, New York
27. Gallagher RH (1975) Finite element analysis — fundamentals. Prentice Hall Inc, Englewood Cliffs, NJ, USA
28. Szilard R (1990) Finite Berechnungsmethoden der Strukturmechanik, Band 2: Flächentragwerke im Bauwesen, Ernst & Sohn, Berlin

29. Wunderlich W, Redanz W (1995) Die Methode der Finiten Elemente, Der Ingenieurbau – Rechnerorientierte Baumechanik. Ernst & Sohn, Berlin
30. Thieme D (1990) Einführung in die Finite-Element-Methode für Bauingenieure. Verlag für Bauwesen, Berlin
31. Strang G, Fix GJ (1973) An analysis of the finite element method. Prentice Hall Inc, Englewood Cliffs, NJ, USA
32. Steinke P (2015) Finite-elemente-methode, 5th edn. Springer, Wiesbaden
33. Irons BM, Razzaque A (1972) Experience with a patch test for convergence of finite elements. The mathematical foundations of the finite element method with application to partial differential equations. Academic Press, New York, 557–582
34. Bathe K-J (2014) Finite element procedures, 2nd edn. K-J Bathe, Watertown, MA (2014). http://meche.mit.edu/people/faculty/kjb@mit.edu; also published by Higher Education Press China 2016
35. Könke C, Krätzig WB, Meskouris K, Petryna YS (2012) Theorie der Tragwerke. In: Zilch K, Diederichs CJ, Katzenbach R, Beckmann KJ (eds) Handbuch für Bauingenieure, 2nd edn. Springer, Heidelberg
36. Betten J (2004) Finite Elemente für Ingenieure 2, 2nd edn. Springer, Berlin
37. Dhatt G, Touzot G et al. (1984) The finite element method displayed. Wiley, Chichester New York
38. Bathe K-J (1996) Solutions manual, finite element procedures. Prentice Hall, Englewood Cliffs, NJ
39. Belytschko T, Liu WK, Moran B, Elkhodary K (2001) Nonlinear finite elements for continua and structures, 2nd edn. Wiley
40. Hughes TJR (2000) The finite element method. Dover, Mineola
41. Wilson EL, Taylor RL, Doherty WP, Ghaboussi J (1973) Incompatible displacement models. Numer Comput Methods Struct Mech 43–57
42. Taylor RL, Beresford PJ, Wilson EL (1976) A non-conforming element for stress analysis. Int J Numer Meth Eng 10:1211–1219
43. Wilson E, Ibrahimbegovic A (1990) Use of incompatible displacement modes for the calculation of element stiffness and stresses. Finite Elem Anal Des 7:229–241
44. Simo JC, Rifai MS (1990) A class of mixed assumed strain methods and the method of incompatible modes. Int J Numer Meth Eng 29:1595–1638
45. Koschnik F (2004) Geometrische Locking-Effekte bei Finiten Elementen und ein allgemeines Konzept zu ihrer Vermeidung, Dissertation, Technische Universität München
46. Lee Y, Lee PS, Bathe K-J (2014) The MITC3 + shell element and its performance. Comput Struct 138:12–23
47. Ko Y, Lee PS, Bathe K-J (2017) A new MITC4 + shell element. Comput Struct 182:404–418
48. Ko Y, Lee Y, Lee PS, Bathe K-J (2017) Performance of the MIT3 + and MIT4 + shell elements in widly-used benchmark problems. Comput Struct 193:187–206
49. Pian THH (1964) Derivation of element stiffness matrices by assumed stress distribution. AIAA J
50. Pian THH, Sumihara K (1984) Rational Approach for Assumed Stress Finite Elements. Int J Numer Methods Eng 20:1685–1695
51. Walder U (1977) Beitrag zur Berechnung von Flächentragwerken nach der Methode der Finiten Elemente, Dissertation, Institut für Baustatik und Konstruktion, ETH Zürich
52. Bergan PG, Felippa CA (1985) A triangular membrane element with rotational degrees of freedom. Computer methods in applied mechanics and engineering, vol 50. Elsevier BV, Amsterdam, Niederlande
53. Alvin K, de la Fuente HM, Haugen B, Felippa CA (1992) Membrane triangles with corner drilling freedoms. Report no CU-CSSC-91-24, Department of Aerospace Engineering Sciences, University of Colorado, Boulder, Colorado
54. Ibrahimbegovic A, Taylor RL, Wilson EL (1990) A robust membrane quadrilateral element with drilling degrees of freedom. Int J Numer Meth Eng 30(3):445–457

55. Ibrahimbegovic A, Wilson EL (1991) A unified formulation for triangular and quadrilateral flat shell finite elements with six nodal degrees of freedom. Commun Appl Numer Methods 7:1–9
56. Babuska I, Szabo BA, Katz IN (1981) The p-version of the finite element method, society for industrial and applied mathematics (SIAM). J Numer Anal 18:512–545
57. Szabo B, Babuschka I (1991) Finite element analysis. Wiley, New York
58. Bellmann J, Rank E (1989) Die p- und hp-Version der Finite-Element-Methode – oder: Lohnen sich höherwertige Elemente? Bauingenieur 64: 67–72
59. Heil W, Sauer R, Schweitzerhof K, Vogel U (1995) Berechnungssoftware für den Konstruktiven Ingenieurbau, Bauingenieur 70. Springer, Berlin, 55–64
60. Hinton E, Owen DRJ, Krause G (1990) Finite Element Programme für Platten und Schalen. Springer, Berlin
61. Hughes TJR, Taylor RL, Kanoknukulchai W (1977) A simple and efficient finite element for plate bending. Int J Numer Methods Eng 11:1529–1543
62. Andelfinger U (1991) Untersuchungen zur Zuverlässigkeit hybrid-gemischter Finiter Elemente für Flächentragwerke, Dissertation, Universität Stuttgart
63. Hughes TJR, Tezduyar TE (1981) Finite elements based upon Mindlin plate theory with particular reference to the four-node bilinear isoparametric element. J Appl Mech 48(3):587–596
64. Bathe K-J, Dvorkin E (1985) A four-node plate bending element based on Mindlin/Reissner plate theory and a mixed interpolation. Int J Numer Methods Eng 21:367–383
65. Bletzinger K-U, Bischoff M, Ramm E (2000) A unified approach for shear-locking-free triangular and rectangular shell finite elements. Comput Struct 75:321–334
66. Batoz J-L, Lardeur P (1989) A discrete shear triangular nine DOF element for the analysis of thick to very thin plates. Int J Numer Methods Eng 28:533–560
67. Przemieniecki JS (1968) Theory of matrix structural analysis. McGraw-Hill, New York
68. Chapelle D, Bathe K-J (2011) The finite element analysis of shells –fundamentals (2nd edn). Springer, Berlin
69. Pian THH (1965) Element stiffness matrices for boundary compatibility and for prescribed boundary stresses. In: Proceedings first conference on matrix methods in structural mechanics, AFFDL TR, vol 66-80
70. Pian THH, Tong P (1969) Basis of finite element methods for solid continua. Int J Numer Methods Eng 1:3–28 (Wiley)
71. Holzer S, Rank E, Werner H (1990) An implementation of the hp-version of the finite element method for Reissner-Mindlin Plate Problems. Int J Numer Methods Eng 30:459–471 (Wiley)
72. Rank E, Bröcker H, Düster A, Rücker M (2001) Integrierte Modellierungs- und Berechnungssoftware für den konstruktiven Ingenieurbau: Die p-Version und geometrische Elemente, Bauingenieur, 76. Springer, Berlin, 53–61
73. Batoz J-L, Bathe K-J, Ho L-W (1980) A study of three-node triangular plate bending elements. Int J Numer Methods Eng 15:1771–1812
74. Bathe K-J, Ho LW (1981) A simple and effective element for analysis of general shell structures. J Comput Struct 13:673–681
75. Bathe K-J (2020) Personal communication, e-mail, 8 Apr 2020
76. Dvorkin EN, Bathe K-J (1984) A continuum mechanics based four-node shell element for general nonlinear analysis. Eng Comput 1
77. Lee P-S, Bathe K-J (2004) Development of MITC isotropic triangular shell finite elements. Comput Struct 11–12, 82:945–962
78. Taylor RL, Simo JC (1985) Bending and membrane elements for analysis of thick and thin shells. In: Proceedings of the NUMEETA 1985 conference, Swansea, Wales
79. Tessler A, Hughes TJR (1983) An improved treatment of transverse shear in the Mindlin-type four-node quadrilateral element. Comput Methods Appl Mech Eng 39:311–335
80. Hartmann F, Katz C (2007) Structural analysis with finite elements, 2nd edn. Springer, Berlin
81. Bischoff M (1999) Theorie und Numerik einer dreidimensionalen Schalenformulierung, Dissertation, Institut für Baustatik, Bericht 30, Universität Stuttgart
82. Cook RD (1989) Finite element modeling for stress analysis. Wiley, New York

83. Grafton PE, Strome DR (1963) Analysis of axi-symmetric shells of revolution by the direct stiffness method. J AIAA 1:2342–2347
84. Nemec I, Kolar V, Sevcik I, Vlk Z, Blaauwendraat J, Bucek J, Teply B, Novak D, Stembera V (2010) Finite element analysis of structures. Shaker Verlag, Aachen
85. Wilson EL (1965) Structural analysis of axisymmetric solids. AIAA J 3(12):2269–2274
86. Buck KE (1973) Rotationskörper unter beliebiger Belastung. In: Finite Elemente in der Statik, Ernst & Sohn, Berlin
87. Werkle H, Friedrich T, Sutter J (2000) Modellierung von Stützen mit Pilzkopfverstärkung bei Flachdecken, 5. FEM/CAD-Tagung, Technische Universität Darmstadt, 2000, Festschrift Bauinformatik, Werner-Verlag, Düsseldorf
88. Werkle H (2002) Analysis of flat slabs by the finite element method, ASEM'02. Pusan, Korea
89. Werkle H (2002) Modelling of connections between dissimilar finite element domains by transition elements. WCCM V, Wien
90. Werkle H (2005) Modeling of connections of beam and shell elements in finite element analysis. NAFEMS world congress, Malta
91. Kugler J (1999) Finite-Element-Modellierung von Starrkörper- und Übergangsbedingungen in der Statik, Dissertation, Universität Karlsruhe, Institut für Baustatik, Bericht 5
92. Wagner W, Gruttmann F (2002) Modeling of shell-beam transitions in the presence of finite rotations. Comput Assist Mech Eng Sci 9:405–418
93. Bathe K-J, Ho LW (1981) Some results in the analysis of thin shell structures. In: Wunderlich et al (eds) Nonlinear finite element analysis in structural mechanics. Springer, Berlin, 123–150
94. Gmür T, Kauten RH (1993) Three-dimensional solid-to-beam transition elements for structural dynamics analysis. Int J Numer Methods Eng 36:1429–1444 (Wiley)
95. Gmür Th, Schorderet A (1993) A set of three-dimensional solid to shell transition elements for structural dynamics. Comput Struct 46:583–591
96. Surana K (1980) Transition finite elements for three-dimensional stress analysis. Int J Numer Methods Eng 7, 15:991–1020 (Wiley)
97. Werkle H (2000) Konsistente Modellierung von Stützen bei der Finite-Element-Berechnung von Flachdecken, Bautechnik. Ernst & Sohn, Berlin
98. Werkle H (1981) Ein Randelement zur dynamischen Finite-Element-Berechnung dreidimensionaler Baugrundmodelle, Dissertation, Universität Karlsruhe
99. Werkle H (1986) Dynamic finite element analysis of three-dimensional soil models with a transmitting element. Earthquake engineering and structural dynamics, vol 14. Wiley, New York, 41–60
100. Schwer LE (2006) An overview of the asme guide for verification and validation in computational solid mechanics, 5. LS-Dyna Anwenderforum, Ulm
101. Schwer LE (2005) Verification and validation in computational solid mechanics and the ASME standards commettee, WIT transactions on the built environment, vol 84. WIT Press
102. VDI 6201 Blatt 1 (2015) Softwaregestützte Tragwerksberechnung - Grundlagen, Anforderungen, Modellbildung, Verein Deutscher Ingenieure eV, Düsseldorf
103. Szabo B, Babuska I (2011) Introduction to finite element analysis. Wiley, Chichester
104. Rombach GA (2011) Finite-Element design of concrete structures, 2nd edn. ICE Publishing, London
105. Pacoste C, Plos M, Johansson M (2012) Recommendations for finite element analysis for the design of reinforced concrete slabs. KTH - Royal Institute of Technology, Stockholm, Sweden
106. Kemmler R, Ramm E (2001) Modellierung mit der Methode der Finiten Elemente. Beton-Kalender 2001, Ernst & Sohn, Berlin, 143–208
107. Barth C, Rustler W (2013) Finite Elemente in der Baustatik-Praxis, 2nd edn., Beuth-Verlag, Berlin
108. Timoshenko SP, Goodier JN (1985) Theory of elasticity. McGraw-Hill, Cambridge
109. Girkmann K (1986) Flächentragwerke. Springer, Wien
110. Johnson KL (1985) Contact mechanics. Cambridge University Press, Cambridge
111. Lubarda VA (2013) Circular loads on the surface of a half-space: displacement and stress discontinuities under the load. Int J Solids Struct 50, Elsevier, 1–14

112. Werkle H, Avak R (2003) Finite Elemente. In: Stahlbetonbau aktuell 2003. Bauwerk Verlag, Berlin
113. Babuska I, Hae-Soo O (1987) Pollution problem of the p- and h–p versions of the finite element method. Communications in applied numerical methods, vol 3, 6. Wiley, 553–561
114. Rank E, Krause R, Schweinegruber M (1992) Netzadaption durch intelligentes Pre- und Post-processing, Finite Elemente, Anwendungen in der Baupraxis (Karlsruhe 1991). Verlag Ernst & Sohn, Berlin, 541–551
115. Hughes TJ, Akin JE (1980) Techniques for developing special finite element shape functions with particular reference to singularities. Int J Numer Methods Eng 15:733–751
116. Welsch M (2016) Bewertung von Spannungsspitzen und Singularitäten in FEM- Rechnungen. In: D. Pieper: 11. Norddeutsches Simulationsforum 2015, Diplomica Verlag, Hamburg
117. Robinson J (1968) The cre method of testing and the Jakobian shape parameters, reliability of methods for engineering analysis. In: Proceedings of the international conference on Swansea 1986. Pineridge Press, 407–424
118. Macneal R, Harder RL (1985) A proposed standard set of problems to test finite element accuracy, finite elements in analysis and design. Elsevier, (North-Holland), Amsterdam
119. Rothe D (2019) Querkräfte und Drillmomente schubweicher Platten, Personal Communication, Sept 2019
120. Thompson JF (1985) Numerical grid generation. Elsevier Science Ltd
121. Zienkiewicz OC, Phillips DV (1971) An Automatic Mesh Generation Scheme for Plane and Curved Surfaces by Isoparametric Coordinates. Int J Numer Meth Eng 3:519–528
122. Werkle H, Gong D (1993) CAD-unterstützte Generierung ebener Finite-Element-Netze auf der Grundlage von Makroelementen, Bauingenieur 68. Springer, Berlin, 351–358
123. Owen SJ (1998) A survey of unstructured mesh generation technology. In: Proceedings of the 7th international meshing roundtable, 239–267
124. Liseikin VD (2010) Grid generation methods, 2nd edn. Springer, Dordrecht
125. Sorger CG (2012) Generierung von Netzen für Finite Elemente hoher Ordnung in zwei und drei Raumdimensionen, Dissertation, Technische Universität München
126. Lo SH (1985) A new Mesh generator scheme for arbitrary planar domains. Int J Numer Methods Eng 21:1403–1426
127. Zhu JZ, Zienkiewicz O, Hinton E, Wu J (1991) A new approach to the development of automatic quadrilateral Mesh generation. Int J Numer Methods Eng 32:849–866
128. Rehle N, Ramm E (1995) Generieren von FE-Netzen für ebene und gekrümmte Flächentragwerke, Bauingenieur 70. Springer, Berlin, 357–364
129. Rank E, Rücker M, Schweingruber M (1994) Automatische Generierung von Finite-Element-Netzen, Bauingenieur 69. Springer, Berlin
130. Rank E, Schweingruber M, Sommer M (1993) Adaptive mesh generation and transformation of triangular to quadrilateral meshes. Commun Numer Methods Eng 9:121–129
131. Rank E, Krause R, Schweingruber M (1992) Netzadaption durch intelligentes Pre- und Post-processing, Finite Elemente - Anwendungen in der Bautechnik (Karlsruhe 1991). Ernst & Sohn, Berlin, 541–551
132. Rehle N, Ramm E (1995) Adaptive Vernetzung bei mehreren Lastfällen, Finite Elemente in der Baupraxis (Stuttgart 1995). Ernst & Sohn, Berlin
133. Ramm E, Rehle N (1996) Qualitätssicherung durch angepasste FE-Netze – Möglichkeiten und Grenzen, Baustatik-Baupraxis 6, Weimar
134. Rust W, Stein E (1992) 2D-Finite-Element Mesh adaption in structural mechanics including shell analysis and non-linear calculations. In: Ladeveze P, Zienkiewicz OC (eds) Proceedigns of the conferences europe sur les nouvelles advancées en calcul des structures
135. Münzner D (2004) Freiformflächen in der Tragwerks- und Objektplanung, Sofistik- Seminar 2004, Sofistik Oberschleißheim
136. Lo SH (1991) Volume Discretization into Tetrahedra-I. Verification and orientation of boundary surfaces. Comput Struct 39(5):493–500
137. Lo SH (1991) Volume Discretization into Tetrahedra-II. 3D triangulation by advancing front approach. Comput Struct 39(5):501–511

138. Tilander S (1994) A finite element mesh generation program, Interner Bericht, Fachhochschule Konstanz, Lehrgebiet Baustatik (Prof Dr-Ing H Werkle)
139. Schaper G, Spiess G (2000) Nachvollziehbare Beispiele aus der Baupraxis – Berechnungen mit verschiedenen FE-Programmen, Festschrift Bauinformatik zu Ehren von Univ-Prof Dr-Ing habil Udo F. Meißner zum 60. Geburtstag, Hrsg.: U Rüppel, K Wassermann, Werner-Verlag, Düsseldorf
140. Williams ML (1952) Stress Singularities resulting from various boundary conditions in angular corners of plates in extension. J Appl Mech: 526–528
141. Rössle A (2000) Corner singularities and regularity of weak solutions for the two-dimensional Lamé equations on domains with angular corners. J Elast 60:57–77
142. Schneider M (1999) Finite-Element-Berechnungen von Scheibentragwerken, Diplomarbeit, HTWG Konstanz, Fachbereich Bauingenieurwesen, Lehrgebiet Baustatik (Prof Dr-Ing Horst Werkle), Konstanz
143. Tompert K (1995) Einige Problemfälle bei FE-Berechnungen aus der Sicht eines Prüfingenieurs, Finite Elemente in der Baupraxis, (Stuttgart 1995). Ernst & Sohn, Berlin
144. Schäfer K (1995) FE-Berechnung oder Stabwerkmodelle?, Finite Elemente in der Baupraxis, (Stuttgart 1995). Ernst & Sohn, Berlin
145. Thieme D (1992) Erfahrungen aus 100 Scheibenberechnungen, Finite Elemente, Anwendungen in der Baupraxis (Karlsruhe 1991). Ernst & Sohn, Berlin
146. Blaauewendraad J (2010) Plates and FEM. Springer, Dortrecht
147. Melzer H, Rannacher R (1980) Spannungskonzentrationen in Eckpunkten der Kirchhoffschen Platte, Bauingenieur 55. Springer, Berlin, 181–184
148. Williams ML (1951) Surface stress singularities resulting from various boundary conditions in angular corners of plates under bending. First US congress of applied mechanics. Illinois Institute of Technology, New York/Chicago, 325–329
149. Rössle A, Sändig A-M (2011) Corner singularities and regularity results for the Reissner/Mindlin plate model. J Elast 103:113–135
150. Häggblad B, Bathe K-J (1990) Specifications of boundary conditions for Reissner/Mindlin plate bending elements. Int J Numer Meth Eng 30:981–1011
151. Beyer M, Scharpf D (1985) Zur Frage der Drillmomente an freien Rändern dünner Platten, Finite Elemente - Anwendungen in der Baupraxis (München 1984). Ernst & Sohn, Berlin
152. Franz G (1983) Konstruktionslehre des Stahlbetonbaus, Band I – Grundlagen und Bauelemente, Teil B Die Bauelemente und ihre Bemessung. Springer, Berlin
153. Czerny F (1990) Tafeln für Rechteckplatten, Betonkalender 1990. Ernst & Sohn, Berlin
154. Babuška I, Pitkaranta J (1990) The plate paradox for hard and soft simple Support. SIAM J Math Anal 21(3):551–576
155. Grasser E, Thielen G (1988) Hilfsmittel zur Berechnung der Schnittgrößen und Formänderungen von Stahlbetontragwerken nach DIN 1045, Deutscher Ausschuß für Stahlbetonbau, H. 240. Ernst & Sohn, Berlin
156. Morley LSD (1963) Skew plates and structures. Pergamon Press, Oxford
157. Ramm E, Stander N, Stegmüller H (1988) Gegenwärtiger Stand der Methode der Finiten Elemente, Finite Elemente, Anwendungen in der Baupraxis (Bochum 1988). Ernst & Sohn, Berlin
158. Wegner R (1978) Einfluß der Stützensteifigkeit auf die Zustandsgrößen von Flachdecken, Bauingenieur 53. Springer, Berlin, 169–178
159. Gerold F (2004) 3D-Modellierung von Gebäuden mit der Methode der Finiten Elemente, Diplomarbeit, Fachhochschule Konstanz, Fachbereich Bauingenieurwesen, Lehrgebiet Baustatik (Prof Dr-Ing H Werkle), Konstanz
160. Werkle H, Gerold F (2005) Modellierung stabartiger Bauteile bei Flächentragwerken, Sofistik Seminar 2005, Nürnberg, Sofistik GmbH Oberschleißheim
161. Stein E, Niekamp R (1998) Ein objekt-orientiertes Programm für adaptive parallele Berechnungen von zwei- und dreidimensionalen Problemen der Strukturmechanik, Finite Elemente in der Baupraxis, Technische Universität Darmstadt, 1998. Ernst & Sohn, Berlin

162. Berger H, Gabbert U, Graeff-Weinberg K, Grochla J, Köppe H (1996) Über eine Technik zur lokalen Netzverdichtung in der FEM und ihre Anwendung im Bauwesen, TH Darmstadt, VDI-Verlag, Düsseldorf, 4. FEM/CAD-Tagung
163. Cramer H (1999) Anwendungsorientierte Finite-Element-Modelle für die Kopplung verschiedenartiger Tragwerksteile, 7. Fachtagung Baustatik-Baupraxis RWTH Aachen, AA Balkema, Rotterdam (1999)
164. Glahn H (1973) Ein Kraftgrößenverfahren zur Behandlung von Pilz- und Flachdecken über rechteckigem Grundriß, Dissertation, TH Aachen
165. Glahn H, Trost H (1974) Zur Berechnung von Pilzdecken. Der Bauingenieur, Springer, Berlin
166. Holzer S (1998) Was kann ein Finite-Element-Programm für Platten in der Praxis leisten?, Finite Elemente in der Baupraxis (Darmstadt 1998). Ernst & Sohn, Berlin
167. Ramm E, Müller J (1985) Flachdecken und Finite Elemente - Einfluß des Rechenmodells im Stützenbereich, Finite Elemente - Anwendungen in der Baupraxis, München 1984. Ernst & Sohn, Berlin
168. Konrad A, Wunderlich W (1995) Erfahrungen bei der baupraktischen Anwendung der FE-Methode bei Platten- und Scheibentragwerken, Finite Elemente in der Baupraxis (Stuttgart 1995). Ernst & Sohn, Berlin
169. Scharpf D (1985) Anwendungsorientierte Grundlagen Finiter Elemente, Landesvereinigung der Prüfingenieure für Baustatik, Baden-Württemberg, Tagungsbericht 10, Freudenstadt
170. Sutter J (2000) Finite-Element-Berechnung von Flachdecken mit Pilzkopfverstärkung, Diplomarbeit, Fachhochschule Konstanz, Fachbereich Bauingenieurwesen, Lehrgebiet Baustatik (Prof Dr-Ing H Werkle)
171. Zimmermann S (1989) Parameterstudie an Platte mit Unterzug, 1. FEM-Tagung Kaiserslautern
172. Katz C, Stieda J (1992) Praktische FE-Berechnung mit Plattenbalken, Bauinformatik 1/1992
173. Wunderlich W, Kiener G, Ostermann W (1994) Modellierung und Berechnung von Deckenplatten mit Unterzügen, Bauingenieur 89. Springer, Berlin
174. Katz C (1994) Neues zu Plattenbalken, Sofistik-Seminar, Nürnberg 1994, Sofistik GmbH
175. Rothe H (1994) Anwendungen von FE-Programmen und ihre Grenzen, Seminar Tragwerksplanung der Vereinigung der Prüfingenieure für Baustatik in Hessen
176. Bachmaier T (1994) Verfahren zur Bemessung von Unter- und Überzügen in FEM-Platten, 3. FEM/CAD-Tagung, TH Darmstadt
177. Bechert H, Bechert A (1995) Kopplung von Platten-, Scheiben- und Balkenelementen - das Problem der voll mittragenden Breite, Finite Elemente in der Baupraxis (Stuttgart 1995). Ernst & Sohn, Berlin
178. DIN EN (2011) 1992-1-1, Eurocode 2: Bemessung und Konstruktion von Stahlbeton- und Spannbetontragwerken – Teil 1-1: Allgemeine Bemessungsregeln und Regeln für den Hochbau, Januar 2011
179. Gupta Ma (1977) Error in eccentric beam formulation. Int J Numer Methods Eng 11:1473–1477
180. Wassermann K (1995) Unter- und Überzug in MicroFe und PlaTo, mb-news, Oktober 1995, mb-Programme, Hameln
181. Schöck Isokorb® (2020) Technical Information, Schöck Isokorb® T for reinforced concrete structures, Bicester, UK
182. Stiglat K, Wippel H (1976) Die vierseitig frei drehbar gelagerte Platte mit abhebenden Ecken unter Gleichlast, Beton- und Stahlbetonbau 8, vol 71. Ernst & Sohn, Berlin, 207
183. Marques JMMC, Owen DRJ (1984) Infinite elements in quasi-static materially nonlinear problems. Comput Struct 18(4):739–751
184. Zienkiewicz OC, Emson C, Bettess P (1983) A novel boundary infinite element. Int J Numer Methods Eng 19:393–404
185. v Wolffersdorff P-A, Kimmich S (1999) Erstellung von Bauwerk/Bodenmodellen für komplexe Flächengründungen. In: Meskouris K: Baustatik-Baupraxis 7. Balkema, Rotterdam
186. NN (1990) Hafenbautechnische Gesellschaft und deutsche Gesellschaft für Erd- und Grundbau/Arbeitsausschuss Ufereinfassungen: Empfehlungen des Arbeitsausschusses Ufereinfassungen EAU 1990, Ernst & Sohn, Berlin

187. Arslan U (2009) Bodenmechanik und Felsmechanik. Institut für Werkstoffe und Mechanik im Bauwesen, Technische Universität Darmstadt, Darmstadt
188. Waas G, Riggs R, Werkle H (1985) Displacement solutions for dynamic loads in transversly-isotropic stratified media. Earthquake Engineering and Design, vol 13. Wiley, 173–193
189. Smoltczyk U (1980) Grundbau Taschenbuch. Ernst & Sohn, Berlin
190. HASE NN (2019) Halbraumanalyse für statische Boden-Struktur-Interaktion, Manual Version 2018 Aug, Sofistik AG, Oberschleißheim, Germany
191. Bellmann J, Katz C (1994) Bauwerk-Boden-Wechselwirkungen, 3. FEM/CAD-Tagung Darmstadt, TH Darmstadt
192. Tsudik E (2013) Analysis of structures on elastic foundations. J Ross Publishing, Cengage Learning, Delhi
193. Kany M (1974) Berechnung von Flächengründungen, 2nd edn. Ernst & Sohn, Berlin
194. Hahn J (1970) Durchlaufträger, Rahmen, Platten und Balken auf elastischer Bettung, Werner-Verlag, Düsseldorf
195. Bares R (1979) Berechnungstafeln für Platten und Wandscheiben — tables for the analysis of plates, slabs and diaphragms. Bauverlag, Wiesbaden
196. Hirschfeld K (1969) Baustatik. Springer, Berlin
197. Buczkowski R, Torbacki W (2001) Finite element modelling of thick plates on two-parameter elastic foundation. Int J Numer Anal Meth Geomech 25:1409–1427
198. Buczkowski R, Taczaka M, Kleiber M (2015) A 16-node locking-free Mindlin plate resting on a two-parameter elastic foundation — static and eigenvalue analysis. Comput Assist Methods Eng Sci 22:99–114
199. Waas G (1980) Dynamisch belastete Fundamente auf geschichtetem Baugrund. VDI-Berichte Nr 381:185–189
200. Petersen C, Werkle H (2018) Dynamik der Baukonstruktionen, 2nd edn. Springer Vieweg, Wiesbaden
201. DIN-Fachbericht 130 (2003) Wechselwirkung Baugrund/Bauwerk bei Flachgründungen. Beuth-Verlag, Berlin
202. Fischer D (2009) Interaktion zwischen Baugrund und Bauwerk, Schriftenreihe Geotechnik. H 21, Universität Kassel
203. Morgen K, Bente S (2012) Flachgründungen. In: Boley C, Handbuch der Geotechnik. Vieweg + Teubner, Wiesbaden
204. Ahrens H, Winselmann D (1984) Eine iterative Berechnung von Flachgründungen nach dem Steifemodulverfahren, Finite-Elemente – Anwendungen in der Baupraxis – FEM'84. Ernst & Sohn, Berlin
205. Caselunge A, Eriksson J (2012) Structural element approaches for soil-structure interaction. Masterthesis, Department of Civil and Environmental Engineering, Chalmers University of Technology, Schweden
206. Barth Ch, Margraf E (2004) Untersuchung verschiedener Bodenmodelle zur Berechnung von Fundamentplatten im Rahmen von FEM-Lösungen, Bautechnik 81, H5, Ernst & Sohn, Berlin, 337–343
207. Pasternak PL (1954) New method of calculation for flexible substructures on two-parameter elastic foundation. Gosudarstvennoe Izdatelstoo. Literatury po Stroitelstvu i Architekture, Moskau, 1–56 (in Russian)
208. Barwaschow WA (1977) Setzungsberechnungen von unterschiedlichen Modellen, Osnowania, fundamenti i mechanika gruntow, H. 4/77, Moskau (in Russian)
209. Vlasov VZ, Leont'ev NN (1966) Beams, plates and shells on elastic foundations, Israel Program for Scientific Translations, Jerusalem (Translated from Russian; original Russian version published in 1960)
210. Jones R, Xenophontos J (1977) The Vlasov foundation model. Int J Mech Sci 19(1977):317–323
211. Vallabhan CVG, Daloglu AT (1999) Consistent FEM-Vlasov model for plates on layered soil. J Struct Eng ASCE 108–113

212. Kolar V, Nemec I (1986) Modelling of soil–structure interaction. Elsevier Publishers, Amsterdam
213. Tanahashi H (2007) Pasternak model formulation of elastic displacements in the case of a rigid circular foundation. J Asian Arch Buil Eng 6(1):167–173
214. Tanahashi H (2004) Formulas for an infinitely long Bernoulli-Euler beam on the Pasternak model, soils and foundations, vol 44, No 5. Japanese Geotechnical Society, Tokyo
215. Feng Z, Cook RD (1983) Beam elements on two-parameter elastic foundations. J Eng Mech 6 109:1390–1402
216. Iancu-Bogdan T, Vasile M (2010) The modified Vlasov foundation model: an attractive approach for beams resting on elastic supports. EJGE, vol 15
217. Buczkowski R, Torbacki W (2009) Finite element analysis of plate on layered tensionless foundation. Archives in Civil Engineering, LVI 3
218. Straughan WT (1990) Analysis of plates on elastic foundations, Texas Tech University, Dissertation
219. Werkle H, Slongo L (2018) Modellierung des Baugrunds bei der Finite-Element-Berechnung von Bodenplatten, Bautechnik 95, H 8. Ernst & Sohn, Berlin
220. Slongo L (2017) Modellierung der Boden–Bauwerk–Wechselwirkung bei der statischen Berechnung von Fundamentplatten, Bachelorthesis, HTWG Konstanz, Fakultät Bauingenieurwesen, Lehrgebiet Baustatik (Prof Dr-Ing H Werkle), Konstanz
221. Rothe D (2019) Vergleich-2-Parameter-Modelle, personal communication, Hochschule Darmstadt, 28 Aug 2019
222. Zilch K, Katzenbach R (2002) Baugrund–Tragwerk–Interaktion. In: Handbuch für Bauingenieure, Springer, Berlin, 3-462–3-480
223. DIN 4018 (1974) Berechnung der Sohldruckverteilung unter Flächengründungen, einschl. Beiblatt 1 (Erläuterungen und Berechnungsbeispiele) Beuth-Verlag, Berlin, 1974 (DIN 4018) und 1981 (Beiblatt)
224. Meyerhof G (1953) Some recent foundation research and application to design. Struct Eng 31(6):151–167
225. Werkle H, Bock P (1996) Zur Anwendung der Finite-Elemente-Methode in der Praxis – Fehlerquellen bei der Modellbildung und Ergebnisinterpretation, 4. FEM/CAD-Tagung, TH Darmstadt, Darmstadt
226. Laggner TM, Schlicke D, Tue NV, Denk W-D (2021) Statische Analyse mit linear elastischen 3D-Gebäudemodellen, Auswirkungen unterschiedlicher Modellierungsarten auf den vertikalen Lastfluss, Beton- und Stahlbetonbau 116, Ernst & Sohn, Berlin
227. Ramm E, Fleischmann N, Burmeister A (1993) Modellierung mit Faltwerkselementen. Tagung Universtät München, Baustatik-Baupraxis
228. Tompert K (1997) Anwendungen der FE-Methode bei der Berechnung ebener und räumlicher Tragwerke im Hochbau, Probleme beim Aufstellen und Prüfen, Landesvereinigung der Prüfingenieure für Baustatik Baden-Württemberg, Tagungsbericht 24, Freudenstadt
229. Tompert K (1998) Probleme beim Aufstellen und Prüfen von FE-Berechnungen, 2. Technische Universität Dresden, Oktober, Dresdener Baustatik-Seminar
230. Dressel B (1998) Zuverlässigkeit von Standsicherheitsnachweisen aus der Sicht des Prüfingenieurs, 2. Technische Universität Dresden, Oktober, Dresdener Baustatik-Seminar
231. Krebs A, Pellar A, Runte Th (1995) Einsatz der Methode der Finiten Elemente am Beispiel eines schiefwinkligen Rahmenbauwerks, Finite Elemente in der Baupraxis, (Stuttgart 1995). Ernst & Sohn, Berlin
232. Pflüger A (1960) Elementare Schalenstatik. Springer, Berlin
233. Flügge W (1981) Statik und Dynamik der Schalen. Springer, Berlin
234. Schaper G, Helter E (1998) Erfahrungen mit verschiedenen FEM-Programmen bei der Berechnung eines teilweise vorgespannten Faulbehälters, Finite Elemente in der Baupraxis (Darmstadt 1998). Ernst & Sohn, Berlin
235. Pitkäranta J, Babuska I, Szabo B (2008) The Girkmann problem. IACM Expr 22:28
236. Pitkäranta J, Babuska I, Szabo B (2009) The problem of verification with reference to the Girkmann problem. IACM Expr 24:14–15

237. Szabo BA, Babuska I, Pitkäranta J (2010) The problem of verification with reference to the Girkmann problem. Engineering with computers, vol 26. Springer, 171–183
238. Fastabend M, Schäfers T, Albert M, Lommen HG (2009) Zur sinnvollen Anwendung ganzheitlicher Gebäudemodelle in der Tragwerksplanung von Hochbauten, Beton- und Stahlbetonbau 104, H 10. Ernst & Sohn, Berlin
239. Enseleit J (1999) Strukturmechanische Analyse des Entstehens von Bauwerken. Werner-Verlag, Düsseldorf
240. Bischoff M (2010) Statik am Gesamtmodell: Modellierung, Berechnung und Kontrolle, Der Prüfingenieur 36. Zeitschrift der Bundesvereinigung der Prüfingenieure für Bautechnik, Berlin
241. Staller M (2014) Bautechnische Prüfung von Gesamtmodellen im Massivbau, Baustatik-Baupraxis 12, TU München
242. Fischer O, Reinhardt J (2008) Tragwerksplanung mit Gesamtmodellen aus der Sicht einer Baufirma, Baustatik-Baupraxis 10, Universität Karlsruhe (TU)
243. Löwenstein JG (2014) Rechnen am Gesamtsystem, mb-news, 4/2014, mb AEC Software GmbH, Kaiserslautern
244. Bischoff M (2015) Computer und Tragwerksmodellierung – Vorschläge und Impulse für eine moderne universitäre Baustatiklehre. Der Prüfingenieur, Zeitschrift der Bundesvereinigung der Prüfingenieure für Bautechnik, Mai 2015, Berlin
245. Häßler M (2014) Finite-Element-Berechnungen von Bauwerken mit Gesamtmodellen, Bachelorthesis, HTWG Konstanz, Fakultät Bauingenieurwesen, Lehrgebiet Baustatik (Prof Dr-Ing Horst Werkle), Konstanz
246. Fastabend M (2014) Statische Berechnung mit Gesamtmodellen – bautechnische Prüfung und Qualitätssicherung, Baustatik-Baupraxis 12, TU München
247. Rank E, Roßmann A (1987) Fehlerschätzung und automatische Netzanpassung bei Finite-Element-Berechnungen, Bauingenieur 62. Springer, Berlin, 449–454
248. Stein E, Ohnimus S, Seifert B, Mahnken R (1993) Adaptive Finite-Element-Methoden im konstruktiven Ingenieurbau. Tagung Universtät München, Baustatik-Baupraxis
249. Zienkiewicz OC, Zhu Z (1987) A simple error estimator and adaptive procedure for practical engineering analysis. Int J Numer Methods Eng 24:337–357
250. Grätsch T, Bathe K-J (2005) Entwicklung von Finite-Element-Modellen – Stand und Tendenzen, Baustatik-Baupraxis 9, TU Dresden
251. Grätsch T, Bathe K-J (2005) A posteriori estimation techniques in practical finite element analysis. Computers & structures, vol 83. Elsevier, 235–265
252. Zienkiewicz OC, Zhu Z (1989) Error estimates and adaptive refinement for plate bending problems. Int J Numer Methods Eng 28:2853–2893
253. Katz C (1989) Fehlerabschätzungen Finiter Element Berechnungen, 1. FEM-Tagung Kaiserslautern, Tagungsband
254. Hartmann F, Pickardt S (1985) Der Fehler bei finiten Elementen, Bauingenieur 60. Springer, Berlin, 463–468
255. Grätsch T, Hartmann F (2001) Über ein Fehlerbild bei der Schnittgrößenermittlung mit finiten Elementen – Teil 1: Scheiben, Bautechnik 78, H 5. Ernst & Sohn, Berlin
256. Grätsch T, Hartmann F (2005) Über ein Fehlerbild bei der Schnittgrößenermittlung mit finiten Elementen – Teil 2: Platten, Bautechnik 80, H 3. Ernst & Sohn, Berlin
257. Holzer S, Rank E, Werner H (1990) An implementation of the hp-version of the finite element method for Reissner-Mindlin plate problems. Int J Numer Methods Eng 30:459–471
258. Stein E, Ohnimus S (1995) Zuverlässigkeit und Effizienz von Finite-Element-Berechnungen durch adaptive Methoden, Finite Elemente in der Baupraxis (Stuttgart 1995). Ernst & Sohn, Berlin
259. Pravida JM (1999) Zur nichtlinearen adaptiven Finite-Element-Analyse von Stahlbetonscheiben, Dissertation, TU München
260. Duda H (1998) FEM – ein verführerisches Hilfsmittel, Finite Elemente in der Baupraxis (Darmstadt 1998). Ernst & Sohn, Berlin
261. Göttlicher M (2000) Anwendung der FEM in der Baupraxis, Festschrift Bauinformatik zu Ehren von Univ.-Prof Dr-Ing habil. Udo F. Meißner zum 60. Geburtstag, Hrsg: U Rüppel, K Wassermann, Werner-Verlag, Düsseldorf

262. Wörner J-D (2000) Informatik und Ingenieurwesen, Festschrift Bauinformatik zu Ehren von Univ-Prof Dr-Ing. habil. Udo F. Meißner zum 60. Geburtstag, Hrsg: U Rüppel, K Wassermann, Werner-Verlag, Düsseldorf

263. Reineck K-H (1995) Der Schadensfall Sleipner und die Folgerungen für den computerunterstützten Entwurf von Tragwerken aus Konstruktionsbeton, Finite Elemente in der Baupraxis (Stuttgart 1995). Ernst & Sohn, Berlin

264. N N (2002) Anforderungen für das Aufstellen EDV-unterstützter Standsicherheitsnachweise, Fachkommission der Vereinigung der Prüfingenieure für Baustatik in Niedersachsen, Bremen und Hamburg

265. VDI 6201 Blatt 2 (2017) Softwaregestützte Tragwerksberechnung - Verifikationsbeispiele, Verein Deutscher Ingenieure eV, Düsseldorf

266. Eisfeld M, Struss P (2008) Qualitätssicherung im Konstruktiven Ingenieurbau, Jahrbuch Bautechnik 2009, VDI-Verlag

267. Polónyi S, Reyer E (1977) Zuverlässigkeitsbetrachtungen und Kontrollmöglichkeiten (Prüfung) zur praktischen Berechnung mit der Finite-Element-Methode, Bautechnik 11. Ernst & Sohn, Berlin

268. Meißner U, Heller M (1989) Zur Beurteilung von Finite-Element-Ergebnissen, 1. Universität Kaiserslautern, FEM-Tagung

269. Sofistik (2018) Mechanical Benchmarks, verification manual, SOFiSTiK AG, Oberschleissheim

270. Henke P, Rapolder M (2008) Tragwerksplanung mit Gesamtmodellen – Auswirkungen auf die bautechnische Prüfung, Baustatik-Baupraxis 10, Universität Karlsruhe (TU)

271. Henke P, Rapolder M (2011) Erfahrungen und Forderungen aus der bautechnischen Prüfung, Baustatik-Baupraxis 11, Innsbruck

272. Rombach G (2007) Probleme bei der Berechnung von Stahlbetonkonstruktionen mittels dreidimensionaler Gesamtmodelle. Beton- und Stahlbetonbau 102:207–214

273. Rombach G (2010) Statik am Gesamtmodell: Modellierung. Berechnung, Kontrolle, Der Prüfingenieur 36:27–34

274. Minnert J (2013) Black-Box EDV-gestützte Berechnungen?! Notwendigkeit und Möglichkeiten der Ergebniskontrolle, 10. Gießener Bauforum 2013, Shaker-Verlag, Aachen

275. Eisenbiegler G (1988) Gleichgewichtskontrollen bei punktgestützten Platten, Beton- und Stahlbetonbau 83. Ernst & Sohn, Berlin

276. Mestrovic M (2016) An application of Richardson extrapolation on FEM solutions. Int J Math Comput Methods

277. Losch S, Butenweg C (1999) Qualitätsgesteuerte Bemessung von Stahlbetontragwerken mit der FE-Methode, 11. Forum Bauinformatik, Junge Wissenschaftler forschen, Darmstadt, 208–215

278. Ri-EDV-AP-2001 (2001) Richtlinie für das Aufstellen und Prüfen EDV-unterstützter Standsicherheitsnachweise, Ausgabe April 2001, Bundesvereinigung der Prüfingenieure für Bautechnik eV, 2001, Hamburg

279. Bathe K-J (2007) To enrich life. Klaus-Jürgen Bathe, USA

280. Madenci E, Guven I (2015) The finite element method and applications in engineering using ANSYS. Springer International Publishing, New York

281. Wilson EL (1970) SAP — a general structural analysis program, report no UC SESM 70–20. Structural Engineering Laboratory, University of California, Berkeley, USA

282. Bathe K-J, Wilson EL, Peterson FE (1974) SAP IV — a structural analysis program for static and dynamic response of linear systems, report no. EERC 73-11, Earthquake Engineering Research Center, University of California, Berkeley, USA

283. Werner H, Axhausen K, Katz C (1979) Programmaufbau und Datenstrukturen in entwurfsunterstützenden Programmketten. In: Pahl PJ, Stein E, Wunderlich W (eds) Finite Elemente in der Baupraxis. Springer, Berlin

284. Kimmich S (2020) Numerische Mathematik und Anwendung in der Finite Element Methode. Skriptum, HS Stuttgart

285. Batoz J-L (1982) An explicit formulation for an efficient triangular plate-bending element. Int J Numer Methods Eng 18:1077–1089
286. Batoz J-L, Tahar MB (1982) Evaluation of a new quadrilateral thin plate bending element. Int J Numer Methods Eng 18:1655–1677
287. Bucalem ML, Bathe K-J (2011) The mechanics of solids and structures—hierarchical modeling and the finite element solution. Springer, Berlin

Internet References

288. WEB 1 (2019) http://www.adina.com/company.shtml. visited at 18.2.2019, 14:33
289. WEB 2 (2019) https://en.wikipedia.org/wiki/Ansys. visited at 18.2.2019, 15:03
290. WEB 3 (2019) https://www.csiamerica.com/products/sap2000. visited at 18.2.2019, 16:00
291. WEB 4 (2019) https://www.infograph.de/de/geschichte. visited at 18.2.2019, 16:10
292. WEB 5 (2019) https://www.dlubal.com/de/unternehmen/ueber-uns/historie-und-zahlen https://www.dlubal.com/de/unternehmen/ueber-uns/firmenphilosophie. visited at 18.2.2019, 16:30
293. WEB 6 (2019) https://www.sofistik.de/unternehmen/ueber-uns/geschichte/. visited at 18.2.2019, 16:40
294. WEB 7 (2019) https://www.rib-software.com/group/ueber-rib/historie/. Visited at 18.2.2019, 16:50

Finite Element Software

295. SOF 1 (2012) ADINA, Theory and Modeling Guide, Volume I: ADINA, Report ARD 12-8, ADINA R&D Inc, Watertown USA, Dec 2012
296. SOF 2 (2013) ANSYS, ANSYS mechanical APDL element reference, release 15.0, ANSYS Inc, Canonsburg, USA, Nov 2013
297. SOF 3 (2018) InfoCad, InfoCad 18.1 – User manual, InfoGraph GmbH, Aachen, Germany
298. SOF 3a (1997) InfoCad – FEM Programmsystem, InfoGraph GmbH, Aachen, Germany
299. SOF 4 (2016) Dlubal RFEM, RFEM 5 Programmbeschreibung. Dlubal Software GmbH, Tiefenbach, Deutschland
300. SOF 5 (2016) SAP2000, CSI Analysis Reference Manual For SAP2000®, ETABS®, and SAFE®, Computers and Structures (CSI) Berkeley, Calfornia, USA
301. SOF 6 (2014) SOFiSTiK Finite-Element-Software, Version 2014-9 ASE – Allgemeine Statik Finiter Element Strukturen, ASE Manual, Version 2014-9, Software Version, Sofistik 2014, SOFiSTiK AG, Oberschleissheim, Germany, 2015 TAPLA – 2D Finite Elemente in der Geotechnik, TALPA Manual, Version 2014-9, Software Version SOFiSTiK 2014, SOFiSTiK AG, Oberschleissheim, Germany
302. SOF 6a SOFiSTiK Finite-Element-Software, Version 2.0-93, SOFiSTiK GmbH, Oberschleißheim
303. SOF 7 (2019) iTWO structure fem/TRIMAS TRIMAS/ PONTI, Grundlagen Benutzerhandbuch, RIB Software AG, Stuttgart-Möhringen
304. SOF 8 MicroFe, Version 2019, mb AEC Software GmbH, Kaiserslautern
305. SOF 8a (1992) MicroFe, Version 5.21, mb-Programme, Software im Bauwesen GmbH, Hameln, Kopmanshof 69 (aktuelle Programmversion enthält eine anderes Plattenelement)
306. SOF 9 (2019) s3d/microSnap, Detlef Rothe Hochschule Darmstadt & Thomas Schmidt Hochschule Magedeburg-Stendal, Version 0.68

Chapter 5
Dynamic Analysis of Structures

Abstract This chapter deals with the dynamics of structures in the context of the finite element method. It starts with some basic concepts as the kinematics of dynamic systems, mass, and damping forces and leads to the equations of motion of a finite element system. The solution of the equations of motion is treated for free vibrations as well as for vibrations excited by forces or base motion. For free undamped vibration, the concept of eigenmodes and eigenfrequencies is discussed. Forced vibrations are treated for harmonic and for general dynamic excitation. As computational methods, the direct integration in time domain, the solution in frequency domain, and the modal analysis are outlined. A special focus is on earthquake-excited vibrations. In addition to the computational methods mentioned before, the response spectrum analysis is examined. The modeling of structures for dynamic analysis is treated in the last section. In this context, different types of damping such as viscous, hysteretic, modal, and Rayleigh damping are discussed. The influence of the discretization in time and frequency domains is examined. Methods to consider dynamic soil–structure interaction are presented. Different types of modeling buildings for earthquake analyses are examined and demonstrated on an example of a beam model and a three-dimensional building model with and without dynamic soil–structure interaction. A final example demonstrates the accuracy achievable with a finite element analysis by validation of the previously computed vibrations of a building due to an impact in a double-blind test.

5.1 Introduction

Actions on a structure cannot be assumed to be static in all cases. Actions whose variation over time has a significant effect on the stresses and displacements of the structure are called dynamic actions. They cause a time-dependent structural behavior with time-dependent displacements and internal forces. Examples for time-dependent actions are the earthquake excitation of structures, dynamic loads caused by wind, traffic, or machines as well as accident loads [1].

From the point of view of mechanics, the existence of inertial forces is substantial for dynamic problems. Inertial forces are caused by the acceleration of masses. They

© Springer Nature Switzerland AG 2021
H. Werkle, *Finite Elements in Structural Analysis*, Springer Tracts
in Civil Engineering, https://doi.org/10.1007/978-3-030-49840-5_5

are of importance if the temporal variation of the actions occurs particularly fast. Otherwise, for slow motions, the inertial forces are negligible and one gets a static problem. Furthermore, damping forces are practically always present in dynamic processes. These extract energy from the system and cause a motion to decay when free oscillations occur.

In the following, some essential basic concepts of dynamics are first explained and their representation in the finite element method is dealt with. Then the equilibrium conditions to be solved in dynamic systems, the so-called equations of motion, are established, and various methods for their solution are discussed. An introduction to structural dynamics is also given in [2–12].

5.2 Basic Concepts of Dynamics

5.2.1 Kinematics

In dynamic problems, the displacements are no longer described by constant values but by functions dependent on time, the so-called displacement time histories. In addition to the time histories of the displacements, the temporal changes of the displacements, namely the velocities, are important. For a displacement time history $u(t)$ of a point, the time history of the *velocity* is obtained by differentiation as

$$v(t) = \dot{u}(t) = \frac{du}{dt}. \tag{5.1a}$$

The *acceleration* is defined as the change of the velocity. It is obtained by differentiating the velocity to time or as a second derivative of the displacement time history as

$$b(t) = \ddot{u}(t) = \frac{dv}{dt} = \frac{d^2u}{dt^2}. \tag{5.1b}$$

The unit of velocity is usually m/s, that of acceleration m/s^2. Rotations of nodes of a finite element model are angles and thus are unitless. The associated angular velocity has the unit 1/s, the angular acceleration 1/s^2.

The equations are the basis of analytical solutions. For example, for a constant acceleration b_0 from (5.1a), the velocity follows to

$$v(t) = v_0 + \int_0^t b_0 \, d\bar{t} = v_0 + b_0 \cdot t \tag{5.2a}$$

and with (5.1b), the displacement to

$$u(t) = u_0 + \int\limits_0^t v(\bar{t})\, d\bar{t} = u_0 + v_0 \cdot t + \frac{1}{2} b_0 \cdot t^2 \qquad (5.2b)$$

where u_0 is the displacement and v_0 the velocity at the time $t = 0$. These relationships are the basis of numerical integration methods in which the acceleration is assumed to be constant in each time step (see Sect. 5.6.2).

Displacement time histories can be described only in simple cases by analytical functions, such as sine functions. Usually, time histories are expressed numerically in time steps.

In finite element systems, each degree of freedom possesses a time history. In the general three-dimensional case, six time histories are obtained at each node for the three displacements and the three rotations. The amount of data is thus considerably larger for time history analyses than for static problems. The velocities and accelerations are expressed by vectors as the nodal displacements.

5.2.2 Inertial Forces

When a force F is acting on a point-like mass m, the mass is accelerated with $\ddot{u}$ in the direction of the x-axis (Fig. 5.1). According to *Newton's law,* one obtains

$$F = m \cdot \ddot{u}. \qquad (5.3)$$

By introducing the *inertial force* F_T as

$$F_T = -m \cdot \ddot{u}. \qquad (5.3a)$$

(5.3) can be written as an equilibrium condition in statics as

$$F + F_T = 0. \qquad (5.3b)$$

The inertial force has the magnitude $m \cdot \ddot{u}$ and is thus proportional to the mass and the acceleration. Its direction is opposite to the acceleration. It is also known as *d'Alembert force* or *mass force*.

Since forces and masses appear both in dynamic analyses, attention must be paid to the system of units. In the SI system, mass is the basic unit, while force is a

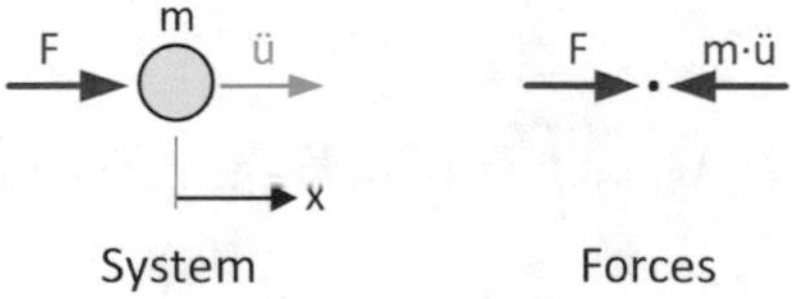

Fig. 5.1 Inertial force at a point mass

System   Forces

derived quantity. For example, 1 m^3 of concrete has the mass $2500 \text{ kg} = 2.5 \text{ t (tons)}$. The force corresponding to this mass is obtained with the acceleration of gravity of approximately $g = 10 \text{ m/s}^2$ as $m \cdot g = 2.5 \cdot 10 \text{ tm/s}^2 = 25 \text{ tm/s}^2 = 25 \text{ kN}$, i.e.

$$1 \text{ t} \frac{\text{m}}{\text{s}^2} = 1 \text{ kN}.$$

The force unit kN can be written as $1 \text{ kN} = 1 \text{ t} \cdot \text{m/s}^2$. For practical reasons, in structural analysis the unit of force is selected as the basic unit and the mass is expressed in this unit. The unit of mass t (ton) is, therefore, expressed as $1 \text{ t} = 1 \text{ kN} \cdot \text{s}^2/\text{m}$. For the mass unit kg, the conversion $1 \text{ kg} = 1 \text{ N} \cdot \text{s}^2/\text{m}$ applies.

If a rigid body is represented in the structural model as a point mass, the mass forces act at the center of gravity (or center of mass) of the body. In the general three-dimensional case, a mass point has three translational and three rotational degrees of freedom. The inertial forces in the three degrees of freedom of displacements in the Cartesian coordinate system result from the acceleration of the mass m in x-, y-, and z-directions. One receives

$$F_x = -m_x \cdot \ddot{u} \quad \text{with} \quad m_x = m \tag{5.4a}$$

$$F_y = -m_y \cdot \ddot{v} \quad \text{with} \quad m_y = m \tag{5.4b}$$

$$F_z = -m_z \cdot \ddot{w} \quad \text{with} \quad m_z = m. \tag{5.4c}$$

In the degrees of freedom of rotation, the inertial forces correspond to moments caused by the angular acceleration. The associated mass quantities are denoted as rotational masses. Thus, three translational masses and three rotational masses are assigned to a mass point. While the translational masses are usually the same in all directions, different values are obtained for the rotational masses. The determination of the rotational mass of a three-dimensional body is shown below exemplarily for the rotational mass θ_x about the x-axis.

An infinitesimal mass element of a rigid body is considered. The Cartesian coordinate system is located in the center of gravity in the direction of the principal axes (Fig. 5.2). The body is accelerated with the angular acceleration $\ddot{\varphi}_x = \mathrm{d}^2\varphi_x/\mathrm{d}t^2$. The infinitesimal mass element $\mathrm{d}m$ is thus accelerated with $\ddot{w} = \ddot{\varphi}_x \cdot y$ in the z-direction and with $\ddot{v} = -\ddot{\varphi}_x \cdot z$ in the y-direction.

They result in the inertial forces $-\ddot{w} \cdot \mathrm{d}m = -\ddot{\varphi}_x \cdot y \cdot \mathrm{d}m$, $-\ddot{v} \cdot \mathrm{d}m = \ddot{\varphi}_x \cdot z \cdot \mathrm{d}m$, and the moments $-\ddot{w} \cdot y \cdot \mathrm{d}m = -\ddot{\varphi}_x \cdot y^2 \cdot \mathrm{d}m$ and $\ddot{v} \cdot z \cdot \mathrm{d}m = -\ddot{\varphi}_x \cdot z^2 \cdot \mathrm{d}m$ on the x-axis. The total moment M_x is obtained by integration over the volume V of the body as

$$M_x = \int (-\ddot{\varphi}_x \cdot y^2 - \ddot{\varphi}_x \cdot z^2) \cdot \mathrm{d}m = -\ddot{\varphi}_x \cdot \int (y^2 + z^2) \cdot \mathrm{d}m = -\Theta_x \cdot \ddot{\varphi}_x.$$

Fig. 5.2 Infinitesimal mass element dm of a body idealized as rigid mass

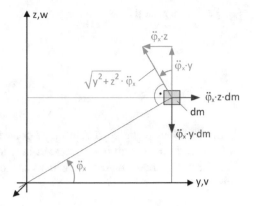

This is the relationship between the moment and the angular acceleration during a rotation about the x-axis corresponds to (5.3a) for displacements. The resulting inertial forces do not occur during an angular acceleration around the x-axis, because the axis system is assumed to be at the center of gravity of the body. Hence, the following applies to angular accelerations around the principal axes:

$$M_x = -\Theta_x \cdot \ddot{\varphi}_x \quad \text{with} \quad \Theta_x = \int (y^2 + z^2) \cdot dm \tag{5.5a}$$

$$M_y = -\Theta_y \cdot \ddot{\varphi}_y \quad \text{with} \quad \Theta_y = \int (x^2 + z^2) \cdot dm \tag{5.5b}$$

$$M_z = -\Theta_z \cdot \ddot{\varphi}_z \quad \text{with} \quad \Theta_z = \int (x^2 + y^2) \cdot dm. \tag{5.5c}$$

The rotational masses are thus the moments of inertia of the body around the principal axes.

Equations (5.4a–5.4c) and (5.5a–5.5c) for the inertial forces and moments of a mass can be written in matrix notation as

$$
\begin{bmatrix} F_x \\ F_y \\ F_z \\ M_x \\ M_y \\ M_z \end{bmatrix}
= -
\begin{bmatrix}
m_x & 0 & 0 & 0 & 0 & 0 \\
0 & m_y & 0 & 0 & 0 & 0 \\
0 & 0 & m_z & 0 & 0 & 0 \\
0 & 0 & 0 & \Theta_x & 0 & 0 \\
0 & 0 & 0 & 0 & \Theta_y & 0 \\
0 & 0 & 0 & 0 & 0 & \Theta_z
\end{bmatrix}
\cdot
\begin{bmatrix} \ddot{u} \\ \ddot{v} \\ \ddot{w} \\ \ddot{\varphi}_x \\ \ddot{\varphi}_y \\ \ddot{\varphi}_z \end{bmatrix}
\tag{5.6}
$$

$$\underline{F}_T = -\underline{M} \cdot \underline{\ddot{u}}. \tag{5.6a}$$

The matrix $\underline{M}$ is the mass matrix of a point.

Fig. 5.3 Rectangular slab

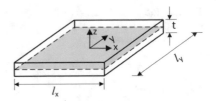

Example 5.1 A rectangular reinforced concrete slab with sizes $l_x = 18.0$ m, $l_y = 14.0$ m, and $t = 0.2$ m is represented in a finite element analysis as a rigid body (Fig. 5.3). The mass matrix related to the center of gravity of the slab shall be determined.

Reinforced concrete has a density of $\rho = 2.5$ t/m^3. The mass of the slab is thus obtained as

$$m = 2.5 \cdot 18.0 \cdot 14 \cdot 0.20 = 126.0 \text{ t} = 126.0 \text{ kNs}^2/\text{m}.$$

For the rotational masses, the following applies according to (5.5a–5.5c) with $dm = \rho \cdot dx \, dy \, dz$:

$$\Theta_x = \int_V (y^2 + z^2) \cdot dm = \rho \cdot t \cdot \iint_A y^2 \, dx \, dy + \rho \cdot A \cdot \int_{-d/2}^{d/2} z^2 \, dz$$

$$= \rho \cdot t \cdot I_x + \rho \cdot A \cdot \frac{t^3}{12} \approx \rho \cdot t \cdot I_x$$

$$\Theta_y = \int_V (x^2 + z^2) \cdot dm$$

$$= \rho \cdot t \cdot \iint_A x^2 \, dx \, dy + \rho \cdot A \cdot \int_{-d/2}^{d/2} z^2 \, dz = \rho \cdot t \cdot I_y + \rho \cdot A \cdot \frac{t^3}{12} \approx \rho \cdot t \cdot I_y$$

$$\Theta_z = \int_V (x^2 + y^2) \cdot dm$$

$$= \rho \cdot t \cdot \left(\int_A x^2 dA + \int_A y^2 dA \right) = \rho \cdot t \cdot (I_x + I_y) = \rho \cdot t \cdot I_p$$

where A is the surface area of the plate, I_x and I_y are the second moments of area about the x- and y-axes, respectively, and I_p is the polar moment of inertia of the area. For the given dimensions, one obtains the rotational masses with

$$I_x = \frac{l_x \cdot l_y^3}{12}, \quad I_y = \frac{l_y \cdot l_x^3}{12}, \quad I_p = I_x + I_y$$

and neglecting the term

$$\rho \cdot A \cdot \frac{t^3}{12} \approx 2.5 \cdot 18.0 \cdot 14.0 \cdot \frac{0.2^3}{12} = 0.4 \text{ kNms}^2$$

as

$$\Theta_x \approx \rho \cdot t \cdot I_x = 2.5 \cdot 0.20 \cdot \frac{18 \cdot 14^3}{12} \text{ kNms}^2 = 2058 \text{ kNms}^2 = 2058 \text{ tm}^2$$

$$\Theta_y \approx \rho \cdot t \cdot I_y = 2.5 \cdot 0.20 \cdot \frac{14 \cdot 18^3}{12} \text{ kNms}^2 = 3402 \text{ kNms}^2 = 3402 \text{ tm}^2$$

$$\Theta_z \approx \rho \cdot t \cdot (I_x + I_y) = 2058 + 3402 = 5460 \text{ kNms}^2 = 5460 \text{ tm}^2.$$

The mass matrix at the point hence is obtained in the units kN, m, and s as

$$
\begin{bmatrix} F_x \\ F_y \\ F_z \\ M_x \\ M_y \\ M_z \end{bmatrix}
= -
\begin{bmatrix}
126 & 0 & 0 & 0 & 0 & 0 \\
0 & 126 & 0 & 0 & 0 & 0 \\
0 & 0 & 126 & 0 & 0 & 0 \\
0 & 0 & 0 & 2058 & 0 & 0 \\
0 & 0 & 0 & 0 & 3402 & 0 \\
0 & 0 & 0 & 0 & 0 & 5460
\end{bmatrix}
\cdot
\begin{bmatrix} \ddot{u} \\ \ddot{v} \\ \ddot{w} \\ \ddot{\varphi}_x \\ \ddot{\varphi}_y \\ \ddot{\varphi}_z \end{bmatrix}.
$$

If the plate is additionally occupied by a distributed mass $\bar{m}$, this can be taken into account in the calculation by increasing the density by a factor $(1 + \bar{m}/(\rho \cdot t))$. For a distributed mass $\bar{m} = 0.976$ t/m^2, one obtains $(1 + \bar{m}/(\rho \cdot t)) = 1 + 0.976/(2.5 \cdot 0.2) = 1.933$, and thus, $m = 1.933 \cdot 126.0 = 244$ t, $\Theta_x = 1.933 \cdot 2058 = 3978$ tm^2, $\Theta_y = 1.933 \cdot 3402 = 6576$ tm^2, and $\Theta_z = 1.933 \cdot 5460 = 10554$ tm^2. ◀

For the calculation of moments of inertia of bodies with other elementary shapes and with arbitrary geometry, see [1].

In finite element computations, the mass forces are determined element by element and represented as element mass matrices. Using the same method as for assembling the global stiffness matrix from element stiffness matrices, the global mass matrix is constructed from the element mass matrices. One then obtains the inertial forces as

$$\underline{F}_{\text{T}} = -\underline{M} \cdot \underline{\ddot{u}} \tag{5.7}$$

where $\underline{M}$ is the mass matrix and $\underline{\ddot{u}}$ the vector of accelerations of the system.

Mass forces are volume forces whose magnitude depends on the acceleration at the point of the element being considered. For the distribution of the accelerations within a finite element, the same shape functions are chosen as in the derivation of the element stiffness matrix. The resulting element mass matrix is called *consistent mass matrix*. The derivation of consistent mass matrices is shown below using the example of a truss member with a linearly variable cross-sectional area.

Fig. 5.4 Truss element with linearly varying cross section

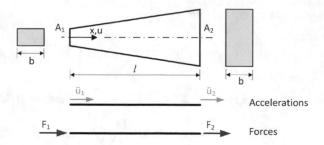

For the truss element with linearly variable cross-sectional area, the stiffness matrix is derived in Sect. 4.3.3 using the principle of virtual displacements. In the following, the consistent mass matrix of the element is determined (Fig. 5.4). For the accelerations as well as for the virtual displacements and the real displacements, linear shape functions are assumed. The virtual displacements are then

$$\bar{u}(x) = \bar{u}_1 + \frac{x}{l} \cdot (\bar{u}_2 - \bar{u}_1) \tag{5.8a}$$

and the accelerations

$$\ddot{u}(x) = \ddot{u}_1 + \frac{x}{l} \cdot (\ddot{u}_2 - \ddot{u}_1). \tag{5.8b}$$

The mass force at an infinitesimal truss element with the length dx is

$$-\rho \cdot \ddot{u}(x) \cdot A(x) \cdot dx$$

where ρ is the density and

$$A(x) = A_1 + \frac{x}{l} \cdot (A_2 - A_1)$$

the linearly varying cross-sectional area. Applying the principle of virtual displacements gives

$$F_1 \cdot \bar{u}_1 + F_2 \cdot \bar{u}_2 = -\int_0^l \rho \cdot \ddot{u}(x) \cdot A(x) \cdot \bar{u}(x) \, dx.$$

After performing the integration and solving the equations for F_2 with $\bar{u}_1 = 0$ and for F_1 with $\bar{u}_2 = 0$, one obtains

$$\begin{bmatrix} F_1 \\ F_2 \end{bmatrix} = -\frac{\rho \cdot l}{12} \begin{bmatrix} 3 \cdot A_1 + A_2 & A_1 + A_2 \\ A_1 + A_2 & A_1 + 3 \cdot A_2 \end{bmatrix} \cdot \begin{bmatrix} \ddot{u}_1 \\ \ddot{u}_2 \end{bmatrix} \tag{5.9}$$

$$\underline{F}_{me} = -\underline{M}^{(e)} \cdot \underline{\ddot{u}}_e. \tag{5.9a}$$

The matrix $\underline{M}^{(e)}$ is the element mass matrix of the truss element. For an element with constant cross-sectional area $A = A_1 = A_2$, Eq. (5.9) simplifies to

$$\begin{bmatrix} F_1 \\ F_2 \end{bmatrix} = -\frac{\rho \cdot A \cdot l}{6} \begin{bmatrix} 2 & 1 \\ 1 & 2 \end{bmatrix} \cdot \begin{bmatrix} \ddot{u}_1 \\ \ddot{u}_2 \end{bmatrix}. \tag{5.10}$$

In general, the element mass matrix of plane and three-dimensional elements with the displacement shape functions $\underline{N}$ is obtained by integration over the volume of the element to

$$\underline{M}^{(e)} = \rho \cdot \int \underline{N}^{\mathrm{T}} \cdot \underline{N} \cdot \mathrm{d}V. \tag{5.11}$$

For example, the element mass matrix of the rectangular plane stress element described in Sect. 4.4 can be determined with (5.11), (Fig. 4.14). With $\underline{N}$ according to (4.27b–4.27e) and $\mathrm{d}V = t \cdot \mathrm{d}x \cdot \mathrm{d}y$ ($t = $ wall thickness), one obtains

$$\underline{M}^{(e)} = \rho \cdot \frac{t \cdot a \cdot b}{36} \cdot \begin{bmatrix} 4 & 0 & 2 & 0 & 1 & 0 & 2 & 0 \\ 0 & 4 & 0 & 2 & 0 & 1 & 0 & 2 \\ 2 & 0 & 4 & 0 & 2 & 0 & 1 & 0 \\ 0 & 2 & 0 & 4 & 0 & 2 & 0 & 1 \\ 1 & 0 & 2 & 0 & 4 & 0 & 2 & 0 \\ 0 & 1 & 0 & 2 & 0 & 4 & 0 & 2 \\ 2 & 0 & 1 & 0 & 2 & 0 & 4 & 0 \\ 0 & 2 & 0 & 1 & 0 & 2 & 0 & 4 \end{bmatrix}. \tag{5.12a}$$

The consistent mass matrix of a beam in bending with the degrees of freedom according to Fig. 3.15 is according to [2]

$$\underline{M}^{(e)} = \frac{\bar{m} \cdot l}{420} \cdot \begin{bmatrix} 156 & 22 \cdot l & 54 & -13 \cdot l \\ 22 \cdot l & 4 \cdot l^2 & 13 \cdot l & -3 \cdot l^2 \\ 54 & 13 \cdot l & 156 & -22 \cdot l \\ -13 \cdot l & -3 \cdot l^2 & -22 \cdot l & 4 \cdot l^2 \end{bmatrix}. \tag{5.12b}$$

Here, l is the beam element length and $\bar{m}$ the mass per unit of length.

For the rectangular element of a shear rigid plate in bending described in Sect. 4.7.2 with the stiffness matrix according to (4.83), the consistent element mass matrix is according to [13]:

$$\underline{M}^{(e)} = \frac{\bar{m} \cdot a \cdot b}{176400} \cdot$$

$$
\begin{bmatrix}
24178 & 3227b & -3227a & 8582 & 1393b & 1918a \\
3227b & 560b^2 & -441ab & 1393b & 280b^2 & 294ab \\
-3227a & -441ab & 560a^2 & -1918a & -294ab & -420a^2 \\
8582 & 1393b & -1918a & 24178 & 3227b & 3227a \\
1393b & 280b^2 & -294ab & 3227b & 560b^2 & 441ab \\
1918a & 294ab & -420a^2 & 3227a & 441ab & 560a^2 \\
2758 & 812b & -812a & 8582 & 1918b & 1393a \\
-812b & -210b^2 & 196ab & -1918b & -420b^2 & -294ab \\
812a & 196ab & -210a^2 & 1393a & 294ab & 280a^2 \\
8582 & 1918b & -1393a & 2758 & 812b & 812a \\
-1918b & -420b^2 & 294ab & -812b & -210b^2 & -196ab \\
-1393a & -294ab & 280a^2 & -812a & -196ab & -210a^2
\end{bmatrix} \cdots\cdots
$$

$$
\cdots
\begin{bmatrix}
2758 & -812b & 812a & 8582 & -1918b & -1393a \\
812b & -210b^2 & 196ab & 1918b & -420b^2 & -294ab \\
-812a & 196ab & -210a^2 & -1393a & 294ab & 280a^2 \\
8582 & -1918b & 1393a & 2758 & -812b & -812a \\
1918b & -420b^2 & 294ab & 812b & -210b^2 & -196ab \\
1393a & -294ab & 280a^2 & 812a & -196ab & -210a^2 \\
24178 & -3227b & 3227a & 8582 & -1393b & -1918a \\
-3227b & 560b^2 & -441ab & -1393b & 280b^2 & 294ab \\
3227a & -441ab & 560a^2 & 1918a & -294ab & -420a^2 \\
8582 & -1393b & 1918a & 24178 & -3227b & -3227a \\
-1393b & 280b^2 & -294ab & -3227b & 560b^2 & 441ab \\
-1918a & 294ab & -420a^2 & -3227a & 441ab & 560a^2
\end{bmatrix}.
\tag{5.12c}
$$

Consistent mass matrices of other elements are given in [14, 15]. From (5.11), it follows that consistent mass matrices are symmetrical matrices.

Instead of consistent mass matrices, element mass matrices are often constructed approximately on the basis of single masses in the nodes, the sum of which corresponds to the total mass of the element. The mass matrix is then a diagonal matrix. For example, the total mass $\rho \cdot A \cdot l$ of a truss element can be distributed evenly to the two nodes. One then gets for the element mass matrix instead of (5.10)

$$
\begin{bmatrix} F_1 \\ F_2 \end{bmatrix} = -\frac{\rho \cdot A \cdot l}{2} \begin{bmatrix} 1 & 0 \\ 0 & 1 \end{bmatrix} \cdot \begin{bmatrix} \ddot{u}_1 \\ \ddot{u}_2 \end{bmatrix}.
\tag{5.13}
$$

In the same way, for a rectangular plane stress element, a diagonal matrix is obtained as a mass matrix, in which all diagonal elements have the value $0.25 \cdot \rho \cdot t \cdot a \cdot b$. If all nodes are accelerated with the same value, the same inertial forces result as with consistent mass matrix. Usually, mass matrices with node masses are used, since the number of calculation operations can be significantly reduced during dynamic analysis with diagonal mass matrices. Methods for determining nodal mass matrices from consistent mass matrices are described in [14].

5.2.3 Damping Forces

Spring forces are proportional to displacements and mass forces are proportional to accelerations. Forces that depend on velocity are called damping forces. If one simply assumes a linear relationship, the damping force F_D is obtained as

$$F_D = -c \cdot \dot{u}, \tag{5.14}$$

where c is denoted as damping constant. It depends on material and system parameters.

The damping proportional to the velocity is called *viscous damping*. Other types of damping are *Coulomb damping*, which is caused by frictional forces, and *hysteretic damping*, which can be determined from the hysteresis loop in the stress–strain diagram of a material under harmonic vibration [1, 3, 4]. Damping processes are very complex and have different reasons. The easiest way to describe them approximately in a mathematical way is by the model of velocity proportional damping. Therefore, normally all damping effects are summarized in an equivalent viscous damping measure.

First, the damping of a single degree of freedom system consisting of a spring, a mass, and a damper is considered. Damping is defined by the *damping coefficient* ξ. It is a related quantity. The damping coefficient is defined with the spring constant k, the mass m, and the damping constant c as

$$\xi = \frac{c}{c_{cr}} = \frac{c}{2\sqrt{k \cdot m}}. \tag{5.15}$$

The damping constant c_{cr} is called *critical damping*. It can be shown that the vibration of a single degree of freedom system changes into a creep movement at $c_{cr} = 1$. The damping coefficient ξ thus indicates the ratio of the damper constant and the critical damping constant. It is also denoted as *damping ratio* and normally given as a percentage. If damping is less than critical, i.e. if $\xi < 1$, the system is called underdamped.

In finite element analysis, the damping forces are expressed by a damping matrix $\underline{C}$ and a velocity vector $\underline{\dot{u}}$ as

$$\underline{F}_D = -\underline{C} \cdot \underline{\dot{u}}. \tag{5.16}$$

Only in simple cases the damping matrix can be determined explicitly. For example, for a viscous damper according to Fig. 5.5, it is

$$\begin{bmatrix} F_1 \\ F_2 \end{bmatrix} = -c \cdot \begin{bmatrix} 1 & -1 \\ -1 & 1 \end{bmatrix} \cdot \begin{bmatrix} \dot{u}_1 \\ \dot{u}_2 \end{bmatrix}. \tag{5.17}$$

The relationship is similar to the stiffness relationship of the truss member (cf. 3.3) and can be transformed with the transformation matrix T according to (3.5) with (3.8b) to any other coordinate system.

Fig. 5.5 Viscous damper

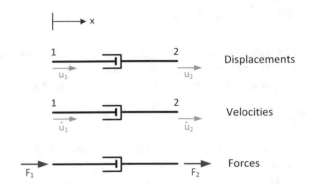

In practice, however, the damping matrix $\underline{C}$ can only be determined in special cases. In the direct integration methods, however, it is explicitly required. Therefore, it is often defined as a linear combination of the stiffness matrix and the mass matrix by

$$\underline{C} = \alpha \cdot \underline{M} + \beta \cdot \underline{K}. \tag{5.18}$$

This type of damping is denoted as *Rayleigh damping*. The damping coefficient ξ can be specified as a function of a given circular frequency ω as

$$\xi = \frac{1}{2} \cdot \left(\frac{\alpha}{\omega} + \beta \cdot \omega \right). \tag{5.18a}$$

The weighting factors α and β are selected so that the damping coefficients for two given circular frequencies ω_1 and ω_2 assume the values ξ_1 and ξ_2, respectively. If one introduces ω_1 and ξ_1 as well as ω_2 and ξ_2 into (5.18a) and solves the two equations obtained for α and β, the following equations are obtained:

$$\alpha = 2 \cdot \omega_1 \cdot \omega_2 \frac{\xi_1 \cdot \omega_2 - \xi_2 \cdot \omega_1}{\omega_2^2 - \omega_1^2} \tag{5.18b}$$

$$\beta = 2 \cdot \frac{\xi_2 \cdot \omega_2 - \xi_1 \cdot \omega_1}{\omega_2^2 - \omega_1^2}. \tag{5.18c}$$

The reference frequencies ω_1 and ω_2 must be selected in such a way that they reflect the desired damping characteristics as closely as possible. In many cases, the two lowest eigenfrequencies are chosen for this purpose. The determination of the natural frequencies is explained in Sect. 5.4. It is also possible to construct damping matrices which have a given damping factor in any number of natural frequencies (see, e.g. [4, 16]).

Example 5.2 The following natural frequencies of a structure were determined:

$$f_1 = 4.0 \text{ Hz}, \quad f_2 = 6.0 \text{ Hz}, \quad f_3 = 13.0 \text{ Hz}, \quad f_4 = 24.0 \text{ Hz}.$$

Two cases are considered. In case "a", the parameters of the Rayleigh damping are to be determined in such a way that in the first two eigenmodes, the damping values

$$\xi_1 = 1\%, \quad \xi_2 = 1\%$$

are fulfilled.

With the natural frequencies $\omega_1 = 2 \cdot \pi \cdot 4.0 = 25.1 \cdot 1/\text{s}$ and $\omega_2 = 2 \cdot \pi \cdot 6.0 = 37.7 \cdot 1/\text{s}$, one obtains with (5.18b, 5.18c)

$$\alpha = 2 \cdot 25.1 \cdot 37.7 \cdot \frac{0.01 \cdot 37.7 - 0.01 \cdot 25.2}{37.7^2 - 25.1^2} = 0.302$$

$$\beta = 2 \cdot \frac{0.01 \cdot 37.7 - 0.01 \cdot 25.2}{37.7^2 - 25.1^2} = 0.00032.$$

With these parameters, the Rayleigh damping matrix is calculated according to (5.18). The damping coefficient depends on the frequency. According to (5.18a), one obtains

$$\xi = \frac{1}{2} \cdot \left(\frac{\alpha}{\omega} + \beta \cdot \omega \right) = \frac{1}{2} \cdot \left(\frac{0.302}{2 \cdot \pi \cdot f} + 0.00032 \cdot 2 \cdot \pi \cdot f \right)$$

$$= \frac{0.02400}{f} + 0.0010 \cdot f.$$

For high frequencies, the damping factor increases. At low frequencies, the mass proportional part of the damping in (5.18a) predominates, at high frequencies the stiffness proportional part (Fig. 5.6). For the natural frequencies f_3 and f_4 given above, one obtains

$$\xi_3 = \frac{0.02400}{13.0} + 0.0010 \cdot 13.0 = 1.5\%, \quad \xi_4 = \frac{0.02400}{24.0} + 0.0010 \cdot 24.0 = 2.5\%.$$

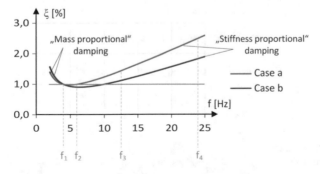

Fig. 5.6 Damping coefficient with Rayleigh damping

It can be seen that with Rayleigh's damping, the specified damping coefficient is only maintained at the two lowest natural frequencies. At the higher natural frequencies, one obtains significantly higher damping coefficients.

In case "b", the parameters should be determined in such a way that a damping of 1% is obtained for the natural frequencies f_1, f_2 and f_3 as accurately as possible. For this purpose, the damping and the two frequencies $\bar{f}_1 = f_1 = 4.0$ Hz and $\bar{f}_2 = 10.0$ Hz are investigated. With the coefficients $\bar{\alpha}_1 = 0.359$ and $\bar{\beta}_1 = 0.00023$ according to (5.18b, 5.18c), one obtains with (5.18a)

$$\bar{\xi} = \frac{1}{2} \cdot \left(\frac{\bar{\alpha}}{\omega} + \bar{\beta} \cdot \omega \right) = \frac{0.028571}{f} + 0.000714 \cdot f.$$

In the frequency range between the two reference frequencies of 4 and 10 Hz, the damping coefficient is approximately constant, then it increases again. At the natural frequencies, one gets now

$$f_1 = 4 \text{ Hz} \rightarrow \xi_1 = 1.0\%, \qquad f_2 = 6 \text{ Hz} \rightarrow \xi_2 = 0.9\%,$$
$$f_3 = 13 \text{ Hz} \rightarrow \xi_3 = 1.1\%, \quad f_4 = 24 \text{ Hz} \rightarrow \xi_4 = 1.8\%.$$

Hence, the approximation can be improved by an appropriate choice of the reference frequencies.   ◀

5.3 Equations of Motion

The equilibrium conditions for a dynamic system with displacements as unknowns are denoted as equations of motion. At first, a *single degree of freedom system* (SDOF system) is considered. It consists of a mass, a spring, and a damper (Fig. 5.7). In the case of undamped vibration, the damper is omitted. On the mass an external force F, a mass force F_T, a spring force F_K, and a damper force F_D are acting. The equilibrium condition at time t is

$$F_K + F_D + F_T + F = 0.$$

With $F_K = -k \cdot u$, F_T according to (5.3a), and F_D according to (5.14), one obtains

$$k \cdot u + c \cdot \dot{u} + m \cdot \ddot{u} = F. \qquad (5.19)$$

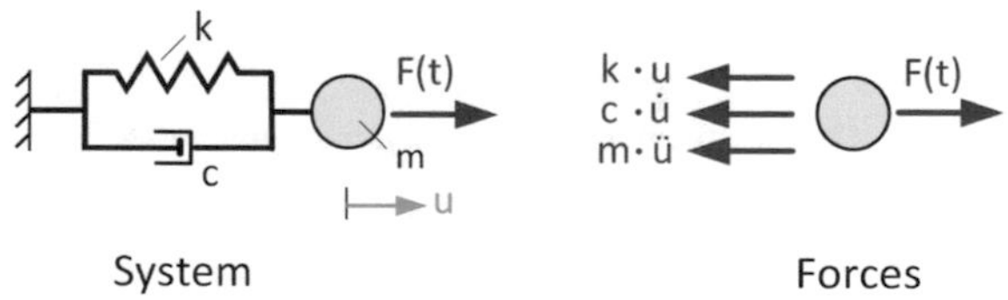

Fig. 5.7 Single degree of freedom system with external force

System   Forces

The excitation force F, the displacement u, the velocity $\dot{u}$, and the acceleration $\ddot{u}$ depend on the time t.

A finite element system is a *multi degree of freedom system* (MDOF system). The equilibrium conditions at a time t are

$$\underline{F}_K + \underline{F}_D + \underline{F}_T + \underline{F} = 0.$$

With the stiffness forces $\underline{F}_K = -\underline{K} \cdot \underline{u}$, the mass forces F_T according to (5.7), and the damping forces $\underline{F}_D$ according to (5.16), the equations of motion are obtained as

$$\underline{K} \cdot \underline{u} + \underline{C} \cdot \dot{\underline{u}} + \underline{M} \cdot \ddot{\underline{u}} = \underline{F}. \tag{5.20}$$

Herein, $\underline{K}$ represents the stiffness matrix, $\underline{C}$ the damping matrix, and $\underline{M}$ the mass matrix of the system. The external forces $\underline{F}$ as well as the vectors $\underline{u}$, $\dot{\underline{u}}$, and $\ddot{\underline{u}}$ of the nodal displacements, velocities, and accelerations vary with time t. The solution of these equations is discussed in the following sections.

Another type of action is base excitation as caused by the ground motion of an earthquake. For base excitation, the equations of motion have to be adapted. In this case, a given displacement time history is applied at the base of the supports of the finite element model. For a purely static displacement, no forces would occur, since the displacements of all supports are identical. However, in case of a dynamic excitation of the finite element model, vibrations occur due to the mass and damping forces.

At first, the single degree of freedom system is considered again. The time history of the base displacement is $u_b(t)$. The displacement of the point mass u_{ges} is then composed of the base displacement u_b and the relative displacement u of the point (Fig. 5.8). For the displacements, velocities, and accelerations, the following applies:

$$u_{ges} = u + u_b \tag{5.21a}$$

$$\dot{u}_{ges} = \dot{u} + \dot{u}_b \tag{5.21b}$$

$$\ddot{u}_{ges} = \ddot{u} + \ddot{u}_b. \tag{5.21c}$$

Fig. 5.8 Single degree of freedom system with base excitation

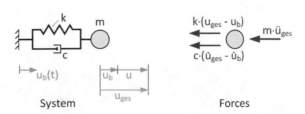

In the equilibrium conditions, the spring and damper forces depend on the relative displacement $u = u_{\text{ges}} - u_{\text{b}}$, while the inertia force results from the total acceleration $\ddot{u}_{\text{ges}}$. Thus, one obtains

$$k \cdot u + c \cdot \dot{u} + m \cdot (\ddot{u} + \ddot{u}_{\text{b}}) = 0$$

or

$$k \cdot u + c \cdot \dot{u} + m \cdot \ddot{u} = -m \cdot \ddot{u}_{\text{b}}. \qquad (5.22)$$

Comparing (5.22) with the equation of motion (5.19) for force excitation, it can be seen that the excitation by base motion corresponds to an external force $-m \cdot \ddot{u}_{\text{b}}$. The solution results in the relative displacements $u(t)$. The total displacements are obtained with (5.21a).

Finite element models can be excited by base motions in the three coordinate directions. The base excitation is different in the x-, y-, and z-directions and is normally zero in the rotational degrees of freedom. To take it into account, the topology vectors $\underline{I}_x$, $\underline{I}_y$, and $\underline{I}_z$ are introduced. The vector $\underline{I}_x$ has 1 in all degrees of freedom corresponding to a displacement in the x-direction and 0 in the remaining degrees of freedom. In the same way, $\underline{I}_y$ and $\underline{I}_z$ are defined for the y- and z-directions, respectively.

For a base excitation $u_{\text{b},x}$ in the x-direction, $u_{\text{b},y}$ in the y-direction, and $u_{\text{b},z}$ in the z-direction, one obtains (Fig. 5.9)

$$\underline{u}_{\text{ges}} = \underline{u} + \left(\underline{I}_x \cdot u_{\text{b},x} + \underline{I}_y \cdot u_{\text{b},y} + \underline{I}_z \cdot u_{\text{b},z} \right), \qquad (5.23\text{a})$$

$$\underline{\dot{u}}_{\text{ges}} = \underline{\dot{u}} + \left(\underline{I}_x \cdot \dot{u}_{\text{b},x} + \underline{I}_y \cdot \dot{u}_{\text{b},y} + \underline{I}_z \cdot \dot{u}_{\text{b},z} \right), \qquad (5.23\text{b})$$

$$\underline{\ddot{u}}_{\text{ges}} = \underline{\ddot{u}} + \left(\underline{I}_x \cdot \ddot{u}_{\text{b},x} + \underline{I}_y \cdot \ddot{u}_{\text{b},y} + \underline{I}_z \cdot \ddot{u}_{\text{b},z} \right). \qquad (5.23\text{c})$$

The stiffness forces $\underline{F}_{\text{K}}$ and the damping forces $\underline{F}_{\text{D}}$ are determined from the relative displacements $\underline{u}$, as with the single degree of freedom system. However, the mass forces $\underline{F}_{\text{T}}$ depend on the total accelerations $\underline{\ddot{u}}_{\text{ges}}$. This gives the following equations of motion:

Fig. 5.9 Finite element model (truss model) with base excitation in the x-direction

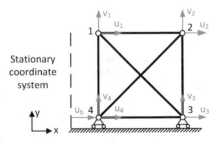

Stationary coordinate system

$$K \cdot \underline{u} + \underline{C} \cdot \dot{\underline{u}} + \underline{M} \cdot \ddot{\underline{u}} = -\underline{M} \cdot (\underline{I}_x \cdot \ddot{u}_{b,x} + \underline{I}_y \cdot \ddot{u}_{b,y} + \underline{I}_z \cdot \ddot{u}_{b,z}). \qquad (5.24)$$

They correspond to the equations of motion (5.20) with an external load vector

$$\underline{F} = -\underline{M} \cdot (\underline{I}_x \cdot \ddot{u}_{b,x} + \underline{I}_y \cdot \ddot{u}_{b,y} + \underline{I}_z \cdot \ddot{u}_{b,z}). \qquad (5.24a)$$

Example 5.3 For the truss model shown in Fig. 5.10 with four nodal masses and the parameters as in Example 3.1, the equations of motion for free vibrations shall be established. The damping will be assumed to be zero. The mass of the members can be neglected. Furthermore, the equations of motion for a base motion $u_{b,x}(t)$ in the x-direction shall be given.

The stiffness matrix of the system has already been determined in Example 3.6. The masses of the degrees of freedom u_1 and v_1 are each 20 t, those of the degrees of freedom u_2 and v_2 each 40 t, and the mass in the degree of freedom u_3 is 10 t. Thus, one receives for free, undamped vibrations with $\underline{C} = \underline{0}$ and $\underline{F} = \underline{0}$ the equations of motion after (5.20):

$$2.80 \cdot 10^5 \cdot \begin{bmatrix} 1.35 & -0.35 & -1.00 & 0 & -0.35 \\ -0.35 & 1.35 & 0 & 0 & 0.35 \\ -1.00 & 0 & 1.35 & 0.35 & 0 \\ 0 & 0 & 0.35 & 1.35 & 0 \\ -0.35 & 0.35 & 0 & 0 & 1.35 \end{bmatrix} \begin{bmatrix} u_1 \\ v_1 \\ u_2 \\ v_2 \\ u_3 \end{bmatrix}$$

$$+ \begin{bmatrix} 20 & 0 & 0 & 0 & 0 \\ 0 & 20 & 0 & 0 & 0 \\ 0 & 0 & 40 & 0 & 0 \\ 0 & 0 & 0 & 40 & 0 \\ 0 & 0 & 0 & 0 & 10 \end{bmatrix} \begin{bmatrix} \ddot{u}_1 \\ \ddot{v}_1 \\ \ddot{u}_2 \\ \ddot{v}_2 \\ \ddot{u}_3 \end{bmatrix} = \begin{bmatrix} 0 \\ 0 \\ 0 \\ 0 \\ 0 \end{bmatrix}$$

or

$$\underline{K} \cdot \underline{u} + \underline{M} \cdot \ddot{\underline{u}} = \underline{0}.$$

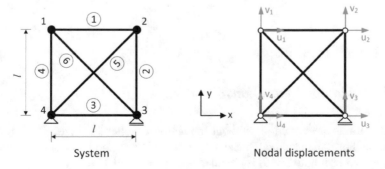

System   Nodal displacements

Fig. 5.10 Truss model with nodal masses

For a base excitation, the topology vectors must first be established. Since the accelerations in the y- and z-directions are zero, i.e. $\ddot{u}_{b,y}(t) = \ddot{u}_{b,z}(t) = 0$, the topology vectors $\underline{I}_y$ and $\underline{I}_z$ can be omitted. In the vector $\underline{I}_x$, the entries in the rows corresponding to the degrees of freedom u_1, u_2, and u_3 are 1 whereas all other entries are zero. The right side of the equations of motion thus is

$$-\underline{M} \cdot \underline{I}_x \cdot \ddot{u}_{b,x}(t) = -\begin{bmatrix} 20 & 0 & 0 & 0 & 0 \\ 0 & 20 & 0 & 0 & 0 \\ 0 & 0 & 40 & 0 & 0 \\ 0 & 0 & 0 & 40 & 0 \\ 0 & 0 & 0 & 0 & 10 \end{bmatrix} \cdot \begin{bmatrix} 1 \\ 0 \\ 1 \\ 0 \\ 1 \end{bmatrix} \cdot \ddot{u}_{b,x}(t)$$

$$= -\begin{bmatrix} 20 \\ 0 \\ 40 \\ 0 \\ 10 \end{bmatrix} \cdot \ddot{u}_{b,x}(t).$$

The scalar function $\ddot{u}_{b,x}(t)$ is the time history of the base or ground motion in the x-direction. The complete equations of motion is according to (5.24) are

$$2.80 \cdot 10^5 \cdot \begin{bmatrix} 1.35 & -0.35 & -1.00 & 0 & -0.35 \\ -0.35 & 1.35 & 0 & 0 & 0.35 \\ -1.00 & 0 & 1.35 & 0.35 & 0 \\ 0 & 0 & 0.35 & 1.35 & 0 \\ -0.35 & 0.35 & 0 & 0 & 1.35 \end{bmatrix} \cdot \begin{bmatrix} u_1 \\ v_1 \\ u_2 \\ v_2 \\ u_3 \end{bmatrix}$$

$$+ \begin{bmatrix} 20 & 0 & 0 & 0 & 0 \\ 0 & 20 & 0 & 0 & 0 \\ 0 & 0 & 40 & 0 & 0 \\ 0 & 0 & 0 & 40 & 0 \\ 0 & 0 & 0 & 0 & 10 \end{bmatrix} \cdot \begin{bmatrix} \ddot{u}_1 \\ \ddot{v}_1 \\ \ddot{u}_2 \\ \ddot{v}_2 \\ \ddot{u}_3 \end{bmatrix} = -\begin{bmatrix} 20 \\ 0 \\ 40 \\ 0 \\ 10 \end{bmatrix} \cdot \ddot{u}_{b,x}(t).$$

◀

5.4   Free Vibrations

5.4.1   Undamped Vibrations

Free vibrations are vibrations without the action of external forces after an initial excitation. For the undamped, linear single degree of freedom system, the equation of motion according to (5.19) is obtained with $c = 0$ and $F = 0$ to

$$k \cdot u + m \cdot \ddot{u} = 0. \tag{5.25}$$

For static system behavior, the equilibrium condition is $k \cdot u = 0$ when the external load is zero, and hence with $u = 0$ the displacement is zero. In the case of dynamic system behavior, however, in addition an inertial force $m \cdot \ddot{u}$ is acting. It is in equilibrium with the spring force $k \cdot u$ at all times. Therefore, time-dependent displacements $u(t)$ can occur without external forces acting on the system.

Equation (5.25) is a homogeneous, linear, second-order differential equation with constant coefficients. Its solution is

$$u = u_a \cdot \sin(\omega \cdot t) + u_s \cdot \cos(\omega \cdot t). \tag{5.26}$$

It can be written with the *amplitude* $u_0 = \sqrt{u_a^2 + u_s^2}$ and *phase angle* $\varphi_0 = -\arctan(u_s/u_a)$ (Fig. 5.11) as

$$u = u_0 \cdot \sin(\omega \cdot t - \varphi_0). \tag{5.27a}$$

This gives the velocity and acceleration as

$$\dot{u} = \omega \cdot u_0 \cdot \cos(\omega \cdot t - \varphi_0), \tag{5.27b}$$

Fig. 5.11 Free undamped vibrations

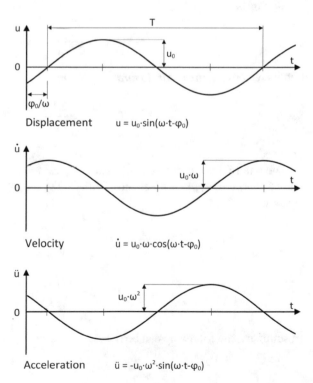

Displacement $u = u_0 \cdot \sin(\omega \cdot t - \varphi_0)$

Velocity $\dot{u} = u_0 \cdot \omega \cdot \cos(\omega \cdot t - \varphi_0)$

Acceleration $\ddot{u} = -u_0 \cdot \omega^2 \cdot \sin(\omega \cdot t - \varphi_0)$

$$\ddot{u} = -\omega^2 \cdot u_0 \cdot \sin(\omega \cdot t - \varphi_0). \tag{5.27c}$$

The validity of the solution can be easily checked by putting it into (5.25). One obtains

$$\left(k - \omega^2 \cdot m\right) \cdot u_0 \cdot \sin(\omega \cdot t - \varphi_0) = 0$$

and with $\sin(\omega \cdot t - \varphi_0) \neq 0$ and $u_0 \neq 0$

$$k - \omega^2 \cdot m = 0.$$

The solution of this equation is

$$\omega = \sqrt{\frac{k}{m}}. \tag{5.28}$$

Free vibrations according to (5.27a–5.27c) are only possible with this circular frequency. It is also referred to as *natural circular frequency*. The phase angle φ_0 can be determined from the displacement or the velocity at $t = 0$.

More often than the natural circular frequency ω, the *natural frequency* or *eigenfrequency*

$$f = \frac{\omega}{2 \cdot \pi} \tag{5.29a}$$

and its reciprocal, the *natural vibration period*

$$T = \frac{1}{f} = \frac{2 \cdot \pi}{\omega} \tag{5.29b}$$

are used.

The unit of a *circular frequency* ω and a *frequency* f is $1/s = $ Hz (1 Hertz); the unit of *vibration period* is s (second). With (5.28), the following applies to the SDOF system:

$$f = \frac{1}{2\pi} \cdot \sqrt{\frac{k}{m}} \tag{5.30a}$$

$$T = 2\pi \cdot \sqrt{\frac{m}{k}} \tag{5.30b}$$

In summary, the following can be noted:

Free vibrations of a SDOF system

The single degree of freedom system has a single natural frequency. It is

$$f = \frac{1}{2\pi} \cdot \sqrt{\frac{k}{m}}.$$

Free undamped vibrations are vibrations with a sinusoidal time history in this frequency. The amplitude and the phase angle of the vibration result from the initial conditions.

Example 5.4 The beam shown in Fig. 5.12 is loaded in the middle with a component with a weight of 50 kN. The natural frequency and natural period of the system are to be determined. The mass of the beam can be neglected.

The spring constant of the system is obtained for a simply supported beam with a point load in the middle as

$$k = \frac{48 \cdot E \cdot I}{l^3} = \frac{48 \cdot 2.1 \cdot 10^8 \cdot 2 \cdot 10^{-4}}{6^3} = 9.333 \cdot 10^3 \text{ kN/m}.$$

The weight of 50 kN corresponds to a mass of

$$m = \frac{50 \text{ kN}}{10 \text{ m/s}^2} = 5 \text{ kN s}^2/\text{m}.$$

The natural frequency is

$$f = \frac{1}{2\pi} \cdot \sqrt{\frac{9.333 \cdot 10^3}{5} \frac{\text{kN/m}}{\text{kN s}^2/\text{m}}} = 6.876 \frac{1}{\text{s}} = 6.876 \text{ Hz}$$

and the natural period

$$T = \frac{1}{f} = \frac{1}{6.876} = 0.145 \text{ s}.$$ ◀

Finite element models are systems with many degrees of freedom and masses. Their equations of motion for undamped, free vibrations are obtained presuming linear system behavior from (5.20) as

$$\underline{K} \cdot \underline{u} + \underline{M} \cdot \underline{\ddot{u}} = \underline{0}. \tag{5.31}$$

Fig. 5.12 Beam with point mass

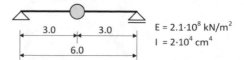

3.0 3.0

6.0

$E = 2.1 \cdot 10^8 \text{ kN/m}^2$
$I = 2 \cdot 10^4 \text{ cm}^4$

The inertial forces $\underline{M} \cdot \underline{\ddot{u}}$ are in equilibrium with the forces $\underline{K} \cdot \underline{u}$ resulting from the stiffness of the system at any time.

To solve the system of equations, as with an SDOF system, a sinusoidal time history is assumed:

$$\underline{u} = \underline{\phi} \cdot \sin(\omega \cdot t). \tag{5.32}$$

Herein, the phase angle in (5.27a) is set to zero, as it is not relevant for the following considerations. The vector $\underline{\phi}$ contains the vibration amplitudes in the individual degrees of freedom; the function $\sin(\omega \cdot t)$ describes the dependence on time. The velocity and acceleration are thus

$$\underline{\dot{u}} = \omega \cdot \underline{\phi} \cdot \cos(\omega \cdot t), \tag{5.32a}$$

$$\underline{\ddot{u}} = -\omega^2 \cdot \underline{\phi} \cdot \sin(\omega \cdot t). \tag{5.32b}$$

Entering this into the equations of motion (5.31), one obtains

$$\left(\underline{K} - \omega^2 \cdot \underline{M}\right) \cdot \underline{\phi} = \underline{0}. \tag{5.33}$$

The equation represents a general eigenvalue problem for the eigenvalues ω^2 and eigenvectors $\underline{\phi}$. Since both, the stiffness matrix and the mass matrix, are real, symmetric, and positive definite, the eigenvalues are real and positive. Solution methods for eigenvalue problems are discussed in Sect. 1.5. The eigenvectors $\underline{\phi}$ of free vibrations are also called *eigenmodes* or *vibration modes,* and the roots ω of the eigenvalues are called *natural circular frequencies* of the vibration. The following applies to them (see Sect. 1.5):

Eigenfrequencies and Eigenmodes of MDOF systems

The eigenfrequencies and modes of a finite element model are the solution of the eigenvalue problem

$$\left(\underline{K} - \omega_i^2 \cdot \underline{M}\right) \cdot \underline{\phi}_i = \underline{0}. \tag{5.34}$$

A finite element model has as many natural circular frequencies ω_i as the degrees of freedom occupied with mass. For each natural circular frequency ω_i exists an eigenmode $\underline{\phi}_i$. The eigenmodes fulfill the orthogonality conditions

$$\underline{\phi}_i^T \cdot \underline{K} \cdot \underline{\phi}_j = 0, \quad \underline{\phi}_i^T \cdot \underline{M} \cdot \underline{\phi}_j = 0 \quad \text{for} \quad i \neq j, \tag{5.34a}$$

$$\underline{\phi}_i^T \cdot \underline{K} \cdot \underline{\phi}_j \neq 0, \quad \underline{\phi}_i^T \cdot \underline{M} \cdot \underline{\phi}_j \neq 0 \quad \text{for} \quad i = j. \tag{5.34b}$$

The natural frequencies are normally ordered by magnitude, so that the following applies:

$$\omega_1 < \omega_2 < \omega_3 \ldots < \omega_j < \ldots \omega_n \tag{5.34c}$$

and for the natural frequencies

$$f_1 < f_2 < f_3 \ldots < f_j < \ldots f_n. \tag{5.34d}$$

As the solution of a homogeneous system of equations, the eigenmodes are defined only up to a constant multiplication factor. Therefore, eigenmodes only describe the shape of a vibrational mode, but do not specify its absolute magnitude. Eigenmodes can be normalized in different ways. One possibility is to set the absolute value of the maximum displacement of an eigenmode to 1. Often, however, eigenmodes are normalized using the *orthogonality relationships*

$$\underline{\phi}_i^{\mathrm{T}} \cdot \underline{M} \cdot \underline{\phi}_i = 1, \tag{5.35a}$$

$$\underline{\phi}_i^{\mathrm{T}} \cdot \underline{K} \cdot \underline{\phi}_i = \omega^2. \tag{5.35b}$$

This normalization has computational benefits when using the eigenmodes to calculate excited vibrations.

The mechanical significance of natural frequencies and the associated mode shapes is similar to that of the single degree of freedom system. If a finite element model is deformed in a mode shape and the bond is released, the system vibrates in the respective mode shape with the associated natural frequency. The ratio of the displacements of different degrees of freedom is preserved at all times during the vibrational process.

Example 5.5 The natural frequencies and mode shapes of the truss system shown in Fig. 5.10 shall be determined.

The stiffness and mass matrix of the system have already been established in Example 5.3. Thus, the eigenvalue problem is

$$
\left(2.80 \cdot 10^5 \cdot
\begin{bmatrix}
1.35 & -0.35 & -1.00 & 0 & -0.35 \\
-0.35 & 1.35 & 0 & 0 & 0.35 \\
-1.00 & 0 & 1.35 & 0.35 & 0 \\
0 & 0 & 0.35 & 1.35 & 0 \\
-0.35 & 0.35 & 0 & 0 & 1.35
\end{bmatrix}
- \omega^2 \cdot
\begin{bmatrix}
20 & 0 & 0 & 0 & 0 \\
0 & 20 & 0 & 0 & 0 \\
0 & 0 & 40 & 0 & 0 \\
0 & 0 & 0 & 40 & 0 \\
0 & 0 & 0 & 0 & 10
\end{bmatrix}
\right) \cdot
\begin{bmatrix}
\hat{u}_1 \\
\hat{v}_1 \\
\hat{u}_2 \\
\hat{v}_2 \\
\hat{u}_3
\end{bmatrix}
=
\begin{bmatrix}
0 \\
0 \\
0 \\
0 \\
0
\end{bmatrix}.
$$

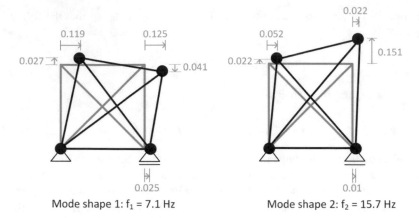

Mode shape 1: f_1 = 7.1 Hz   Mode shape 2: f_2 = 15.7 Hz

Fig. 5.13 Eigenmodes

It can be solved by one of the methods given in Sect. 1.5. Since five degrees of freedom are occupied by masses, the complete solution gives five natural frequencies and mode shapes. They are ordered by size:

$$f_1 = 7.1 \text{ Hz} \quad f_2 = 15.7 \text{ Hz} \quad f_3 = 20.3 \text{ Hz} \quad f_4 = 24.3 \text{ Hz} \quad f_5 = 33.1 \text{ Hz}$$

$$\phi_1 = \begin{bmatrix} 0.119 \\ 0.027 \\ 0.125 \\ -0.041 \\ 0.025 \end{bmatrix} \quad \phi_2 = \begin{bmatrix} 0.052 \\ 0.022 \\ 0.022 \\ 0.151 \\ 0.010 \end{bmatrix} \quad \phi_3 = \begin{bmatrix} 0.032 \\ 0.206 \\ -0.038 \\ -0.014 \\ -0.079 \end{bmatrix}$$

$$\phi_4 = \begin{bmatrix} -0.161 \\ 0.033 \\ 0.084 \\ 0.015 \\ -0.131 \end{bmatrix} \quad \phi_5 = \begin{bmatrix} 0.080 \\ -0.072 \\ -0.017 \\ -0.001 \\ -0.276 \end{bmatrix}.$$

The eigenmodes are normalized with respect to the mass matrix, so that (5.35a) applies. The first and second mode are shown in Fig. 5.13.   ◄

Example 5.6 For the investigation of its horizontal vibrations, an exhaust tower is modeled as a beam fixed at its base. In a simplified finite model, it is discretized into four beam elements (Fig. 5.14). The natural frequencies and modes are to be determined.

First of all, the stiffness matrices of all finite elements are determined as in a static analysis. The element stiffness matrices $\underline{K}_1 = \underline{K}_2 = \underline{K}_3 = \underline{K}_4 = \underline{K}_{el}$ of beams 1–4 are obtained according to (3.24) as

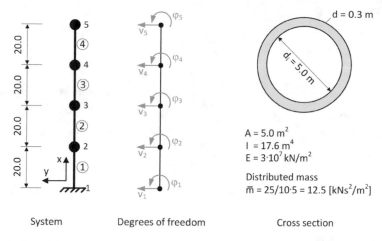

A = 5.0 m²
I = 17.6 m⁴
E = 3·10⁷ kN/m²

Distributed mass
m̄ = 25/10·5 = 12.5 [kNs²/m²]

System   Degrees of freedom   Cross section

Fig. 5.14 Modeling of an exhaust tower as a beam

$$
\underline{K}_{el} = \frac{E \cdot I}{l} \cdot
\begin{bmatrix}
12/l^2 & 6/l & -12/l^2 & 6/l \\
6/l & 4 & -6/l & 2 \\
-12/l^2 & -6/l & 12/l^2 & -6/l \\
6/l & 2 & -6/l & 4
\end{bmatrix}
=
\begin{bmatrix}
0.079 & 0.792 & -0.079 & 0.792 \\
0.792 & 10.56 & -0.792 & 5.28 \\
-0.079 & -0.792 & 0.079 & -0.792 \\
0.792 & 5.28 & -0.792 & 10.56
\end{bmatrix}
\cdot 10^7.
$$

The global stiffness matrix is obtained by assembling the element stiffness matrices as

$$
\underline{K} =
\begin{bmatrix}
0.158 & 0.000 & -0.079 & 0.792 & 0.000 & 0.000 & 0.000 & 0.000 \\
0.000 & 21.120 & -0.792 & 5.280 & 0.000 & 0.000 & 0.000 & 0.000 \\
-0.079 & -0.792 & 0.158 & 0.000 & -0.079 & 0.792 & 0.000 & 0.000 \\
0.792 & 5.280 & 0.000 & 21.120 & -0.792 & 5.280 & 0.000 & 0.000 \\
0.000 & 0.000 & -0.079 & -0.792 & 0.158 & 0.000 & -0.079 & 0.792 \\
0.000 & 0.000 & 0.792 & 5.280 & 0.000 & 21.120 & -0.792 & 5.280 \\
0.000 & 0.000 & 0.000 & 0.000 & -0.079 & -0.792 & 0.079 & -0.792 \\
0.000 & 0.000 & 0.000 & 0.000 & 0.792 & 5.280 & -0.792 & 10.560
\end{bmatrix}
\cdot 10^7, \quad
\underline{u} =
\begin{bmatrix}
v_2 \\
\varphi_2 \\
v_3 \\
\varphi_3 \\
v_4 \\
\varphi_4 \\
v_5 \\
\varphi_5
\end{bmatrix}.
$$

It is related to the vector $\underline{u}$ of the global degrees of freedom. The mass matrix is occupied only in the degrees of freedom of displacements, while it is zero in the degrees of freedom of rotations. It is

$$
\underline{M} =
\begin{bmatrix}
250 & 0 & 0 & 0 & 0 & 0 & 0 & 0 \\
0 & 0 & 0 & 0 & 0 & 0 & 0 & 0 \\
0 & 0 & 250 & 0 & 0 & 0 & 0 & 0 \\
0 & 0 & 0 & 0 & 0 & 0 & 0 & 0 \\
0 & 0 & 0 & 0 & 250 & 0 & 0 & 0 \\
0 & 0 & 0 & 0 & 0 & 0 & 0 & 0 \\
0 & 0 & 0 & 0 & 0 & 0 & 125 & 0 \\
0 & 0 & 0 & 0 & 0 & 0 & 0 & 0
\end{bmatrix}.
$$

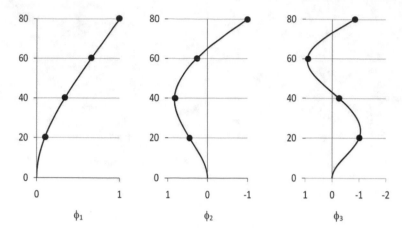

Fig. 5.15 Eigenmodes

The complete solution of the eigenvalue problem $\left(\underline{K} - \omega^2 \cdot \underline{M}\right) \cdot \underline{\phi} = \underline{0}$ consists of four natural frequencies and modes, since only four degrees of freedom are occupied by a mass (Fig. 5.15):

$$\omega_1 = 3.47 \text{ Hz} \qquad \omega_2 = 20.41 \text{ Hz} \qquad \omega_3 = 54.05 \text{ Hz} \qquad \omega_4 = 94.20 \text{ Hz}$$
$$f_1 = 0.55 \text{ Hz} \qquad f_2 = 3.25 \text{ Hz} \qquad f_3 = 8.60 \text{ Hz} \qquad f_4 = 14.99 \text{ Hz}$$
$$T_1 = 1.81 \text{ s}, \qquad T_2 = 0.31 \text{ s}, \qquad T_3 = 0.12 \text{ s} \qquad T_4 = 0.07 \text{ s}$$

$$\underline{\phi}_1 = \begin{bmatrix} 0.592 \\ 0.055 \\ 2.074 \\ 0.089 \\ 4.037 \\ 0.104 \\ 6.172 \\ 0.108 \end{bmatrix} \cdot 10^{-2}, \quad \underline{\phi}_2 = \begin{bmatrix} -2.373 \\ -0.165 \\ -4.325 \\ 0.011 \\ -1.423 \\ 0.264 \\ 5.222 \\ 0.367 \end{bmatrix} \cdot 10^{-2}, \quad \underline{\phi}_3 = \begin{bmatrix} 4.210 \\ 0.127 \\ 1.084 \\ -0.346 \\ -3.862 \\ 0.045 \\ 3.517 \\ 0.531 \end{bmatrix} \cdot 10^{-2}, \quad \underline{\phi}_4 = \begin{bmatrix} -4.037 \\ 0.130 \\ 3.977 \\ 0.075 \\ -2.598 \\ -0.215 \\ 1.502 \\ 0.415 \end{bmatrix} \cdot 10^{-2}.$$

The values above are normalized to a generalized mass one. The eigenmodes normalized to a maximum value of 1 are shown in Fig. 5.15.   ◀

Example 5.7 For the square slab shown in Fig. 4.46a, with the parameters given in Example 4.12, the first natural frequency shall be determined for a discretization into 2 × 2 shear rigid plate elements. The slab has a distributed mass of 1.0 t/m².

For the symmetry reasons, the analysis is carried out for ¼ of the complete system. The stiffness matrix of the symmetrical subsystem of the plate was already determined in Example 4.12. The mass is concentrated at point 4 (Fig. 4.46). Taking into account ¼ of the element surface, i.e. 1.25 m × 1.25 m = 1.5625 m², a mass of 1.5625 t = 1.5625 kNs²/m is obtained.

The equations of motion is

$$\left(10^4 \cdot \begin{bmatrix} 3.200 & 0 & 0.800 & 0 & -0.320 \\ 0 & 3.200 & 0 & 0.800 & 0.320 \\ 0.800 & 0 & 3.200 & 0 & -0.880 \\ 0 & 0.800 & 0 & 3.200 & 0.880 \\ -0.320 & 0.320 & -0.880 & 0.880 & 0.864 \end{bmatrix} \right.$$

$$\left. -\omega^2 \cdot \begin{bmatrix} 0 & 0 & 0 & 0 & 0 \\ 0 & 0 & 0 & 0 & 0 \\ 0 & 0 & 0 & 0 & 0 \\ 0 & 0 & 0 & 0 & 0 \\ 0 & 0 & 0 & 0 & 1.5625 \end{bmatrix} \right) \cdot \begin{bmatrix} \varphi_{x1} \\ \varphi_{y1} \\ \varphi_{x2} \\ \varphi_{y3} \\ w_4 \end{bmatrix} = \begin{bmatrix} 0 \\ 0 \\ 0 \\ 0 \\ 0 \end{bmatrix}.$$

It has 5 degrees of freedom, of which only one degree of freedom possesses a mass. The system, therefore, has only one natural frequency and one mode shape. This is obtained as

$$\omega_1 = 97.76 \text{ Hz}$$

$$f_1 = \frac{\omega_1}{2\pi} = 15.56 \text{ Hz} \quad \text{and} \quad \begin{bmatrix} \varphi_{x1} \\ \varphi_{y1} \\ \varphi_{x2} \\ \varphi_{y3} \\ w_4 \end{bmatrix} = \begin{bmatrix} 0.067 \\ -0.067 \\ 0.533 \\ -0.533 \\ 1.000 \end{bmatrix}.$$

$$T_1 = \frac{1}{f_1} = 0.064 \text{ s}$$

The calculation corresponds to a discretization of the slab in only 2×2 elements. Nevertheless, the natural frequency approximates the exact solution of $f_1 = 17.77$ Hz quite well (cf. Example 5.20). ◀

5.4.2 Damped Vibrations

The equation of motion for free vibrations of a linear single degree of freedom system including damping as given in (5.19) is

$$k \cdot u + c \cdot \dot{u} + m \cdot \ddot{u} = 0. \tag{5.36}$$

The solution of this differential equation can be written for underdamped systems with $\xi < 1$ (Fig. 5.16) as

$$u = u_0 \cdot e^{\xi \omega t} \cdot \sin(\omega_D \cdot t - \psi_0). \tag{5.37a}$$

The natural circular frequency of the damped vibration

Fig. 5.16 Damped vibration

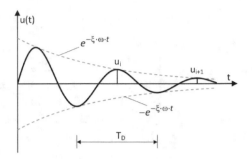

$$\omega_D = \sqrt{1 - \xi^2} \cdot \omega \tag{5.37b}$$

is only slightly different from the natural frequency of the undamped vibration for typical damping values of $\xi = 1-5\%$. This also applies to the period of the vibration

$$T_D = \frac{2\pi}{\omega_D}. \tag{5.37c}$$

The term $e^{-\xi \cdot \omega \cdot t}$ in (5.37a) describes the decay of the vibration (Fig. 5.16). The theoretical limit value $\xi = 1 = 100\%$ is the *critical damping*. At values of $\xi > 1$, the solution of (5.36) yields a decaying function for the displacements (creep movement).

The damping coefficient ξ indicates the ratio of two subsequent amplitudes. From (5.37a), one obtains for two subsequent vibration amplitudes the relationship

$$\xi = \frac{1}{2\pi} \ln \frac{u_i}{u_{i+1}} \tag{5.37d}$$

and for n subsequent amplitudes

$$\xi = \frac{1}{n \cdot 2 \cdot \pi} \ln \frac{u_i}{u_{i+n}}. \tag{5.37e}$$

Example 5.8 For the beam given in Example 5.4, the decaying vibrations for an initial deflection of 1 cm from the static rest position shall be determined. The calculation shall be carried out for the damping coefficients 1%, 2%, and 5%.

The natural circular frequency of the undamped vibration has been obtained in Example 5.4 to be $\omega = 2 \cdot \pi \cdot 6.876 = 43.20$ Hz. First, the phase angle φ_0 and the amplitude u_0 must be determined. It can be shown using (5.37a) that if the initial velocity is zero, the phase angle is $\varphi_0 = -\arctan(1/\xi)$. After that, for a damping factor $\xi = 5\%$, one obtains

$$\varphi_0 = -1.52 \qquad\qquad \text{(with} \quad \xi = 0 : -\pi/2 = -1.57)$$

$$\omega_D = \sqrt{1 - 0.05^2} \cdot \omega = 43.15 \text{ Hz (with} \quad \xi = 0 : \omega = 43.20 \text{ Hz)}.$$

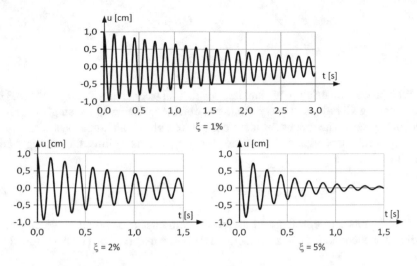

Fig. 5.17 Free damped vibrations of a beam

The difference to the natural frequency of the undamped vibration is practically negligible.

From the initial condition for the displacement at $t = 0$, one obtains with (5.37a)

$$u_0 = 0.01 \text{ m/sin}(-\varphi_0) = 0.01 \text{ m}.$$

Hence, the displacement time history is obtained with (5.37a) as $u(t) = 0.01 \cdot e^{-0.05 \cdot 43.20 \cdot t} \cdot \sin(43.15 \cdot t + 1.52)$ with t in [s] and u in [m].

Correspondingly, one obtains for the other damping coefficients:

$$u(t) = 0.01 \cdot e^{-0.02 \cdot 43.20 \cdot t} \cdot \sin(43.20 \cdot t + 1.55) \text{ with } \xi = 2\%$$
$$u(t) = 0.01 \cdot e^{-0.01 \cdot 43.20 \cdot t} \cdot \sin(43.20 \cdot t + 1.56) \text{ with } \xi = 1\%.$$

The time histories are shown in Fig. 5.17.   ◄

The equations of motion of the free, damped vibrations of a finite element model are according to (5.20)

$$\underline{K} \cdot \underline{u} + \underline{C} \cdot \underline{\dot{u}} + \underline{M} \cdot \underline{\ddot{u}} = \underline{0}. \tag{5.38}$$

It can be shown that real eigenvalues and eigenmodes are only obtained if the damping matrix $\underline{C}$ meets certain conditions. This is the case with proportional damping. One refers to proportional damping when $\underline{C}$ can be represented as a linear combination of the stiffness and mass matrix, as with Rayleigh damping according to (5.18). The damping matrix then fulfills the orthogonality conditions corresponding to (5.34a) and (5.34b) as

$$\underline{\phi}_i^T \cdot \underline{C} \cdot \underline{\phi}_j = 0 \quad \text{for} \quad i \neq j, \tag{5.39a}$$

$$\underline{\phi}_i^T \cdot \underline{C} \cdot \underline{\phi}_j \neq 0 \quad \text{for} \quad i = j. \tag{5.39b}$$

With proportional damping, real damped natural frequencies analogous to (5.37b) and decaying vibrations in every mode are obtained, similar to the single degree of freedom system. The modes are identical to those of the undamped system.

In general, it is assumed that the damping matrix fulfills the orthogonality conditions and that

$$\underline{\phi}_i^T \cdot \underline{C} \cdot \underline{\phi}_i = 2 \cdot \xi_i \cdot \omega_i \cdot \underline{\phi}_i^T \cdot \underline{M} \cdot \underline{\phi}_i \tag{5.40}$$

applies. In this concept, denoted as modal damping, a damping value ξ_i can be assigned at each mode i and an associated damping matrix $\underline{C}$ can be determined (cf. [17]).

Even if the damping matrix does not fulfill the orthogonality conditions, the corresponding eigenvalue problem can be solved. However, the eigenvalues and eigenmodes are then complex. This means that when an eigenmode is excited, the individual degrees of freedom vibrate at the same frequency, but no longer in phase. Each degree of freedom of an eigenmode then has a phase angle to the other degrees of freedom (cf. [3]). However, general damping matrices are not important in practice, since the damping parameters can only be determined with a certain degree of uncertainty anyway and simpler damping approaches are therefore sufficient.

5.5 Forced Vibrations with Harmonic Excitation

Harmonic excitation is understood to be an excitation by forces or base (ground) motions whose time history corresponds to a sine or cosine function. Such excitations can be caused by machine vibrations, for example. Only the case of force excitation is considered here. The excitation by base motion can be treated analogously.

For harmonic excitation, a periodic vibration in the frequency of the excitation takes place after the so-called *transient response*. This state of the system is denoted as *steady state response*.

At first, the single degree of freedom system is considered again. Linear system behavior is assumed. The excitation takes place in the circular frequency Ω due to the load $F_0 \cdot \sin(\Omega \cdot t)$. This results in a steady state motion

$$u = u_0 \cdot \sin(\Omega \cdot t - \varphi_0). \tag{5.41}$$

Herein, u_0 is the amplitude of the vibration and φ_0 the phase angle between the excitation and the response of the system. If one inserts this equation into the equation of motion

$$k \cdot u + c \cdot \dot{u} + m \cdot \ddot{u} = F_0 \cdot \sin(\Omega \cdot t) \tag{5.42}$$

the amplitude and the phase angle of the *harmonic vibration* can be determined. They are [2]

$$u_0 = u_{st} \cdot \frac{1}{\sqrt{\left(1 - \eta^2\right)^2 + 4 \cdot \xi^2 \cdot \eta^2}} \tag{5.43a}$$

$$\varphi_0 = \arctan\left(\frac{2 \cdot \xi \cdot \eta}{1 - \eta^2}\right). \tag{5.43b}$$

Herein, ξ is the damping coefficient according to (5.15),

$$u_{st} = \frac{F_0}{k} \tag{5.43c}$$

is the static displacement due to a force F_0, and

$$\eta = \frac{\Omega}{\omega} \tag{5.43d}$$

the ratio between the frequency of excitation and the natural frequency of the system. The ratio between the dynamic and static displacements is denoted as *magnification factor*. If the excitation frequency and the natural frequency coincide, resonance occurs. It is then $\eta = 1$ and the magnification factor

$$\frac{u_0}{u_{st}} = \frac{1}{2 \cdot \xi} \tag{5.44}$$

depends on the damping coefficient ξ only. Outside the resonance range, on the other hand, damping has little influence on amplitudes.

The phase angle according to (5.43b) is positive for $\eta < 1$ and negative for $\eta > 1$. This means that the displacement and the force are directed at the same time in the same direction for an excitation below resonance and in opposite direction for an excitation above resonance.

Example 5.9 For the beam in Example 5.4, the frequency response curves for excitation by a vertical harmonic load in the middle of the beam are to be determined (Fig. 5.18). The damping coefficients are to be assumed to be 1%, 2%, and 5%.

The beam with a point mass represents a single degree of freedom system with harmonic excitation, whose magnification factor is given by (5.43a). If the excitation frequency $f_F = \Omega/(2 \cdot \pi)$ and the natural frequency as given in Example 5.4 with

$$f = \frac{1}{2 \cdot \pi} \sqrt{\frac{9.333 \cdot 10^3}{5} \frac{kN/m}{kN \, s^2/m}} = 6.876 \text{ Hz}$$

Fig. 5.18 Beam with
harmonic loading

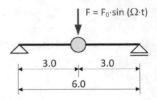

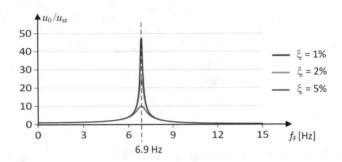

Fig. 5.19 Variation of dynamic magnification factor with excitation frequency and damping

are inserted into (5.43a), one obtains the magnification factor as

$$\frac{u_0}{u_{st}} = \frac{1}{\sqrt{\left(1 - \left(\frac{2 \cdot \pi \cdot f_F}{2 \cdot \pi \cdot f}\right)^2\right)^2 + 4 \cdot \xi^2 \cdot \left(\frac{2 \cdot \pi \cdot f_F}{2 \cdot \pi \cdot f}\right)^2}}$$

$$= \frac{1}{\sqrt{\left(1 - \left(\frac{f_F}{6.876}\right)^2\right)^2 + 4 \cdot \xi^2 \cdot \left(\frac{f_F}{6.876}\right)^2}}.$$

The variation of the dynamic magnification factor with the excitation frequency
and damping are shown in Fig. 5.19. The peak values at resonance are for damping
coefficients of 1, 2, and 5%:

$$\xi = 1\% : \quad u_0 = 50 \cdot u_{st}$$
$$\xi = 2\% : \quad u_0 = 25 \cdot u_{st}.$$
$$\xi = 5\% : \quad u_0 = 10 \cdot u_{st}$$

◄

For finite element models, the loading for harmonic vibrations can be written as

$$\underline{F} = \underline{F}_0 \cdot \sin(\Omega \cdot t). \tag{5.45}$$

Herein, $\underline{F}_0$ represents the vector of the loads acting on the system with the circular frequency Ω. Thus, the equations of motion are

$$\underline{K} \cdot \underline{u} + \underline{C} \cdot \underline{\dot{u}} + \underline{M} \cdot \underline{\ddot{u}} = \underline{F}_0 \cdot \sin(\Omega \cdot t). \tag{5.46}$$

At first, undamped vibrations are treated. For steady state response, the displacements are written as

$$\underline{u} = \underline{u}_0 \cdot \sin(\Omega \cdot t) \tag{5.47}$$

where $\underline{u}_0$ is the vector of the amplitudes in the individual degrees of freedom. The displacements are in phase or—with negative values—in counterphase to the force. With (5.47), the accelerations are $\underline{\ddot{u}} = -\Omega^2 \cdot \underline{u}$. The equations of motion are thus

$$\left(\underline{K} - \Omega^2 \cdot \underline{M}\right) \cdot \underline{u}_0 \cdot \sin(\Omega \cdot t) = \underline{F}_0 \cdot \sin(\Omega \cdot t)$$

or

$$\left(\underline{K} - \Omega^2 \cdot \underline{M}\right) \cdot \underline{u}_0 = \underline{F}_0. \tag{5.48}$$

The amplitudes $\underline{u}_0$ of the displacements are obtained from the solution of a linear system of equations with real coefficients. In the resonance case, the excitation frequency Ω coincides with one of the natural frequencies ω_i of the system, and the left side of (5.48) becomes identical with the left side of the eigenvalue problem (5.34). Equation (5.34) is a homogeneous equation system, which has a solution only if the coefficient matrix is singular. Therefore, in the resonance case, with $\Omega = \omega_i$ the matrix $\left(\underline{K} - \Omega^2 \cdot \underline{M}\right)$ is singular and (5.48) has no solution.

From (5.48), it is interesting to note that for harmonic vibrations without damping, the mass forces correspond to a negative stiffness $-\Omega^2 \cdot \underline{M}$. The displacements in the case of harmonic vibrations are obtained as the solution of an equivalent static problem with the stiffness reduced by $\Omega^2 \cdot \underline{M}$. For a diagonal matrix $\underline{M}$ with point masses, the expression $-\Omega^2 \cdot \underline{M}$ can be interpreted as a stiffness matrix with negative spring constants $k_{j,j} = -\Omega^2 \cdot m_{j,j}$ on the diagonal. Similarly, analogies between the equations of the continuously supported beam and the harmonically vibrating beam and between the continuously supported plate and the harmonically vibrating plate can be established.

In harmonic vibrations of damped systems, each time-varying quantity has not only an amplitude but also a phase shift. Therefore, two quantities, amplitude and phase angle, instead of one, the amplitude, must be determined. In general, complex notation is used for harmonic vibrations. It will be briefly introduced here, but not explained in detail (cf. [3, 17]).

With Euler's formula

$$e^{\mathrm{i}\cdot\Omega\cdot t} = \cos(\Omega \cdot t) + \mathrm{i} \cdot \sin(\Omega \cdot t) \tag{5.49}$$

all time-dependent quantities of harmonic vibrations can be written as complex quantities. The displacement of a single degree of freedom system according to (5.41)

$$u = u_0 \cdot \sin(\Omega \cdot t - \varphi_0),$$

for example, can be written as in complex notation as

$$\hat{u} = u_0 \cdot e^{i \cdot (\Omega \cdot t - \varphi_0)} = u_0 \cdot \cos(\Omega \cdot t - \varphi_0) + i \cdot u_0 \cdot \sin(\Omega \cdot t - \varphi_0). \qquad (5.50)$$

Complex quantities are marked with ^. The imaginary part of the displacement $\hat{u}$ corresponds to the instantaneous value (5.41) of the displacement time history of the harmonic vibration.

To calculate the amplitude and phase angle, the equation is reformulated. One receives

$$\hat{u} = u_0 \cdot e^{i \cdot (\Omega \cdot t - \varphi_0)} = u_0 \cdot e^{-i \cdot \varphi_0} \cdot e^{i \cdot \Omega \cdot t}$$

or

$$\hat{u} = \hat{u}_0 \cdot e^{i \cdot \Omega \cdot t} \qquad (5.51)$$

with

$$\hat{u}_0 = u_0 \cdot e^{-i \cdot \varphi_0} = u_0 \cdot \cos(-\varphi_0) + i \cdot u_0 \cdot \sin(-\varphi_0) = \mathrm{Re}(\hat{u}_0) + i \cdot \mathrm{Im}(\hat{u}_0).$$

From the real and imaginary parts of the complex amplitude, one gets

$$\mathrm{Re}(\hat{u}_0)^2 + \mathrm{Im}(\hat{u}_0)^2 = (u_0 \cdot \cos(-\varphi_0))^2 + (u_0 \cdot \sin(-\varphi_0))^2$$
$$= u_0^2 \cdot \left((\cos(-\varphi_0))^2 + (\sin(-\varphi_0))^2\right) = u_0^2.$$

The amplitude of the vibration

$$u_0 = \sqrt{\mathrm{Re}(\hat{u}_0)^2 + \mathrm{Im}(\hat{u}_0)^2} \qquad (5.51a)$$

is the absolute value of the complex amplitude $\hat{u}_0$. With

$$\tan(\varphi_0) = \frac{\sin(\varphi_0)}{\cos(\varphi_0)} = -\frac{u_0 \cdot \sin(-\varphi_0)}{u_0 \cdot \cos(-\varphi_0)} = -\frac{\mathrm{Im}(\hat{u}_0)}{\mathrm{Re}(\hat{u}_0)}$$

the phase angle is

$$\varphi_0 = -\arctan\left(\frac{\mathrm{Im}(\hat{u}_0)}{\mathrm{Re}(\hat{u}_0)}\right). \tag{5.51b}$$

If $\hat{u}_0$ is real and therefore the imaginary part of $\hat{u}_0$ is zero, the phase angle is zero.

The velocity and acceleration can be easily determined by differentiating (5.51) with respect to time as to $\dot{\hat{u}} = \mathrm{i} \cdot \Omega \cdot \hat{u}$ and $\ddot{\hat{u}} = -\Omega^2 \cdot \hat{u}$.

Writing the load in complex notation, one gets

$$\hat{\underline{F}} = \underline{F}_0 \cdot e^{\mathrm{i}\cdot\Omega\cdot t} = \underline{F}_0 \cdot \cos(\Omega \cdot t) + \mathrm{i} \cdot \underline{F}_0 \cdot \sin(\Omega \cdot t). \tag{5.52}$$

The imaginary part of $\hat{\underline{F}}$ corresponds to the load $\underline{F}$ according to (5.45).

Inserting the expressions for force, displacement, velocity, and acceleration into the equations of motion yields

$$\left(\underline{K} + \mathrm{i} \cdot \Omega \cdot \underline{C} - \Omega^2 \cdot \underline{M}\right) \cdot \hat{\underline{u}}_0 = \underline{F}_0. \tag{5.53}$$

It is a linear system of equations with complex coefficients. Its solution gives the complex displacements $\hat{\underline{u}}_0$. The stiffness and damping matrix can be combined into the *complex stiffness matrix*

$$\hat{\underline{K}} = \underline{K} + \mathrm{i} \cdot \Omega \cdot \underline{C} \tag{5.53a}$$

so that the equations of motion are

$$\left(\hat{\underline{K}} - \Omega^2 \cdot \underline{M}\right) \cdot \hat{\underline{u}}_0 = \underline{F}_0. \tag{5.53b}$$

The complex stiffness matrix $\hat{\underline{K}}$ can also be constructed element by element from complex element stiffness matrices. To each individual element, a different damping coefficient can be assigned. The element matrices are then determined using the *complex modulus of elasticity*

$$\hat{E} = E \cdot (1 + 2 \cdot \mathrm{i} \cdot \xi_\mathrm{h}). \tag{5.54}$$

The damping coefficient ξ_h is the so-called *hysteretic damping coefficient*. It can be assumed to be equal to the viscous damping coefficient according to (5.15). In case of resonance, the magnification factor is the same for the hysteretic and viscous damping ratio. Minor differences in the frequency response curves occur only near the resonance peak.

Example 5.10 For the truss system in Example 5.3, the steady state response due to a harmonic force shall be determined. The sinusoidal force with a magnitude of 10 kN acts in horizontal direction at node 2 with a frequency of $f_\mathrm{E} = 8$ Hz. The horizontal displacement of node 2 is to be computed. The hysteretic damping coefficient is 2% for all truss elements.

The stiffness and mass matrix have already been given in Example 5.3. Since the damping factor is the same in all members, the complex stiffness matrix is obtained by multiplying the real stiffness matrix with the factor $(1 + 2 \cdot i \cdot \xi_h) = (1 + i \cdot 0.04)$. The load vector has the value 10 kN in the excited degree of freedom. Thus, the equations of motion are

$$
\left(\hat{k} \cdot \begin{bmatrix} 1.35 & -0.35 & -1.00 & 0 & -0.35 \\ -0.35 & 1.35 & 0 & 0 & 0.35 \\ -1.00 & 0 & 1.35 & 0.35 & 0 \\ 0 & 0 & 0.35 & 1.35 & 0 \\ -0.35 & 0.35 & 0 & 0 & 1.35 \end{bmatrix} - \Omega^2 \cdot \begin{bmatrix} 20 & 0 & 0 & 0 & 0 \\ 0 & 20 & 0 & 0 & 0 \\ 0 & 0 & 40 & 0 & 0 \\ 0 & 0 & 0 & 40 & 0 \\ 0 & 0 & 0 & 0 & 10 \end{bmatrix} \right)
$$

$$
\cdot \begin{bmatrix} \hat{u}_1 \\ \hat{v}_1 \\ \hat{u}_2 \\ \hat{v}_2 \\ \hat{u}_3 \end{bmatrix} = \begin{bmatrix} 0 \\ 0 \\ 10 \\ 0 \\ 0 \end{bmatrix}
$$

with $\hat{k} = 2.80 \cdot 10^5 \cdot (1 + i \cdot 0.04)$.

The solution for the complex system of equations for $\Omega = 2 \cdot \pi \cdot 8$ Hz $= 50.3$ Hz is

$$
\underline{\hat{u}}_0 = \begin{bmatrix} -2.821 - 0.415 \cdot i \\ -0.665 - 0.094 \cdot i \\ -2.855 - 0.443 \cdot i \\ 1.012 + 0.142 \cdot i \\ -0.599 - 0.088 \cdot i \end{bmatrix} \cdot 10^{-4}.
$$

The horizontal displacement of node 2 corresponds to the third degree of freedom. For this, the amplitude and the phase angle of the vibration are obtained as

$$
u_0 = \sqrt{(-2.855)^2 + (-0.443)^2} \cdot 10^{-4} = 2.889 \cdot 10^{-4} \text{ m}
$$

$$
\varphi_0 = -\arctan\left(\frac{-0.443 \cdot 10^{-4}}{-0.2855 \cdot 10^{-4}}\right) = -0.155 \hat{=} -8.88°.
$$

The phase angle is negative because the excitation frequency of 8 Hz is higher than the first natural frequency of 7.1 Hz (cf. Example 5.5). The force and displacement at the same instant of time are in opposite directions.   ◀

If the calculations of a system excited by a point force of magnitude 1 are carried out for a series of excitation frequencies, and the amplitudes and phases of the response are plotted in a diagram, the so-called *frequency response* is obtained, related to an excitation point and a force direction.

When formulating the equations of motion, the stiffness matrices are usually set up as for static problems, and the mass matrices are determined for point masses or as consistent mass matrices. However, for the calculation of harmonic vibrations, the stiffness matrices of beams with a continuous mass distribution can also be derived from the analytical solution [1, 18–20]. Therefore, a simplified discretization of distributed masses as point masses is not necessary. Numerical difficulties that may arise due to transcendental functions contained in the analytical solution can be avoided by using a formulation with series expansions, as given in [21].

5.6 Forced Vibrations with General Dynamic Excitation

5.6.1 Generals

The solution of the equations of motion of a finite element system can be done by different methods (Fig. 5.20). If the equations of motion (5.20) are solved directly with numerical methods this is called *direct integration in the time domain*. In this case, the damping matrix $\underline{C}$ must be explicitly known. Normally, Rayleigh damping is assumed (cf. Sect. 5.2.3). However, the structural model may also contain specific damping elements. The procedure is computationally intensive. However, it is the only one that allows the consideration of nonlinear material and system behavior.

A particular method for solving the equations of motion uses the transformation of all time-dependent quantities into the frequency domain. For each frequency, the equations of motion for harmonic vibrations are solved (Sect. 5.5). This is called

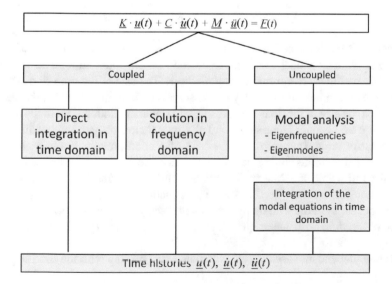

Fig. 5.20 Computational methods for force excited vibrations

a *solution in the frequency domain*. Subsequently, the time-dependent quantities of interest have to be transformed back into the time domain. The advantage of this method is its ability to take into account frequency-dependent spring and damping values, such as those that occur in dynamic soil–structure interaction problems. The definition of the material damping as hysteretic damping by means of a complex modulus of elasticity allows considering individual damping coefficients in each finite element. For large systems, this method is computationally intensive.

In a *modal analysis*, the equations of motion are not explicitly solved as coupled equations. The first step is to determine the eigenfrequencies and eigenmodes of the system. Since a few eigenmodes are often able to represent the dynamic behavior of the system, the computation can be limited to the lowest eigenmodes. By transforming the equations of motion into the space of the eigenmodes, the so-called *modal equations* are obtained. These are integrated into the time domain. In contrast to direct integration, however, these are scalar equations. The computational effort is, therefore, significantly lower than that for direct integration. The damping is introduced as a damping coefficient ξ. The damping coefficient can be chosen differently in each eigenmode. This is then referred to as *modal damping*. The modal equations can also be solved in the frequency domain instead of the time domain in order to compute the frequency response of a system.

5.6.2 Direct Numerical Integration

Direct numerical integration methods are calculation methods in which the equations of motion (5.20)

$$\underline{K} \cdot \underline{u}(t) + \underline{C} \cdot \underline{\dot{u}}(t) + \underline{M} \cdot \underline{\ddot{u}}(t) = \underline{F}(t)$$

are integrated step by step. One speaks of direct integration as the equations of motion of a finite element model are not transformed before integration. The numerical integration methods are approximation methods. The size of the time step Δt of the integration determines the accuracy of the solution. For some methods, the stability depends on the size of the time step. Absolutely stable methods are methods in which the solution $u(t)$ cannot increase beyond all limits for arbitrary initial conditions and for arbitrarily large time steps. In the case of a conditionally stable method, the solution remains limited only if the time step does not exceed a critical value.

In the following, the *Newmark method*, which is frequently used in dynamic computations, is treated [16, 22, 23]. The equations are given for finite element models, i.e. for systems with many degrees of freedom. In the formulation for a one degree of freedom with scalar quantities, they apply to a single degree of freedom system.

All time-dependent variables are calculated step by step with the time step

$$\Delta t = t_{n+1} - t_n. \tag{5.55}$$

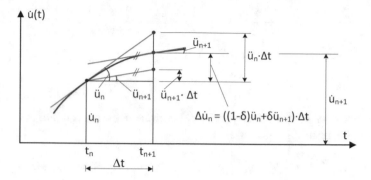

Fig. 5.21 Newmark method

The accelerations at time t_n are referred to as $\ddot{\underline{u}}_n$ and at time t_{n+1} as $\ddot{\underline{u}}_{n+1}$. Accelerations are assumed to be constant in each time step, as a weighted average of $\ddot{\underline{u}}_n$ and $\ddot{\underline{u}}_{n+1}$. In the Newmark method, weighting factors are the integration parameter δ or the complementary expression $(1 - \delta)$ with $0 < \delta < 1$. According to Fig. 5.21, (5.2a) gives the following expression for the velocity $\dot{\underline{u}}_{n+1}$ at time t_{n+1}:

$$\dot{\underline{u}}_{n+1} = \dot{\underline{u}}_n + \left((1 - \delta) \cdot \ddot{\underline{u}}_n + \delta \cdot \ddot{\underline{u}}_{n+1}\right) \cdot \Delta t. \tag{5.55a}$$

It is composed of the sum of the velocity $\dot{\underline{u}}_n$ at the time t_n and the acceleration multiplied by the time step Δt. The acceleration is introduced as weighted average of $\ddot{\underline{u}}_n$ and $\ddot{\underline{u}}_{n+1}$.

The expression for displacements is based on an averaged acceleration too. With (5.2b), one obtains the displacement at time t_{n+1} as

$$\underline{u}_{n+1} = \underline{u}_n + \dot{\underline{u}}_n \cdot \Delta t + \left((0.5 - \alpha) \cdot \ddot{\underline{u}}_n + \alpha \cdot \ddot{\underline{u}}_{n+1}\right) \cdot \Delta t^2 \tag{5.55b}$$

with the weighted average acceleration

$$\left((1 - 2\alpha) \cdot \ddot{\underline{u}}_n + 2\alpha \cdot \ddot{\underline{u}}_{n+1}\right). \tag{5.55c}$$

Here, $\underline{u}_n$ is the displacement at time t_n and α with $0 < \alpha < 0.5$ is again a parameter for weighting the accelerations $\ddot{\underline{u}}_n$ and $\ddot{\underline{u}}_{n+1}$. Equations (5.55a) and (5.55b) can also be written as

$$\ddot{\underline{u}}_{n+1} = a_0 \cdot \left(\underline{u}_{n+1} - \underline{u}_n\right) - a_2 \cdot \dot{\underline{u}}_n - a_3 \cdot \ddot{\underline{u}}_n \tag{5.56a}$$

$$\dot{\underline{u}}_{n+1} = \dot{\underline{u}}_n + a_6 \cdot \ddot{\underline{u}}_n + a_7 \cdot \ddot{\underline{u}}_{n+1} \tag{5.56b}$$

with

$$a_0 = \frac{1}{\alpha \cdot \Delta t^2}, \quad a_1 = \frac{\delta}{\alpha \cdot \Delta t}, \qquad a_2 = \frac{1}{\alpha \cdot \Delta t}, \qquad a_3 = \frac{1}{2 \cdot \alpha} - 1$$

$$a_4 = \frac{\delta}{\alpha} - 1, \quad a_5 = \frac{\Delta t}{2} \cdot \left(\frac{\delta}{\alpha} - 2\right), \quad a_6 = (1 - \delta) \cdot \Delta t, \quad a_7 = \delta \cdot \Delta t.$$

$$(5.56c)$$

If the two expressions for velocity and acceleration are introduced into the equations of motion, the resulting equation is

$$\left(\underline{K} + a_1 \cdot \underline{C} + a_0 \cdot \underline{M}\right) \cdot \underline{u}_{n+1} = \underline{F}_{n+1} + \underline{M} \cdot \left(a_0 \cdot \underline{u}_n + a_2 \cdot \underline{\dot{u}}_n + a_3 \cdot \underline{\ddot{u}}_n\right)$$
$$+ \underline{C} \cdot \left(a_1 \cdot \underline{u}_n + a_4 \cdot \underline{\dot{u}}_n + a_5 \cdot \underline{\ddot{u}}_n\right). \quad (5.56d)$$

The equation represents the equilibrium conditions at time t_{n+1}. The right side of the equation is formulated with the external forces at time t_{n+1} and the displacements, velocities, and accelerations at time t_n known from the previous time step.

The stability of the solution depends on the choice of integration parameters α and δ. According to [16], the method is absolutely stable if the following conditions are fulfilled:

$$\delta \geq \frac{1}{2}, \quad \alpha \geq \frac{1}{4}\left(\frac{1}{2} + \delta\right)^2. \quad (5.57)$$

When $\delta = 1/2$ and $\alpha = 1/4$, the average $(\underline{\ddot{u}}_n + \underline{\ddot{u}}_{n+1})/2$ is taken as constant acceleration, for both the displacements in (5.55b) and the velocities in (5.55a). This method is also known as Newmark's constant-average-acceleration method or trapezoidal rule. In [16], $\delta = 1/2$ and $\alpha = 1/4$ are recommended for reasons of accuracy. In this case, the coefficients simplify to

$$a_0 = \frac{4}{\Delta t^2}, \quad a_1 = \frac{2}{\Delta t}, \quad a_2 = \frac{4}{\Delta t}, \quad a_3 = a_4 = 1, \quad a_5 = 0, \quad a_6 = a_7 = \frac{\Delta t}{2}.$$

$$(5.58)$$

The equation to be solved is

$$\left(\underline{K} + \frac{2}{\Delta t} \cdot \underline{C} + \frac{4}{\Delta t^2} \cdot \underline{M}\right) \cdot \underline{u}_{n+1}$$
$$= \underline{F}_{n+1} + \underline{M} \cdot \left(\frac{4}{\Delta t^2} \cdot \underline{u}_n + \frac{4}{\Delta t} \cdot \underline{\dot{u}}_n + \underline{\ddot{u}}_n\right) + \underline{C} \cdot \left(\frac{2}{\Delta t} \cdot \underline{u}_n + \underline{\dot{u}}_n\right) \quad (5.58a)$$

with

$$\underline{\ddot{u}}_{n+1} = \left(\underline{u}_{n+1} - \underline{u}_n\right) \cdot \frac{4}{\Delta t^2} - \underline{\dot{u}}_n \cdot \frac{4}{\Delta t} - \underline{\ddot{u}}_n \quad (5.58b)$$

$$\underline{\dot{u}}_{n+1} = \underline{\dot{u}}_n + \left(\underline{\ddot{u}}_n + \underline{\ddot{u}}_{n+1}\right) \cdot \frac{\Delta t}{2}. \quad (5.58c)$$

The displacements $\underline{u}_0 = \underline{u}(0)$ and velocities $\underline{\dot{u}}_0 = \underline{\dot{u}}(0)$ at time $t = 0$ are chosen as initial conditions. The initial acceleration $\underline{\ddot{u}}_0$ at time $t = 0$ is thus obtained according to (5.20) as

$$\underline{\ddot{u}}_0 = \underline{M}^{-1} \cdot \left(\underline{F}(0) - \underline{K} \cdot \underline{u}_0 - \underline{C} \cdot \underline{\dot{u}}_0 \right). \tag{5.58d}$$

The method can be summarized as:

Newmark method

Start

1. Determination of the initial values $\underline{u}_0$, $\underline{\dot{u}}_0$, and $\underline{\ddot{u}}_0$ with

$$\underline{\ddot{u}}_0 = \underline{M}^{-1} \cdot \left(\underline{F}(0) - \underline{K} \cdot \underline{u}_0 - \underline{C} \cdot \underline{\dot{u}}_0 \right).$$

2. Selection of the time step Δt and the integration parameters α and δ with

$$\delta \geq 0.5, \quad \alpha \geq 0.25 \cdot (0.5 + \delta)^2$$

3. Determination of the constants a_0 to a_7

$$a_0 = \frac{1}{\alpha \cdot \Delta t^2}, \ a_1 = \frac{\delta}{\alpha \cdot \Delta t}, \qquad a_2 = \frac{1}{\alpha \cdot \Delta t}, \qquad a_3 = \frac{1}{2 \cdot \alpha} - 1$$

$$a_4 = \frac{\delta}{\alpha} - 1, \ a_5 = \frac{\Delta t}{2} \cdot \left(\frac{\delta}{\alpha} - 2 \right), \ a_6 = (1 - \delta) \cdot \Delta t, \ a_7 = \delta \cdot \Delta t.$$

4. Computation of the effective stiffness matrix

$$\underline{\tilde{K}} = \underline{K} + a_1 \cdot \underline{C} + a_0 \cdot \underline{M}.$$

5. Triangularization of $\underline{\tilde{K}}$

$$\underline{\tilde{K}} = \underline{L} \cdot \underline{D} \cdot \underline{L}^{\mathrm{T}}.$$

Computation for each time step

1. Determination of effective loads at time t_{n+1}

$$\underline{\tilde{F}}_{n+1} = \underline{F}_{n+1} + \underline{M} \cdot \left(a_0 \cdot \underline{u}_n + a_2 \cdot \underline{\dot{u}}_n + a_3 \cdot \underline{\ddot{u}}_n \right)$$
$$+ \underline{C} \cdot \left(a_1 \cdot \underline{u}_n + a_4 \cdot \underline{\dot{u}}_n + a_5 \cdot \underline{\ddot{u}}_n \right).$$

2. Computation of the displacements at time t_{n+1}

$$\underline{L} \cdot \underline{D} \cdot \underline{L}^{\mathrm{T}} \cdot \underline{u}_{n+1} = \underline{\tilde{F}}_{n+1}.$$

3. Determination of the accelerations and velocities at time t_{n+1}

$$\underline{\ddot{u}}_{n+1} = a_0 \cdot \left(\underline{u}_{n+1} - \underline{u}_n\right) - a_2 \cdot \underline{\dot{u}}_n - a_3 \cdot \underline{\ddot{u}}_n$$
$$\underline{\dot{u}}_{n+1} = \underline{\dot{u}}_n + a_6 \cdot \underline{\ddot{u}}_n + a_7 \cdot \underline{\ddot{u}}_{n+1}.$$

With the Newmark method, only the displacements, velocities, and accelerations of the previous time step are required to calculate their values in the next time step. Such procedures are also called *one-step methods*. In *multi-step methods*, the values of more than one previous time step are considered. In the context of finite element calculations, one-step methods are mostly used. In addition to the Newmark method, the *Wilson-θ-method* is frequently employed [16]. It assumes a linear variation of the acceleration within a time step. However, the equilibrium conditions are not formulated at time $t = t_{n+1} = t_n + \Delta t$, but at time $t = t_n + \theta \cdot \Delta t$. This can cause undesired effects in nonlinear analyses. The method is absolutely stable for $\theta > 1.37$. In general, $\theta = 1.4$ is chosen.

The numerical error caused by the time discretization can appear in the calculation result by a decaying amplitude and an extension of the vibration period. Since the decay in amplitude is similar to an artificial damping, it is also called numerical damping. Examples of Wilson-θ-method and other methods are given in [16]. With the Newmark method, no numerical damping occurs for $\delta = 1/2$ and $\alpha = 1/4$. However, if one chooses $\delta > 1/2$, an artificial amplitude decay also appears with the Newmark method. The influence of the numerical damping depends on the ratio of the time step to the vibration period considered. Short periods or high frequencies are damped more than low frequencies. This effect can be desirable to filter out higher natural frequencies from the vibrational response.

The Newmark method and the Wilson method belong to the so-called *implicit integration methods* based on equilibrium conditions at time t_{n+1}. The displacements are determined by solving a linear system of equations. In addition, there are the *explicit integration methods* which are based on the equilibrium conditions at time t. In the *central difference method*, the acceleration and velocity are approximated by the finite difference expressions

$$\underline{\ddot{u}}_n = \frac{1}{\Delta t^2} \cdot \left(\underline{u}_{n-1} - 2 \cdot \underline{u}_n + \underline{u}_{n+1}\right) \tag{5.59a}$$

$$\underline{\dot{u}}_n = \frac{1}{2 \cdot \Delta t} \cdot \left(-\underline{u}_{n-1} + \underline{u}_{n+1}\right). \tag{5.59b}$$

The initial values $\underline{u}_0$ and $\underline{\dot{u}}_0$ are preset. The initial acceleration $\underline{\ddot{u}}_0$ is obtained after (5.58d) and

$$\underline{u}_{-1} = \underline{u}_0 - \Delta t \cdot \underline{\dot{u}}_0 + \frac{\Delta t^2}{2} \cdot \underline{\ddot{u}}_0 \tag{5.59c}$$

with (5.59a, 5.59b) for $n = 0$. If (5.59a, 5.59b) are inserted into the equations of motion (5.20) for $t = t_n$, one obtains

$$\left(\frac{1}{2\Delta t} \cdot \underline{C} + \frac{1}{\Delta t^2} \cdot \underline{M} \right) \cdot \underline{u}_n = \underline{F}_n + \left(\underline{K} - \frac{2}{\Delta t^2} \cdot \underline{M} \right) \cdot \underline{u}_n$$
$$- \left(\frac{1}{2\Delta t} \cdot \underline{C} + \frac{1}{\Delta t^2} \cdot \underline{M} \right) \cdot \underline{u}_{n-1} \tag{5.59d}$$

and with $\underline{C} = 0$ for systems without damping

$$\left(\frac{1}{\Delta t^2} \cdot \underline{M} \right) \cdot \underline{u}_n = \underline{F}_n + \left(\underline{K} - \frac{2}{\Delta t^2} \cdot \underline{M} \right) \cdot \underline{u}_n - \left(\frac{1}{\Delta t^2} \cdot \underline{M} \right) \cdot \underline{u}_{n-1}. \tag{5.59e}$$

For point masses in all degrees of freedom, the mass matrix is a diagonal matrix that can be inverted easily. The time-consuming solution of a linear system of equations is, therefore, not necessary. The method can thus be faster and more efficient than implicit methods, but has disadvantages with regard to stability. According to [16], for the central difference method, the time step must be selected so that

$$\Delta t \le \Delta t_{\text{crit}} = \frac{T_{\min}}{\pi} \tag{5.59f}$$

applies. The critical time step Δt_{crit} depends on the smallest period $T_{\min}$ of the finite element model and thus on the fineness of the finite element discretization. For small element sizes, this requires extraordinarily small time steps. Explicit methods are, therefore, only used in special problems such as impact problems.

Example 5.11 The beam in Example 5.4 is suddenly loaded by a vertical force of 10 kN in the middle of the beam. Afterwards, the load remains constant. The response vibration is to be determined using the Newmark method for a damping coefficient of $\xi = 2\%$.

The natural period according to Example 5.4 is $T = \sqrt{9333/5}/(2 \cdot \pi) = 0.145$ s. The time step is set to $\Delta t = 0.01$ s $< T/10$, the integration parameters $\delta = 1/2$ and $\alpha = 1/4$ are chosen. With (5.56c), this results in the integration constants

$$a_0 = 18910, \quad a_1 = 137.5, \quad a_2 = 275.0,$$
$$a_3 = a_4 = 1, \quad a_5 = 0, \quad a_6 = a_7 = 0.007.$$

The effective stiffness is obtained as $\tilde{k} = k + a_1 \cdot c + a_0 \cdot m = 1.051 \cdot 10^5$ kN/m with $c = 2 \cdot \xi \cdot \sqrt{k \cdot m} = 2 \cdot 0.02 \cdot \sqrt{9.33 \cdot 10^3 \cdot 5} = 8.64$ kN s/m, $m = 5$ kN s^2/m, and $k = 9.333 \cdot 10^3$ kN/m.

The load is $F_0 = F_1 = F_2 = \ldots = F_{n,\text{max}} = 10$ kN and, therefore, independent of time in this example. The calculation procedure for the step-by-step integration is

$$\tilde{F}_{n+1} = F_{n+1} + m \cdot (a_0 \cdot u_n + a_2 \cdot \dot{u}_n + a_3 \cdot \ddot{u}_n) + c \cdot (a_1 \cdot u_n + a_4 \cdot \dot{u}_n + a_5 \cdot \ddot{u}_n)$$

$$u_{n+1} = \frac{\tilde{F}_{n+1}}{\tilde{k}}, \quad \ddot{u}_{n+1} = a_0 \cdot (u_{n+1} - u_n) - a_2 \cdot \dot{u}_n - a_3 \cdot \ddot{u}_n,$$

$$\dot{u}_{n+1} = \dot{u}_n + a_6 \cdot \ddot{u}_n + a_7 \cdot \ddot{u}_{n+1}.$$

The initial values of the displacement and the velocity are $u_0 = 0$ and $\dot{u}_0 = 0$. The initial value of the acceleration is obtained with (5.19) as $\ddot{u}_0 = F_0/m = 10/5 = 2.0$ m/s^2. The results for the first 10 time steps are given in Table 5.1.

The time histories shown in Fig. 5.22 show a decaying vibration. The displacements tend toward the static displacement $u_{\text{st}} = F_0/k = 1.071$ mm, the velocity and acceleration toward zero. The maximum displacement of 2.066 mm is slightly lower

Table 5.1 Displacements, velocities, and accelerations with the Newmark method

n	t [s]	u_n [mm]	$\dot{u}_n$ [mm/s]	$\ddot{u}_n$ $\left[\text{m/s}^2\right]$
0	0	0.000	0.000	2.000
1	0.01	0.095	18.952	1.790
2	0.02	0.361	34.241	1.267
3	0.03	0.748	43.224	0.529
4	0.04	1.186	44.414	−0.291
5	0.05	1.597	37.729	−1.046
6	0.06	1.908	24.479	−1.604
7	0.07	2.066	7.117	−1.869
8	0.08	2.045	−11.219	−1.799
9	0.09	1.853	−27.271	−1.412
10	0.10	1.525	−38.236	−0.781

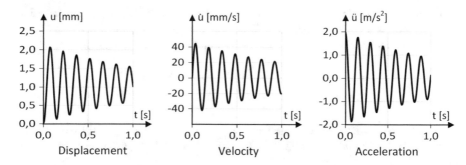

Fig. 5.22 Displacement, velocity, and acceleration time histories for $\xi = 2\%$

Fig. 5.23 Displacement
time histories for $\xi = 0$

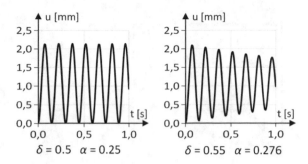

$\delta = 0.5 \quad \alpha = 0.25$ $\delta = 0.55 \quad \alpha = 0.276$

than the solution for $\xi = 0$, which is (exactly) $2 \cdot u_{st} = 2.142$ mm. The small difference is mainly due to the damping and only to a lesser extent due to the numerical integration error. The small difference of $(2.14 - 2.07)/2.14 = 3\%$ shows that for impact problems damping is only of minor importance.

In order to investigate the influence of numerical damping, the calculation is repeated with the damping coefficient $\xi = 0$, whereby the time step $\Delta t = 0.01$ s is maintained. The displacement time histories are given for two parameter combinations of α and δ in Fig. 5.23. For $\delta = 0.5$ and $\alpha = 0.25$, the amplitudes remain constant, i.e. no numerical damping occurs. If, however, $\delta = 0.55$ and $\alpha = 0.25(0.5 + \delta)^2 = 0.276$ are selected, the amplitudes decay due to the numerical damping, although the damping coefficient is $\xi = 0$. ◀

5.6.3 Modal Analysis

In modal analysis, the natural frequencies and modes of a structure are used to represent its response to a time-dependent loading. While direct numerical integration is also applicable to nonlinear systems, modal analysis requires linear system behavior. The mass and stiffness matrices must, therefore, be constant. For the damping matrix, there are restrictions which will be discussed later.

The approach is explained by means of an example. Figure 5.24 shows a system with three masses. It is a frame with rigid horizontal beams. Since the rotations due to the assumption of the rigid beams are fixed, the 3 horizontal storey displacements are the only degrees of freedom of the system.

The system possesses three eigenfrequencies and mode shapes. Any displacement state of the frame is described by the three displacement values u_1, u_2 and u_3 or $\begin{bmatrix} u_1 & u_2 & u_3 \end{bmatrix}^{T}$. Alternatively, it can be represented by the superposition of the three mode shapes $\underline{\phi}_1, \underline{\phi}_2$, and $\underline{\phi}_3$ as

$$\begin{bmatrix} u_1 \\ u_2 \\ u_3 \end{bmatrix} = \alpha_1 \cdot \underline{\phi}_1 + \alpha_2 \cdot \underline{\phi}_2 + \alpha_3 \cdot \underline{\phi}_3 = \sum \alpha_i \cdot \underline{\phi}_i.$$

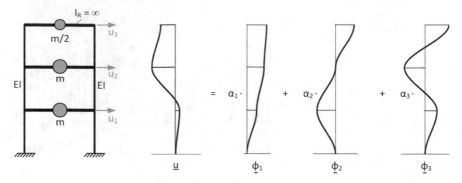

Fig. 5.24 Mode shapes of a system with 3 degrees of freedom

Instead of the three displacements u_1, u_2, and u_3, the variables α_1, α_2, and α_3 are used to describe the displacements of the structure. Mathematically, this can be understood as a change of basis. From the basis of the geometric coordinates of the finite element model, one switches to the basis of the eigenmodes. In general, the superposition of the eigenmodes can be written as $\underline{u} = \sum \alpha_i \cdot \underline{\phi}_i$ or, since the displacements depend on time, as

$$\underline{u}(t) = \sum \alpha_i(t) \cdot \underline{\phi}_i. \tag{5.60}$$

The functions $\alpha_i(t)$ are called generalized coordinates, normal coordinates, or modal coordinates.

Considering m eigenmodes of a system with $n > m$ degrees of freedom occupied by mass, the following applies:

$$\underline{u}(t) = \sum_{i=1}^{m} \alpha_i(t) \cdot \underline{\phi}_i = \underline{\Phi} \cdot \underline{A} \tag{5.61}$$

with the modal matrix

$$\underline{\Phi} = \begin{bmatrix} \underline{\phi}_1 & \underline{\phi}_2 & \underline{\phi}_3 & \cdots & \underline{\phi}_m \end{bmatrix} \tag{5.61a}$$

and the vector of the generalized coordinates

$$\underline{A} = \begin{bmatrix} \alpha_1 \\ \alpha_2 \\ . \\ . \\ \alpha_m \end{bmatrix}. \tag{5.61b}$$

If the modal matrix contains all n eigenmodes of the system, it is quadratic and regular, since the eigenvectors are linearly independent. It can then be inverted to determine the generalized coordinates as $\underline{A} = \underline{\Phi}^{-1} \cdot \underline{u}$. In general, however, fewer eigenmodes are used than the number of degrees of freedom of the system occupied by mass. In this case, the modal matrix is rectangular and there is no inverse relationship. Therefore, $\underline{u}(t)$ cannot be represented by the generalized coordinates exactly. Instead, an approximation of the displacements $\underline{u}(t)$ is obtained, but the accuracy in the range of the considered eigenmodes is maintained. In most practical cases, the displacements can be described by a few eigenmodes and the contribution of higher eigenmodes is small. The approximation becomes the better, the less the eigenmodes which have been omitted participate in the vibrations of the structure.

To determine the generalized coordinates, one starts from the equations of motion. If (5.61) is inserted into the equations of motion (5.20), one obtains for a single eigenmode $\underline{\phi}_i$

$$\underline{K} \cdot \underline{\phi}_i \cdot \alpha_i(t) + \underline{C} \cdot \underline{\phi}_i \cdot \dot{\alpha}_i(t) + \underline{M} \cdot \underline{\phi}_i \cdot \ddot{\alpha}_i(t) = \underline{F}(t).$$

Now, the equation is multiplied by the transposed jth eigenmode $\underline{\phi}_j^T$ and one obtains

$$\underline{\phi}_j^T \cdot \underline{K} \cdot \underline{\phi}_i \cdot \alpha_i(t) + \underline{\phi}_j^T \cdot \underline{C} \cdot \underline{\phi}_i \cdot \dot{\alpha}_i(t) + \underline{\phi}_j^T \cdot \underline{M} \cdot \underline{\phi}_i \cdot \ddot{\alpha}_i(t) = \underline{\phi}_j^T \cdot \underline{F}(t).$$

Due to the orthogonality relations (5.34a, 5.34b) of the eigenmodes and assuming proportional damping according to (5.40), the expression leads to generalized coordinates $\alpha_i(t) \neq 0$ only for $i = j$. The equation thus becomes the scalar equation, which is also called the *modal equation of motion*

$$k_i^* \cdot \alpha_i(t) + c_i^* \cdot \dot{\alpha}_i(t) + m_i^* \cdot \ddot{\alpha}_i(t) = f_i^* \tag{5.62}$$

with the definitions

$$k_i^* = \underline{\phi}_i^T \cdot \underline{K} \cdot \underline{\phi}_i \qquad \textit{generalized stiffness} \tag{5.62a}$$

$$c_i^* = \underline{\phi}_i^T \cdot \underline{C} \cdot \underline{\phi}_i \qquad \textit{generalized damping} \tag{5.62b}$$

$$m_i^* = \underline{\phi}_i^T \cdot \underline{M} \cdot \underline{\phi}_i \qquad \textit{generalized mass} \tag{5.62c}$$

$$f_i^*(t) = \underline{\phi}_i^T \cdot \underline{F}(t) \qquad \textit{generalized load.} \tag{5.62d}$$

With (5.62), the solution of the equations of motion (5.20) of the finite element model is reduced to the solution of the m scalar equations of motion, which correspond to those of a single degree of freedom system.

Damping can be written for any eigenmode as

$$c_i^* = 2 \cdot \xi_i \cdot \sqrt{k_i^* \cdot m_i^*} = 2 \cdot \xi_i \cdot \omega_i \cdot m_i^*. \tag{5.63}$$

In this concept, which is denoted as *modal damping*, an individual damping coefficient ξ_i is assigned to each eigenmode.

With (5.63) and the circular eigenfrequency $\omega_i = \sqrt{k_i^*/m_i^*}$, (5.62) can be expressed as

$$\omega_i^2 \cdot \alpha_i(t) + 2 \cdot \xi_i \cdot \omega_i \cdot \dot{\alpha}_i(t) + \ddot{\alpha}_i(t) = f_i^*(t)/m_i^*. \tag{5.64}$$

If the eigenmodes are already normalized according to (5.35a) with respect to the generalized mass with $m_i^* = 1$, the practical calculation simplifies.

The equation of motion (5.19) of a single degree of freedom system can be solved analytically or numerically with the so-called *Duhamel integral* or convolution integral [1, 2]

$$u(t) = \frac{1}{m \cdot \omega_D} \int_0^t F(\tau) \cdot e^{-\xi \cdot \omega \cdot (t-\tau)} \cdot \sin(\omega_D \cdot (t-\tau)) \, d\tau. \tag{5.65}$$

Herein, $\omega = \sqrt{k/m}$ is the circular frequency of the undamped single degree of freedom system, ω_D the circular frequency of the damped single degree of freedom system according to (5.37b), and ξ the damping coefficient according to (5.15). The Duhamel integral can also be used to solve (5.62), where

$$\alpha_i(t) \equiv u(t), \quad k_i^* \equiv k, \quad c_i^* = 2 \cdot \xi_i \cdot \sqrt{k_i^* \cdot m_i^*} \equiv c,$$
$$m_i^* \equiv m, \quad f_i^*(t) \equiv F(t). \tag{5.65a}$$

For elementary load-time functions, (5.65) can be evaluated analytically. For a general load-time function, the modal equations of motion (5.62) are solved by direct numerical integration, for example, with the Newmark method (Sect. 5.6.2). The time step must be chosen the same for all modal equations. The superposition to the complete solution is then performed with (5.61).

The method is summarized below:

Modal analysis

Start
1. Selection of the number m of natural frequencies to be considered and of the time step Δt.
2. Establishment and solution of the eigenvalue problem of undamped vibration for m eigenmodes

$$\left(\underline{K} - \omega_i^2 \cdot \underline{M} \right) \cdot \underline{\phi}_i = \underline{0}.$$

The results are the natural circular frequencies ω_i and the corresponding eigenmodes $\underline{\phi}_i$ for $i = 1, 2, \ldots, m$.

Computation for each natural vibration

1. Computation of the generalized mass and the generalized load-time history

$$m_i^* = \underline{\phi}_i^T \cdot \underline{M} \cdot \underline{\phi}_i, \; f_i^*(t) = \underline{\phi}_i^T \cdot \underline{F}(t).$$

2. Selection of the modal damping coefficient ξ_i.
3. Stepwise integration of the modal equation of motion with the time Δt

$$\omega_i^2 \cdot \alpha_i(t) + 2 \cdot \xi_i \cdot \omega_i \cdot \dot{\alpha}_i(t) + \ddot{\alpha}_i(t) = f_i^*(t)/m_i^*.$$

Computation of time histories

1. Displacement time histories by superposition of the natural vibrations

$$\underline{u}(t) = \sum \alpha_i(t) \cdot \underline{\phi}_i.$$

2. Time histories of internal forces and stresses using section force or stress matrices.

A benefit of a modal analysis is that often a small number of natural vibrations is sufficient to represent the motion of a structure. The number of eigenmodes to be taken into account is determined by the frequency content of the time history of the excitation and the spatial distribution of the load quantities. Assuming the same accuracy, for the determination of internal forces and stresses, more eigenmodes may be required than for displacements, because these are quantities which are obtained by differentiation. Higher eigenmodes are, therefore, more important for the internal forces and stresses than for the displacements.

Example 5.12 The truss system in Example 5.3 is loaded in node 2 by a horizontal force with the impact-like load-time function as shown in Fig. 5.25. The response of the undamped system with $\xi = 0$ is to be determined by modal analysis. In addition to the displacements, the time histories of the normal forces are requested in all members.

The eigenmodes of the truss have been determined in Example 5.5. They are normalized with respect to the generalized mass, so that $m_i^* = 1$ applies to each eigenmode. Since the load $\underline{F}(t)$ is described by a single time function $\tilde{F}(t)$, it can be written as $\underline{F}(t) = \underline{I} \cdot \tilde{F}(t)$. The vector $\underline{I} = \begin{bmatrix} 0 & 0 & 1 & 0 & 0 \end{bmatrix}^T$ specifies the spatial distribution of the load and contains a 1 in the excited degree of freedom while all

Fig. 5.25 Load-time function $\tilde{F}(t)$

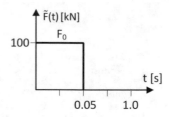

other elements are zero. The generalized load according to (5.62a) and (5.64) thus is

$$\frac{f_i^*(t)}{m_i^*} = \frac{1}{m_i^*} \cdot \underline{\phi}_i^{\mathrm{T}} \cdot \underline{F}(t) = \frac{\Gamma_i}{m_i^*} \cdot \tilde{F}(t) \quad \text{with} \quad \Gamma_i = \underline{\phi}_i^{\mathrm{T}} \cdot \underline{I}.$$

The parameters Γ_i are called *participation factors*. The participation factors of all five natural vibrations are obtained with the eigenmodes according to Example 5.5 as

$$\Gamma_1 = 0.125, \quad \Gamma_2 = 0.022, \quad \Gamma_3 = -0.038, \quad \Gamma_4 = 0.084, \quad \Gamma_5 = -0.017.$$

Due to the spatial distribution of the load (horizontal force at node 2), in particular, the first and fourth modes are excited.

Here, the integration of the equation of motion (5.64) can be performed analytically because the load-time function is rather elementary. With

$$\begin{aligned} f_i^*(t) &= \Gamma_i \cdot \tilde{F}(t) = \Gamma_i \cdot F_0 \quad &\text{if} \quad t \leq t_{\mathrm{F}} \\ &= 0 \quad &\text{if} \quad t > t_{\mathrm{F}} \end{aligned}$$

the solution of (5.64) for the undamped system with $\xi_i = 0$ is obtained applying the Duhamel integral (5.65) as

$$\begin{aligned} \alpha_i(t) &= \frac{f_i^*(t)}{\omega_i^2 \cdot m_i} \cdot (1 - \cos(\omega_i \cdot t)) \quad &\text{if} \quad t \leq t_{\mathrm{F}} \\ &= \frac{f_i^*(t)}{\omega_i^2 \cdot m_i} \cdot (\cos(\omega_i \cdot (t_{\mathrm{F}} - t)) - \cos(\omega_i \cdot t)) \quad &\text{if} \quad t > t_{\mathrm{F}} \end{aligned}$$

The functions are evaluated with the time step $\Delta t = 0.005$ s. Table 5.2 gives the generalized coordinates $\alpha_i(t)$ related to the respective load factor $(\Gamma_i \cdot F_0)$ for the first 20 time steps. The functions are shown for the time up to 0.3 s in Fig. 5.26.

Table 5.2 Time histories of the generalized coordinates from 0 to 0.1 s

t [s]	$\dfrac{\alpha_1(t)}{\Gamma_1 \cdot F_0}$	$\dfrac{\alpha_2(t)}{\Gamma_2 \cdot F_0}$	$\dfrac{\alpha_3(t)}{\Gamma_3 \cdot F_0}$	$\dfrac{\alpha_4(t)}{\Gamma_4 \cdot F_0}$	$\dfrac{\alpha_5(t)}{\Gamma_5 \cdot F_0}$
0	0	0	0	0	0
0.005	0.124	0.122	0.121	0.119	0.114
0.010	0.492	0.460	0.436	0.410	0.344
0.015	1.083	0.933	0.821	0.712	0.463
0.020	1.870	1.426	1.126	0.858	0.354
0.025	2.813	1.821	1.230	0.767	0.124
0.030	3.865	2.024	1.093	0.489	0.000
0.035	4.973	1.985	0.768	0.178	0.104
0.040	6.083	1.715	0.383	0.007	0.334
0.045	7.139	1.277	0.090	0.070	0.463
0.050	8.089	0.777	0.003	0.333	0.363
0.055	8.761	0.213	0.035	0.530	0.020
0.060	8.997	−0.402	0.054	0.434	−0.343
0.065	8.785	−0.921	0.052	0.097	−0.368
0.070	8.136	−1.218	0.029	−0.293	−0.030
0.075	7.082	−1.223	−0.005	−0.521	0.337
0.080	5.676	−0.934	−0.037	−0.460	0.372
0.085	3.988	−0.421	−0.055	−0.145	0.040
0.090	2.101	0.194	−0.051	0.251	−0.331
0.095	0.109	0.762	−0.027	0.508	−0.376
0.100	−0.189	1.147	0.008	0.483	−0.050

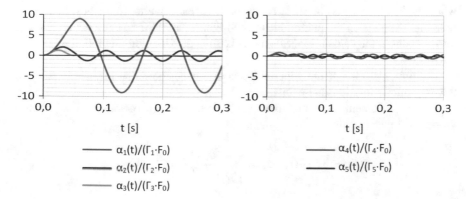

Fig. 5.26 Generalized coordinates

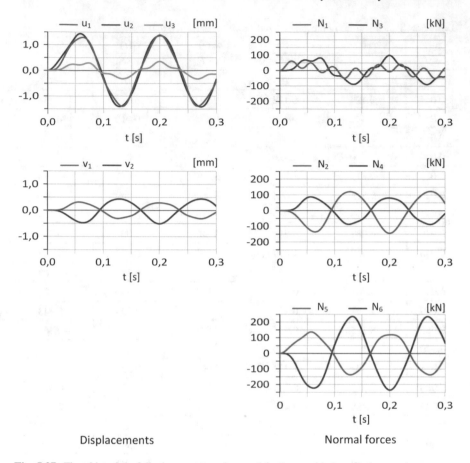

Fig. 5.27 Time histories of displacements and normal forces considering all eigenmodes

The values are exact because they are based on an analytical solution. In general, however, the load-time functions are more complicated so that they cannot be integrated analytically. The modal equations of motion are then solved numerically, whereby the time step is chosen to be the same for each natural frequency. Its magnitude thus depends on the magnitude of the highest natural frequency which is taken into account in the calculation. Since this is an approximation method, a numerical error occurs.

The time histories of the generalized coordinates show that especially the two lowest natural vibrations are excited.

The displacement time histories result from the superposition of m natural vibrations in

$$\underline{u}(t) = \sum_{i=1}^{m} \alpha_i(t) \cdot \underline{\phi}_i = \sum_{i=1}^{m} \frac{\alpha_i(t)}{\Gamma_i \cdot F_0} \cdot (\Gamma_i \cdot F_0) \cdot \underline{\phi}_i.$$

They are shown in Fig. 5.27, taking into account all natural vibrations, i.e. with $m = 5$. From this, the corresponding element stresses and internal forces are obtained using the sectional force matrices of the elements. This will be shown for the time history of the normal force in element 1. The sectional force matrices of the truss elements are given in Example 3.3. For element 1, one gets

$$N_1 = 2.80 \cdot 10^5 \cdot \begin{bmatrix} -1 & 1 \end{bmatrix} \cdot \begin{bmatrix} u_1 \\ u_2 \end{bmatrix} \quad \text{with} \quad u_1 \text{ and } u_2 \text{ in [m] and } N_1 \text{ in [kN].}$$

Since the displacements u_1 and u_2 are time-dependent variables, this results in a time history that is for the normal force N_1.

The displacements of nodes 1 and 2 and the normal force in element 1 are summarized in Table 5.3 up to $t = 0.1$ s. The time histories of the displacements and normal forces are shown in Fig. 5.27.

So far, all eigenmodes have been included in the superposition. The solution is therefore exact. Now, it shall be investigated how the result changes if not all eigenmodes are taken into account. Table 5.4 shows the maxima of displacements and normal force N_1 that are obtained for $t \leq 0.3$ s if the superposition is only

Table 5.3 Time histories

t [s]	u_1 [mm]	v_1 [mm]	u_2 [mm]	v_2 [mm]	N_1 [kN]
0	0.000	0.000	0.000	0.000	0.000
0.005	0.001	0.000	0.031	0.000	8.331
0.010	0.013	0.000	0.116	−0.002	28.707
0.015	0.060	0.002	0.238	−0.011	50.001
0.020	0.161	0.010	0.379	−0.032	61.053
0.025	0.320	0.031	0.523	−0.068	56.923
0.030	0.520	0.069	0.667	−0.120	41.301
0.035	0.729	0.125	0.815	−0.184	23.971
0.040	0.916	0.190	0.970	−0.254	15.197
0.045	1.062	0.250	1.135	−0.324	20.323
0.050	1.165	0.293	1.298	−0.387	37.338
0.055	1.236	0.312	1.414	−0.438	49.927
0.060	1.282	0.309	1.441	−0.472	44.568
0.065	1.290	0.289	1.381	−0.483	25.575
0.070	1.238	0.260	1.251	−0.464	3.575
0.075	1.107	0.225	1.070	−0.413	−10.486
0.080	0.893	0.183	0.855	−0.330	−10.752
0.085	0.609	0.135	0.613	−0.222	1.075
0.090	0.287	0.080	0.347	−0.099	16.875
0.095	−0.038	0.019	0.055	0.026	26.153
0.100	−0.333	−0.046	−0.257	0.142	21.324

Table 5.4 Influence of the number of eigenmodes considered on the structural response

Modes	max $|u_1|$ [mm]	max $|v_1|$ [mm]	max $|u_2|$ [mm]	max $|v_2|$ [mm]	max N_1 [kN]	min N_1 [kN]
1	1.341	0.307	1.412	0.464	20.01	−20.01
1, 2	1.353	0.312	1.417	0.484	21.24	−18.36
1, 2, 3	1.352	0.309	1.418	0.484	21.63	−18.44
1, 2, 3, 4	1.390	0.313	1.442	0.483	59.47	−46.62
1, 2, 3, 4, 5	1.394	0.312	1.441	0.483	61.05	−48.05

performed up to the eigenmode n_{max} and higher eigenmodes are neglected. It can be seen that the first eigenmode dominates the displacements. The differences with additional consideration of higher eigenmodes are about 3%. For the normal force N_1, however, the fourth eigenmode proves to be significant. If only the first eigenmode is taken into account, the normal force N_1 is underestimated by 67% compared to the complete solution.

Figure 5.28 shows the time histories of the displacements and normal forces obtained by considering only the first eigenmode. The displacement time histories are essentially the same as with the complete solution, even if the slight harmonics—in particular for u_3—are missing. This also applies to the normal forces N_2, N_4, N_5, and N_6, but not to N_1 and N_3, where higher eigenmodes are significantly involved. Basically, the error of internal forces is greater for a lower number of eigenmodes than for displacements, because they are calculated as difference of erroneous (displacement) quantities. ◄

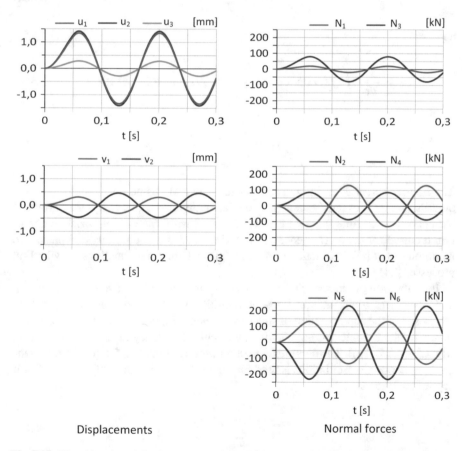

Displacements   Normal forces

Fig. 5.28 Time histories of displacements and normal forces considering the first eigenmode

5.6.4 *Fourier Transformation*

The equations of motion of linear systems can be solved by a Fourier transformation of all displacement and force quantities. The loads and the vibrations excited thereby can be periodic or nonperiodic.

Under certain conditions, a periodic function can be represented by a so-called Fourier series. For a function $f(t)$ and a period T_0 with which the function is repeated, the Fourier series is

$$f(t) = a_0 + 2 \cdot \sum_{k=1}^{\infty} (a_k \cdot \cos(\Omega_k \cdot t) + b_k \cdot \sin(\Omega_k \cdot t)) \tag{5.66}$$

with

$$\Omega_k = \frac{2 \cdot \pi}{T_0} \cdot k \tag{5.66a}$$

and

$$a_0 = \frac{1}{T_0} \cdot \int_{t=0}^{T_0} f(t)\, dt, \quad a_k = \frac{1}{T_0} \cdot \int_{t=0}^{T_0} f(t) \cdot \cos(\Omega_k \cdot t)\, dt, \tag{5.66b}$$

$$b_k = \frac{1}{T_0} \cdot \int_{t=0}^{T_0} f(t) \cdot \sin(\Omega_k \cdot t)\, dt. \tag{5.66c}$$

The Fourier series is an infinite series with sine and cosine terms and represents the function $f(t)$ exactly except at jump discontinuities. For example, an arbitrary but periodic load-time function can be described with (5.66). The Dirichlet condition required for the convergence of the series (piecewise continuity and piecewise monotony of $f(t)$) is practically always fulfilled for time functions of vibration processes [24].

For vibration processes, $f(t)$ describes the time history of a static quantity, e.g. a displacement or an internal force. According to (5.66), the time history is thus composed of the superposition of sine and cosine oscillations. Each Fourier element stands for a harmonic oscillation with the circular frequency Ω_k (comparing (5.41) with (5.26), (5.27a)). The complex notation introduced in Sect. 5.5 for the representation of harmonic oscillations is used for the representation of the Fourier series. It applies as follows:

$$f(t) = \sum_{k=-\infty}^{\infty} \hat{c}_k \cdot e^{i \cdot \Omega_k \cdot t} \tag{5.67}$$

with

$$
\hat{c}_k =
\begin{cases}
a_k - i \cdot b_k & \text{for } k > 0 \\
a_0 & \text{for } k = 0 \\
a_k + i \cdot b_k & \text{for } k < 0
\end{cases}
\tag{5.67a}
$$

The summation is done from $-\infty$ to ∞. Using Euler's formula (5.49), it can be shown that (5.66) and (5.67) are identical. Since in (5.66) the summation begins at 1, in (5.67), after the insertion of (5.67a), the terms for k and $-k$ must be combined. In doing so, it has to be considered that $a_{-k} = a_k$ and $b_{-k} = -b_k$ follow from $\sin(\Omega_{-k} \cdot t) = -\sin(\Omega_k \cdot t)$ and $\cos(\Omega_{-k} \cdot t) = \cos(\Omega_k \cdot t)$, respectively, according to (5.66b, 5.66c).

For nonperiodic functions, the so-called *Fourier transform* is obtained from (5.66) or (5.67) as a limit for $T_0 \to \infty$. The sum in (5.66) or (5.67) then becomes an integral and instead of discrete values Ω_k, a function $\Omega(t)$ is obtained. This form of the Fourier transform is of importance for analytical solutions and will not be pursued further here. For numerical calculations, nonperiodic functions can be represented approximately as a series. However, the period T_0 must then be chosen sufficiently large. With the so-called *Discrete Fourier Transform* or DFT, the period T_0 is also discretized in N time steps. The time step is obtained as

$$
\Delta t = \frac{T_0}{N}
\tag{5.68}
$$

and the function $f(t)$ is available at the points $t = t_j = j \cdot \Delta t$ with $j = 0, 1, \ldots, (N-1)$. Because of $T_0 = N \cdot \Delta t$, the relationship $\Omega_k \cdot t_j = \frac{2 \cdot \pi}{N \cdot \Delta t} \cdot k \cdot j \cdot \Delta t = 2 \cdot \pi \frac{k \cdot j}{N}$ applies.

The coefficients a_k and b_k are determined numerically according to (5.66b, 5.66c) as

$$
a_k = \frac{1}{N} \cdot \sum_{j=0}^{N-1} f(t_j) \cdot \cos\left(2 \cdot \pi \cdot \frac{k \cdot j}{N}\right),
\tag{5.69a}
$$

$$
b_k = \frac{1}{N} \cdot \sum_{j=0}^{N-1} f(t_j) \cdot \sin\left(2 \cdot \pi \cdot \frac{k \cdot j}{N}\right).
\tag{5.69b}
$$

In complex notation, with (5.67a) and Euler's formula

$$
e^{-2 \cdot \pi \cdot i \frac{k \cdot j}{N}} = \cos\left(2 \cdot \pi \cdot \frac{k \cdot j}{N}\right) - i \cdot \sin\left(2 \cdot \pi \cdot \frac{k \cdot j}{N}\right)
$$

the complex Fourier coefficients

$$\hat{c}_k = \frac{1}{N} \cdot \sum_{j=0}^{N-1} f(t_j) \cdot e^{-2\cdot\pi\cdot\mathrm{i}\,\frac{k\cdot j}{N}} \tag{5.70a}$$

are obtained. For the pair of numbers $k = m$ and $k = N - m$ with $e^{-2\cdot\pi\cdot\mathrm{i}} = \cos(2\cdot\pi) - \mathrm{i}\cdot\sin(2\cdot\pi) = 1$, the relationship

$$e^{-2\cdot\pi\cdot\mathrm{i}\cdot(N-m)\cdot j/N} = e^{-2\cdot\pi\cdot\mathrm{i}} \cdot e^{2\cdot\pi\cdot\mathrm{i}\cdot m\cdot j/N} = e^{2\cdot\pi\cdot\mathrm{i}\cdot m\cdot j/N}$$

applies. With Euler's formula, it can be shown that the coefficients $\hat{c}_k$ for $k > N/2$ are repeated in the sense that the real parts are equal, while the imaginary parts have the same absolute value but opposite signs, i.e. are conjugated complex values [2]. In this sense, for example, the value for $j = 1$ corresponds to the value $j = N - 1$.

The function $f(t)$ is obtained at the discretization points with (5.67) as

$$f(t_j) = \sum_{k=0}^{N-1} \hat{c}_k \cdot e^{2\cdot\pi\cdot\mathrm{i}\,\frac{k\cdot j}{N}}. \tag{5.70b}$$

The summation over k is carried out from 0 to $(N - 1)$, which corresponds to the summation from $-N/2$ to $N/2$ due to the above mentioned property of the Fourier coefficients [4]. This means that it is sufficient to determine the values for $0 \geq k \geq N/2$. The Fourier terms for $k > N/2$ then result as conjugated complex values:

$$\hat{c}_k = \mathrm{Re}(\hat{c}_{N-k}) - \mathrm{i}\cdot\mathrm{Im}(\hat{c}_{N-k}) \quad \text{for} \quad N/2 + 1 \leq k \leq N - 1. \tag{5.70c}$$

Equation (5.70b) describes the *Inverse Discrete Fourier Transform* called IDFT.

With (5.70a), the Fourier coefficients of a function $f(t)$ are determined in dependence of the frequency $\Omega_k = 2\pi \cdot k/T_0$. This is called a transformation of $f(t)$ from the time to the frequency domain. With the so-called inverse Fourier transformation, the function is transformed with (5.70b) from the frequency domain into the time domain. Both the frequency domain and the time domain are discretized in the DFT, the frequency domain into the circular frequencies $\Omega_k = 2 \cdot \pi \cdot k/(N \cdot \Delta t)$ or the frequencies $f_k = \Omega_k/(2 \cdot \pi) = k/(N \cdot \Delta t)$ with $k = 0, 1 \ldots, (N - 1)$ and the time domain into the points $t_j = j \cdot \Delta t$ with $j = 0, 1, \ldots, (N - 1)$.

In practice, a particularly efficient numerical technique is used to calculate the Discrete Fourier Transform, the so-called *Fast Fourier Transform* (FFT) by Cooley–Tukey [25]. In the FFT, the number N of discretization points must be an integer power of 2, for example, $2^{10} = 1024$ [4]. This can easily be achieved by adding a missing number of discretization points with a function value of zero to the function.

The resolution in the frequency domain is limited due to the discretization. Since the Fourier coefficients for $k > N/2$ are repeated in the abovementioned sense, the maximum frequency that can be represented is for $k = N/2$. For a discretization in

and $\hat{\underline{v}}_k$ for $k > N/2$ as conjugated complex values (5.70c).

$$\ddot{\underline{u}}(t_j) = \sum_{k=0}^{N-1} \hat{\underline{a}}_k \cdot e^{2 \cdot \pi \cdot i \frac{k \cdot j}{N}} \quad \text{with} \quad \hat{\underline{a}}_k = -\Omega_k^2 \cdot \underline{u}_k$$

and $\hat{\underline{a}}_k$ for $k > N/2$ as conjugated complex values (5.70c).

2. Retransformation of internal forces (coefficients for $k > N/2$ as conjugated complex values (5.70c)).

Example 5.14 The truss system in Example 5.12 is to be examined for an impact load of 100 kN with the time function $F(t) = \tilde{F}(t)$ shown in Fig. 5.25. The hysteretic material damping is $\xi = 2\%$. The influence of an additional damper with the damper constant $c_{damp} = 500$ kNs/m which is attached at node point 1 in the horizontal direction shall be examined (Fig. 5.32). The analysis has to be performed by a Discrete Fourier Transform.

First, the time step and period are specified. The time step must be selected in such a way that both the load and the vibration behavior of the system are represented with sufficient accuracy. Discretizing the duration of the impact load of 0.05 s in 10 time steps results in $\Delta t \leq 0.05/10 = 0.005$ s. The vibration behavior of the system is characterized by the natural frequencies between $f_1 = 7.1$ Hz and $f_5 = 33.1$ Hz which have been determined in Example 5.5. To represent the frequency of 33.1 Hz in the analysis, a time step of at least $\Delta t = 1/(2 \cdot f_{max}) = 1/(2 \cdot 33.1) = 0.015$ s is required. This corresponds to the Nyquist frequency and thus is a very rough representation of a vibration with the frequency f_5. The time step is therefore selected as $\Delta t \approx 0.005$ s. The selection of the period T_0 must ensure that the vibration has practically decayed to zero after its expiration in order to avoid an influence of the subsequent periodic repetition of the vibration function. It will be assumed conservatively that the first natural vibration after time T_0 should have decayed to 1% of the initial amplitude. After (5.37e), one gets

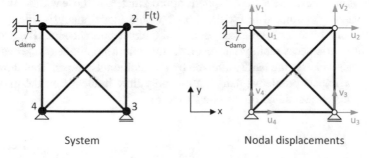

System   Nodal displacements

Fig. 5.32 Truss system with a damper

$$n = \frac{1}{\xi \cdot 2 \cdot \pi} \ln \frac{u_i}{u_{i+n}} = \frac{1}{0.02 \cdot 2 \cdot \pi} \ln 100 = 37 \quad \text{and with it}$$

$$T_0 \geq n \cdot T_1 = 37 \cdot \frac{1}{7.1} \approx 5 \text{ s.}$$

This corresponds to $N = T_0/\Delta t = 5/0.005 = 1000$ time steps. Hence, a period of $T_0 = 5.0$ s and $N = 2^{10} = 1024$ time steps with a size of $\Delta t = 5/1024 = 0.00488$ s is chosen.

According to (5.73) and Example 5.3, the equations of motion the kth term of the Fourier series are obtained as

$$
\left(
2.80 \cdot 10^5 \cdot
\begin{bmatrix}
1.35 & -0.35 & -1.00 & 0 & -0.35 \\
-0.35 & 1.35 & 0 & 0 & 0.35 \\
-1.00 & 0 & 1.35 & 0.35 & 0 \\
0 & 0 & 0.35 & 1.35 & 0 \\
-0.35 & 0.35 & 0 & 0 & 1.35
\end{bmatrix}
\cdot (1 + 2 \cdot i \cdot \xi)
\right.
$$

$$
\left.
+ i \cdot \Omega_k
\begin{bmatrix}
c_{\text{damp}} & 0 & 0 & 0 & 0 \\
0 & 0 & 0 & 0 & 0 \\
0 & 0 & 0 & 0 & 0 \\
0 & 0 & 0 & 0 & 0 \\
0 & 0 & 0 & 0 & 0
\end{bmatrix}
- \Omega_k^2 \cdot
\begin{bmatrix}
20 & 0 & 0 & 0 & 0 \\
0 & 20 & 0 & 0 & 0 \\
0 & 0 & 40 & 0 & 0 \\
0 & 0 & 0 & 40 & 0 \\
0 & 0 & 0 & 0 & 10
\end{bmatrix}
\right)
\cdot
\begin{bmatrix}
\hat{u}_{k,1} \\
\hat{v}_{k,1} \\
\hat{u}_{k,2} \\
\hat{v}_{k,2} \\
\hat{u}_{k,3}
\end{bmatrix}
=
\begin{bmatrix}
0 \\
0 \\
\hat{F}_{k,x,2} \\
0 \\
0
\end{bmatrix} .
$$

Herein, $\Omega_k = k \cdot 2 \cdot \pi/T_0$ represents the circular frequency of the harmonic vibration for the kth coefficient of the Fourier series. On the right side are the complex Fourier coefficients of the force at node 2, which are determined as shown in Example 5.13. The solution of the complex system of equations gives the displacements in frequency domain. From the displacements, other quantities like the accelerations in the frequency domain can be determined. The transformation in time domain results in the time histories of the investigated quantities.

The time histories of the horizontal displacement and acceleration of node 2 are shown in Fig. 5.33a, taking into account the material damping of $\xi = 2\%$ but without considering the additional damper. Due to the material damping, the amplitudes of the displacements and accelerations decay as expected. The maximum vibration amplitude, however, corresponds approximately to those without damping, i.e. with values determined in Example 5.12 with a modal analysis. Figure 5.33b shows the time histories taking the additional damper into account. The vibration time histories now decay rapidly. However, the maximum amplitudes at the beginning of the vibration history are hardly affected by the additional damper. This shows that for shock or impact problems, damping has only little influence on the maximum displacements and sectional forces of a structure.   ◀

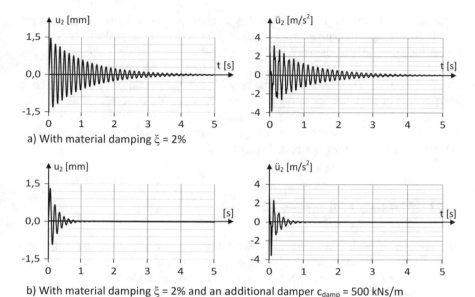

a) With material damping ξ = 2%

b) With material damping ξ = 2% and an additional damper c$_{damp}$ = 500 kNs/m

Fig. 5.33 Time histories of the horizontal displacements and accelerations of nodal point 2

5.7 Earthquake Excitation

5.7.1 Generals

Earthquakes are caused by a sudden rupture in the earth's crust. The rupture occurs when the strength limits of the rock are exceeded in a fault after previous slow displacements and stress redistributions. From the point of rupture, waves propagate in the ground, which manifest themselves as vibrations on the earth's surface. These are measured as acceleration time histories and have one vertical and two horizontal components. They are called free field accelerations and are applied to the foundation of a building as ground motion in dynamic analyses. For larger buildings, the free field acceleration at different points of a foundation slab can be quite different (cf. [3]). Rigid foundations, however, compensate locally different ground displacements and filter out short-wave components of the ground waves that correspond to higher excitation frequencies [26–28]. Therefore, the effect of different multi-point seismic excitations is mostly neglected.

For the seismic design of structures, the vertical seismic acceleration is generally of minor importance. On the one hand, it is usually lower than the horizontal acceleration and on the other hand, a certain degree of seismic safety against vertical forces is already achieved by designing the structure for the vertical dead weight loads taking a safety factor into account. For the seismic design of structures, therefore, only the horizontal earthquake acceleration is considered in general, even if there are cases where the vertical component can have a significant influence [29].

Load assumptions on earthquake loads to be expected at a site possess a large uncertainty compared to other types of loads. This concerns the probability of occurrence, the duration, the intensity, and characteristic time histories of an earthquake. In addition, the structural behavior during an earthquake can only be described with some uncertainty. This applies, for example, to the interaction of non-load-bearing components or the behavior in the nonlinear range. A full coverage of all possible earthquake risks is hardly possible and economically not justifiable. The aim of the design is, therefore, an earthquake design in which some damage is allowed, but a collapse of the supporting structure is avoided.

5.7.2 Methods of Analysis

The analysis of the internal forces and displacements due to seismic excitation can be performed following different methods. These result from the solution of the equations of motion (5.24) of a system excited by a ground motion (Fig. 5.34).

In a *time history analysis*, an acceleration time history is assumed as ground motion. This results in time histories of internal forces and displacements whose maxima are decisive in design. A time history analysis is the most general method to investigate the seismic response of a structure. It is also applicable to nonlinear analyses.

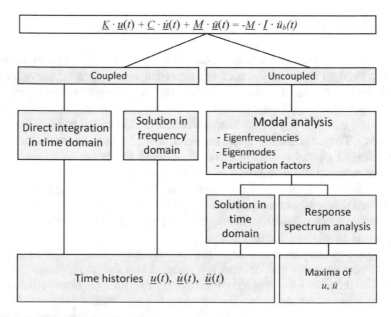

Fig. 5.34 Computational methods for seismic excitation

A *response spectrum analysis* is based on a so-called response spectrum, which comprises the characteristics of a large number of earthquakes. A modal analysis provides the maxima of the earthquake internal forces and displacements in each eigenmode, which are then superimposed using probabilistically based approximation methods (Fig. 5.34). The time histories and the temporal correlation of the internal forces are lost. On the other hand, the computational effort is low and some inaccuracies of the procedure can be accepted because of the uncertainty inherent in the load assumptions.

A different approach is followed with *probabilistic methods*. Here, the uncertainties in the analysis of the vibrational response are described by probability distributions. These uncertainties consist of the geometric and material-related uncertainties of the structural system and the uncertainties of the seismic load. Since the uncertainties of the structural system are orders of magnitude smaller than those of the seismic excitation, they are usually neglected and the analysis is done with deterministic system characteristics. The uncertainty of the seismic characteristics (reference value of the ground acceleration, duration of the earthquake, frequency content, etc.) is, however, described by probabilities of occurrence and distribution functions. As a result of the dynamic analysis, probability distributions of the internal forces and displacements are obtained. In practice, probabilistic methods are rarely used, since the computational effort is high and the required information on the seismic parameters is difficult to obtain [4].

5.7.3 Time History Analysis

For time history analysis, the response of a structure is computed for a particular earthquake time history (Fig. 5.35). For this purpose, natural earthquake time histories obtained from measurements or artificial earthquake time histories generated on the basis of given earthquake characteristics are used. The artificial generation of time histories is based on random vibrations. The random vibrations are generated as a stationary random process of unlimited length and multiplied by a time-dependent envelope curve to obtain a time history of finite duration. The frequency content is adapted to soil conditions at the site by a filtering process. With different methods, it is possible to generate time histories which correspond to a given response spectrum for a given damping coefficient [30, 31].

For time history analyses, the equations of motion according to (5.24)

$$\underline{K} \cdot \underline{u} + \underline{C} \cdot \underline{\dot{u}} + \underline{M} \cdot \underline{\ddot{u}} = -\underline{M} \cdot \left(\underline{I}_x \cdot \ddot{u}_{b,x} + \underline{I}_y \cdot \ddot{u}_{b,y} + \underline{I}_z \cdot \ddot{u}_{b,z} \right)$$

are considered. They can be solved with direct numerical integration methods for both linear and nonlinear systems. For linear systems, however, a modal analysis is often preferred, since its computational effort is considerably lower, provided that the dynamic response can be described by a few eigenmodes.

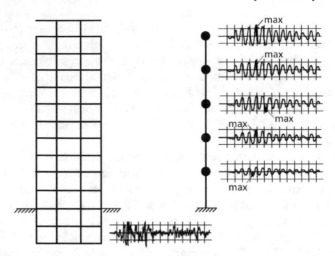

Fig. 5.35 Time history analysis

In the modal analysis, the natural frequencies and eigenmodes are first determined. Thus, one receives the modal equations of motion (5.62) as

$$k_i^* \cdot \alpha_i(t) + c_i^* \cdot \dot{\alpha}_i(t) + m_i^* \cdot \ddot{\alpha}_i(t) = f_i^* \tag{5.74}$$

with

$$f_i^*(t) = \underline{\phi}_i^{\mathrm{T}} \cdot \underline{F}(t) = -\underline{\phi}_i^{\mathrm{T}} \cdot \underline{M} \cdot \left(\underline{I}_x \cdot \ddot{u}_{\mathrm{b},x} + \underline{I}_y \cdot \ddot{u}_{\mathrm{b},y} + \underline{I}_z \cdot \ddot{u}_{\mathrm{b},z} \right). \tag{5.74a}$$

For a single direction of excitation, (5.74) is simplified to

$$f_i^*(t) = \underline{\phi}_i^{\mathrm{T}} \cdot \underline{F}(t) = -\underline{\phi}_i^{\mathrm{T}} \cdot \underline{M} \cdot \underline{I} \cdot \ddot{u}_{\mathrm{b}}(t) = -\Gamma_i \cdot \ddot{u}_{\mathrm{b}}(t) \tag{5.75}$$

where $\underline{I}$ is a topology vector (in x-, y-, or z-direction) and $\ddot{u}_{\mathrm{b}}(t)$ is the ground acceleration time history in the direction of the excitation. The value

$$\Gamma_i = \underline{\phi}_i^{\mathrm{T}} \cdot \underline{M} \cdot \underline{I} \tag{5.75a}$$

is denoted as *participation factor* of the i-ten eigenmode. The solution of the modal equations of motion (5.74) is obtained with numerical integration methods. An analytical solution can be written with the Duhamel integral (5.65) as

$$\alpha_i(t) = -\frac{\Gamma_i}{m_i^* \cdot \omega_{\mathrm{D},i}} \int_0^t \ddot{u}_{\mathrm{b}}(\tau) \cdot e^{-\xi_i \cdot \omega_i \cdot (t-\tau)} \cdot \sin\left(\omega_{\mathrm{D},i} \cdot (t-\tau)\right) \mathrm{d}\tau. \tag{5.76}$$

The time histories of the displacements are obtained by superimposing the eigenmodes according to (5.60). The corresponding internal force time histories are determined by means of the stress or section force matrices of the finite elements.

Example 5.15 For the exhaust tower in Example 5.6, an earthquake time history analysis shall be performed. For this purpose, an acceleration time history recorded during the Albstadt earthquake of January 16, 1978, is employed (Fig. 5.36). It is the North-South component of the recording at the Jungingen station, approximately 11 km away from the epicenter of the earthquake [32, 33]. The earthquake has the magnitude $M_L = 4.6$.

The structural response is computed using modal analysis. The exhaust tower is again represented by a beam structure with 4 masses (Fig. 5.14). In Example 5.6, the lowest three natural frequencies and the eigenmodes normalized with respect to the mass matrix were determined as

$$
\begin{array}{lll}
\omega_1 = 3.471 \text{ Hz} & \omega_2 = 20.40 \text{ Hz} & \omega_3 = 54.03 \text{ Hz} \\
f_1 = 0.552 \text{ Hz} & f_2 = 3.25 \text{ Hz} & f_3 = 8.60 \text{ Hz} \\
T_1 = 1.81 \text{ s} & T_2 = 0.31 \text{ s} & T_3 = 0.12 \text{ s}
\end{array}
$$

and

$$
\underline{\phi}_1 = \begin{bmatrix} 0.592 \\ 0.055 \\ 2.074 \\ 0.089 \\ 4.037 \\ 0.104 \\ 6.172 \\ 0.108 \end{bmatrix} \cdot 10^{-2} \quad
\underline{\phi}_2 = \begin{bmatrix} -2.373 \\ -0.165 \\ -4.325 \\ 0.011 \\ -1.423 \\ 0.264 \\ 5.222 \\ 0.367 \end{bmatrix} \cdot 10^{-2} \quad
\underline{\phi}_3 = \begin{bmatrix} 4.210 \\ 0.127 \\ 1.084 \\ -0.346 \\ -3.862 \\ 0.045 \\ 3.517 \\ 0.531 \end{bmatrix} \cdot 10^{-2}.
$$

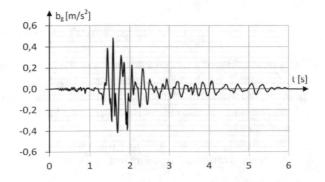

Fig. 5.36 Earthquake acceleration time history (Albstadt, Germany, January 16, 1978, Jungingen station, NS component)

The topology vector $\underline{I}$ contains a 1 in all degrees of freedom in the direction of horizontal ground motion u_b. These are rows 1, 3, 5, and 7, which correspond to the degrees of freedom of displacement v_2, v_3, v_4, and v_5 (Fig. 5.14). All other elements of $\underline{I}$ are zero:

$$\underline{I} = \begin{bmatrix} 1 \\ 0 \\ 1 \\ 0 \\ 1 \\ 0 \\ 1 \\ 0 \end{bmatrix}.$$

First the modal equations are established. The participation factors $\Gamma_i = -\underline{\phi}_i^T \cdot \underline{M} \cdot \underline{I}$ of the eigenmodes according to (5.75a) are

$$\Gamma_1 = -24.47, \quad \Gamma_2 = 13.78, \quad \Gamma_3 = -7.98.$$

The generalized load of the ith eigenmode for earthquake excitation is obtained with (5.75) as

$$f_i^*(t) = \underline{\phi}_i^T \cdot \underline{F}(t) = -\underline{\phi}_i^T \cdot \underline{M} \cdot \underline{I} \cdot \ddot{u}_b(t) = -\Gamma_i \cdot \ddot{u}_b(t)$$

where $\ddot{u}_b(t)$ represents the acceleration time history given in Fig. 5.36.

Thus the modal equations according to (5.74) and (5.64)

$$\omega_i^2 \cdot \alpha_i(t) + 2 \cdot \xi_i \cdot \omega_i \cdot \dot{\alpha}_i(t) + \ddot{\alpha}_i(t) = f_i^*(t)/m_i^*$$

can be solved by numerical integration. In all modes the generalized mass is $m_i^* = 1$ and the damping coefficient is $\xi_i = 0.05$. The solutions $\alpha_i(t)$ are shown in Fig. 5.37. They were determined numerically according to (5.76) with a time step of 0.01 s.

The displacement time histories of all degrees of freedom are obtained with (5.61) as

$$\underline{u}(t) = \alpha_1(t) \cdot \underline{\phi}_1 + \alpha_2(t) \cdot \underline{\phi}_2 + \alpha_3(t) \cdot \underline{\phi}_3.$$

The time histories of the displacements of nodal points 2 and 5 as well as the rotation of nodal point 2 are shown as an example in Fig. 5.38. While at the top of the model (point 5), the first natural frequency dominates in the vibration response, the second natural frequency is also clearly represented in the vibrational response at the lowest point of the model.

The internal forces are obtained from the displacement time histories and the internal force matrices. According to (3.25),

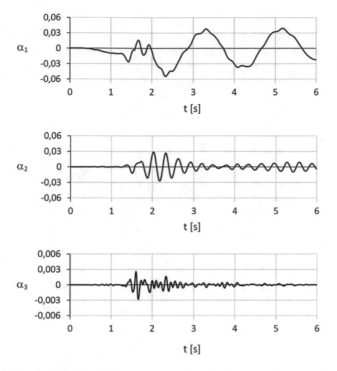

Fig. 5.37 Generalized coordinates $\alpha_1(t)$, $\alpha_2(t)$, and $\alpha_3(t)$

$$
\begin{bmatrix} V_1 \\ M_1 \\ V_2 \\ M_2 \end{bmatrix} = \frac{E \cdot I}{l} \cdot \begin{bmatrix} 12/l^2 & 6/l & -12/l^2 & 6/l \\ -6/l & -4 & 6/l & -2 \\ 12/l^2 & 6/l & -12/l^2 & 6/l \\ 6/l & 2 & -6/l & 4 \end{bmatrix} \cdot \begin{bmatrix} v_1 \\ \varphi_1 \\ v_2 \\ \varphi_2 \end{bmatrix}
$$

applies.

The internal forces of beam element 1 shall be determined at the fixing point. Due to the restraint, $v_1 = \varphi_1 = 0$ applies. Introducing the time histories $u_2(t)$ and $\varphi_2(t)$ shown in Fig. 5.38 for v_2 and φ_2, one obtains (E, I, l, cf. Example 5.6):

$$
\begin{bmatrix} V_1 \\ M_1 \\ V_2 \\ M_2 \end{bmatrix} = \frac{E \cdot I}{l} \cdot \begin{bmatrix} -12/l^2 & 6/l \\ 6/l & -2 \\ -12/l^2 & 6/l \\ 6/l & 4 \end{bmatrix} \cdot \begin{bmatrix} u_2(t) \\ \varphi_2(t) \end{bmatrix}
$$

$$
= \begin{bmatrix} -0.079 & 0.792 \\ 0.792 & -5.280 \\ -0.079 & 0.792 \\ -0.792 & 10.560 \end{bmatrix} \cdot 10^7 \cdot \begin{bmatrix} u_2(t) \\ \varphi_2(t) \end{bmatrix}.
$$

The time histories of the shear force $V_1(t)$ and the bending moment $M_1(t)$ are shown in Fig. 5.39. For both internal forces, the influence of the second mode is apparent.

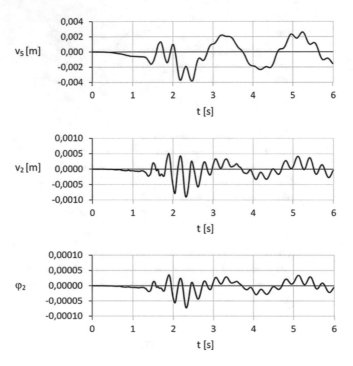

Fig. 5.38 Time histories of the displacements at nodal points 2 and 5 and the rotation at point 2

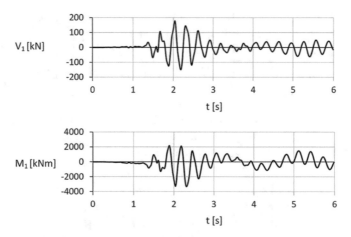

Fig. 5.39 Time histories of the shear force and the bending moment at the fixing point

Their maxima are

$$V_{1,\max} = 179 \text{ kN}, \quad M_{1,\max} = 3307 \text{ kNm.}$$ ◄

5.7.4 Response Spectrum Analysis

The response spectrum method is commonly used for earthquake design of buildings. It is based on modal analysis. The vibration response is thus represented in modal coordinates, i.e. by the eigenmodes of the system. However, unlike in time history analysis, not every modal equation of motion is solved for a specific earthquake time history. Instead, only the maximum of the vibrational response in the respective natural vibration is determined using the so-called response spectrum. The maxima of the eigenmodes are then superimposed according to probabilistic considerations [1, 4, 26, 30].

An important advantage of a response spectrum analysis compared to a time history analysis is that a large number of earthquake time histories are covered by a single response spectrum. The consideration of a multitude of earthquakes can be done with a low computational effort. However, due to the superposition rule of maxima, only approximate solutions are obtained and the method is limited to linear systems. Nonlinear effects can only be considered indirectly by increasing the damping.

First of all, a single degree of freedom system is considered, which is excited by a ground motion. The acceleration time history $\ddot{u}_b(t)$ of a natural earthquake is used as the ground motion. The equation of motion of the system is then according to (5.22)

$$k \cdot u + c \cdot \dot{u} + m \cdot \ddot{u} = -m \cdot \ddot{u}_b$$

where u denotes the relative displacement of the mass to the ground. Dividing the equation by the mass m gives

$$\frac{k}{m} \cdot u + \frac{c}{m} \cdot \dot{u} + \ddot{u} = -\ddot{u}_b$$

and with the natural circular frequency $\omega = \sqrt{k/m}$ and the damping coefficient

$$\xi = \frac{c}{2 \cdot \sqrt{k \cdot m}} = \frac{c}{2 \cdot m \cdot \omega}$$

after (5.15)

$$\omega^2 \cdot u + 2 \cdot \omega \cdot \xi \cdot \dot{u} + \ddot{u} = -\ddot{u}_b. \tag{5.77}$$

The solution of the modal equation then depends only on the natural frequency of the system, the damping coefficient ξ, and the given time history $\ddot{u}_b(t)$. It can be determined numerically by direct integration or with the Duhamel integral according to (5.65) as

$$u(t) = -\frac{1}{\omega_D} \int_0^t \ddot{u}_b(\tau) \cdot e^{-\xi \cdot \omega \cdot (t-\tau)} \cdot \sin(\omega_D \cdot (t - \tau)) \, d\tau. \qquad (5.78)$$

From the relative displacements $u(t)$, one obtains the relative accelerations by differentiating twice with respect to time and the absolute accelerations by adding the ground accelerations. The maximum of the absolute accelerations is

$$S_a(\omega, \xi) = \max|\ddot{u}(\omega, \xi, t) + \ddot{u}_b(t)|. \qquad (5.79a)$$

If the calculation is carried out for different natural frequencies and the maxima of the response acceleration are plotted against the natural frequency $f = \omega/(2 \cdot \pi)$ or the period of vibration $T = 1/f = 2 \cdot \pi/\omega$, a so-called *acceleration response spectrum* is obtained for a given earthquake and a given damping coefficient. An acceleration response spectrum thus shows the maximum response acceleration of a single degree of freedom system as a function of its period of vibration (Fig. 5.40).

Response spectra can also be evaluated for the relative velocity and the relative displacement by the corresponding maximum values

$$S_v(\omega, \xi) = \max|\dot{u}(\omega, \xi, t)| \qquad (5.79b)$$

$$S_d(\omega, \xi) = \max|u(\omega, \xi, t)|. \qquad (5.79c)$$

The relationship between the (relative) displacement, (relative) velocity, and (absolute) acceleration response spectra of an earthquake is defined as

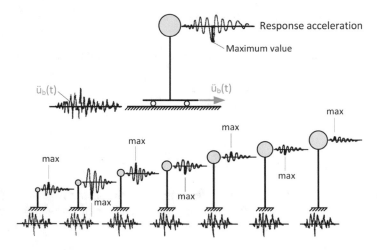

Fig. 5.40 Determination of a response spectrum for seismic excitation [26]

$$S_a = \omega^2 \cdot S_d \qquad (5.80a)$$

$$S_v = \omega \cdot S_d. \qquad (5.80b)$$

For the undamped single degree of freedom system, the relationship (5.80a) is exact as can be shown by introducing $\xi = 0$ into (5.77). From (5.77) follows $-\ddot{u} - \ddot{u}_b = \omega^2 \cdot u$ or $S_a = \max|\ddot{u} + \ddot{u}_b| = \max|\omega^2 \cdot u| = \omega^2 \cdot \max|u| = \omega^2 \cdot S_d$. For the damped single degree of freedom system (5.80a) and (5.80b) are only valid approximately. Therefore, the spectral velocity S_v and spectral acceleration S_a are referred to as *pseudo velocity* and *pseudo acceleration*, respectively [2, 4].

For a rigid system with $k \to \infty$, the circular frequency tends to $\omega = \sqrt{k/m} \to \infty$ and the period of vibration to $T \to 0$. The motion of the mass is identical with that of the ground, so that for $T \to 0$ the response acceleration is equal to the maximum *free field acceleration* of the ground. The limit value $S_a(T \to 0, \xi) = S_a(T \to 0) = \max|\ddot{u}_b(t)|$ is independent of the damping coefficient.

Example 5.16 For the acceleration time history in Fig. 5.36, which was recorded during the Albstadt earthquake in January 16, 1978, at the Jungingen station (Germany), the acceleration response spectrum for a damping coefficient of 5% is to be determined.

The response spectrum shall be evaluated in the range of the natural vibration periods T between 0 and 1.0 s. As step size on the ordinate $1/100 = 0.010$ s is selected, the calculations are to be performed for 100 natural periods.

For each period, the equation of motion (5.77) is solved with the given time history $\ddot{u}_b(t)$. The solution is carried out with the Newmark method (Sect. 5.6.2). Alternatively, efficient methods for computing the Duhamel integral can be used [2, 4].

The time histories $u(t)$ for $T = 0.2$ s and $T = 0.5$ s are shown as an example in Fig. 5.41. The maximum values of the displacement time histories are $1.407 \cdot 10^{-3}$m

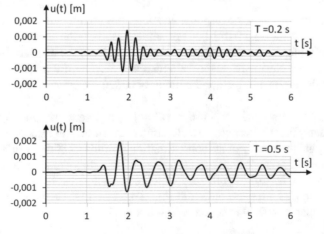

Fig. 5.41 Vibrational response for two single degree of freedom systems with $\xi = 5\%$

Fig. 5.42 Acceleration response spectra of a recorded earthquake acceleration time history for 2, 5, 10, and 15% damping (Albstadt, Germany, January 16, 1978, station Jungingen, NS component)

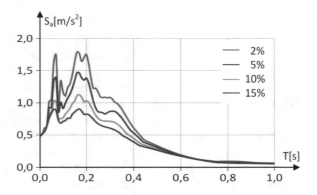

Table 5.7 Maxima of the response acceleration of a single degree of freedom system with different periods and 5% damping

	T [s]					
	0.000	0.100	0.175	0.200	0.300	0.400
S_d [mm]	0.000	0.235	1.111	1.407	1.961	2.140
$\omega^2 \cdot S_d$ [m/s²]	0.482	0.928	1.443	1.388	0.860	0.528
S_a [m/s²]	0.482	0.912	1.441	1.383	0.865	0.534
	T [s]					
	0.500	0.600	1.000	1.500	1.800	2.000
S_d [mm]	1.946	1.658	1.493	2.148	2.229	2.287
$\omega^2 \cdot S_d$ [m/s²]	0.307	0.182	0.059	0.038	0.027	0.023
S_a [m/s²]	0.310	0.184	0.060	0.039	0.028	0.024

for $T = 0.2$ s and $1.946 \cdot 10^{-3}$m for $T = 0.5$ s. Thus, the spectral values of the response acceleration according to (5.80a) are $1.407 \cdot 10^{-3} \cdot (2 \cdot \pi / 0.2)^2 = 1.388$ m/s² for $T = 0.2$ s and $1.946 \cdot 10^{-3} \cdot (2 \cdot \pi / 0.5)^2 = 0.307$ m/s² for $T = 0.5$ s. The response spectrum, where the response accelerations are plotted over the periods T, is shown in Fig. 5.42 from 0 to 1.0 s. For some periods T, the spectral accelerations are given in Table 5.7 for 5% damping. For comparison, the accelerations directly obtained with the Newmark method are also shown. The differences between the spectral accelerations and the accelerations obtained by direct integrations can be neglected.

Figure 5.42 shows the response spectra for different damping ratios. With increasing damping, the response accelerations decrease. The response spectra can as well illustrate the quality of the approximation formula (5.80c) given below. Thus, for a damping of 2% according to (5.80c), $\eta = 1.2$ and thus a maximum acceleration of $1.2 \cdot 1.46 = 1.75$ m/s² is obtained which corresponds well with the maximum acceleration of 1.79 m/s² according to Fig. 5.42. For 15% damping, 1.03 m/s² is obtained compared to 0.90 m/s² according to Fig. 5.42.   ◀

Response spectra of individual earthquakes are significantly influenced by the randomness of the recorded time history and show pronounced peaks and valleys. To compensate for these randomness, the response spectra are determined for a large number of recorded time histories that are averaged and smoothed according to probabilistic criteria. Depending on the smoothing parameters, one obtains response spectra with different exceedance probabilities [26, 27]. The procedure is summarized as follows:

Computation of an earthquake response spectrum

1. Selection of the damping coefficient for which the response spectrum is to be determined, e.g. 5%.
2. Selection of a series of characteristic, recorded acceleration time histories of earthquakes for the region considered.
3. Computation of the time history of absolute acceleration $\ddot{u}(t)$ of the mass of the single degree of freedom system with a period T by numerical integration, e.g. with the Newmark method.
4. Repetition of step 3 for other periods (e.g. $T = 0.01 - 5.00$ s).
5. Plotting of the maximum values $S_a = \max(\ddot{u}(t))$ over the period T of the single degree of freedom system gives the acceleration response spectrum for an individual earthquake and a specific damping coefficient ξ.
6. Repetition of steps 2–5 for other relevant earthquakes.
7. Smoothing of the spectra for a given exceedance probability. The result is a smoothed acceleration response spectrum for several earthquakes and for a specific damping factor ξ. If the response spectra for other damping coefficients ξ are required, steps 1–7 are to be repeated.

In standards for the seismic design of buildings, normalized response spectra are specified. Figure 5.43 shows the horizontal elastic response spectrum according to DIN EN 1998 (Eurocode 8) and the national annex valid in Germany [34]. The corner periods T_B, T_C and T_D and thus their shapes are influenced by the geological subsoil and the ground type at the site (Table 5.8). The design ground acceleration a_{gR} is 0.4 m/s^2 in seismic zone 1, 0.6 m/s^2 in seismic zone 2, and 0.8 m/s^2 in seismic zone 3. The importance factor γ_1 takes values between 0.8 and 1.4. It takes into account the importance of a structure for the protection of the general public. The elastic response spectrum applies to a damping of 5% and can be adapted with the damping correlation factor

$$\eta = \sqrt{10/(5 + \xi)} \geq 0.55 \qquad (5.80c)$$

to other damping coefficients ξ. For further details, reference is made to DIN EN 1998 [34] or [35, 36]. The ductile behavior of structures is approximately taken into account in Eurocode 8 by modifying the elastic response spectrum. For this purpose, in the equations in Fig. 5.43 the damping correction factor η is replaced by $1/q$.

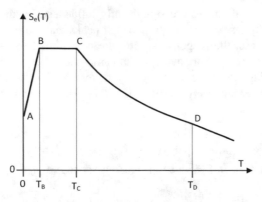

Parameters

a_{gR} : design ground acceleration

$\eta \triangleq 1/q$: damping correlation factor

($\eta = 1$ for 5% damping)

$\beta_0 = 2.5$

γ_I :   Importance factor

S :   Soil factor

$$0 \le T \le T_B\!: S_e(T) = a_{gR} \cdot \gamma_I \cdot S \cdot \left[1 + \frac{T}{T_B} \cdot (\eta \cdot \beta_0 - 1)\right] \qquad T_C \le T \le T_D\!: S_e(T) = a_{gR} \cdot \gamma_I \cdot S \cdot \eta \cdot \beta_0 \cdot \frac{T_C}{T}$$

$$T_B \le T \le T_C\!: S_e(T) = a_{gR} \cdot \gamma_I \cdot S \cdot \eta \cdot \beta_0 \qquad\qquad T_D \le T\!: \qquad S_e(T) = a_{gR} \cdot \gamma_I \cdot S \cdot \eta \cdot \beta_0 \cdot \frac{T_C \cdot T_D}{T^2}$$

Fig. 5.43 Elastic acceleration spectrum according to DIN EN 1998 [34]

Table 5.8 Parameters of the elastic horizontal acceleration spectrum

	A-R	B-R	C-R	B-T	C-T	C-S
S	1.00	1.25	1.50	1.00	1.25	0.75
T_B [s]		0.05			0.10	
T_C [s]	0.20	0.25	0.30	0.30	0.40	0.50
T_D [s]			2.0			
		A	Firm to medium-firm soil $v_s > 800$ m/s			
Building ground classes		B	Loose soil (gravel to coarse sands, marls) 350 m/s $< v_s \le 800$ m/s			
		C	Fine-grained soil (fine sand, loesses) 150 m/s $< v_s \le 350$ m/s			
Subsoil classes		R	Rock			
		T	Transition between R and S			
		S	Sediments			

The behavior coefficient q depends on the ductility of the structure. Such spectra are called *inelastic response spectra*.

Employing the acceleration response spectrum, the seismic load of a single-mass vibrator can be easily determined as

$$H = m \cdot \max|\ddot{u}(\omega, \xi, t) + \ddot{u}_b(t)| = m \cdot S_a(\omega, \xi). \tag{5.81}$$

For finite element systems with multiple degrees of freedom and masses, the natural frequencies and eigenmodes are computed first. With them, one receives the modal equations of motion (5.62) as

$$k_i^* \cdot \alpha_i(t) + c_i^* \cdot \dot{\alpha}_i(t) + m_i^* \cdot \ddot{\alpha}_i(t) = f_i^*(t) \tag{5.82}$$

and for excitation in a single direction

$$f_i^*(t) = \underline{\phi}_i^T \cdot \underline{F}(t) = -\underline{\phi}_i^T \cdot \underline{M} \cdot \underline{I} \cdot \ddot{u}_b(t) = -\Gamma_i \cdot \ddot{u}_b(t) \tag{5.82a}$$

and

$$k_i^* \cdot \alpha_i(t) + c_i^* \cdot \dot{\alpha}_i(t) + m_i^* \cdot \ddot{\alpha}_i(t) = -\Gamma_i \cdot \ddot{u}_b(t). \tag{5.82b}$$

The value

$$\Gamma_i = \underline{\phi}_i^T \cdot \underline{M} \cdot \underline{I} \tag{5.82c}$$

is the participation factor of the ith eigenmode.

If several orthogonal directions of excitation are to be considered simultaneously, the internal forces in the individual directions are determined separately using the response spectrum method and the results are approximately superimposed with suitable superposition formulae, such as those given in DIN EN 1998 [34].

The solutions $\alpha_i(t)$ of the modal equations of motion are obtained in the time history method by direct integration for a given time history of the ground acceleration $\ddot{u}_b(t)$. In the response spectrum method, however, only their maximum values are determined, i.e.

$$\alpha_{i,\max} = \max|\alpha(t)| = \frac{\Gamma_i}{m_i^*} \cdot S_d(\omega_i, \xi) = \frac{\Gamma_i}{m_i^* \cdot \omega_i^2} \cdot S_a(\omega_i, \xi_i). \tag{5.82d}$$

The values S_d and S_a represent the response spectra of the displacements and accelerations according to (5.79a–5.79c) and (5.80a–5.80b), respectively, as it can be checked easily by comparing the coefficients of (5.22) and (5.82b). In general, the acceleration response spectrum is used. Its values S_a depend on the period T_i of the ith eigenmode and the damping coefficient ξ_i. Normally, the damping coefficient ξ_i is the same for all eigenmodes. However, it is also possible to consider a different damping coefficient ξ_i for each natural vibration i. This type of damping is called *modal damping*.

The maximum displacements in the ith mode are thus

$$\underline{u}_{i,\max} = \alpha_{i,\max} \cdot \underline{\phi}_i^T. \tag{5.83}$$

Fig. 5.44 Superposition of
time histories of different
modes [26]

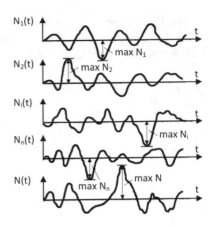

From this, the maximum internal forces can be determined using the sectional force
matrices. The maximum accelerations are

$$a_{i,\max} = \omega_i^2 \cdot u_{i,\max}. \tag{5.84}$$

The response spectrum method provides maxima of the internal forces and
displacements in each mode. However, the times at which the maxima occur in
each eigenmode are not specified. The question arises on how the maxima computed
can be superimposed. Summing up the contributions of each eigenmode is not a
resonable option because the maxima occur at different times (Fig. 5.44). Rather,
superposition rules based on probabilistic theory must be applied.

A classical superposition formula is the *Square Root of Sum of Squares* (SRSS).
Denoting the maximum value of an internal force or a displacement in the ith mode
by N_i, the superimposed value N becomes

$$N = \sqrt{\sum_{i=1}^{m} N_i^2} \tag{5.85}$$

where m is the number of modes considered. This formula assumes that the individual eigenmodes are excited independently of each other. Then the probability
of exceeding the total quantity N is equal to the probability of exceeding a partial
quantity N_i.

It can then be assumed that the individual eigenmodes are excited independently
of each other if the distance between the individual eigenfrequencies is sufficiently
large. If, on the other hand, the eigenfrequencies are located close together, the
conditions for applying the SRSS superposition are no longer fulfilled. For this case,
various superposition rules have been developed on a probabilistic basis. They take
into account the modal interaction between the eigenmodes. With the *Complete
Quadratic Combination method* (CQC), the total quantity N is determined as

$$N = \sqrt{\sum_{i=1}^{m} \sum_{j=1}^{m} N_i \cdot \rho_{ij} \cdot N_j} \tag{5.86}$$

with the cross-modal coefficient

$$\rho_{ij} = \frac{8 \cdot \sqrt{\xi_i \cdot \xi_j} \cdot \left(\xi_i + r_{ij} \cdot \xi_j\right) \cdot r_{ij}^{3/2}}{\left(1 - r_{ij}^2\right)^2 + 4 \cdot \xi_i \cdot \xi_j \cdot r_{ij} \cdot \left(1 + r_{ij}^2\right) + 4 \cdot \left(\xi_i^2 + \xi_j^2\right) \cdot r_{ij}^2}. \tag{5.86a}$$

Here,

$$r_{ij} = \omega_j/\omega_i = f_j/f_i \tag{5.86b}$$

is the ratio of the eigenfrequencies. The parameters ξ_i and ξ_j are the damping coefficients of the modes i and j [37]. If the damping coefficients are the same for all natural vibrations, (5.86a) is simplified with the uniform damping coefficient ξ to

$$\rho_{ij} = \frac{8 \cdot \xi^2 \cdot \left(1 + r_{ij}\right) \cdot r_{ij}^{3/2}}{\left(1 - r_{ij}^2\right)^2 + 4 \cdot \xi^2 \cdot r_{ij} \cdot \left(1 + r_{ij}\right)^2}. \tag{5.86c}$$

The cross-modal coefficients ρ_{ij} are plotted as a function of the ratio $r_{ij} = \omega_j/\omega_i$ of the natural frequencies for various damping coefficients in Fig. 5.45. For clearly separated natural frequencies, the cross-modal coefficient is $\rho_{ij} \approx 0$, i.e. the coupling is negligible. With conventional damping coefficients, an influence exists only in the range $0.5 < f_i/f_j < 2.0$ and especially for closely spaced natural frequencies with $r_{ij} = f_j/f_i \approx 1$. The prerequisites of (5.86a), namely that the earthquake duration is considerably longer than the fundamental period and that the response spectrum is sufficiently smooth, are always fulfilled.

The values obtained after (5.85) and (5.86) are absolute maxima and are to be used as design values in both, positive and negative directions.

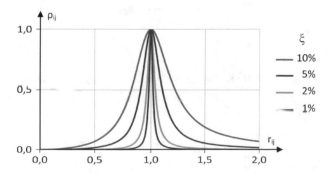

Fig. 5.45 Cross-modal coefficients [37]

Example 5.17 The exhaust tower in Example 5.6 is subjected to an earthquake with the response spectrum shown in Fig. 5.42 for a damping ratio of 5%. The total internal forces have to be determined for the beam model. For comparison, a calculation with the response spectrum according to DIN EN 1998 shall be performed. The computation is done in the units kN, m, and s.

The modal analysis is performed with three eigenmodes. The eigenfrequencies and eigenmodes of the system discretized into four elements and nodal masses have been determined in Example 5.6 and the participation factors in Example 5.15. They are

$$\omega_1 = 3.471 \text{ Hz} \qquad \omega_2 = 20.40 \text{ Hz} \qquad \omega_3 = 54.03 \text{ Hz}$$

$$T_1 = 1.81 \text{ s} \qquad T_2 = 0.31 \text{ s} \qquad T_3 = 0.12 \text{ s}$$

$$\underline{\phi}_1 = \begin{bmatrix} 0.592 \\ 0.055 \\ 2.074 \\ 0.089 \\ 4.037 \\ 0.104 \\ 6.172 \\ 0.108 \end{bmatrix} \cdot 10^{-2} \quad \underline{\phi}_2 = \begin{bmatrix} -2.373 \\ -0.165 \\ -4.325 \\ 0.011 \\ -1.423 \\ 0.264 \\ 5.222 \\ 0.367 \end{bmatrix} \cdot 10^{-2} \quad \underline{\phi}_3 = \begin{bmatrix} 4.210 \\ 0.127 \\ 1.084 \\ -0.346 \\ -3.862 \\ 0.045 \\ 3.517 \\ 0.531 \end{bmatrix} \cdot 10^{-2}$$

$$\Gamma_1 = -24.47 \qquad \Gamma_2 = 13.78 \qquad \Gamma_3 = -7.98.$$

The eigenmodes are normalized in such a way that $m_i^* = 1$ applies to all eigenmodes. From the response spectrum in Fig. 5.42 and Table 5.7, the response accelerations in the three natural periods are obtained as

$$S_a(1.81) = 0.027 \text{ m/s}^2, \quad S_a(0.31) = 0.873 \text{ m/s}^2, \quad S_a(0.12) = 1.040 \text{ m/s}^2.$$

With it, all values are known to determine the maxima of the generalized coordinates with (5.82d) or

$$\alpha_{i,\text{max}} = \frac{\Gamma_i}{m_i^* \cdot \omega_i^2} \cdot S_a(T_i)$$

as

$$\alpha_{1,\text{max}} = -0.055, \quad \alpha_{2,\text{max}} = 0.029, \quad \alpha_{3,\text{max}} = -0.0028.$$

The maximum displacements in the eigenmodes are obtained with (5.83) as

$$
\underline{u}_{1,\max} = \alpha_{1,\max} \cdot \underline{\phi}_1 =
\begin{bmatrix}
-0.325 \\
-0.030 \\
-1.137 \\
-0.049 \\
-2.214 \\
-0.057 \\
-3.385 \\
-0.059
\end{bmatrix} \cdot 10^{-3},
$$

$$
\underline{u}_{2,\max} = \alpha_{2,\max} \cdot \underline{\phi}_2 =
\begin{bmatrix}
-0.686 \\
-0.035 \\
-1.250 \\
0.003 \\
-0.411 \\
0.076 \\
1.509 \\
0.106
\end{bmatrix} \cdot 10^{-3}
$$

$$
\underline{u}_{3,\max} = \alpha_{3,\max} \cdot \underline{\phi}_3 =
\begin{bmatrix}
-0.120 \\
-0.004 \\
-0.031 \\
0.010 \\
0.110 \\
-0.001 \\
-0.100 \\
-0.015
\end{bmatrix} \cdot 10^{-3}.
$$

The superimposed displacements are determined with (5.85). For example, the displacement of point 5 at the upper end of the beam (Fig. 5.14) is obtained as

$$
v_5 = \sqrt{(-3.385)^2 + (1.509)^2 + (-0.100)^2} \cdot 10^{-3} = 3.708 \cdot 10^{-3} \text{ m.}
$$

It can be seen that the first eigenmode is dominant in the vibration response of the uppermost point, while the third eigenmode can be practically neglected. The value corresponds in good approximation to the value of $3.8 \cdot 10^{-3}$ m determined in the time history analysis (Fig. 5.38).

For the other degrees of freedom, the following maxima result:

$$\underline{u}_{\max} = \begin{bmatrix} 0.768 \\ 0.057 \\ 1.690 \\ 0.050 \\ 2.255 \\ 0.095 \\ 3.708 \\ 0.122 \end{bmatrix} \cdot 10^{-3}.$$

Like the displacements, the maximum internal forces are first determined in the individual eigenmodes and then superimposed. The internal forces related to the eigenmodes are obtained using the internal force matrices and the maximum displacements $\underline{u}_{i,\max}$. As an example, the shear force and the bending moment at the base are determined. According to (3.25), the relationship between the displacements and the internal forces of element 1 is (E, I, l, cf. Example 5.6)

$$\begin{bmatrix} V_1 \\ M_1 \\ V_2 \\ M_2 \end{bmatrix} = \frac{E \cdot I}{l} \cdot \begin{bmatrix} 12/l^2 & 6/l & -12/l^2 & 6/l \\ -6/l & -4 & 6/l & -2 \\ 12/l^2 & 6/l & -12/l^2 & 6/l \\ 6/l & 2 & -6/l & 4 \end{bmatrix} \cdot \begin{bmatrix} v_1 \\ \varphi_1 \\ v_2 \\ \varphi_2 \end{bmatrix}.$$

With $v_1 = \varphi_1 = 0$ at nodal point 1 and the maxima $\underline{u}_{i,\max}$ given above for v_2 and φ_2, one obtains

$$\begin{bmatrix} V_{i,1,\max} \\ M_{i,1,\max} \\ V_{i,2,\max} \\ M_{i,2,\max} \end{bmatrix} = \frac{E \cdot I}{l} \cdot \begin{bmatrix} -12/l^2 & 6/l \\ 6/l & -2 \\ -12/l^2 & 6/l \\ -6/l & 4 \end{bmatrix} \cdot \begin{bmatrix} v_{i,2,\max} \\ \varphi_{i,2,\max} \end{bmatrix}$$

$$= \begin{bmatrix} -0.079 & 0.792 \\ 0.792 & -5.280 \\ -0.079 & 0.792 \\ -0.792 & 10.56 \end{bmatrix} \cdot 10^7 \cdot \begin{bmatrix} v_{i,2,\max} \\ \varphi_{i,2,\max} \end{bmatrix}$$

and hence V_1 and M_1 in the three modes are:

Eigenmode 1:

$v_{1,2,\max} = -0.325 \cdot 10^{-3}$ m, $\varphi_{1,2,\max} = -0.03 \cdot 10^{-3}$
$V_{1,1,\max} = 16$ kN $M_{1,1,\max} = -965$ kNm

Eigenmode 2:

$v_{2,2,\max} = -0.686 \cdot 10^{-3}$ m, $\varphi_{2,2,\max} = -0.048 \cdot 10^{-3}$
$V_{2,1,\max} = 166$ kN $M_{2,1,\max} = -2914$ kNm

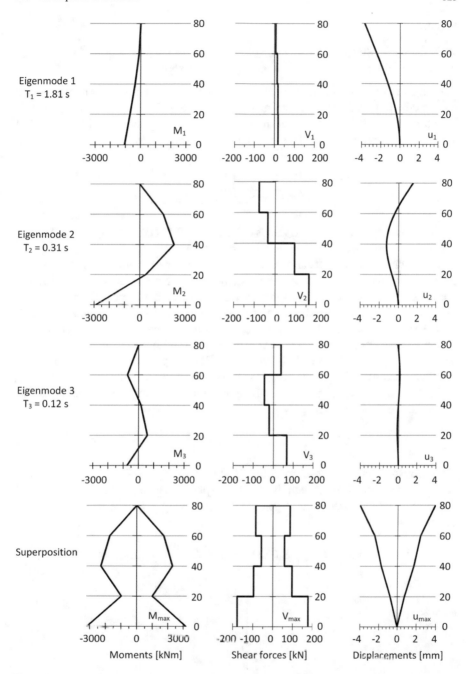

Fig. 5.46 Modal displacements and internal forces with superposition for a recorded earthquake

Eigenmode 3:

$$v_{3,2,\max} = -0.120 \cdot 10^{-3} \text{ m}, \qquad \varphi_{3,2,\max} = -0.004 \cdot 10^{-3}$$
$$V_{3,1,\max} = 66 \text{ kN} \qquad\qquad M_{3,1,\max} = -756 \text{ kNm}$$

The largest contributions to the shear force and moment result from the second eigenmode. In the response spectrum according to Fig. 5.42, the second eigenmode with a period of 0.31 s, is excited more than the first eigenmode with a period of 1.81 s. The superposition of the eigenmodes gives the maxima

$$V_{1,\max} = \sqrt{16^2 + 166^2 + 66^2} = 179 \text{ kN},$$
$$M_{1,\max} = \sqrt{965^2 + 2914^2 + 756^2} = 3162 \text{ kNm}.$$

The maxima agree well with the values of $V_{1,\max} = 179$ kN and $M_{1,\max} = 3307$ kNm calculated in Example 5.15 with the time history method.

All modal displacements and internal forces of the system as well as their super-positions are shown in Fig. 5.46. The superposed displacements and internal forces are to be understood in terms of design values and do not represent internal force distributions that are statically related to each other (e.g. shear forces and moments). As absolute design variables, they are given in positive and negative direction.

For comparison, the exhaust tower is investigated with the elastic response spectrum given in DIN EN 1998 [34] with the following parameters: $a_{gR} = 0.8$ m/s^2 (earthquake zone 3, Jungingen site), importance factor $\gamma_1 = 1$ (importance class II), and subsoil according to category B-R (gravel/rocky subsoil). The damping is 5%. According to Table 5.8, one gets

$$S = 1.25, \quad T_B = 0.05 \text{ s}, \quad T_C = 0.25 \text{ s}, \quad T_D = 2.00 \text{ s}.$$

The corresponding response spectrum is shown in Fig. 5.47. It exhibits significantly higher accelerations than the response spectrum given in Fig. 5.42 (with 5% damping). For the three natural periods, the following response accelerations are obtained:

$$S_a(1.81) = 0.345 \text{ m/s}^2, \quad S_a(0.31) = 2.03 \text{ m/s}^2, \quad S_a(0.12) = 2.50 \text{ m/s}^2$$

Fig. 5.47 Acceleration response spectrum according to DIN EN 1998

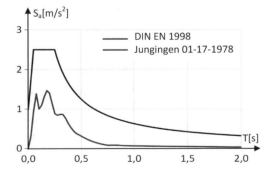

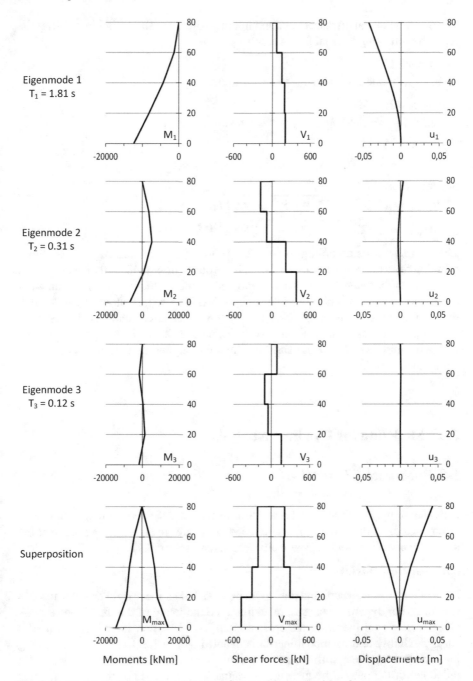

Fig. 5.48 Modal displacements and internal forces with superposition for a response spectrum according to DIN EN 1998

The resulting modal displacements and internal forces are shown in Fig. 5.48. At the base, the following internal forces result:

Eigenmode 1:
$$V_{1,1,max} = 208 \text{ kN} \quad M_{1,1,max} = 12327 \text{ kNm}$$

Eigenmode 2:
$$V_{2,1,max} = 385 \text{ kN} \quad M_{2,1,max} = 6771 \text{ kNm}$$

Eigenmode 3:
$$V_{3,1,max} = 159 \text{ kN} \quad M_{3,1,max} = 1817 \text{ kNm}$$

Superposition:
$$V_{1,max} = \sqrt{207^2 + 385^2 + 159^2} = 465 \text{ kN},$$
$$M_{1,max} = \sqrt{12327^2 + 6771^2 + 1817^2} = 14181 \text{ kNm}$$

The internal forces according to DIN EN 1998 are 3–4.5 times higher than those resulting from the recorded time history. This is due to the significantly lower spectral values of the measured earthquake compared to the earthquake to be assumed at the Jungingen site according to DIN EN 1998 (Fig. 5.47). It should be noted that the earthquake with the magnitude $M_L = 4.6$ measured in January 16, 1978, is not a maximum earthquake event. At the same site, a much stronger earthquake with a magnitude of $M_L = 5.4$ and a considerably higher accelerations occurred in September 3, 1978.   ◀

5.8 Modeling for Dynamic Analysis

5.8.1 Structural Model

In dynamic analyses in addition to the stiffnesses, the masses and damping parameters are modeled. In the following, recommendations are given which supplement the modeling considerations for static analyses given in Sect. 4.11.

Stiffnesses and masses

For static analyses, beams and plates can be modeled as shear rigid or shear flexible elements. For dynamic analyses, in addition the modeling of the translational mass and the rotational inertia of the cross section has to be considered. A beam with shear flexibility and rotational inertia is denoted as the *Timoshenko beam*. Taking into account the shear stiffness and rotational inertia of a cross section results in a reduction of the natural frequencies of beams. The influence of the shear stiffness is noticeable for thick beams, whereas it is small and can be neglected for slender beams. It is greater for higher eigenmodes than for lower ones. Beams with low

Table 5.9 Eigenfrequencies and eigenmodes of shear rigid beams [1, 35]

Support condition	Parameters	Eigenmode
	$\alpha_i = i \cdot \pi$	$\phi_i(x) = \sin\left(\frac{i \cdot \pi \cdot x}{l}\right)$
	$\alpha_1 = 1.88$ $\alpha_2 = 4.69$ $\alpha_3 = 7.85$ $\alpha_4 = 11.00$	$\phi_i(x) = \sin(\alpha_i \cdot x/l) - \sinh(\alpha_i \cdot x/l) + (\cosh(\alpha_i \cdot x/l) - \cos(\alpha_i \cdot x/l)) \cdot \dfrac{\sinh(\alpha_i) + \sin(\alpha_i)}{\cosh(\alpha_i) + \cos(\alpha_i)}$
	$\alpha_1 = 3.93$ $\alpha_2 = 7.07$ $\alpha_3 = 10.21$ $\alpha_4 = 13.35$	$\phi_i(x) = \sinh(\alpha_i \cdot x/l) - \sin(\alpha_i \cdot x/l) - (\cosh(\alpha_i \cdot x/l) - \cos(\alpha_i \cdot x/l)) \cdot \dfrac{\sinh(\alpha_i) - \sin(\alpha_i)}{\cosh(\alpha_i) - \cos(\alpha_i)}$
	$\alpha_1 = 4.73$ $\alpha_2 = 7.85$ $\alpha_3 = 11.00$ $\alpha_4 = 14.14$	$\phi_i(x) = \sin(\alpha_i \cdot x/l) - \sinh(\alpha_i \cdot x/l) + (\cosh(\alpha_i \cdot x/l) - \cos(\alpha_i \cdot x/l)) \cdot \dfrac{\sinh(\alpha_i) - \sin(\alpha_i)}{\cosh(\alpha_i) - \cos(\alpha_i)}$

E Modulus of elasticity,
I Area moment of inertia,
l Length of the beam,
$\bar{m}$ Distributed mass (mass per unit length)

Eigenfrequencies

$$f_i = \frac{1}{2\pi} \cdot \frac{\alpha_i^2}{l^2} \cdot \sqrt{\frac{EI}{\bar{m}}} \qquad i = 1, 2, 3 \ldots$$

Table 5.10 Eigenfrequencies of an elastically clamped beam

$$f_i = \frac{1}{2\pi} \cdot \frac{\alpha_i^2}{l^2} \cdot \sqrt{\frac{EI}{\bar{m}}} \qquad i = 1, 2, 3 \ldots \qquad k_b = \frac{EI}{k_\varphi \cdot l}$$

k_b	0.00	0.10	0.20	0.40	0.60	0.80	1.00	1.50	2.00	3.00	4.00	5.00	$k_b > 5$
α_1	1.88	1.72	1.62	1.47	1.38	1.30	1.25	1.15	1.08	0.98	0.92	0.87	$\approx \sqrt[4]{\dfrac{3}{k_b}}$
α_2	4.69	4.40	4.27	4.14	4.09	4.05	4.03	4.00	3.98	3.96	3.96	3.95	≈ 3.93
α_3	7.85	7.45	7.32	7.22	7.17	7.15	7.13	7.11	7.10	7.09	7.09	7.08	≈ 7.07
α_4	11.00	10.52	10.40	10.32	10.29	10.27	10.26	10.24	10.23	10.23	10.22	10.22	≈ 10.21

Table 5.11 Eigenfrequencies and eigenmodes of rectangular shear rigid plates in bending [1, 35]

Support conditions	Eigenfrequencies and eigenmodes
 Simply supported plate (all edges)	$$\alpha_{i,j} = \frac{\pi}{2} \cdot \left(i^2 + \frac{a^2}{b^2} \cdot j^2 \right)$$ Eigenmodes $$\phi_{i,j} = \sin\left(\frac{i \cdot \pi \cdot x}{a} \right) \cdot \sin\left(\frac{j \cdot \pi \cdot y}{b} \right)$$
 Clamped plate (all edges)	$$\alpha_{1,1} = 1{,}57 \cdot \sqrt{5{,}14 + 3{,}13 \cdot (a/b)^2 + 5{,}14 \cdot (a/b)^4}$$ $$\alpha_{2,1} = 9{,}82 \cdot \sqrt{1 + 0{,}298 \cdot (a/b)^2 + 0{,}132 \cdot (a/b)^4}$$ $$\alpha_{12} = 1{,}57 \cdot \sqrt{5{,}14 + 11{,}65 \cdot (a/b)^2 + 39{,}06 \cdot (a/b)^4}$$
E Modulus of elasticity, t Plate thickness μ Poisson ratio $\tilde{m}$ Distributed mass (mass per unit area)	Eigenfrequencies $$f_{i,j} = \frac{\alpha_{i,j}}{a^2} \cdot \sqrt{\frac{E \cdot t^3}{12 \cdot (1 - \mu^2) \cdot \tilde{m}}}$$ with $i, j = 1, 2, 3 \ldots$

slenderness should be modeled with shear flexible beam elements. The influence of the rotational mass of a cross section can usually be neglected, except at very high frequencies.

The natural frequencies of a system depend on its stiffness and mass. For simple systems, the formulas for natural frequencies and natural modes given in Tables 5.9, 5.10, and 5.11 are useful.

Example 5.18 The influence of the shear stiffness and the rotational inertia on the natural frequencies is to be investigated for a simply supported beam.

The eigenfrequencies of a simply supported, shear rigid Euler–Bernoulli beam are given in Table 5.9 as

$$f_i = \frac{\pi}{2} \cdot \frac{i^2}{l^2} \cdot \sqrt{\frac{E \cdot I}{\tilde{m}}} \quad \text{with} \quad i = 1, 2, 3 \ldots . \tag{5.87}$$

Figure 5.49a shows the corresponding eigenmodes. The natural frequencies of the simply supported shear flexible beam with rotational inertia are given in [1, 5]. Thereafter, one obtains:

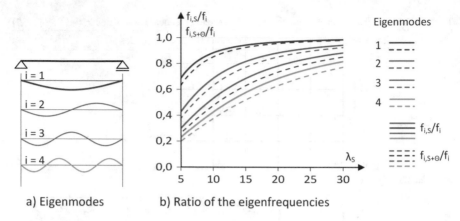

a) Eigenmodes   b) Ratio of the eigenfrequencies

Fig. 5.49 Influence of the shear flexibility and the rotational inertia on the eigenfrequencies of a simply supported beam

Shear flexible beam without rotational inertia:

$$f_{i,\mathrm{S}} = \frac{1}{\sqrt{1 + \dfrac{2 \cdot \pi^2}{\lambda_s^2} \cdot \dfrac{1+\mu}{\kappa} i^2}} \cdot f_i. \tag{5.87a}$$

Shear flexible beam with rotational inertia:

$$f_{i,\mathrm{S}+\Theta} = \frac{1}{\sqrt{1 + \dfrac{\pi^2}{\lambda_s^2} \cdot \left(1 + 2 \cdot \dfrac{1+\mu}{\kappa}\right) \cdot i^2}} \cdot f_i. \tag{5.87b}$$

The parameter

$$\lambda_s = \frac{l}{\sqrt{I/A}} \tag{5.87c}$$

is the slenderness ratio, μ the Poisson ratio, and $\kappa = A_s/A$ the ratio of the shear area A_s to the cross-sectional area A. For rectangular cross sections, the slenderness can also be written as $\lambda_s = \sqrt{12} \cdot l/h \approx 3.5 \cdot l/h$ with h as the height of the cross section. The ratios of the natural frequencies of the Timoshenko and the Euler–Bernoulli beam are plotted versus the slenderness ratio in Fig. 5.49b for the Poisson's ratio $\mu = 0.2$ and the ratio $\kappa = 5/6$ of a rectangular cross section. It can be noted that the influence of the shear stiffness and rotational mass of the cross section is of particular importance for short and deep (thick) beams and increases with the height of the eigenmode.   ◀

Table 5.12 Typical damping coefficients [106]

Material	Damping coefficient ξ [%]	
	Elastic range	Elastoplastic range
Reinforced concrete	1–2	7
Prestressed concrete	0.8	5
Steel, screwed construction	1	7
Steel, welded construction	0.4	4
Timber	1–3	–
Masonry	1–2	7

Damping

Damping is assumed to be proportional to velocity as given in (5.14) and (5.16), mainly for computational reasons. Its actual physical behavior, however, is much more complex [1, 27, 38]. The term damping covers such different influences as material damping in components, friction effects in connections, crack formation, local plastic deformations, energy radiation in the soil, and others. The damping is strongly dependent on the magnitude of a load and normally increases with increasing load. Measured damping coefficients are, therefore, by nature subjected to large scattering. Characteristic damping coefficients are defined for different building materials and types of construction [1]. Typical coefficients depending on strain level are given in Table 5.12.

Actions

For the modeling of dynamic actions, reference is made to [1].

5.8.2 Finite Element Model

Discretization in finite elements

As with static analyses, the finite element modeling is based on the structural model and the properties of the finite elements. In addition, however, the discretization of the continuous mass distribution must be taken into account. When vibrations occur, structures usually deform in a "wave-like" manner. For the half (sinusoidal) wavelength, at least three to four elements should be provided to represent the mass discretization, or at least six to eight elements for the whole wavelength or sinusoidal period. This applies not only to bending vibrations of beams and plates, but also to wave propagation processes in two- and three-dimensional solids such as in soils (Fig. 5.50). The length of waves in an elastic solid is for shear waves

$$\lambda_S = \frac{v_S}{f} = \frac{1}{f} \cdot \sqrt{\frac{G}{\rho}} \qquad (5.88a)$$

Fig. 5.50 Discretization of a wavelength λ

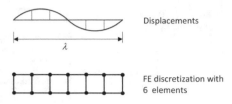

and for compressional waves

$$\lambda_P = \frac{v_P}{f} = \frac{1}{f} \cdot \sqrt{\frac{G}{\rho} \cdot \frac{1-\mu}{1-2\mu}}. \tag{5.88b}$$

Here, f is the frequency of vibration, G the shear modulus, ρ the density of the material, and v_S, v_P the velocity of the shear wave and the compressional wave, respectively, with

$$v_S = \sqrt{\frac{G}{\rho}} \tag{5.88c}$$

$$v_P = \sqrt{\frac{G}{\rho} \cdot \frac{1-\mu}{1-2\mu}}. \tag{5.88d}$$

Since the wavelength decreases with the frequency of the vibration, finer finite element meshes must be selected for modeling high-frequency vibrations than for low-frequency vibrations. This is true for analyses in the frequency domain, for the determination of eigenfrequencies and modes as well as for direct numerical integration methods.

Example 5.19 The beam in Fig. 5.14, clamped at one end, has been discretized into 4 beam elements in Example 5.6. The results are compared with the exact analytical solution and a modeling with 2–8 elements. Shear deformations can be neglected.

For the clamped beam with continuous mass distribution, the exact analytical solution is given in Table 5.9. The eigenfrequencies and modes are determined with $E = 3 \cdot 10^7$ kN/m^2, $I = 17.6$ m^4, $l = 80$ m, and $\bar{m} = 12.5$ t/m. The eigenfrequencies and modes of the system discretized into 4 elements and node masses have been determined numerically in Example 5.6. These as well as those of the system discretized into 2 and 8 elements are given in Table 5.13. The eigenmodes are normalized to the maximum value of 1 and given at $x = 20, 40, 60$ and 80 m (Fig. 5.51).

The results show that the natural frequency f and the natural period T are not particularly sensitive with regard to the discretization of the system. In the case of a one degree of freedom system, they are calculated with (5.28) from the root of the spring stiffness, so that errors in stiffness have only a minor effect on the natural frequency. The error of the natural frequency of the model with 4 elements is 3.2% in the first eigenfrequency, 8.6% in the second eigenfrequency, 13.6% in the third eigenfrequency, and 23.3% in the fourth eigenfrequency. For an earthquake analysis, the accuracy of the model with four elements appears to be appropriate for

Table 5.13 Eigenfrequencies and modes of a clamped beam with continuous mass distribution with different finite element discretizations

Eigenmode			Eigenmode 1				Eigenmode 2			
Number of elements			2	4	8	∞	2	4	8	∞
f [Hz]			0.510	0.553	0.560	0.571	2.627	3.248	3.480	3.555
T [s]			1.961	1.810	1.772	1.751	0.381	0.308	0.281	0.281
$\phi_{i,k}$	x [m]	80	1.000	1.000	1.000	1.000	0.655	1.000	1.000	1.000
		60	0.644*	0.654	0.657	0.658	−0.511*	−0.272	−0.164	−0.138
		40	0.327	0.336	0.339	0.340	−1.000	−0.828	−0.725	−0.717
		20	0.093*	0.096	0.097	0.097	−0.482*	−0.454	−0.416	−0.419
Eigenmode			Eigenmode 3				Eigenmode 4			
Number of elements			2	4	8	∞	2	4	8	∞
f [Hz]			–	8.602	9.579	9.960	–	14.993	18.47	19.556
T [s]			–	0.116	0.104	0.100	–	0.067	0.054	0.051
$\phi_{i,k}$	x [m]	80	–	0.835	1.000	1.000	–	0.372	1.000	1.000
		60	–	−0.917	−0.670	−0.585	–	−0.644	−0.690	−0.619
		40	–	0.257	0.082	0.021	–	0.985	0.876	0.704
		20	–	1.000	0.707	0.728	–	−1.000	−0.882	−0.682

* Values obtained by interpolation of the bending line

determining the first two natural frequencies and even for a rough estimation of the third natural frequency.

The higher the eigenmode, the more elements are needed to represent it. In the 4-element model, the first mode corresponds to a quarter sine or 16 elements/sine-period

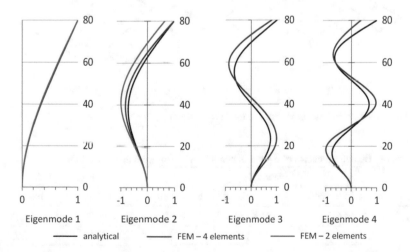

Fig. 5.51 Eigenmodes of a clamped beam

("wavelength"), the second mode a $3/4$ sine or $16/3 \approx 5$ elements/sine-period, and the third eigenmode corresponds to $5/4$ sine or $16/5 \approx 3$ elements/period. According to the recommendation to discretize a sine period into at least 6–8 elements, the model with 4 elements is suitable for representing the first and second eigenmodes. It is not or less suitable for the higher eigenmodes. The comparison of the finite element results with the exact solutions confirms the rule.   ◀

Example 5.20 The first two eigenfrequencies of the square, simply supported slab in Fig. 4.46a, are computed with a finite element model with 4×4 elements and compared with the exact analytical solution. The plate has a thickness of $t = 0.20$ m, a side length of $a = 5.0$ m, a modulus of elasticity $E = 3 \cdot 10^7$ kN/m^2, and a Poisson ratio of $\mu = 0$. The distributed mass is $\bar{m} = 1.0$ t/m^2.

The eigenfrequencies and modes of the continuous system can be determined analytically with the formulae given in Table 5.11. The first eigenfrequency is obtained for $i = j = 1$, the second for $i = 1, j = 2$ or $j = 1, i = 2$. The eigenmodes are described by sinusoidal functions.

The finite element analysis is performed with a shear rigid plate element. Due to the ratio $t/a = 0.2/5.0 = 1/25$, shear deformations can be neglected. The eigenfrequencies given in Table 5.14 are obtained.

The eigenmodes obtained with a finite element discretization in 4×4 elements are shown in Fig. 5.52. Similar to the previous example, the 4×4 discretization proves to be adequate for describing the first eigenmode, where there are 4 elements per half-sine period or 8 elements per sine-period. For the second eigenmode, the description of the eigenmode is coarser, but the natural frequency is still reproduced with good accuracy.   ◀

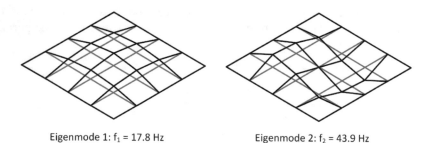

Eigenmode 1: $f_1 = 17.8$ Hz   Eigenmode 2: $f_2 = 43.9$ Hz

Fig. 5.52 Finite element analysis of the modes of a square plate

Table 5.14 Eigenfrequencies of a plate in bending with different discretizations

	Eigenmode 1			Eigenmode 2	
	2×2 elements	4×4 elements	analytically	4×4 elements	analytically
f [Hz]	15.56	17.76	17.77	43.9	44.43
T [s]	0.064	0.056	0.056	0.023	0.023

Damping

To represent damping, there are various models that are closely related to a dynamic analysis method (Fig. 5.53). For direct integration in the time domain, the damping matrix $\underline{C}$ is explicitly required. It is obtained according to (5.18) as Rayleigh damping. The parameters α and β can be determined in such a way that at two reference frequencies, e.g. for two natural frequencies, predefined damping coefficients ξ_1 and ξ_2 are met. The damping at other (natural) frequencies is obtained according to (5.18a), i.e. it can no longer be controlled. Typically, higher natural frequencies are damped more. Whether this can be accepted must be checked in each individual case. In addition, individual dampers can be considered in the model.

When solving the equations of motion in the frequency domain, the material damping is represented as hysteretic damping and introduced into the analysis with a complex modulus of elasticity. Each finite element may have a different damping coefficient. Single dampers can also be included in the damping matrix. Both the hysteretic damping coefficient and individual dampers can be frequency-dependent. Frequency-dependent soil springs and dampers for the representation of the dynamic soil–structure interaction of rigid foundations are a typical field of application. The computational results obtained with a constant, frequency-independent hysteretic damping coefficient ξ_h hardly differ from those obtained with a modal analysis with a constant viscous damping coefficient ξ.

In a modal analysis, damping is introduced in the solution of the modal equations of motion. Each eigenmode i can possess a different viscous damping coefficient ξ_i. If damping coefficients of the individual modes are different, this is called modal damping. Modal damping can also be considered in the response spectrum

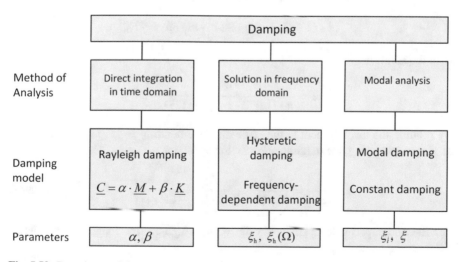

Fig. 5.53 Damping models

method. In this case, response spectra are required for different damping coefficients. Details on the determination of modal damping coefficients for dynamic soil–structure interaction are given in Sect. 5.8.5.

Example 5.21 For the system dealt with in Example 5.9, the vibration amplitude (steady state response) in the middle of the beam under harmonic force excitation is to be determined once with viscous damping and for comparison with a hysteretic damping formulation. The damping is assumed with $\xi = \xi_h = 10\%$ and $\xi = \xi_h = 15\%$, respectively. The dynamic magnification factor shall be plotted versus frequency.

For viscous damping, one obtains with (5.42) and (5.15)

$$k \cdot u + 2 \cdot \sqrt{k \cdot m} \cdot \xi \cdot \dot{u} + m \cdot \ddot{u} = F_0 \cdot \sin(\Omega \cdot t).$$

Applying the solution (5.41) for the steady state response, the magnification factor (5.43a) is obtained with $\omega = \sqrt{k/m}$ and $u_{st} = F_0/k$ as

$$\frac{u_0}{u_{st}} = \frac{1}{\sqrt{\left(1 - \left(\frac{\Omega}{\omega}\right)^2\right)^2 + 4 \cdot \xi^2 \cdot \left(\frac{\Omega}{\omega}\right)^2}}. \tag{5.89a}$$

The analysis with hysteretic damping in frequency domain is done with the equations of motion (5.73). For the single degree of freedom system, one receives with the hysteretic damping according to (5.53a) and (5.54) the complex equation

$$\left(k \cdot (1 + 2 \cdot i \cdot \xi_h) - \Omega^2 \cdot m\right) \cdot \hat{u} = F_0$$

with the solution

$$\hat{u} = \frac{F_0}{k \cdot (1 + 2 \cdot i \cdot \xi_h) - \Omega^2 \cdot m}.$$

By relating the displacement to the static displacement $u_{st} = F_0/k$, as in the case of viscous damping, the magnification factor becomes

$$\left| \frac{\hat{u}}{u_{st}} \right| = \left| \frac{k}{k \cdot (1 + 2 \cdot i \cdot \xi_h) - \Omega^2 \cdot m} \right|. \tag{5.89b}$$

For the undamped natural frequency $\Omega = \omega = \sqrt{k/m}$, the magnification factor assumes the value

$$\left| \frac{\hat{u}}{u_{st}} \right| = \frac{1}{2 \cdot \xi_h} \tag{5.89c}$$

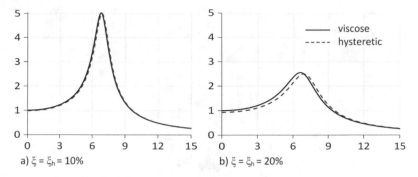

Fig. 5.54 Magnification factors of an SDOF with viscous and hysteretic damping

as for viscous damping in (5.44). Both magnification functions are shown in Fig. 5.54 for a damping coefficient of 10% and 20%. The differences are small and can be neglected. At the undamped natural frequency, both magnification factors are exactly the same. ◀

Mass matrices

The modeling of the masses is usually straightforward. In practice, the permanent loads and an effective part of the traffic loads are used to determine the masses. Mostly, masses are concentrated at nodal points. Instead of nodal masses consistent mass matrices may be used, which then result in fully occupied mass matrices. While nodal masses lead to an overestimation of the natural vibration periods, these are slightly underestimated with consistent mass matrices. In [39], therefore, an averaging of the nodal mass matrix M_{node} and the consistent mass matrix M_{cons} to

$$\underline{M} = \alpha_M \cdot \underline{M}_{cons} + (1 - \alpha_M) \cdot \underline{M}_{node} \qquad (5.90)$$

with $\alpha_M = 0.5$ is proposed. If the analysis is performed in the frequency domain, the additional computational effort required to construct a consistent or an averaged mass matrix is negligible.

Example 5.22 The natural frequencies of the exhaust tower examined in Example 5.6 should be calculated with consistent mass matrices and compared with the analytical solution given in Example 5.19.

The model with four beam elements is investigated. The analysis differs from that in Example 5.6 only by the mass matrix. To determine the mass matrix, the element matrix (5.12b) is used. The uniformly distributed mass is $m = 12.5$ t/m, the element length $l = 20.0$ m. The matrices are constructed element by element and assembled to the global mass matrix analogous to the element stiffness matrices. The consistent mass matrix of the system is obtained as

Table 5.15 Eigenfrequencies and periods with consistent mass matrices

	Eigenmode 1			Eigenmode 2			Eigenmode 3			Eigenmode 4		
	Nodal masses	Consistent matrix	∞	Nodal masses	Consistent matrix	∞	Nodal masses	Consistent matrix	∞	Nodal masses	Consistent matrix	∞
f [Hz]	0.552	0.568	0.563	3.247	3.563	3.501	8.599	10.05	9.808	14.99	19.82	19.26
T [s]	1.811	1.760	1.778	0.308	0.280	0.290	0.116	0.100	0.102	0.067	0.050	0.052

$$M = \begin{bmatrix} 186 & 0 & 32 & -155 & 0 & 0 & 0 & 0 \\ 0 & 1905 & 155 & -714 & 0 & 0 & 0 & 0 \\ 32 & 155 & 186 & 0 & 32 & -155 & 0 & 0 \\ -155 & -714 & 0 & 1905 & 155 & -714 & 0 & 0 \\ 0 & 0 & 32 & 155 & 186 & 0 & 32 & -155 \\ 0 & 0 & -155 & -714 & 0 & 1905 & 155 & -714 \\ 0 & 0 & 0 & 0 & 32 & 155 & 93 & -262 \\ 0 & 0 & 0 & 0 & -155 & -714 & -262 & 952 \end{bmatrix}.$$

The natural vibration periods and natural frequencies obtained with nodal masses and with consistent mass matrices are given in Table 5.15 together with the analytical solution. The results obtained with consistent mass matrices confirm the behavior described above. In particular, they differ less from the exact solution for the higher natural frequencies than the natural frequencies determined with nodal masses. From a practical point of view, however, the differences are small in both cases. ◀

Static and dynamic condensation

For complex finite element models, large systems of equations can arise, the solution of which requires long computing times, especially in dynamic analyses. In order to reduce the number of degrees of freedom, *condensation methods* have been developed. There are two types of condensation, static and dynamic.

Static condensation is a condensation of the global stiffness matrix. For this purpose, the stiffness matrix is divided into the degrees of freedom $\underline{u}_c$ which are to be condensed out and into the degrees of freedom $\underline{u}_a$ which are to remain in the system of equations:

$$\begin{bmatrix} \underline{K}_{aa} & \underline{K}_{ac} \\ \underline{K}_{ca} & \underline{K}_{cc} \end{bmatrix} \cdot \begin{bmatrix} \underline{u}_a \\ \underline{u}_c \end{bmatrix} = \begin{bmatrix} \underline{F}_a \\ \underline{F}_c \end{bmatrix} \tag{5.91}$$

The vectors $\underline{F}_a$ and $\underline{F}_c$ contain the corresponding nodal forces. If the second equation is solved for $\underline{u}_c$, one obtains

$$\underline{u}_c = \underline{K}_{cc}^{-1} \cdot \left(\underline{F}_c - \underline{K}_{ca} \cdot \underline{u}_a \right). \tag{5.91a}$$

This equation is inserted into the first equation of (5.91) and one gets

$$\left(\underline{K}_{aa} - \underline{K}_{ac} \cdot \underline{K}_{cc}^{-1} \cdot \underline{K}_{ca} \right) \cdot \underline{u}_a = \underline{F}_a - \underline{K}_{cc}^{-1} \cdot \underline{F}_c. \tag{5.91b}$$

The matrix

$$\underline{K}_{cond} = \underline{K}_{aa} - \underline{K}_{ac} \cdot \underline{K}_{cc}^{-1} \cdot \underline{K}_{ca} \tag{5.91c}$$

is the condensed stiffness matrix, the vector

$$\underline{F}_{\text{cond}} = \underline{F}_a - \underline{K}_{cc}^{-1} \cdot \underline{F}_c \qquad (5.91\text{d})$$

is the corresponding load vector.

Equation (5.91) can also be understood as the "element equation" of a *macro-element* or a *substructure*. This macro-element can be combined with other elements or macro-elements to form the complete structure. This can be useful for large systems to reduce the computational effort, especially if no forces are acting on the degrees of freedom condensed out, i.e. if $\underline{F}_c = \underline{0}$. An example is a building with identical stories, where one storey is formulated as a substructure. After having computed the entire system, the displacements of the condensed degrees of freedom in the respective substructure are obtained with (5.91a).

Example 5.23 In the equations of motion in Example 5.6, the degrees of freedom not occupied by masses shall be condensed out and the natural frequencies shall be calculated with the reduced system.

First, the degrees of freedom in the global stiffness matrix are rearranged. The degrees of freedom of displacements u_2, u_3, u_4, u_5 are assigned to masses, the rotational degrees of freedom φ_2, φ_3, φ_4, φ_5 are free of masses. The stiffness matrix given in Example 5.6 now has the new structure

$$\underline{K}_{\text{sort}} = \begin{bmatrix} 0.158 & -0.079 & 0 & 0 & 0 & 0.792 & 0 & 0 \\ -0.079 & 0.158 & -0.079 & 0 & -0.792 & 0 & 0.792 & 0 \\ 0 & -0.079 & 0.158 & -0.079 & 0 & -0.792 & 0 & 0.792 \\ 0 & 0 & -0.079 & 0.079 & 0 & 0 & -0.792 & -0.792 \\ 0 & -0.792 & 0 & 0 & 21.114 & 5.279 & 0 & 0 \\ 0.792 & 0 & -0.792 & 0 & 5.279 & 21.114 & 5.279 & 0 \\ 0 & 0.792 & 0 & -0.792 & 0 & 5.279 & 21.114 & 5.279 \\ 0 & 0 & 0.792 & -0.792 & 0 & 0 & 5.279 & 10.557 \end{bmatrix} \cdot 10^7.$$

It is related to the vector

$$\underline{u}_{\text{sort}} = \begin{bmatrix} u_2 \\ u_3 \\ u_4 \\ u_5 \\ \varphi_2 \\ \varphi_3 \\ \varphi_4 \\ \varphi_5 \end{bmatrix} = \begin{bmatrix} \underline{u}_a \\ \underline{u}_c \end{bmatrix} \quad \text{with} \quad \underline{u}_a = \begin{bmatrix} u_2 \\ u_3 \\ u_4 \\ u_5 \end{bmatrix}, \quad \underline{u}_c = \begin{bmatrix} \varphi_2 \\ \varphi_3 \\ \varphi_4 \\ \varphi_5 \end{bmatrix}.$$

The corresponding mass matrix is

$$
M_{\text{sort}} = \begin{bmatrix}
250 & 0 & 0 & 0 & 0 & 0 & 0 & 0 \\
0 & 250 & 0 & 0 & 0 & 0 & 0 & 0 \\
0 & 0 & 250 & 0 & 0 & 0 & 0 & 0 \\
0 & 0 & 0 & 125 & 0 & 0 & 0 & 0 \\
0 & 0 & 0 & 0 & 0 & 0 & 0 & 0 \\
0 & 0 & 0 & 0 & 0 & 0 & 0 & 0 \\
0 & 0 & 0 & 0 & 0 & 0 & 0 & 0 \\
0 & 0 & 0 & 0 & 0 & 0 & 0 & 0
\end{bmatrix}.
$$

The submatrices according to (5.91) are

$$
K_{aa} = \begin{bmatrix}
0.158 & -0.079 & 0 & 0 \\
-0.079 & 0.158 & -0.079 & 0 \\
0 & -0.079 & 0.158 & -0.079 \\
0 & 0 & -0.079 & 0.079
\end{bmatrix} 10^7,
$$

$$
K_{cc} = \begin{bmatrix}
21.114 & 5.279 & 0 & 0 \\
5.279 & 21.114 & 5.279 & 0 \\
0 & 5.279 & 21.114 & 5.279 \\
0 & 0 & 5.279 & 10.557
\end{bmatrix} 10^7,
$$

$$
K_{ac} = K_{ca}^{\text{T}} = \begin{bmatrix}
0 & 0.792 & 0 & 0 \\
-0.792 & 0 & 0.792 & 0 \\
0 & -0.792 & 0 & 0.792 \\
0 & 0 & -0.792 & -0.792
\end{bmatrix} 10^7.
$$

With (5.91c), one obtains the condensed stiffness matrix as

$$
K_{\text{cond}} = \begin{bmatrix}
12.407 & -7.795 & 2.939 & -0.490 \\
-7.795 & 9.469 & -6.326 & 1.714 \\
2.939 & -6.326 & 6.530 & -2.408 \\
-0.49 & 1.714 & -2.408 & 1.061
\end{bmatrix} \cdot 10^5
$$

and the corresponding mass matrix as

$$
M_{aa} = \begin{bmatrix}
250 & 0 & 0 & 0 \\
0 & 250 & 0 & 0 \\
0 & 0 & 250 & 0 \\
0 & 0 & 0 & 125
\end{bmatrix}.
$$

Thus the condensed eigenvalue problem is

$$\left(\begin{bmatrix} 12.407 & -7.795 & 2.939 & -0.490 \\ -7.795 & 9.469 & -6.326 & 1.714 \\ 2.939 & -6.326 & 6.530 & -2.408 \\ -0.49 & 1.714 & -2.408 & 1.061 \end{bmatrix} \cdot 10^5 - \omega^2 \cdot \begin{bmatrix} 250 & 0 & 0 & 0 \\ 0 & 250 & 0 & 0 \\ 0 & 0 & 250 & 0 \\ 0 & 0 & 0 & 125 \end{bmatrix} \right)$$

$$\cdot \begin{bmatrix} v_1 \\ v_2 \\ v_3 \\ v_4 \end{bmatrix} = \begin{bmatrix} 0 \\ 0 \\ 0 \\ 0 \end{bmatrix}.$$

Its eigenvalues are obtained as

$$\omega_1 = 3.5 \text{ Hz}, \quad \omega_2 = 20.4 \text{ Hz}, \quad \omega_3 = 54.0 \text{ Hz}, \quad \omega_4 = 94.2 \text{ Hz}.$$

The natural frequencies correspond to the values determined in Example 5.6 with the fully occupied system of equations. The same applies to the eigenmodes, whereby with the condensed system, obviously only the values for the degrees of freedom of displacements are obtained. The values of the rotational degrees of freedom can be calculated with (5.91a).   ◀

In this context it should be pointed out that the (condensed) stiffness matrix can also be derived by means of the flexibility matrix. This can be effective for small systems to get the stiffness matrix by a hand calculation (see examples in [1]).

Example 5.24 The stiffness matrix related to the degrees of freedom of displacements and the equations of motion in Example 5.6 shall be derived by means of the flexibility matrix (Fig. 5.14).

The bending line of a clamped bar of length L, which is loaded by a force F at a distance x from the clamping point, is described by

$$v(x, L) = \frac{F \cdot L^3}{6 \cdot E \cdot I} \cdot \left(2 - 3 \cdot \frac{L - x}{L} + \left(\frac{L - x}{L} \right)^3 \right).$$

The flexibility matrix describes the displacements of the beam in Fig. 5.14 at points 2–5 due to unit forces applied at points 2–5. With the coordinates of the points in Fig. 5.14, the flexibility matrix can be determined. For example, the displacement of point 2 at $x = 20$ m due to a force $F = 1$ at point 4 with $L = 60$ m is given as $v(20, 60) = 2.021 \cdot 10^{-5}$ (in units kN and m). Taking into account the symmetry of the flexibility matrix, one obtains

$$\begin{bmatrix} v_2 \\ v_3 \\ v_4 \\ v_5 \end{bmatrix} = \begin{bmatrix} v(20, 20) & v(20, 40) & v(20, 60) & v(20, 80) \\ v(20, 40) & v(40, 40) & v(40, 60) & v(40, 80) \\ v(20, 60) & v(40, 60) & v(60, 60) & v(60, 80) \\ v(20, 80) & v(40, 80) & v(60, 80) & v(80, 80) \end{bmatrix} \cdot \begin{bmatrix} F_2 \\ F_3 \\ F_4 \\ F_5 \end{bmatrix}$$

or

$$\underline{v} = \underline{N} \cdot \underline{F}$$

with

$$\underline{N} = \begin{bmatrix} 0.505 & 1.263 & 2.021 & 2.779 \\ 1.263 & 4.041 & 7.073 & 10.104 \\ 2.021 & 7.073 & 13.640 & 20.460 \\ 2.779 & 10.104 & 20.460 & 32.332 \end{bmatrix} \cdot 10^{-5}.$$

The stiffness matrix is obtained with

$$\underline{F} = \underline{N}^{-1} \cdot \underline{v} = \underline{K} \cdot \underline{v}$$

as

$$\underline{K} = \underline{N}^{-1} = \begin{bmatrix} 12.407 & -7.795 & 2.939 & -0.490 \\ -7.795 & 9.469 & -6.326 & 1.714 \\ 2.939 & -6.326 & 6.530 & -2.408 \\ -0.49 & 1.714 & -2.408 & 1.061 \end{bmatrix} \cdot 10^{5}.$$

Thus, the eigenvalue problem is the same as in Example 5.23:

$$\left(\begin{bmatrix} 12.407 & -7.795 & 2.939 & -0.490 \\ -7.795 & 9.469 & -6.326 & 1.714 \\ 2.939 & -6.326 & 6.530 & -2.408 \\ -0.49 & 1.714 & -2.408 & 1.061 \end{bmatrix} \cdot 10^{5} - \omega^{2} \cdot \begin{bmatrix} 250 & 0 & 0 & 0 \\ 0 & 250 & 0 & 0 \\ 0 & 0 & 250 & 0 \\ 0 & 0 & 0 & 125 \end{bmatrix} \right)$$
$$\cdot \begin{bmatrix} v_1 \\ v_2 \\ v_3 \\ v_4 \end{bmatrix} = \begin{bmatrix} 0 \\ 0 \\ 0 \\ 0 \end{bmatrix}.$$

◀

In Example 5.23, the eigenvalue problem of the condensed equations of motion gives the same solution as the one without condensation. The reason is that in Example 5.23, the condensed-out degrees of freedom are not occupied by mass and a static condensation of the stiffness matrix is completely sufficient. If this is not the case, in addition to the stiffness matrix, the mass matrix must be condensed. This is denoted as *dynamic condensation*. A method for this is the so-called *Guyans reduction*. The mass matrix is subdivided as the stiffness matrix in (5.91) with

$$\begin{bmatrix} \underline{K}_{aa} & \underline{K}_{ac} \\ \underline{K}_{ca} & \underline{K}_{cc} \end{bmatrix} \cdot \begin{bmatrix} \underline{u}_a \\ \underline{u}_c \end{bmatrix} = \begin{bmatrix} \underline{F}_a \\ \underline{F}_c \end{bmatrix}$$

into submatrices as

$$\underline{M} = \begin{bmatrix} \underline{M}_{aa} & \underline{M}_{ac} \\ \underline{M}_{ca} & \underline{M}_{cc} \end{bmatrix} \quad \text{with} \quad \underline{M}_{ca} = \underline{M}_{ac}^{\mathrm{T}}. \tag{5.92a}$$

The condensed mass matrix is obtained according to [40] as

$$\underline{M}_{\mathrm{cond}} = \underline{M}_{aa} - \underline{M}_{ac} \cdot \underline{K}_{cc}^{-1} \cdot \underline{K}_{ca} - \left(\underline{K}_{cc}^{-1} \cdot \underline{K}_{ca}\right)^{\mathrm{T}} \cdot \left(\underline{M}_{ca} - \underline{M}_{cc} \cdot \underline{K}_{cc}^{-1} \cdot \underline{K}_{ca}\right).$$
$$\tag{5.92b}$$

As this is an approximation, the natural frequencies of the condensed and non-condensed equations of motion no longer match exactly. Another method for dynamic condensation is the *Craig–Bampton method*. It is more complex and also suited for the formulation of substructures in dynamic analyses [41, 42].

5.8.3   Discretization in Time and Frequency Domain

Dynamic analyses require not only the spatial discretization of the structure into finite elements but also a temporal discretization. For this purpose, a time step must be selected to represent all time-dependent variables. Decisive for the choice of the time step is the frequency content of the load and the dynamic model. The selected time step must be able to represent both, the time history of the load and the vibrational behavior of the finite element model.

The required frequency content of a vibrational response and the modeling of the structure in the structural and finite element models must be mutually adapted. A realistic determination of high frequencies of a complex structure also requires a quite detailed structural and finite element model. Simple models, which are suited to analyze the global vibrational behavior, are not sufficient to represent the structural response in detail (e.g. in Example 5.32). In fact, a structural model is only suitable for representing vibrations up to a certain frequency. The realistic representation of higher natural frequencies and modes also requires a more detailed modeling of a structure or a component.

With regard to the accuracy of the displacements and stresses, the same applies as for static finite element analyses. Stresses and internal forces react as differentiated quantities sensitively to errors in the displacements. Therefore, small errors in the displacements can lead to significantly larger errors in the internal forces and stresses. For example, in beam models for buildings, the additional consideration of higher eigenmodes has a greater influence on the shear forces than on the bending moments or displacements.

Choice of the computational method

Time history analyses can be carried out by different methods. Modal analysis with subsequent superposition of the eigenmodes, direct numerical integration and analyses in frequency domain with Fourier transform are available (Fig. 5.20). For systems whose vibration behavior can be described by only a few eigenmodes, it is recommended to choose modal analysis. Direct numerical integration is effective for shock-like processes where the relevant maximum response of the system occurs shortly after the load application. An analysis in frequency domain is useful for periodic processes such as machine vibrations and for computational models with frequency-dependent finite elements such as soil springs and dampers or boundary elements. The analysis in the frequency domain is also the basis of the computation of random vibrations with probabilistic methods.

Modal analysis

Modal analysis is often used for the vibration analysis of structures. It is very efficient if only a few eigenmodes are required to determine the dynamic response. This is normally the case for beam structures. The dynamic response of finite element models for large plates in plane stress or bending or of three-dimensional solids, however, can only be described with a large number of eigenmodes (cf. Example 5.32). For these models, one should therefore check whether other methods are more suitable.

The modal analysis is mathematically exact if all eigenmodes of the computational model are considered in the superposition. For a finite element model with n degrees of freedom occupied by mass, these are n eigenmodes. In practice, however, a few eigenmodes are normally used to describe the vibration response. In every modal analysis, the question arises on how many eigenmodes are required to obtain results with sufficient accuracy. The required number of eigenmodes must take two influences into account: Firstly, it must be able to reflect the frequency content of the excitation and secondly, the local distribution of the load. Both influences are now to be assessed separately. For this purpose, the case is considered where all loads act with the same time history but different magnitudes at the nodes of the finite element model. It shall therefore apply

$$\underline{F}(t) = \underline{\Lambda} \cdot f(t). \tag{5.93}$$

Herein, $\underline{\Lambda}$ represents a vector, which gives the distribution of the forces at the nodes, and $f(t)$ a scalar function, which describes the time history of the load. For earthquake excitation, in a single direction according to (5.24)

$$\underline{\Lambda} = -\underline{M} \cdot \underline{I} \tag{5.94}$$

applies. With regard to the time-dependent function $f(t)$, the range of eigenfrequencies and eigenmodes must be sufficiently large in order to cover all the relevant excitation frequencies.

To investigate the distribution requirements, described by the vector $\underline{\Lambda}$, the static equilibrium of the finite element system with

$$\underline{K} \cdot \underline{u} = \underline{F} \tag{5.95}$$

is considered. The load vector is

$$F = \underline{\Lambda} \cdot f_{\text{st}} \tag{5.95a}$$

where f_{st} represents an arbitrary reference value of the static load. The displacements in (5.95) are now expressed by the eigenmodes of the system as

$$\underline{u} = \sum \alpha_i \cdot \underline{\phi}_i \tag{5.96}$$

and one gets

$$\sum \left(\underline{K} \cdot \underline{\phi}_i \cdot \alpha_i \right) = \underline{\Lambda} \cdot f_{\text{st}}. \tag{5.97}$$

If the summation is done over all n eigenmodes, it is a mathematically complete set of eigenvectors that are able to describe any displacement vector $\underline{u}$. If fewer eigenmodes are taken into account, an approximate solution is obtained for the loads on the right-hand side of the equation. The error, i.e. the difference between the approximate solution and the exact solution, is

$$\underline{E} = \underline{\Lambda} \cdot f_{\text{st}} - \sum \left(\alpha_i \cdot \underline{K} \cdot \underline{\phi}_i \right). \tag{5.98}$$

This expression will now be transformed. From the eigenvalue problem (5.33) follows

$$\underline{K} \cdot \underline{\phi}_i = \omega_i^2 \cdot \underline{M} \cdot \underline{\phi}_i$$

and so

$$\underline{E} = \underline{\Lambda} \cdot f_{\text{st}} - \sum \left(\alpha_i \cdot \underline{M} \cdot \underline{\phi}_i \cdot \omega_i^2 \right).$$

The factors α_i are obtained from (5.64) with (5.62d) and $\dot{\alpha} = 0$, $\ddot{\alpha} = 0$ in the static case. One gets

$$\alpha_i = \frac{1}{m_i^* \cdot \omega_i^2} \cdot \underline{\phi}_i^{\mathrm{T}} \cdot \underline{F} = \frac{1}{m_i^* \cdot \omega_i^2} \cdot \underline{\phi}_i^{\mathrm{T}} \cdot \underline{\Lambda} \cdot f_{\text{st}}.$$

Thus with (5.98), the vector of error forces is

$$\underline{E} = \underline{\Lambda} \cdot f_{\text{st}} - f_{\text{st}} \cdot \sum \left(\frac{1}{m_i^*} \cdot \underline{\phi}_i^{\mathrm{T}} \cdot \underline{\Lambda} \cdot \underline{M} \cdot \underline{\phi}_i \right). \tag{5.99}$$

It indicates the extent to which the load effective in the computation differs from the applied load $\underline{\Lambda} \cdot f_{\text{st}}$.

For an earthquake excitation, one obtains with (5.94), (5.75a), and $f_{\text{st}} = -1$

$$\underline{E} = \underline{M} \cdot \underline{I} - \sum \left(\frac{1}{m_i^*} \cdot \underline{\phi}_i^{\mathrm{T}} \cdot \underline{M} \cdot \underline{I} \cdot \underline{M} \cdot \underline{\phi}_i \right) = \underline{M} \cdot \underline{I} - \sum \left(\frac{\Gamma_i}{m_i^*} \cdot \underline{M} \cdot \underline{\phi}_i \right).$$
$$(5.100)$$

Here, the applied load consists of the masses acting in the degrees of freedom in the direction of the earthquake acceleration. The error vector $\underline{E}$ indicates to what extent the computationally effective masses differ from the real masses.

The magnitude of the error caused by limiting the number of eigenmodes will be measured by a norm of the vector $\underline{E}$. More suitable than the Euclidean vector norm is here the absolute norm (1.10)

$$\|\underline{E}\| = \sum |E_j|, \qquad (5.101)$$

where E_j must be summed over all vector elements. Therewith (assuming that the sum in (5.100) positive),

$$\|\underline{E}\| = \left\| \underline{M} \cdot \underline{I} - \sum \left(\frac{\Gamma_i}{m_i^*} \cdot \underline{M} \cdot \underline{\phi}_i \right) \right\| = \|\underline{M} \cdot \underline{I}\| - \left\| \sum \left(\frac{\Gamma_i}{m_i^*} \cdot \underline{M} \cdot \underline{\phi}_i \right) \right\|.$$
$$(5.101a)$$

Considering only the error in the masses excited by the earthquake, with

$$\Gamma_i = \underline{\phi}_i^{\mathrm{T}} \cdot \underline{M} \cdot \underline{I} = \underline{I}^{\mathrm{T}} \cdot \underline{M} \cdot \underline{\phi}_i$$

one gets

$$\|\underline{E}\| = \|\underline{M} \cdot \underline{I}\| - \sum \frac{\Gamma_i^2}{m_i^*}. \qquad (5.102)$$

The error $\|\underline{E}\|$ is now related to the magnitude of the load, i.e. to $\|\underline{M} \cdot \underline{I}\|$. This gives the relative error to

$$e = \frac{\|\underline{E}\|}{\|\underline{M} \cdot \underline{I}\|} = \frac{\|\underline{M} \cdot \underline{I}\| - \sum \frac{\Gamma_i^2}{m_i^*}}{\|\underline{M} \cdot \underline{I}\|} = 1 - \frac{\sum \frac{\Gamma_i^2}{m_i^*}}{\|\underline{M} \cdot \underline{I}\|} \qquad (5.103)$$

or

$$e = 1 - \sum \varepsilon_i \qquad (5.104)$$

with

$$m_{\mathrm{ges}} = \|\underline{M} \cdot \underline{I}\| \qquad (5.104a)$$

$$m_{\text{eff},i} = \frac{\Gamma_i^2}{m_i^*} \qquad (5.104\text{b})$$

and

$$\varepsilon_i = \frac{m_{\text{eff},i}}{m_i^*} = \frac{\Gamma_i^2}{m_i^* \cdot m_{\text{ges}}}. \qquad (5.104\text{c})$$

The relative error e indicates the proportion (or percentage) by which the computationally effective masses differ from the real masses in the degrees of freedom of earthquake excitation. The expression is denoted as *effective mass factor*. It gives the ratio between the *effective mass* $m_{\text{eff},i}$ in mode i and the total mass m_{ges} of the system excited by an earthquake (in a given direction). It can be shown that

$$\varepsilon_1 > \varepsilon_2 > \ldots > \varepsilon_i > \ldots > \varepsilon_n \qquad (5.105)$$

and

$$\sum_{i=1}^{n} \varepsilon_i = 1. \qquad (5.106)$$

The formulae allow to estimate how many eigenmodes are required to represent the earthquake loading of a structure. If the error e according to (5.104) is sufficiently small or the sum of the effective mass factors of the k eigenmodes considered

$$s_e = \sum_{i=1}^{k} \varepsilon_i \qquad (5.106\text{a})$$

is close to 1, the contribution of the higher eigenmodes can be skipped, unless they have to be included because of the frequency content of the excitation.

Example 5.25 For the model of an exhaust tower examined in Examples 5.6, 5.15, and 5.17, the effective masses for earthquake excitation shall be determined when considering 1, 2, 3, and 4 eigenmodes.

The vectors of the masses and of the effective masses in (5.100) are

$$\underline{M} \cdot \underline{I} = \begin{bmatrix} 250 \\ 0 \\ 250 \\ 0 \\ 250 \\ 0 \\ 125 \\ 0 \end{bmatrix}, \quad \frac{\Gamma_1}{m_1^*} \cdot \underline{M} \cdot \underline{\phi}_1 = \begin{bmatrix} 36 \\ 0 \\ 127 \\ 0 \\ 247 \\ 0 \\ 189 \\ 0 \end{bmatrix}, \quad \sum_1^2 \frac{\Gamma_i}{m_i^*} \cdot \underline{M} \cdot \underline{\phi}_i = \begin{bmatrix} 118 \\ 0 \\ 276 \\ 0 \\ 296 \\ 0 \\ 99 \\ 0 \end{bmatrix},$$

Table 5.16 Influence of the number of modes considered in modal analysis

i	Γ_i	$\sum_i m_{\text{eff},i}$	ε_i	$\sum_i \varepsilon_i$	$e\ (\%)$
1	-24.47	599	0.684	0.684	31.6
2	13.78	789	0.217	0.901	9.9
3	-7.98	852	0.073	0.974	2.6
4	-4.77	875	0.026	1.000	0

$$\sum_1^3 \frac{\Gamma_i}{m_i^*} \cdot \underline{M} \cdot \underline{\phi}_i = \begin{bmatrix} 202 \\ 0 \\ 298 \\ 0 \\ 219 \\ 0 \\ 134 \\ 0 \end{bmatrix}, \quad \sum_1^4 \frac{\Gamma_i}{m_i^*} \cdot \underline{M} \cdot \underline{\phi}_i = \begin{bmatrix} 250 \\ 0 \\ 250 \\ 0 \\ 250 \\ 0 \\ 125 \\ 0 \end{bmatrix}.$$

The norm $\left\| \underline{M} \cdot \underline{1} \right\| = 875$ according to (5.101) gives the total mass m_{ges} of the system in the degrees of freedom in the direction of the earthquake excitation. The participation factors Γ_i have been given in Example (5.15). The calculation of the effective mass factors ε_i and the relative errors e according to (5.104) is shown in Table 5.16.

In the example, all masses are assigned to horizontal degrees of freedom and are thus in the direction of the earthquake excitation. The total mass of the system is equal to the real mass of $\left\| \underline{M} \cdot \underline{1} \right\| = 875$ t. If only one eigenmode is taken into account, the effective mass (equivalent mass) is 599 t compared to 875 t total mass. The error e, which indicates the difference between the effective mass and its exact value, is $(875 - 599)/875 = 31.6\%$. This demonstrates that additional eigenmodes must be included in the computation. In general, the error e should not exceed 10%. This is achieved here when the second eigenmode is included.

In Example 5.17, in which the system is calculated with the response spectrum method, according to Figs. 5.46 and 5.48, the shear forces at the clamping point in the second eigenmode and in Fig. 5.46, even in the third eigenmode, are significantly higher than in the first eigenmode. This is due to the second criterion to be considered, namely the frequency content of the load. The acceleration values $S_a(T)$ of the response spectrum are considerably lower in the first eigenmode than in the other eigenmodes due to the long natural vibration period $T_1 = 1.81$ s. When comparing with the response spectrum calculation in Example 5.17, it should also be noted that the errors e apply to a response acceleration S_a that is identical in all eigenmodes and that in (5.101) is based on the sum of the absolute values of the contributions of all eigenmodes. The SRSS superposition of the sum of all shear forces in the individual eigenmodes corresponds to an Euclidean norm and, therefore, provides different values than the superposition corresponding to the absolute norm in (5.101). ◄

Numerical time integration

The numerical integration methods require the selection of a time step. Its size depends on the frequency content of the load and the natural vibration behavior of the structure. In addition, the selected numerical integration method is also important. From a mathematical point of view, accuracy and stability are characteristic for a method. For numerically unstable methods, there is a critical time step Δt_{crit}, beyond which the solution grows without bound for mathematical reasons (cf. (5.59f)). Numerically stable methods lead to less precise solutions with large time steps, but these solutions remain bounded. Finite element models with many degrees of freedom occupied by mass possess a high number of natural frequencies. The vibration period associated with the highest natural frequency is therefore extremely small and would require an even smaller time step in numerically unstable methods in order to ensure the boundedness of the solution. For finite element models, therefore, numerically stable integration methods are used almost exclusively. An exception is the analysis of impact processes where the investigation of short time histories is sufficiently long to determine the relevant maximum values, and implicit methods with extremely short time steps can be efficient.

Analyses in the frequency domain

When solving the equations of motion with the discrete Fourier transform, vibrations are represented as a Fourier series. It is thus a process which is repeated with the period T_0. However, this does not result in a limitation of the generality. For the treatment of nonperiodic processes of limited duration, T_0 is chosen so large that the oscillation decays within T_0 and overlapping effects are thus avoided. The smallest time step still to be represented is obtained due to the frequency content of the load and the vibrational response of the system. This results in (5.68) as the minimum required number N of steps in the time domain. The related questions have already been discussed in Sect. 5.6.4.

5.8.4   Building Models

Dynamic investigations of buildings are done in particular with regard to earthquake-resistant designs. For slender and high structures, dynamic actions caused by wind loads may become relevant as well. In both cases, mainly horizontal loads act on a building. Due to the different deformation behavior under horizontal loads, different types of structures can be distinguished (Fig. 5.55). Tower-like structures are mainly subjected to bending. Concrete frame constructions can be modeled, for simplification, as shear beam models with pure shear deformations [26]. In diaphragm buildings, both bending and shear deformations must be taken into account. In addition, there are combinations of these types of structures, such as frame constructions with a building core [26].

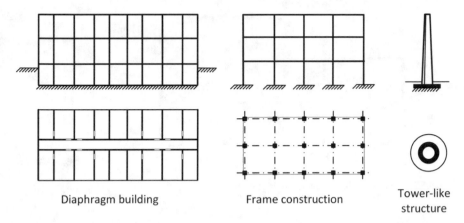

Diaphragm building   Frame construction   Tower-like structure

Fig. 5.55 Types of structures

For the analysis, various types of structural models may be employed. These are discussed in the following for reinforced concrete buildings with regard to their seismic loads.

Beam models

For the analysis of earthquake response vibrations, buildings can be modeled as beams flexible in shear and bending, which are rigidly or elastically clamped in the ground and vibrate horizontally. The model is particularly applicable for tower-like structures and diaphragm buildings. A beam also denoted as an equivalent beam represents the stiffness of the building against horizontal loads. Floor slabs are assumed to be rigid in their plane and infinitely flexible perpendicular to it, i.e. the bending stiffness of floor slabs is neglected. Masses are represented as discrete floor masses at the height of the floor slabs. The masses of the vertical elements such as walls, columns, and facades are proportionally distributed between the floors above and below. They can be eccentric to the axis of the equivalent beam. Rotational inertia about horizontal axes is normally neglected [43].

The floor plan of a 5-storey building with a basement is shown in Fig. 5.56. The basement walls create a laterally stiff box, above which the shear walls are continuous over the height of the building. The x- and y-axes lie in a horizontal plane, the z-axis is oriented upwards, i.e. in the longitudinal direction of the beam. The center of mass is marked M, the center of stiffness S.

If the floor plan is nearly symmetrical with respect to two orthogonal axes in terms of horizontal stiffness and mass distribution, two separate, plane beam models can be analyzed. Further requirements for splitting into two separate models are formulated in standards [34]. With these two models, the investigation of the structure for earthquakes acting in the x- and y-directions is performed separately. Otherwise, a computation with a single, spatial model is required (Fig. 5.56).

In the most basic case of double-symmetrical, tower-like structures and diaphragm buildings, the bending and shear stiffnesses of the individual stiffening components

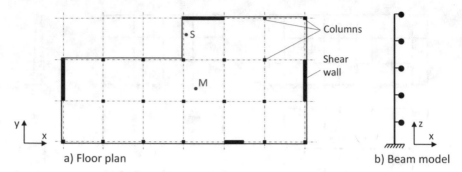

a) Floor plan   b) Beam model

Fig. 5.56 Building with asymmetrical floor plan

(shear walls) are added together in the direction of loading. This gives the second
moment of area of the beam

$$I_x = \sum_i I_{x,i}, \quad I_y = \sum_i I_{y,i} \tag{5.107}$$

where $I_{x,i}$, $I_{y,i}$ are the moments of area of the ith component (shear wall) with respect
to the x- and y-axes. The center of rigidity (center of stiffness) and the center of mass
are located at the center of gravity of the floor plan area. The masses are summarized
to single (lumped) masses at the height of the floor slabs. The shear forces V_x and
V_y are distributed to the individual stiffening components (shear walls) in the ratio
of the bending stiffnesses

$$V_{x,i} = V_x \cdot \frac{I_{y,i}}{I_y}, \quad V_{y,i} = V_y \cdot \frac{I_{x,i}}{I_x}. \tag{5.107a}$$

With asymmetrical floor plans, the center of mass and the center of stiffness of the
beam cross section do not coincide. Therefore, there is an additional torsional action
on the cross section. For cross sections with a single axis of symmetry, the product
moment of area is $I_{xy} = 0$. Then the center of rigidity S (shear center) according to
[43, 44] is

$$x_S = \frac{\sum_i I_{x,i} \cdot x_i}{I_x}, \quad y_S = \frac{\sum_i I_{y,i} \cdot y_i}{I_y}. \tag{5.108}$$

It has been assumed that $I_{xy,i} = 0$ for all bracing elements and shear stiffness can
be neglected.

Due to the torsional stresses, the cross section does not remain plane. In addition
to the St. Venant's torsion, warping torsion occurs. According to [44], the cross-
sectional parameters of the beam, including the torsional constant I_T and the warping
constant I_ω, can be calculated as

$$I_x = \sum_i I_{x,i}, \quad I_y = \sum_i I_{y,i}, \quad I_T = \sum_i I_{T,i} \tag{5.108a}$$

$$I_\omega = \sum_i \left(I_{x,i} \cdot x_{S,i}^2 + I_{y,i} \cdot y_{S,i}^2 + I_{\omega,i} \right). \tag{5.108b}$$

In (5.108b), $x_{S,i}$ and $y_{S,i}$ are the distances between the center of rigidity S and the center of gravity of the ith component in the x- and y-directions, respectively. When distributing the horizontal loads to the stiffening components, the torsional moments shall be taken into account in addition to the shear forces. One obtains

$$V_{x,i} = V_x \cdot \frac{I_{y,i}}{I_y} - M_z \cdot \frac{I_{y,i} \cdot y_{S,i}}{I_\omega} \tag{5.109a}$$

$$V_{y,i} = V_y \cdot \frac{I_{x,i}}{I_x} + M_z \cdot \frac{I_{x,i} \cdot x_{S,i}}{I_\omega}. \tag{5.109b}$$

The shear forces V_x and V_y are related to the center of rigidity, M_z is the secondary torsional moment of the beam. The warping moment in a component i is obtained as

$$M_{\omega,i} = M_\omega \cdot \frac{I_{\omega,i}}{I_\omega}. \tag{5.109c}$$

In general, the shear stiffness of the walls should be taken into account. For a detailed treatment of the modeling of buildings as equivalent beams, please refer to [26, 43–46]. For finite element modeling of nonsymmetrical structures, beam elements with warping torsion are required. Special attention should be paid to the transfer of the warping degree of freedom across the individual floors and its fixing at the foundation.

The structure can also be modeled as a multi-beam model instead of a single equivalent beam. In a multi-beam model, the shear walls and other horizontally stiffening components are each modeled individually as beams whose horizontal displacements are coupled by floor slabs at the height of the floors. The horizontal stiffness of the floor slabs should be taken into account similarly as for 3D building models. It is also important to consider the shear deformations of the shear walls, i.e. the modeling should be carried out with shear flexible beam elements. The slab masses can be assigned to the nodes of the individual walls on the basis of load distribution areas. For multi-beam models, the internal forces of the shear walls are obtained directly, so that a redistribution as for the equivalent beam model is not necessary. Figure 5.57 shows a building ground plan with several shear walls and bracing components. Columns can usually be neglected due to their significantly lower horizontal stiffness compared to shear walls.

Three-dimensional building models

Three-dimensional finite element models of buildings are used in statics for structural design. They are also well suited for dynamic earthquake analyses [47]. With regard

Fig. 5.57 Floor plan of a building with several stiffening walls of a multi-beam model

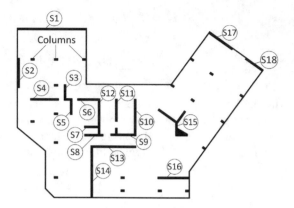

to modeling, the same aspects apply as for the static models, especially with regard to horizontal loads. In addition, there are aspects for dynamic modeling, for example, with regard to the number of eigenmodes to be considered in a modal analysis or the response spectrum method. In a modal analysis, local eigenmodes with vibrations of individual components or parts of the structure can appear, whose modal mass and their contribution to the global vibration behavior, however, is small. Stresses must be combined separately for all eigenmodes to resultant internal forces, which can then be superimposed. The computational effort of a dynamic analysis of three-dimensional building models is significantly higher than that for a static analysis and requires powerful hardware and software, especially for large high-rise buildings. As with the static analysis of three-dimensional building models in structural design, dynamic analyses require higher expertise and demands on the interpretation of the results [48]. A comparative analysis with a simplified beam model may be worthwhile.

Example 5.26 The building with the floor plans shown in Fig. 5.58 has a basement and 3 stories. It is to be examined as a beam model and as a three-dimensional building model. All natural vibrations below 20 Hz are to be determined. The basement is also modeled here for better comparability with the following examples. Normally, modeling the basement is not required for buildings with a stiff box-type basement without soil–structure interaction, i.e. the model is fixed at the level of the ground floor [43].

The floor heights are 3.00 m each in the basement and on the second and third floors, and 4.00 m on the ground floor. The wall thickness of the shear walls is 0.30 m, the columns have the dimensions 0.30 m × 0.30 m. The floor slab thickness is 0.20 m, the foundation slab has a thickness of 0.40 m. The modulus of elasticity is $E = 28300$ MN/m^2, the Poisson's ratio is $\mu = 0.2$. Openings in the floor slabs are not taken into account for simplification. The masses result from the dead weight of the slabs and the walls. Furthermore, a mass occupancy of 0.23 t/m^2 (0.26 t/m^2 in the top floor) is assumed for the slabs due to attributable traffic loads. The building is symmetrical with respect to an axis parallel to the y-axis (cf. [49]).

First, the beam model is examined. For this purpose, the stiffness parameters for the components W1–W4 and U1 are determined. In the upper floors, one obtains for

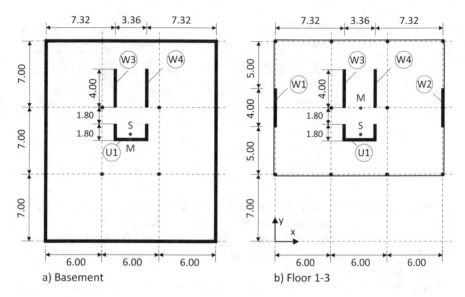

Fig. 5.58 Building plan

the shear walls W1–W4

$$I_x = 1.600 \text{ m}^4, \quad I_y = 0.009 \text{ m}^4, \quad I_T = 0.036 \text{ m}^4,$$
$$I_\omega = 0, \quad A_{Sx} = A_{Sy} = 1.000 \text{ m}^2,$$

and for the cross section U1 with the warping constant according to Table 3.11

$$I_x = 0.577 \text{ m}^4, \quad I_y = 3.75 \text{ m}^4, \quad I_T = 0.045 \text{ m}^4, \quad I_\omega = 1.11 \text{ m}^6,$$
$$A_{Sx} = 1.01 \text{ m}^2, \quad A_{Sy} = 0.83 \text{ m}^2.$$

The stiffness parameters of the beam model are determined with (5.108)–(5.109b). They are compiled in Table 5.17. In the basement, a load propagation of the vertical forces of the overlying walls of 45° was assumed in order to determine the moments of inertia, i.e. walls W1 and W2 are assumed to continue in the basement with an average length of 7.0 m. For the basement, the shear areas and the torsional constant were computed for the full hollow box section, since the transmission of horizontal

Table 5.17 Parameters of the beam model

	x_S [m]	y_S [m]	I_x [m^4]	I_y [m^4]	I_T [m^4]	I_ω [m^6]	A_{Sx} [m^2]	A_{Sy} [m^2]
Floors 1–3	9.00	11.00	6.98	3.79	0.19	270.0	5.08	5.00
Basement	9.00	11.01	20.95	3.86	2198	–	26.41	26.23

Table 5.18 Floor masses of the beam model

Coordinate z [m]	i	Coordinates of the center of gravity			Masses	Rotational inertia
		$x_{M,i}$ [m]	$y_{M,i}$ [m]	$z_{M,i}$ [m]	m_i [t]	$\Theta_{z,i}$ [t·m²]
0	1	9.00	10.50	0	568	36194
3	2	9.00	10.50	3.00	413	26234
7	3	9.00	14.00	7.00	244	10554
10	4	9.00	14.00	10.00	235	10184
13	5	9.00	14.00	13.00	217	9413

forces is ensured by the slab. The small difference in the location of the stiffness centers S in the upper floors and in the basement is neglected.

Assuming a warping-free basement, the warping degree of freedom is fixed at the level of the floor slab above the basement at $z = 3.0$ m. All other degrees of freedom are fixed at the base of the model.

The floor masses consist of the mass of the slabs and the proportionate masses of the walls. Furthermore, the rotational inertia of the slabs is evaluated (cf. Example 5.1). The masses and coordinates of the centers of gravity M are listed in Table 5.18. The masses, which act in the x- and y-directions, were connected with beam elements of large stiffness to the nodal points of the equivalent beam. The beam model is shown in Fig. 5.59.

The analysis is carried out with shear flexible, three-dimensional beam elements and a warping degree of freedom with [SOF 6]. The element describes the warping torsion with shape functions. In order to take into account the discretization error associated with this, the equivalent beam on each floor was divided into 5 beam elements. The model has 166 degrees of freedom.

The three-dimensional building model is shown in Fig. 5.60. Walls and floor slabs are modeled as a folded structure with shear flexible 4-node shell elements. The columns are modeled with hinges at the ends. The additional masses are applied as distributed masses. All degrees of freedom of the foundation slab are fixed. The model has 22296 degrees of freedom. The analyses were performed with [50, SOF 6].

Fig. 5.59 Beam model

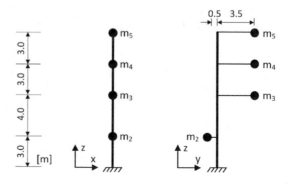

Fig. 5.60 Three-dimensional building model [50]

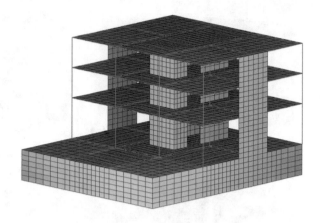

The natural frequencies below 20 Hz determined with both models are summarized in Table 5.19. The three lowest natural frequencies of both models are in good agreement. The eigenmodes are shown in Figs. 5.61 and 5.62. Due to the lower bracing of the building in this direction, the first eigenmode is a vibration in the x-direction, which additionally shows a small torsional component. The second

Table 5.19 Eigenfrequencies

Eigenmode i	Eigenfrequency f_i [Hz]		Description
	Beam model	3D building model	
1	3.1	3.3	Translation in the x-direction with torsion
2	5.5	5.9	Translation in the y-direction
3	6.8	6.2	Torsion
4	18.7	12.6	Translation in the x-direction

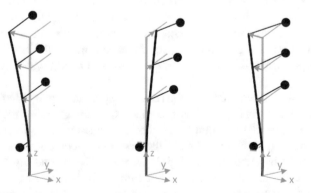

Eigenmode 1: 3.1 Hz   Eigenmode 2: 5.5 Hz   Eigenmode 3: 6.8 Hz

Fig. 5.61 Eigenmodes of the beam model

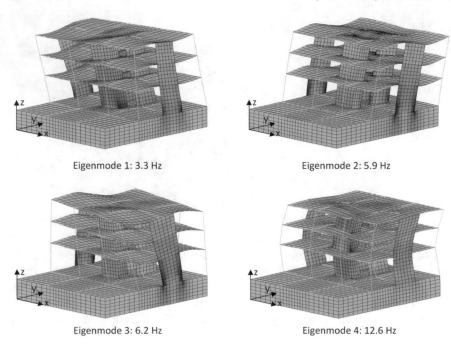

Fig. 5.62 Eigenmodes of the building model [50]

eigenmode is a pure translational vibration in the y-direction. A torsional effect is excluded because of the symmetry of the building. The third eigenmode describes a torsional vibration of the building. For the higher eigenmodes, there are major differences in both models. However, the eigenfrequencies and modes in the range of seismic excitation are represented equally well by both, the beam model and the three-dimensional building model.   ◀

Example 5.27 The beam model of the building discussed in Example 5.26 is to be examined with the response spectrum method for earthquake excitation in the x- and y-directions (Fig. 5.59). The respective response spectrum given in Fig. 5.63 corresponds to a horizontal response spectrum according to DIN EN 1998 with the parameters $a_{gR} = 0.8$ m/s^2, $\gamma_1 = 1$, and $\eta = 1$ for soil conditions of type C-S with $S = 0.75$.

The internal forces obtained with the response spectrum method in the first three eigenmodes are shown in Fig. 5.64. For seismic excitation in the x-direction, the first and the third eigenmodes are significant. The comparison of the internal forces in both modes shows that the influence of the first eigenmode dominates by far. The fourth eigenmode is practically negligible as well. The second eigenmode is not excited and thus V_y and M_x are zero. If an earthquake is acting in the y-direction, however, only the second eigenmode is excited due to the symmetry of the building, so that solely the internal forces V_y and M_x occur.

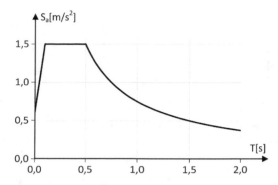

Fig. 5.63 Acceleration response spectrum

From the internal forces of the beam model, the internal forces in the individual bracing components can be calculated. This will be shown for the internal forces of the shear wall W1 in a section at $z = 3.0$ m above the slab of the ground floor. If the seismic excitation is acting in the y-direction, the distribution of internal forces can be evaluated according to (5.107a) due to the symmetry of the building. With the total shear force of 917 kN of the beam model in the second eigenmode, the shear force in the wall W1 is obtained as

$$V_{y,\text{W1}} = V_y \cdot \frac{I_{x,\text{W1}}}{I_x} = 917 \cdot \frac{1.6}{7.0} = 210 \text{ kN}$$

and accordingly, the bending moment for a total moment of 7266 kNm as

$$M_{x,\text{W1}} = M_x \cdot \frac{I_{x,\text{W1}}}{I_x} = 7266 \cdot \frac{1.6}{7.0} = 1660 \text{ kNm}.$$

Since no other, higher eigenmodes are relevant for excitation in the y-direction, the superposition according to (5.85) or (5.86) is not necessary. The maximum internal forces for excitation in the y-direction are thus

$$V_{\text{W1,max}} = 210 \text{ kN}, \quad M_{\text{W1,max}} = 1660 \text{ kNm}.$$

For earthquake excitation in the x-direction, torsional vibrations with warping torsion are excited in addition to bending vibrations. In the first eigenmode, one obtains with (5.109b) and $M_{z,\text{s}}^{(1)} = 4128$ kNm

$$V_{y,\text{W1}}^{(1)} = V_y^{(1)} \cdot \frac{I_{x,\text{W1}}}{I_x} + M_{z,\text{s}}^{(1)} \cdot \frac{I_{x,\text{W1}} \cdot x_{\text{S,W1}}}{I_\omega} = 0 + 4128 \cdot \frac{1.6 \cdot 9.0}{270} = 220 \text{ kN}$$

and in the third eigenmode with $M_{z,\text{s}}^{(3)} = 1088$ kNm

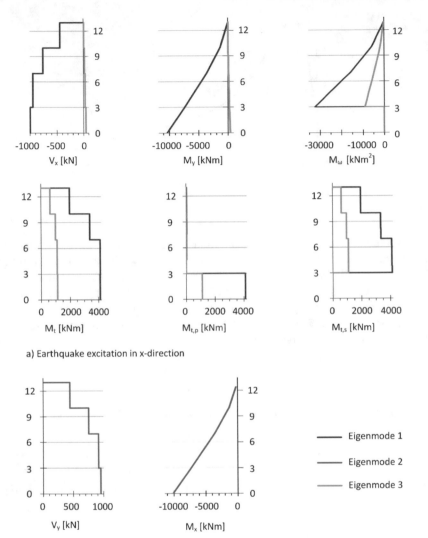

a) Earthquake excitation in x-direction

b) Earthquake excitation in y-direction

Fig. 5.64 Internal forces of the beam model

$$V_{y,\text{W1}}^{(3)} = V_y^{(3)} \cdot \frac{I_{x,\text{W1}}}{I_x} + M_{z,\text{s}}^{(3)} \cdot \frac{I_{x,\text{W1}} \cdot x_{S,\text{W1}}}{I_\omega} = 0 + 1088 \cdot \frac{1.6 \cdot 9.0}{270} = 58 \text{ kN}.$$

The resulting shear force for seismic excitation in the x-direction is obtained according to (5.85) as

$$V_{\text{W1,max}} = \sqrt{220^2 + 58^2} = 228 \text{ kN}.$$

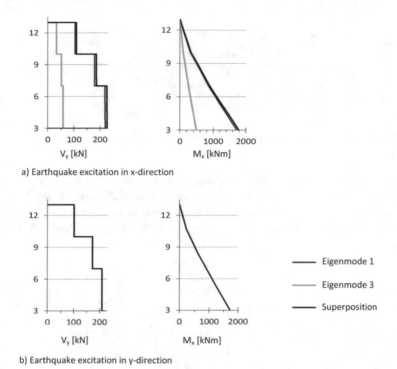

a) Earthquake excitation in x-direction

b) Earthquake excitation in y-direction

Fig. 5.65 Internal forces of the shear wall W1 of the beam model

The bending moments are calculated from the shear forces in the shear wall W1. Figure 5.65 shows the internal forces of the shear wall W1 for earthquake excitation in the *x*- and *y*-directions. ◀

Example 5.28 The three-dimensional building model in Example 5.26 is to be examined for earthquake excitation with the response spectrum method. The earthquake parameters as specified in Example 5.27 shall be applied.

The response spectrum method gives the stresses or the internal forces in the shell elements in each eigenmode. These can be taken for the static design or they can be summed up to stress resultants. Figure 5.66 shows, for example, the shear stresses τ_{yz} at the base of the shear wall W1 in the second eigenmode for earthquake excitation in the *y*-direction. The resulting shear force is obtained as

$$V_{y,\mathrm{W1}} = (390 + 114 + 69 + 69 + 120 + 388) \cdot 0.30 \cdot \frac{4.00}{6} = 230\,\mathrm{kN}.$$

Under consideration of the different model assumptions, the value agrees well with that of the beam model of 210 kN.

The building model, however, provides additional information on internal forces in structural components that are not represented by the beam model. Figure 5.67 shows the principal stresses (membrane stresses) in the floor slab at $z = 10$ m in

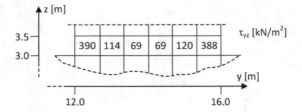

Fig. 5.66 Shear stresses in the shear wall W1 for seismic excitation in the y-direction

the first and second eigenmode for earthquake excitation in the x- and y-directions, respectively. They can serve as a basis for the earthquake design of the floor slab. For the beam model, a two-dimensional finite element model for the floor slab or a simplified strut-and-tie model would have to be investigated for the slab design additionally. ◀

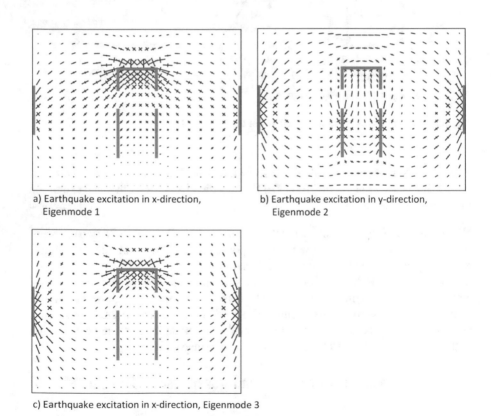

a) Earthquake excitation in x-direction,
   Eigenmode 1

b) Earthquake excitation in y-direction,
   Eigenmode 2

c) Earthquake excitation in x-direction, Eigenmode 3

Fig. 5.67 Principal stresses in the floor slab at $z = 10$ m

5.8.5 Soil–Structure Interaction

General

For the dynamic analysis of buildings and earthquake investigations, it is frequently assumed that the building is rigidly fixed to the ground. This is justified for soft structures on stiff soil. In the case of stiff structures on soft soil, however, the flexibility of the soil influences the natural frequencies of the system. In particular, the rotational stiffness of a foundation resting on the soil is important. For an elastically clamped beam with a distributed mass, the natural frequencies can be determined according to Table 5.10. This allows for simple cases estimates on the influence of soil flexibility to be made.

The soil influences not only the stiffness but also the damping of the soil–structure system. Due to the radiation of waves in soil, the vibrations of a foundation are damped, which can significantly reduce its displacements. This effect is denoted as *radiation damping*, which is added to the *material damping* of the soil. The so-called *soil–structure interaction* (SSI) can be modeled not only with spring-damper models but also with complex soil models [1, 51–54].

Spring-damper models

When modeling buildings as an equivalent beam, the flexibility of the subsoil can be represented by a damped elastic restraint of the foundation nodal point (Fig. 5.68). This requires that a stiff box-type basement exists so that the basement can be assumed to be a rigid body. For this purpose, soil springs and dampers are used, which are determined for a rigid plate on a homogeneous or inhomogeneous elastic half-space. Table 5.20 gives the spring constants of a rigid circular foundation on an elastic soil, the stiffness of which increases linearly with depth according to Fig. 5.69 as

$$G(z) = G_0 + g_0 \cdot z = G_0 \cdot \left(1 + \alpha \cdot \frac{z}{r}\right), \tag{5.110}$$

Fig. 5.68 Spring-damper model for soil–structure interaction

Table 5.20 Spring constants of rigid foundations on an inhomogeneous soil

Degree of freedom	Spring constant	Equivalent radius	
		Foundation with arbitrary geometry	Rectangular foundation
Vertical displacement	$k_z = \dfrac{4 \cdot r_z \cdot G(r_z)}{1-\mu} = \dfrac{4 \cdot r_z}{1-\mu} \cdot (G_0 + r_z \cdot g_0)$	$r_x = r_y = r_z = \sqrt{\dfrac{A}{\pi}}$	$r_x = r_y = r_z = \sqrt{\dfrac{a \cdot b}{\pi}}$
Horizontal displacement	$k_x = k_y = \dfrac{8 \cdot r_x \cdot G(0.5 \cdot r_x)}{2-\mu} = \dfrac{8 \cdot r_x}{2-\mu} \cdot (G_0 + 0.5 \cdot r_x \cdot g_0)$		
Rotation about the x-axis	$k_{\varphi x} = \dfrac{8 \cdot r_{\varphi x}^3 \cdot G(0.4 \cdot r_{\varphi x})}{3 \cdot (1-\mu)} = \dfrac{8 \cdot r_{\varphi x}^3}{3 \cdot (1-\mu)} \cdot (G_0 + 0.4 \cdot r_{\varphi x} \cdot g_0)$	$r_{\varphi x} = \sqrt[4]{\dfrac{4 \cdot I_x}{\pi}}$	$r_{\varphi x} = \sqrt[4]{\dfrac{a \cdot b^3}{3 \cdot \pi}}$
Rotation about the y-axis	$k_{\varphi y} = \dfrac{8 \cdot r_{\varphi y}^3 \cdot G(0.4 \cdot r_{\varphi y})}{3 \cdot (1-\mu)} = \dfrac{8 \cdot r_{\varphi y}^3}{3 \cdot (1-\mu)} \cdot (G_0 + 0.4 \cdot r_{\varphi y} \cdot g_0)$	$r_{\varphi y} = \sqrt[4]{\dfrac{4 \cdot I_y}{\pi}}$	$r_{\varphi y} = \sqrt[4]{\dfrac{a^3 \cdot b}{3 \cdot \pi}}$
Rotation about the z-axis	$k_{\varphi z} = \dfrac{16 \cdot r_{\varphi z}^3 \cdot G(r_{\varphi z})}{3} = \dfrac{16 \cdot r_{\varphi z}^3}{3} \cdot (G_0 + 0.2 \cdot r_{\varphi z} \cdot g_0)$	$r_{\varphi z} = \sqrt[4]{\dfrac{2 \cdot I_p}{\pi}}$	$r_{\varphi z} = \sqrt[4]{\dfrac{a \cdot b \cdot (a^2 + b^2)}{6 \cdot \pi}}$

Soil

$G(z) = G_0 + g_0 \cdot z$ Shear modulus

μ Poisson ratio

Foundation

Arbitrary geometry:

A Surface area of the foundation

$I_x, I_y, I_p = I_x + I_y$ Areal moments second order

Rectangular foundations:

a, b Side lengths

Circular foundations:

r Radius, $r_x = r_y = r_z = r_{\varphi x} = r_{\varphi y} = r_{\varphi z} = r$

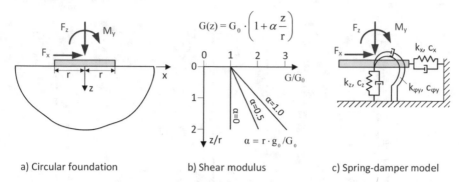

a) Circular foundation   b) Shear modulus   c) Spring-damper model

Fig. 5.69 Soil model of an inhomogeneous half-space with linearly increasing shear modulus

$$\alpha = \frac{g_0}{G_0} \cdot r. \tag{5.110a}$$

For a homogeneous soil, with $\alpha = 0$, the formulae represent the analytical solution for a homogeneous, elastic half-space. The equations of inhomogeneous soil are based on numerical analyses with a semi-finite method and the subsequent adjustment of the parameters [55, 56]. The calculations were performed with the Poisson ratios $\mu = 0.2, 0.3, 0.45$ and the increase parameters $g_0 \cdot r/G_0 = 1$ and $g_0 \cdot r/G_0 = 2$ so the method is valid for $g_0 \cdot r/G_0 \leq 2$ or $G(z = r) \leq 3 \cdot G_0$, and $\mu \leq 0.45$.

The stiffnesses of an inhomogeneous half-space correspond to those of a homogeneous half-space with a shear modulus at a representative depth of the inhomogeneous soil. The representative depth is different in the individual degrees of freedom, since the stress distribution in the soil differs for motions in the different degrees of freedom. The representative depths are obtained for vertical displacement as $\bar{z} = r$, for horizontal displacement as $\bar{z} = 0.5 \cdot r$, for rotation about a horizontal axis (rocking) as $\bar{z} = 0.4 \cdot r$, and for rotation about the z-axis (torsion) as $\bar{z} = 0.2 \cdot r$.

The equations have been derived for circular foundations. Foundations with other geometry are replaced by equivalent circular foundations. For this purpose, in the degrees of freedom of displacement the areas are equated. In the degrees of freedom of rotation the second-order area moments around the respective axes are equated. This gives the equivalent radii given in Table 5.20.

The values of the dampers in the individual degrees of freedom are given in Table 5.21. The associated damping coefficients of the foundation are

$$\xi_{frg} = \frac{a_{0,frg}}{2} \cdot \frac{\bar{c}_{frg}}{\bar{k}_{frg}} + \xi_{\text{Mat}} = \frac{a_{0,frg}}{2} \cdot \bar{c}_{frg} + \xi_{\text{Mat}} \tag{5.111}$$

where for the index "*frg*", the respective degree of freedom according to Table 5.21 is to be inserted. The coefficients k_{frg} are assumed to be frequency-independent, so that $k_{frg} = 1$ applies for all degrees of freedom. The damping coefficient ξ_{Mat} is the material damping of the soil, while the first part of the expression in (5.93) stands

Table 5.21 Damper coefficients of rigid foundations on an inhomogeneous soil

Degree of freedom	Parameter	Dimensionless reference frequency	Reference wave length	Reference frequency
Vertical displacement	$\bar{c}_z = 0.85$	$a_{0,z} = 2 \cdot \pi \cdot f_z \cdot r_z \cdot \sqrt{\dfrac{\rho}{G_0 + 1.5 \cdot \lambda_{sz} \cdot g_0}}$	$\lambda_{sz} = \dfrac{v_{S0}}{f_z}$	f_z
Horizontal displacement in the x-direction	$\bar{c}_x = 0.57$	$a_{0,x} = 2 \cdot \pi \cdot f_x \cdot r_x \cdot \sqrt{\dfrac{\rho}{G_0 + 0.75 \cdot \lambda_{sx} \cdot g_0}}$	$\lambda_{sx} = \dfrac{v_{S0}}{f_x}$	f_x
Horizontal displacement in the y-direction	$\bar{c}_y = 0.57$	$a_{0,y} = 2 \cdot \pi \cdot f_y \cdot r_y \cdot \sqrt{\dfrac{\rho}{G_0 + 0.75 \cdot \lambda_{sy} \cdot g_0}}$	$\lambda_{sy} = \dfrac{v_{S0}}{f_y}$	f_y
Rotation about the x-axis	$\bar{c}_{\varphi x} = 0.3 \cdot \dfrac{a_{0,\varphi x}^2}{1 + a_{0,\varphi x}^2}$	$a_{0,\varphi x} = 2 \cdot \pi \cdot f_{\varphi x} \cdot r_{\varphi x} \cdot \sqrt{\dfrac{\rho}{G_0 + 0.75 \cdot \lambda_{s\varphi x} \cdot g_0}}$	$\lambda_{s\varphi x} = \dfrac{v_{S0}}{f_{\varphi x}}$	$f_{\varphi x}$
Rotation about the y-axis	$\bar{c}_{\varphi y} = 0.3 \cdot \dfrac{a_{0,\varphi y}^2}{1 + a_{0,\varphi y}^2}$	$a_{0,\varphi y} = 2 \cdot \pi \cdot f_{\varphi y} \cdot r_{\varphi y} \cdot \sqrt{\dfrac{\rho}{G_0 + 0.75 \cdot \lambda_{s\varphi y} \cdot g_0}}$	$\lambda_{s\varphi y} = \dfrac{v_{S0}}{f_{\varphi y}}$	$f_{\varphi y}$
Rotation about the z-axis	$\bar{c}_{\varphi z} = 0.3 \cdot \dfrac{a_{0,\varphi z}^2}{2 + a_{0,\varphi z}^2}$	$a_{0,\varphi z} = 2 \cdot \pi \cdot f_{\varphi z} \cdot r_{\varphi z} \cdot \sqrt{\dfrac{\rho}{G_0 + 0.25 \cdot \lambda_{s\varphi z} \cdot g_0}}$	$\lambda_{s\varphi z} = \dfrac{v_{S0}}{f_{\varphi z}}$	$f_{\varphi z}$

$v_{S0} = \sqrt{\dfrac{G_0}{\rho}}$ Shear wave velocity at the soil surface at $z = 0$

ρ Density of soil

for the radiation damping. The latter results from the propagation of elastic waves from the foundation into the ground and the energy radiation associated with it. The damping according to (5.111) and Table 5.21 depends on the vibration frequency $\omega = 2 \cdot \pi \cdot f$. With $g_0 = 0$, the coefficients $\bar{c}_{frg}$ are valid for the elastic half-space and can be approximated by the functions specified in [57]. In calculations in the frequency domain, the frequency-dependent damping is directly included in the calculation according to (5.111). In modal analysis, it is related to the corresponding natural frequencies, as shown below.

Other formulas for springs and dampers of rigid foundations, for example, for embedded foundations or foundations on an elastic soil layer, are given in [27, 57, 58] and in the summary in [1].

Example 5.29 For the foundation shown in Fig. 5.70, the spring and damper coefficients for vibrations in the horizontal direction are to be determined. The foundation is stiffened by basement walls and can be assumed to be rigid. Two homogeneous soils are investigated with the parameters according to Table 5.22. Soil A is a dense gravel material and soil B consists of a stiff loam or clay material.

First, the surface area and the second-order area moments of the foundation are determined:

$$A = 378 \text{ m}^2, \quad I_x = 13892 \text{ m}^4, \quad I_y = 10206 \text{ m}^4, \quad I_p = 24098 \text{ m}^4.$$

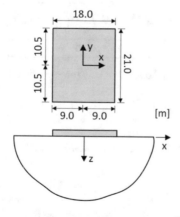

Fig. 5.70 Foundation and soil model

Table 5.22 Soil parameter

Soil type	G_0 [kN/m^2]	g_0 [kN/m^3]	μ	ρ [t/m^3]	v_{S0} [m/s]	ξ_{Mat} [%]
A	162000	8100	0.33	1.8	300	5
B	30000	0	0.33	1.8	129	5

According to Table 5.20, the equivalent radii are

$$r_x = r_y = \sqrt{\frac{378}{\pi}} = 10.97 \text{ m}, \quad r_{\varphi x} = 11.53 \text{ m}, \quad r_{\varphi y} = \sqrt[4]{\frac{4 \cdot 10206}{\pi}} = 10.68 \text{ m},$$

$$r_{\varphi z} = 11.13 \text{ m}.$$

The soil springs according to Table 5.20 are obtained, for example, for soil A and displacements in the y-direction and rotation around the x-axis as

$$k_y = \frac{8 \cdot 10.97}{2 - 0.33} \cdot (162000 + 0.5 \cdot 10.97 \cdot 8100) = 1.085 \cdot 10^7 \frac{\text{kN}}{\text{m}},$$

$$k_{\varphi x} = \frac{8 \cdot 11.53^3}{3 \cdot (1 - 0.33)} \cdot (162000 + 0.4 \cdot 11.53 \cdot 8100) = 1.217 \cdot 10^9 \text{ kNm}.$$

The soil springs are listed together with the corresponding values for soil B in Table 5.23.

To determine the damping, the reference frequencies must first be selected. They are taken from Example 5.30 for the vibrations in the respective degrees of freedom. With this, the dimensionless frequencies $a_{0,frg}$, the coefficients $\bar{c}_{frg}$, and the damping values in the respective degrees of freedom are determined. For example, for displacements in the y-direction and rotation around the x-axis, one obtains

$$a_{0,y} = 2 \cdot \pi \cdot 5.1 \cdot 10.97 \cdot \sqrt{\frac{1.8}{162000 + 0.75 \cdot \frac{300}{5.1} \cdot 8100}} = 0.65, \quad \bar{c}_y = 0.57$$

$$a_{0,\varphi x} = 2 \cdot \pi \cdot 5.1 \cdot 11.53 \cdot \sqrt{\frac{1.8}{162000 + 0.75 \cdot \frac{300}{5.1} \cdot 8100}} = 0.69,$$

$$\bar{c}_{\varphi x} = 0.3 \cdot \frac{0.69^2}{1 + 0.69^2} = 0.10$$

and according to (5.111), the damping coefficients ξ_{frg} as

Table 5.23 Spring and damper values

	Spring values [kN/m], [kNm]			Damping ratios [%]			Reference frequencies [Hz]	
	Soil A	Soil B		Soil A	Soil B		Soil A	Soil B
k_x	$1.085 \cdot 10^7$	$1.576 \cdot 10^6$	ξ_x	14	51		3.0	2.6
k_y	$1.085 \cdot 10^7$	$1.576 \cdot 10^6$	ξ_y	24	83		5.1	3.8
$k_{\varphi x}$	$1.217 \cdot 10^9$	$1.831 \cdot 10^8$	$\xi_{\varphi x}$	8	43		5.1	3.8
$k_{\varphi y}$	$9.523 \cdot 10^8$	$1.453 \cdot 10^8$	$\xi_{\varphi y}$	5	21		3.0	2.6
$k_{\varphi z}$	$1.324 \cdot 10^9$	$2.206 \cdot 10^8$	$\xi_{\varphi z}$	14	53		3.0	2.6

$$\xi_y = \frac{0.65}{2} \cdot 0.57 + 0.05 = 23.7 \ \%, \quad \xi_{\varphi x} = \frac{0.69}{2} \cdot 0.10 + 0.05 = 8.3 \ \%.$$

The values for the other degrees of freedom and for soil B are given in Table 5.23. The damping coefficients are larger in the degrees of freedom of displacements than in the degrees of freedom of rotations around the x- and y-axes. ◀

Modal damping

In principle, different damping coefficients can appear in the individual elements of a finite element model. This is the case when modeling the soil–structure interaction, where soil possesses a damping different from the structure. The question arises on how these different damping coefficients can be considered in an analysis. An arbitrary damping matrix can be treated in time history analyses, which are performed with numerical integration methods. When solving eigenvalue problems, an arbitrary, non-orthogonal damping matrix would lead to complex eigenvalues and eigenmodes, which can be used for time history analyses. For earthquake analyses, however, the simpler response spectrum method is generally preferred. For this reason and due to the natural fluctuation range of the damping coefficients, simplified heuristic methods have been developed in order to be able to include different damping coefficients of individual elements in a modal analysis.

In a modal analysis, a different damping coefficients ξ_i in each eigenmode i can be adopted when solving the modal equations (5.64). In order to take a different damping value for each element into account, appropriate modal damping coefficients are determined as a weighted sum of the damping coefficients of the individual elements. The ratio of the deformation energy stored in the respective element to the deformation energy of the total system is used as weighting factor.

The elastic strain energy of the total system with the system stiffness matrix $\underline{K}$, undergoing a displacement $\underline{u}$, is

$$E_{\text{ges}} = \frac{1}{2} \cdot \underline{u}^{\text{T}} \cdot \underline{K} \cdot \underline{u}. \tag{5.112}$$

For a displacement with the ith eigenmode $\underline{\phi}_i$, one obtains

$$E_{\text{ges},i} = \frac{1}{2} \cdot \underline{\phi}_i^{\text{T}} \cdot \underline{K} \cdot \underline{\phi}_i. \tag{5.113}$$

If the eigenmodes are normalized so that the generalized mass is 1, then according to (5.35b) it is

$$\underline{\phi}_i^{\text{T}} \cdot \underline{K} \cdot \underline{\phi}_i = \omega_i^2.$$

The energy in the kth element is

$$E_{k,i} = \frac{1}{2} \cdot \underline{\phi}_{k,i}^{\text{T}} \cdot \underline{K}_e \cdot \underline{\phi}_{k,i}. \tag{5.114}$$

Here, $\underline{K}_e$ is the element stiffness matrix and $\underline{\phi}_{k,i}$ are the displacements of the element k in the ith eigenmode. The weighting factors in the ith eigenmode, therefore, are $E_{k,i}/E_{\text{ges},i}$. If the damping in element k is designated by $\xi_{e,k}$, the damping coefficient of the ith eigenmode is obtained as

$$
\xi_i = \sum \frac{E_{k,i}}{E_{\text{ges},i}} \cdot \xi_{e,k} = \frac{\sum E_{k,i} \cdot \xi_{e,k}}{E_{\text{ges},i}} = \frac{\sum \underline{\phi}_{k,i}^{\mathrm{T}} \cdot \underline{K}_e \cdot \underline{\phi}_{k,i}}{\underline{\phi}_i^{\mathrm{T}} \cdot \underline{K} \cdot \underline{\phi}_i}. \tag{5.115}
$$

The summation is to be performed over all finite elements. In practice, the modal damping coefficients obtained in this way are limited by codes to values of, e.g. $\xi_i \leq 15\%$, because the method is based approximately on the real eigenmodes and frequencies of the undamped system.

Other methods for determining weighted damping factors are given in [59–62].

Example 5.30 The building treated in Examples 5.26 and 5.27 is investigated as a beam model, taking into account the soil–structure interaction (Fig. 5.71). For this purpose, the building is examined with the spring-damper model determined in Example 5.29 for the two soil types in Table 5.22.

The spring values were already determined in Example 5.29 and are given in Table 5.23. With them, the natural frequencies given in Table 5.24 are obtained. For the stiff soil A, they differ only slightly from those of the rigidly clamped beam model. In the case of soft soil B, however, significant differences occur, especially in the second natural frequency. Due to the elastic support, the natural frequencies decrease compared to those of the fixed base.

To determine the strain energy in the springs, their displacements in the eigenmodes are required. The eigenmodes are here normalized to the mass matrix so that (5.35a, 5.35b) apply as

$$
m_i^* = \underline{\phi}_i^{\mathrm{T}} \cdot \underline{M} \cdot \underline{\phi}_i = 1, \quad \underline{\phi}_i^{\mathrm{T}} \cdot \underline{K} \cdot \underline{\phi}_i = \omega_i^2 \quad \text{for} \quad i = 1, 2, 3 \ldots.
$$

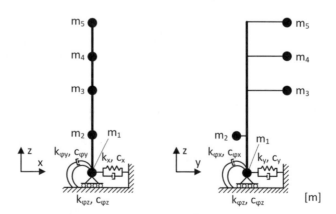

Fig. 5.71 Beam model of a building with soil springs and damper

Table 5.24 Eigenfrequencies and modal damping coefficients of the beam model

Eigenmode i	Eigenfrequency f_i [Hz]			Modal damping coefficient ξ_i [%]		Description
	Rigid	Soil A	Soil B	Soil A	Soil B	
1	3.1	3.0	2.6	5.5	15.3	Translation in the x-direction with torsion
2	5.5	5.1	3.6	6.6	28 (<47.3)	Translation in the y-direction
3	6.8	6.7	5.8	5.1	25.9	Torsion
4	18.7	14.4	7.1	10.0	28 (<37.0)	Translation in the x-direction

Thus, the expression for the strain energy in the ith eigenmode according to (5.113) is

$$E_{\text{ges},i} = \frac{1}{2} \cdot \underline{\phi}_i^{\text{T}} \cdot \underline{K} \cdot \underline{\phi}_i = \frac{1}{2} \cdot \omega_i^2 \quad \text{for} \quad i = 1, 2, 3 \ldots.$$

The strain energy in the horizontal and rotational spring is obtained from (5.114) to

$$E_{x,i} = \frac{1}{2} \cdot \phi_{kx,i} \cdot k_x \cdot \phi_{kx,i} \quad \text{and} \quad E_{\varphi x,i} = \frac{1}{2} \cdot \phi_{k\varphi x,i} \cdot k_{\varphi x} \cdot \phi_{k\varphi x,i}.$$

Here, $\phi_{kx,i}$ and $\phi_{k\varphi x,i}$ are the displacements and rotations, respectively, of the springs in the ith mode. For the model in Fig. 5.71, the displacements and rotations of node 1 are listed in Table 5.25. In the following, the model with soil B will be discussed.

For the first eigenmode, with the units kN, s, and m, the following numerical values are obtained:

Table 5.25 Displacements and rotations of nodal point 1 in the eigenmodes 1–4

i	u	v	φ_x	φ_y	φ_z
			Soil A		
1	$0.868 \cdot 10^{-3}$	0	0	$-0.099 \cdot 10^{-3}$	$0.028 \cdot 10^{-3}$
2	0	$2.659 \cdot 10^{-3}$	$0.228 \cdot 10^{-3}$	0	0
3	$1.035 \cdot 10^{-3}$	0	0	$-0.079 \cdot 10^{-3}$	$-0.207 \cdot 10^{-3}$
4	$20.422 \cdot 10^{-3}$	0	0	$-0.279 \cdot 10^{-3}$	$0.304 \cdot 10^{-3}$
			Soil B		
1	$5.236 \cdot 10^{-3}$	0	0	$-0.492 \cdot 10^{-3}$	$0.132 \cdot 10^{-3}$
2	0	$-12.050 \cdot 10^{-3}$	$-0.746 \cdot 10^{-3}$	0	0
3	$11.699 \cdot 10^{-3}$	0	0	$-0.221 \cdot 10^{-3}$	$-1.296 \cdot 10^{-3}$
4	$-27.817 \cdot 10^{-3}$	0	0	$-0.371 \cdot 10^{-3}$	$-0.851 \cdot 10^{-3}$

$$E_{ges,1} = \frac{1}{2} \cdot \omega_1^2 = \frac{1}{2} \cdot (2 \cdot \pi \cdot 2.6)^2 = 133.4$$

$$E_{x,1} = \frac{1}{2} \cdot 1.576 \cdot 10^6 \cdot 0.005235^2 = 21.6$$

$$E_{\varphi y,1} = \frac{1}{2} \cdot 1.453 \cdot 10^8 \cdot 0.000492^2 = 17.6$$

$$E_{\varphi z,1} = \frac{1}{2} \cdot 2.206 \cdot 10^8 \cdot 0.000132^2 = 1.9.$$

For the strain energy of the structure, the following portion of the energy remains:

$$E_{build,1} = 133.4 - 21.6 - 17.6 - 1.9 = 92.3.$$

The damping coefficients for $f_1 = 2.6$ Hz have been determined in Example 5.29 as $\xi_x = 51\%$, $\xi_{\varphi y} = 21\%$, and $\xi_{\varphi z} = 53\%$ for the degrees of freedom "horizontal displacement in the x-direction", "rotation around the y-axis", and "rotation around the z-axis" (torsion), respectively. The material damping of the structure is 5%. According to (5.115), the weighted modal damping in the first eigenmode is

$$\xi_1 = \frac{92.3}{133.4} \cdot 0.05 + \frac{21.6}{133.4} \cdot 0.51 + \frac{17.6}{133.4} \cdot 0.21 + \frac{1.9}{133.4} \cdot 0.53$$
$$= 0.153 = 15.3\%.$$

The modal damping coefficients of the other eigenmodes are determined in the same way. The modal damping coefficients of all eigenmodes are summarized in Table 5.24. For soil B, the modal damping increases with higher eigenmodes. In the case of stiff soil A, the additional damping contribution due to soil–structure interaction is small due to the low foundation displacements and rotations. Hence, the modal damping coefficients mainly reflect the material damping of the structure.

The seismic loads on the building are determined using the response spectrum method. In order to take modal damping values into account, the response spectrum in Fig. 5.63, which applies to 5% damping, is adapted to modified damping values with (5.80c). After (5.80c), the coefficient η must not be greater than 0.55. This means that the damping coefficient is limited to a maximum of $\xi = 10/\eta^2 - 5 = 10/0.55^2 - 5 \approx 28\%$.

The internal forces of the beam model superposed according to (5.85) for earthquake excitation in the y-direction are shown in Fig. 5.72. The second eigenmode is dominant here. The additional consideration of higher eigenmodes with deformations in the y-direction has only a negligible influence on the internal forces. For comparison, the internal forces of the model with a rigid restraint at the base are also given (cf. Example 5.27). For soft soil B, the bending moments and shear forces are reduced due to the soil–structure interaction. For stiff soil A, the differences between the internal forces of the two models with and without soil–structure interaction are

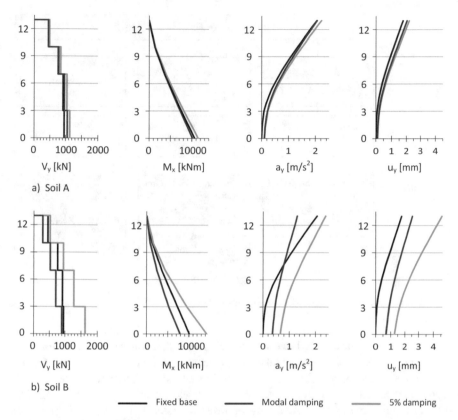

a) Soil A

b) Soil B

———— Fixed base   ———— Modal damping   ———— 5% damping

Fig. 5.72 Displacements, accelerations, and internal forces for earthquake excitation in the y-direction

small. For stiff soils, the soil–structure interaction can, therefore, be neglected and the building can be represented as rigidly clamped on the foundation level.

Taking into account the soil springs and assuming a constant damping of 5%, which corresponds to the material damping of the structure as well as the soil, the internal forces are overestimated. This means that for models with soil–structure interaction, the radiation damping must not be neglected. When modeling the elastic flexibility of the soil, the soil damping should be taken into account as well.   ◀

Finite element modeling of soil

Modeling the foundation as a rigid body and modeling the soil as a spring-damper model is the easiest way to deal with the dynamic soil–structure interaction. However, soil may also be represented as a three-dimensional finite element solid model together with the structure. In this case, the size of the soil model has to be chosen in such a way that elastic waves, which originate from the foundation and propagate in the soil, are not reflected at the arbitrarily introduced boundaries of the finite element model and affect the foundation vibrations. For short, shock-like processes, this can

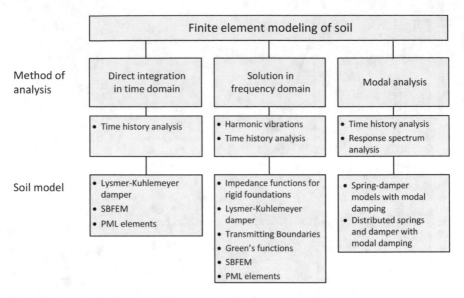

Fig. 5.73 Models for dynamic soil–structure interaction

be achieved by a sufficiently large soil model. This is no longer practicable for long-lasting processes, such as earthquakes. In this case, special elements are arranged at the boundaries of the soil region discretized in finite elements, which mechanically reproduce the static and dynamic properties of an elastic soil region extending "into infinity". Figure 5.73 gives an overview of different methods, distinguishing between methods suitable for analyses in the time domain, in the frequency domain, and for modal analysis [63].

In soil dynamics, the decay behavior of a wave can be described by an exponential function similar to that of a damped single degree of freedom system [30]. The decay of the amplitudes at time t is described by

$$u_{\max}(t) = u_0 \cdot e^{-\xi \cdot \omega \cdot t} = u_0 \cdot e^{-2 \cdot \pi \cdot f \cdot \xi \cdot t}.$$

A shear wave with the shear wave velocity v_S travels the distance $x = v_S \cdot t$ in time t. Thus, one obtains for a shear wave

$$u_{\max}(x) = u_0 \cdot e^{-2 \cdot \pi \cdot f \cdot \xi \cdot x / v_S}. \tag{5.116}$$

The amplitude decay is determined by the distance x and the damping coefficient ξ. With harmonic vibration, the amplitudes of waves with higher frequencies decrease more rapidly than those with lower frequencies. When representing soils as three-dimensional solids, extraordinarily large finite element models are thus required, unless measures are taken at the boundaries of the model to transmit waves.

In the *Lysmer–Kuhlemeyer model*, dampers are arranged at the boundaries of the finite element model in order to enable the transmission of P- and S-waves (Fig. 5.74) [64]. In the case of shear and compression waves, there is a linear relationship between

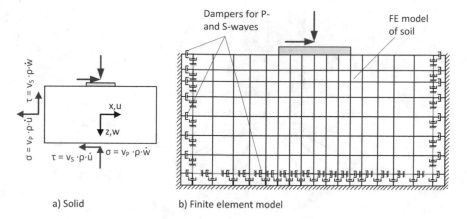

Fig. 5.74 Model with Lysmer–Kuhlemeyer dampers at the boundaries of the finite element model

the stress and the velocity which can be written as $\tau = v_S \cdot \rho \cdot \dot{w}$ or $\sigma = v_P \cdot \rho \cdot \dot{u}$, respectively, with the wave velocities v_S and v_P as given in (5.88c, 5.88d). With dampers per unit area

$$\bar{c}_\perp = v_P \cdot \rho, \tag{5.117a}$$

$$\bar{c}_\| = v_S \cdot \rho, \tag{5.117b}$$

orthogonal or parallel to the edge, compression and shear waves incident perpendicular to the edge are absorbed. However, since obliquely incident waves are partially reflected, this is an approximate method. Therefore, the edges of the finite element model must be sufficiently far away from the region to be investigated—as a foundation—so that reflected waves have already substantially decayed. The model does not require any calculation in the frequency domain and can be applied with standard finite element software.

In *Scaled Boundary Finite Element Method* (SBFEM), a contact surface is defined between an irregular region discretized into finite elements and the homogeneous linear elastic soil region extending into infinity (Fig. 5.75). Starting from a scaling point, the soil region is discretized into segmented elements. In each segment, shape functions of the displacements are defined, namely on the boundary surface as polynomials as in the finite element method and in the radial direction extending into infinity as analytical functions which fulfill the associated, ordinary differential equation. The wave radiation condition is thus fulfilled. The method was published first by Wolf and Song [65–69] and subsequently improved especially with regard to its efficiency [70–73]. It is available in the frequency and time domain, but only represents a homogeneous elastic half-space.

For finite element models of the soil, which are bounded at the bottom by a rigid subsoil, so-called *transmitting boundaries* have been developed [39, 74–80]. They

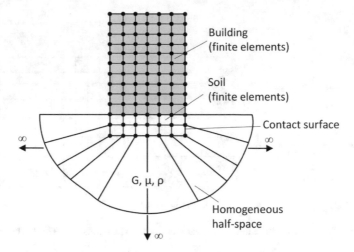

Fig. 5.75 SBFEM model with a building on a homogeneous half-space

are arranged along the vertical boundaries of the model (Fig. 5.76a). Transmitting boundaries have been derived in the frequency domain. They are semi-finite elements which, like the finite elements, are based on the shape functions of the displacements. However, the soil is only discretized in the vertical direction by piecewise linear or square shape functions. In the horizontal direction, the exact analytical solution is fulfilled, so that no discretization is required in this direction. A dynamic stiffness matrix of the soil at the vertical boundary, extending to infinity is obtained. The analyses are carried out in the frequency domain. Any horizontal stratification of the soil can be modeled.

In addition, attention should also be drawn to the so-called *PML elements* (*Perfectly Matched Layer*), which are arranged at the boundaries of the discretized

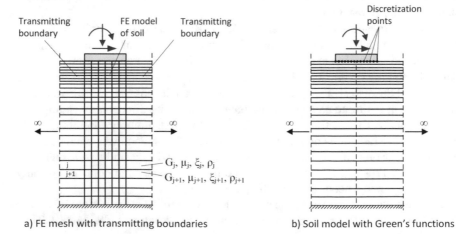

a) FE mesh with transmitting boundaries   b) Soil model with Green's functions

Fig. 5.76 Semi-finite elements of soil

soil region similar to transmitting boundaries to absorb incoming waves. They are based on stretching functions of the coordinates introduced in the derivation of the mass and stiffness matrices of finite continuum elements and can be specified in both frequency and time domain [81–86].

Green's functions describe the displacements at the surface (and inside) of a homogeneous or layered elastic half-space due to a single force at the surface of the half-space in the frequency domain. They can be used to construct soil flexibility matrices to be inverted to stiffness matrices. Green's functions for a layered half-space can be derived with semi-analytical solution procedures (*thin-layer method*) or analytically [80, 87, 88]. Solutions are also possible in the time domain for the half-space [89, 90].

Solutions based on the *Boundary Element Method* (BEM) start from an analytical solution for the elastic continuum and develop a dynamic stiffness matrix of the soil, which is related to the boundary of the model. Here, too, the computations usually require a calculation in the frequency domain. An overview of this development is given in [91–101].

Spring-damper models presume a rigid foundation (Fig. 5.69). Solutions in the frequency domain are available for various soil conditions and foundation types [1, 58]. Their frequency dependence is described by impedance functions of stiffness and damping. They can also be determined for any type of foundation and layered soils using numerical methods such as the thin-layer method based on Green's functions [55]. Impedance functions can be used for analyses in the frequency domain. In a modal analysis, the frequency dependence can be considered approximately with appropriate modal damping coefficients. In three-dimensional building models, springs and dampers can be assumed to be distributed over the foundation surface area as continuous elastic damped support. This allows the modeling of flexible foundation slabs. The coupling between the vertical degrees of freedom of displacements and the degrees of freedom of rotation around the two horizontal axes can be modeled using the EST model (Sect. 4.10.4) [102].

For the practical application of the abovementioned methods, their availability in commercial programs and efficiency are decisive. For simple beam models, the dynamic soil–structure interaction can be captured with spring-damper models. The calculation can be performed with a modal analysis and modal damping. For three-dimensional building models, springs and dampers can be distributed over the foundation surface and modeled as continuous elastic support with damping. More complex soil and building models can be investigated with SBFEM. It is implemented in [SOF 6]. Special software has also been developed for the analysis of complex soil and structure models using the thin-layer method in the frequency domain [76, 103]. For complex soil models, the radiation damping can be approximated with Lysmer–Kuhlemeyer dampers. The model is simple, but can be applied to nonlinear problems in time history analyses too. It can be used with general finite element software packages like [SOF 2], but it is also implemented in programs specialized for soil dynamics like [SOF 10].

Example 5.31 The three-dimensional building model in Examples 5.26 and 5.28 shall be examined under consideration of soil–structure interaction for an earthquake excitation in the *y*-direction. The soil properties of soil B given in Table 5.22 will be applied.

The analyses are performed with an SBFEM model with [SOF 6]. In addition to modeling the building, the excavation volume of the soil has to be modeled with solid elements (cf. Fig. 5.75). Figure 5.77 shows the building model with the soil elements up to the contact surface with the SBFEM element [50]. The SBFEM

Fig. 5.77 Finite element model of the building and the soil [50]

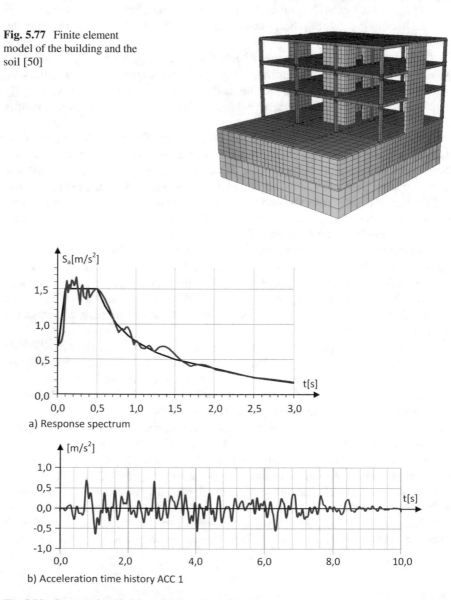

a) Response spectrum

b) Acceleration time history ACC 1

Fig. 5.78 Generated artificial acceleration time history

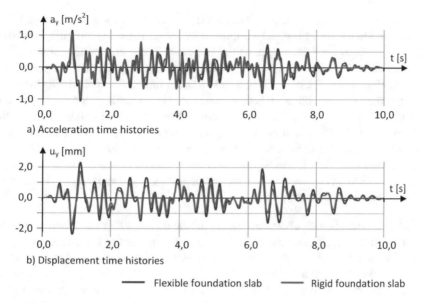

Fig. 5.79 Time history of the displacements and accelerations in the top floor [50]

element describes the radiation damping but does not contain any material damping of the soil.

For the acceleration time history of the ground, several artificial earthquake time histories compatible with the response spectrum according to Fig. 5.63 are generated. An acceleration time history generated with [SOF 11] according to [30] is given in Fig. 5.78b. Figure 5.78a shows the response spectrum of the generated artificial time history and the original response spectrum [50]. The calculations were performed with a total of 4 artificially generated time histories.

In the following, the displacements and accelerations in the center of rigidity of the building determined with the building model are examined. The time histories

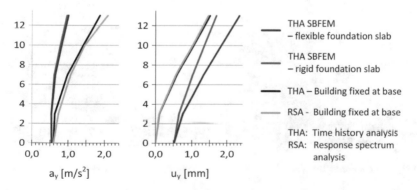

Fig. 5.80 Distribution of maximum displacements and accelerations over height [50]

of the displacements u_y and accelerations a_y in the top floor at $z = 13$ m in the y-direction are shown in Fig. 5.79. Figure 5.80 shows the maximum displacements and accelerations over the height of the building. These are the maximum values averaged from 4 earthquakes. For comparison, the accelerations and displacements of the model fixed at the base are also shown according to Example 5.28. Due to the radiation damping, the building accelerations are significantly reduced compared to the rigidly fixed model without soil–structure interaction. The maximum accelerations correspond well with those of the beam model in Fig. 5.72. The foundation slab has a thickness of 40 cm. For comparison, the calculations were also carried out with a rigid foundation slab. It can be seen that the influence of the deformations of the foundation slab and the basement on the accelerations is low and thus the assumption of a rigid basement is appropriate.

The maximum (relative) displacements for the model with soil–structure interaction include the displacements of the foundation compared to the ground in the far field. They are thus greater than the displacements of the rigidly fixed model, where the foundation does not move relative to the ground. If a rigid foundation is assumed, they correspond well to those of the beam model in Fig. 5.72. They increase slightly if the deformability of the foundation slab is taken into account (Fig. 5.79b).

Similar results are obtained with a building model with continuous elastic support and modal damping. The subgrade reaction moduli are obtained by means of the spring constants of the beam model for soil B in Table 5.23 in vertical and horizontal directions as

$$k_{S,z} = \frac{k_{\varphi x}}{I_x} = \frac{1.83 \cdot 10^8}{13892} = 13183 \ \frac{\text{kN}}{\text{m}^3}, \quad k_{S,y} = \frac{k_y}{A} = \frac{1.576 \cdot 10^6}{378} = 4170 \ \frac{\text{kN}}{\text{m}^3}.$$

The calculations were carried out for a rigid foundation slab and a flexible foundation slab with a slab thickness of 40 cm. The lowest natural frequency for translation in the y-direction of 3.9 Hz for the rigid and 3.4 Hz for the flexible foundation slabs corresponds to that of the beam model of 3.6 Hz (cf. Example 5.30, Table 5.24). The corresponding eigenmode in Fig. 5.81 illustrates the deformations of the base-

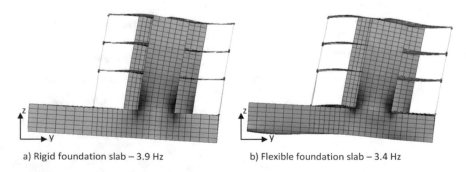

a) Rigid foundation slab – 3.9 Hz   b) Flexible foundation slab – 3.4 Hz

Fig. 5.81 Lowest eigenmodes for translation of the building in the y-direction

ment. The accelerations determined by a time history analysis and a response spectrum analysis with modal damping are given in [50, 102]. They correspond in good approximation to those in Fig. 5.80. ◀

5.8.6 Modeling and Validation

Problems in structural dynamics are usually characterized by a greater uncertainty than in structural analysis. Load assumptions often show a larger variation than in static analyses. This applies, for example, to the load assumptions for earthquakes and the assumed distributions and time histories of wind pressures. Variations in the stiffness parameters of structural elements and the stiffness of non-load-bearing walls, which can be neglected in a static structural analysis can have a significant influence on the dynamic behavior in the resonance range.

In structural dynamics, systems can often be modeled with the simple model of the single degree of freedom system when the first natural frequency is prevailing. However, the realistic modeling of complex structures for dynamic loads can also require very detailed models. For example, when calculating the horizontal load transfer of a structure statically, the internal forces normally can be determined floor by floor. The calculation of the horizontal vibrations of the structure, however, requires a model for the entire structure, including all floors and, if necessary, of soil–structure interaction. Interactions between the individual floors, which are often insignificant in the static case, are the basis of the vibration processes occurring. In general, in dynamic processes mass and stiffness forces of different structural components interact with each other as a function of time and thus the structural components cannot be treated separately in the analysis, as is often the case in statics. The reason for this lies in an important property of vibrational and wave propagation processes, namely their ability to preserve or transport energy. Wave processes are found in nature to transport small amounts of energy over long distances, as is the case with sound waves in acoustics. A similar situation is found in soil dynamics. If a soil is statically loaded with a rigid foundation slab, the displacements on its surface quickly decay with the distance from the foundation. Under dynamic loading, however, waves propagate which, despite the amplitude reduction caused by radiation and material damping, have a considerably greater range of influence than static deformations. When calculating soil with finite element models the models must be sufficiently large to avoid unintended wave reflections at the edges. In the case of building analysis a calculation in the high-frequency range requires a detailed modeling of the building structure.

In the resonance range, the response vibrations of structural models can react sensitively to minor changes of the structural model or the material parameters. Often there is no "safe side" of the model assumptions in an analysis like it is in static problems. This can be taken into account by parameter variation of the affected values.

The required degree of discretization of a structure or component is influenced by many factors. The size of the frequency range to be considered as well as the type and location of the load and the required results play an important role. Due to the difficulties described for complex vibration models, simple models are preferable, at least for comparative calculations. Due to the small number of parameters, they offer the advantage of being more manageable and easier to interpret in their behavior.

Some important aspects of modeling are summarized below.

Modeling in structural dynamics

- In a modal analysis, the representation of higher eigenmodes requires a finer discretization than that of lower eigenmodes.

    - For beams, a sinusoidal part (spacial period) of a mode shape must be discretized into at least 6 elements.
    - For plane or spatial solids, a wavelength must be discretized into at least 6 elements with linear shape functions.
    - The determination of internal forces, especially shear forces, requires a finer discretization than that of displacements.

- In a modal analysis, a sufficient number of eigenmodes is to be considered.

    - In earthquake analyses, the verification can be done by means of the effective masses or the effective mass factors.
    - The determination of internal forces usually requires more eigenmodes than that of deformations with the same degree of accuracy.

- In the case of Rayleigh damping, in addition to the damping at the two reference frequencies, the effects on damping in the entire frequency range of interest should be considered.
- With numerical time integration, the size of the time step should not exceed 1/10 of the period of the highest natural frequency to be considered.
- For analyses in the frequency domain, the fundamental period of the Fourier transform must be selected in such a way that the vibrations completely decay. The time step should be selected in such a way that the period of the highest frequency to be considered is represented with 6 time steps at least.

Example 5.32 The accuracy achievable in building modeling shall be demonstrated by the validation of a calculation carried out in the context of an experimental investigation. It concerns a large-scale test on a 64 m high power plant building with a diameter of 22 m [104, 105]. The building consists of an external structure made up of a hemispherical shell with an adjoining cylindrical shell and an internal structure of walls and ceilings on several floors. The outer and inner structure are located on a common foundation slab (Fig. 5.82). During the test, the reinforced concrete shell of the outer structure was subjected to an impact load. This was realized by a

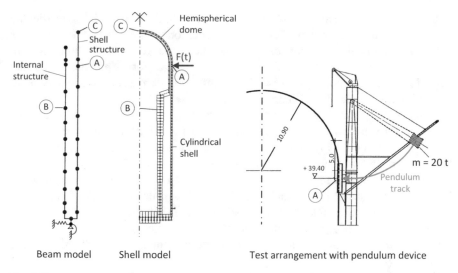

Fig. 5.82 Large-scale tests on a building

pendulum device with an impact body with a mass of 20 t and a drop height of 5 m. The force-time histories of the impact load assumed in the analysis in advance and measured later in the test are shown in Fig. 5.83. They are in good agreement.

Prior to the tests, calculations were carried out and the results were deposited with the institution responsible for the tests. After the tests had been carried out, they were compared with the test results. Later adjustments of calculation parameters were thus excluded. Neither the calculation engineers nor the test supervisors knew the results of the calculation and the subsequent test. Hence, it was a double-blind test.

For comparison, two structural models, namely a beam model and a shell model, were examined. For the beam model, the inner and outer structure was represented as beams and the soil as springs. It should be able to represent the global vibration behavior of the structure.

In the shell model the outer structure, i.e. the dome and the adjoining cylinder, were modeled as a shell. This model was intended to represent both, the high-frequency shell vibrations with their short wavelengths as well as the overall vibration behavior

Fig. 5.83 Load-time history

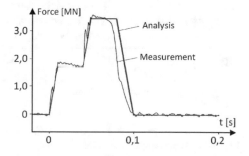

of the structure, which requires small element sizes. In order to be able to restrict the fine discretization required to one dimension, rotationally symmetric shell elements with non-rotationally symmetric displacement fields were used for the finite element model (cf. Sect. 4.8.3). The shape functions are linear for membrane action and cubic for bending action. In the direction of the circumference, all space-dependent quantities are described by a Fourier series.

The dynamic analysis was carried out as a modal analysis. All natural frequencies below 80 Hz were considered. The natural frequencies for the Fourier term $n = 1$ to $n = 9$ are shown in Fig. 5.84. For $n > 9$, there are no natural frequencies below 80 Hz. Figure 5.84 shows two typical eigenmodes.

The time histories of the horizontal accelerations at selected points are shown in Fig. 5.85. First, the results at point A, which is located on the outer shell directly at the point of impact, are considered. The vibrations of the shell model at the beginning of the time history with large amplitudes and high frequencies are local shell vibrations. The subsequent vibrations with low amplitudes represent the global vibrations of the structure.

The shell model is able to reproduce both the local shell vibrations and the global structural vibrations with high accuracy. The global, low-frequency building response, which is governed by the lowest eigenmodes, can also be reliably reproduced by the beam model, whereas this is inherently not the case for the local shell vibrations.

The accelerations of points B and C are essentially determined by the global building vibration. Therefore, the results of the beam model are nearly identical to those of the shell model. Both agree well with the measured acceleration time histories.

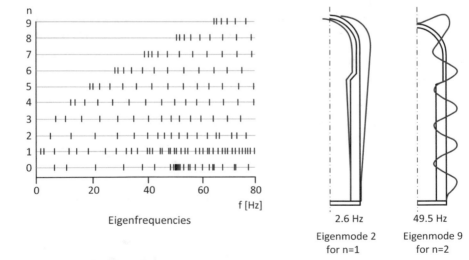

Fig. 5.84 Modal analysis of the shell model

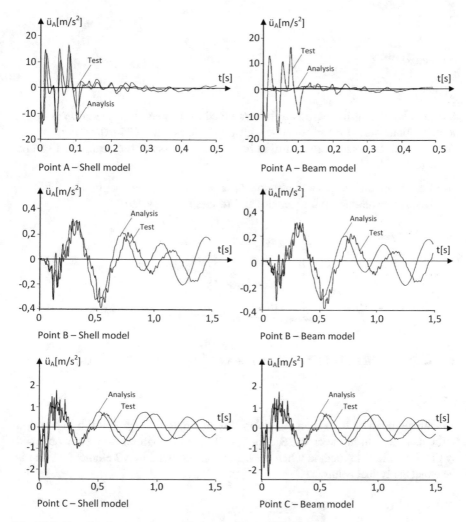

Fig. 5.85 Computed versus measured horizontal accelerations of the beam and shell models

As assumed in the analysis, the stresses in the structure remained within the linear range of material behavior. The comparison of the test and calculation results shows that even for complex structures the analysis can accurately predict the dynamic behavior within the scope of the model assumptions. ◄

Exercises

Problems

5.1

A beam clamped at the base is to be examined as a three-dimensional system. At point 2, a plate of reinforced concrete with a specific weight of 25 kN/m^3 is arranged. The plate can be considered as a rigid body and the mass of the beam itself can be neglected.

(a) Determine the masses and rotational masses of the system at point 2.
(b) Determine all natural frequencies and modes of the system.

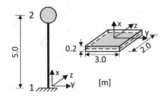

$A = 0.015\,\text{m}^2$, $I_T = 9.244 \cdot 10^{-4}\,\text{m}^4$, $I_y = I_z = 4.622 \cdot 10^{-4}\,\text{m}^4$, $E = 2.1 \cdot 10^8\,\text{kN/m}^2$, $\mu = 0.2$

5.2

A beam clamped on both sides is equipped with a mass of $\bar{m} = 0.8$ t/m in addition to the dead weight of the cross section. Specify the first natural frequency obtained by the finite element method when the beam is discretized into 2 elements. What is the exact analytical solution?

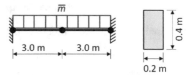

$E = 3.0 \cdot 10^7\,\text{kN/m}^2$ $\rho = 2.5\,\text{t/m}^3$.

5.3

The beam shown is discretized into 3 beam elements.

(a) Formulate the eigenvalue problem to determine the eigenfrequencies and modes. How many eigenfrequencies does the finite element model possess?

(b) Set up the equations of motion for an earthquake excitation in vertical direction
     with the earthquake acceleration $a_{g,v}(t)$.

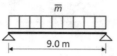

$$E = 2.1 \cdot 10^8 \text{ kN/m}^2 \quad I = 33740 \text{ cm}^4, \quad \bar{m} = 0.5 \text{ t/m}.$$

5.4

A non-sway frame with infinitely large extensional stiffness is discretized as shown
in beam elements. The distributed mass $\bar{m}$ is effective on the beam; the mass of the
columns can be neglected.

(a) Determine the stiffness and mass matrices of the system exploiting the symmetry
     of the system.
(b) Calculate the natural frequency determined by a finite element program.

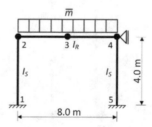

$$E = 2.1 \cdot 10^8 \text{ kN/m}^2 \quad \bar{m} = 2.0 \text{ t/m}, \quad I_R = 6 \cdot 10^{-4} \text{m}^4 \quad I_S = 2 \cdot 10^{-4} \text{m}^4.$$

5.5

Model the frame as a two degrees of freedom system and determine the natural
frequencies and modes.

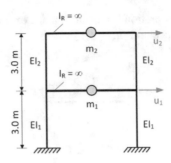

$m_1 = 250$ t, $m_2 = 125$ t, $I_1 = 32143$ cm^4, $I_2 = 21429$ cm^4, $E = 2.1 \cdot 10^8$ kN/m^2.

5.6

Establish the flexibility matrix of the system and determine the stiffness matrix from it. Calculate the natural frequencies and eigenmodes of the system. Normalize the eigenmodes so that the generalized mass is 1.

$E = 2.1 \cdot 10^8$ kN/m^2, $I = 20000$ cm^4, $m_1 = 3.0$ t, $m_2 = 2.0$ t

5.7

The eigenmodes of the frame shown are given to

$$\underline{\phi}_1 = \begin{bmatrix} 0.366325 \\ 0.796587 \\ 1.000000 \end{bmatrix}, \quad \underline{\phi}_2 = \begin{bmatrix} 1.000000 \\ 0.338228 \\ -0.963319 \end{bmatrix}, \quad \underline{\phi}_3 = \begin{bmatrix} 0.715061 \\ -1.000000 \\ 0.696111 \end{bmatrix}.$$

(a) Prove that the eigenmodes fulfill the orthogonality conditions with an accuracy of 3 decimals.
(b) Determine the generalized stiffnesses and masses in the three eigenmodes.
(c) How are the three natural frequencies of the system?

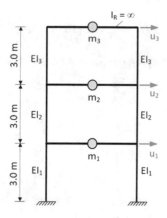

$$E = 2.1 \cdot 10^8 \text{ kN/m}^2, \quad I_1 = 32143 \text{ cm}^4, \quad I_2 = I_3 = 21429 \text{ cm}^4,$$
$$m_1 = 400 \text{ t}, \quad m_2 = 350 \text{ t}, \quad m_3 = 250 \text{ t}.$$

5.8

Determine the natural frequency of a plate clamped on all sides. The calculation is to be performed with a single shear rigid plate element using the symmetry of the plate. Compare the result with the analytical solution.

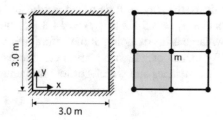

$$E = 2.1 \cdot 10^8 \text{ kN/m}^2, \quad \mu = 0.2, \quad t = 2.0 \text{ cm}.$$

5.9

The first eigenmode and the corresponding natural vibration period of a mast are

$$\varphi_1 = \begin{bmatrix} 0.0118 \\ 0.2429 \\ 0.5912 \\ 1.0000 \end{bmatrix}, \quad T_1 = 0.51 \text{ s}.$$

(a) Normalize the eigenvector in such a way that the generalized mass is 1.
(b) Establish the modal equation of motion for the first eigenmode if the load

$$F(t) = 2 \cdot \sin(10 \cdot t) \ [kN]$$

is applied at node 3 for a damping coefficient of $\xi = 5\%$.

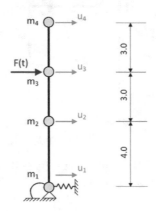

$m_1 = 4 \ t, \quad m_2 = 7 \ t, \quad m_3 = 5 \ t, \quad m_4 = 3 \ t$

5.10

For a time domain analysis, the Rayleigh damping is adjusted in such a way that at the two natural frequencies $f_1 = 3.5$ Hz and $f_3 = 14.8$ Hz, the damping coefficient of 5% is obtained. How large is the damping at the natural frequencies $f_2 = 14.8$ Hz and $f_4 = 24.6$ Hz? By means of which equation can the damping coefficient be approximated for high frequencies and which for low frequencies?

5.11

A damped single degree of freedom system is in rest at the time $t = 0$. Determine the displacement, acceleration, and the spring force at the time $t = 0.12$ s by the Newmark method with the time step $\Delta t = 0.03$ s for a time-varying force $F(t) = F_0 \cdot \cos(20 \cdot t)$ with t in s and $F_0 = 2.0$ kN.

$k = 500$ kN/m, $\quad m = 1.0$ t, $\quad \xi = 10\%$.

5.12

Determine the displacement, acceleration, and spring force of the single degree of freedom system in Exercise 5.10 at time $t = 0.12$ s by the Newmark method with the time step $\Delta t = 0.03$ s for an initial displacement $u_0 = 0.5$ cm at time $t = 0$.

5.13

The frame in Exercise 5.4 is loaded at the level of the beam by the harmonic horizontal force $F(t) = F_0 \cdot \sin(\Omega \cdot t)$ with $F_0 = 10$ kN. How large are the amplitudes of the horizontal displacements of the beam in steady state response with an excitation frequency of $f = 0$ (static force), $f = 1.80$ Hz, $f = 2.00$ Hz, and $f = 5.00$ Hz, when a damping coefficient $\xi = 0$ is assumed.

5.14

The system specified in Exercise 5.1 is excited in the y-direction by an earthquake. Determine the displacements of the system by the response spectrum method. Only the lowest eigenmode for vibrations in the y-direction is to be considered. Use the internal forces matrix of the beam to determine the internal forces due to earthquakes. The response spectrum according to DIN EN 1998 shall be based on the following parameters:

$a_{gR} = 0.6$ m/s^2, $T_B = 0.1$ s, $T_C = 0.4$ s, $T_D = 2.0$ s,
$S = 1.25$, $\gamma_I = 1.2$, $q = 1.5$, $\beta_0 = 2.5$.

5.15

The clamped beam in Exercise 5.6 shall be examined by the response spectrum method. Determine the maximum displacements and accelerations in each eigenmode and their superposition. Give the horizontal forces acting on the beam in each eigenmode. Both eigenmodes are to be considered. The response spectrum according to DIN EN 1998 shall be based on the following parameters:

$a_{gR} = 0.6$ m/s^2, $T_B = 0.1$ s, $T_C = 0.4$ s, $T_D = 2.0$ s,
$S = 1.25$, $\gamma_I = 1.2$, $q = 1.5$, $\beta_0 = 2.5$.

Solutions

5.1

(a) $m_x = m_y = m_z = 3.0$ t, $\Theta_x = 3.25$ kNms^2,
     $\Theta_y = 1.00$ kNms^2, $\Theta_z = 2.25$ kNms^2

(b)   With $\underline{u} = \begin{bmatrix} u_2 \\ v_2 \\ w_2 \\ \varphi_{x,2} \\ \varphi_{y,2} \\ \varphi_{z,2} \end{bmatrix}$ one obtains:

$$f_1 = 4.29 \text{ Hz}, \quad \underline{\phi}_1 = \begin{bmatrix} 0 \\ 1 \\ 0 \\ 0 \\ 0 \\ 0.306 \end{bmatrix}, \quad f_2 = 4.37 \text{ Hz}, \quad \underline{\phi}_2 = \begin{bmatrix} 0 \\ 0 \\ 1 \\ 0 \\ -0.303 \\ 0 \end{bmatrix},$$

$$f_3 = 11.23 \text{ Hz}, \quad \underline{\phi}_1 = \begin{bmatrix} 0 \\ 0 \\ 0 \\ 1 \\ 0 \\ 0 \end{bmatrix}, f_4 = 30.57 \text{ Hz}, \quad \underline{\phi}_4 = \begin{bmatrix} 0 \\ -0.23 \\ 0 \\ 0 \\ 0 \\ 1 \end{bmatrix},$$

$$f_5 = 45.02 \text{ Hz}, \quad \underline{\phi}_5 = \begin{bmatrix} 0 \\ 0 \\ 0.101 \\ 0 \\ 1 \\ 0 \end{bmatrix}, \quad f_6 = 4.29 \text{ Hz}, \quad \underline{\phi}_6 = \begin{bmatrix} 1 \\ 0 \\ 0 \\ 0 \\ 0 \\ 0 \end{bmatrix}.$$

5.2

$f_{\text{FE}} = 15.5 \text{ Hz}, \quad f_{\text{exact}} = 17.7 \text{ Hz}.$

5.3

(a)   System:

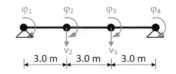

Eigenvalue problem:

$$\left(\begin{bmatrix} 0.945 & -0.472 & 0.472 & 0 & 0 & 0 \\ -0.472 & 0.630 & 0 & -0.315 & 0.472 & 0 \\ 0.472 & 0 & 1.889 & -0.472 & 0.472 & 0 \\ 0 & -0.315 & -0.472 & 0.630 & 0 & 0.472 \\ 0 & 0.472 & 0.472 & 0 & 1.889 & 0.472 \\ 0 & 0 & 0 & 0.472 & 0.472 & 0.945 \end{bmatrix} \cdot 10^5 \right.$$

$$\left. -\omega^2 \cdot \begin{bmatrix} 0 & 0 & 0 & 0 & 0 & 0 \\ 0 & 1.5 & 0 & 0 & 0 & 0 \\ 0 & 0 & 0 & 0 & 0 & 0 \\ 0 & 0 & 0 & 1.5 & 0 & 0 \\ 0 & 0 & 0 & 0 & 0 & 0 \\ 0 & 0 & 0 & 0 & 0 & 0 \end{bmatrix}\right) \cdot \begin{bmatrix} \bar{\varphi}_1 \\ \bar{v}_2 \\ \bar{\varphi}_2 \\ \bar{v}_3 \\ \bar{\varphi}_3 \\ \bar{\varphi}_4 \end{bmatrix} = \begin{bmatrix} 0 \\ 0 \\ 0 \\ 0 \\ 0 \\ 0 \end{bmatrix}$$

The system possesses 2 eigenfrequencies.

(b) Equations of motion for vertical seismic excitation:

$$\begin{bmatrix} 0.945 & -0.472 & 0.472 & 0 & 0 & 0 \\ -0.472 & 0.630 & 0 & -0.315 & 0.472 & 0 \\ 0.472 & 0 & 1.889 & -0.472 & 0.472 & 0 \\ 0 & -0.315 & -0.472 & 0.630 & 0 & 0.472 \\ 0 & 0.472 & 0.472 & 0 & 1.889 & 0.472 \\ 0 & 0 & 0 & 0.472 & 0.472 & 0.945 \end{bmatrix} \cdot 10^5 \cdot \begin{bmatrix} \varphi_1 \\ v_2 \\ \varphi_2 \\ v_3 \\ \varphi_3 \\ \varphi_4 \end{bmatrix}$$

$$+ \begin{bmatrix} 0 & 0 & 0 & 0 & 0 & 0 \\ 0 & 1.5 & 0 & 0 & 0 & 0 \\ 0 & 0 & 0 & 0 & 0 & 0 \\ 0 & 0 & 0 & 1.5 & 0 & 0 \\ 0 & 0 & 0 & 0 & 0 & 0 \\ 0 & 0 & 0 & 0 & 0 & 0 \end{bmatrix} \cdot \begin{bmatrix} \ddot{\varphi}_1 \\ \ddot{v}_2 \\ \ddot{\varphi}_2 \\ \ddot{v}_3 \\ \ddot{\varphi}_3 \\ \ddot{\varphi}_4 \end{bmatrix} = - \begin{bmatrix} 0 \\ 1.5 \\ 0 \\ 1.5 \\ 0 \\ 0 \end{bmatrix} \cdot a_{g,v}(t)$$

5.4

(a) $\underline{K} = \begin{bmatrix} 16.800 & -4.725 \\ -4.725 & 2.363 \end{bmatrix} \cdot 10^4$, $\quad \underline{M} = \begin{bmatrix} 0 & 0 \\ 0 & 4 \end{bmatrix}$, $\quad \underline{u} = \begin{bmatrix} \varphi_2 \\ v_3 \end{bmatrix}$,

(b) $f_1 = 8.1$ Hz, $\quad T_1 = 0.124$ s.

5.5

$$\underline{K} = \begin{bmatrix} 10 & -4 \\ -4 & 4 \end{bmatrix} \cdot 10^4, \quad \underline{M} = \begin{bmatrix} 250 & 0 \\ 0 & 125 \end{bmatrix}, \quad \underline{u} = \begin{bmatrix} u_1 \\ u_2 \end{bmatrix}$$

$$f_1 = 1.82 \text{ Hz}, \quad f_2 = 3.87 \text{ Hz}, \quad \underline{\phi}_1 = \begin{bmatrix} -0.843 \\ 1 \end{bmatrix}, \quad \underline{\phi}_2 = \begin{bmatrix} 0.593 \\ 1 \end{bmatrix}.$$

5.6

$$\underline{K} = \begin{bmatrix} 9.000 & -2.813 \\ -2.813 & 1.125 \end{bmatrix} \cdot 10^3, \quad \underline{M} = \begin{bmatrix} 3000 & 0 \\ 0 & 2000 \end{bmatrix}, \quad \underline{u} = \begin{bmatrix} u_1 \\ u_2 \end{bmatrix}.$$

$$f_1 = 1.65 \text{ Hz}, \quad f_2 = 9.36 \text{ Hz}, \quad \underline{\phi}_1 = \begin{bmatrix} 0.324 \\ 1 \end{bmatrix}, \quad \underline{\phi}_2 = \begin{bmatrix} 1 \\ -0.486 \end{bmatrix}.$$

Normalization with respect to the generalized mass:

$$\underline{\phi}_1 = \begin{bmatrix} 0.673 \\ 2.078 \end{bmatrix} \cdot 10^{-2}, \quad \underline{\phi}_2 = \begin{bmatrix} 1.697 \\ -0.825 \end{bmatrix} \cdot 10^{-2}.$$

5.7

(a) $\left| \underline{\phi}_1^T \cdot \underline{M} \cdot \underline{\phi}_2 \right| < 10^{-4}, \quad \left| \underline{\phi}_1^T \cdot \underline{M} \cdot \underline{\phi}_3 \right| < 10^{-3}, \quad \left| \underline{\phi}_3^T \cdot \underline{M} \cdot \underline{\phi}_3 \right| < 10^{-3}$

(b) $\underline{\phi}_1^T \cdot \underline{M} \cdot \underline{\phi}_1 = 526, \quad \underline{\phi}_2^T \cdot \underline{M} \cdot \underline{\phi}_2 = 672, \quad \underline{\phi}_3^T \cdot \underline{M} \cdot \underline{\phi}_3 = 676$

$\underline{\phi}_1^T \cdot \underline{K} \cdot \underline{\phi}_1 = 17112, \quad \underline{\phi}_2^T \cdot \underline{K} \cdot \underline{\phi}_2 = 145279, \quad \underline{\phi}_3^T \cdot \underline{K} \cdot \underline{\phi}_3 = 263408.$

(c) $f_1 = 0.91 \text{ Hz}, \quad f_2 = 2.34 \text{ Hz}, \quad f_3 = 3.14 \text{ Hz}$

$T_1 = 1.10 \text{ s}, \quad T_2 = 0.43 \text{ s}, \quad T_3 = 0.32 \text{ s}$

5.8

$f_{\text{FE}} = 14.0 \text{ Hz}, \quad f_{\text{exact}} = 19.4 \text{ Hz}.$

5.9

(a) $\phi_1 = \begin{bmatrix} 0.0052 \\ 0.1069 \\ 0.2602 \\ 0.4402 \end{bmatrix},$

(b) $151.78 \cdot \alpha_1(t) + 1.232 \cdot \dot{\alpha}_1(t) + \ddot{\alpha}_1(t) = 0.52 \cdot \sin(10 \cdot t).$

5.10

$\xi(f_2) = 4.0\%, \quad \xi(f_4) = 7.3\%.$

Low frequencies: mass proportional damping: $\xi(f) = \dfrac{0.142}{f}$

High frequencies: stiffness proportional damping: $\xi(f) = 0.002732 \cdot f$

5.11

$t = 0: \quad u = 0, \qquad\qquad \dot{u} = 0, \qquad\qquad \ddot{u} = 2.000 \text{ m/s}, \quad F_k = 0$

$t = 0.12 \text{ s}: u = 2.542 \text{ mm}, \ \dot{u} = -5.6 \text{ cm/s}, \ \ddot{u} = -2.494 \text{ m/s}, \ F_k = 1.27 \text{ kN}$

5.12

$t = 0: \quad u = 5.000 \text{ mm}, \qquad \dot{u} = 0, \qquad\qquad \ddot{u} = -2.500 \text{ m/s}, \ F_k = 2.5 \text{ kN}$

$t = 0.12 \text{ s}: u = -3.112 \text{ mm}, \ \dot{u} = -4.7 \text{ cm/s}, \ \ddot{u} = 1.767 \text{ m/s}, \quad F_k = -16.3 \text{ kN}$

5.13

$f = 0: \qquad u_1 = 0.17 \text{ mm}, \qquad u_2 = 0.17 \text{ mm}$

$f = 1.8 \text{ Hz}: \quad u_1 = 2.16 \text{ mm}, \qquad u_2 = 3.60 \text{ mm}$

$f = 2.0 \text{ Hz}: \quad u_1 = -0.16 \text{ mm}, \quad u_2 = -0.32 \text{ mm}$

$f = 5.0 \text{ Hz}: \quad u_1 = -0.02 \text{ mm}, \quad u_2 = 0.01 \text{ mm}$

5.14

$$\underline{u}_2 = \begin{bmatrix} v_2 \\ \varphi_{z2} \end{bmatrix} = \begin{bmatrix} 1.9 \text{ mm} \\ 0.59 \cdot 10^{-3} \end{bmatrix},$$

$$V_1 = \begin{bmatrix} V_{y1,1} \\ M_{z1,1} \\ V_{y2,1} \\ M_{z2,1} \end{bmatrix} = \begin{bmatrix} -4.20 \\ 21.99 \\ -4.20 \\ 0.97 \end{bmatrix}, \ V_2 = \begin{bmatrix} V_{y1,2} \\ M_{z1,2} \\ V_{y2,2} \\ M_{z2,2} \end{bmatrix} = \begin{bmatrix} -0.22 \\ 0.38 \\ -0.22 \\ -0.71 \end{bmatrix},$$

$$V_{max} = \begin{bmatrix} V_{y1,max} \\ M_{z1,max} \\ V_{y2,max} \\ M_{z2,max} \end{bmatrix} = \begin{bmatrix} 4.21 \\ 21.99 \\ 4.21 \\ 1.20 \end{bmatrix}.$$

5.15

Eigenmode 1:

$f = 1.65 \text{ Hz}: u_1 = 3.844 \text{ mm}, \ u_2 = 11.862 \text{ mm}, \ a_1 = 0.411 \text{ m/s}^2, \ a_2 = 1.267 \text{ m/s}^2$
$$F_1 = 1.232 \text{ kN}, \quad F_2 = 2.534 \text{ kN}.$$

Eigenmode 2:

$f = 9.36 \text{ Hz}: u_1 = 0.253 \text{ mm}, \ u_2 = -0.123 \text{ mm}, \ a_1 = 0.876 \text{ m/s}^2, \ a_2 = -0.426 \text{ m/s}^2$
$$F_1 = 2.63 \text{ kN}, \quad F_2 = -0.85 \text{ kN}.$$

Superposition:

$$u_1 = 3.85 \text{ mm}, \quad u_2 = 11.86 \text{ mm}, \quad a_1 = 9.67 \text{ m/s}^2, \quad a_2 = 1.34 \text{ m/s}^2.$$

References

1. Petersen C, Werkle H (2018) Dynamik der Baukonstruktionen, 2nd edn. Springer Vieweg, Wiesbaden
2. Paz M (1991) Structural dynamics. Chapman & Hall, New York
3. Pfaffinger D (1989) Tragwerksdynamik. Springer, Wien
4. Clough RW, Penzien J (1975) Dynamics of structures, McGraw-Hill
5. Chopra AK (2012) Dynamics of structures, 4th edn. Prentice Hall, Englewood Cliffs
6. Marguerre K, Wölfel H (1979) Technische Schwingungslehre, Bibliographisches Institut Mannheim
7. Bucher C, Zabel V (2008) Dynamische Modellbildung und Analyse von Tragwerken, Beton-Kalender 2008, Ernst & Sohn, Berlin
8. Müller G, Buchschmid M (2014) Modellierung und Berechnung in der Baudynamik, Stahlbau-Kalender 2014. Ernst & Sohn, Berlin
9. Westerhagen-Freitag C, Wörner J-D (2015) Baudynamik, Beton-Kalender 2015. Ernst & Sohn, Berlin
10. Strommen EN (2014) Structural dynamics. Springer International, Cham, Switzerland
11. Meskouris K, Butenweg C, Hinzen K-G, Höffer R (2019) Structural dynamics with applications in earthquake and wind engineering, 2nd edn. Springer, Berlin
12. Kausel E (2017) Advanced structural dynamics. Cambridge University Press, Cambridge (USA)
13. Przemieniecki JS (1968) Theory of matrix structural analysis. McGraw-Hill, New York
14. Cook RD, Malkus DS, Plesha ME (1989) Concepts and applications of finite element analysis. Wiley, New York
15. Chandrupatla TR, Belegundu AD (1997) Introduction into finite elements in engineering. Prentice Hall of India, New Delhi
16. Bathe K-J (ed) (2014) Finite element procedures, 2nd edn. Watertown MA
17. Weaver W, Timoshenko SP, Young D (1990) Vibration problems in engineering. Wiley, Singapore
18. Waller H, Krings K (1975) Matrizenmethoden in der Maschinen- und Bauwerksdynamik. Bibliographisches Institut, Zürich
19. Müller G, Buchschmid M (2014) Modellierung und Berechnung in der Baudynamik. In: Stahlbau-Kalender 2014, Ernst & Sohn, Berlin
20. Kolousek V (1973) Dynamics of engineering structures. Butterworths, London
21. Rubin H (1988) Eine einheitliche Formulierung des ebenen Stabproblems bei Berücksichtigung der M- und Q-Verformungen, Theorie I. und II. Ordnung, elastischer Bettung sowie harmonischer Schwingungen, Bauingenieur 63, Spinger, 195–204
22. Zienkiewicz OC (1984) Methode der Finiten Elemente. Carl Hanser Verlag, München
23. Newmark MM (1959) A method of computation for structural dynamics. A.S.C.E. J Eng Mech Div 85:67–94
24. Bronstein IN, Semendjajew KA, Musiol G, Mühlig H (2008) Taschenbuch der Mathematik, 7th edn. Verlag Harri Deutsch, Frankfurt am Main
25. Cooley PM, Tukey JW (1965) An algorithm for the machine computation of complex fourier series. Math Comput 19:297–301
26. Müller FP, Keintzel E (1984) Erdbebensicherung von Hochbauten. Ernst & Sohn, Berlin

27. Flesch R (1993) Baudynamik praxisgerecht, Bd. 1. Berechnungsgrundlagen. Bauverlag, Wiesbaden
28. Werkle H (1984) Kinematic interaction of rigid circular foundations on layered soil under surface wave excitation. In: Proceeding 8WCEE, vol 3. San Francisco, 945–952
29. Werkle H, Zahn F (2008) Bauwerksschäden infolge des Bebens in Yogyakarta, Java, am 27. Mai 2006, Bautechnik 11, Ernst & Sohn, Berlin
30. Meskouris K, Hinzen K-G, Butenweg C, Mistler M (2011) Bauwerke und Erdbeben, 3rd edn., Vieweg+Teubner Verlag | Springer Fachmedien Wiesbaden GmbH
31. Ferreira F, Moutinho C, Cunha A, Caetano E (2020) An artificial accelerogram generator code written in Matlab, Engineering Reports, Wiley
32. Schneider G, Wiek J (1979) Herdnahe Messungen der seismischen Bodenbeschleunigung in Südwestdeutschland, Mitteilungen des Instituts für Bautechnik, 2/1979, Berlin
33. Hosser D (1987) Realistische seismische Lastannahmen für Bauwerke, Bauingenieur 62. Springer, Berlin, 567–574
34. DIN EN 1998-1, Eurocode 8: Design of structures for earthquake resistance – Part 1: General rules, seismic actions and rules for buildings, 2010-12; DIN EN 1998-1/NA: Nationally determined parameters to DIN EN 1998-1, 2011-01
35. Werkle H (2020) Baudynamik. In: Schneider-Bautabellen für Ingenieure, 24th edn. Reguvis Verlag, Köln
36. Albert A (2020) Bauten in deutschen Erdbebengebieten. In: Schneider-Bautabellen für Ingenieure, 24th edn. Reguvis Verlag, Köln
37. Wilson EL, Der Kiuregian A, Bayo EP (1981) A replacement for the SRSS method in seismic analysis. Earthq Eng Struct Dyn 9:187–194
38. Waas G (1987) Dämpfung von Bauwerksschwingungen, 4. Jahrestagung der Deutschen Gesellschaft für Erdbeben-Ingenieurwesen und Baudynamik, Hrsg: Dolling H-J, Berlin 1987, DGEB-Publikation Nr. 2
39. Lysmer J et al. (1974) LUSH–a computer program for complex response analysis of soil-structure-systems, Earthquake Engineering Research Center, Report EERC 74-4, University of California, Berkeley
40. Guyan RJ (1965) Reduction of stiffness and mass matrices. AIAA J Am Inst Aeronaut Astronaut 3(2):380
41. Hurty WC (1965) Dynamic analysis of structural systems using component modes. AIAA J 3(4)
42. Craig Jr RR, Bampton MCC (1968) Coupling of substructures for dynamic analyses. AIAA J 6(7), 1313-1319
43. Bachmann H (2002) Erdbebensicherung von Bauwerken. Birkhäuser Verlag, Basel
44. Goris A (2020) Stahl- und Spannbetonbau nach EC2. In: Schneider-Bautabellen für Ingenieure, 24th edn. Reguvis Verlag, Köln
45. König G, Liphardt S (2003) Hochhäuser aus Stahlbeton, Betonkalender 2003. Ernst & Sohn, Berlin
46. Beck H, Schäfer H (1969) Die Berechnung von Hochhäusern durch Zusammenfassung aller aussteifenden Bauteile zu einem Balken, Der Bauingenieur, H. 3, Springer, Berlin
47. Fastabend M, Schäfers T, Albert M, Lommen H-G (2009) Zur sinnvollen Anwendung ganzheitlicher Gebäudemodelle in der Tragwerksplanung von Hochbauten, Beton- und Stahlbetonbau 104 , H. 10, Ernst & Sohn, Berlin
48. Findeiß R (2014) Räumliches Stabtragwerk vs. 3D-FEM-Faltwerksystem–Berechnungs-modelle für das Hotel Intercontinental in Davos, Baustatik-Baupraxis 12, TU München
49. Fajfar P, Kreslin M (2011) Modelling and analysis, Eurocode 8 – background and applications, workshop by the European Commission/CEN/LNEC, Lisbon, Portugal
50. Volarevic J (2013) Boden-Bauwerk-Wechselwirkung bei der dynamischen Finite-Element-Berechnung von Gesamtmodellen, Masterthesis, HTWG Konstanz, Fakultät Bauingenieur-wesen, Lehrgebiet Baustatik (Prof. Dr.-Ing. H. Werkle), Konstanz
51. NIST (2012) Soil–structure interaction for building structures, NIST GCR 12-917-21, US Department of Commerce

52. Kausel E (2010) Early history of soil–structure interaction. Soil Dyn Earthq Eng 30:822–832
53. Lou M, Wang H, Chen X, Zhai Y (2011) Structure-soil–structure interaction: literature review. Soil Dyn Earthq Eng 31:1724–1731
54. Wolf JP (1985) Dynamic soil–structure interaction. Prentice Hall, Englewood Cliffs N.J
55. Waas G, Werkle H (1984) Schwingungen von Fundamenten auf inhomogenem Baugrund, VDI-Schwingungstagung, Bad Soden, VDI-Berichte
56. Werkle H (1988) Steifigkeit und Dämpfung von Fundamenten auf inhomogenem Baugrund, 3. Jahrestagung der Deutschen Gesellschaft für Erdbeben-Ingenieurwesen und Baudynamik, Trans Tech Publications, Clausthal
57. Richardt FE, Hall JR, Woods JD (1970) Vibrations of soils and foundations. Prentice-Hall, Englewood Cliffs N.J
58. Studer J, Ziegler A (1986) Bodendynamik. Springer, Berlin
59. Roesset JM, Whitman RV, Dobry R (1973) Modal analysis for structures with foundation interaction. ASCE J Struct Div 99:339–416
60. Novak M (1974) Effect of soil on structural response to wind and earthquake. Earthq Eng Struct Dyn 3:79–96
61. Bielak J (1976) Modal analysis for building–soil–interaction. ASCE, J Eng Mech Div 102:771–786
62. Tsai NC (1974) Modal damping for soil–structure–interaction. ASCE, J Eng Mech Div 105:323–341
63. Lehmann L (2007) Wave propagation in infinite domains – with application to structure interaction. Springer, Berlin, New York
64. Lysmer J, Kuhlemeyer RL (1969) Finite dynamic model for infinite media. ASCE, J Eng Mech Div 95:859–878
65. Wolf JP, Song C (1996) Finite element modeling of unbounded media. Wiley, New York
66. Bazyar MH, Song C (2006) Time harmonic response of non-homogeneous elastic unbounded domains by using the scaled boundary finite-element method. Earthq Eng Struct Dyn 35:377–383
67. Bazyar MH, Song C (2006) Transient analysis of wave propagation in non-homogeneous elastic unbounded domains by using the scaled boundary finite-element method. Earthq Eng Struct Dyn 35:1787–1806
68. Wolf JP (2003) The scaled boundary finite element method. Wiley, Chichester, England
69. Wolf JP, Song C (2000) The scaled boundary finite-element method–a primer: derivations. Comput Struct 78:191–210
70. Zhang X, Wegner JL, Haddow JB (1999) Three-dimensional dynamic soil–structure interaction analysis in time domain. Earthq Eng Struct Dyn 28:1501–1524
71. Yan J, Zhang X, Jing F (2004) A coupling procedure of FE and SBFE for soil–structure interaction in time domain. Int J Numer Methods Eng 59:1453–1471
72. Radmanovic B (2009) Evaluation of dynamic soil–structure interaction in frequency and time domain. Technische Universität München, Masterthesis
73. Radmanovic B, Katz C (2010) High performance SBFEM. In: Proceedings of the world congress of computational mechanics (WCCM), Sydney, Australia
74. Waas G (1972) Linear two-dimensional analysis of soil dynamics problems in semi-infinite media. Dissertation, University of Californiy, Berkeley
75. Kausel E (1974) Forced vibrations of circular foundations on layered media, MIT Research Report R74-11, MIT, Cambridge Mass
76. Waas G (1984) Berechnung dynamischer Boden-Bauwerk-Wechselwirkungen, Finite Elemente – Anwendungen in der Bautechnik, Ernst & Sohn, Berlin
77. Werkle H (1986) Dynamic finite element analysis of three-dimensional soil models with a transmitting element. Earthq Eng Design 14:41–60 (Wiley)
78. Werkle H (1987) A transmitting boundary for the dynamic finite element analysis of cross anisotropic soils. Earthq Eng Design 15:831–838 (Wiley)
79. Lin H, Thassoulas JL (1986) A hybrid method for three-dimensional problems of dynamics of foundations, Earthq Eng Design 14:61–74 (Wiley)

80. Waas G, Riggs R, Werkle H (1985) Displacement solutions for dynamic loads in transversly-isotropic stratified media. Earthq Eng Design 13:173–193 (Wiley)
81. Kausel E, de Oliveira Barbosa JM (2012) PML's: a direct approach. Int J Numer Methods Eng 90:343–352
82. Basu U, Chopra A (2003) Perfectly matched layers for time-harmonic elastodynamics of unbounded domains: theory and finite-element implementation. Comput Methods Appl Mech Eng 192:1337–1375 (Elsevier BV)
83. Basu U, Chorpa AK (2004) Perfectly matched layers for transient elastodynamics of unbounded domains. Int J Numer Methods Eng 59:1039–1074
84. Feldbusch A, Sadegh-Azar H (2020) Nicht-lineare Untersuchungen zur dynamischen Boden–Bauwerk–Interaktion, Bauingenieur 95, Nr. 4, Springer, Berlin
85. Nguyen CT, Tassoulas JL (2017) Reciprocal absorbing boundary condition for the time-domain numerical analysis of wave motion in unbounded layered media. Proc R Soc A Math Phys Eng Sci 1–25
86. Nguyen CT, Tassoulas JL (2020) Transient analysis of full-space unbounded domains by reciprocal absorbing boundaries. Int J Numer Anal Methods Geomech 44:3–18
87. Kausel E (1981) An explicit solution for the Green functions for dynamic loads in layered media, Report R81-13 with Errata, No. 699, Department of Civil Engineering, MIT, Cambridge, Massachusetts
88. Barbosa O, Kausel E (2012) The thin-layer method in a cross-anisotropic 3-D-space. Int J Numer Methods Eng 537–560
89. Bode C (2000) Numerische Verfahren zur Berechnung von Baugrund–Bauwerk–Interaktionen im Zeitbereich mittels Greenscher Funktionen für den Halbraum. Dissertation, TU Berlin
90. Bode C, Hirschauer R, Savidis S (2002) Soil–structure interaction in time domain using halfspace Green's functions. Soil Dyn Earthq Eng 22:283–295 (Elsevier BV)
91. Kausel E, Peek R (1982) Boundary integral method for stratified soils, MIT report R82-50, Massachusetts Institute of Technology, Cambridge, Massachusetts
92. Wolf JP (1985) Dynamic soil–structure–interaction. Prentice Hall Inc., Englewood Cliffs, NJ
93. Karabalis DL, Beskos DE (1985) Dynamic response of 3-D flexible foundations by time domain BEM and FEM. Soil Dyn Earthq Eng 4(2):91–101
94. Antes H, Estroff OV (1987) Erschütterungsausbreitung im Boden und dynamische Interaktionseffekte – Untersuchungen mit einer Randelementmethode im Zeitbereich, Bauingenieur 62. Springer, Berlin, 201–208
95. Wolf JP (1988) Soil–structure–interaction analysis in time domain. Prentice Hall Inc., Englewood Cliffs, NJ
96. Hub Y, Schmid G (1988) Dynamische Bauwerk-Baugrund-Wechselwirkung im Frequenzbereich mit Hilfe der Randelemente, Bauingenieur 63. Springer, Berlin, 125–131
97. Estroff O, Kausel E (1989) Coupling of boundary and finite elements for soil–structure interaction problems. Earthq Engi Struct Dyn 18:1065–1075
98. Müller G (1993) Ein Verfahren zur Kopplung der Randelementmethode mit analytischen Lösungsansätzen. Habilitation, TU München
99. Hirschauer R (2001) Kopplung von Finiten Elementen mit Randelementen zur Berechnung der dynamischen Baugrund-Bauwerk-Interaktion, Dissertation, TU Berlin
100. Schepers W, Savidis SA (2009) Kopplung von Finiten Elementen und Randelementen zur Lösung ausgedehnter dynamischer Boden–Bauwerk–Interaktionsprobleme im Frequenzbereich. In: VDI-Berichte Nr. 2063, Tagungsband Baudynamik 2009, 14.-15. Mai, Kassel, VDI Verlag, 371–390
101. Romero A, Galvi P, Dominguez J (2013) 3D non-linear time domain FEM BEM approach to soil–structure interaction problems. Eng Anal Bound Elem 37:501–512
102. Werkle H (2016) Modeling of soil–foundation–structure interaction for earthquake analysis of 3D BIM models. International workshop on seismic performance of soil–foundation–structure systems, University of Auckland, Neuseeland, 2016. In: Chouw N, Odense RP, Larkin T (2017) Seismic performance of soil–foundation–structure–interaction. CRS Press/Balkema, Taylor & Francis, London

103. Lysmer J, Tabatabaie-Raissi M, Tajirian FF, Vahdani S, Ostadan F (1981) SASSI–a system for analysis of soil-structure interaction, Report No. UCB/GT/81–02, Geotechnical Engineering, University of California, Berkeley, CA, April 1981
104. Werkle H (1985) HDR-Sicherheitsprogramm – Vorausberechnungen der Versuchsgruppe STO, Hochtief AG, Frankfurt, August 1985 (unpublished)
105. Werkle H, Waas G (1987) Computed versus measured response of HDR reactor building in impact test, SMIRT 9, Lausanne
106. Müller FP (1978) Baudynamik, Betonkalender 1978. Ernst & Sohn, Berlin

Finite Element Software

107. SOF 6 (2014) SOFiSTiK Finite-Element-Software, Version 2014-9 ASE–Allgemeine Statik Finiter Element Strukturen, ASE Manual, Version 2014-9, Software Version, Sofistik 2014, SOFiSTiK AG, Oberschleissheim, Germany, 2015 TAPLA–2D Finite Elemente in der Geotechnik, TALPA Manual, Version 2014-9, Software Version SOFiSTiK 2014, SOFiSTiK AG, Oberschleissheim, Germany
108. SOF 2 (2013) ANSYS ANSYS Mechanical APDL Element Reference, Release 15.0, ANSYS Inc., Canonsburg, USA
109. SOF 10 (2020) PLAXIS Dynamics 2D/3D, Bentley, Dublin, Ireland
110. SOF 11 (2003) SYNTH–Ermittlung eines Synthetischen spektrumskompatiblen Beschleunigungszeitverlaufs. In: Meskouris K, Hinzen K-G, Bauwerke und Erdbeben, 1. Auflage, Vieweg, Wiesbaden

The Book's Homepage

The book has its own homepage:

www.Finite-Elemente-in-der-Baustatik.de

There you will find the latest info and updates about the book.

© Springer Nature Switzerland AG 2021
H. Werkle, *Finite Elements in Structural Analysis*, Springer Tracts
in Civil Engineering, https://doi.org/10.1007/978-3-030-49840-5

ADINA is a registered trademark of K.-J. Bathe/ADINA R&D, Inc.
ANSYS is a registered trademark of ANSYS, Inc.
InfoGraph is a registered trademark of InfoGraph GmbH, Aachen, Germany
PLAXIS is a registered trademark of the PLAXIS company (Plaxis bv)
RFEM™ is a registered trademark of the Dlubal Software GmbH
SAP 2000 is a registered trademark of Computers and Structures, Inc.
SOFISTIK is a trademark of SOFiSTiK AG
TRIMAS is a trademark of RIB SOFTWARE SE

© Springer Nature Switzerland AG 2021
H. Werkle, *Finite Elements in Structural Analysis*, Springer Tracts
in Civil Engineering, https://doi.org/10.1007/978-3-030-49840-5

Index

© Springer Nature Switzerland AG 2021
H. Werkle, *Finite Elements in Structural Analysis*, Springer Tracts
in Civil Engineering, https://doi.org/10.1007/978-3-030-49840-5